Lecture Notes in Networks and Systems

Volume 350

AF870025

The series "Lecture Notes in Networks and Systems" publishes the latest developments in Networks and Systems—quickly, informally and with high quality. Original research reported in proceedings and post-proceedings represents the core of LNNS.

Volumes published in LNNS embrace all aspects and subfields of, as well as new challenges in, Networks and Systems.

The series contains proceedings and edited volumes in systems and networks, spanning the areas of Cyber-Physical Systems, Autonomous Systems, Sensor Networks, Control Systems, Energy Systems, Automotive Systems, Biological Systems, Vehicular Networking and Connected Vehicles, Aerospace Systems, Automation, Manufacturing, Smart Grids, Nonlinear Systems, Power Systems, Robotics, Social Systems, Economic Systems and other. Of particular value to both the contributors and the readership are the short publication timeframe and the world-wide distribution and exposure which enable both a wide and rapid dissemination of research output.

The series covers the theory, applications, and perspectives on the state of the art and future developments relevant to systems and networks, decision making, control, complex processes and related areas, as embedded in the fields of interdisciplinary and applied sciences, engineering, computer science, physics, economics, social, and life sciences, as well as the paradigms and methodologies behind them.

Indexed by SCOPUS, INSPEC, WTI Frankfurt eG, zbMATH, SCImago.

All books published in the series are submitted for consideration in Web of Science.

For proposals from Asia please contact Aninda Bose (aninda.bose@springer.com).

More information about this series at https://link.springer.com/bookseries/15179

Abrar Ullah · Sajid Anwar · Álvaro Rocha ·
Steve Gill
Editors

Proceedings of International Conference on Information Technology and Applications

ICITA 2021

Springer

Editors
Abrar Ullah
School of Mathematical and Computer Science
Heriot-Watt University
Dubai, United Arab Emirates

Álvaro Rocha
Lisbon School of Economics and Management
University of Lisbon
Portugal, Portugal

Sajid Anwar
Institute of Management Sciences
Center of Excellence in Information Technology
Peshawar, Pakistan

Steve Gill
School of Mathematical and Computer Science
Heriot-Watt University
Dubai, United Arab Emirates

ISSN 2367-3370 ISSN 2367-3389 (electronic)
Lecture Notes in Networks and Systems
ISBN 978-981-16-7617-8 ISBN 978-981-16-7618-5 (eBook)
https://doi.org/10.1007/978-981-16-7618-5

This Springer imprint is published by the registered company Springer Nature Singapore Pte Ltd.
The registered company address is: 152 Beach Road, #21-01/04 Gateway East, Singapore 189721, Singapore

Conference Organization

Honorary Chairs

Steve Gill, Associate Professor, Head of School of Mathematical and Computer Sciences, Heriot-Watt University, Dubai, United Arab Emirates
Prof. Álvaro Rocha, Professor, University of Lisbon, Portugal, President of AISTI (Iberian Association for Information Systems and Technologies), Chair of IEEE SMC Portugal Section Society Chapter

General Chair

Dr. Abrar Ullah, Associate Professor, School of Mathematical and Computer Sciences, Heriot-Watt University, Dubai, United Arab Emirates

General Co-chairs

Dr. Ryad Soobhany, Assistant Professor, School of Mathematical and Computer Sciences, Heriot-Watt University, Dubai, United Arab Emirates
Dr. Imran Razzak, Senior Lecturer, School of Information Technology, Deakin University, Victoria, Australia
Talal Shaikh, Associate Professor, School of Mathematical and Computer Sciences, Heriot-Watt University, United Arab Emirates

International Chair

Dr. Sajid Anwar, Associate Professor, Institute of Management Sciences, Peshawar, Pakistan

International Advisor

Dr. David Tien, Vice Chairman, IEEE Computer Chapter, NSW, Australia

Workshop Chairs

Prof. Ibrahim A. Hameed, Professor, Norwegian University of Science, and Technology (NTNU), Chair of the IEEE Computational Intelligence Society (CIS), Norway
Dr. B. B. Gupta, Assistant Professor, National Institute of Technology Kurukshetra, India

Special Session Chairs

Prof. Fernando Moreira, Professor Catedrático, Diretor do Departamento de Ciência e Tecnologia, Universidade Portucalense, Porto, Portugal
Dr. Abdul Rauf, RISE Research Institute of Sweden

Publicity Chairs

Prof. Maria José Sousa, Professor, University Institute of Lisbon, Portugal
Dr. Salma Noor, Assistant Professor, Shaheed Benazir Bhutto Women University, Pakistan

Programme Committee Chairs

Dr. M. Tanveer, Associate Professor Indian Institute of Technology Indore, India
Dr. Abdul Qayyum, Lecturer, University of Burgundy, France

Preface

This conference addresses the importance that IT professionals, academics and researchers stretch across narrowly defined subject areas and constantly acquire a global technical and social perspective. ICITA 2021 offers such an opportunity to facilitate cross-disciplinary and social gatherings. Due to breadth and depth of the topics, it is challenging to class them into specific categories; however, for the convenience of readers, the conference covers a wide range of topics which are broadly split into software engineering, machine learning, network security and digital media and education.

The need for novel software engineering (SE) tools and techniques which are highly reliable and greatly robust are order of the day. There is a greater understanding that design and evolution of the software systems and tools must be "smart" if it is to remain efficient and effective. The nature of artifacts, from specifications through to delivery, produced during construction of software systems can be very convoluted and difficult to manage. A software engineer cannot find all its intricacies by examining these artifacts manually. Automated tools and techniques are required to reflect over business knowledge to identify what is missing or could be effectively changed while producing and evolving these artifacts. There is an agreed belief among researchers that SE provides an ideal platform to apply and test the recent advances in artificial intelligence (AI) tools and techniques. More and more SE problems are now resolved through the application of AI, such as through tool automation and machine learning algorithms.

Machine learning is a broad subfield of computational intelligence that is concerned with the development of techniques that allow computers to "learn." With an increased and effective use of machine learning techniques, there has been rising demand for the use of this approach in different fields of life. There is a wider application of machine learning in different domains of computer science including ecommerce, software engineering, robotics, digital media and education and computer security. Given the opportunities and challenges of the emerging machine learning applications, this area has a great research potential for further investigation.

The growth of data has revolutionized the production of knowledge within and beyond science, by creating efficient ways to plan, conduct, disseminate and assess high-quality novel research. The past decade has witnessed the creation of innovative approaches to produce, store and analyze data, culminating in the emergence of the field of data science, which brings together computational, algorithmic, statistical and mathematical techniques toward extrapolating knowledge from ever growing data sources. This area of research is continuously growing and attracts a lot of interest.

Computer security is a process of protecting computer software, hardware and networks against harm. The application of computer security has a wider scope, including hardware, software and network security. In the wake of rising security threats, it is eminent to improve security postures. This is an ongoing and active research area which attracts a lot of interests from researchers and practitioners.

With the advent of the Internet and technology, the traditional teaching and learning has largely transformed into digital education. Teachers and students are significantly reliant upon the use of digital media in face-to-face classrooms and remote online learning. The adoption of digital media in education profoundly modifies the landscape of education, particularly with regards to online learning, e-learning, blended learning and face-to-face digital-assisted learning, offering new possibilities but also challenges that need to be explored and assessed.

The International Conference on Information Technology and Applications (ICITA) is an initiative to consider the above-mentioned considerations and challenges. Besides the above topics, International Workshop on Information and Knowledge in the Internet of Things (IKIT) 2021 was run in conjunction with ICITA 2021 with a focus on Internet of Things (IoT).

ICITA 2021 was able to attract 112 submissions from 30 different countries across the world. From the 132 submissions, we accepted 65 submissions, which represents an acceptance rate of 49.2%. Out of 65, IKIT 2021 received 20 submissions with 4 accepted papers. Out of all submissions, 60 were selected to be published in this volume. The accepted papers under this volume were categorized under four different themes, including software engineering, machine learning and data science, network security and digital media and education. Each submission is reviewed by at least two reviewers, who are considered experts in the related submitted paper. The evaluation criteria include several issues, such as correctness, originality, technical strength, significance, quality of presentation, interest and relevance to the conference scope. This volume is published in Lecture Notes in Networks and Systems Series by Springer, which has a high SJR impact.

We would like to thank all program committee members as well as the additional reviewers for their effort in reviewing the papers. We hope that the topics covered in ICITA proceedings will help the readers to understand the intricacies involving the methods and tools of software engineering that have become an important element of nearly every branch of computer science. A special thank you to Teresa Guarda, Director of the CIST Research and Innovation Center, Universidad Estatal Península de Santa Elena, La Libertad, Ecuador, for organizing the International Workshop on Information and Knowledge in the Internet of Things (IKIT) 2021.

We would like to extend our special thanks to the keynote speakers, David Tien Senior Lecturer, Charles Sturt University and Vice Chairman, IEEE Computer

Chapter, NSW, Australia, Anthony Lewis Brooks, Associate Professor, Department of Architecture, Design and Media Technology, Aalborg University, Denmark, and Mohamed Quafafou, Professor of Computer Science at Aix-Marseille University, France.

Dubai, United Arab Emirates — Abrar Ullah, Ph.D.
Peshawar, Pakistan — Sajid Anwar, Ph.D.

Contents

Editors and Contributors

About the Editors

Dr. Abrar Ullah is an Associate Professor and Director of Postgraduate Studies at the School of Mathematical and Computer Science, Heriot Watt University, Dubai Campus. Abrar received the M.Sc. (Computer Science, 2000) from University of Peshawar. Abrar received the Ph.D. (Security and Usability) from University of Hertfordshire, United Kingdom. Abrar has been working in industry and academia for over 20 years. He has vast experience in teaching and development of enterprise systems. Abrar started his teaching career in 2002 as lecturer at the University of Peshawar and Provincial Health Services Academy Peshawar. In 2008, Abrar joined the ABMU NHS UK as Lead Developer and contributed to a number of key systems in the NHS. In 2011, Abrar joined professional services at Cardiff University as "Team Lead & Senior Systems Analyst" and led a number of successful strategic and national level projects. In the same period, besides his professional role, he also worked as lecturer of "Digital Media Design" for School of Medicine, Cardiff University. In 2017, Abrar held the role of lecturer at school of management and computer science, Cardiff Metropolitan University. He also held the role of "Lead Developer" at the NHS - Health Education and Improvement Wales (HEIW) until 2019. Abrar is General Chair of the 15th ICITA conference to be held in 13–14 Noverumber 2021 at Dubai UAE. His research interests are cross-disciplinary & industry focused. Abrar has research interest in Security Engineering, Information Security, Usability, Usable Security, Online Examinations and Collusion Detection, Applying Machine Learning techniques to solve real world security problems. Abrar has published over 16 research articles in prestigious conferences and journals.

Dr. Sajid Anwar is an Associate Professor at the Center of Excellence in Information Technology Institute of Management Sciences (IMSciences), Peshawar, Pakistan. He received his MS (Computer Science, 2007) and Ph.D. degrees (Software Engineering, 2011) from NUCES-FAST, Islamabad. Previously, he was head of Undergraduate Program in Software Engineering at IMSciences. Dr. Sajid Anwar is leading expert in Software architecture engineering and Software maintenance prediction. His research interests are cross-disciplinary & industry focused and includes: Search based Software Engineering; Prudent based Expert Systems; Customer Analytics, Active Learning and applying Data Mining and Machine Learning techniques to solve real world problems. Dr. Sajid Anwar is Associate editor of Expert Systems Journal Wiley. He has been a Guest Editor of numerous journals, such as *Neural Computing and Applications*, *Cluster Computing* Journal Springer, *Grid Computing* Journal Springer, *Expert Systems* Journal Wiley, *Transactions on Emerging Telecommunications Technologies* Wiley, and *Computational and Mathematical Organization Theory* Journal Springer. He is also Member Board Committee Institute of Creative Advanced Technologies, Science and Engineering, Korea (iCatse.org) . He has supervised to completion many MS research students. He has conducted and led collaborative research with Government organizations and academia and has published over 45 research articles in prestigious conferences and journals.

Dr. Álvaro Rocha holds the title of Honorary Professor, and holds a D.Sc. in Information Science, Ph.D. in Information Systems and Technologies, M.Sc. in Information Management, and BCs in Computer Science. He is a Professor of Information Systems at the ISEG—Lisbon School of Economics & Management, University of Lisbon. He is also President of AISTI (the Iberian Association for Information Systems and Technologies), Chair of the IEEE Portugal Section Systems, Man, and Cybernetics Society Chapter, and Editor-in-Chief of both JISEM (*Journal of Information Systems Engineering & Management*) and RISTI (*Iberian Journal of Information Systems and Technologies*). Moreover, he has served as Vice-Chair of Experts for the European Commission's Horizon 2020 program, and as an Expert at the COST—intergovernmental framework for European Cooperation in Science and Technology, at the Government of Italy's Ministry of Education, Universities and Research, at the Government of Latvia's Ministry of Finance, at the Government of Mexico's National Council of Science and Technology, and at the Government of Polish's National Science Centre.

Dr. Steve Gill is Academic Head for the School of Mathematical and Computer Sciences at Heriot-Watt University's Dubai campus. He teaches on a range of computer science subjects including information systems development, use of methodologies for developing software systems and on professional development skills. His research interests are in data mining particular interestingness measures He is also interested in student learning difficulties particularly in academic writing.

Contributors

Muhammad Aamir Department of Computer Science, Huanggang Normal University, Huangzhou, Hubei, China

Muhammad Zahid Abbas COMSATS University Islamabad, Vehari Campus, Pakistan

Qandeel Abbas Department of Computer Science and Information Technology, University of Sargodha, Sargodha, Pakistan

Sohail Abbas Department of Computer Science, College of Computing and Informatics, University of Sharjah, Sharjah, UAE

Faima Abbasi Department of Computer Science, Bahria University Islamabad, Islamabad, Pakistan

Mohammed Hamzah Abed Computer Science Department, College of Computer Science and Information Technology, Al-Qadisiyah University, Al Diwaniyah, Iraq

Shiemaa Adlan School of Mathematical and Computer Science, Heriot-Watt University Knowledge Park, Dubai, UAE

Zunaira Afzal Department of Computer Science and Information Technology, University of Sargodha, Sargodha, Pakistan

Abdul Ahad Department of AI, Anurag University, Hyderabad, Telangana, India

Maaz Bin Ahmad Karachi Institute of Economics and Technology (KIET), Karachi, Pakistan

Mudassar Ahmad Department of Computer Science, National Textile University (NTU), Faisalabad, Pakistan

Mukhtiar Ahmad Institute of Computer Sciences and IT, The University of Agriculture, Peshawar, Pakistan

Waleed Ahmed Department of Computer Science and Information Technology, University of Sargodha, Sargodha, Pakistan

Sahar Ajmal Riphah International University, Faisalabad Campus, Rawalpindi, Pakistan

Raghad Al Barghash College of Computing and Informatics, University of Sharjah, Sharjah, UAE

Rouaa Al Zoubi Department of Computer Science, College of Computing and Informatics, University of Sharjah, Sharjah, UAE

Muntasir Al-Asfoor Computer Science Department, College of Computer Science and Information Technology, Al-Qadisiyah University, Al Diwaniyah, Iraq

Feras Al-Obeidat College of Technological Innovation, Zayed University, Abu Dhabi, United Arab Emirates

Khubaib Amjad Alam Department of Computer Science, National University of Computer and Emerging Science (NUCES-FAST), Islamabad, Pakistan

Saleh Abdel Afou Alaliyat Department of Information and Communication Technology and Natural Sciences, Norwegian University of Science and Technology, Ålesund, Norway

Mohanad Alayedi Scientific Instrumentation Laboratory (LIS), Department of Electronics, Faculty of Technology, Ferhat Abbas University of Setif 1, Setif, Algeria

Asmaa Mohamed Ali Eabayed Aldhanhani Department of Computer & Information Sciences (CIS), Higher Colleges of Technology, Fujairah Women's College, Fujairah, United Arab Emirates

Abdihamid Ali United States International University Africa, Nairobi, Kenya

Haider Ali School of Information Science and Technology, Northwest University, Xian, China

Shaik Siddiq Ali Department of Electronics and Communication Engineering, Hindustan Institute of Technology and Science, Chennai, India

Shaukat Ali Department of Computer Science, University of Peshawar, Peshawar, Pakistan

Meera Abdalla Mohamed Qusoom Alnaqbi Department of Computer & Information Sciences (CIS), Higher Colleges of Technology, Fujairah Women's College, Fujairah, United Arab Emirates

Eiman Abdulla Obaid Salem Alyammahi Department of Computer & Information Sciences (CIS), Higher Colleges of Technology, Fujairah Women's College, Fujairah, United Arab Emirates

Khadija Mubarak Mohamed Awasiya Alzaabi Department of Computer & Information Sciences (CIS), Higher Colleges of Technology, Fujairah Women's College, Fujairah, United Arab Emirates

Adnan Amin Center for Excellence in Information Technology, Institute of Management Sciences, Peshawar, Pakistan

V. S. Anoop Rajagiri College of Social Sciences (Autonomous), Kochi, Kerala, India

Mohammad Samar Ansari Aligarh Muslim University, Aligarh, India

M. Hammad Anwar Department of Computer Science and Information Technology, University of Sargodha, Sargodha, Pakistan

Sajid Anwar Centre of Excellence in IT, Institute of Management Sciences (IMSciences), Peshawar, Pakistan

Muhammad Arif Department of Software Engineering, The Superior University, Lahore, Pakistan

Mamoona Naveed Asghar Athlone Institute of Technology, Athlone, Ireland; The Islamia University of Bahawalpur, Bahawalpur, Pakistan

Madnia Ashraf Department of Computer Science and IT, University of Sargodha, Sargodha, Pakistan;
Punjab Information Technology Board, Lahore, Pakistan

Muhammad Asif Department of Computer Science, Lahore Garrison University, Lahore, Pakistan

Noor Ayesha School of Clinical Medicine, Zhengzhou University, Zhengzhou, Henan, PR China

João Azevedo Algoritmi Research Centre, University of Minho, Braga, Portugal

Javed Iqbal Bangash Institute of Computer Sciences and IT, The University of Agriculture, Peshawar, Pakistan

Larissa Cristina Barbieri IFRO, Vilhena, Brazil

Raul Afonso Pommer Barbosa Fundação Getúlio Vargas, Rio de Janeiro, Brazil;
Fundação Getulio Vargas FGV-EAESP, São Paulo, Brazil

Hadj Batatia MACS School, Heriot-Watt University, Dubai, United Arab Emirates

Mughair Aslam Bhatti School of Geography, Nanjing Normal University, Nanjing, China;
Key Laboratory of Virtual Geographic Environment, Ministry of Education, Nanjing Normal University, Nanjing, China

Uzair Aslam Bhatti School of Geography, Nanjing Normal University, Nanjing, China;
Key Laboratory of Virtual Geographic Environment, Ministry of Education, Nanjing Normal University, Nanjing, China

Tetiana Biloborodova G.E. Pukhov Institute for Modelling in Energy Engineering, Kyiv, Ukraine

Bassem Bouaziz MIRACL Laboratory, University of Sfax, Sfax, Tunisia

Boubakar Seddik Bouazza Technology of Communication Laboratory (LTC), Electronics department, Faculty of Technology, Dr. Tahar Moulay University of Saida, Saida, Algeria

Amal Boudaya University of Sfax, MIRACL Laboratory, CRNS, Sfax, Tunisia

Zied Bouraoui CRIL, Université d'Artois and CNRS, Arras, France

Eliana Paula Calegari Instituto Benjamin Constant, Rio de Janeiro, Brazil

Siwar Chaabene University of Sfax, MIRACL Laboratory, CRNS, Sfax, Tunisia

Lotfi Chaari University of Toulouse, Dubai, United Arab Emirates;
IRIT Laboratory, University of Toulouse, INP, Labège, France

Anthony Chan Charles Sturt University, Wagga Wagga, New South Wales, Australia

Bhanu Chander Department of Computer Science and Engineering, Pondicherry University, Pondicherry, India

K. Chandrasekaran National Institute of Technology Karnataka, Surathkal, Karnataka, India

Ngan-Khanh Chau An Giang University, VNU-HCM, Ho Chi Minh City, Vietnam

Asma Chebli Computer Science Department, LRI Laboratory, University of Badji Mokhtar Annaba, Annaba, Algeria

Abdelhamid Cherifi Technology of Communication Laboratory (LTC), Electronics department, Faculty of Technology, Dr. Tahar Moulay University of Saida, Saida, Algeria

Ismael Costa ISLA Santarém, Largo Cândido Dos Reis, Santarém, Portugal

Luciana Aparecida Barbieri Da Rosa IFRO, Vilhena, Brazil

Luciana Aparecida Barbieri da Rosa Instituto Federal de Rondônia, Vilhena, Brazil

Gertrudes Aparecida Dandolini Engineering and Knowledge Management Department, Federal University of Santa Catarina, Florianópolis, Brazil

Renata Danielienė Vilnius University, Vilnius, Lithuania;
Information Technologies Institute, Vilnius, Lithuania

Tarim Dar University of Engineering and Technology, Taxila, Pakistan

Andreia de Bem Machado Engineering and Knowledge Management Department, Federal University of Santa Catarina, Florianópolis, Brazil

Maria Jose de Sousa Instituto Universitário de Lisboa, Lisbon, Portugal;
ISCTE, Lisbon, Portugal

Lavadya Nirmala Devi Department of ECE, University College of Engineering, Osmania University, Hyderabad, India

Usha Divakarla NMAM Institute of Technology, Nitte, Karkala, Karnataka, India

Akila Djebbar Computer Science Department, LRI Laboratory, University of Badji Mokhtar Annaba, Annaba, Algeria

Thanh-Nghi Do Can Tho University, Can Tho, Vietnam

Alcione Félix do Nascimento IFRO, Vilhena, Brazil

João Rodrigues dos Santos Economics and Management Department, IADE/Universidade Europeia, Lisbon, Portugal

Muhammad Ejaz Department of Computer Science and Information Technology, University of Sargodha, Sargodha, Pakistan

Mohamed Fakhfakh MIRACL Laboratory, University of Sfax, Sfax, Tunisia

Enda Fallon Athlone Institute of Technology, Athlone, Ireland

Abdelhak Ferhat Hamida Laboratory of Optoelectronics and Components (LOC), Electronics department, Faculty of Technology, Ferhat Abbas University of Setif 1, Setif, Algeria

Hafiz Yasir Ghafoor The Superior University, Lahore, Pakistan

Jeet Gohil United States International University Africa, Nairobi, Kenya

Teresa Guarda ISLA Santarém, Largo Cândido Dos Reis, Santarém, Portugal;
CIST—Centro de Investigación en Sistemas y Telecomunicaciones, Universidad Estatal Península de Santa Elena, La Libertad, Ecuador;
Algoritmi Centre, Minho University, Guimarães, Portugal

Haji Gul Center for Excellence in Information Technology, Institute of Management Sciences, Peshawar, Pakistan

Khushi Gupta United States International University Africa, Nairobi, Kenya

Priyam Gupta Delhi Technological University, New Delhi, Delhi, India

Siddharth Gupta Computer Science and Engineering Department, Graphic Era Deemed to Be University, Dehradun, Uttarakhand, India

Muhammad Asif Habib Department of Computer Science, National Textile University (NTU), Faisalabad, Pakistan

Ibrahim A. Hameed Department of Information and Communication Technology and Natural Sciences, Norwegian University of Science and Technology, Ålesund, Norway

Zeeshan Hameed Department of Computer & Information Sciences (CIS), Higher Colleges of Technology, Fujairah Women's College, Fujairah, United Arab Emirates

Muhammad Umair Hassan Department of Information and Communication Technology and Natural Sciences, Norwegian University of Science and Technology, Ålesund, Norway

Diarmuid Healy Athlone Institute of Technology, Athlone, Ireland

Abrar Hussain Department of Computer Science, Lahore Garrison University, Lahore, Pakistan

Hafsa Ilyaas University of Engineering and Technology, Taxila, Pakistan

Muhammad Ilyas Department of Computer Science and Information Technology, University of Sargodha, Sargodha, Pakistan

Florentin Ipate Department of Computer Science, University of Bucharest, Bucharest, Romania

Muhammad Waseem Iqbal The Superior University, Lahore, Pakistan

Saba Iqbal School of Mathematical and Computer Science, Heriot-Watt University Knowledge Park, Dubai, UAE

Umer Iqbal Riphah International University, Faisalabad Campus, Rawalpindi, Pakistan

Muhammad Ishaq Institute of Computer Sciences and IT, The University of Agriculture, Peshawar, Pakistan

Arfan Jaffar The Superior University, Lahore, Pakistan

Rashid Jahangir COMSATS University Islamabad, Vehari Campus, Pakistan

Ali Javed University of Engineering and Technology, Taxila, Pakistan

Shahzeb Javed Riphah International University, Faisalabad Campus, Rawalpindi, Pakistan

Sana ben Jemea Department of Physiology, Faculty of Medicine of Sfax, Sfax, Tunisia

Y. Jennifer Department of Electronics and Communication Engineering, Hindustan Institute of Technology and Science, Chennai, India

Aneesah Abdul Kadhar Heriot-Watt, Dubai, UAE

Nadia Kanwal Athlone Institute of Technology, Athlone, Ireland;
Lahore College for Women University, Lahore, Pakistan

Sreenath Kashyap Department of CSE, Vidya Jyothi Institute of Technology, Hyderabad, Telangana, India

Aaishwarya Khalane Heriot-Watt University Dubai, Dubai, UAE

Malika Khalil Intelligent Criminology Research Lab National Center of Artificial Intelligence, Al Khawarizmi Institute of Computer Science, University of Engineering and Technology, Lahore, Pakistan

Abdullah Khan Institute of Computer Sciences and IT, The University of Agriculture, Peshawar, Pakistan;
Department of Computer Science, University of Swabi, Swabi, Pakistan

Ahmed Khan COMSATS University, Abbotabad, Pakistan

Asfandyar Khan Institute of Computer Sciences and IT, The University of Agriculture, Peshawar, Pakistan

Muhammad Usman Ghani Khan Intelligent Criminology Research Lab National Center of Artificial Intelligence, Al Khawarizmi Institute of Computer Science, University of Engineering and Technology, Lahore, Pakistan

Muhammad Zeeshan Khan Intelligent Criminology Research Lab National Center of Artificial Intelligence, Al Khawarizmi Institute of Computer Science, University of Engineering and Technology, Lahore, Pakistan

Shah Khusro Department of Computer Science, University of Peshawar, Peshawar, Pakistan

Ayesha Komal Department of Computer Science, National College of Business Administration and Economics Lahore, Sub Campus Multan, Multan, Pakistan

Mark Koverha Department of Computer Science and Engineering, Volodymyr Dahl East Ukrainian National University, Severodonetsk, Ukraine

Smitha S. Kumar Heriot-Watt, Dubai, UAE

Kumaravelan Department of Computer Science and Engineering, Pondicherry University, Pondicherry, India

Fátima Leal REMIT, Universidade Portucalense, Porto, Portugal

Truong-Thanh Ma CRIL, Université d'Artois and CNRS, Arras, France

Bayan Mahfood Department of Computer Science, College of Computing and Informatics, University of Sharjah, Sharjah, UAE

Nasir Mahmood Department of Computer Science, National Textile University, Faisalabad, Pakistan

Toqeer Mahmood Department of Computer Science, National Textile University, Faisalabad, Pakistan

Rikesh Makwana Heriot-Watt University Dubai, Dubai, UAE

Benedita Malheiro INESC TEC, Porto, Portugal;
School of Engineering, Polytechnic Institute of Porto, Porto, Portugal

Hassaan Malik Department of Computer Science, National College of Business Administration and Economics Lahore, Sub Campus Multan, Multan, Pakistan;
School of Systems and Technology, University of Management and Technology, Lahore, Pakistan

Célio Gonçalo Marques TECHN&ART | LIED.IPT, Polytechnic Institute of Tomar, Tomar, Portugal

Robert Mathew Anugraham Neurocare, Thiruvananthapuram, India

Hayet Farida Merouani Computer Science Department, LRI Laboratory, University of Badji Mokhtar Annaba, Annaba, Algeria

Zaid Mohammed Athlone Institute of Technology, Athlone, Ireland

Fernando Moreira REMIT, IJP, Universidade Portucalense, Porto, Portugal;
IEETA, Universidade de Aveiro, Aveiro, Portugal

Teresa Mourão ISCTE—University Institute of Lisbon, Lisbon, Portugal

Leah Mutanu United States International University Africa, Nairobi, Kenya

Muhammad Muzammal Department of Computer Science, Bahria University Islamabad, Islamabad, Pakistan

Saqib Ali Nawaz College of Information and Communication Engineering, Hainan University, Haikou, China

Anam Naz Department of Computer Science and Information Technology, University of Sargodha, Sargodha, Pakistan

Saeeda Naz Govt. Girls Postgraduate College, Abbotabad, Pakistan

Kashif Nazir Department of Computer Science, National Textile University (NTU), Faisalabad, Pakistan

Shah Nazir Department of Computer Science, University of Swabi, Swabi, Pakistan

Mariam Orabi College of Computing and Informatics, University of Sharjah, Sharjah, UAE

Avnish Panwar Computer Science and Engineering Department, Graphic Era Hill University, Dehradun, Uttarakhand, India

Hélder Pestana TECHN&ART | LIED.IPT, Polytechnic Institute of Tomar, Tomar, Portugal

Andrew Phelan Athlone Institute of Technology, Athlone, Ireland

Carla Pires CBIOS - Universidade Lusófona's Research Center for Biosciences and Health Technologies, Lisbon, Portugal

Filipe Portela Algoritmi Research Centre, University of Minho, Braga, Portugal; IOTECH—Innovation on Technology, Trofa, Portugal

A. S. Poyyamozhi Department of Electronics and Communication Engineering, Hindustan Institute of Technology and Science, Chennai, India

Shanza Rasham Department of Computer Science and Information Technology, University of Sargodha, Sargodha, Pakistan

C. B. M. Rashidi Advanced Communication Engineering, Center of Excellence School of Computer and Communication Engineering (ACE-Co-SCCE), Universiti Malaysia Perlis, (UniMAP), Perlis, Malaysia

Ali Raza Department of Computer Science and IT, University of Sargodha, Sargodha, Pakistan

M. Arslan Raza Department of Computer Science, Lahore Garrison University, Lahore, Pakistan

Imran Razzak Deakin University, Geelong, Australia

Amjad Rehman Artificial Intelligence & Data Analytics Lab CCIS, Prince Sultan University, Riyadh, Saudi Arabia

Arshia Rehman COMSATS University, Abbotabad, Pakistan

Benjamin Jacob Reji Heriot-Watt University, Dubai, UAE

Tanees Riaz University of Engineering and Technology, Taxila, Pakistan

Maria Carolina Martins Rodrigues Universidade de Algarve, Faro, Portugal

Giedrius Romeika Vilnius University, Vilnius, Lithuania

Tanzila Saba Artificial Intelligence & Data Analytics Lab CCIS, Prince Sultan University, Riyadh, Saudi Arabia

K. Sakthidasan Sankaran Department of Electronics and Communication Engineering, Hindustan Institute of Technology and Science, Chennai, India

Sercan Sari Department of Computer Engineering, Yeditepe University, Istanbul, Turkey

Akilan Selvacoumar Heriot-Watt University, Dubai, UAE

Agustin Servat Athlone Institute of Technology, Athlone, Ireland

Babar Shah College of Technological Innovation, Zayed University, Abu Dhabi, UAE

Peer Azmat Shah Department of Computer & Information Sciences (CIS), Higher Colleges of Technology, Fujairah Women's College, Fujairah, United Arab Emirates

Talal Shaikh Heriot-Watt University Dubai, Dubai, UAE

Raj Kumar Singh Delhi Technological University, New Delhi, Delhi, India

Inna Skarga-Bandurova School of Engineering, Computing and Mathematics, Oxford Brookes University, Oxford, UK

Illia Skarha-Bandurov Luhansk State Medical University, Rubizhne, Ukraine

Ahmad Ryad Soobhany School of Mathematical and Computer Science, Heriot-Watt University Knowledge Park, Dubai, UAE

Maria José Sousa ISCTE-IUL—Instituto Universitário de Lisboa, Lisbon, Portugal;
Political Science and Public Policy Department, ISCTE—University Institute of Lisbon, Lisbon, Portugal

S. Sreelakshmi Indian Institute of Information Technology and Management-Kerala (IIITM-K), Thiruvananthapuram, India

Sreenithi Sridharan Heriot-Watt University, Dubai, UAE

Putty Srividya Department of ECE, University College of Engineering, Osmania University, Hyderabad, India

Muhammad Tahir College of Computing and Informatics, Saudi Electronic University, Riyadh, Saudi Arabia

Asadullah Tariq Department of Computer Science, The Superior University, Lahore, Pakistan

Maria Emília Teixeira IJP, Universidade Portucalense, Porto, Portugal

David Tien Charles Sturt University, Wagga Wagga, New South Wales, Australia

Krzysztof Trzcinski Athlone Institute of Technology, Athlone, Ireland

Abdallah Tubaishat College of Technological Innovation, Zayed University, Abu Dhabi, United Arab Emirates

Adrian Turcanu School of Mathematical and Computer Sciences, Heriot-Watt University, Dubai, UAE

Cristina Nicoleta Turcanu University of Pitesti, Pitești, Romania

Prince Tyagi Delhi Technological University, New Delhi, Delhi, India

Muhammad Talha Ubaid Intelligent Criminology Research Lab National Center of Artificial Intelligence, Al Khawarizmi Institute of Computer Science, University of Engineering and Technology, Lahore, Pakistan

Abrar Ullah School of Mathematical and Computer Science, Heriot-Watt University Knowledge Park, Dubai, UAE;
School of Mathematical and Computer Science, Heriot-Wat University, Edinburgh, UK

Milan Varghese Rajagiri College of Social Sciences (Autonomous), Kochi, Kerala, India

Bruno Veloso INESC TEC, Porto, Portugal;
Universidade Portucalense, Porto, Portugal

Marlene Grace Verghese Department of CSE, Vidya Jyothi Institute of Technology, Hyderabad, Telangana, India

Irfan Yaqoob Department of Electrical and Computer Engineering, Clarkson University, Potsdam, USA

Yelyzaveta Yevsieieva School of Medicine, V. N. Karazin, Kharkiv National University, Kharkiv, Ukraine

Linwang Yuan School of Geography, Nanjing Normal University, Nanjing, China;
Key Laboratory of Virtual Geographic Environment, Ministry of Education, Nanjing Normal University, Nanjing, China

Zhaoyuan Yu Key Laboratory of Virtual Geographic Environment, Ministry of Education, Nanjing Normal University, Nanjing, China

Numan Zafar Department of Electrical and Computer Engineering, Clarkson University, Potsdam, USA

Ahmad Zaib Women Medical College, DHQHospital, Abbotabad, Pakistan

Hela Zouari Department of Physiology, Faculty of Medicine of Sfax, Sfax, Tunisia

Machine Learning and Data Science

A Convolutional Neural Network for Artifacts Detection in EEG Data

Amal Boudaya, Siwar Chaabene, Bassem Bouaziz, Hadj Batatia, Hela Zouari, Sana ben Jemea, and Lotfi Chaari

Abstract Electroencephalography (EEG) is an effective tool for neurological disorders diagnosis such as seizures, chronic fatigue, sleep disorders, and behavioral abnormalities. Various artifacts types may impact EEG signals regardless the used, resulting in an erroneous diagnosis. Various data analysis tools have therefore been developed in the biomedical engineering literature to detect and/or remove these artifacts. In this sense, deep learning (DL) is one of the most promising methods. In this paper, we develop a novel method based on artifacts detection using a convolutional neural network (CNN) architecture. The available EEG data was collected using 32 channels from the *Nihon Kohden Neurofax EEG-1200*. The data are preprocessed and analyzed using our CNN to perform binary artifact detection. The suggested method highlights the best classification results with a maximal accuracy up to 99.20%.

1 Introduction

Artifacts detection has become a topic of interest in various biomedical signals such as electroencephalogram (EEG) [1], electromyogram (EMG) [2], electrocardiogram

A. Boudaya (✉) · S. Chaabene · B. Bouaziz
University of Sfax, MIRACL Laboratory, CRNS, Sfax, Tunisia

B. Bouaziz
e-mail: bassem.bouaziz@isims.usf.tn

H. Batatia
MACS School, Heriot-Watt University, Dubai-Campus, Dubai, United Arab Emirates
e-mail: H.Batatia@hw.ac.uk

H. Zouari · S. Jemea
Department of Physiology, Faculty of Medicine of Sfax, Sfax, Tunisia

L. Chaari
IRIT Laboratory, University of Toulouse, INP, Labège, France
e-mail: lotfi.chaari@toulouse-inp.fr

A. Ullah et al. (eds.), *Proceedings of International Conference on Information Technology and Applications*, Lecture Notes in Networks and Systems 350,
https://doi.org/10.1007/978-981-16-7618-5_1

(ECG) [3], and electrooculogram (EOG) [4]. EEG is a non-invasive recording technology which is the most used for the analysis of electrical brain activity for diagnosis [5].

In addition to standard clinicals such as Nihon Kohden and Micromed, there are a variety of wearable EEG headsets like Emotiv EPOC + which are being widely used for applications such as drowsiness detection [6]. Nevertheless, EEG recordings are frequently affected by a number of artifact types [7, 8], which may affect the analysis and processing results based on the recorded signals. Specifically, physiological artifacts [9] such as eye blink, pulse, or muscular activity can affect the quality of EEG signals. Other non-physiological artifacts [10] like faulty signal recording, line frequency, lights, and external movements are produced outside the body by electromagnetic fields and may alter the accuracy of the recorded data. Removing these artifacts by discarding affected data segments or applying a restoration technique [11] needs an automatic method to detect them.

In this sense, various techniques are used for binary artifacts classification. Since defining a clear model for these artifacts is complicated due to artifacts variability, most methods rely on machine learning (ML) techniques [12]. However, most of the methods request the intervention of an expert who define explicit features to summarize relevant data segments. Indeed, the authors in [12] define a set of statistical features that are presented to classifiers for the artifacts task. To avoid this features extraction step, artificial neural networks (ANNs) may be considered as an alternative, especially in the case of deep convolutional networks [13]. Indeed, convolutional neural networks (CNNs) use a set of convolutional filters to allow implicit features extraction. The depth of the network helps to achieve high abstraction levels and hence to reach better detection and classification performances [13, 14].

In this study, we propose a CNN based method to detect artifacts in EEG segments. The proposed methods rely on a binary detection scheme: *artifacts* and *No-artifacts.* This classification will help physicians to avoid extra time when reading all EEG signal and let them concentrate only on the No-artifacts segments and consequently will lead to a better diagnosis. A specific architecture is proposed to handle data segments. In this sense, various segment lengths are tested in order to identify the best configuration in terms of detection accuracy. Comparisons are performed with the state of the art, showing promising results.

The rest of the paper is described below. Section 2 describes the background and formulates the problem of artifact detection in EEG segments. Section 3 introduces the proposed method for binary classification as well as the used architecture. An experimental validation is then conducted in Sect. 4. Finally, conclusions and future work are drawn in Sect. 5.

2 Problem Formulation

2.1 EEG Signals

The brain is divided into two hemispheres, and as illustrated in Fig. 1, each hemisphere is composed of four lobes: frontal, temporal, parietal, and occipital.

Clinical EEG signals collect brain data using multiple electrodes, generally ranging from 20 to 32.

EEG data analysis generally involves a preprocessing step in order to extract typical frequency waves [15], and each of them reflects a specific brain behavior as explained in Table 1.

A relevant analysis of these waves must exclude data segments distorted by different types of recording artifacts. Specifically, we can define artifacts as anything that reduces signal-to-noise ratio and obstruct accurate data acquisition in the brain [16]. Different examples of physiological and non-physiological artifacts are reported in Table 2.

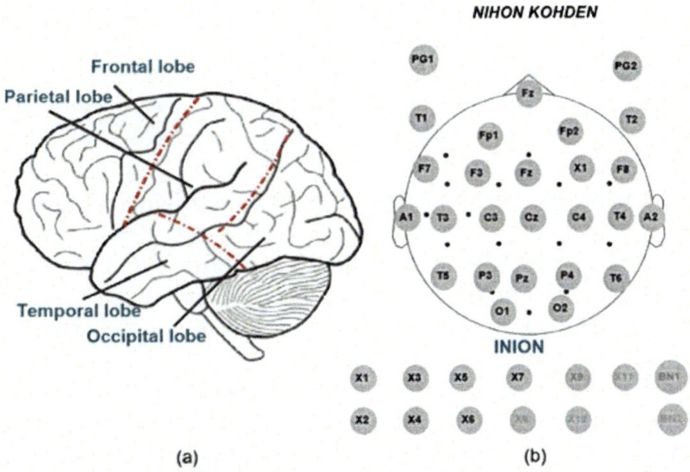

Fig. 1 **a** The four brain NIHON KOHDEN lobes; **b** reference electrode NASION positions

Table 1 Fundamental brain waves

Band	Frequency	Description
Delta	< 4 Hz	Deep sleep
Theta	4–8 Hz	Relaxed state
Alpha	8–13 Hz	Consciousness
Beta	13–30 Hz	24–26
Gamma	> 30 Hz	Vigilance

Table 2 Fundamental brain waves

Artifact types	Sources	Frequency range	Amplitude	Characteristics
Muscle	Body movements	$\leq$35 Hz	Low	Beta frequency
Ocular	Eye	0.5–3 Hz	100 mV	Delta wave
Electrode	Electrode contact	Very low	High	Various morphology
Cardiac	Heart	$\geq$1 Hz	1-10mv	Epilepsy

2.2 Related Works

In the recent literature, various studies have been proposed to classify artifacts [8, 10, 16, 17]. In [18], the authors suggest the use of an auto-regressive as a feature with a support vector machine (SVM) classifier to detect the contamination of EEG signals artifacts. The accuracy value is approximately equal to 94%.

As developed in [12], three classifiers are used as support vector machine (SVM), decision tree (DT), and k-nearest neighbors (KNN) to classify various muscular and ocular artifacts sources. The data are divided here into overlapping segments. Binary classification as *artifacts* and *No-artifacts* is done with a high accuracy detection equal to 98.78% using DT algorithm.

In [8], an ANN architecture is used to detect a physiological artifact of eye movement in EEG signals. A binary classification of *artifacts* and *No-artifacts* accuracy up to 97.50% is achieved. Most ML methods need a large volume of data.

Based on the literature, deep learning (DL) networks have been employed to denoise EEG recordings, and their efficiencies have been compared with traditional approaches.

In [19], a novel CNN model is proposed to avoid over-fitting by removing the muscle artifacts from EEG signals. The model contains fourteen convolutional layers including six average pooling, one flatten, and one fully connected layers. The reported average accuracy rate is equal to 86.3%.

As proposed in [17], a CNN architecture is implemented for EEG artifact identification. The model is composed by convolutional, max-pooling, and softmax layers. An accuracy rate of 74% is reached for *artifacts* and *No-artifacts* classification.

In [16], the authors propose a DL model for artifacts classification during signals collection to help clinicians in artifacts resolving problems. Various architectures are used as recurrent neural network (RNN), deep CNN (DCNN), and CNN. An accuracy rate of 80% is given by the CNN model. In this study, a DL method based on CNN architecture is developed to detect artifacts in EEG signals. The proposed method gives the best results compared to other previous studies.

3 Proposed Method

In this section, we propose a new method based on a CNN architecture for *artifacts* and *No-artifacts* classification. Our data are collected from the *Nihon Kohden* using 32 channels located around the patient's brain.

The suggested approach pipeline is proposed in Fig. 2. Our strategy revolves around two main parts as data collection and model evaluation. Each step is described in the following two subsections.

3.1 Data Collection

The signals were recorded from four patients for approximately forty minutes using the *Nihon Kohden* equipment at the hospital. Our data are annotated by neurologists who detect various artifacts types. Based on binary classification, we made the fusion of different types of artifacts for *artifacts* and *No-artifacts* classes. The different characteristics of *Nihon Kohden* machine are presented in the following Table 3.

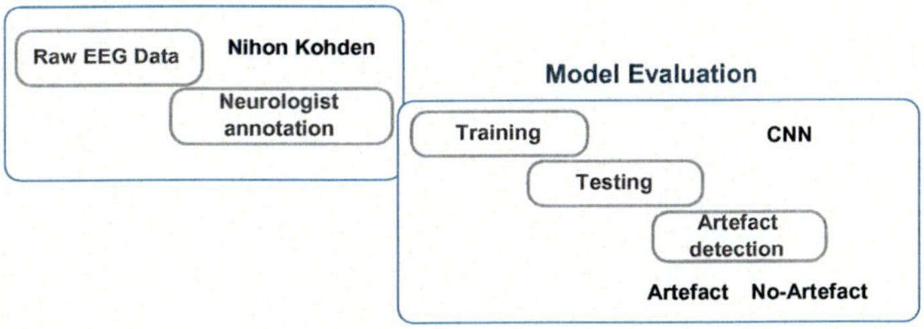

Fig. 2 Pipeline of the pro-Data Collection posed method

Table 3 Characteristics of the *Nihon Kohden*

Characteristics	*Nihon Kohden*
32 EEG channels	Fp1, Fp2, F3, F4, C3, C4, P3, P4, O1, O2, F7, F8, T3, T4, T5, T6, Fz, Cz, Pz, PG1, PG2, A1, A2, T1, T2, X1, X2, X3, X4, X5, X6, X7
DC channels	DC SpO2, DC EtCO2, DC DC03, DC DC04, DC DC05, DC DC06, DC pulse, DC CO2 wave
STIM channels	STIM events
Sampling rate	500 Hz

3.2 Model Evaluation

Figure 3 shows the suggested CNN employed in our artifact detection. Various segments duration with and without overlap are the input of our architecture. An example of overlapping interval is shown in Fig. 4.

The proposed CNN model contains 10 layers which are two batch-normalization, three convolutional, three max-pooling, one flatten, and one dense layers. However, the batch-normalization layers are used to avoid over-fitting and to accelerate the learning step. The convolutional layers are integrated to extract features of the input signal. The max-pooling layers are included to reduce the dimensionality of the

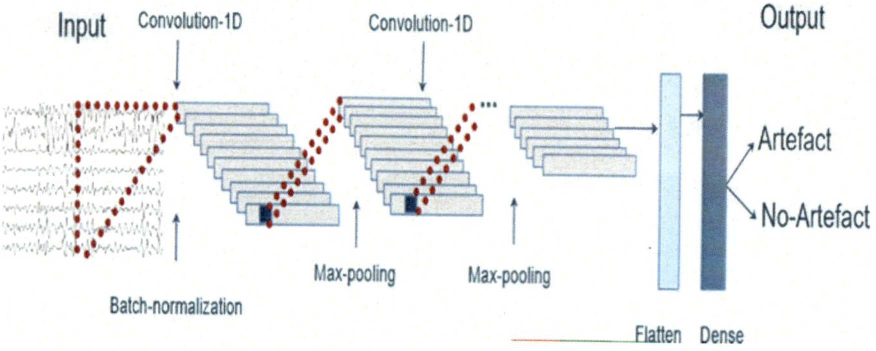

Fig. 3 Proposed CNN architecture

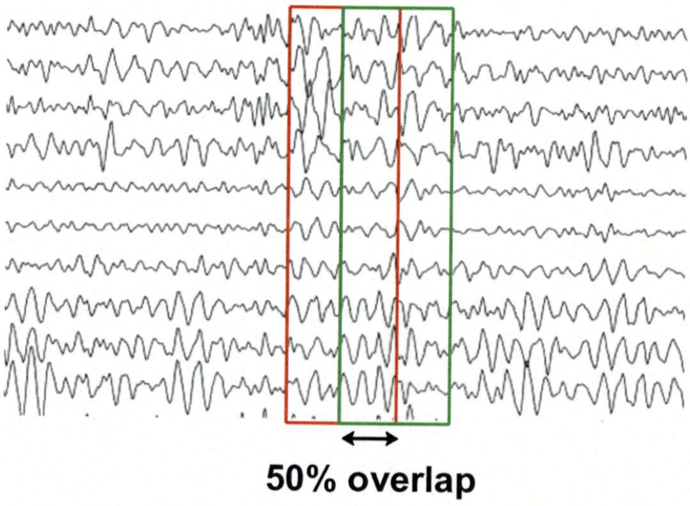

Fig. 4 Overlap interval

input layers by sub-sampling them. A flatten layer is used to turn data into a one-dimensional array that is connected to the dense layer. A binary classification is developed, which means that a single neuron is enough to denote class "1" for the *No-artifact* class and class "0" for the *artifact* class.

4 Results Analysis

4.1 Evaluation Tools

Our dataset is divided into 70% for the training set and 30% for the test set. For the learning model, we used 20% from the training set for the validation set in order to avoid over-fitting and increase the accuracy rate. The different layers, the output shapes, and the parameters of our model are presented in Table 4.

The network is trained with a stochastic gradient descent (SGD) optimizer which is widely used in DL. Using the binary cross-entropy for binary classification, the proposed CNN architecture is trained to 100th epochs.

For the first step, we segment our EEG signals into various duration as 1 s, 2 s, and 3 s with 50% overlap. The different experimental results associated with five runs are indicated in Table 5.

According to the above results, we can assume that the best results of the various segments are obtained without overlap based on the various results in Table 5. Besides, the high accuracy for the testing set is obtained using 1 s with overlap equal to 78.09%, while the best accuracy is obtained using 2 s without overlap with 92.42%. To increase the accuracy test value, we made others experiments with 0.1 s and 0.03 s without overlap. Table 6 presents the average experimental results for 0.1 s and 0.03 s.

Table 4 Model summary

Layer	Output shape	Param #
batch_normalization_10	(None, 1, 32)	4
conv1d_29 (Conv1D)	(None, 1, 32)	4
batch_normalization_10	(None, 1, 32)	1056
max_pooling1d_28 (MaxPooling)	(None, 1, 32)	0
conv1d_30 (Conv1D)	(None, 1, 64)	2112
max_pooling1d_29 (MaxPooling)	(None, 1, 64)	0
conv1d_31 (Conv1D)	(None, 1, 128)	8320
max_pooling1d_30 (MaxPooling)	(None, 1, 128)	0
flatten_8 (Flatten)	(None, 128)	0
dense_8 (Dense)	(None, 1)	129

Table 5 Experimental results for 1 s, 2 s, and 3 s segments with and without overlap

Duration	1 s		2 s		3 s	
	With (%)	Without (%)	With (%)	Without (%)	With (%)	Without (%)
Overlap	99.82	99.81	96.79	99.86	97.03	99.74
Accuracy train validation	99.82	99.85	96.48	99.94	97.02	99.84
Accuracy train test	78.09	77.55	56.34	92.42	56.45	88.03

Table 6 Average experimental results for 0.1 s and 0.03 s segments without overlap

Duration	0.1 s (%)	0.03 s (%)
Accuracy train	99.84	99.44
Accuracy validation	99.78	98.95
Accuracy test	96.91	98.60

The average accuracy test is equal to 98.60% which is obtained by using 0.03 s segment without overlap. In this way, our method converges with achieving stability after 50 epochs. As shown in Figs. 5 and 6, the accuracy curves of training and validation sets increase while their loss curves decrease.

To measure the performance of the classification problem, we need to use a confusion matrix to make various combinations of actual and predicted values. The used confusion matrix is presented in Table 7.

where TP, TN, FP, and FN are, respectively, true positives, true negatives, false positives, and false negatives. Table 8 indicates the accuracy, precision, recall, and F1-score results based on the confusion matrix. Our model reach a maximal predicted accuracy equal to 99.20%.

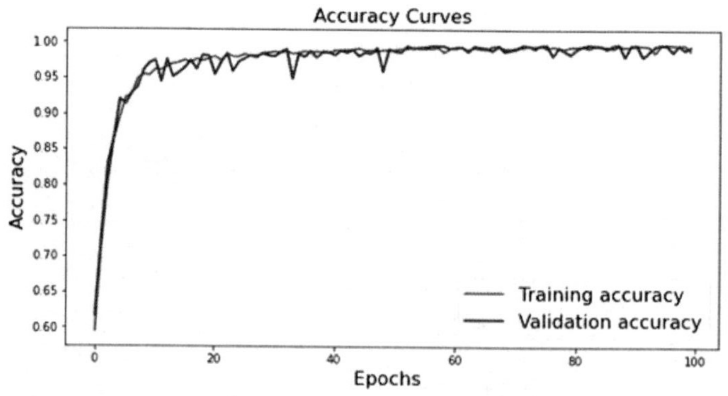

Fig. 5 Accuracy curves

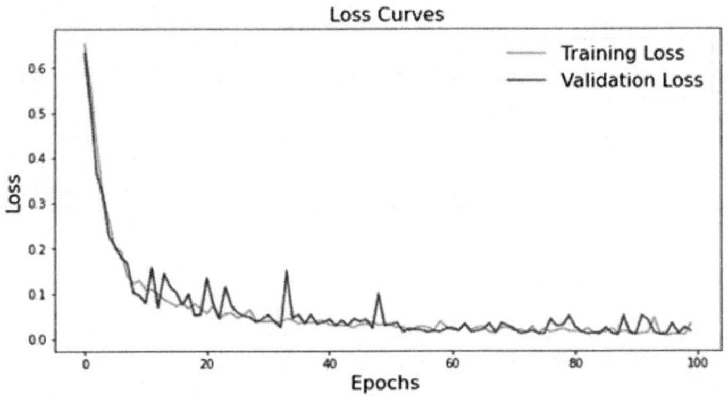

Fig. 6 Loss curves

Table 7 Confusion matrix results

Types	*Artifact*	*No-Artifact*
Artifact	4152/4200 = 98.85% (TP)	48/4200 = 1.14% (FN)
No-Artifact	19/4200 = 0.45% (FP)	4181/4200 = 99.54% (TN)

Table 8 Evaluation of the best results of the proposed method

Accuracy	(TP + TN)/(TP + TN + FP + FN)	99.20%
Precision (P)	TP/(TP + FP)	98.86%
Recall (R)	TP/(TP + FN)	99.54%
F1-score	(2 × P × R)/(P + R)	99.20%

4.2 Comparison with Existing Approach

Based on the recent literature, we compare our best accuracy result with existed research works for binary classification artifacts in EEG signals. In [8], the authors propose an ANN architecture for eye movement artifact detection. An accuracy rate of 97.50% is achieved. According to [16], the authors use a fast artifact annotator based on various DL architectures as RNN, DCNN, and CNN for binary artifacts detection in EEG signals. The best accuracy rate is equal to 80% using the CNN architecture which contains one input layer, six convolutional layers, six batch-normalization layers, five max-pooling layers, one flatten layer, and two dense layers. As developed in [17], the authors propose an energy-efficient CNN architecture to detect artifacts in EEG signals. The proposed model contains five layers in which two convolutional layers, two max pool layers, and one softmax layer. The experimental results reach a 74% accuracy rate. The results of the different architectures compared to our method are presented in Table 9.

Table 9 Comparison with other detection artifacts methods

Approach	Accuracy test (%)	Used method
M. N. Tibdewal et al. [8]	82.44	ANN
D. Kim et al. [16]	93	CNN
M. Khatwani et al. [17]	98.61	CNN
Proposed method	99.20	CNN

5 Conclusion and Further Directions

An EEG method based on artifacts detection is proposed in this paper using a DL technique. Our data is collected at hospital from the *Nihon Kohden* machine and the annotation step is done by neurologists. The suggested method divides the signal into segments and applies a CNN algorithm for binary artifacts classification. The maximal classification accuracy achieves 99.20%. In the coming approach, we can use the second method based on removing intervals for binary classification of EEG signals.

References

1. Sadiya SS, Chantland E, Alhanai T, Liu T, Ghassemi MM (2021) Unsupervised EEG artifact detection and correction. Frontiers Digital Health 2:57
2. Mandrile F, Farina D, Pozzo M, Merletti R (2003) stimulation artifact in surface emg signal: effect of the stimulation waveform, detection system, and current amplitude using hybrid stimulation technique. IEEE Trans Neural Syst Rehab Eng 11:407–415
3. Moeyersons J, Smets E, Morales J, Villaa A, Raedt WD, Testelmans D, Buysec B, Hoof CV, Willems R, Huffel SV, Varona C (2019) Artefact detection and quality assessment of ambulatory ECG signals. Comput Methods Programs Biomed 182:105050
4. Tandle A, Jog N, D'cunha P, Chheta M (2016) Classification of artefacts in EEG signal recordings and EOG artefact removal using EOG subtraction. Commun Appl Electron 4:12–19
5. Abreu R, Leal A, Figueiredo P (February 2018) EEG-informed fmri: a review of data analysis methods. Frontiers Human Neurosci 12:29
6. Chaabene S, Bouaziz B, Boudaya A, Hökelmann A, Ammar A, Chaari L (2021) Convolutional neural network for drowsiness detection using eeg signals. Sensors 21:1734
7. Skupch AM, Dollfuß P, Fürbaß F, Hartmann M, Perko H, Pataraia E, Lindinger G, Kluge T (2013) "EEG artifact detection using spatial distribution of rhythmicity." APCBEE Procedia, the 3rd international conference on biomedical engineering and technology ICBET, vol. 7, pp 16–20
8. Tibdewal MN, Fate RR, Mahadevappa M, Ray AK, Malokar M (2017) Classification of artifactual EEG signal and detection of multiple eye movement artifact zones using novel Time-amplitude algorithm. SIViP 11:333–340
9. Jiang X, Bian G, Tian Z (2019) Removal of Artifacts from EEG Signals. Sensors 19:5
10. Rincón AQ, D'Giano C, Batatia H (2021) Artefacts detection in eeg signals. Adv Signal Process: Rev (Book Series) 2:413–441
11. Chaari L, Batatia H, Dobigeon N, Tourneret JY (May 2014) "A hierarchical sparsity-smoothness Bayesian model for $\ell 0+ \ell 1+ \ell 2$ regularization." IEEE International Conference on Acoustics, Speech and Signal Processing (ICASSP), pp 1901–1905

12. Nedelcu E, Portase R, Tolas R, Muresan R, Dinsoreanu M, Potolea R (Nov 2017) "Artifact detection in eegusing machine learning." 13th IEEE international conference on intelligent computer communication and processing (ICCP), pp 77–83
13. Zorgui S, Chaabene S, Bouaziz B, Batatia H, Chaari L (June 2020) "A convolutional neural network for lentigo diagnosis." International conference on smart living and public health (ICOST), pp 89–99
14. Boudaya A, Bouaziz B, Chaabene S, Chaari L, Ammar A, Hökelmann A (Jun 2020) "EEG-based hypo-vigilance detection using convolutional neural network." International conference on smart living and public health (ICOST),pp 69–78
15. Teplan M (2002) Fundamentals of eeg measurement. Psychology 2:1–11
16. Kim D, Keene S (Dec 2019) "Fast automatic artifact annotator for eeg signals using deep learning." IEEE Signal Processing in Medicine and Biology Symposium (SPMB), pp 1–5
17. Khatwani M, Hosseini M, Paneliya H, Hairston WD, Waytowich N, Mohsenin T (Oct 2018) "Energy efficient convolutional neural networks for eeg artifact detection." IEEE biomedical circuits and systems conference (BioCAS), pp 1–4
18. Lawherna V, Hairstonb WD, McDowell K, Westerfieldc M, Robbins K (2012) Detection and classification of subject-generated artifacts in EEG signals using autoregressive models. J Neurosci Methods 208:181–189
19. Zhang H, Wei C, Zhao M, Wu H, Liu Q (May 2021) "A novel convolutional neural network model to remove muscle artifacts from EEG." arXiv, pp 1265–1269

Real-World Protein Particle Network Reconstruction Based on Advanced Hybrid Features

Haji Gul, Feras Al-Obeidat, Fernando Moreira, Muhammad Tahir, and Adnan Amin

Abstract Biological network proteins are key operational particles that substantially and operationally cooperate to bring out cellular progressions. Protein links with some other biological network proteins to accomplish their purposes. Physical collaborations are commonly referred to by the relationships of domain-level. The interaction among proteins and biological network reconstruction can be predicted based on various methods such as social theory, similarity, and topological features. Operational particles of proteins collaboration can be indirect among proteins based on mutual fields, subsequently particles of proteins involved in an identical biological progression be likely to harbor similar fields. To reconstruct the real-world network of proteins particles, some methods need only the notations of proteins domain, and then, it can be utilized to multiple species. A novel method we have introduced will analyze and reconstruct the real-world network of protein particles. The proposed technique works based on protein closeness, algebraic connectivity, and mutual proteins. Our proposed method was practically tested over different data sets and reported the results. Experimental results clearly show that the proposed technique worked best as compared to other state-of-the-art algorithms.

H. Gul · A. Amin
Center for Excellence in Information Technology, Institute of Management Sciences, Peshawar, Pakistan
e-mail: ms171606772@imsciences.edu.pk

A. Amin
e-mail: adnan.amin@imsciences.edu.pk

F. Al-Obeidat
Zayed University, Abu Dhabi, UAE
e-mail: feras.al-obeidat@zu.ac.ae

F. Moreira (✉)
REMIT, IJP, Universidade Portucalense, Porto, Portugal
e-mail: fmoreira@uportu.pt

IEETA, Universidade de Aveiro, Aveiro, Portugal

M. Tahir
College of Computing and Informatics, Saudi Electronic University, Riyadh 11673, Saudi Arabia
e-mail: m.tahir@seu.edu.sa

A. Ullah et al. (eds.), *Proceedings of International Conference on Information Technology and Applications*, Lecture Notes in Networks and Systems 350,
https://doi.org/10.1007/978-981-16-7618-5_2

Keywords Reconstruction biological network · Protein–protein interaction · Real-world entity relationship prediction · Complex networks

1 Introduction

The field of protein particles is a valuable structure feature for the analysis of particle purposes and natural progressions since a field is an essential and well-designed part that facilitates numerous fundamental biological progressions and collaborations among proteins. It is comparatively less complex to reproduction of the real-world protein network purely based on repeated protein particle structures. Many databases of protein particles consist of numerous unique items expressing the structure of repeated proteins [1]. This huge freely available source of domain facts has been employed for analyzing the purposes of specific proteins, reconstruction, and also collaborations among proteins.

In this paper, we concentrate on the issue of real-world biological network reconstruction. The biological system can be represented by a graph where nodes and edges are expressed by proteins/genes and the interaction or relationships among proteins/genes. The edges of these networks may be directed, undirected, weighted, or un-weighted express numerous biological characteristics. Though, for several systems, the data access to, is not a straight sketch of the causal complex network. Additionally, the determination is to analyze the biological networks that have been topological structure reconstructed or topological inferred from protein particles. Specifically, consuming the data of time series from the vertices in a real-world network system is to infer expected associates among them [2, 3].

Choosing the best suitable procedure for this problem is a disputed issue in a real-world proteins network. Many real-world network reconstruction methods commonly have dissimilar assumptions of features, and in the real-world network system, their accuracy is also dissimilar over different systems. This is due to the structure and complexity of real-world networks. A usual procedure for this problem is to utilize several diverse network reconstruction methods and associate with each other the experimental results. On the other hand, the comparisons of the real-world system are as well not a common trick.

The proposed algorithm analyzed experimentally over seven publicly available datasets and compared them with five other state-of-the-art network reconstruction algorithms. For comparisons, we used area under the curve (AUC) evaluation matrix.

2 Overview of the Problem

Real-world network systems are made out of mixtures (nodes) and responses (links). At the point when numerous responses are realized that share reactants and items, they can be graphically connected. Organization remaking is a long and exhausting interaction that includes building an enormous organization in a bit-by-bit style by recognizing each response in turn. It is basic to the bottom-up perspective to the field of biological science. Due to some issues, if any part of the real-world network is damaged or crashed, the interaction among proteins can be reconstructed by the proposed algorithm. The prediction accuracy of the proposed algorithm is also compared with state-of-the-art algorithms which clearly shows that the proposed method outperforms as compared to others.

3 Related Work

In various articles, the researcher has combined the geography technique with social hypothesis extra local area data like client conduct and interest to improve the time complexity and accuracy of network completion by link mining [4, 5]. Numerous frameworks were proposed in the new years for the biomedical field network analysis and information retrieval. For the conduct of bio-inspiration and algorithmic, a novel technique was created by [6]. These frameworks follow various instruments, for example, neural organization, etc. The neural organization is one of the significant techniques that give answers for different issues in numerous fields. This organization contains various layers, the significance of neural organization clarified in [7, 8].We have enumerated below some state-of-the-art complex network reconstruction procedures.

3.1 Real-World Structure-Based Similarities Methods

There are heaps of frameworks that are present to calculate similarity among pair of nodes / protein even lacking protein characteristics. This framework utilized real-world proteins network information and identified as topological/structure-based procedures. These procedures are further considered into global and local similarities frameworks.

Leicht Lome Newman Index (LHN) [9]:

$$LHN(x, y) = \frac{|\boldsymbol{\Gamma}(\mathbf{x}) \cap \boldsymbol{\Gamma}(\mathbf{y})|}{Kx * Ky} \quad (1)$$

Parameter Dependent:

$$PD(x, y) = \frac{|\boldsymbol{\Gamma}(\mathbf{x}) \cap \boldsymbol{\Gamma}(\mathbf{y})|}{(|\boldsymbol{\Gamma}(\mathbf{x}) \cap \boldsymbol{\Gamma}(\mathbf{y})|)} \quad (2)$$

Adamic Adar (AA) [10]:

$$AA(x, y) = \sum_{w \int \{\Gamma(x) \cap \Gamma(y)\}} \frac{1}{\log|\Gamma(\omega)|'} \quad (3)$$

Common Neighbor (CN) [11]:

$$\mathrm{CN}(x, y) = |\Gamma(x) \cap \Gamma(y)| \quad (4)$$

Hub Promoted (HP) [12]:

$$S_{uv} = \frac{|\boldsymbol{\Gamma}(\boldsymbol{x}) \cap \boldsymbol{\Gamma}(\mathbf{y})|}{\mathbf{min}\{\boldsymbol{k}(\mathbf{x}), \boldsymbol{k}(\mathbf{y})\}} \quad (5)$$

4 Propose Work

We introduce a novel framework based on real-world network structure features. Multiple advanced features are combined in the proposed method to improve the accuracy of real-world biomedical proteins network reconstruction. Structure-based similarity of real-world complex networks for protein particle network reconstruction usually is determined by the topological organization of a real-world network. Main goal of such a procedure is to predict the interaction / association among proteins / nodes utilizing their resident association structure. The resident structure-based similarity methods work based on the information of proteins, their mutual relationships, and features, such as Adamic Adar structure-based proteins / nodes similarity technique allocates greater probability to mutual proteins particles with a lesser number of link or degree.

The proposed algorithm work based on biomedical complex network graph algebraic connectivity, proteins mutual relationship, and closeness centrality. Mathematically, representation of the proposed work can be defined as:

$$\text{Proposed Method}(x, y) = \left[\sum_{(i,j)\in E} C_{i,j}\left(x_i, y_j\right)^2\right] \cdot \left(\frac{n-1}{\sum_{v\in V}(x, y)}\right) . |\Gamma(x) \cap \Gamma(y)|$$

The given equation no.6 consist of different parts to compute the similarity. Complete detail of the proposed equation is given below.

Algebraic Connectivity: or Fiedler eigenvalue of a real-world complex network G considers different eigenvalues independently which is a second-littlest eigenvalue of the Laplacian grid of graph G. This eigenvalue is more noteworthy than 0 if the real-world network graph G is an associated graph. This is a culmination of the way that the occasion 0 shows up as an eigenvalue in the Laplacian is the number of associated parts in the real-world network graph. The size of this worth reflects how all-around associated the whole graph is. It has been utilized in synthesizing the healthiness and synchronizability of a real world.

The algebraic connectivity of a real-world network graph G may be negative/positive. Furthermore, the worth of the arithmetical availability is limited above by the edges network of the real-world graph, AC <= availability. On the off chance, the quantity of vertices of an undirected associated graph with positive vertex loads is n and D is the diameter, the AC is additionally known to be limited beneath by AC > 1. In contrast to traditional connectivity, the AC is reliant on the sum of edges also on the way in which edges are merged.

$$\text{Algebrai Connectivity}(x, y) = \left[\sum_{(i,j)\in E} C_{i,j}\left(x_i, y_j\right)^2 \right] \tag{7}$$

Closeness Centrality: is a proportion of the normal briefest separation from every protein to another protein. In particular, it is backward of the normal briefest distance between the protein and any remaining proteins in the real-world biomedical complex organization. The mathematical representation of closeness centrality is given below.

$$\text{Closeness Centrality}(x, y) = \left(\frac{n-1}{\sum_{v\in V}(x, y)} \right) \tag{8}$$

Also, we say that assuming the amount of the distances is enormous, the closeness is little and the other way around. A vertex with a high closeness centrality would mean it has cozy associations with numerous vertices.

Mutual Neighbor of Protein: Mutual protein neighbor is a similarity method that computes the similarity based on the number of proteins between the source and destination proteins. The higher number of mutual proteins between source and destination proteins also the higher similarity will be assigned to this source and destination proteins/vertices.

$$\text{Mutual Neighbor}(x, y) = |\Gamma(x) \cap \Gamma(y)| \tag{9}$$

5 Experimental Setup

To compare the performance of the proposed algorithm with other state-of-the-art similarity algorithms, we have partitioned the datasets into training E^T and testing / probe set E^P. In the experiment, 10% of data limited in the testing set while the leftover 90% were utilized for the training set. For 100 times, the investigation

Table 1 After the 100-time execution of each algorithm, the average result is reported here

Data sets	LHN	PD	AA	CN	HP	Proposed method
Cerebral Cortex	0.17246	0.28593	0.74465	0.86439	0.86439	**0.86812**
Rhesus Brain 2	0.057516	0.17427	0.74711	0.88308	0.88308	**0.8966**
Visual Cortex 1	0.39938	0.37875	**0.40127**	0.3961	0.39616	0.39331
Eco Everglades	0.24882	0.23347	0.69784	0.69852	0.69852	**0.69907**
Eco Stmarks	0.39915	0.36712	**0.66525**	0.66187	0.66187	0.66229
Eco Foodweb Baywet	0.40182	0.19782	0.61301	0.61441	0.61441	**0.62839**
Eco Foodweb Baydry	0.40316	0.19766	0.60972	0.61198	0.61198	**0.62534**

Area under the curve (AUC) is used for evaluation and comparisons

was repeated, and irregular independent sampling of the predicted construction was accomplished. The average performance of 100 times executions is described in table no.1.

Real-world biological networks can be considered and described by a computable network/graph. In a graph, the proteins of a biomedical network express by vertices, and the interaction among these proteins can be denoted by edges. We have used in our experiments seven popular and mostly utilized biomedical real-world protein particle network data sets. Most of the data sets that we have downloaded from [13] are given below.

- Cerebral Cortex
- Brain 2
- Visual Cortex 1
- Eco Everglades
- Eco Stmarks
- Eco Foodweb Baywet
- Eco Foodweb Baydry (Table 1).

5.1 *Evaluation Measurement*

Area under the curve (AUC) [14] denoted the performance and quality of a method utilized for real-world biomedical network reconstruction issue, which is presented up by the mathematical condition given as

$$\text{AUC} = \frac{n' + 0.5n''}{n} \tag{10}$$

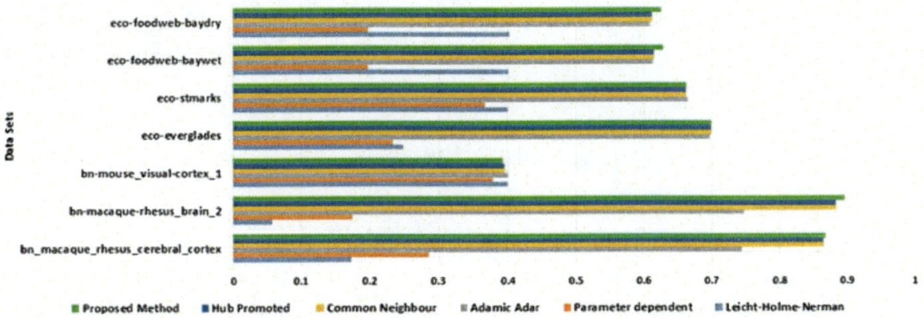

Fig. 1 Results graphical view

In the given mathematical condition, n describes the total number of execution, while n´ indicates the number of executions where influence part of the biomedical network or protein interaction achieved higher accuracy. Finally, n´ presents the number of occasions where the influence part and future shape of the protein network produce similar accuracy. The numerical value 0.5 was used when accuracy produced to be similar and independent distribution.

6 Results and Discussions

The prediction results of different real-world biomedical network reconstruction algorithms are given in Table 1. There are a total of six methods experimentally analyzed and utilized over seven real-world protein data sets. The similarity method Adamic Adar accomplished high accuracy over two data sets only while the proposed algorithm accuracy is best over five data sets as compared to other similarity algorithms.

Experimental results also described in the form of horizontal bar graph in Fig. 1.

7 Conclusion

The real-world biomedical network reconstruction is one of the most important problem in a complex network. The main aim of protein network reconstruction is to build the network and complete the affected part of the biomedical network. The network part or protein interaction can be damaged due to different issues. In this research paper, we have developed a novel algorithm for biomedical proteins network reconstruction. The proposed algorithm worked based on different advance features algebraic connectivity, closeness centrality, and mutual relationship among proteins in the biomedical network. We have compared the proposed algorithm with five other state-of-the-art similarity algorithms over seven different publically available data sets.

The experimental results clearly show that the proposed algorithm performance is best as compared to others. In the future, we can use some other advanced level of feature to improve the accuracy.

References

1. Finn RD, Attwood TK, Babbitt PC, Bateman A, Bork P, Bridge AJ, Chang H-Y, Oszt´anyi Z, El-Gebali S, Fraser M et al (2017) Interpro in 2017—beyond protein family and domain annotations. Nucleic Acids Res 45:D190–D199
2. Runge J (2018) Causal network reconstruction from time series: from theoreticalassumptions to practical estimation. Chaos: An Interdisciplinary J Nonlinear Sci 28:075310
3. Brugere I, Gallagher B, Berger-Wolf TY (2018) Network structure infer- ence, a survey: Motivations, methods, and applications. ACM Comput Surveys (CSUR) 51:1–39
4. Gul H, Amin A, Adnan A, Huang K (2021) A systematic analysis of link prediction in complex network. IEEE Access 9:20531–20541
5. Gul H, Amin A, Nasir F, Ahmad S, Wasim M (2021) Link prediction using double degree equation with mutual and popular nodes, pp 328–337
6. Molina D, Poyatos J, Del Ser J, Garc´ıa S, Hussain A, Herrera F, (2020) Comprehensive taxonomies of nature-and bio-inspired optimization: inspiration versus algorithmic behavior, critical analysis recommendations. Cognitive Comput 12:897–939
7. Scardapane S, Scarpiniti M, Baccarelli E, Uncini A (2020) Why should we add early exits to neural networks? Cogn Comput 12:954–966
8. Jiang X, Yan T, Zhu J, He B, Li W, Du H, Sun S (2020) Densely connected deep extreme learning machine algorithm. Cogn Comput 12:979–990
9. Leicht EA, Holme P, Newman ME (2006) Vertex similarity in networks. Phys Rev E 73:026120
10. Adamic LA, Adar E (2003) Friends and neighbors on the web. Social networks 25:211–230
11. Newman ME (2001) Clustering and preferential attachment in growing networks. Phys Rev E 64:025102
12. Ravasz E, Somera AL, Mongru DA, Oltvai ZN, Barab´asi A-L (2002) Hierarchical organization of modularity in metabolic networks. Science 297:1551–1555
13. Rossi RA, Ahmed NK (2015) The network data repository with interactive graph analytics and visualization, in: AAAI,. URL: http://networkrepository.com
14. Hanley JA, McNeil BJ (1982) The meaning and use of the area under a receiver operating characteristic (roc) curve. Radiology 143:29–36

Improving Coronavirus (COVID-19) Diagnosis Using Deep Transfer Learning

Arshia Rehman, Saeeda Naz, Ahmed Khan, Ahmad Zaib, and Imran Razzak

Abstract *Background*: Coronavirus disease (COVID-19) is an infectious dis- ease caused by a new virus. Exponential growth is not only threatening lives, but also impacting businesses and disrupting travel around the world. *Aim*: The aim of this work is to develop an efficient diagnosis of COVID-19 disease by differentiating it from viral pneumonia, bacterial pneumonia, and healthy cases using deep learning techniques. *Method*: In this work, we have used pre-trained knowledge to improve the diagnostic performance using transfer learning techniques and compared the performance of different CNN architectures. *Results*: Evaluation results using K-fold (10) showed that we have achieved state-of-the-art performance with overall accuracy of **98.75%** on the perspective of CT and X-ray cases as a whole. *Conclusion*: Quantitative evaluation showed high accuracy for automatic diagnosis of COVID-19. Pre-trained deep learning models developed in this study could be used for early screening of coronavirus; however, it calls for extensive need to CT or X-rays dataset to develop a reliable application.

A. Rehman · A. Khan
COMSATS University, Abbotabad, Pakistan
e-mail: Ahmed.Khan@ciit.edu

S. Naz (✉)
Govt. Girls Postgraduate College, No.1, Abbotabad, Pakistan

A. Zaib
Women Medical College, DHQHospital, Abbotabad, Pakistan
e-mail: ahmed.zaib@pmc.med

I. Razzak (✉)
Deakin University, Geelong, Australia
e-mail: imran.razzak@ieee.org

A. Ullah et al. (eds.), *Proceedings of International Conference on Information Technology and Applications*, Lecture Notes in Networks and Systems 350,
https://doi.org/10.1007/978-981-16-7618-5_3

1 Introduction

The revolution of atypical and individual-to-individual transmissible pneumonia brought about by the serious intense respiratory disorder, coronavirus (SARS-COV-2, also known as COVID-19 and 2019-nCov), has caused a global alarm [1]. In December 2019, Wuhan, Hubei Province, China, turned into the focal point of an outbreak of unknown cause of pneumonia, which got attention not only in China but also internationally [2, 3]. Later, Chinese Center for Disease Control and Prevention (China CDC) determined a non-SARS novel coronavirus in a patient of Wuhan at January 7, 2020. As of 3 April, more than 1 million people have been infected in 180 countries, according to the Johns Hopkins University Center for Systems Science and Engineering. There have been over 53,000 deaths globally. Just over 3000 of those deaths have occurred in mainland China. More than 211,000 people are recorded as having recovered from the coronavirus. [4, 5].

Coronaviruses (CoV) are a large family of viruses that cause sickness ranging from the common cold to more severe diseases such as Middle East respiratory syndrome (MERS-CoV) and severe acute respiratory syndrome (SARS-CoV). Coronaviruses caused diseases in mammals and birds, a zoonotic virus that is transmitted between animals and people. Detailed investigations found that SARS-CoV was transmitted from civet cats to humans and MERS-CoV from dromedary camels to humans. When a virus spread circulating animal populations affects human population, this is termed as spillover event. It is speculated that the 2019 novel coronavirus was originated in bats and was transmitted in humans, possibly with pangolin as an intermediary host. Coronavirus disease (COVID-19) is a new strain that was discovered in 2019 (not previously recognized in humans). Unfortunately, COVID-19 is now spreading from humans to humans.

COVID-19 diagnosis relies on clinical symptoms, positive CT images or positive pathogenic testing, and epidemiological history. Common clinical characteristics of COVID-19 include respiratory symptoms, fever, cough, shortness of breath, and breathing difficulties. In more severe cases, infection can cause pneumonia, severe acute respiratory syndrome, kidney failure, and even death [6–9]. However, these symptoms may be related to isolated cases; e.g., infected CT scan of chest indicated pneumonia, while pathogenic test came back positive for the virus. If a person is identified under investigation, samples of lower respiratory tract will be collected for pathogenic test like sputum, tracheal aspirate, and bronchoalveolar lavage. The laboratory technology used sequencing of nucleic acid and real-time reverse-transcription polymerase chain reaction (RT-PCR) for the detection of COVID-19 [10, 11]. The detection rates of nucleic acid is low estimates (between 30 and 50%), due to the factors like availability, quality, quantity, reproducibility, and stability of testing and detection kits [10–12]. Therefore, tests need to be repeated many times after confirmation in many cases.

Two well-known radiological imaging modalities are used for the detection of COVID-19, i.e., X-Rays and computed tomography (CT) scans. These two modalities are frequently used by the clinicians to diagnose pneumonia, respiratory tract infection, lung inflammation, enlarged lymph nodes, and abscesses. Since COVID-19 affects the epithelial cells of respiratory tract, X-rays are used to analyze the health of a patient's lungs. However, most of COVID-19 cases have similar features on CT images like ground-glass opacity in the early stage and pulmonary consolidation in the late stage. Therefore, CT scans play an important role in the diagnosis of COVID-19 as an advanced imaging modality.

Machine learning researchers and computer scientists play a vital role in the era when COVID-19 spreads all over the world. One of the breakthroughs of AI is deep learning that extracts the detailed features from the images [13, 14]. This will help the doctors and healthcare practitioners in real-time assistance, less costly, time effectively, and accurately. In this work, we have used pre-trained knowledge to improve the diagnostic performance using transfer learning techniques and compared the performance different transfer learning methods. Evaluation results using K-fold (10) showed that we have achieved state-of-the-art performance. The contributions of researchers using machine leaning and deep learning-based techniques for the detection of COVID-19 are presented in Sect. 2. The datasets information deployed in our study is detailed in Sect. 2. Then, we present our proposed COVID-19 detection system based on transfer learning technique in Sect. 3. The evaluated results are discussed in Sect. 4. Finally, we draw conclusion of our study in Sect. 5.

2 Dataset

One of the prerequisites in image recognition and computer vision research is of finding appropriate dataset. Massive amount of training dataset is one of the essential requirements of deep learning; however, large medical imaging data is an obstacle in the success of deep learning. COVID-19 lung image datasets are currently limited as it is recent emerging disease. Dr. Joseph Cohen established the public GitHub repository where X-Ray and CT images of COVID-19 are collected as confirmed cases are reported. The dataset contains images of ARDS (acute respiratory distress syndrome), MERS (Middle East respiratory syndrome), pneumocystis, SARS (severe acute respiratory syndrome), and streptococcus. We have collected 200 COVID-19 images from Covid2019 chestXray dataset of GitHub repository [15] accessed on April 01, 2020. The metadata.csv provided with the dataset is parsed to select the positive samples of COVID-19 images. Furthermore, we collect 200 images of bacterial and viral pneumonia from Kaggle repository entitled by Chest X-Ray Images (Pneumonia) [16]. In this way, dataset of 200 X-Ray and CT images of COVID-19, 200 images of healthy subjects, 200 images of patients affected from bacterial pneumonia, and 200 images of patients affected from viral pneumonia, respectively.

3 Proposed COVID-19 Detection System

Recently, several deep learning methods have been proposed for the diagnosis of COVID-19. In order to improve the diagnostic performance, we have used transfer learning for diagnosis of coronavirus through the transfer of already learnt knowledge. For this purpose, we used ImageNet as a source domain denoted by Ds where the source task Ts is to classify 1000 classes of natural images. The target domain denoted by Dt is the Covid2019 Chest X-Ray dataset where the target task Tt is to classify two classes, i.e., Covid2019 and Healthy in binary classification and four classes, i.e., Covid2019, healthy, bacterial pneumonia, and viral pneumonia in multi-classification. The source domain consists of two components:

$$Ds = \{Xi, P(Xi)\} \tag{1}$$

where Xi is the feature space of ImageNet and $P\ (Xi)$ is the marginal distribution such that $Xi = \{\times 1, \ldots, xn\}$, $xi \in Xi$. The target domain also consists of two components:

$$Dt = \{Xc, P(Xc)\} \tag{2}$$

where Xc is the feature space of Covid2019 Chest X-Ray dataset and $P\ (Xi)$ is the marginal distribution such that $Xc = \{\times 1, \ldots, xn\}$, $xj \in Xc$. On the other hand, source task Ts is the tuple of two elements defined as:

$$Ts = \{\gamma i, P(Xi|X)\} = \{\gamma, \eta\} \tag{3}$$

where γ is the label space of source data and η is the objective function that learned from the pair of feature vector and label space, i.e., (xi, yi) $xi \in Xi$, $yi \in \gamma i$.

The target task Tt is the tuple of two elements defined as:

$$Ts = \{\gamma j, P(Xj|X)\} = \{\gamma, \eta\} \tag{4}$$

where γ is the label space of source data and η is the objective function that learned from the pair of feature vector and label space i.e.(x_j, y_i) $x_j \in X_j$, $y_j \in \gamma_j$.

The objective of transfer learning is to train a base network on large source dataset and then transfer learned features to the small target dataset. For the given source domain Ds and target domain Dt with their corresponding source task Ts and target task Tt, the aim of transfer learning is to learn conditional probability distribution of target domain, i.e., $(PYT\ |XT)$ with respect to the information gain of source domain Ds and source task Ts.

The proposed system employs different powerful CNN architectures (AlexNet, SqueezeNet, GoogLeNet, VGG, MobileNet, ResNet18, ResNet50, ResNet101, and DenseNet) coupled with transfer learning techniques. The system consists of three

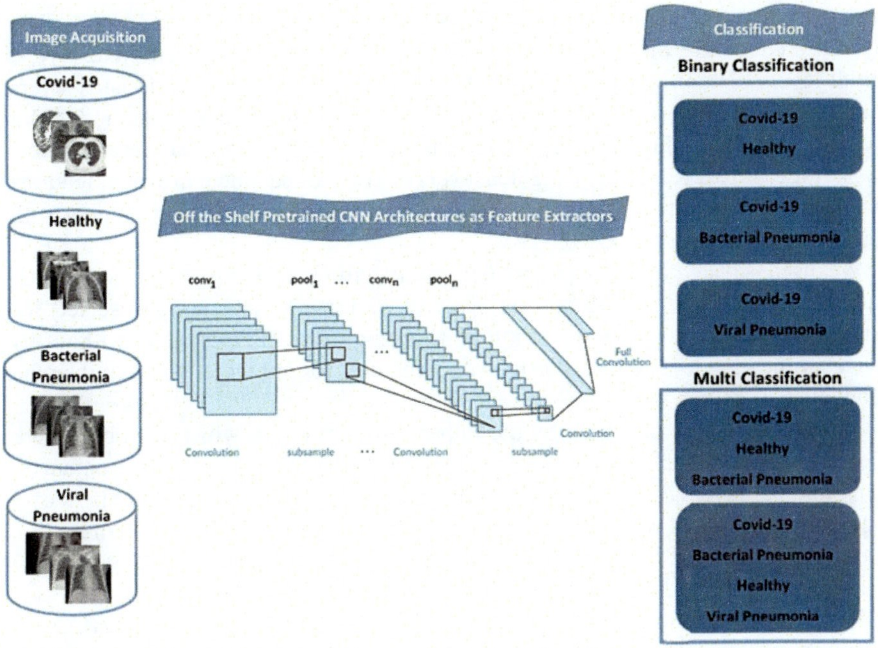

Fig. 1 Proposed system for detection and classification of COVID-19

main steps: image acquisition, off-the-shelf pre-trained models as feature extractors, and classification as illustrated in Fig. 1. In the first step, images are acquired from two well-known datasets: Covid2019 Chest X-Ray downloaded from GitHub repository and Chest X- Ray Images (Pneumonia) from Kaggle repository. The next step is to extract the automated features using off-the-shelf pre-trained architectures of CNN. In the last step, two types of classification are performed: binary classification and classification. In binary classification, three scenarios are investigated: COVID-19 and healthy classification, COVID-19 and bacterial pneumonia classification, and COVID-19 and viral pneumonia classification, respectively. In multi-class problem, two scenarios are explored using three classes, i.e., COVID-19, healthy, and bacterial pneumonia. Furthermore, four classes are also used in the experimentation of multi-class classification, i.e., COVID-19, healthy, bacterial pneumonia, and viral pneumonia. The detail of steps is described in the coming sections.

3.1 Off-The-Shelf Pre-Trained Models as Feature Extractors

Deep learning models have got significant interest in the last few decades due to its capability of extracting automated features [17, 18]. The automated features extracted by CNN models have the capability of parameter sharing and local connectivity;

thus, these features are more robust and discriminative than traditional features. The layered architecture of CNN is used to extract the features from different layers. Initial layers contain generic features (lines, edges, blobs etc.,), while later layers contain specific features. The purpose of fine-tuning is to extract the features from pre-trained CNN architectures and then train a shallow network on these features. The key idea is to leverage the weighted layers of pre-trained model to extract features of ImageNet but not to update the weights of the models layers during training with data for the detection of COVID-19.

We used seven powerful pre-trained architectures of CNN (AlexNet, VGG, SqueezeNet, GoogLeNet, MobileNet, ResNet with its variants, and DenseNet) for extracting the robust automatic features. Utilizing these CNN models without its own classification layer enables us to extract features for our target task based on the knowledge of source task.

AlexNet [19] was proposed by the Supervision group, in which members were Alex Krizhevsky, Geoffrey Hinton, and Ilya Sutskever. The architecture contains 5 convolutional layers (Conv1 with 96 filters of 11 × 11, Conv2 with 256 filters of 5 × 5, Conv3 and Cov4 with 384 filters of 3 × 3, and Conv5 with 256 filters of 3 × 3), three fully connected layers (FC6 and FC7 with 4096 neurons and FC8 with 1000 classes of ImageNet dataset).

VGG [20] has a uniform architecture designed by Simonyan and Zisserman. It comprises 16 convolutional layers with 3 × 3 convolutions (but with lot of filters) and three fully connected layers like AlexNet.

One of the micro-architectures of CNN was released in 2016 named as **SqueezeNet** [21]. It contains fire modules with squeezed convolutional layer (with 1 × 1 filters) fallow with an expanded layer (with a combination of 1 × 1 and 3 × 3 filters).

GoogLeNet [22] was proposed by Szegedy et al. in 2014. The concept of inception module was introduced in GoogLeNet in which different convolutional and pooling layers are concatenated for improving learning. GoogLeNet contains nine inception modules, 4 max-pooling layers, 2 convolutional layers, 1 convolutional layer for dimension reduction, 1 average pooling, 2 normalization layers, 1 fully connected layer, and finally a linear layer with softmax activation in the output.

ResNet [23, 24] is short name of residual network in which the idea of skip connection was introduced. Skip connection means stacking the convolution layers together one after the other to mitigate the problem of vanishing gradient. Three variants of ResNet used in this study are: ResNet18, ResNet50, and ResNet101. **ResNet18** contains 5 convolution blocks, each containing 2 residual blocks. Each residual block contains 2 convolution layers with the same number of 3 × 3 filters. The **ResNet50** contains 5 residual blocks each with a convolution and identity block. The convolution blocks and identity blocks have 3 convolution layers. Similarly, **ResNet101** contains 3 residual blocks with 3 convolution and identity blocks.

The idea of skip connections has been extended to connect the later blocks of densely connected layers in **DenseNet** [25] model. The dense blocks contain 1 × 1 convolutional filters and max-pooling layers in order to reduce the number of tunable parameters. Contrary to the skip connections in the ResNets, the output of the dense

Table 1 Summary of pre-trained CNN architectures employed in proposed system gray Pre-trained Model Input Size Layers Size (MB) Parameters (M)

AlexNet	227 × 227	8	240	60
lightgray SqueezeNet	227 × 227	18	4.8	5
GoogLeNet	224 × 224	22	96	6.8
lightgray VGG16	224 × 224	16	528	138
MobileNetv2	224 × 224	53	16	3.5
lightgray ResNet18	224 × 224	18	45	11.174
ResNet50	224 × 224	50	98	23.52
lightgray ResNet101	224 × 224	101	170	42.512
DenseNet201	224 × 224	201	80	20.2

block is not added but instead concatenated. **MobileNetV2** [26] have a new CNN layer, the inverted residual and linear bottleneck layer. The new CNN layer builds on the depth-wise separable convolutions. The detail of pre-trained architectures employed in this study is presented in Table 1.

3.2 Classification and Detection

After extracting the automated features, we again employ pre-trained CNN architectures for classification and detection of COVID-19. The pre-trained architectures used in our study for classification are: AlexNet, VGG, SqueezeNet, GoogLeNet, MobileNet, ResNet with its variants, and DenseNet. Each of these networks has three fully connected (FC) layers, where the last fully connected is used for the classification purpose.

We initialize the number of neurons of last fully connected layer according to the target dataset.

We train the network on Caffee library 7 with GPU (NVIDIA CUDA) having the multiple processors of 2.80 GHz, 16 GB DDR4-SDRAM, 1 TB HDD, 128 GB SSD. The parameters of fine-tuning method are not set by the network itself, and it is essential to set and optimize these parameters according to the results of training the images in improving the performance. In our case, each network is trained with Adam optimizer in maximum 30 epochs. The value of batch size is set to 11 with the initial learn rate of $3e^{-4}$. The number of best epochs is varied according to validation criteria with validation frequency.

Two types of classification have been performed in our study: binary classification and multi-class classification. In binary classification, the dataset is split into two categories: COVID-19 and healthy, COVID-19 and bacterial pneumonia, and COVID-19 and viral pneumonia. The images of healthy lungs are acquired from Kaggle repository, while images of lungs affected from COVID-19 are taken from GitHub repository. We will not only classify the healthy lungs from COVID-19 but also classify different causes of pneumonia either caused by some bacteria, virus, or COVID-19. Thus, in multi-class classification, dataset is split into three categories: healthy, COVID-19, pneumonia bacterial and four categories: healthy, COVID-19, pneumonia bacterial, and Pneumonia viral. Like binary classification, images of COVID-19 are taken from GitHub repository, while images of healthy, pneumonia bacterial, and pneumonia viral are acquaint from Kaggle repository. The detail of both scenarios along with results is presented in coming section.

4 Result Analysis and Discussions

In this section, we analyze the effectiveness of our proposed framework in light of results of experiments conducted. As discussed earlier, the experimental study is con- ducted using two publicly available repositories. The statistics of datasets are presented in Table 2. We deploy CNN architectures coupled with fine-tuning strategy by splitting 80% dataset in training set, 20% of training data in validation set, and 20% in testing.

The effectiveness of proposed system is evaluated by computing four major outcomes of evaluation measures: true positives (tp), false positives (fp), true negatives (tn), and false negatives (fn). We use a well-known evaluation measure, i.e., accuracy. **Accuracy** is used to determine the classes of proposed system correctly. To evaluate the accuracy of a test set, we compute the proportion of true positive and true negative in all evaluated cases. We first compute the accuracy of the proposed system for all classes either binary class or multi-class. Then, we compute mean classification accuracy using K-fold measure. Mean classification accuracy is calculated by taking average of the accuracy achieved by each of the tenfold.

We have conducted two studies of classification in our proposed system: *Binary classification* and *Multi-classification.* In binary classification, three scenarios are used: COVID-19 and healthy, COVID-19 and bacterial pneumonia, COVID-19 and viral pneumonia. Series of experiments are conducted in each scenario using nine pre-trained CNN networks, i.e., AlexNet, SqueezeNet, GoogLeNet, VGG16,

Table 2 Dataset split statistics

	Binaryclass dataset			Multiclass dataset		
Class	Train set	Validation set	Test set	Train set	Validation set	Test set
COVID-2019	160	40	40	160	40	40
Healthy	160	40	40	160	40	40
Bacterial pneumonia	–	–	–	160	40	40
Viral pneumonia	–	–	–	160	40	40

MobileNet, ResNet18, ResNet50, ResNet101, and DenseNet. Our initial claim is to explore the fine-tuning technique of transfer learning by extracting the features of pre-trained CNN networks. We have achieved highest accuracy of 98.75% using VGG16, ResNet18, ResNet50, ResNet101, and DenseNet in Scenario A: COVID-19 and healthy. Similarly, we attained 98.75% using all pre-trained models in Scenario B: COVID-19 and bacterial pneumonia. The accuracy of 98.75% is achieved using SqueezeNet, GoogLeNet, MobileNet, ResNet50, and ResNet101 in Scenario C: COVID-19 and viral pneumonia. For a deeper insight on Table 3, it is observed that the performance of the automated features extracted from each pre-trained networks used in this study is comparable with one another. This supports our initial claim that fine-tuning the pre-trained CNN networks can be successfully deployed to a limited class dataset even without augmentation.

The study of multi-class classification is further divided into two scenarios. In Scenario D, dataset is split into three classes: COVID-19, healthy, and bacterial pneumonia. We achieve 97.20% in this scenario using MobileNet. In Scenario E, dataset is

Table 3 Experimental results of binary classification and multi-classification

Gray scenario	Model	Train set (%)	Validation set (%)	Test set (%)	Covid19	Healthy (%)	Bacterial pneumonia (%)	Viral Pneumonia (%)
9*A	AlexNet	98.44	97.20	97.04	92.59	100	–	–
	SqueezeNet	100	98.80	98.89	100	97.5	–	–
	GoogLeNet	100	100	98.15	96.29	100	–	–
	VGG16	100	100	98.75	100	97.5	–	–
	MobileNet	98.80	96.30	96.30	100	97.5	–	–
	ResNet18	100	100	98.75	100	97.5	–	–
	ResNet50	100	100	98.75	100	97.5	–	–
	ResNet101	100	100	**98.75**	100	97.5	–	–
	DenseNet	100	100	98.75	100	97.5	–	–
9*B	AlexNet	100	100	98.75	100	–	97.5	–
	SqueezeNet	100	100	98.75	100	–	97.5	–
	GoogLeNet	100	100	98.75	100	–	97.5	–
	VGG16100	100	100	98.75	100	–	97.5	–
	MobileNet	100	100	98.75	100	–	97.5	–
	ResNet18	100	100	98.75	100	–	97.5	–
	ResNet50	100	100	98.75	100	–	97.5	–
	ResNet101	100	100	**98.75**	100	–	97.5	–
	DenseNet	100	100	98.75	100	–	97.5	–
9*C	AlexNet	98.14	97.01	92.54	81.48	–	–	100

(continued)

Table 3 (continued)

Gray scenario	Model	Train set (%)	Validation set (%)	Test set (%)	Covid19	Healthy (%)	Bacterial pneumonia (%)	Viral Pneumonia (%)
	SqueezeNet	100	100	98.51	100	–	–	97.5
	GoogLeNet	100	100	98.51	100	–	–	97.5
	VGG16	100	89.55	88.06	96.55	–	–	82.5
	MobileNet	100	100	98.51	97.5			100
	ResNet18	100	100	95.52	92.59			97.5
	ResNet50	100	100	98.75	100			97.5
	ResNet101	100	100	**98.75**	100			97.5
	DenseNet	100	100	97.01	96.29			97.5
9*D	AlexNet	100	100	87.85	96.29	77.5	92.5	–
	SqueezeNet	100	100	94.39	100	97.5	87.5	–
	GoogLeNet	100	100	91.59	96.29	85	95	–
	VGG16	98.37	98.37	92.52	100	95	85	–
	MobileNet	100	100	**97.20**	100	100	92.5	–
	ResNet18	99.77	99.77	95.33	100	92.5	80	–
	ResNet50	100	100	92.52	100	90	90	–
	ResNet101	100	100	89.72	100	92.5	80	–
	DenseNet	100	99.77	93.46	100	87.5	95	–
9*E	AlexNet	99.81	99.32	63.27	92.59	55	55	60
	SqueezeNet	96.26	94.56	70.07	92.59	87.5	45	57.5
	GoogLeNet	100	100	75.51	100	85	75	50
	VGG	97.96	97.11	63.95	92.59	62.5	65	45
	MobileNet	99.83	100	**80.95**	96.29	87.5	75	70
	ResNet18	100	100	74.15	100	70	80	55
	ResNet50	99.32	99.32	72.79	100	70	77.5	52.5
	ResNet101	98.98	97.96	70.75	92.59	77.5	60	60
	DenseNet	96.26	94.56	70.07	92.59	80	70	45

split into four categories: COVID-19, healthy, bacterial pneumonia, and viral pneumonia. We achieve the highest accuracy of 80.95% using MobileNet architecture of CNN on test set. It can be observed from Table 3 that accuracy on multi-class problem ranging from 63.95 to 80.95. To get deeper insight, class-wise accuracy is reported in Table 3. It can be noted from Table 3 that some of bacterial and viral pneumonia lead to misclassification. The proposed system is classifying most of the viral pneumonia scans as healthy person.

For more illustration, confusion matrix of each scenario with model of highest accuracy is presented in Fig. 1. Scenario A accurately classifies Covid19 with 100% classification rate and misclassifies one instance of healthy subject with overall test accuracy of 98.75%. Scenario B and C classify Covid19 from bacterial pneumonia and viral pneumonia with 98.75% classification rate using ResNet101. In multi-classification scenarios, it is noted that viral pneumonia and bacterial pneumonia are classified as healthy subject. It is due to the reason that COVID-19 is caused by SARS-COV-2 virus and pneumonia is also cause by virus or bacteria; thus, we need a strong overlap in the feature space between these two classes. The main reason is the immunity system of human being has got adaptation toward pre-existing viruses. So, the human being when get infected by old viral pneumonia, they did not have so many in the chest X-Rays or CT scans. Number of viral pneumonia is misclassified as a bacterial pneumonia. Another reason is the pediatric X-Ray images and virus attacked the child for the first time after the birth. Sometimes, the proposed system misclassified some images of bacterial pneumonia as COVID-19 and vice versa. In case of COVID-19, when a person gets pneumonia, then it shows a lot of changes in the X-Rays or CT like bacteria pneumonia or worst because human being has not yet developed the immunity against this novel corona virus. The COVID-19 images are classified accurately by number of architectures of CNN. The accuracy will be increased with increased data, more deep mesoscopic model, and long fine-tuning transfer learning process.

In addition with accuracy, we further evaluate our proposed system with four evaluation measures: sensitivity, specificity, precision, and F-score. Sensitivity or recall measures the ability of system to correctly classify the classes and is calculated from the proportion of true positives.

Overall accuracy, sensitivity, specificity, precision, and F-score of proposed system achieved by employment of each CNN architecture using fine-tuned features with 20% test set in both scenarios are presented in Table 4.

For the effectiveness of any proposed system, comparative analysis with the state of the art gives notable insight. We have compared the evaluation of our proposed system with existing techniques discussed in Table 4. A meaningful comparison is possible with Wang et al. [1] and Xu et al. [27] presented in Table 5. Wang et al. [1] deployed transfer learning techniques using Inception model. They conducted experiment on 453 CT scans of pathogen confirmed COVID-19 cases and attained 89.5% accuracy. Another deep learning-based system was proposed by Xu et al. [27]. They differentiate COVID-19 pneumonia from influenza-A viral pneumonia. They used ResNet with location-attention classification on 618 CT samples and attained 86.7% accuracy. Our proposed system significantly differentiate the COVID-19 from bacterial pneumonia with 98.75% and COVID-19 from viral pneumonia with 98.51% accuracy, respectively. We have deployed nine pre-trained CNN networks to investigate the transfer learning techniques and conclude that fine-tuning the pre-trained CNN networks can be successfully deployed to a limited class dataset even without augmentation.

5 Conclusion

CT imaging is an efficient tool to diagnose COVID-19 and assess its. In this work, we used pre-trained knowledge to improve the diagnostic performance of COVID-19. In conclusion, our proposed study revealed the feasibility of fine-tuning transfer learning technique of deep learning to assist doctors to detect the COVID-19. Our

Table 4 Overall accuracy of COVID-19 detection system achieved by employment of each CNN architecture using fine-tuned features (20% test set)

Gray Exp	Model	Accuracy	Specificity	Sensitivity	Precision	F-Score
1	AlexaNet	96.30	0.9259	100	9310	96.42
Lightgray 2	SqueezeNet	98.15	100	96.43	100	96.30
3	GoogLeNet	98.15	100	96.43	100	96.30
lightgray 4	VGG	98.75	97.50	100	96.43	98.18
5	MobileNet	96.30	100	92.59	100	96.15
lightgray 6	ResNet18	98.75	97.50	100	96.43	98.18
7	ResNet50	98.75	97.50	100	96.43	98.18
lightgray 8	ResNet101	98.75	97.50	100	96.43	98.18
8	DenseNet	98.75	97.50	100	96.43	98.18
B. Scenario COVID-19 Bacterial Pneumonia						
Gray Exp	Model	Accuracy	Specificity	Sensitivity	Precision	F-Score
1	AlexaNet	98.75	97.50	100	96.43	98.18
lightgray 2	SqueezeNet	98.75	97.50	100	96.43	98.18
3	GoogLeNet	98.75	97.50	100	96.43	98.18
lightgray 4	VGG	98.75	97.50	100	96.43	98.18
5	MobileNet	98.75	97.50	100	96.43	98.8
6	ResNet18	98.75	97.50	100	96.43	98.18
lightgray 7	ResNet50	98.75	97.50	100	96.43	98.18
8	ResNet101	98.75	97.50	100	96.43	98.18
lightgray 9	DenseNet	98.75	97.50	100	96.43	98.18
C. Scenario COVID-19 Viral Pneumonia						
Gray Exp	Model	Accuracy	Specificity	Sensitivity	Precision	F-Score
1	AlexaNet	92.54	100	81.48	100	89.80
Lightgray 2	SqueezeNet	98.51	97.50	100	96.43	98.18
3	GoogLeNet	98.51	97.50	100	96.43	98.18
lightgray 4	VGG	88.06	82.50	96.30	78.79	86.67
5	MobileNet	98.51	97.50	100	96.43	98.18
6	ResNet18	95.52	97.50	92.59	96.15	94.34
lightgray 7	ResNet50	98.75	97.50	100	96.43	98.18

(continued)

Table 4 (continued)

Gray Exp	Model	Accuracy	Specificity	Sensitivity	Precision	F-Score
8	ResNet101	98.75	97.50	100	96.43	98.18
lightgray 9	DenseNet	97.01	97.50	96.30	96.30	96.30
D. Scenario Three Classes						
Gray Exp	Model	Accuracy	Specificity	Sensitivity	Precision	F-Score
1	AlexaNet	87.85	93.53	88.77	89.67	88.94
Lightgray 2	SqueezeNet	94.39	97.01	95.00	95.29	94.99
3	GoogLeNet	91.59	95.52	92.10	92.96	92.33
Lightgray 4	VGG	92.52	96.02	93.33	93.60	93.32
5	MobileNet	97.20	98.51	97.50	97.67	97.50
6	ResNet18	95.33	97.51	95.83	95.85	98.83
lightgray 7	ResNet50	92.52	96.02	93.33	93.60	93.32
8	ResNet101	89.72	94.53	90.83	91.22	90.80
lightgray 9	DenseNet	93.46	96.52	94.17	94.32	94.16
E. Scenario Four Classes						
Gray Exp	Model	Accuracy	Specificity	Sensitivity	Precision	F-Score
1	AlexaNet	74.15	91.25	75.95	76.67	74.33
lightgray 2	SqueezeNet	70.07	89.30	70.65	71.61	70.10
3	GoogLeNet	75.51	91.61	77.50	76.52	76.81
lightgray 4	VGG	63.95	87.72	66.27	66.25	66.11
5	MobileNet	80.95	93.48	82.20	82.52	82.23
6	ResNet18	74.15	91.25	75.95	76.67	74.33
lightgray 7	ResNet50	72.79	90.10	74.90	73.23	72.43
8	ResNet101	70.75	90.00	72.52	73.70	72.64
lightgray 9	DenseNet	70.07	89.77	71.90	71.23	71.43

Table 5 Performance comparison of performance with recent works using machine learning techniques for CT and X-ray images

Gray methods	Features	Model	Data samples performance	
Wang et al. [1]	Automated	Inception model	453	89.5%
Lightgray Xu et al. [27] Automated	Automated	ResNet	618	86.7%
Proposed system	Automated	ResNet101	728	98.75

system differentiate the COVID-19 from viral pneumonia and bacterial pneumonia. Our system holds great potential to improve the efficiency of diagnosis, isolation, and treatment of COVID-19 patients, relieve the pressure of radiologists, and give control of epidemic. The proposed system 100% classifies COVID-19 from healthy, COVID-19 from bacterial pneumonia, and COVID-19 from viral pneumonia using

ResNet in binary classification. In multi-class classification, 97.20% is achieved on three classes and 80.95% on four classes using MobileNet, respectively.

References

1. Wang S, Kang B, Ma J, Zeng X, Xiao M, Guo J, Cai M, Yang J, Li Y, Meng X et al (2020) A deep learning algorithm using ct images to screen for corona virus disease (covid-19). medRxiv
2. Wang C, Horby PW, Hayden FG, Gao GF (2020) A novel coronavirus outbreak of global health concern. The Lancet 395(10223):470–473
3. Naseem U, Razzak I, Khushi M, Eklund PW, Kim J (2021) Covid-senti: a large-scale benchmark twitter data set for covid-19 sentiment analysis. IEEE Trans Comput Social Syst
4. https://www.theguardian.com/world/2020/mar/17/coronavirus-symptoms-should-i-see-doctor-covid-19. Accessed 03 Apr 2020
5. Qayyum A, Razzak I, Tanveer M, Kumar A (2021) Depth-wise dense neural network for automatic covid19 infection detection and diagnosis. Ann Oper Res 1–21
6. Wang D, Hu B, Hu C, Zhu F, Liu X, Zhang J, Wang B, Xiang H, Cheng Z, Xiong Y et al (2020) Clinical characteristics of 138 hospitalized patients with 2019 novel coronavirus–infected pneumonia in wuhan, china. JAMA
7. Chen N, Zhou M, Dong X, Jieming Q, Gong F, Han Y, Qiu Y, Wang J, Liu Y, Wei Y et al (2020) Epidemiological and clinical characteristics of 99 cases of 2019 novel coronavirus pneumonia in wuhan, china: a descriptive study. The Lancet 395(10223):507–513
8. Li Q, Guan X, Wu P, Wang X, Zhou L, Tong Y, Ren R, Leung KSM, Lau EHY, Wong JY et al (2020) Early transmission dynamics in wuhan, china, of novel coronavirus–infected pneumonia. New England J Med
9. Huang C, Wang Y, Li X, Ren L, Zhao J, Yi H, Zhang L, Fan G, Jiuyang X, Xiaoying G et al (2020) Clinical features of patients infected with 2019 novel coronavirus in Wuhan, China. The Lancet 395(10223):497–506
10. Corman VM, Landt O, Kaiser M, Molenkamp R, Meijer A, Chu DKW, Bleicker T, Bru¨ nink S, Schneider J, Schmidt ML et al (2020) Detection of 2019 novel coronavirus (2019-ncov) by real-time rt-pcr. Eurosurveillance 25(3)
11. Chu DKW, Pan Y, Cheng S, Hui KPY, Krishnan P, Liu Y, Ng DYM, Wan CKC, Yang P, Wang Q, et al (2020) Molecular diagnosis of a novel coronavirus (2019-ncov) causing an outbreak of pneumonia. Clin Chem
12. Zhang N, Wang L, Deng X, Liang R, Meng S, He C, Lanfang H, Yudan S, Ren J, Fei Y et al (2020) Recent advances in the detection of respiratory virus infection in humans. J Med Virol 92(4):408–417
13. Razzak MI, Naz S, Zaib A (2018) Deep learning for medical image processing: Overview, challenges and the future. In Classification in BioApps, Springer, pp 323–350
14. Rehman A, Naz S, Razzak I (2021) Leveraging big data analytics in healthcare enhancement: trends, challenges and opportunities. Multimedia Syst, pp 1–33
15. https://github.com/ieee8023/covid-chestxray-dataset. Accessed: 01 Apr 2020
16. https://www.kaggle.com/paultimothymooney/chest-xray-pneumonia. Accessed 01 Apr 2020
17. Naseer A, Rani M, Naz S, Razzak MI, Imran M, Xu G (2020) Refining parkinson's neurological disorder identification through deep transfer learning. Neural Comput Appl 32(3):839–854
18. Rehman A, Naz S, Razzak MI, Akram F, Imran M (2020) A deep learning-based framework for automatic brain tumors classification using transfer learning. Circuits Syst Signal Process 39(2):757–775
19. Krizhevsky A, Sutskever I, Hinton GE (2012) Imagenet classification with deep convolutional neural networks. In Advances in neural information processing systems, pp 1097–1105
20. Simonyan K, Zisserman A (2014) Very deep convolutional networks for large-scale image recognition. arXiv preprint arXiv:1409.1556

21. Iandola FN, Han S, Moskewicz MW, Ashraf K, Dally, WJ, Keutzer K (2016) Squeezenet: Alexnet-level accuracy with 50x fewer parameters and¡ 0.5 mb model size. arXiv preprint arXiv:1602.07360
22. Szegedy C, Liu W, Jia Y, Sermanet P, Reed S, Anguelov D, Erhan D, Vanhoucke V, Rabinovich A (2015) Going deeper with convolutions. In Proceedings of the IEEE conference on computer vision and pattern recognition, pp 1–9
23. He K, Zhang X, Ren S, Sun J (2016) Deep residual learning for image recognition. In Proceedings of the IEEE conference on computer vision and pattern recognition, pp 770–778
24. Targ S, Almeida D, Lyman K (2016) Resnet in resnet: generalizing residual architectures. arXiv preprint arXiv:1603.08029
25. Huang G, Liu Z, Van Der Maaten L, Weinberger KQ (2017) Densely connected convolutional networks. In Proceedings of the IEEE conference on computer vision and pattern recognition, pp 4700–4708
26. Sandler M, Howard A, Zhu M, Zhmoginov A, Chen LC (2018) Mobilenetv2: Inverted residuals and linear bottlenecks. In Proceedings of the IEEE confer- ence on computer vision and pattern recognition, pp 4510–4520
27. Xu X, Jiang X, Ma C, Du P, Li X, Lv S, Yu L, Chen Y, Su J, Lang G et al (2020) Deep learning system to screen coronavirus disease 2019 pneumonia. arXiv preprint arXiv:2002.09334

Bayesian Optimization for Sparse Artificial Neural Networks: Application to Change Detection in Remote Sensing

Mohamed Fakhfakh, Bassem Bouaziz, Hadj Batatia, and Lotfi Chaari

Abstract Artificial neural networks (ANNs) are today the most popular machine learning algorithms. ANNs are widely applied in various fields such as medical imaging and remote sensing. One of the main challenges related to the use of ANNs is the inherent optimization problem to be solved during the training phase. This optimization step is generally performed using a gradient-based approach with a backpropagation strategy. For the sake of efficiency, regularization is generally used. When non-smooth regularizers are used to promote sparse networks, this optimization becomes challenging. Classical gradient-based optimizers cannot be used due to differentiability issues. In this paper, we propose an efficient optimization scheme formulated in a Bayesian framework. Hamiltonian dynamics are used to design an efficient sampling scheme. Promising results show the usefulness of the proposed method to allow ANNs with low complexity levels reaching high accuracy rates while performing faster that with other optimizers.

Keywords Artificial neural networks · Machine learning · Optimization · Deep learning · MCMC · Hamiltonian dynamics

M. Fakhfakh (✉) · B. Bouaziz
MIRACL Laboratory, University of Sfax, Sfax, Tunisia
e-mail: mohamed.fakhfakh.doc@enetcom.usf.tn

B. Bouaziz
e-mail: bassem.bouaziz@isims.usf.tn

H. Batatia
MACS School, Heriot-Watt University, Dubai-Campus, Dubai, United Arab Emirates
e-mail: H.Batatia@hw.ac.uk

L. Chaari
University of Toulouse, IRIT-ENSEEIHT, Dubai, United Arab Emirates
e-mail: lotfi.chaari@toulouse-inp.fr

A. Ullah et al. (eds.), *Proceedings of International Conference on Information Technology and Applications*, Lecture Notes in Networks and Systems 350,
https://doi.org/10.1007/978-981-16-7618-5_4

1 Introduction

Deep learning (DL) [1] has grown at a remarkable rate, attracting a great number of researchers and practitioners. It is today one of the most attractive research directions in many applications such as recognition, medical diagnoses, self-driving cars, and recommendation systems. The performance of a DL algorithm depends on several factors, especially the optimization procedure used during the learning process. The essence of most architectures is to build an optimization model and learn the parameters from the given data.

Indeed, optimization methods can be divided into three categories [2, 3]: (1) first-order optimization methods such as stochastic gradient; (2) high-order optimization methods, mainly Newton's algorithm; and (3) heuristic derivative-free optimization methods.

For gradient-based methods, the weights are updated in the opposite direction of the objective function gradient. When non-convex or non-differentiable functions are used, these methods may suffer from slow convergence and local minima issues. The stochastic gradient descent (SGD) method [4, 5] is one of the core techniques behind the success of deep neural networks since it alleviates the abovementioned difficulties. Its main limitation lies in the use equal-sized steps for all parameters, irrespective of gradient behavior. Hence, an efficient way of deep network optimization is to make adaptive step sizes for each parameter. Other variants have been widely used in recent years such as Adam optimizer [6, 7]. Compared with their first-order counterpart, high-order methods [8, 9] provide convergence at a faster speed at the expense of more memory load especially for the storage of the inverse Hessian matrix. As regards heuristic optimization methods, they are being widely used for complex problems [10, 11].

In parallel, Bayesian techniques have demonstrated their ability to provide efficient optimization algorithms with better convergence guarantees than variational techniques. In a Bayesian framework, these techniques assume that all parameters are realizations of random variables. Likelihood and prior distributions are formulated to model the available information on the target parameters. An estimator for these parameters is generally derived using a maximum *a posteriori* (MAP) framework. However, the main difficulty is to derive analytical closed-form expressions of the estimators because of the posterior distribution form which can be complex if sophisticated priors are use, such as those promoting sparsity. In this case, Markov chain Monte Carlo (MCMC) techniques are generally used to sample coefficients from the target posterior [12–14]. The main limitation of such techniques lies in the high complexity level, especially when multidimensional data are handled. For such cases, efficient sampling methods have been proposed in the literature such as the random walk Metropolis–Hastings (MH) algorithm [12] or the Metropolis-adjusted Langevin algorithm (MALA) [15, 16]. Recently, sampling using Hamiltonian dynamics [17, 18] has been investigated developing the so-called Hamiltonian Monte Carlo (HMC) sampling. A more sophisticated algorithm has been proposed in [19–21] called non-smooth Hamiltonian Monte Carlo (ns-HMC) sampling. This method solves the problem of HMC schemes that cannot be used in the case of exponential distributions with non-differentiable energy function.

In this paper, we propose a Bayesian optimization method to minimize the target cost function and derive the optimal weights vector. The proposed method targets regularization schemes promoting sparse networks [22]. Indeed, gradient-based optimization methods in this case are not very efficient due to differentiability and convergence issues. Learning performances can, therefore, be altered. We demonstrate that using the proposed method leads to high accuracy results with simple architectures, which cannot be reached using standard optimizers.

We apply our ns-HMC optimizer to change detection [23] in bitemporal remotely sensed images. Indeed, change detection is nowadays a very active research area mainly due to the availability of free and highly resoluted time series. Several applications are studies such as in agriculture [24] or ocean monitoring [25].

The rest of this paper is organized as follows. The addressed problem is formulated in Section 2. The proposed efficient Bayesian optimization scheme is developed in Section 3 and validated in Section 4. Finally, conclusions and future work are drawn in Section 5.

2 Problem Formulation

In this paper, we propose a method to allow weights optimization under non-smooth regularizations. Let us denote by x an input to be presented to the ANN. The estimated label will be denoted by $\hat{y}(x; W)$ as a nonlinear function of the input x and the weights vector $W \in \mathbb{R}^N$, while the ground truth label will be denoted by y. Using a quadratic error with an ℓ_1 regularization with M input data for the learning step, the weights vector can be estimated as:

$$\hat{W} = \arg\min_{W} \mathcal{L}(W) = \arg\min_{W} \sum_{m=1}^{M} \| \hat{y}\big(x^m; W\big) - y^{(m)} \|_2^2 + \lambda W_1 \qquad (1)$$

where λ is a regularization parameter balancing the solution between the data fidelity and regularization terms, and M is the number of learning data.

Since the optimization problem in (1) is not differentiable, the use of gradient-based algorithms with backpropagation is not possible. In this case, the learning process is costly and very complicated.

In Section 3, we present a method to efficiently estimate the weights vector without increase of learning complexity. The optimization problem in (1) is formulated and solved in a Bayesian framework.

3 Bayesian Optimization

As stated above, the weights optimization problem is formulated in a Bayesian framework. In this sense, the problem parameters and hyperparameters are assumed to follow probability distributions. More specifically, a likelihood distribution is defined to model the link between the target weights vector and the data, while a prior distribution is defined to model the prior knowledge about the target weights.

3.1 *Hierarchical Bayesian Model*

According to the principle of minimizing the error between the reference label y and the estimated one $\hat{y}$, and assuming a quadratic error (first term in (1)), we define the likelihood distribution as

$$f(y; W, \sigma) \propto \prod_{m=1}^{M} \exp\left(-\frac{1}{2\sigma^2} \|\hat{y}(x^m; W) - y^{(m)}\|^2\right), \tag{2}$$

where σ^2 is a positive parameter to be set.

As regards the prior knowledge on the weights vector W, we propose the use of a Laplace distribution in order to promote the sparsity of the neural network:

$$f(W; \lambda) \propto \prod_{k=1}^{N} \exp\left(-\frac{\|w^{[k]}\|_1}{\lambda}\right), \tag{3}$$

where λ is a hyperparameter to be fixed or estimated.

By adopting a maximum a posteriori (MAP) approach, we first need to express the posterior distribution. Based on the defined likelihood and prior, this posterior writes:

$$\begin{aligned} f(W; y, \sigma, \lambda) &\propto f(y; W, \sigma) \propto f(W; \lambda) \\ &\propto \prod_{m=1}^{M} \exp\left(-\frac{1}{2\sigma^2} \|\hat{y}(x^m; W) - y^{(m)}\|^2\right) \\ &\quad x \prod_{k=1}^{N} \exp\left(-\frac{\|W^{[k]}\|_1}{\lambda}\right). \end{aligned} \tag{4}$$

It is clear that this posterior is not straightforward to handle in order to derive a closed-form expression of the estimate $\hat{W}$. For this reason, we resort to a stochastic sampling approach in order to numerically approximate the posterior and, hence, to calculate an estimator for $\hat{W}$. The following section details the adopted sampling procedure.

3.2 Hamiltonian Sampling

Let us denote $\alpha = \frac{\lambda}{\sigma^2}$ and $\theta = \{\sigma^2, \lambda\}$. For a weight W^k, we define the following energy function

$$E_\theta^k(W^k) = \frac{\alpha}{2}\sum_{m=1}^{M} \|\hat{y}(x^m; W) - y^{(m)}\|_2^2 + \|W^k\|_1. \quad (5)$$

The posterior in (4) can, therefore, be reformulated as

$$f(W; y, \theta) \propto \exp\left(-\sum_{k=1}^{N} E_\theta^k(W^k)\right). \quad (6)$$

To sample according to this exponential posterior, and since direct sampling is not possible due to the form of the energy function E_θ^k, Hamiltonian sampling is adopted. Indeed, Hamiltonian dynamics [17] strategy has been widely used in the literature to sample from high-dimensional vectors. However, sampling using Hamiltonian dynamics requires computing the gradient of the energy function, which is not possible in our case due to the ℓ_1 term. To overcome this difficulty, we resort to a non-smooth Hamiltonian Monte Carlo (ns-HMC) strategy as proposed in [19]. More specifically, we use the plug and play procedure developed in [21]. Indeed, this strategy requires to calculate the proximity operator only at an initial point and uses the shift property [26, 27] to deduce the proximity operator during the iterative procedure [21], Algorithm 1.

As regards the proximity operator calculation, let us denote by $G_{\mathcal{L}}(W^k)$ the gradient of the quadratic term of the loss function $\mathscr{L}$ with respect to the weight W^k. Let us also denote by $\varphi(W^k) = {W^k}_1$. Following the standard definition of the proximity operator [26, 27], we can write for a point z

$$\text{prox}_{E_\theta^k}(z) = p \Leftrightarrow z - p \in \partial E_\theta^k(p). \quad (7)$$

Straightforward calculations lead to the following expression of the proximity operator:

$$\text{prox}_{E_\theta^k}(z) = \text{prox}_\varphi\left(z - \frac{\alpha}{2} G_{\mathcal{L}}(W^k)\right). \quad (8)$$

Since prox_{φ} is nothing but the soft thresholding operator [27], the proximity operator in (8) can be easily calculated once a single gradient step is applied (back propagation) to calculate $G_{\mathcal{L}}(W^k)$.

The main steps of the proposed method are detailed in Algorithm 1.

After convergence, Algorithm 1 provides chains of coefficients sampled according to the target distribution of each W^k. These chains can be used to compute an MMSE (minimum mean square error) estimator (after discarding the samples corresponding to the burn-in period).

Algorithm 1 Main steps of the proposed Bayesian optimization.

- Fix the hyperparameters λ and ;
- Initialize with some W_0;
- Perform one backpropagation step to provide an initialization for $G_{\mathcal{L}}(W_0)$;
- Compute $\text{prox}_{E_\theta}(W_0)$ according to (8);
- Use the Gibbs sampler in [21], Algorithm 1 until convergence;

It is worth noting that hyperprior distributions can be put on λ and σ in order to integrate them in the hierarchical Bayesian model. These hyperparameters can, therefore, be estimated from the data at the expense of some additional complexity.

4 Experimental Validation

In order to validate the proposed method, we experiment it on change detection in bitemporal remotely sensed images. The first step of our change detection methodology is to generate difference images from acquisitions at times T1 and T2 [28]. Relevant change areas, therefore, appear as main information in the difference image. Change detection is formulated as a binary classification problem: change and no change. This experiment is conducted on an open database extracted from Google Earth (DigitalGlobe).[1] This database contains pairs of season-varying images: seven and four pairs with spatial resolutions of 4725×2700 and 1900×1000, respectively. The dataset was generated by cropping 256×256 randomly rotated fragments $(0-2\pi)$ with at least a part of target object. Therefore, object center coordinates were unique, and distance between object centers for each axis was 32 pixels. Therefore, in our experiment, the dataset contained more than 280 image sets with image size 256×256 pixels: 220 train sets and 66 test sets.

[1] kttpB://drive.google.com/file/d/1GX656JqqOyBiEf0w65kDGYto−nHrNB9.

4.1 Loss and Accuracy Assessment

To perform binary classification of difference images, we used a CNN architecture similar to LeNet [29] with three convolutional (Conv-32, Conv-64, and Conv-128) and two fully connected (FC-128 and FC-softamx) layers. Each convolutional layer includes filters with 3 x 3 kernels followed by 2×2 maxpooling with stride size equal to 1. In order to avoid overfitting, dropout (with a dropout rate equal to 0.45) and batch normalization were applied for the sake of regularization. As regards coding, we used Python programming language with Keras and TensorFlow libraries on an Intel(R) Core(TM) i7 3630QM CPU 2.40GHZ architecture with 8 Gb memory.

The used CNN is trained using the proposed Bayesian optimization method. For the sake of comparison, two other families of optimizers are used : *i)* MCMC-based method, precisely the standard Metropolis–Hastings (MH) algorithm and the random walk Metropolis–Hastings (rw-MH) [12] and *ii)* Adam [7], the most popular optimization technique used in DL, with a learning rate equal to 10^{-3}.

Table 1 reports accuracy, loss, sensitivity, and specificity scores for all optimizers, in addition to the computational time. The reported scores indicate that the proposed method clearly outperforms the competing optimizers in terms of learning precision and, hence, classification performance. This is confirmed by Fig. 1 which displays accuracy and loss curves for the training and test phases.

The displayed curves clearly indicate a convergence with a relatively low test accuracy rate for some optimizers, indicating potential overfitting and low generalization capacity of the model when trained using these optimizers. This can be explained by potential local minima issued. One can also notice that the Bayesian formulation can help to alleviate this problem. The proposed optimizer (ns-HMC) performs well for this case with an accuracy score up to 95%. The competing optimizers also require longer computational time in comparison to the proposed method. It is worth noting that curves irregularity for Bayesian techniques (proposed method, MH, and rw-MH) are due to the random sampling effect. No monotonic behavior is expected.

Table 1 Results for the change detection in bitemporal remotely sensed images with the proposed ns-HMC

Optimizers	Comp. time (minutes)	Accuracy	Loss	Sens	Spec
Proposed method	**20.16**	**0.95**	**0.15**	**0.93**	**0.91**
MH	29.04	0.84	0.24	0.82	0.77
rw-MH	30.93	0.87	0.20	0.85	0.82
Adam	25.21	0.89	0.31	0.87	0.86

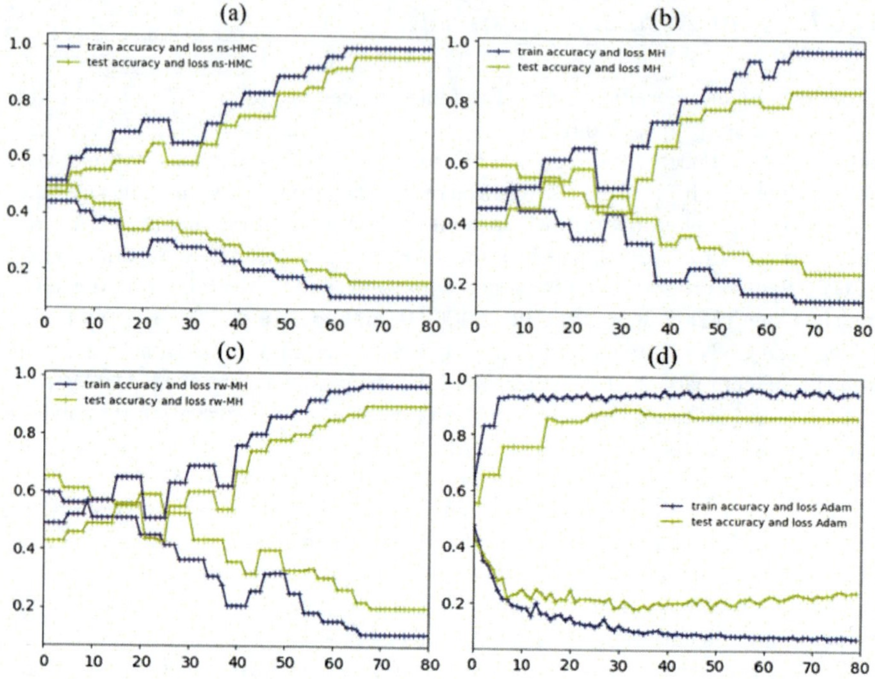

Fig. 1 Train and test curves for the proposed HMC method **a**, MH **b**, rw-MH **c**, and Adam **d**

4.2 *Qualitative Analysis*

In this section, we illustrate some representative results obtained on pairs of images. These time series are made up of two images acquired at times T_1 and T_2, with ground truth and the difference images. Figure 2 displays four couples of images, their difference image, and the binary change map as a ground truth. Detection scores obtained using the proposed method are also provided. The two examples on the top of the figure correspond to scenes with change, while the two others correspond to stable scenes. The reported detection score indicates the probability of change between images ate T_1 and T_2. When analyzing the ground truth change map, it is clear that the performed change detection is also relevant for moderate change scenarios.

5 Conclusion

In this paper, we proposed a novel Bayesian optimization method to fit weights for artificial neural networks. The proposed method relies on Hamiltonian dynamics and solves optimization issues related to the use of sparse regularizations. Our results

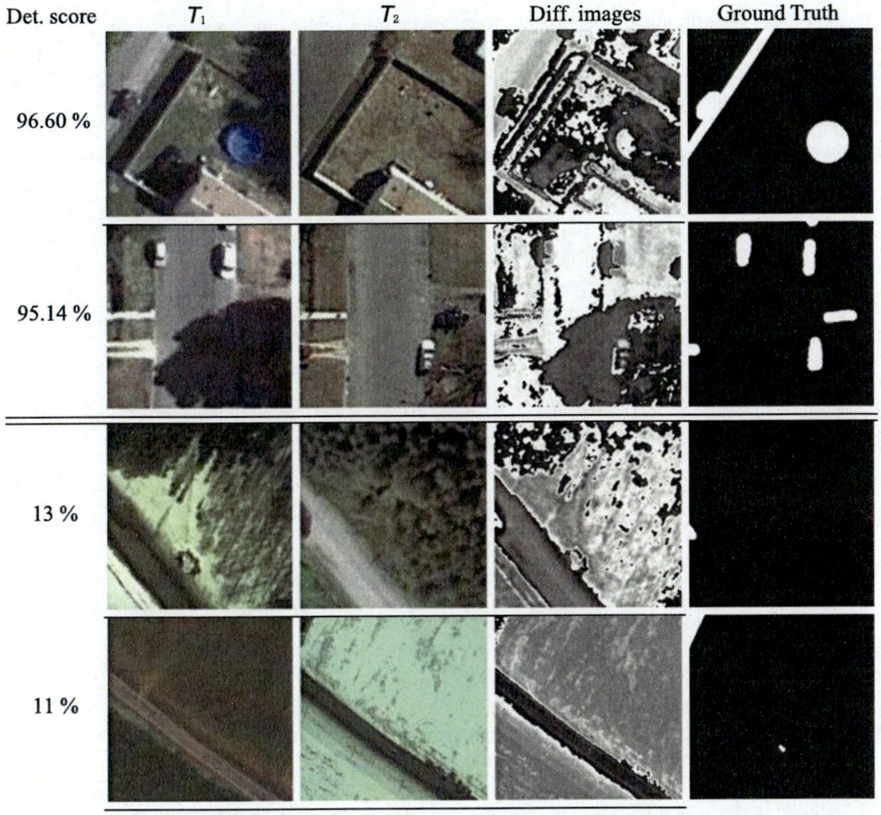

Fig. 2 Four examples of image couples (T_1, T_2) with change (two top examples) and non-change (two bottom examples). The reported detection score indicates the change probability for each example

demonstrated the good performance of the proposed method in comparison with standard optimizers. Moreover, the proposed technique allows simple networks to enjoy high accuracy and generalization properties. Future work will focus on further reducing the computational time by resorting to parallel GPU architectures.

References

1. Litjens G, Kooi T, Ehteshami BB, Setio AAA, Ciompi F, Ghafoorian M, Laak VD, Jeroen AWM, Ginneken BV, Sánchez C (2017) A survey on deep learning in medical image analysis. Medical Image Anal 42:60–88
2. Zaheer R, Shaziya H (2019) A study of the optimization algorithms in deep learning. In 2019 third international conference on inventive systems and control (ICISC), pp 536–539. IEEE
3. Dogo EM, Afolabi OJ, Nwulu NI, Twala B, Aigbavboa CO (2018) A comparativeanalysis of gradient descent-based optimization algorithms on convolutional neural networks. In 2018

international conference on computational techniques, electronics and mechanical systems (CTEMS), pp 92–99
4. Robbins H, Monro S (1951) A stochastic approximation method. Ann Math Statistics, pp 400–407
5. Jain P, Kakade S, Kidambi R,Netrapalli P, Sidford A (2018) Parallelizing stochastic gradient descent for least squares regression: mini-batching, averaging, and model misspecification. J Mach Learning Res 18
6. Kingma DP, Ba J (2014) Adam: A method for stochastic optimization. arXiv preprint arXiv: 1412.6980
7. Sun S, Cao Z, Zhu H, Zhao J (2019) A survey of optimization methods from a machine learning perspective. IEEE Trans Cybernetics 50(8):3668–3681
8. Shanno DF (1970) Conditioning of quasi-newton methods for function minimization. Math Comput 24(111):647–656
9. Hu J, Jiang B, Lin L, Wen Z, Yuan YX (2019) Structured quasi-newton methods for optimization with orthogonality constraints. SIAM J Sci Comput 41(4):A2239–A2269
10. Rios LM, Sahinidis NV (2013) Derivative-free optimization: a review of algorithms and comparison of software implementations. J Global Optim 56(3):1247–1293
11. Berahas AS, Byrd RH, Nocedal J (2019) Derivative-free optimization of noisy functions via quasi-newton methods. SIAM J Optim 29(2):965–993
12. Robert C, Casella G (2013) Monte Carlo statistical methods.Springer Science and Business Media
13. Chaari L, Batatia H, Dobigeon N, Tourneret J (2014) A hierarchical sparsity-smoothness bayesian model for l0+l1+l2 regularization. In 2014 IEEE international conference on acous tics, speech and signal processing (ICASSP), pp 1901–1905
14. Chaari L (2019) A bayesian grouplet transform. SIViP 13:871–878
15. Roberts GO, Tweedie RL (1996) Exponential convergence of langevin distributions and their discrete approximations. Bernoulli 2(4):341–363
16. Girolami M, Calderhead B (2011) Riemann manifold langevin and Hamiltonian Monte Carlo methods. J Royal Stat Soc: Series B (Statistical Methodology) 73(2):123–214
17. Hanson KM (2001) Markov Chain Monte Carlo posterior sampling with the hamiltonian method. InMedical imaging 2001: image processing, International Society for Optics and Photonics, vol 4322, pp 456–467
18. Brooks S, Gelman A, Jones G, Meng XL (2011) Handbook of Markov Chain Monte Carlo. CRC press
19. Chaari L, Tourneret J-Y, Chaux C, Batatia H (2016) A Hamiltonian Monte Carlo method for non-smooth energy sampling. IEEE Trans Signal Process 64(21):5585–5594
20. Chaari L, Tourneret J-Y, Batatia H (Sept2017) A general non-smooth Hamiltonian Monte Carlo scheme using Bayesian proximity operator calculation. European signal processing conference EUSIPCO, 1260–1264
21. Chaari L, Tourneret JY, Batatia H (Sept 2018) A plug and play Bayesian algorithm for solving myope inverse problems. European signal processing conference EUSIPCO 742–746
22. Mocanu DC, Mocanu E, Stone P, Nguyen PH, Gibescu M, Liotta A (2018) Scalable training of artificial neural networks with adaptive sparse connectivity inspired by network science. Nat Commun 9:1–12
23. Si Salah H, Goldin SE, Rezgui A, Nour El Islam B, Ait-Aoudia S (2020) What is a remote sensing change detection technique? towards a conceptual framework. Int J Remote Sens 41(5):1788–1812
24. Addink E (2001) Change detection with remote sensing: relating NOAA-AVHRR to environmental impact of agriculture in Europe
25. Faghmous J, Chamber Y, Boriah S, Vikebø F, Liess S, dos Santos Mesquita M, Kumar V (2012) A novel and scalable spatio-temporal technique for ocean eddy monitoring. In Proceedings of the AAAI Conference on Artificial Intelligence 26
26. Moreau JJ (1965) Proximité et dualité dans un espace hilbertien. Bull Soc Math France 93:273–299

27. Chaux C, Combettes PL, Pesquet JC, Wajs VR (2007) A variational formulation for frame-based inverse problems. Inverse Prob 23(4):1495
28. Benazza-Benyahia A, Gharbi W, Chaari L (2020) Unsupervised bayesian change detection for remotely sensed images. Signal Image Video Process 15:205–213
29. LeCun Y, Bottou L, Bengio Y, Haffner P (1998) Gradient-based learning applied to document recognition. Proc IEEE 86(11):2278–2324

Context-Aware Multimodal Emotion Recognition

Aaishwarya Khalane and Talal Shaikh

Abstract Making human–computer interaction more organic and personalized for users essentially demands advancement in human emotion recognition. Emotions are perceived by humans considering multiple factors such as facial expressions, voice tonality, and information context. Although significant research has been conducted in the area of unimodal/multimodal emotion recognition in videos using acoustic/visual features, few papers have explored the potential of textual information obtained from the video utterances. Humans experience emotions through their audio-visual and linguistic senses, making it quintessential to take the latter into account. This paper outlines two different algorithms for recognizing multimodal emotional expressions in online videos. In addition to acoustic (speech), visual (facial), and textual (utterances) feature extraction using BERT, we utilize bidirectional LSTMs to capture the context between utterances. To obtain richer sequential information, we also implement a multi-head self-attention mechanism. Our analysis utilizes the benchmarking CMU multimodal opinion sentiment and emotion intensity (CMU-MOSEI) dataset, which is the largest dataset for sentiment analysis and emotion recognition to date. Our experiments result in improved F1 scores in comparison to the baseline models.

Keyword Multimodal · Emotion · Recognition · CMU-MOSEI · Multi-head attention · BERT · Context-aware

1 Introduction

Modern technology demands that machines be smarter than ever while understanding and responding to the needs of humans. In today's day and age, it is vital to make the interaction between humans and machines, commonly known as human–computer interaction (HCI), fluid, and hassle-free. In recent years, a whole new interdisciplinary

A. Khalane (✉) · T. Shaikh
Heriot-Watt University Dubai, Dubai, UAE

T. Shaikh
e-mail: t.a.g.shaikh@hw.ac.uk

A. Ullah et al. (eds.), *Proceedings of International Conference on Information Technology and Applications*, Lecture Notes in Networks and Systems 350,
https://doi.org/10.1007/978-981-16-7618-5_5

field called affective computing has emerged [8]. It is concerned with machine recognition, perception, and interpretation of human emotions. Affective computing and HCI are closely associated with emotion recognition (ER), which finds applications in numerous fields including tutoring, healthcare, customer satisfaction, gaming, chat-bots, smart home solutions, and robotics. For modeling emotions as close to human perception as possible, training models over videos shot in the wild is imperative. By leveraging such natural videos showcasing spontaneous and real reactions, it is possible to add real-life nuances to the ER model. With a multitude of natural videos being posted on social media and streaming services, there is an abundance of conversational/reaction-based video data that is readily available online and is an excellent source of natural emotion data for ER models. Despite its complexity, it is certainly not impossible to achieve accurate human ER. Bringing intelligent machines as close as possible to interpreting emotions like humans is being extensively researched. This paper outlines a few of the current methods used in ER and proposes two models for classifying human emotions.

2 Background

Human emotions are complex and hence need to be perceived using a fusion of various cues like facial expressions (visual), voice modulation (acoustic), and words spoken (textual). A smile could be a sarcastic expression of "disappointment" but classified as "happy" if the only focus is the visual cue. In ER, most research has traditionally used unimodal (single cue) methods, which are insubstantial, insufficient, and unrealistic, since this human experience is fundamentally multimodal [8]. As a result, it is not ideal to use a single modality for ER, such as facial or speech ER. Consequently, multimodal approaches (fusion of two or more cues) prove to be more effective. The multimodal ER technique reflects nuances of real emotional perception, as the different methods complement each other and together make the system more robust and reliable [9]. In spite of the fact that the majority of a message is conveyed by facial expressions (55%) and voice modulations (38%), around 7% is conveyed through the words spoken, implying that textual aspects should not be ignored [10]. Despite the considerable amount of research on audio-visual ER in videos, we discovered only a few studies that explored the utterances in these clips, which could help enhance audio-visual ER. Therefore, this paper explores the use of the contextual information provided by the utterances in conjunction with audio-visual methods. Using deep learning techniques discussed in the upcoming sections, we propose a trimodal approach using visual, acoustic, and textual features to train our deep neural network for reliably identifying emotions.

2.1 CMU-MOSEI Dataset

CMU-MOSEI (Multimodal Opinion Sentiment and Emotion Intensity Dataset) created in 2018 [11] is presently the largest multimodal dataset for sentiment analysis and emotion recognition. It comprises 23,453 video segments (3228 videos) across 250 topics with 1000 distinct subject matters. The videos have been procured from video sharing and social media platforms such as YouTube. Each video features an individual looking into the camera and sharing their review about a particular theme. Quality-checking of the videos has been performed by 14 judges across all three modalities. Using Ekman's method, emotions have been categorized on a [0,3] Likert scale, with 0 indicating no evidence of an emotion and 3 indicating strong evidence, by a panel of three crowdsourced judges from Amazon mechanical turk. This dataset was suitable for our experiment because:

1. The dataset is open-source.
2. The video content is natural, not staged.
3. It contains a diversity of topics and people, increasing our ability to generalize our model and use it under real-world scenarios.

2.2 Feature Extraction

An individual's facial expressions serve as the primary means of assessing their emotional state. Apart from the handcrafted methods like the facial action coding system (FACS) developed by Friesen and Ekman and active appearance models (AAMs) used in computer vision, 2D CNNs like AlexNet and 3D CNNs like C3D have been profoundly used for automatic facial feature extraction.

Audio features can be categorized into three main types—voice quality features, spectrum-based relevant features such as mel frequency cepstral coefficients (MFCCs) and linear prediction cepstral coefficients (LPCCs), and prosodic features (pitch, energy, intensity). Researchers have discovered that the variation in these features correlates with their emotional states [7]. The combination of these audio features and the inference of relationships between them was used to design a speech emotion recognition system [4].

Researchers in the area of textual ER have focused primarily on the use of emotional keywords (words that convey emotion). By not taking into account the syntax and semantics of a sentence, they are at the risk of being unable to classify sentences that have no emotional keywords. The bidirectional encoder representations from transformer (BERT) utilize transformers to generate models for pre-training representations of texts/languages [6]. Pre-trained models can then be fine-tuned for specialized tasks, such as classification, translation, or extraction of high-quality textual features. Word2Vec, GloVe embeddings traditionally extract a single word vector per word in a sentence, ignoring the context, while BERT utilizes its bidirectional attention mechanism to extract contextual information. Moreover, the pre-trained BERT model contains pre-encoded language information which translates into quicker development and smaller training data requirements saving additional computation costs.

2.3 *Feature Fusion*

A suitable fusion methodology can increase the amount of information extracted from various modalities and thus contribute to an improved understanding of how emotions are categorized.

Early Fusion or Feature-Level Fusion is the process of combining the different features into a single feature vector which is input to the classifier. By combining various features, one can establish correlations between them and potentially improve results. In order to concatenate all the features into a single vector, all of their formats need to be matched.

Decision-Level or Late Fusion uses independent prediction methods to classify features and combine the resulting classifications into a new decision vector. Using this method, all features do not have to be converted to the same format, which reduces overhead. Consequently, the results can be improved by utilizing the best classifiers suited to each of the modalities.

2.4 *Related Research*

Our baseline paper used a fusion mechanism called DFG—dynamic fusion graph to evaluate the CMU-MOSEI dataset [11]. The graphs dynamically modified their structure, thereby accommodating n modalities (n belongs to 1, 2, 3) depending on the combination of modalities giving the best results. The memory fusion network (MFN) architecture used was composed of series of parallel LSTMs, each representing a single modality. Often, audio and text modalities were fused. Video was only fused if meaningful information was discovered. Unlike text, audio was found to merge with both other modalities.

DAR-DRNN [3] mimicked the capacity of CNNs and RNNs to integrate spatiotemporal dimensionality alongside previously acquired information using uni- and bidirectional LSTMs. The feature extraction methods used by the baseline at CMU-MOSEI were combined with an early fusion strategy to simplify and scale the analysis. The self-attention mechanism was combined with residual neural networks for using contextual information from previous layers which improved the accuracy by around 1.5% compared to the baseline LSTM.

Multi-attention LSTM (MALM) [1] introduced a novel feature-level fusion method based on a multi-utterance bimodal attention (MMMU-BA) mechanism developed with LSTM, which assessed each modality separately before fusion. Attention mechanism calculated the importance of the segments at each step. Following their attention scores, the modalities were then integrated, prioritizing those that understood emotions better, before they were fused together for improving accuracy (priority fusion). Text features were classified using stacked CNNs (TextCNNs), and visual features classified using 3D CNNs. OpenSmile was used to extract 6373 audio features including pitch, MFCC, and sound intensity (LLDs). According to the research, multimodal multi-attention outperformed the unimodal uni-attention, unimodal multi-attention, and multimodal uni-attention methods by around 2%.

The authors of [5] used a faster end-to-end training process for feature extraction from raw features and feature classification. They used a cross-modal attention system with CNNs to introduce two distinct models—fully end-to-end multimodal model (FE2E) and multimodal end-to-end sparse model (MESM). MESM had a similar structure to FE2E except that its CNN layers were replaced with sparse CNN blocks to reduce computational costs. Both models were evaluated in terms of weighted accuracy and F1 scores.

3 Our Experiments

Pre-extracted audio and video features used by our baseline paper [11] were utilized. Our baseline had extracted visual features by extracting frames at 30 Hz from the videos. The facial action units were extracted using facial action coding system. MultiComp OpenFace was used to extract head pose and orientation, several facial landmarks, facial shape parameters, and eye gaze features. The final visual embeddings were then extracted by utilizing deep learning facial feature extraction frameworks such as FaceNet, DeepFace, and SphereFace. Using static facial images; Emotient FACET was employed to extract the six emotions (happy, sad, angry, surprise, disgust, fear). Covarep was employed to extract audio features such as 12 MFCCs, pitch, maxima dispersion quotients (MDQs—which represent tense-lax contrasts in voice patterns) to portray emotions described by speech tonality. Consequently, 35 video features and 74 audio features per time step as extracted by our baseline [11] were used for our experiments.

Extracting BERT textual embeddings instead of GloVe embeddings (used by our baseline) was a crucial step. The textual features were obtained by fine-tuning the pre-trained base BERT model (provided in the Huggingface Transformers library) for emotion recognition. In addition to the 12 stacked transformer encoder layers, the model had 768 hidden nodes along with 12 attention heads within the feed-forward network. The BERT base model took in sequences of words which were then passed through the stack of transformers as new sequences continued to be input. This resulted in a compressed vector of 768 bytes. The following sections describe our two models (early and late fusion) for classifying emotions with varying model architectures.

The code for the above experiments was based on the following sources.[1,2,3]

3.1 *Model 1-Early Fusion with Bidirectional LSTMs*

The three feature vectors (text, audio, video) were concatenated using early fusion, then classified using bidirectional LSTM (Fig. 1). Due to their ability to understand the context by remembering important information, LSTMs have proven effective for training models on sequential data, thereby solving the vanishing gradient drawback that RNNs face. Bidirectional LSTMs have two networks (for capturing the past and the future) instead of one and thus can accommodate contextual information better. A two-layered LSTM having a hidden layer with 256 neurons was used. For loss calculation, the output from the LSTM was then passed through a fully connected layer followed by sigmoid activation.

3.2 *Model 2-Late Fusion with Multi-headed Self-Attention and BiLSTMs*

The early fusion model simply averaged the video and audio vectors. This model used the attention mechanism rather than a simple average of feature vectors, partly inspired by the [1] architecture (Fig. 2). Multi-head self-attention was applied to each feature vector separately (text, audio, video). In multi-head self-attention, multiple heads or agents work together in parallel to obtain multiple representations for the input, then combine them. This enables a greater understanding of different contexts and subspaces in the same sequence, which isn't possible with a single head. The initial step multiplied input vectors with the square root of input dimensions to enlarge all values in the vectors so that the added positional encoding was relatively small,

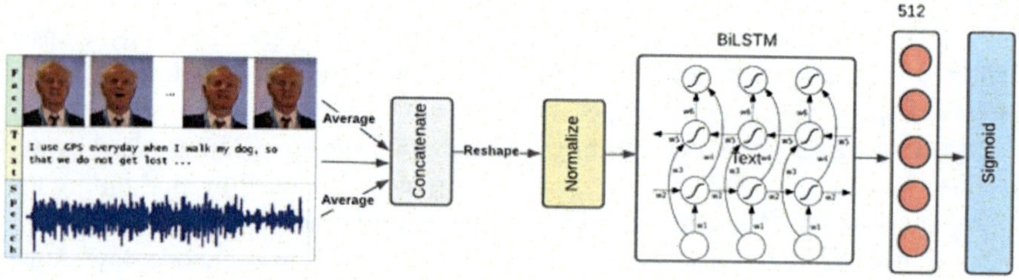

Fig. 1 Early fusion with BiLSTM model

[1] https://github.com/A2Zadeh/CMU-MultimodalSDK.

[2] https://mccormickml.com/2019/07/22/BERT-fine-tuning/.

[3] https://github.com/Ighina/MultiModalSA.

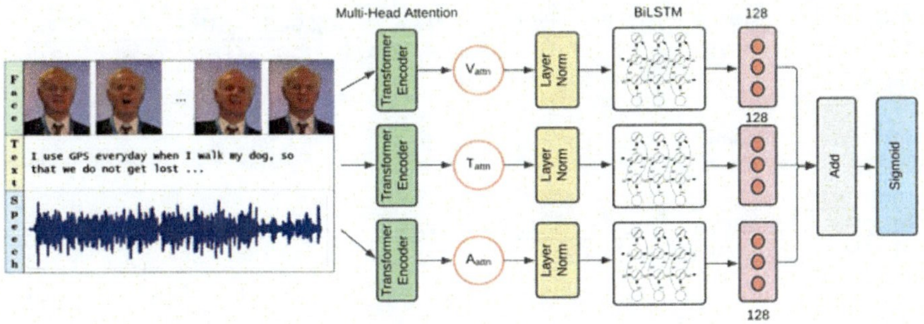

Fig. 2 Late fusion with multi-headed self-attention and BiLSTMs model

to avoid loss of the actual vector values. Positional encoding was then applied to these enlarged vectors. Next, the encoded input for each feature vector was passed into Pytorch 3.3.3's transformer encoder classes to perform multi-head attention. For each vector, the encoder produced an attention representation that was passed on to the decoder. To generate the final attention vectors and corresponding attention weights (required for model evaluation), an attention decoder class was created, instead of using the traditional transformer decoder. A score function was used to compute attention scores, whereby the encoder output was batch multiplied with the target input and scaled by dividing it with d_k where d_k = Number of input dimension/Number of attention heads.

The attention weights were then transformed into probabilities using softmax layer. The output predictions were obtained by passing each of the three attention feature vectors through a parallel BiLSTM (each modeling a single modality). Summing up the three output predictions and running them through sigmoidal activation yielded the final prediction.

4 Results

Since our dataset is imbalanced, the authors of the baseline paper evaluate it using F1 scores as well as the weighted accuracy metric described by the formula in Fig. 3, where N is the total number of negative labels, while P is the total number of positive labels. TP represents true positives, while TN represents true negatives. The results evaluated from the two models—bidirectional LSTM (BiLSTM) and multi-head attention (MHA)—were compared against the baseline Graph-MFN model [11], the

$$\text{WAcc.} = \frac{TP \times N/P + TN}{2N}$$

Fig. 3 Weighted accuracy formula

MODEL	ANGRY	HAPPY	SAD	FEAR	DISGUST	SURPRISE	AVERAGE
BiLSTM	75.053	**71.873**	71.506	86.025	**84.145**	**85.646**	79.041
MHA	**75.401**	71.772	**72.150**	**86.196**	84.084	85.084	**79.115**

Fig.4 Precision scores for text, audio, visual modalities

MODEL	ANGRY	HAPPY	SAD	FEAR	DISGUST	SURPRISE	AVERAGE
BiLSTM	78.335	**71.879**	**74.903**	**91.634**	84.106	**90.176**	81.839
MHA	**78.507**	71.815	75.397	91.506	**84.341**	89.897	**81.910**

Fig. 5 Recall scores for text, audio, visual modalities

state-of-the-art DAR-DRNN [3], FE2E [5], and MESM [5] models described in Sect. 2.4.

4.1 *Precision and Recall*

Overall, the average precision obtained in all classes was 79% (Fig. 4). Fear achieved the most precision, followed by Surprise, while Sadness and Happiness achieved the least. Real-time ER for criminal activity surveillance and depression detection requires high precision scores for fear and surprise, as false alarms in such scenarios could be hazardous. Fear might sometimes be misinterpreted as sadness, which indeed affects the precision of sadness.

The average recall across all classes was around 82% (Fig. 5). In natural videos, (not staged/enacted), emotions like fear, surprise, and disgust are vividly noticeable in terms of audio and visual features compared to happiness or sadness.

This justifies the high recall values for fear, surprise, and disgust. Humans who label the data can better understand instances of sarcasm, which is difficult to be captured by the ML algorithm, justifying lower recall for happiness, while sadness might often be mistaken as a neutral emotion. Comprehensively, our models obtained higher recall values than precision, implying that they are able to identify most instances within a class correctly even though they could perform better at reducing misclassification.

4.2 *F1 Scores*

Figure 6 compares the F1 scores obtained by our models with those obtained by the other state-of-the-art-models. The high F1 values can be attributed to high precision and recall scores (Fig. 6).

MODEL	ANGRY	HAPPY	SAD	FEAR	DISGUST	SURPRISE	AVERAGE
BiLSTM [our]	73.919	71.718	72.269	87.819	84.125	**86.636**	79.414
MHA [our]	**74.771**	71.779	**72.807**	87.871	**84.207**	86.508	**79.657**
Graph-MFN	72.8	66.3	66.9	**89.9**	76.6	85.5	76.333
DAR-DRNN	45.68	72.39	46.58	23.21	45.64	22.87	42.728
FE2E	49.6	**72.600**	29.100	26.8	57.1	49	47.367
MESM	49.3	72.300	46.600	28.9	56.4	27.2	46.783

Fig. 6 Comparison of F1 scores for text, audio, visual modalities

In 3/6 classes (Angry, Sad, Disgust), our multi-head attention LSTM model received the highest F1, while for Surprise, our simple early fusion BiLSTM received the highest score. In addition, the MHA model achieved an average F1 score of 79.66%, a 3% increase over the baseline model while being over 30 + % above the DAR-DRNN, MESM, and FE2E models. In five out of six classes, our models outperformed the graph-MFN baseline model on F1. The significantly high F1 for the Happy class of FE2E seems peculiar compared to the other classes which have lower F1 scores. It might be because the class Happy was dominant in our imbalanced dataset, so the authors of FE2E were able to get better recall as well as precision, contributing to the higher F1 score.

4.3 Weighted Accuracies

Our MHA model achieved higher weighted accuracies than the baseline GraphMFN for class Happy and Disgust (Fig. 7). Furthermore, it outperformed the DAR-DRNN model in terms of average weighted accuracy over all classes, while being quite close to the graph-MFN model.

The BiLSTM model, with its simpler architecture, was still able to attain comparable results to all the other complex models. A possible explanation would be the usage of BERT for extracting the textual features unlike other complex models.

MODEL	ANGRY	HAPPY	SAD	FEAR	DISGUST	SURPRISE	AVERAGE
BiLSTM [our]	57.729	71.372	58.490	50.178	72.307	51.931	60.335
MHA [our]	59.086	**71.550**	59.147	50.462	71.762	51.878	60.648
Graph-MFN	62.6	66.3	60.4	62	69.1	53.7	62.35
DAR-DRNN	61.458	63.372	58.805	59.318	64.332	56.359	60.607
FE2E	67	65.400	**66.700**	63.800	**77.7**	65.200	**67.633**
MESM	**66.8**	64.100	63.000	**65.800**	75.6	**65.700**	66.833

Fig. 7 Comparison of weighted accuracies for text, audio, visual modalities

Overfitting is a prominent cause of lower accuracy. Our models were fine-tuned to avoid model overfitting by implementing mechanisms like early stopping, dropout layers, learning rate decay, and hyperparameters tuning. The accuracy could be improved by using a different feature extraction mechanism to extract richer, more effective features or by changing the model classifier architecture or by balancing the dataset for equal class distribution.

5 Conclusion and Future Work

In order to achieve the best results, we experimented with several features fusion and classification methods. Using comparatively simpler model architectures, we were able to achieve similar results to the baseline paper [11], while outperforming other models like [3, 5] in terms of F1. Furthermore, we outperformed [5] in the overall weighted average. Compared to the baseline model, our models were able to achieve better precision, recall and F1 scores, but our weighted accuracy scores did not improve significantly, which is a limitation we plan to improve in our future work. We would also need to optimize our complex MHA model for quicker evaluation. We would also explore different feature extraction techniques, such as 3D CNN and facial feature focused extraction [2]. Since the complexity of the model architecture is proportional to its computation time, our ultimate aim would be to strike an appropriate balance between performance efficacy and real-time deployment for various applications including online schooling, health monitoring, and pervasive computing.

References

1. Wang A, Baoshan Sun RJ (2019) An improved model of multi-attention lstmfor multimodal sentiment analysis—proceedings of the 2019 3rd international conference on computer science and artificial intelligence 2019. Acm.org, https://dl.acm.org/doi/abs/https://doi.org/10.1145/3374587.3374606
2. Avots E, Sapin´ski T, Bachmann M, Kamin´ska D (Jul 2018) Audiovisual emotionrecognition in wild. Mach Vision Appl 30(5):975–985. https://doi.org/10.1007/s00138-018-0960-9, https://link.springer.com/article/https://doi.org/10.1007/s00138-018-0960-9
3. Chandra E, Hsu JYJ (Nov 2019) Deep learning for multimodal emotion recognitionattentive residual disconnected rnn. 2019 International Conference on Technologies and Applications of Artificial Intelligence (TAAI). https://doi.org/10.1109/taai48200.2019.8959913, https://ieeexplore.ieee.org/document/8959913
4. Chatterjee J, Mukesh V, Hsu HH, Vyas G, Liu Z (Aug 2018) Speech emotion recognition using cross-correlation and acoustic features. 2018 IEEE 16th Intl Conf on Dependable, Autonomic and Secure Computing, 16th Intl Conf on pervasive intelligence and computing, 4th Intl Conf on Big Data Intelligence and Computing and Cyber Science and Technology Congress(DASC/PiCom/DataCom/CyberSciTech). https://doi.org/10.1109/dasc/picom/datacom/cyberscitec.2018.00050, https://ieeexplore.ieee.org/document/8511893

5. Dai W, Cahyawijaya S, Liu Z, Fung P Multimodal end-to-end sparse modelfor emotion recognition. https://arxiv.org/pdf/2103.09666.pdf
6. Devlin J, Chang MW, Lee K, Google K, Language A (2019) BERT: pre-trainingof deep bidirectional transformers for language understanding, https://arxiv.org/pdf/1810.04805.pdf
7. Khalil RA, Jones E, Babar MI, Jan T, Zafar MH, Alhussain T (2019) Speechemotion recognition using deep learning techniques: a review. IEEE Access 7:117327–117345 . https://doi.org/10.1109/access.2019.2936124, https:// ieeexplore.ieee.org/document/8805181
8. Poria S, Cambria E, Bajpai R, Hussain A (Sep 2017) A review of affective computing: From unimodal analysis to multimodal fusion. Inf Fusion 37:98–125 . https://doi.org/10.1016/j.inffus.2017.02.003, https://www.sciencedirect.com/science/article/pii/S1566253517300738
9. Qi H, Wang X, Hall F, Sitharama S, Cs I, Hall C, Chakrabarty K, Hall H Multisensor data fusion in distributed sensor networks using mobile agents. http://users.cis.fiu.edu/~iyengar/images/publications/data_ fusion_mobile_agents.pdf
10. Rao KS, Koolagudi SG (Jul 2013) Recognition of emotions from video using acousticand facial features. Signal Image Video Process 9(5):1029–1045. https://doi.org/10.1007/s11760-013-0522-6, https://link.springer.com/article/https://doi.org/10.1007/s11760-013-0522-6
11. Zadeh A, Liang P, Vanbriesen J, Poria S, Tong E, Cambria E, Chen M, Morency LP (2018) Multimodal language analysis in the wild: cmu-mosei dataset and interpretable dynamic fusion graph, pp 2236–2246. https://www.aclweb.org/anthology/P18-1208.pdf

Beard and Hair Detection, Segmentation and Changing Color Using Mask R-CNN

Muhammad Talha Ubaid, Malika Khalil, Muhammad Usman Ghani Khan, Tanzila Saba, and Amjad Rehman

Abstract Beard and hair detection and segmentation have a significant role in gender identification, age assessment, and facial recognition. Due to the variability of their forms, colors, and intensities and the impact of shadows and light objects. This paper has proposed an efficient state-of-the-art system for beard and hair detection and segmentation for changing color in challenging facial images. After segmentation, the color of hair and beard can be changed. We have used a modified version of a Mask R-CNN model for hair and beard detection and segmentation. We have collected and prepared a dataset of 1500 images equally divided into both hair and beard images. This dataset is online available on the NCAI1 website. Finally, we have retrained a modified version of Mask R-CNN through transfer learning on our dataset to detect and segment out hair and beard on any given image. Mask R-CNN has outperformed compared to earlier systems designed for the same task with an accuracy of 91.2%.

Keywords Segmentation · Mask R-CNN · Transfer learning · Facial recognition · Body image

M. T. Ubaid · M. Khalil · M. U. G. Khan
Intelligent Criminology Research Lab National Center of Artificial Intelligence, Al Khawarizmi Institute of Computer Science, University of Engineering and Technology, Lahore, Pakistan
e-mail: Talha.Ubaid@kics.edu.pk

M. Khalil
e-mail: Malika.khalil@kics.edu.pk

M. U. G. Khan
e-mail: Usman.ghani@kics.edu.pk

T. Saba · A. Rehman (✉)
Artificial Intelligence & Data Analytics Lab CCIS, Prince Sultan University, Riyadh 11586, Saudi Arabia

A. Ullah et al. (eds.), *Proceedings of International Conference on Information Technology and Applications*, Lecture Notes in Networks and Systems 350,
https://doi.org/10.1007/978-981-16-7618-5_6

1 Introduction

Image filtering apps are picture-morphic applications and use artificial intelligence and neural facial transformations to make faces creepy, funny, weird, and sometimes fascinating. The app can use images from your library. Using artificial intelligence, the app creates realistic transformations of user faces using various filters and features. FaceApp [1] became viral for the first time in 2017, but the app has become even more practical with several improvements since then, rendering it viral again. The free version includes a limited choice of plugins. You also have options to change your appearance, from turning your lips to a grin, changing the color of your eyes, and changing your gender. The app has been reviewed almost 200,000 times in the App Store and is rated 4.7 times.

FaceApp offers its users plenty of features to cheer. One of its abilities is to transform user's photos to make them look younger by changing their hairstyle color like if a person has some white hair, it will change its color to black or whatever color they like. It will enhance the personality of a person to make him look younger.

As of June 2019, Snapchat has exceeded 210 million complex clients every day (Statista, 2019). By way of comparison, in the prior quarter of 2019, Snapchat had 190 million dynamic users daily and 188 million dynamic customers in the second quarter of 2018. It suggests that diverse users have increased by 8% year-over-year daily and that each person uses facial filters.

Image segmentation is a preprocess whenever a specific object in an image is processed, i.e., changing color or removing the background, etc. In image segmentation, each object's pixel-wise mask in an image is created. Figure 1 is showing image segmentation. There are two different types of image segmentation which are object segmentation and instance segmentation.

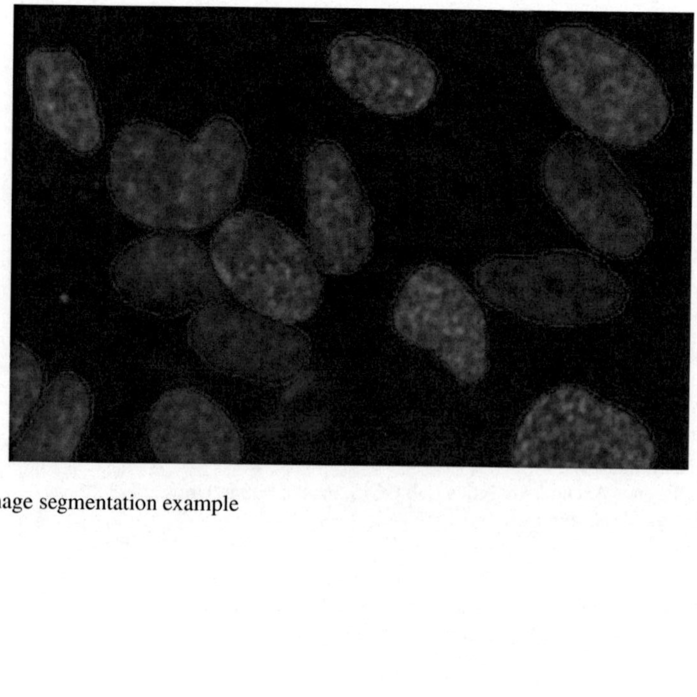

Fig. 1 Image segmentation example

In object segmentation, all objects of the same class are classified as a single object. In contrast, in an instance, segmentation, each object of the same type is classified individually, as shown in Fig. 2. This article has applied a state-of-the-art image segmentation technique based on the Mask R-CNN model to segment instances in any given image. Mask R-CNN is fundamentally a Faster R-CNN extension. Faster R-CNN is widely used for object detection tasks. It returns the bounding box and class label coordinates for each object in the image for a given image. The mask R-CNN architecture is built on the top of Faster R-CNN. Therefore, it returns the class labels and bounding box coordinates and masks each instance in the given image.

We have explored mask R-CNN, a state-of-the-art model to use for hair and beard detection and segmentation. The working of mask R-CNN can be understood in Fig. 3. Initially, this model is trained on the MS-COCO dataset to detect and segment 81 different objects (classes), including bicycle, car, jug, traffic signal, etc. We have collected a local dataset of 1500 images. Furthermore, we have annotated

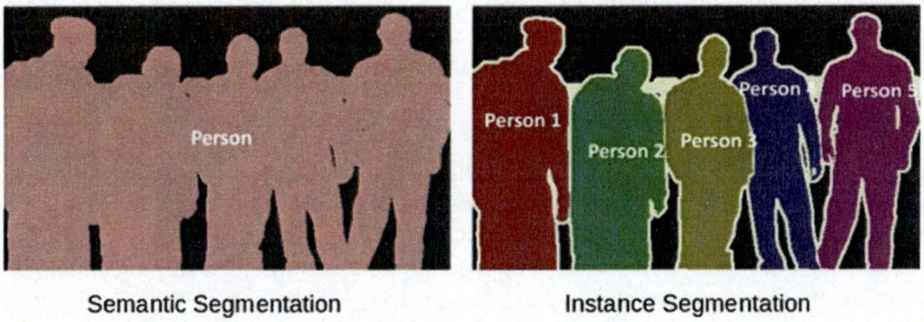

Fig. 2 Object (Semantic) segmentation versus instance segmentation

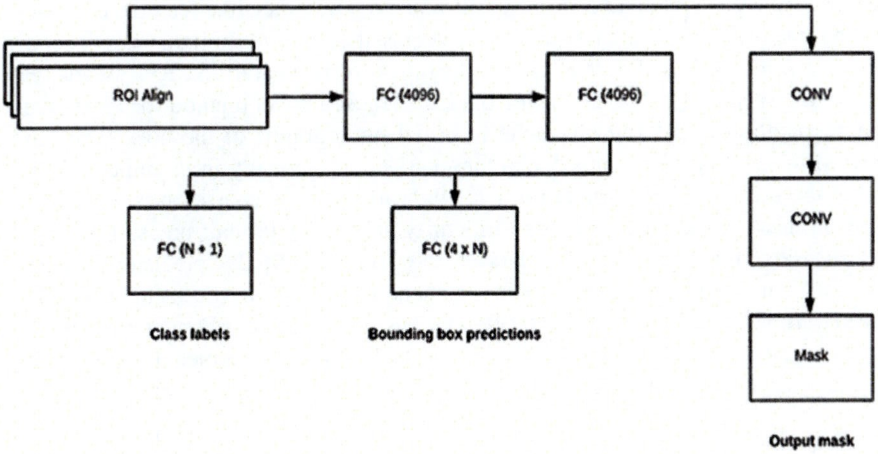

Fig. 3 Working of Mask R-CNN

this dataset against beard and hair using a well-known state-of-the-art tool named VIA annotation tool developed by Stanford. This dataset is and published online on NCAI's website. Finally, we have retrained the existing mask R-CNN model using transfer learning. The four output layers in the current model were excluded, which are "mrcnn class logits," "mrcnn bbox FC," "mrcnn bbox," and "mrcnn mask." These layers are retrained using our dataset to segmentation two objects, i.e., hair and beard, instead of 81 classes.

2 Literature Survey

The use of neural networks for facial identification has a long history [2–4]. A few of the research studies are illustrated here. The critical earlier work we will learn about hair identification is designed by Yacoob and Davis [5]. Their approach constructs a basic shading paradigm and makes use of it to view the hair image. In any event, their invention can only operate in a regulated state of base and minor variation of hair shading. Then again, hair presentation, amalgamation, and action have just become complex testing topics [6].

Currently, several researchers work on face generation as well as on the generation of face-related attributes. It has a wide range of applications in video surveillance, intelligent criminology, and automatic forensic analysis [7]. By Fusing Features, deepfakes and facial alterations were detected in the picture in [8]. A novel multi-stream deep learning method is used to detect false faces in video frames, in which three streams are combined at the feature level using the fusion layer. Rowley et al. [9] use a sequence of neural network-based filters to detect the multi-scale presence of faces and merge detections from different filters. Detectors based on cascade [10] and deformable part frameworks have also dominated the face detection approaches over the past decades. Viola and Jones [11] introduced quick-hair-like features computation by integral picture and boosted cascade classification. A similar system accompanies various studies after that. One of the top performers at the SURF cascade was among the variants [12]. T. Hoang et al. [13] suggest a novel method for identifying and segmenting beard and mustache in brutal photographs of the nose. They use MASM to identify 79 primary facial landmarks automatically in the input picture. Using the self-quotient formula removes light artifacts. A coarse classifier is then used on these self-quotient photos to identify skin or facial hair areas. They used two datasets, and on the MBGC dataset, they achieved 96.2% precision in beard and 98.8% in the mustache. 95.8% in mustache and 97.0% in mustache in FERET dataset. Yoon et al. [14] utilized Mobile-Unet using the U-Net segmentation model, which includes MobileNetV2's optimization methods. Hair segmentation accuracy was tested on people of various genders and ethnicities, and the average accuracy was 89.9%.

Nguyen et al. [15] also developed a system for facial beard synthesis and editing. Nonetheless, this method only works on images of high resolution to indicate that facial hair still exists. Also, there was a need for the Graph cuts-based initial seed method. Their approach indicated was the failure of conclusive experimental results.

Hoang et al. [16] developed a system that concurrently senses and segments the beard/mustache. Both the pre-trained model and the self-trained model are used to benefit. They had been focusing on superpixels. Propose an incremental search approach to surmount landmark limitations. Propose a new feature that can accentuate the high-frequency facial hair detail. They use databases such as MBGC, FERET, and PINELLAS. PINELLAS has the smallest amount of time to extract the features.

Hoang et al. [17] perform detailed and varied network research to check our proposed facial hair self-detection and segmentation algorithm. The photos used for our experimental analysis are drawn from the NIST Multiple Biometric Grand Challenge-2008 (MBGC), including 34,696 front pictures of 810 people with different facial expressions and lighting conditions. In addition, the NIST Color Facial Recognition Technology (FERET) includes 989 color images of 989 objects under different illuminations. There is a slight improvement in precision on these three databases, but MBGC has higher accuracy.

3 Methodology

3.1 Proposed Solution

In this proposed method, firstly, an image is passed to the system as input, for which we need to change the color of hair or beard. In the next step, we detect the face from this image using a well-known library named dlib2. Next, this face image is passed to our trained model, i.e., Mask R-CNN for segmentation of beard and hair from it and to generate a mask for both of these objects in the given image. Mask is a binary image with the exact dimensions as the input image, in which all pixels are set to 1, where beard or hair are detected, and all other pixels are 0 and considered background. With this mask image, a region of interest (ROI) is segmented in the original image. This is a subset in the image representing the beard and hair portion in the image. Also, the model assigns an appropriate class to each segmented region in the image, i.e., beard, hair, or background. The model assigns scores to each object concerning three classes based on the probability of matching each object with all three classes. A class having the highest probability is considered a class label. Finally, we can change the color of these segmented objects with the help of ROI. The user is provided with an option of 12 different colors from where any of the colors can be chosen, and an enhanced image is an output of the system. Figure 4 represents our proposed methodology.

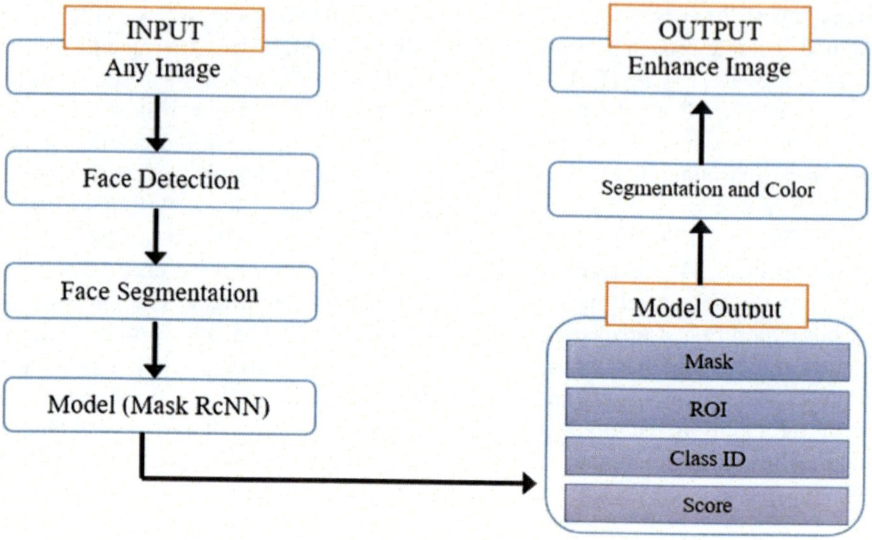

Fig. 4 The proposed methodology for beard and hair segmentation using Mask R-CNN

3.2 Backbone Model

In the mask R-CNN model, we have used ResNet 101 for extracting the features from the images. This is a similar approach to Faster R-CNN, where ConvNet is used for the same task. Thus, the first move is to use ResNet 101 architecture to take a picture and extract its features. These characteristics serve as a reference for the next layer.

3.3 Region Proposal Network (RPN)

In this step, a Region proposed network (RPN) is applied to the feature map obtained from the previous step. As a result, we predict that either an object is present or not in this region. Finally, we take only those regions or feature maps of the image to predict an object.

3.4 Region of Interest (RoI)

The feature maps or regions we have selected in the previous step using RPN may have different shapes. However, we need to have the same shape as all these regions for further processing. Therefore, a pooling layer is applied to equalize the shapes of all the selected regions. In the next step, to predict bounding box coordinates and class labels, these regions or feature maps are passed through an FCN (fully connected network).

The working mechanism for both models is almost the same up to this point. Now, the working mechanism or difference between both frameworks (models) is started. Mask R-CNN also produces masks for each object in the image for segmentation. We measure the area of concern for that first so that the processing period can be minimized. We calculate the Intersection over Union (IoU) with ground truth boxes for all predicted regions. So we can calculate IoU using the following Eq. 1.

$$\mathrm{IoU} = \frac{\mathrm{Areaofintersection}}{\mathrm{AreaofUnion}} \tag{1}$$

3.5 Mask Segmentation

We will add a mask division to the current framework once the ROIs3 is centered on the IoU4 values. A segmentation mask (28 $\times$ 28) is returned for all regions where an object is present. This size is scaled up at the time of inference.

4 Dataset

We have collected a dataset of 1500 images for both categories' hair and beard. The dataset contains 750 images for the beard and 750 against the hair (Head). We have used google images API to download images automatically for different styles of beard and hair. To apply different beards and hairstyles, we have downloaded 20 different images for each of the categories. All these hair and beard style images are in PNG format to avoid the background as the PNG background is transparent.

We manually annotated all of these images using a well-known VGG Image Annotator (VIA) tool at Oxford university to train our modified Mask R-CNN model. Using this tool, we got annotations of all images in a JSON file that was later fed to a network model for training. Figure 5 shows how we annotate each image.

5 Results and Experiments

For experimentation, we have used the NVIDIA Geforce GTX-1080Ti graphical processing unit (GPU). These experiments took 290 min to complete the training of the model on the GPU mentioned above. We have used Keras, OpenCV libraries with Tensorflow backend. The model was trained on 200 epochs. For the evaluation of the model, we have used accuracy metrics.

Fig. 5 Example of image annotation for the training of Mask R-CNN

Fig. 6 Results of segmentation of hair using mask R-CNN

We have used the Mask R-CNN model, which is an extension of Faster R-NN. The main reason for using this model is that Faster R-CNN can only generate bounding boxes around an object in the image. In contrast, Mask R-CNN generates a bounding box and generates a mask against each object in the image, as shown in Fig. 6. Therefore, using the Mask R-CNN model, we can easily segment an object in any given image. Our research needs to change the color of beard and hair so that the bounding box cannot be used for object segmentation because it will change the color of neighboring pixels of beard and hair.

We have used a dataset of 1500 images manually annotated using the VIA annotation tool for training and testing the Mask R-CNN model. We have used 70% of the dataset to train the model and the remaining 30% to test the model—Fig. 6 shown the results of our proposed method for beard and hair detection and segmentation.

We have trained our model on 200 epochs and 3000 iterations per step. Our proposed method Mask R-CNN has achieved 91% accuracy on the testing dataset for detection and segmentation of beard and hair. In [14], we achieved 89% accuracy on the hair segmentation while we achieved more in our experiments. Figures 8 and 9 show the result statistics of this model, i.e., accuracy and loss, respectively. As in Fig. 9, the loss drops drastically from almost 6 to 0.5. After that, it continues to

Fig. 7 Changing of beard and hair color after segmentation

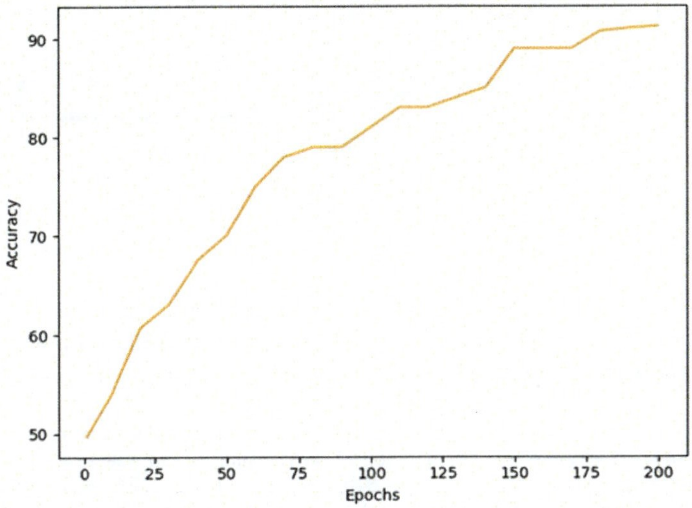

Fig. 8 Accuracy graph of proposed model

decrease with little change. This model outputs a mask of two sub-images containing beard and hair in the original image. We can easily then change the color of these pixels as per user choice, as shown in Fig. 7.

6 Conclusion

Hair and beard style plays an essential role in the personality of any human being. However, the detection and segmentation of hair and beard from an image are never has been an easy task. It contains a lot of complexities in the segmentation of these objects to change the color. We have proposed a solution for this task using the Mask R-CNN model, which is a state-of-the-art model for object segmentation and, for instance, segmentation. It creates masks for each object present in the image. It also returns RoI against each object in the image. We have trained an existing Mask

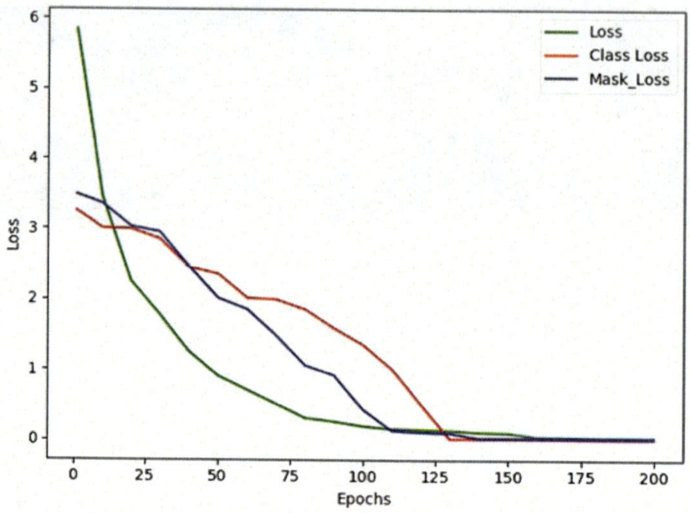

Fig. 9 Error loss graph of proposed model

R-CNN model (trained on 81 classes) on three classes: hair, beard, and background using transfer learning. We have achieved state-of-the-art results by this model for hair and beard segmentation.

Acknowledgements We want to offer our sincere thanks to the National Center of Artificial Intelligence for ultimately supporting our exploration work. The authors also want to acknowledge the appreciation of all group members (associates & the executives) and association (KICS) for their help, commitment, specialized meetings, and information sharing exertion for this. Without their support and encouragement, this research work could not be accomplished.

References

1. https://apps.apple.com/us/app/faceapp-ai-face-editor/id1180884341 access at: Aug 17, 2021
2. Meethongjan K, Dzulkifli M, Rehman A, Altameem A, Saba T (2013) An intelligent fused approach for face recognition. J Intell Syst 22(2):197–212
3. Sharif M, Naz F, Yasmin M, Shahid MA, Rehman A (2017) Face recognition: a survey. J Eng Sci Technol Rev 10(2):166–177
4. Saba T, Kashif M, Afzal E (2021) "Facial expression recognition using patch-based lbps in an unconstrained environment", Proc. IEEE first international conference on artificial intelligence and data analytics (CAIDA), pp 105–108
5. Yacoob Y, Davis LS (2006) Detection and analysis of hair. IEEE Trans Pattern Anal Mach Intell 28(7):1164–1169. https://doi.org/10.1109/TPAMI.2006.139
6. Raza M, Sharif M, Yasmin M, Khan MA, Saba T, Fernandes SL (2018) Appearance based pedestrians' gender recognition by employing stacked auto encoders in deep learning. Futur Gener Comput Syst 88:28–39

7. Khan MZ, Jabeen S, Khan MU, Saba T, Rehmat A, Rehman A, Tariq U (2020) A realistic image generation of face from text description using the fully trained generative adversarial networks. IEEE Access 10(9):1250–1260
8. Yavuzkilic S, Sengur A, Akhtar Z, Siddique K (2021) Spotting deepfakes and face manipulations by fusing features from multi-stream cnns models. Symmetry 13(8):1352
9. Rowley HA, Baluja S, Kanade T (Jan 1998) "Neural network-based face detection", IEEE Trans Pattern Anal Mach Intell 20(1):23–38
10. Chen D, Ren S, Wei Y, Cao X, Sun J (2014) "Joint cascade face detection and alignment". Proc Eur Conf Comput Vis, pp 109–122
11. Viola P, Jones MJ (2004) Robust real-time face detection. Int J Comput Vis 57(2):137–154
12. Li J, Zhang Y (2013) "Learning surf cascade for fast and accurate object detection". Proc IEEE Conf Comput Vis Pattern Recog, pp 3468–3475
13. Le THN, Luu K, Seshadri K, Savvides M (2012) "Beard and moustache segmentation using sparse classifiers on self-quotient images," in Proceedings—international conference on image processing, ICIP, pp 165–168. https://doi.org/10.1109/ICIP.2012.6466821
14. Yoon H-S, Park S-W, Yoo J-H (2021) Real-time hair segmentation using mobile-unet. Electronics 10(2):99
15. Nguyen MH, Lalonde JF, Efros AA, De La Torre F (2008) Image—based shaving. Comput Graph Forum 27(2):627–635. https://doi.org/10.1111/j.1467-8659.2008.01160.x
16. Le THN, Luu K, Zhu C, Savvides M (2017) Semi self- training beard/moustache detection and segmentation simultaneously. Image Vis Comput 58:214–223. https://doi.org/10.1016/j.imavis.2016.07.009
17. Shen Y, Peng Z, Zhang Y (2014) "Image based hair segmentation algorithm for the application of automatic facial caricature synthesis." Sci World J 2014. https://doi.org/10.1155/2014/748634

A Robust Remote Sensing Image Watermarking Algorithm Based on Region-Specific SURF

Uzair Aslam Bhatti, Zhaoyuan Yu, Linwang Yuan, Saqib Ali Nawaz, Muhammad Aamir, and Mughair Aslam Bhatti

Abstract Remote sensing image watermarking algorithm has weak resistance to geometric attacks; therefore, paper proposes a reversible watermarking algorithm based on SURF (Speeded Up Robust Features) feature points to select ROI and embed mid and low-frequency subbands, respectively, which can effectively resist geometric attack. The algorithm first extracts the SURF feature points of the carrier and then performs an inverse wavelet transform on the carrier image to filter out the low-frequency coefficients of the ROI and the intermediate frequency coefficients of the non-interest area (ROB). With sampling pyramid decomposition, the near subband after watermark decomposition is embedded in the low-frequency subband of the region of interest, and the residual subband is embedded in the intermediate frequency coefficient of the non-interesting region. Experimental data show that the algorithm can resist conventional geometric attacks. The similarity of the watermark is high, and the NC value is kept above 0.89, which has good reversibility and robustness.

Keywords SURF feature detection · Reversible watermark · Remote sensing image

U. A. Bhatti · Z. Yu · L. Yuan (✉) · M. A. Bhatti
School of Geography, Nanjing Normal University, Nanjing 210023, China

Key Laboratory of Virtual Geographic Environment, Ministry of Education, Nanjing Normal University, No. 1 Wenyuan Road, Nanjing, China

S. A. Nawaz
College of Information and Communication Engineering, Hainan University, Haikou 570228, China
e-mail: 90472@njnu.edu.cn

M. Aamir
Department of Computer Science, Huanggang Normal University, Huangzhou 438000, Hubei, China

A. Ullah et al. (eds.), *Proceedings of International Conference on Information Technology and Applications*, Lecture Notes in Networks and Systems 350,
https://doi.org/10.1007/978-981-16-7618-5_7

1 Introduction

In computer network information security, data encryption technology is a very secure computer network information protection measure. The fractal theory of natural geometry has been widely used in many fields of natural science and social science. It is not only a new branch of modern mathematics but also very active in nonlinear scientific research. Since the most important feature of fractals is self-similarity, fractals are highly favored by information scientists [1]. Especially in the past ten years, fractals have played an increasingly important role in computer graphics, computer vision, and image processing and analysis [2]. To play a role in its copyright protection and data security maintenance. At present, digital watermarking technology has achieved great development, and various algorithms based on the spatial domain, transform domain, and compression domain has been proposed [3]. At present, the development and application of remote sensing technology have become a hot field. Remote sensing images are widely used in agriculture and forestry, water conservancy, environment, meteorology, military, and other departments, and have played a role in promoting the construction and development of the national economy [4]. The rapid development of computer multimedia technology and network technology has not only brought convenience to the transmission and application of remote sensing images, but also enabled some malicious attackers to edit, copy, and distribute remote sensing images without restrictions, greatly harming the ownership of remote sensing images [5]. The interests of the owner. It can be seen that the copyright protection of remote sensing images has become very important. At present, the development and application of remote sensing technology have become a hot field. Remote sensing images are widely used in agriculture and forestry, water conservancy, environment, meteorology, military, and other departments, and have played a role in promoting the construction and development of the national economy [6].

ROI (Region of Interest) and ROB (Region of Background) regions, and embed them as matched watermark images. Functions used for modification it cannot be embedded separately, and it is not qualified to restore the image carrier after the watermark extracted is not high. Embed two watermarks in the JPEG2000 image and embed both in the ROI and background area of the JPEG2000 image to digital watermarks [7, 8, 9]. This algorithm is of guiding significance for designing ROI watermarks. Unfortunately, ROI data cannot be recovered after watermark extraction, and the high quality of ROI cannot be guaranteed. They designed an ROI reversible watermarking algorithm in the Contourlet domain [10]. The algorithm embeds the watermark into the subband of the Contourlet of the image ROI through differential technology. After the extraction of the watermark, the data of ROI can be recovered without loss. The reversible watermarking algorithm has important reference significance for the watermarking algorithm in the region of interest [11]. Block watermarks are included in the low-frequency subband components of interest. Once the extracted watermark exists, the data algorithm cannot retrieve the field of interest. Each value represents the main feature of the image. If the image is subjected to a general attack after the watermark is embedded, the single value is practically unchanged [12, 13]. The differential histogram technique can ensure the high quality of the ROI after watermark extraction. Scrambling can enhance the security effect of watermarking. They implemented three algorithms for pixel scrambling, row scrambling, and color saturation scrambling, and obtained the best encryption effect for pixel scrambling. In response to the above problems, this paper proposes a reversible watermarking algorithm based on the combination of ROI and SURF, which can resist signal attacks and geometric attacks and can restore the original image with high quality.

2 Related Theories

2.1 SURF Feature Detection

SURF (speeded up robust features) is a fast-robust local feature detection algorithm proposed based on SIFT operator. In general, the standard SURF operator is several times faster than the SIFT operator and has better robustness under multiple images [14]. This paper uses ROI selection based on SURF features. The basic idea is as follows: first, calculate the integral image and traverse the image once to get the sum of all pixels. Then construct the Hessian matrix [15] and perform Gaussian filtering on the image. After filtering, the Hessian matrix expression is:

$$H = \begin{bmatrix} L_{XX}((x, y), \sigma) & L_{XY}((x, y), \sigma) \\ L_{YX}((x, y), \sigma) & L_{YY}((x, y), \sigma) \end{bmatrix} \tag{1}$$

If the Hessian matrix discriminator has an extreme value, the current point will be brighter or darker than the surrounding points, and the candidate object may be divided by the extreme value. To increase the speed, SURF uses a box filter to approximate a Gaussian filter. If the endpoint is a physical endpoint, it is very important to calculate the Hessian discriminant for each pixel. If it is a positive number, the pixel is a local extreme point, otherwise, it is not. The extreme point is obtained is used as a candidate feature point. Then, the non-maximum suppression of the 3 * 3 * 3 cube neighborhood adjacent to this point [16], that is, the candidate extremum point is related to 8 extremum points of the same scale neighborhood and 18 extremum points of the adjacent scale. In comparison, the higher the significance of the pixel and the greater the contribution to the ROI selection. The feature point contribution is defined as

$$w_p = \left| \frac{v_p - \mu_{N(P)}}{\mu_{N(P)}} \right| \tag{2}$$

where v_p represents the d (H) value of the feature point p, and $\mu_{N(P)}$ is the average of the d (H) values of 26 points around the point p. The matrix composed of the contribution of feature points is the contribution matrix. Using the idea of dynamic programming to determine the largest sub-matrix, the matrix is the part with the largest contribution of feature points, that is, ROI.

SURF feature point correction: Let (X_i, Y_i) and $\left(X_j, Y_j\right)$ be any two feature points in the original image feature points, $\left(X'_i, Y'_i\right)$ and $\left(X'_j, Y'_j\right)$ are Feature points of image matching after suffering a geometric attack.

Rotation correction: If the number of matching feature points is N, then the angle between the vectors of the matching feature points of the two images is the angle of rotation. From the vector angle formula (3), the maximum rotation angle is removed. The obtained angle is averaged to obtain the rotation angle β.

$$\beta_i = \frac{(x_i - x_j)\left(x'_i - x'_j\right) + (y_i - y_j)\left(y'_i - y'_j\right)}{\sqrt{(x_i - x_j)^2 + \left(x'_i - x'_j\right)^2}\sqrt{(y_i - y_j)^2 + \left(y'_i - y'_j\right)^2}} \tag{3}$$

$$\beta = \frac{1}{n}\sum_{i=1}^{n}\beta_i \quad i \le N - 1, n \le N - 1 \tag{4}$$

Scaling correction: According to the matching feature points of the two images, the scaling ratio of the image length and width can be estimated, and the points with larger errors can be removed, and the scaling ratio of length and width can be obtained by averaging.

$$\alpha_x = \frac{\left|x_i - x_j\right|}{\left|x'_i - x'_j\right|}, \quad \alpha_y = \frac{\left|y_i - y_j\right|}{\left|y'_i - y'_j\right|} \tag{5}$$

Translation correction: Calculate the difference between the abscissa and ordinate of each pair of matching feature points of the two images, remove the larger error value, and calculate the average value to get the translation distance.

$$\begin{cases} \Delta x = \left|x_i - x'_i\right| \\ \Delta y = \left|y_i - y'_i\right| \end{cases} \tag{6}$$

2.2 *Sampling Pyramid Decomposition*

The digital watermark is sampled, and the residual subband is calculated to generate a sampling golden tower composed of a series of residual subbands and an approximate subband. The image of this golden tower structure has scalable characteristics. Set the original image G_0 as the bottom layer of the sampling pyramid (layer 0), down-sampling G_0 to obtain the first layer G_1 of the sampling golden tower, and then fill the G_1 with the interpolation method to form the same as the original image The size of the image G_0^*. Then the difference between G_0 and G_0^* is used to construct the residual image L_0. After decomposing the sampled golden tower in one layer, an approximate image G_1 and a residual image L_0 are generated. If the sampling gold tower decomposition needs to be continued, a similar operation is performed on the approximate subband image G_1 to generate an approximate image G_2 and a residual image L_1.

$$G_0^*(i, j) = 4 \sum_{m=-2}^{2} \sum_{n=-2}^{2} w(m, n) G_0\left(\frac{i+m}{2}, \frac{j+n}{2}\right) \tag{7}$$

Among them:

The image is composed of 5/3 3-level IWT to decompose the extract of wavelet coefficients of ROI and ROB.

(1) Arnold scrambling of watermark and three-level sampling pyramid decomposition to obtain four subband data: G_3, L_2, L_1, and L_0, where G_3 is approximate subband, L_2, L_1, and L_0 are the third level, second level, and the first level residual subband.
(2) The approximate G_3 subband watermark is decomposed into a sampling pyramid, and then ROI is embedded in the LL_3 subband using the reversible watermark histogram algorithm.
(5) The residual subbands L_2, L_1, and L_0 of the watermark are embedded into LH_3, LH_2, and LH_1 of ROB through singular value decomposition. Embedding method: After each h–h block is divided into h × h blocks, SVD decomposition is performed, $A = USV^T$, and Q = round (S (1,1) / Q) is calculated. S (1,1) represents the first singular value after singular value decomposition of each block, q is the embedding strength, and round is rounding.

Embed the watermark according to Eq. (10).

$$G_0^*\left(\frac{i+m}{2}, \frac{j+n}{2}\right) = \begin{cases} G_0\left(\frac{i+m}{2}, \frac{j+n}{2}\right) & \frac{i+m}{2}, \frac{j+n}{2} \text{is an integer} \\ 0 & \text{else} \end{cases} \tag{8}$$

When reconstructing the image sampling pyramid, from the top to the bottom of the sampling golden tower, the following formula is used to restore layer by layer, and then the original image is obtained.

When reconstructing the image sampling pyramid from the top to the bottom of the tower, use the following formula to copy the sampled gold layer by layer, and then save the original image.

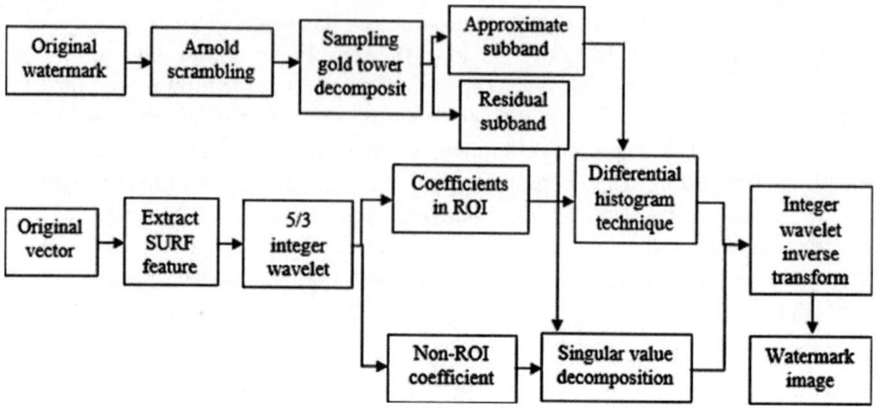

Fig. 1 Watermark embedding algorithm block diagram

$$\begin{cases} G_1 = L_1 + G^*_{l+1} & 0 \leq l \leq N \\ G_N = L_N & l = N \end{cases} \tag{9}$$

3 Watermark Embedding and Extraction

3.1 *Watermark Embedding*

The specific steps of watermark embedding are shown in Fig. 1. Extract the SURF feature points of the carrier image, as described in Sect. 1.1, select the image ROI according to the part with a large contribution of feature points:

$$S'(1.1) = \begin{cases} (Q - 0.5) \times q \text{ if } T_i = 1 \\ (Q + 0.5) \times q \text{ if } T_i = 0 \end{cases} \tag{10}$$

After the singular value is modified, an inverse SVD transformation is performed.

3.2 *Watermark Extraction*

The specific steps of watermark extraction are shown in Fig. 2.

(1) Use the SURF feature point S' and the original carrier SURF feature point S after the geometric attack on the watermark carrier to correct the geometric attack on the watermark carrier.

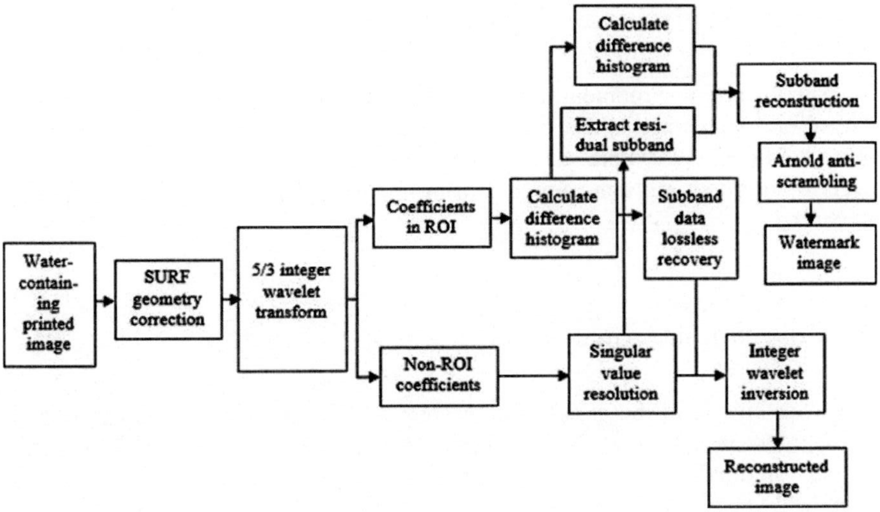

Fig. 2 Watermark extraction algorithm block diagram

(2) The corrected image undergoes a three-level integer wavelet transform to extract the coefficients in ROI and ROB respectively.

(3) Extract the approximate subband information of the watermark in the LL_3 subband of the region of interest using the differential histogram reversible watermarking algorithm and restore the wavelet subband data of the ROI.

(4) Using the singular value decomposition algorithm to extract the watermark residual subband information L_2, L_1. , L_0 of LH_3, LH_2, LH_1 in ROB. The extraction method is similar to the embedding method, and SVD

decomposition is performed on each h × h block, $A' = US'V^T$ and calculates $d = \text{floor}\big(S'(1,1)/q\big)$, where floor is rounded down and $S'(1,1)$ is the first singular value of each sub-block. Calculate the value of mod(d,2), and use parity discriminant (11), to extract the subband information of each resolution watermark.

$$W = \begin{cases} 1 \text{ if } \bmod (d, 2) = 1 \\ 0 \text{ if } \bmod (d, 2) = 0 \end{cases} \tag{11}$$

Perform sampling pyramid reconstruction on the watermark subband information extracted in st (3) and step (4). Then, the inverse Arnold transformation is performed on the reconstructed image to obtain the extracted watermark.

4 Results Analysis and Discussion

The experimental environment is MATLAB2018, which performs invisibility test, multi-resolution extraction test, and robustness test, respectively. The experimental carrier is 512 × 512 remote sensing image, and the watermark is 32 × 32 binary image. Figure 3 shows that remote sensing image and the method of watermark embedding.

4.1 Conventional Attacks

The carrier images of the experiment are remote sensing images. The embedding intensity of the watermark in ROB and remote sensing image is obtained after embedding the watermark, as shown in Fig. 3, and the peak signal-to-noise ratio (PSNR) is shown in Table 1. The Gaussian noise 4% and JPEG 20% shows that NC is 1 after extraction, and the robustness is very good.

Fig. 3 Original remote sensing image and watermark

Table 1 The NC PSNR under conventional attacks

Conventional attack	Gaussian noise			JPEG compression		
	2%	4%	6%	10%	20%	50%
PSNR (db)	17.37	14.69	13.23	23.7	25.69	28.66
NC	1	1	1	1	1	1

Table 2 PSNR and NC under geometric attacks

Geometric Attacks	Attack strength	PSNR (dB)	NC
Rotation (clockwise)	10°	12.71	0.86
	30°	11.57	0.96
	50°	11.43	0.86
Rotation (Anticlockwise)	10°	12.64	1
	30°	11.58	0.96
	50°	11.41	0.96
Scaling	× 0.6	–	0.82
	× 0.8	–	0.86
Translation (Right)	10%	11.71	1
	20%	10.77	0.82
	30%	10.18	0.85
Translation (down)	10%	12.46	0.80
	30%	10.87	0.90
	50%	10.05	0.92
Clipping (Y direction)	10%	–	0.96
	30%	–	0.96
Clipping (X direction)	10%	–	0.96
	30%	–	0.96

4.2 Geometric Attack

According to the algorithm proposed in this paper, the image under attack is corrected and then the watermark is extracted, and the image is only rotated, respectively. The rotation angle of the test is set to 10°–50°. The algorithm first calculates the difference of the original image, calculates the difference histogram of the image and finds the peak value, and embeds the watermark through the peak value.

Table 2 and Fig. 4 show that the results against different attacks: For geometric attacks such as translation, rotation, and scaling, the NC values extracted by the algorithm in this paper are all above 0.82, and the NC value can be 1 when the rotation angle anticlockwise is 10° and translation right 10%. It can be seen that all attack results of NC value are good.

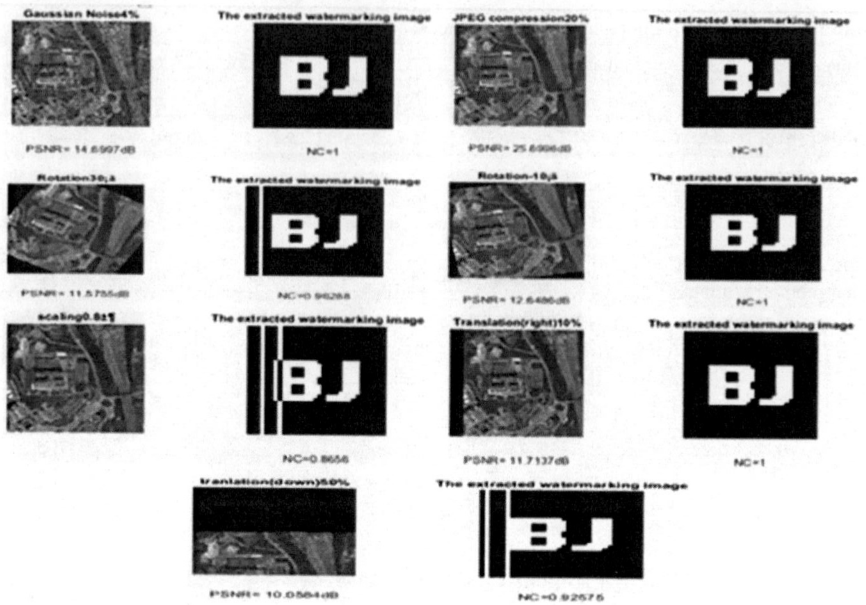

Fig. 4 Different attacks on remote sensing image

5 Conclusion

Based on the remote sensing image, this paper proposes a new digital watermarking technology that uses the local self-similarity in the image to embed the watermark. This paper uses the invariance of SURF feature points to propose a reversible remote sensing image watermarking algorithm based on SURF feature points to select the ROI region. The carrier image is subjected to SURF feature point extraction and then subjected to three-level integer wavelet decomposition, and the watermark is subjected to three-level sampling golden tower decomposition. The approximate subband of the watermark is embedded in the low-frequency subband of the ROI, and the residual subband is embedded in the LH_3, LH_2, and LH_1 in ROB, a retractable structure is formed. Experimental results show that the algorithm has strong robustness against JPEG compression, noise addition, and sharpening attacks, and the watermark embedding has little effect on image classification.

References

1. Bhatti UA, Yu Z, Li J, Nawaz SA, Mehmood A, Zhang K, Yuan L (2020) Hybrid watermarking algorithm using Clifford algebra with Arnold scrambling and chaotic encryption. IEEE Access 8:76386–76398

2. Barni M, Bartolini F, Cappellini V, Magli E, Olmo G (Jun 2002) Near-lossless digital watermarking for copyright protection of remote sensing images. In IEEE international geoscience and remote sensing symposium, vol 3, pp 1447–1449. IEEE
3. Nawaz SA, Li J, Bhatti UA, Mehmood A, Shoukat MU, Bhatti MA (2020) Advance hybrid medical watermarking algorithm using speeded up robust features and discrete cosine transform. Plos One 15(6):e0232902
4. Barni M, Bartolini F, Cappellini V, Magli E, Olmo G (Dec 2001) Watermarking-based protection of remote sensing images: requirements and possible solutions. In Mathematics of data/image coding, compression, and encryption IV, with applications, vol 4475, pp 191–202, International society for optics and photonics
5. Jing L, Zhang Y, Chen G (Oct 2008) Zero-watermarking for copyright protection of remote sensing image. In 2008 9th international conference on signal processing, pp 1083–1086. IEEE
6. Nawaz SA, Li J, Liu J, Bhatti UA, Zhou J, Ahmad RM (Jul 2019) A Feature-based hybrid medical image watermarking algorithm based on SURF-DCT. In The international conference on natural computation, fuzzy systems and knowledge discovery, pp. 1080–1090. Springer, Cham
7. Serra-Ruiz J, Megías D (2011) A novel semi-fragile forensic watermarking scheme for remote sensing images. Int J Remote Sens 32(19):5583–5606
8. Zhu P, Jia F, Zhang J (2013) A copyright protection watermarking algorithm for remote sensing image based on binary image watermark. Optik 124(20):4177–4181
9. Zhou J, Li J, Li H, Liu J, Liu J, Dai Q, Nawaz SA (Dec 2019) Multi-watermarking algorithm for medical image based on NSCT-RDWT-DCT. In International symposium on cyberspace safety and security, pp. 501–515. Springer, Cham
10. Nawaz SA, Li J, Bhatti UA, Shoukat MU, Mehmood A (2020) Advance watermarking algorithm using SURF with DWT and DCT for CT images. In Innovation in medicine and healthcare, Springer, Singapore, pp 47–55
11. Bhatti UA, Yuan L, Yu Z, Li J, Nawaz SA, Mehmood A, Zhang K (2021) New watermarking algorithm utilizing quaternion Fourier transform with advanced scrambling and secure encryption. Multimedia Tools Appl 1–21
12. Qin F, Li J, Li H, Liu J, Nawaz SA, Liu Y (Jul 2020) A robust zero-watermarking algorithm for medical images using curvelet-dct and RSA pseudo-random sequences. In International conference on artificial intelligence and security, Springer, Cham, pp 179–190
13. Li LL, Sun JG (2012) A watermarking algorithm for remote sensing image based on DFT and watermarking segmentation. In Advanced Materials Research. Trans Tech Publications Ltd, vol 433, pp 2504–2508
14. Sweldens W (1996) The lifting scheme: a custom-design construction of biorthogonal wavelets. Appl Comput Harmon Anal 3(2):186–200
15. Lei B, Soon Y, Zhou F, Li Z, Lei H (2012) A robust audio watermarking scheme based on lifting wavelet transform and singular value decomposition. Signal Process 92(9):1985–2001
16. Liu J, Li J, Ma J, Sadiq N, Bhatti UA, Ai Y (2019) A robust multi-watermarking algorithm for medical images based on DTCWT-DCT and Henon map. Appl Sci 9(4):700

Fake News Identification on Social Media Using Machine Learning Techniques

Hafiz Yasir Ghafoor, Arfan Jaffar, Rashid Jahangir, Muhammad Waseem Iqbal, and Muhammad Zahid Abbas

Abstract The devastating effect of spreading fake news related to politics, health, and customer reviews cannot be neglected over social media on the decision-making approach of an individual. The problem of fake news needs the attention of social media administrators, law enforcement agencies, and academic researchers. To handle this issue, researchers suggested various artificial intelligence techniques. However, most of the studies used only a specific type of news that leads to dataset biases. This study used three different standard datasets collected from Kaggle and GitHub. Preprocessed the datasets to remove unwanted text. Then these preprocessed datasets are applied on three classifiers: passive aggressive, machine learning, and naïve Bayes of 30–70, 40–60, 50–50, 60–40, and 70–30, respectively. To evaluate the performance accuracy, precision and recall are used. Results clearly show that this study outperforms the state-of-the-art techniques.

Keywords Fake news detection · Machine learning · Classification · Social media · Twitter

1 Introduction

With the invention of web 2.0, social media has become so popular that everyone wants to use it and has access to diffuse large amounts of information. Social media is a novel platform where millions of users create their profiles and act as a community regardless of their physical availability and geographical location [3]. Social media

H. Y. Ghafoor · A. Jaffar · M. W. Iqbal
The Superior University, Lahore, Pakistan
e-mail: arfan.jaffar@superior.edu.pk

M. W. Iqbal
e-mail: waseem.iqbal@superior.edu.pk

R. Jahangir (✉) · M. Z. Abbas
COMSATS University Islamabad, Vehari Campus, Pakistan

M. Z. Abbas
e-mail: zahidabbas@cuivehari.edu.pk

A. Ullah et al. (eds.), *Proceedings of International Conference on Information Technology and Applications*, Lecture Notes in Networks and Systems 350,
https://doi.org/10.1007/978-981-16-7618-5_8

applications, like Facebook, Twitter, Instagram, YouTube, etc., are online platforms where people create their profiles for sharing their opinions, posting photos, videos, and establish a connection with people across the globe. According to an individual's profile and interests, these applications suggest further links and generate an idea about an individual or product to make decisions.

As there is no mechanism of news authentication, the growing trend of fake news spreading on social media is alarming and needs to be cater. Social media has become a platform for cybercrimes like phishing [1], spamming [35], and malware spreading [34]. The most compromised areas of fake news are politics [4, 17], medical [11], [13, 20], and E-Commerce [23].

Firstly, fake political news puts the people away from politics [29] and polarizes the supporters of politicians into groups [29]. Bessi and Ferrara [5] employed a bot detection algorithm to collect election data using streaming API and reported that, during the 2016 US presidential election campaign, one-fifth of bot accounts were active for the Trump election campaign. The study conducted by Howard and Kollanyi [18] reported that bots played an active role in the Pro-Trump election debate and hated speech during the 2016 US election campaign. Moreover, [14] concluded that those bot accounts that helped President Trump win the 2016 US elections were reactivated during the 2017 French elections. Secondly, in the medical field, many medical professionals interact with people on social media for consultancy [32]. It plays a vital role in developing the online medical center to improve healthcare service [32]. However, quality control and posted information credibility are crucial [12]. For instance, a study by researchers found that half of the students could not discriminate between peer review health solutions and non-peer review solutions [20]. In addition, a study by [13] analyzed that 40% of the shared link on health are fake. Lastly, social media also plays a significant role in e-business through customers' opinions and product feedback. Companies utilize these opinions and input for further decision-making to enhance revenue [16]. However, [23] detected spam reviews on social media blogs and categorize them into three groups namely, untruthful reviews, reviews on brans, and non-reviews [8].

As discussed earlier, fake news can be spread intentionally to deceive people, and it is becoming prominent as there is no mechanism for news authenticity and verification of data. The verification process for news reporters is not accessible due to the amount and speed of data propagation [6]. Thus, the publication of misleading content can defame publishing agencies. In January 2017, a German government spokesperson stated that they "are dealing with a phenomenon of a dimension that [They] have not seen before," referring to the spread of fake news [26]. Similarly, one group of researchers found that 75% accuracy found when high school students were having trouble determining whether or not an article was fake [9, 10]. Fake news diffusion is faster, deeper, and has more impact than factual news [33]. There is much hidden information inside comments, text, images, and videos, and it was not easy to find human behavior [22].

Nowadays, human behavior is analyzed by a social network analysis (SNA) which is the qualitative and quantitative analysis of a social media network [27]. Currently, two main approaches are used for SNA, namely ML and DL. In ML techniques, convenient features, like user, content, URL, and network-based [7], are generated from selected datasets to form a master feature vector (MFV). These MFVs train ML classifiers to build classification models [19]. This classification model is evaluated using different evaluation metrics with a test dataset. However, the discriminative features can enhance the performance of ML techniques. DL techniques were introduced to overcome the fatigue feature generation and biasness on feature types [2].

This type of news is intentionally written to misguide the reader because a typical reader has no tool to differentiate between true and false information [28]. Most of the studies in fake news detection used twitter-based datasets. Twitter is a microblog on which people communicate using a short text (140 characters) called a tweet. This short text can include a link to a website. Twitter users follow other users, and this follower can see all the posts of the followers timeline. New tweets can be created or retweet the existing tweets. Every tweet has its text, number of likes, and number of retweets (responses). It is easy to access its public data using series of authentications and APIs. It has more than 327 million monthly active users. It gained recognition over the years as a powerful and quicker news broadcasting platform. All the organizations, political parties, celebrities, and even government offices have their Twitter accounts to share news quickly.

2 Propoced Methodology

At the very first stage, three datasets are collected from Kaggle and GitHub. Then this data is preprocessed to remove unwanted text. Then the information is divided into two parts training and test data, followed by features extraction using python APIs. After features extraction, three different classifiers are trained and then tested. The proposed methodology employed in this study is shown in Fig. 1. The detail of each phase is presented in subsequent subsections.

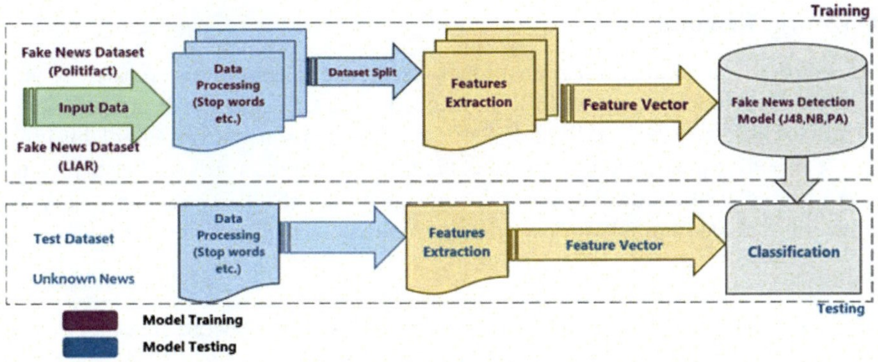

Fig. 1 Proposed research methodology

2.1 *Datasets for Fake News Detection*

Three datasets are used in this study. Detail is given below:

Dataset1: This dataset is collected from "Tencent's Fact Platform" and "People Daily Online," comprised of 3 columns, namely title, content, and label. The contents collected from "People Daily Online" are Real, while others are Fake from 15,825 records. In addition, the last column has two classes, namely Real and Fake.

Dataset2: This dataset has 7796 records with four attributes. These attributes are id, title text, and label. It is the unique identification number of each record. Title attribute contains the title of given news, and text has a detail of each information. Label column has a fake and real label.

Dataset3: This is an entire dataset. The contents of these datasets are collected from various other datasets. It also has the same column as in dataset two with the same number of features.

2.2 *Data Processing*

Preprocessing involves the process of removing unnecessary noise and meaningless data from collected raw data. This raw data contains high-level noise, like complex vocabulary, slang words, emoticons, sparsity, grammatical and spelling mistakes. For better fake news classification results, detailed preprocessing of collected data is required, which involves stopping words removal, punctuation and particular symbol removal, empty spaces removal, spell correction, case conversion, tokenizing, stemming, lemmatization, and normalization. However, preprocessing of data is usually not required for the public datasets. Most of the studies employed publicly available benchmark datasets to avoid preprocessing fatigue, biasness, easy comparison with existing models, and availability of processed data. In contrast, few articles [15, 33] discussed basic preprocessing techniques, including removing stop words, stemming, and splitting/tokenization. Detail is given below:

Stop Words Removal

Stop words are commonly utilized words in a sentence. These words are useless and affect the classification process [30]. In English language, stop words are pronouns and articles like "a," "the," "is," "and," "are," etc. The possible reason behind this is to remove less informative words and select the most important words for better results of fake news classification. The natural language toolkit (NLTK) has a dictionary of stop words and a commonly used stop word removal in Python language.

Stemming

This method is utilized to discover the stem or root of each word like the words connect, connection, and attached are stem to its core as connect [25]. They are converting the morphological form of the term to its stem, considering that each one is semantically the same [30]. This suffixes removal aims to save processing time and memory, reduce corpus size, and find standard stems. Two types of errors can be occurred in stemming: (i) under-stemming and (ii) over-stemming. In under-stemming, two words contain the same root that is not of different stems, and in over-stemming, two different words stem from the same word. NLTK toolkit is used for stemming.

Splitting/Tokenization

In this phase, a stream of text is converted into sentences, words, and meaningful symbols [21]. The purpose of tokenization is to explore the sentence and find keywords, and this group of tokens becomes an input for the next parser or fake news classification. After tokenization, there is a challenge to tackle abbreviations and acronyms that need to convert into meaningful words. Word tokenization with Python NLTK, Mila tokenizer, NLTK word tokenizer [31] tools can be used for tokenization.

2.3 Machine Learning Techniques for Fake News Detection

ML is a field of artificial intelligence where the system can learn from features and improve based on experience. There are three types of ML approaches, including unsupervised, supervised, and reinforcement learning. In this study, passive aggressive, decision tree, and naïve Bayes classifiers are used. All these algorithms are run machine learning algorithms. In this approach, the machine learns under guidance from training data and explicitly tells the device about the class of specific input data. A complete overview of the machine learning algorithm is given in Fig. 2.

Passive Aggressive

This classifier is a part of the machine learning family. This classifier is used where data size is huge and very useful for real-time data. As in real-time, data comes in sequence, whereas in dataset sequence is not maintained. In this method, the classifier uses an example for training, update itself, and through it. If data is actual, it is put into a passive list, otherwise in an aggressive list. Experimental results show that this classifier outperformed as compared to DT and NB.

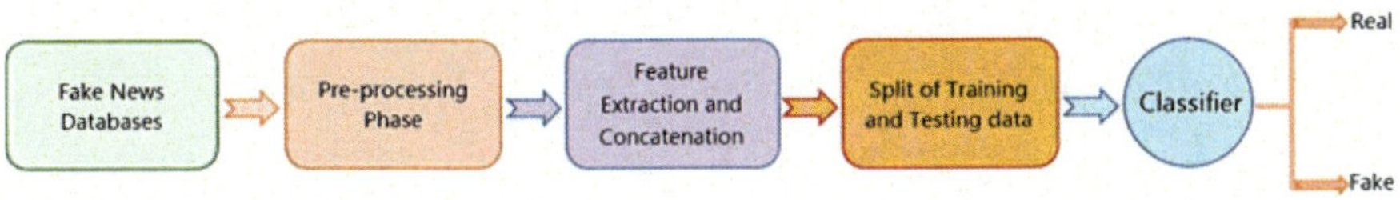

Fig. 2 Classical machine learning process

Decision Tree

A decision tree (DT) is a tree shape having edges, paths, decision nodes, and leaf nodes used to decide the course of action. Every branch of the tree denotes a possible decision, reaction, or occurrence. DT is simple to understand and easy to interpret; hence, the DT is a core algorithm that can be utilized to analyze data and construct a graphical classification model. Attribute test in DT is represented in core node; test results are described in the form of branches, and every leaf node represents a class. The advantages of using DT are simple to understand, easy to visualize and prepare data, and handling both categorical and numerical data. However, slight variations in training samples can result in significant variations in decision logic, making it easy to overfit and challenging to interpret the large decision trees. Various studies employed the DT algorithm for fake news detection. Experimental results show that DT gave better results as compare to NB.

Naïve Bayes

Naive Bayes (NB) is a widely used ML classifier for various applications such as face recognition, weather prediction, medical diagnosis, and news classification. It works on Bays theorem that operates on the assumption that every data pair is independent and equivalent in calculating predictive features. This theorem gives the conditional probability of an event X shown another event Y has occurred. Mathematically, this can be written as:

$$P(Y|X) = \frac{P(X|Y)P(Y)}{P(X)}$$

The advantages of NB on other classifiers are: easy to implement, needs less training data, handle both continuous and discrete data, highly saleable with the number of predictors and data points, highly effective for real-time prediction, and not sensitive for irrelevant features. However, NB cannot learn the relationship between the elements and cause data distribution and scarcity [1]. NB assumes that the text is generated by a parametric model and utilizes training data to compute Bayes-optimal estimates of the model parameters. With these approximations, it categorizes generated test data [24]. NB classifier can deal with an Arbitrary Number of continuous or categorical independent features, and a high dimensional density estimation task is reduced to one-dimensional kernel density estimation [7].

2.4 *Evaluation Metrics*

In this step, the classifier's performance is measured to evaluate whether it classifies data right or wrong. The accuracy of the classifier can be calculated by calculating:

- True Positive (TP) is used to measure the exact prediction

Table 1 Confusion matrix

	Actual instances	
	Yes	No
Predicted Instances		
Yes	TP	FN
No	FP	TN

Table 2 Performance evaluation metrics

Metric	Description	Formula
Accuracy	Accuracy is how much the data is labeled correctly. It computes total appropriately known occurrences	$\text{Accuracy} = \frac{(\text{TP+TN})}{(\text{TN+TP+FN+FP})}$
Precision	It is the ratio between correct optimistic guesses over total positive guesses. It minimizes mistakes in guessing positive labels	$\text{Precision} = \frac{\text{TP}}{\text{FP+TP}}$
Recall	It is the ratio between correct positive guesses overall positive labels. The other name is an actual positive rate (TPR)	$\text{Recall} = \frac{\text{TP}}{\text{FN+TP}}$

- True Negative (TN) is a measure in which the classifier predicts it is true, but in actual it is not true.
- False Positive (FP) is the inaccurate prediction of an object
- False Negative (FN) is an instance that is not precisely predicted.

These four members from the confusion matrix for binary classification are as shown in Table 1.

The performance metrics used to evaluate the machine as mentioned above learning classifiers are presented in Table 2.

3 Experimental Setup

This section presents the experimental setup of the construction of the fake news detection models using the extracted features and three machine learning algorithms. An extensive set of experiments was performed to measure the constructed models' performance and compare them with baseline fake news detection models. To evaluate the performance of the constructed affected news detection models, experiments were performed systematically in three different settings. All experiments were conducted in Jupyter Notebook (Python 3.8.5) environment on a PC with 64-bit Windows 10 operating system, 8 GB RAM, and Intel(R) Core(TM) i5-3210 M CPU.

Table 3 Classification results on dataset 1

	Accuracy			Recall			Precision		
Data size	PA	J48	NB	PA	J48	NB	PA	J48	NB
30–70	92.3	78.1	78.8	91.6	77.8	80.3	92.0	78.1	83.1
40–60	92.5	78.5	82.2	93.0	79.9	79.5	92.5	79.8	86.6
50–50	93.3	79.9	78.4	93.1	80.1	84.2	94.0	80.1	85.5
60–40	93.6	79.7	81.7	93.3	80.4	82.8	93.5	80.4	85.3
70–30	**93.7**	**81.7**	**83.3**	**93.6**	**82.4**	**83.3**	**93.1**	**79.1**	**86.2**

4 Experimental Results

This section presents the results of all the experiments conducted in this study. The results are offered based on three experimental settings where three machine learning algorithms, including decision tree (J48), naïve Bayes (NB) and passive aggressive (PA), were trained and evaluated on three different fake news datasets.

4.1 Results of Experimental Setting I

In this experiment, three classifications, as mentioned earlier, were applied on a publicly available dataset1. The dataset was divided into various ratios (30–70, 40–60, 50–50, 60–40, 70–30) for classifier training and testing and to validate the robustness of each classifier. The experimental results revealed that a passive-aggressive algorithm achieved the best performance on all data ratios. The best performance was obtained 93.70% when the dataset was divided by 70–30% ratio (70% data was used for training and 30% data was used for testing). Moreover, the decision tree and naïve Bayes also achieved reasonably good classification accuracy by 81% and 83%, respectively, as shown in Table 3. In addition, all three classification algorithms achieved the lowest accuracy when data was divided by a 30–70% ratio. This is because a small amount of data was used for training the machine learning classifiers.

4.2 Results of Experimental Setting II

In this experiment, three classification algorithms, namely passive aggressive, naïve Bayes, and decision tree, were applied on a publicly available dataset2. The dataset was divided into various ratios (30–70, 40–60, 50–50, 60–40, 70–30) for classifier training and testing and to validate the robustness of each classifier. The experimental results revealed that a passive-aggressive algorithm achieved the best performance

Table 4 Classification results on dataset 2

	Accuracy			Recall			Precision		
Data size	PA	J48	NB	PA	J48	NB	PA	J48	NB
30–70	92.2	77.9	86.8	92.2	78.9	85.0	92.4	78.0	86.5
40–60	92.2	78.7	81.8	92.5	78.5	82.5	92.1	79.0	84.7
50–50	92.8	80.3	80.7	93.5	80.1	82.5	92.8	78.9	86.8
60–40	93.9	83.4	84.2	94.1	82.2	83.0	93.3	81.0	84.6
70–30	**93.9**	**80.3**	**81.5**	**93.9**	**81.2**	**79.9**	**93.9**	**83.0**	**86.1**

on all data ratios. The best performance was obtained 93.90% when the dataset was divided by 70–30% ratio (70% data was used for training and 30% data was used for testing). Moreover, the decision tree and naïve Bayes also achieved reasonably good classification accuracy by 81% and 83%, respectively, as shown in Table 4. In addition, all three classification algorithms achieved the lowest accuracy when data was divided by a 30–70% ratio. This is because a small amount of data was used for training the machine learning classifiers.

4.3 Results of Experimental Setting III

In this experiment, three classification algorithms, namely passive aggressive, naïve Bayes, and decision tree, were applied on a publicly available dataset3. The dataset was divided into various ratios (30–70, 40–60, 50–50, 60–40, 70–30) for classifier training and testing and to validate the robustness of each classifier. The experimental results revealed that a passive-aggressive algorithm achieved the best performance on all data ratios. The best performance was obtained 99% when the dataset was divided by a 70–30% ratio (70% data was used for training, and 30% data was used for testing). Moreover, the decision tree and naïve Bayes also achieved reasonably good classification accuracy by % and 83%, respectively, as shown in Table 5. In

Table 5 Classification results on dataset 3

	Accuracy			Recall			Precision		
Data size	PA	J48	NB	PA	J48	NB	PA	J48	NB
30–70	99.6	99.8	98.5	99.7	99.9	98.6	99.6	99.8	98.5
40–60	99.7	99.8	98.3	99.7	99.8	98.7	99.7	99.8	98.8
50–50	99.8	99.8	98.7	99.7	99.8	98.6	99.7	99.8	98.7
60–40	99.8	100	98.8	99.8	99.8	98.6	99.9	99.8	98.9
70–30	**99.9**	**99.9**	**98.8**	**99.8**	**99.8**	**98.7**	**99.8**	**100**	**98.8**

Bold indicates that overall best results are achieved when the 70–30 data ratio was used for training and testing of model respectively addition, all three classification algorithms achieved the lowest accuracy when data was divided by a 30–70% ratio. This is because a small amount of data was used for training the machine learning classifiers. This dataset is exclusive and unbalance.

5 Discussion

This section presents the theoretical analysis of the fake news classifiers employed in this research. This research reveals that the proposed features coupled with machine learning classifiers can detect fake news with an accuracy between 79 and 99%. The results show that the proposed features with a training and testing ratio of 70–30% provided the maximum accuracy using the passive-aggressive machine learning algorithm on all three datasets. The possible reason for the excellent performance of the passive-aggressive algorithm is that it can handle large scale of data coming from various social media applications like Twitter and Facebook, etc. Moreover, as this unstructured data contains slang words, emojis, emoticons, punctuation marks, and spelling mistakes, preprocessing techniques were applied to remove this unwanted text and improve the classification accuracy. Furthermore, the features extracted performed better because of the stemming and tokenization practices.

6 Conclusion

In this study, we presented a method to classify fake news from social media platforms. As shown in the result section, our classification method gave 99% accurate accuracy, precision, and recall. The passive-aggressive classifier produced excellent results as compared to the decision tree and naïve Bayes classifiers. For future work, we will use real-time data for this purpose.

References

1. Aggarwal A, Rajadesingan A, Kumaraguru P (2012) PhishAri: automatic realtime phishing detection on twitter. eCrime Researchers Summit (eCrime)
2. Ajao O, Bhowmik D, Zargari S (2018) Fake news identification on twitter with hybrid cnn and rnn models. Proceedings of the 9th International Conference on Social Media and Society
3. Al-garadi MA, Varathan KD, Ravana SD (2016) Cybercrime detection in online communications: the experimental case of cyberbullying detection in the Twitter network. Comput Hum Behav 63:433–443
4. Bessi A, Ferrara E (2016) Social bots distort the 2016 US Presidential election online discussion
5. Bessi A, Ferrara E (2016) Social bots distort the 2016 US Presidential election online discussion. First Monday 21(11–7)

6. Boididou C, Papadopoulos S, Zampoglou M, Apostolidis L, Papadopoulou O, Kompatsiaris Y (2018) Detection and visualization of misleading content on Twitter. Int J Multimedia Inf Retrieval 7(1):71–86. https://doi.org/10.1007/s13735-017-0143-x
7. Buczak AL, Baugher B, Guven E, Ramac-Thomas LC, Elbert Y, Babin SM, Lewis SH (2015) Fuzzy association rule mining and classification for the prediction of malaria in South Korea. BMC Med Inform Decis Mak 15(1):47
8. Ding X, Liu B, Yu PS (2008) A holistic lexicon-based approach to opinion mining. Proceedings of the 2008 international conference on web search and data mining
9. Domonoske C (2016) Students have 'dismaying'inability to tell fake news from real, study finds. National Public Radio 23
10. Edkins B (2016) Americans Believe ey Can Detect Fake News. Studies Show ey Can't.(December 2016). In
11. Eysenbach G (2008) Credibility of health information and digital media: new perspectives and implications for youth. MacArthur foundation digital media and learning initiative
12. Eysenbach G (2008) Credibility of health information and digital media: new perspectives and implications for youth. Digital Media Youth Credibility 123–154
13. Fernández-Luque L, Bau T (2015) Health and social media: perfect storm of information. Healthcare Inf Res 21(2):67–73
14. Ferrara E (2017) Disinformation and social bot operations in the run up to the 2017 French presidential election
15. Girgis S, Amer E, Gadallah M (2018) Deep learning algorithms for detecting fake news in online text. 2018 13th international conference on computer engineering and systems (ICCES)
16. Heydari A, Ali Tavakoli M, Salim N, Heydari Z (2015) Detection of review spam: a survey. Expert Syst Appl 42(7):3634–3642
17. Howard PN, Kollanyi B (2016) Bots,# StrongerIn, and# Brexit: computational propaganda during the UK-EU referendum. Available at SSRN 2798311
18. Howard, P. N., & Kollanyi, B. (2016). Bots,# strongerin, and# brexit: Computational propaganda during the uk-eu referendum. *Browser Download This Paper.*
19. Ishtiaq U, Kareem SA, Abdullah ERMF, Mujtaba G, Jahangir R, Ghafoor HY (2019) Diabetic retinopathy detection through artificial intelligent techniques: a review and open issues. Multimedia Tools Appl 1–44
20. Ivanitskaya L, Boyle IO, Casey AM (2006) Health information literacy and competencies of information age students: results from the interactive online Research Readiness Self-Assessment (RRSA). J Med Internet Res 8(2):e6
21. Kannan S, Gurusamy V (2014) Preprocessing techniques for text mining. Conference Paper. India
22. Lauw H, Shafer JC, Agrawal R, Ntoulas A (2010) Homophily in the digital world: A LiveJournal case study. IEEE Internet Comput 14(2):15–23
23. Liu, B. (2007). *Web data mining: exploring hyperlinks, contents, and usage data.* Springer Science & Business Media.
24. McCallum A, Nigam K (1998) A comparison of event models for naive bayes text classification. AAAI-98 workshop on learning for text categorization
25. Ramasubramanian C, Ramya R (2013) Effective pre-processing activities in text mining using improved porter's stemming algorithm. Int J Adv Res Comput Commun Eng 2(12):4536–4538
26. Ruchansky N, Seo S, Liu Y (2017) Csi: a hybrid deep model for fake news detection. Proceedings of the 2017 ACM on conference on information and knowledge management
27. Scott J (2017) Social network analysis. Sage
28. Shu K, Sliva A, Wang S, Tang J, Liu H (2017) Fake news detection on social media: a data mining perspective. ACM SIGKDD Explorations Newsl 19(1):22–36
29. Tucker J, Guess A, Barberá P, Vaccari C, Siegel A, Sanovich S, Stukal D, Nyhan B (2018) Social media, political polarization, and political disinformation: a review of the scientific literature
30. Vijayarani S, Ilamathi MJ, Nithya M (2015) Preprocessing techniques for text mining-an overview. Int J Comput Sci Commun Netw 5(1):7–16

31. Vijayarani S, Janani R (2016) Text mining: open source tokenization tools-an analysis. Adv Comput Intell: Int J (ACII) 3(1):37–47
32. Viviani M, Pasi G (2017) Credibility in social media: opinions, news, and health information—a survey. Wiley Interdisciplinary Rev: Data Mining and Knowl Discovery 7(5)
33. Vosoughi S, Roy D, Aral S (2018) The spread of true and false news online. Science 359(6380):1146–1151
34. Yang C, Harkreader R, Zhang J, Shin S, Gu G (2012) Analyzing spammers' social networks for fun and profit: a case study of cyber criminal ecosystem on twitter. Proceedings of the 21st international conference on World Wide Web
35. Yardi S, Romero D, Schoenebeck G (2009) Detecting spam in a twitter network. First Monday 15(1)

Utility of Deep Learning Model to Prioritize the A&E Patients Admission Criteria

Krzysztof Trzcinski, Mamoona Naveed Asghar, Andrew Phelan, Agustin Servat, Nadia Kanwal, Mohammad Samar Ansari, and Enda Fallon

Abstract Overcrowding in hospital emergency departments is a rudimentary issue due to patients who are presenting for treatment, but do not require admission or could be treated by their own general practitioner or over-the-counter remedies. This research work analyses the existing process of patient triage admission in accident and emergency departments and attempts to apply deep learning techniques to automate, improve and evaluate the triage process. This research proposed to utilize a deep learning model for efficiency and reducing the requirement for specialized triage professionals when evaluating and determining admission, treatment in accident and emergency departments. Automating the triage process could potentially be developed into an online application which a patient or less specialized medical practitioner could potentially perform prior to presenting at emergency departments, reducing the overall inflow to emergency departments and freeing up resources to better treat those who do require admission or treatment. The core areas to be considered are the use of emergency health records (EHR), as a suitable data source for performing the triage process in emergency departments and the application of deep learning methods using the said EHR dataset(s).

Keywords Deep learning · Densely connected network · Electronic health records (EHR) · A&E admissions · Triage

K. Trzcinski · M. N. Asghar (✉) · A. Phelan · A. Servat · N. Kanwal · E. Fallon
Athlone Institute of Technology, Athlone, Ireland
e-mail: masghar@ait.ie

M. N. Asghar
The Islamia University of Bahawalpur, Bahawalpur, Pakistan

N. Kanwal
Lahore College for Women University, Lahore, Pakistan

M. S. Ansari
Aligarh Muslim University, Aligarh, India

A. Ullah et al. (eds.), *Proceedings of International Conference on Information Technology and Applications*, Lecture Notes in Networks and Systems 350,
https://doi.org/10.1007/978-981-16-7618-5_9

1 Introduction

Accurate triage is a crucial element that is required to ensure smooth and efficient operation of any hospital's Accident & Emergency (A&E) department around the world. It helps to prioritize admissions and allows to distinguish between urgent and less-urgent cases. Unfortunately, in real-life scenarios, triage and prioritization in emergency departments are a complex task and optimal accuracy is hard to achieve due to several reasons. This can be a factor that leads to bottlenecks and overcrowding which is particularly dangerous for patients that require immediate attention. Overcrowding is one of the biggest problems in emergency departments across the world and can result in longer wait times for serious patients, ambulance diversions, increased mortality and lowered personnel morale.

A&E departments must often balance a demand on health services and limited resources available to them at the time. Hence, accurate assessment and prediction of such requirements is crucial. Early identification of patients who need to be prioritized could allow more efficient planning and as a result, it could help with mitigation of impact and prevention of overcrowding.

In this research paper, use of Artificial Intelligence (AI) algorithms in combination with Electronic Health Records (EHR) has been evaluated in relation to automation of triage in A&E settings. Towards that end, first a literature review was conducted which helped build understanding of attempts of predicting admissions using machine learning models, such as logistic regression or natural language processing via neural network models. However, in most cases, it is argued that these methods are challenging to adopt due to complexity in their application. The aim of this article is to contribute to up-to-date research around use of AI in Emergency Departments, specifically by:

- Investigation on prioritization of the Accident & Emergency (A&E) patients around the world that helps to distinguish between urgent and less-urgent cases,
- Investigate the best possible lightweight deep learning for modelling the A&E admissions and patient flow,
- Implementation of Densely Connected Deep Learning model to prioritize A&E admissions by predicting Emergency Severity Index (ESI).

Rest of the paper is organized as follows. Section 2 presents an overview of the most related works. Section 3 presents the proposed methodology with pertinent details of test configurations. Section 4 contains a discussion on the results and their applicability along with the limitations of the presented work. Section 5 contains concluding remarks and scope for future improvements.

2 Literature Review

This section discusses recent related works on triage admission system based on machine learning models. The study in [14] was conducted to determine sensitivity and specificity in ability to predict hospital admissions at the time of triage. The findings of this study have shown that physicians scored poorly (51.8%) when it came to predict whether a patient should be admitted to a hospital (sensitivity) even if their confidence levels were high. A review article [4] with the objective to assess how intelligent clinical decision support systems (CDSS) for triage have been contributing to the improvement of quality of care in the emergency department triage. The top three intelligent techniques most used for the development of the triage models were: Logistic regression, Classification and Regression Tree (CART) and Deep Neural Networks. The most used performance measures were accuracy and 'area under the curve' (AUC).

AI and ML applications that were discussed by Jonathon Stewart et al. [12] include clinical image analysis where a labelled image datasets and DL algorithms are used to detect conditions such as pneumonia from X-ray images or to assess a risk of a stroke in patients, using Computed Tomography (CT) scans. In a study conducted by Jae Yong Yu et al. [17], four ML classifiers have been developed using logistic regression and deep learning models. The primary goal of the models was to predict mortality or ICU admission for patients visiting A&E department. It can be observed that logistic regression and deep learning classifiers achieved the most promising results with accuracy values of 87.2% and 87.6%, respectively.

A similar study was conducted by Scott Levin et al. [10], to evaluate ML-based e-triage and its prediction of severe outcomes and to compare it against Emergency Severity Index (ESI) conventional method. In this case, the AUC values range between 0.73 and 0.92 which proves that the classifier matches or even out-performs more conventional ESI methods. The aim of the study in [3] was to compare the performance of Machine Learning with human triage. The automated software used for the study selects different subsequent survey questions based on patient's answers. They found that inter-rater and intrarater reliability among physicians was low. These results conclude that Machine Learning is better than humans in triage decisions.

In recent years, deep learning has shown promising diagnostic performance in different medical specialties, such as ophthalmology, radiology and dermatology [13]. Despite these positive findings, there are still many limitations in terms of its safe integration into practice. The accuracy of the training datasets ground truth is the most important consideration in a study. The study in [7] is a retrospective cross-sectional study that examined if deep-learning approaches were able to identify critically ill patients only using data immediately available at triage. In this study, neural network and gradient-boosting models demonstrated significantly higher accuracy than traditional methods of triage, suggesting that these models have the potential to significantly enhance the triage process.

In another research paper [6], the authors attempt to train and test four distinct machine learning models, to demonstrate the usage of machine learning to apply the Chinese Emergency Triage Scale (CETS) method accurately and efficiently. The research concludes with the assertion that all, or any of the four applied machine learning techniques can be used to accurately apply the CETS triage classification process and highlighted Extreme Gradient Boosting as the most efficient. In [5], the goal of the research is to apply a mix of machine learning techniques to train and evaluate the models when attempting to predict admission rates using the ESI system. Similar conclusions are found that machine learning can be applied to robustly predict hospital admissions when trained to apply triage classification, with an additional conclusion that the inclusion of historical patient data increases the accuracy of the model(s).

The focus of [9] is towards looking at paediatric emergency departments admissions using the Korean triage and acuity system (KTAS) and paediatric early warning system (PEWS). The authors deduced that the implementation of machine learning techniques can, in some areas, provide significant improvement to existing triage process(es). Shiyao Wang et al. [15] published a method that demonstrates the benefits of using Densely Connected CNN over other techniques for text classification. In this paper, the author suggests a Densely Connected CNN model to be used for text classification to overcome multi-scale features limitations.

Clemens Kruse et al. [8] conducted a systematic literature review to assess EHR use by identifying facilitators and barriers to its adoption. It was concluded that facilitators outweigh barriers by 3:2. The top three facilitators were: increased productivity, improved data and care quality and improved data management. The top three barriers were: errors and missing data, lack of standard for interoperability and loss of productivity due to training activities. In research article [16], the researchers analysed existing research papers in deep learning using EHR, while attempting to highlight challenges in the space of deep learning models of EHR datasets. The most common analytical tasks were found to be disease detection or classification, sequential prediction of medical events, data privacy and concept embedding and data augmentation. The core subject of [11] was an analysis of the existing research in the application of deep learning in medical data, and appropriateness of the usage of deep learning when applied to medical data, including electronic health records (EHR).

3 Methodology

In this section, the proposed test configuration for the prioritization of A&E submission is presented. The proposed architecture is based on Python version 3.7.9, TensorFlow version 2.3.0 and implementation of Dense (Fully Connected) Neural Network model using the TensorFlow framework. The implementation of the classifier is based on Rampurawala [1]. In densely (fully) connected models, all the neurons in a layer are connected to neurons in the next layer. The methodology is shown in Fig. 1.

Dataset used for classification was downloaded from Kaggle website [2] and sourced from the article published by Hong et al. [5]. Original dataset consists of

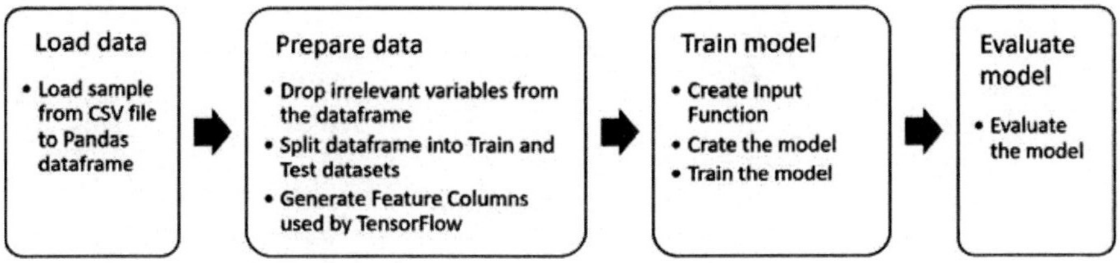

Fig. 1 Test configuration diagram

560,486 observations and 972 variables of Emergency Department visits between March 2014 and July 2017. For this study, different samples of the dataset have been extracted and tested against the DNN Classifier. ESI (Emergency Severity Index) variable has been selected as the dependency variable that consists of 5 classes ranging from 1 to 5 (1 being the highest severity and 5 being the lowest). Figure 1 shows the steps performed while evaluating the model. In the first step, sample dataset is being loaded from the previously prepared CSV into a Pandas dataframe. Pandas dataframes are tabular data structures (rows and columns) that are two dimensional and heterogeneous. They are commonly used in machine learning in combination with Python and its packages, such as NumPy or Sklearn.

To split the sample data into train and test datasets, the Sklearn function called train_test_split was used. Then, dataframe columns are mapped into feature columns which are used by the DNN model as inputs. Because hundreds of different columns are being used in this scenario, a for loop has been created to automatically map these columns to the correct type (numeric, categorical, etc.). To train the model, an input function was created. Input functions are used to feed Pandas dataframes into models provided by TensorFlow Estimators. Like Keras, Estimators provide model-level abstraction that simplify building models within TensorFlow framework. The model used for this study is DNN Classifier from the Estimator API and it was trained using the input function created in the previous step.

4 Results

Initially in this research, the DNN model was trained using approximately 4,985 observations and 215 variables (including the dependant variable–ESI) which contains 5 classes ranging from 1 to 5 (1 being the highest severity and 5 being the lowest). Majority of variables from the original dataset have been removed for the purpose of this study, including variables representing historical labs, historical vitals, imaging, outpatient medications and past medical history. Disposition variable (which was used to hold information on whether patients was discharged or admitted) was also removed from the sample dataset. Some observations were also removed from the sample as they contained ESI index 0 which is outside of ESI scale. The

Table 1 Model optimization tests

Test	Learning R	Batch Size	Epochs	Accuracy	Loss
1	0.001	50	50	0.61	9.52
2	0.005	50	50	0.60	9.19
3	0.01	50	50	0.45	12.22
4	0.0001	50	50	0.62	8.77
5	0.00001	50	50	0.47	11.40
6	0.0001	100	50	0.63	8.65
7	0.0001	200	50	0.64	8.55
8	0.0001	300	50	0.62	8.84
9	0.0001	200	100	0.64	8.51
10	0.0001	200	500	0.65	8.54

tests were conducted using variables representing chief complaints, demographics and triage information.

Several tests were conducted using the above data sample. For each test, different optimization techniques were evaluated (see Table 1): (i) adjusting learning rate of the Adam Optimizer which can reduce local minima (i.e., function minimum values), (ii) adjusting the batch size of the input function which can improve the model's pattern recognition and (iii) adjusting the number of epochs of the input function which determines how effective the model is when it comes to fitting the training data.

Evaluation functions provide two metrics that will allow to measure performance of the model—accuracy and loss. Accuracy represents a percentage of how many observations were accurately predicted vs. the total observations to predict. Loss function is a sum of errors the model made in each iteration. Later in this research, further analysis has been conducted using larger data samples and TensorBoard visualization tool. Regarding the data samples, patient's past medical history variables have also been added to the dataset. The tests were conducted using variables representing chief complaints, past medical history, demographics and triage information.

As per the initial analysis, models have been trained using optimal parameters, i.e. Learning Rate: 0.0001, Batch Size: 200 and Number of Epochs: 500. Two data samples have been used in the experiment that contain 50,000 and 100,000 observations. However, after data pre-processing that involved removing records with missing values, the final number of records reduced to 20,546 and 40,402, respectively.

Table 2 shows the data distribution per class. It can be observed that data distribution is non-uniform, i.e. there is a large discrepancy between the number of records for each class within the datasets. For example, in experiment 1 (Sample 1), there are 18,198 records with the ESI value of 3 (Medium Severity), but only 41 records

Table 2 Records distribution per class

Class	Severity	Description	[S1, S2]
1	Highest	Immediate life-saving intervention is required	30, 41
2	High	High Risk situation, disorientation, severe pain, or vitals in danger zone	5260, 10,129
3	Medium	Multiple resources required, vitals not in danger zone	9404, 18,198
4	Low	Single resource required to stabilize the patient	5198, 10,604
5	Lowest	No resources required to stabilize the patient	654, 1430

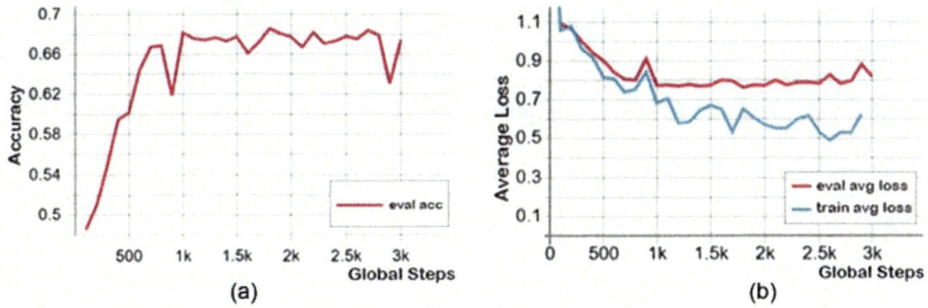

Fig. 2 **a** Evaluation accuracy per 100 global steps (data sample 1), **b** Training (blue line) and Evaluation (red line) average loss per 100 global steps (data sample 1)

with the ESI value of 1 (Highest Severity). This will likely have a negative impact on the accuracy of the model, and it will require further research.

Figure 2a represents accuracy score of the model per each 100 global steps of the model learning process. In contrast with epoch, one global step is one iteration of the model learning process through a single batch of data, where epoch is equal to one iteration through the entire dataset. Improved accuracy can be observed at around 1000 step and it continuous throughout the learning process, up until approx. 2800 steps. The highest accuracy point recorded for the data sample 1 is at 1800 step and it equals 0.6856. Figure 2b shows average loss for training (blue line) and evaluation/testing (red line) datasets per 100 steps in the smaller data sample. It can be observed that discrepancy between the blue and red lines increases significantly after approx. 1000 global steps which can be an indication of over-fitting. At 1000 global steps, the evaluation average loss value is equal to 0.774.

Slightly better results were observed with the larger data sample 2. For this experiment, only 2000 global steps were used instead of 3000 to prevent model over-fitting. Figure 3a visualizes evaluation accuracy with the highest accuracy point at 1,800 step and it equals 0.697. In terms of the average loss, as it can be observed on Fig. 3b, both lines, i.e., validation (green line) and evaluation (grey line) are more tightly coupled than in the previous experiment. This is positive observation as it indicates no over-fitting in this scenario. Also, the lowest average loss can be located at step

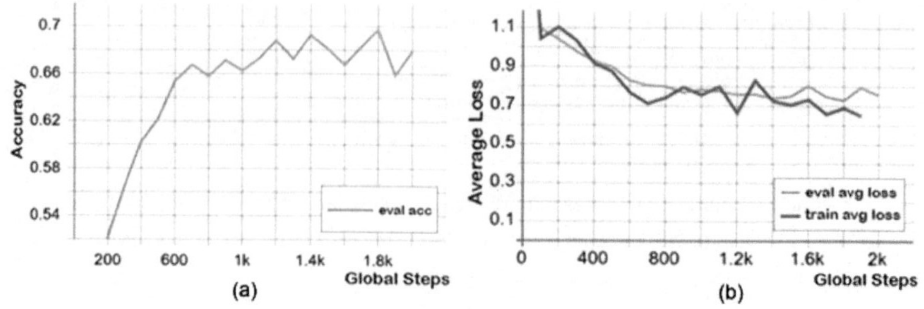

Fig. 3 **a** Evaluation accuracy per 100 global steps (data sample 2), **b** training (green line) and evaluation (grey line) average loss per 100 steps (data sample 2)

Table 3 Precision and recall summary for the experiments

Class	Label	Sample	Precision	Recall
1	Highest Severity	1	0.000	0.000
		2	0.000	0.000
2	High Severity	1	0.689	0.655
		2	0.871	0.386
3	Medium Severity	1	0.648	0.772
		2	0.616	0.875
4	Low Severity	1	0.717	0.543
		2	0.708	0.668
5	Lowest Severity	1	0.361	0.262
		2	0.640	0.034

1800 step and it equals 0.7288. Lastly, the Precision and Recall metrics for the two set of data samples are presented in Table 3.

Figure 4a, b show the confusion matrices for the experiment with the smaller and the larger data samples, respectively. It is another way of visualizing performance of the classifier, and it conveniently shows breakdown of error by classes. Figure 5a, b show the true positive, false positive, false negative and true negative results for each severity level, for data sample 1 and data sample 2, respectively.

5 Conclusion

This research article showed the utility of deep learning model for the A&E admission criteria in hospitals on the basis of five classes. Initially in this research, the optimal performance achieved on the data sample was 0.6512758 accuracy and 8.548257 loss. It was concluded that the optimal hyper-parameters to control the learning process are

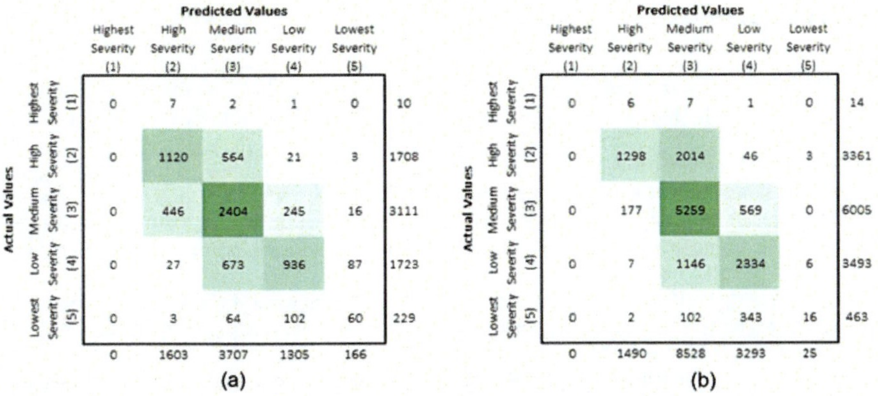

Fig. 4 **a** Confusion matrix (data sample 1), **b** confusion matrix (data sample 2)

		Data Sample 1				Data Sample 2			
Severity	Class	TP	FP	FN	TN	TP	FP	FN	TN
Highest	1	0	0	10	6771	0	0	8	13328
High	2	1120	483	588	4590	1298	192	2063	9783
Medium	3	2404	1303	707	2367	5259	3269	746	4062
Low	4	936	369	787	4689	2334	959	1159	8884
Lowest	5	60	106	169	6446	16	9	447	12864

Fig. 5 True positive, false positive, true negative and false negative values for the two data samples

as follows: Learning Rate: 0.0001, Batch Size: 200 and Number of Epochs: 500. Later in this research, the model performance was improved by using larger datasets (20 K and 40 K records) and additional variables for patients' medical history. Achieved accuracy on the larger evaluation set was 0.697 and average loss was 0.7288.

Recommendations for the future work involve the use of even larger data samples and further optimization using other relevant hyper-parameters. Also, more even data distribution across classes needs to be considered. Due to the imbalance form of data as given in data distribution table (Table 2), the deep learning model accuracy is only at around 70%. Having more evenly distributed data in classes should help to improve the accuracy. Finally, further review of variables available in the original dataset is recommended to identify any additional value adding data.

References

1. "Classification with TensorFlow and dense neural networks," Heartbeat, 8 February 2019. [online]. https://heartbeat.fritz.ai/classification-with-tensorflowand-dense-neural-networks-8299327a818a, Accessed 27th Feb 2021
2. "Hospital triage and patient history data," Kaggle, 3 June 2019. [online]. https://www.kaggle.com/maalona/hospital-triage-and-patient-history-data. Accessed 27 Feb 2021
3. Entezarjou A, Bonamy AKE, Benjaminsson S, Herman P, Midlo¨v P (2020) Human-versus machine learning–based triage using digitalized patient histories in primary care: comparative study. JMIR Med Inf 8(9):e18930
4. Fernandes M, Vieira SM, Leite F, Palos C, Finkelstein S, Sousa JM (2020) Clinical decision support systems for triage in the emergency department using intelligent systems: a review. Artificial Intell Med 102:101762
5. Hong WS, Haimovich AD, Taylor RA (2018) Predicting hospital admission at emergency department triage using machine learning. PLoS ONE 13(7):e0201016
6. Jiang H, Mao H, Lu H, Lin P, Garry W, Lu H, Yang G, Rainer TH, Chen X (2021) Machine learning-based models to support decision-making in emergency department triage for patients with suspected cardiovascular disease. Int J Med Inf 145:104326
7. Joseph JW, Leventhal EL, Grossestreuer AV, Wong ML, Joseph LJ, Nathanson LA, Donnino MW, Elhadad N, Sanchez LD (2020) Deep-learning approaches to identify critically ill patients at emergency department triage using limited information. J Am College Emergency Phys Open 1(5):773–781
8. Kruse CS, Stein A, Thomas H, Kaur H (2018) The use of electronic health recordsto support population health: a systematic review of the literature. J Med Syst 42(11):1–16
9. Kwon JM, Jeon KH, Lee M, Kim KH, Park J, Oh BH (2019) Deep learningalgorithm to predict need for critical care in pediatric emergency departments. Pediatric Emergency Care
10. Levin S, Toerper M, Hamrock E, Hinson JS, Barnes S, Gardner H, Dugas A, Linton B, Kirsch T, Kelen G (2018) Machine-learning-based electronic triage more accurately differentiates patients with respect to clinical outcomes compared with the emergency severity index. Ann Emerg Med 71(5):565–574
11. Miotto R, Wang F, Wang S, Jiang X, Dudley JT (2018) Deep learning for healthcare: review, opportunities and challenges. Brief Bioinform 19(6):1236–1246
12. Stewart J, Sprivulis P, Dwivedi G (2018) Artificial intelligence and machine learningin emergency medicine. Emerg Med Australas 30(6):870–874
13. Ting DS, Rim TH, Choi YS, Ledsam JR (2019) Deep learning in medicine. Are We Ready?
14. Vlodaver ZK, Anderson JP, Brown BE, Zwank MD (2019) Emergency medicinephysicians' ability to predict hospital admission at the time of triage. Am J Emerg Med 37(3):478–481
15. Wang S, Huang M, Deng Z (2018) Densely connected cnn with multi-scale featureattention for text classification. In: IJCAI, pp 4468–4474
16. Xiao C, Choi E, Sun J (2018) Opportunities and challenges in developing deep learning models using electronic health records data: a systematic review. J Am Med Inform Assoc 25(10):1419–1428
17. Yu JY, Jeong GY, Jeong OS, Chang DK, Cha WC (2020) Machine learning andinitial nursing assessment-based triage system for emergency department. Healthcare Inf Res 26(1):13

A Conceptual and Effective Scheme for Brain Tumor Identification Using Robust Random Forest Classifier

K. Sakthidasan Sankaran, A. S. Poyyamozhi, Shaik Siddiq Ali, and Y. Jennifer

Abstract The diagnosis of brain tumor identification is an essential and challenging process by extracting the MRI brain images. The various assortments as brain image shapes, the intensity of images, regions of an image, illuminations, contrast are the conceivable factors that define the brain tumor identification. MRI images play a vital role in finding a brain tumor. That needs to attain a better efficiency of performance, which enhances the computer-aided diagnosis (CAD) method performance. The MRI brain images addressed the limitations of brain tumor radiance, the contrast of the brain images. In this approach, classification of robust random forest (RRF) is employed to enhance MRI brain images' above inabilities. Here, the MRI input process for the preprocessing stage to remove the noise present in the image, edge detection using high-pass filter (HPF). The preprocessed image moves to the image segmentation and then moves to the feature extraction for extracting and selecting the features of the MRI images using adaptive independent component analysis (AICA) for non-radiance of MRI images. The obtained image moves to the robust random forest (RRF) classifier to enhance the accurate prediction of brain tumor detection (normal/abnormal MRI images). Finally, the brain tumor area is to be detected accurately. The performance of the brain tumor identification experiments carried out using MATLAB and evaluates the better performance. The outcome of the results compared with the other conventional brain tumor identification techniques.

Keywords Brain tumor detection · MRI images · HPF · ICA · RF classifier

1 Introduction and Background

In recent years, brain tumor detection can evolve human life, whether benign or malignant. A brain tumor is a bunch of abnormal/affected cells in the brain, forming an uncontrolled condition to build a grouping of brain cells in the brain and nearby the cells. The brain tumor is an irregular shape of cells that might be filled by the

---

K. S. Sankaran (✉) · A. S. Poyyamozhi · S. S. Ali · Y. Jennifer
Department of Electronics and Communication Engineering, Hindustan Institute of Technology and Science, Chennai, India

A. Ullah et al. (eds.), *Proceedings of International Conference on Information Technology and Applications*, Lecture Notes in Networks and Systems 350,
https://doi.org/10.1007/978-981-16-7618-5_10

solid/liquid state. When a specialist evolves, the brain tumors that are never changing the brain of the lesion but due to the tumor size, which may change. The brain tumors may have in large/small lumps. However, any type of brain tumor cells can form into a tumor [1]. There are multiple types of tumors, such as origin, shape, and cell types. A brain tumor is an irregularity in the brain caused by brain cells developing and partitioning in an uncontrolled way. The brain tumor leads to generate a life-threatening problem; there is an early stage of prediction that enables the vital key component [2]. In medical image processing, brain tumors' identification plays a crucial role in processing the exact outcome. Brain tumor detection has some challenging factors while diagnosing the MRI brain images. The brain image clarifies the benign and malignant of the MRI images.

One of the most significant reasons for increasing morality like the reduced lifespan of the human leads to various diseases. Recent studies show that brain tumor detection has the contests [3]. Machine learning algorithms such as neural networks, support vector machine, and convolution neural network enable brain tumor classification. Here, the random forest (RF) classifier evolves the non-radiance, low contrast of brain tumor images. The optimized features allow the ICA to approach [4]. The researcher is typically supposed to recognize the brain tumor and in turn track their progress of treatment procedure [5–7]. By consider the effective classification performance of brain tumor identification, the proposed robust random forest (RF) classifier performs better than other conventional classification algorithms. RF classifier exploits the less radiation, higher contrast levels, and spatial resolution with the accurate classification prediction of MRI brain images. From this paper, the improved and enhanced performance of brain tumor detection is using MRI brain images with the efficient classification of RF classifier approach.

The respite of the manuscript has been structured as follows: Sect. 2 offers a thorough analysis of the related works. Section 3 discusses the paper's problem statement; Sect. 4 shows the proposed methodology with the detailed explanation, and Sect. 5 offers the results and discussion of the paper.

2 Related Works

In [8], the authors stated the machine learning (ML) through DNA methylation of brain tumor classification to retain the accurate results. The high-quality molecular data through the DNA methylation from the specimens of formalin-fixed paraffin-embedded pathology. The additions of molecular testing enable the copy number analysis and identify the pediatric brain tumors by considering the DNA methylation process.

In [9], the authors proposed the brain tumor images are extracted through the radiomic features. The radiomic model consists of ML-based feature extraction and linear model classifier. In [10], the authors proposed the two unique feature extraction methods like nLBP and αLBP, to achieve the classification of brain MRI images. The formation of nLBP enables the brain tumor image pixels around the neighbors to allow the consecutive distance. The operator of αLBP depends on each pixel with its angle value with the high performance of classification and feature extraction using various classification algorithms.

In [11], the authors stated the Internet of medical things (IoMT) computation method used to detect the brain tumor-based images CT scans/MRI images. Here, the approach of partial tree (PART) for the usage of a rule-based association learner with the high accuracy with the less time consideration to attain the brain tumor detection.

In [12], the authors stated an efficient brain pathology identification (BPI) algorithm that performs the human error reduction during the diagnosis of tumor detection. The rapid growth of the brain tumor was identified through CAD tool.

In [13], the authors proposed an intelligent-based medical imaging approach to evolve the leading-edge software packages, metrics validation, glioma evaluation. The medical images and the tumor diagnosis enable the exact accuracy of prediction.

In [14], the authors proposed the multiparametric deep learning model (DLM) for imaging meningiomas. The planning of therapy and monitoring in meningiomas were done with the distinct MRI data including images. The MRI datasets are [T1/T2-weighted, T1-weighted contrast-enhanced (T1CE), FLAIR] of meningiomas with the several brain tumor cases. The segmentation of DLM performance is compared with the conventional segmentation methods.

In [15], the authors proposed the brain tumor segmentation challenge (BraTS) of ensemble learning methods. To enable the reliable automatic segmentation and decision support algorithms helps to a strong development of brain tumor detection and intensification MRI by the concern of segmentation algorithms.

The segmentation of manual abnormal brain tissues from the brain cell that is healthy in turn devours massive time and could yield imprecise outcomes. So as to offer the resolution for this also to aid the medical specialists for the brain tumor regions segmentation from MRI images, on the way to produce a novel CAD method to systematize this procedure using deep learning processes. In this broadside, a new plan was presented which in turn resolves the issue of medical image processing for recognizing the brain tumor through more accurateness.

3 Proposed Method

The proposed approach's performance enables the radiance of brain images to attain the contrast levels with better methodologies such as preprocessing, segmentation, feature extraction, and classification of MRI brain images. The schematic diagram of the proposed method is shown in Fig. 1.

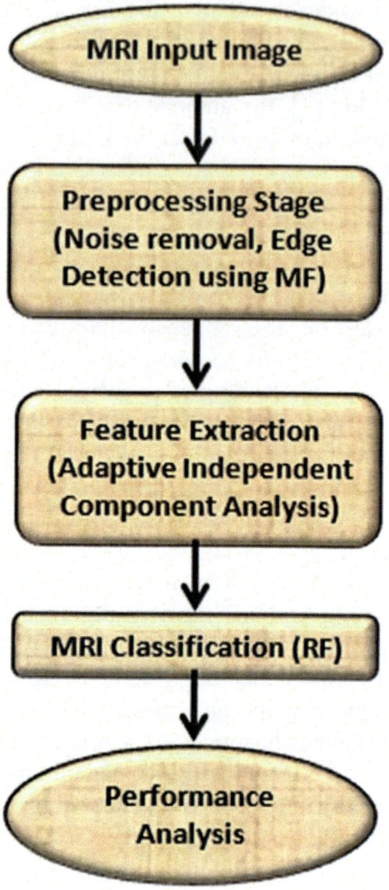

Fig. 1 Flow of the proposed strategy

3.1 *Preprocessing*

The brain tumor-based images took from magnetic resonance imaging (MRI). Before the image segmentation, the preprocessing stage of evolving. The preprocessing stage includes the filtering process. Namely, as median filter (MF) evolves the noise reduction of the graylevel images in the brain images, fine details of the image, missing data replacement, and intensity variation by reducing the amount, thereby the variation in the neighboring pixels. The noise removal is an essential process in MRI pictures. Here, for noise removal median filter is examined. A median filter is a non-straight approach used for the removal of noise present in the MR images. The median is evaluated by ordering all the pixel approval, and after that, displace the pixel consideration with the center of the pixel value. The mean filter incorporates the pixel to pixel variation by replacing each pixel value with the average of neighboring pixels includes itself also. The filter helps to evaluate the noise in the original MR images without sharpness diminishment conditions. Consider the input image I (a, b) size of M N, which consists of noise. Next, to separate the image into n number of levels L and each level has one center of the pixel value. Now, remove the noise and replace the center pixel through the use of median value evaluations. The equation of the median filter is given

$$\mathrm{I(a,\ b)} = \mathrm{median}_{(\mathrm{s,\ t}) \in C_{xy}}\{\mathrm{H\ (u,\ v)}\} \tag{1}$$

where C_{xy} is the set of rectangular sub-image levels, centered at the point (x, y). This process is repeated for the entire level present in the input image when the speckle noise to be removed. Finally, to obtain the noise-free image in the preprocessing stage. The preprocessed image can evolve the image segmentation to attain the normal/abnormal MR images according to its features.

3.2 *Feature Extraction Using Adaptive ICA*

Adaptive independent component analysis (ICA) helps extract the MRI brain images for the feature extraction stage. ICA incorporated the accumulation of brain image features contribution. ICA is an unsupervised learning technique because its performance results as the amount of maximum independent component features (fa, fb), which involves the uniform distribution such as [−1, 1] for binary feature extraction and the output performance (P) results denoted as given below:

$$\mathrm{P} = \mathrm{O} = \{0if\, fa + fb < 01if\, fa + fb \geq 0 \tag{2}$$

P is the binary feature extraction value to the nearby pixels (fa, fB) referred to as the grayscale values on the centralized pixel value. The obtained results show the decimal formation into the feature vectors. This ICA technique is connected to the dispersal contribution. The feature extraction pattern can enable independent components features of the obtained data pattern. ICA is mainly used to extract "pure" details of images from the impure sources of brain MRI features. ICA degrades the obtained signal mixtures as an M × N matrix. The number of the signal source is well-known and less than the number of signal mixtures matrix, then the number of signals-based MRI images get extracted by using independent component analysis, which can be reduced by preprocessing signal source mixtures. For accurate detection of brain tumors through MRI, the MR image should define clearly between the multiple brain cells, such as white detail tracts and gray details. The traditional MRI images never display multiple tumor cells in detail. But, it is possible to use different MRI settings such that every setting captures a multiple image mixture of the source signals additional with the multiple brain tumor cells. where M describes the measured mixture signals, N describes the number of variables. In matrix formation,

$$X = A.S \tag{3}$$

where

X is a nonlinear vector with M*N elements.

S is independent component vectors.

A gets concealed mixing matrix.

The A-1 can calculate, and it is referred to as W. The independent components vectors are evaluated by finding an individual matrix W. The individual matrix W, which is also called as demixing of the component matrix, can be calculated through an algorithm that optimizes the statistical independence of the component vectors of the mixture. If consider the two features for feature extraction.

3.3 *Robust Random Forest (RF)-Based Classification*

The classification of brain tumor is the final stage as well as the important stage in the prediction of whether the brain tumor images are normal or abnormal images. Here, the robust random forest (RRF) helps to specify the benign/malignant images from the previous stage as adaptive thresholding (AT) segmented images. An ideal hyperplane separation separates the MR images of two classes that also maximizes the hyperplane margin. For efficient prediction and classification of brain tumor MR images, the RF classifier performs better classification results. Random forest classifier helps to classify the brain tumor features. Random forest classifier is an ensemble classifier that suits several decision trees on multiple subsamples of the MRI brain tumor images. Random forest performs better than other conventional classifiers, and also, there is no necessity for the number of classes to respond to the precise nonlinear segment features. The ensemble method is based on each tree prepared to enable the nonlinearly inspected information (features) with the equivalent alternatives from training samples of brain tumor images amid a specific time of learning. RF is a potential analysis that provides an evaluation of errors and significant variables, etc. Random forest is carried into a functional brain image classification within an association of time slots.

In the RF-enabled classification process, the features of MRI brain images used for training each association tree calculated the original MR images, not all the samples of training MR images. The nonlinear selection of brain tumor images is essential for the classifier, namely RF classifier. When separating the nodes of each decision tree, the feature subspace is extracted instead of the better features of MR images in all the train sets. It separates the input MRI brain tumor image into subsequent nodes. The nonlinear feature selection relieves the overfitting problems of RF classifier. By putting to the test data, the MRI image-based pixel vector is extracted in the similar method and contribution to the proficient model. Classification probabilities of all trees “t” are defined as:

$$P(y/F) = \frac{1}{T}\sum_{t=1}^{T} P_t(y/f\,(F,\ A)),\ t\ \in (1,\ T) \tag{4}$$

where

P is referred to as the performance of classification.

F defines image intensity, and Y defines the label set.

An essential tunable parameter of random forest is the association rule-based tree depth, which can carry out to minimize the errors that occurred in the technique.

4 Performance Analysis

The performance analysis of the brain tumor identification evolves through the MATLAB simulation.

Here the true positive (TP) is images that are appropriately acknowledged, and false negative (FN) is the image incorrectly eliminated. True negatives (TN) are the circumstances adequately reduced, and false positives (FP) are the MR images that are recognized wrongly, as brain tumor-based MR Images.

Sensitivity: The proportion of true positives (TP) among the complete discharge of MR images.

$$\text{Sensitivity} = \frac{TP}{TP + FN} \tag{8}$$

Specificity: The capability to detect the portion of the brain tumor cells spreading that does not have discharges that is true negatives (TN), and it is defined as the proportion of TN among the complete non-discharges of MRI.

$$\text{Specificity} = \frac{TN}{TN + FP} \tag{9}$$

Accuracy: The measurement approximation degree that conforms to the corrected value.

$$\text{Accuracy} = \frac{TP + TN}{TP + TN + FP + FN} \tag{10}$$

Table 1 is the representation of comparative analysis of the proposed and existing techniques in terms of accuracy, sensitivity, specificity, TN, TP, FN, and FP.

The comparative analysis of accuracy is provided in Fig. 2, sensitivity in Fig. 3, and specificity comparison in Fig. 4. From the analysis, it was evident that the proposed mechanism is better than existing technique. The overall comparative analysis of sensitivity, specificity, and accuracy analysis is shown in Fig. 5. The overall comparative analysis of TN, TP, FN, FP is shown in Fig. 6.

Table 1 Comparison of accuracy with various parameters

Obtained factors	Existing method (17)	Proposed method
Sensitivity (%)	98.48	99.02
Specificity (%)	94.28	95.4
Accuracy (%)	97.02	98.8
True Negative (TN)	66	72
True Positive (TP)	130	146
False Negative (FN)	2	1
False Positive (TP)	4	2

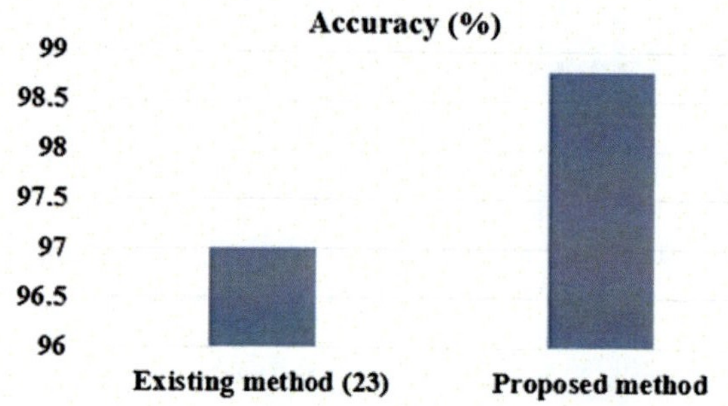

Fig. 2 Comparative analysis of accuracy

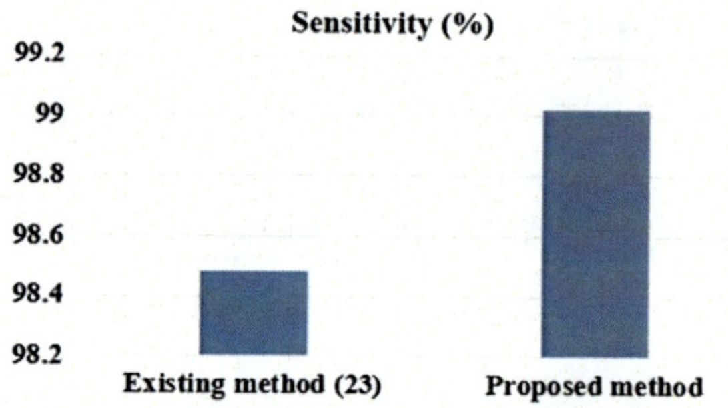

Fig. 3 Comparative analysis of sensitivity

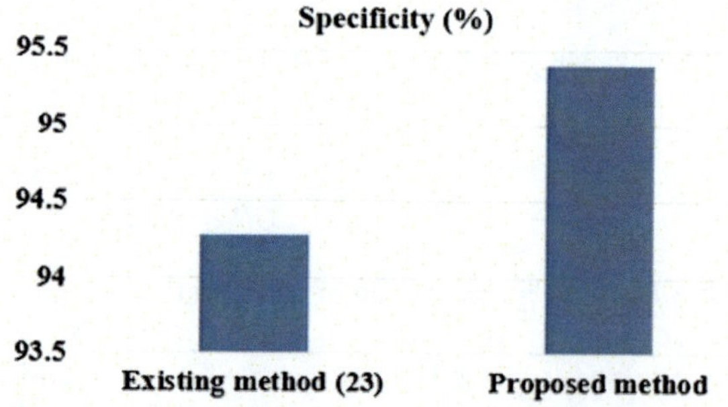

Fig. 4 Comparative analysis of specificity

Fig. 5 Overall comparative analysis of Sensitivity, Specificity, and Accuracy analysis

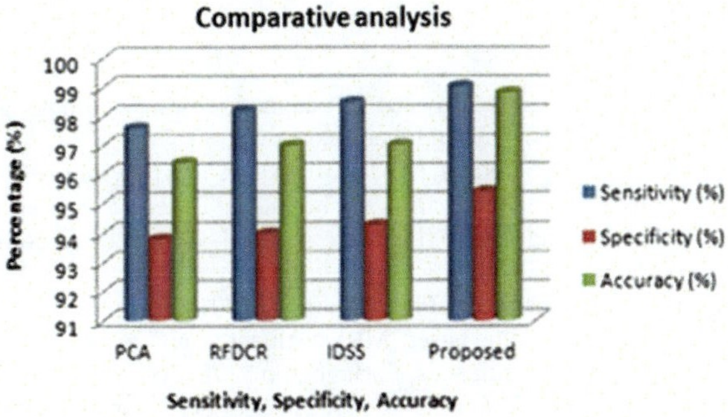

Fig. 6 Comparative analyses of TN, TP, FN, FP

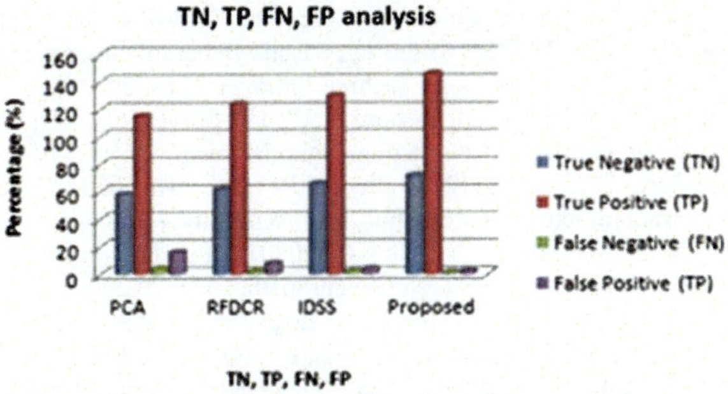

5 Conclusion

The brain tumor leads to the high-risk factor for humans, which is a life-threatening factor. The brain tumor detection in the image processing technique employs perfectly with the accurate detection of results whether the tumor is normal or abnormal through the MR images. In this paper, the RF classifier employed the accurate prediction of a brain tumor in an early-stage identification. Here, the various stages like preprocessing through HPF to remove the speckle noise in the grayscale level, feature extraction through the AICA by extracting the images according to its features, classification is done through RF classifier by classifying the normal/abnormal MR Image of the brain tumor cells. The binary patterns of the MRI to be filtered and retain the perfect edges get extracted in the extraction stage. The experiment was carried out by using the simulation of MATLAB. The accuracy of the RF classification is 98.8%, which is more efficient than the existing IDSS-based SVM classifier (97.02%). The sensitivity, specificity, TN, TP, FN, FP values get improved by comparison to the existing approach. By analyzing the result, the RF classifier-based brain tumor detection actively enables the efficient brain tumor identification process.

References

1. Jalab HA, Hasan AM (2019) "Magnetic resonance imaging segmentation techniques of brain tumors: a review." Archives Neurosci 6
2. Sakthidasan Sankaran K, Manishankar P, Teja KR, Reddy PK, Kumar TP (2020) "Digital image de-noising and restoration method using differential filters for improving the image quality." in Proceedings of 9th IEEE international conference on communication and signal processing, pp 1377–1380
3. Panda A, Mishra TK, Phaniharam VG (2019) "Automated brain tumor detection using discriminative clustering based MRI segmentation," in Smart innovations in communication and computational sciences, Springer, pp 117–126
4. Rajagopal R (2019) Glioma brain tumor detection and segmentation using weighting random forest classifier with optimized ant colony features. Int J Imaging Syst Technol 29:353–359
5. Mittal M, Goyal LM, Kaur S, Kaur I, Verma A, Hemanth DJ (2019) Deep learning based enhanced tumor segmentation approach for MR brain images. Appl Soft Comput 78:346–354
6. Amarapur B (2019) Cognition-based MRI brain tumor segmentation technique using modified level set method. Cogn Technol Work 21:357–369
7. Hameurlaine M, Moussaoui A (2019) Survey of brain tumor segmentation techniques on magnetic resonance imaging. Nano Biomed. Eng 11:178–191
8. Perez E, Capper D (2020) Invited Review: DNA methylation-based classification of paediatric brain tumours. Neuropathol Appl Neurobiol 46:28–47
9. Kim B, Tabori U, Hawkins C (2020) An update on the CNS manifestations of brain tumor polyposis syndromes. Acta Neuropathol 139:703–715
10. Kaplan K, Kaya Y, Kuncan M, Ertunç HM (2020) Brain tumor classification using modified local binary patterns (LBP) feature extraction methods. Medical Hypotheses 139:109696
11. Hussain UN, Khan MA, Lali IU, Javed K, Ashraf I, Tariq J et al (2020) A Unified design of ACO and skewness based brain tumor segmentation and classification from MRI scans. J Control Eng Appl Inf 22:43–55
12. Gudigar A, Raghavendra U, Hegde A, Kalyani M, Ciaccio EJ, Acharya UR (2020) Brain pathology identification using computer aided diagnostic tool: a systematic review. Comput Methods Programs Biomed 187:105205
13. Mohan G, Subashini MM (2019) "Medical imaging with intelligent systems: a review." Deep learning and parallel computing environment for bioengineering systems, pp 53–73
14. Laukamp KR, Thiele F, Shakirin G, Zopfs D, Faymonville A, Timmer M et al (2019) Fully automated detection and segmentation of meningiomas using deep learning on routine multiparametric MRI. Eur Radiol 29:124–132
15. Győrfi Á, Kovács L, Szilágyi L (2019) "Brain tumour segmentation from multispectral mr image data using ensemble learning methods." in Iberoamerican Congress Patt Recogn, pp 326–335

A Machine Learning Approach for Load Balancing in a Multi-cloud Environment

Usha Divakarla and K. Chandrasekaran

Abstract A multi-cloud environment makes use of two or more cloud computing services from different cloud vendors. A typical multi-cloud environment can consist of either only private clouds or only public clouds or a combination of both. Load balancing mechanism is essential in such a computing environment to distribute user requests or network load efficiently across multiple servers or virtual machines, ensuring high availability and reliability. Scalability is also achieved by sending requests only to those servers that are healthy and available to take up the computing workload and thus providing the flexibility to scale up and scale down to satisfy QoS requirements as well, in order to save costs. In our proposed model, a time series-based approach as well as predictive load balancing has been experimented and the results are presented.

Keywords Multi-cloud · Load balancing · Virtual machine · Predictive load balancing · Time series modeling

1 Introduction

Load balancing refers to efficiently distributing incoming user requests and network traffic across several backend servers. High-traffic websites in modern times must serve thousands of concurrent requests from users and return the appropriate application data in a fast and reliable manner. To scale in order to meet these requests, it is general practice to scale horizontally by adding more servers. This can be achieved via load balancing. Many load balancing solutions have been proposed so far. Notable ones among these are: round-robin algorithm, genetic algorithm, Max-Min load balancing algorithm, and the honeybee foraging load balancing algorithm.

U. Divakarla (✉)
NMAM Institute of Technology, Nitte, Karkala, Karnataka, India
e-mail: ushad@nitte.edu.in

K. Chandrasekaran
National Institute of Technology Karnataka, Surathkal, Karnataka, India
e-mail: kch@nitk.edu.in

A. Ullah et al. (eds.), *Proceedings of International Conference on Information Technology and Applications*, Lecture Notes in Networks and Systems 350,
https://doi.org/10.1007/978-981-16-7618-5_11

The goal of a hybrid or multi-cloud strategy is to meet the requirements with a plan that describes which workloads should run in or migrated to which cloud computing environment. Also, it is to decide how the clouds will be assigned weights and with what algorithms intra-cloud load balancing can be achieved. The load balancer has the additional functionality of saving costs by using the resources judiciously and scaling down (shutting off servers) immediately when possible and feasible. This is also known as elasticity.

The aim of a load balancing is to improve the efficiency and performance of the system by means of distributing the workload uniformly to all compute elements. Also, the resources should be utilized to the better extent while the QoS metric is to be improved. The services provided by cloud environment should remain unaffected due to the increase in the numbers of the user tasks, and thus, the scalability is expected to be supported. The system should also be fault tolerant such that whenever a compute element node fails the other nodes should share the workload.

The aim of this paper is to increase the efficiency and performance of system by efficiently distributing the workload using load balancing algorithm. This paper contributes to load balancing by effectively using time series concept in the load balancing algorithm which in turn increases the system performance.

2 Literature Review

Nikolay Grozev et al.[1] focused on an augmented architecture over the traditional three-tier cloud architecture where there are two additional layers—an entry point layer and a data center control layer. The entry point layer checks all the servers in the multi-cloud environment for availability and routes the user to the appropriate data center. The data center control layer is responsible for admission control, load balancing, and scaling up/ scaling down resources. The typical cloud architecture has three layers—presentation layer, domain layer (with application servers), and the data layer (with data servers). The paper explains that there are two types of application servers—stateful and stateless. A stateful application is one that keeps session data in the memory of the assigned server and hence requires sticky load balancing, which ensures that all requests of a session are routed to the same server. A stateless application is one which does not keep any such data in memory, and hence, this type of an application can be dealt with using elastic load balancing. The paper provides a load balancing scheme/solution that provides high availability, scalability, and fault tolerance in both cases. To evaluate the performance of their algorithm, they have used the CloudSim discrete event simulator, which has been used in both the industry and academia for performance evaluation of cloud environments and applications. They have used CPU utilization, RAM utilization, and number of requests to the application as input parameters for load balancing.

Sanjaya K. Panda et al. [2] focused on three task scheduling algorithms: MCC, MEMAX, CMMN for heterogeneous multi-cloud. They have included makespan and average cloud utilization as performance metrics to evaluate the performance of the system. Viewing application as collection of multiple tasks as a directed acyclic graph (DAG). They have explained that scheduling tasks for execution with minimum makespan are NP-complete, and hence, an optimal algorithm takes exponential time to complete. So various heuristics have been given to provide a near-optimal solution.

Fularani Vhansure et al.[3] focused on a specific load balancing algorithm called genetic algorithm. Combination of genetic algorithm and ant colony optimization algorithm is used. The ant colony optimization algorithm picks the shortest efficient path and reduces the makespan.

Chenhao Qu et al.[4] have proposed a system that supplements and enhances autoscalers in handling short-term overload situations for multi-cloud applications, a decentralized load balancing framework that detects and handles short-term overload events, and an optimal load distribution algorithm and admission control mechanism that quickly adapts to the overload situations. They also have a prototype implementation of the proposed system.

Sanjaya K. Panda et al. [5] have proposed three allocation-aware task scheduling algorithms for a multi-cloud environment. The algorithms are based on the traditional Min–Min and Max–Min algorithm and extended for multi-cloud environments. All the algorithms undergo three common phases, i.e., matching, allocating, and scheduling to fit them in the multi-cloud environment.

Bo Zhang et al. [6] have proposed a multi-cloud architecture to address the particular requirements in CWA. They have designed a novel resource scheduling algorithm to minimize the entire system cost. They also have designed an optimal user scheduling strategy based on the MFDL of system load paths.

Cui et al. [7] have designed learning-based approach to route requests in order to balance the load. In their model, the performance of microservice is modeled explicitly through machine learning models and the model can derive the response time from request volume, route decision, and other cloud metrics. Then the balanced route decision is obtained from optimizing the model with Bayesian optimization.

3 Details of the Proposed System

At a high level, the load balancer has the following components:

- Client—end users of the tenant
- End-user Server—Server that intercepts the client (end user) requests
- Private Cloud Agent—Agent in charge of the private cloud of the tenant
- Overall Public Cloud Agent—Agent in charge of the entire public cloud composition of the tenant's application.
- Internal Public Agent—Agent in charge of the individual public cloud servers
- Virtual machines (VM).
- Health checkers—checking the health of each VM for ensuring resilience.

The components mentioned above can be brought together to realize a cohesive system. The following diagram (Fig. 1) describes the structure of the system and

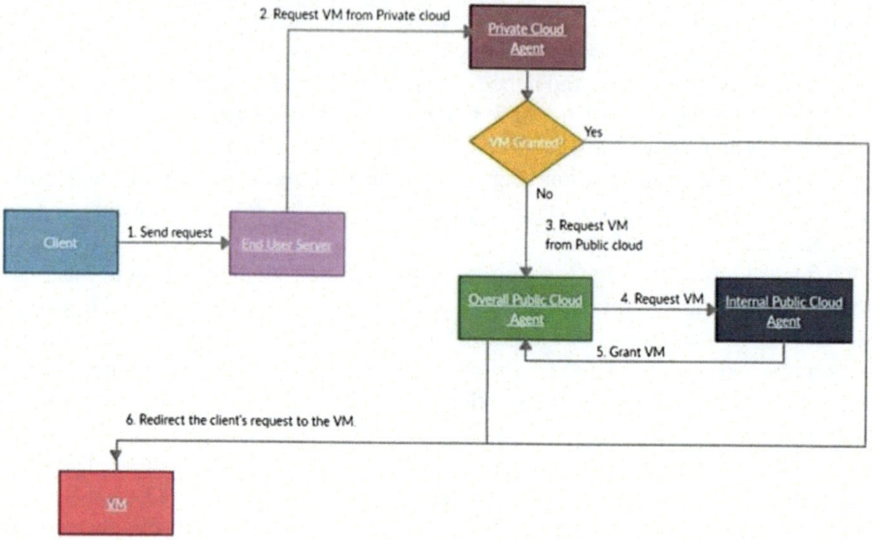

Fig. 1 Modules of the system and their interactions—logical design

the interactions between the interfaces and the modules that achieve the working described in the preceding subsections. This can be termed as the logical design of the system.

3.1 *Detailed Description of the System Components Server for the End User*

This server has a public static IP address. The users of the application using the load balancer will be using this address. Thus, this server intercepts all the user requests. It will send requests for VMs to the private cloud agent which will then check whether it is possible to provision the load among the private cloud VMs. If it is not possible, then the server sends requests for VMs to the overall public cloud agent. One of these agents will respond back with a response containing the IP address and port of the VM that can handle the incoming user request. The server then redirects the user request to this new IP address and port.

Private Cloud Agent

This agent is a server that performs intra-cloud load balancing for the private cloud. It will provide the load among the virtual machines in a modified round-robin fashion while also performing health checks on the VMs. The agent will maintain a queue of the virtual machines such that the pop() function returns the IP address of the VM that is healthy and can run the user requests. Internally, if the pop() function returns an unhealthy VM, it will recursively call itself until a healthy VM's address and port is obtained and will then send that tuple over to the end-user server in response. During the execution of the pop function if the queue becomes empty (there is no VM that is both healthy and underloaded), then the agent sends an appropriate response to the end-user server. After the responses are sent, all the popped VMs are pushed back into the queue in the order in which they are popped.

The round-robin is best suited for the private cloud agent because it ensures that all the servers are used as much as possible (since these are on-premises infrastructure there is no cost-saving needed by scaling up or down).

Overall Public Cloud Agent

This agent will perform inter-cloud load balancing among the public clouds. Based on the developer's settings for the configurations (such as the SLA specifications and geographical proximity), it will decide which public cloud should provision the incoming load.

Internal Public Cloud Agent

This agent will perform the intra-cloud load balancing for the public cloud. In case of non-sticky persistence of session data or in the condition that all the VMs are treated equally, algorithms that should be considered are:

1. Least Connection Method—This method sends traffic to the VM with the least active connections.
2. Least Bandwidth Method—This method selects the VM that is currently serving the least amount of traffic.
3. Least Response Time Method—This algorithm forwards traffic to the VM with the least number of active connections and the lowest average response time.
4. Weighted Round-Robin Method—The weighted round-robin scheduling is designed to handle servers with different capacities in a better manner. Servers are weighted based on their capability, and requests are received in the order of higher to lower weighted servers. Servers with higher weights receive new connections before those with lower weights.
5. IP Hash—A hash of the client's IP address is calculated to redirect the request to a VM.
6. Geographical Load Balancing—It considers geographical proximity of the servers to the users to provision the load.

In the next section, we have proposed a predictive algorithm for elastic load balancing using time series modeling. One main consideration here is the elasticity. The above algorithms are for provisioning on existing spun-up VMs. However, a separate algorithm will be considered for spinning up and spinning down the VMs and thereby maintaining flexibility and elasticity.

If the session data is maintained on the VM itself, the load balancer has to be configured in a way that the requests from the same client are directed to the same VM. Sticky sessions can also be used when the application is dependent on local web server resources (such as file system). Using sticky sessions helps improve user

Fig. 2 Load balancer for sticky sessions

experience. In this case, load balancer creates an affinity between the end-user and a specific network server for the duration of a session, (i.e., the time a specific IP spends on a website).

To handle the sticky sessions as shown in Fig. 2, a load balancer assigns an identifying attribute to each user, traditionally by issuing a cookie or by tracking the IP details. Then, according to the tracking attribute, the load balancer will route all future requests of that user to one specific server for the duration of the session. This is very helpful because the HTTP protocol is stateless and was not devised with session persistence in mind. Nevertheless, many web applications do need to serve personal user data (e.g., keeping logs of items in the cart of an e-commerce website or chat conversations) over the course of a session. Without session persistence, the web application itself would have to maintain that data across many servers which can be inefficient especially for big networks.

Health Checkers

These modules will constantly check the operational status of each VM in the clouds they are in-charge of and will maintain a set of healthy VMs. In this way, the cloud agent can check whether a given VM is healthy or not in amortized O(1) time.

Application Server Layer

This layer consists of the actual infrastructure that will run the user requests. Each virtual machine in this layer will create a new thread for every user request that is redirected to it.

Backups

Each agent will have a backup that will run in case the original agent is not running properly. The agent is thus in continuous and periodic communication with the backup agent. Consistent hashing will be used to maintain the same state in the back-up agents. Necessary requests will be made to the end-user server to direct all new requests to the backup agent in case of switching.

4 Predictive Load Balancing Using Time Series Modeling

Time series modeling is a machine learning technique which uses a series of data points indexed in the order of time to predict data for some future point(s) in time.

Some examples for time series are daily closing values of stocks and shares, ocean tides, climate change among many others.

4.1 Terminology

Throughout the text, we have used the following terminologies:

1. Frame—a span of time with some fixed duration. This can be in any units—seconds, minutes, or hours
2. VM—Virtual Machine
3. L t—A 3 × 1 vector in the time frame t. It has the following components:

   a. L t [0]—Total CPU utilization of all deployed AS VMs in time frame t
   b. L t [1]—Total RAM utilization of all deployed AS VMs in time frame t.

4.2 Basic Idea

Our algorithm aims to forecast the total CPU and RAM parameters for all VMs in the public clouds and thereby provision/scale down VMs in the cloud to efficiently handle the incoming demand. To make the predictions, we will be using a time series forecasting method, called the Holt-Winters method, which uses a technique called triple exponential smoothing to make good predictions about future values.

4.3 Standard Algorithm for Scaling Up/Down

Scalability is required to help us achieve elastic scaling up/scaling in our infrastructure, keeping sticky sessions in mind, along with a few additions to handle the forecasted load. The algorithm marks those servers whose CPU and RAM usages exceed certain pre-fixed thresholds or whose TCP network buffers/queues are full, as over-loaded VMs and accordingly provisions an excess of virtual machines in the cloud to cope with sudden spikes in demand. The algorithm will run every Δ seconds. The input parameters for the algorithm are:

- VMAS—list of currently deployed application server (AS) VMs
- N—number of over-provisioned AS VMs to cope with sudden spikes in demand
- Δ—time period between successive algorithm repetitions.

The algorithm has the following steps:

1. Initialize variables:
   a. Set nOverloaded to 0.
   b. Set list FreeVMs to empty list
2. Inspect the status of all VMs. If there is a VM whose CPU and RAM utilizations are greater than the configured values or whose network buffers are overloaded, then increases the number of Overloaded VMs by 1. If there is a VM which is serving no sessions, then add the VM to list FreeVMs.

At the end of this step, we will have the total number of overloaded VMs and the list of free VMs.

3. Set nFree to the length of list FreeVMs
4. Set nAS to the length of VMAS
5. Set all Overloaded to 1 if all VMs are overloaded. Else set it to 0.
6. Provision more VMs if nFree $\leq$ N
7. Else if nFree > N, release VMs

We aim at modifying this algorithm to provide a more efficient way of provisioning VMs to handle the workload on our environment in the next time frame. We will run the modified algorithm once in every time frame. It will predict the value of the load vector L in the next time frame.

4.4 Data Collection

To describe our forecasting model, we first need data. We have considered both short-term and long-term factors to collect the data for our forecasting model. Since the product at hand is a load balancer for a multi-cloud web application, we will naturally have peak hours during the day when the demand for the application is very high and there will be hours when the demand is low. This trend approximately repeats throughout the week, month, and year. The second aspect we have considered is the short-term spikes in demand for our application. For example, if the application we are serving is an online marketing website, and a new popular product is on sale, we can expect to have a lot of users using the website simultaneously to order the product. This can happen at any time during the day. The data for public cloud was collected from the compute instances on Amazon Web Services which were created by us and for private cloud we had created a private cloud at our laboratory and collected the required data from the inhouse cloud.

To account for these observations, our data will be in the form of a rolling window of hourly usage details of all the VMs in the infrastructure. Keeping short-term spikes in mind, the usage details are sampled every 5 minutes. This way, we will have both the details for the current day and the long-term trends in the usage of the application.

4.5 The Algorithm

The standard algorithm for scaling up/down explained above has been modified using Holt-Winters forecasting model [8]. The standard algorithm for scaling up/down does not take into consideration the overall effect of the incoming load on the cloud. To mitigate this, we have considered the total CPU and RAM utilization of all the deployed AS VMs in order to forecast the future load. We accordingly spinned up new VMs or block and calculated the number of VMs. The input parameters for the new algorithm are:

- *VMAS*—list of currently deployed AS VMs
- *N*—number of currently deployed AS VMs
- nBlocked—number of currently blocked AS VM
- Δ—time period between successive algorithm repetitions, or time frame size
- *ThCPU*—Threshold CPU consumption ratio in the range [0, 1]
- *ThRAM*—Threshold RAM consumption ratio in the range [0,1].

We create the following files in the working directory of the project to manage inter-process communication between the forecasting module and the load balancer module:

load_details.dat—It contains the total load/ usage details in each time frame in the rolling window

The following algorithm runs once in every time frame on the load balancer's side:

1. Get predicted load vector L' from forecasting module
2. Calculate M as shown in Sect. 8 using L', *ThCPU,* and *Th RAM*
3. If $M > N$, we need to have $M - N$ new unblocked VMs.
4. Unblock *min* (*nBlocked*, $M - N$) VMs and put these VMs in the active list.

5. If there are still more VMs to be provisioned, spin up the required number of VMs.
6. If $M < N$, block $N - M$ already deployed VMs from getting more sessions.
7. Write the usage details, i.e., sum of consumption of CPU and RAM utilizations of every VM in the current frame to the load_details.dat file.

The forecasting module reads from load_details.dat to build the load vector L. Using the forecasting approach, the load in the next frame is predicted. This load is sent to the load balancer module for processing. Holt-Winters forecasting is a surprisingly powerful tool despite its simplicity. It can handle lots of complicated seasonal patterns by simply finding the central value, then adding in the effects of trend and seasonality. The challenge here is giving it the right parameters. Thus, we have shown a way to predict future load on the cloud environment, keeping in mind the trend and seasonality traits in the usage of the application running on our clouds.

5 Experiments, Results, and Analysis

As a proof of concept for the proposed model, a mini-version of a multi-cloud environment was created. Individual laptops were used for the private cloud and with different sockets listening on different threads to act as virtual machines. For the public cloud, a compute instance with a Linux-based OS was created and the virtual machines are simulated in a similar fashion as the private cloud. In this case, spinning up would be spawning a new thread and spinning down is terminating existing thread(s). Assumptions have been made about the configurations of the public clouds while implementing and that the private cloud is on-premises infrastructure completely owned by the organization using the load balancer.

The Python programming language and its libraries are used for the load balancer system. The tools libraries used are:

- **socket—**We will use this library to create sockets and thus create servers and clients that are constantly listening.
- **threading—**We will use this library to achieve multithreading functionalities and to create multi-client servers (such as the VMs or the end-user server).

For creating the public cloud, virtual machines compute instances on Amazon Web Services (EC2 Linux instances) are created. To simulate the end users, the Apache JMeter tool is used with which test plans have been created. User threads containing http requests are spawned in fixed time intervals and sent to the end-user server.

For the predictive load balancing algorithm, the following libraries were used:

- **pandas** to import data from csv file
- **numpy**
- **statsmodels.tsa.api** for our models

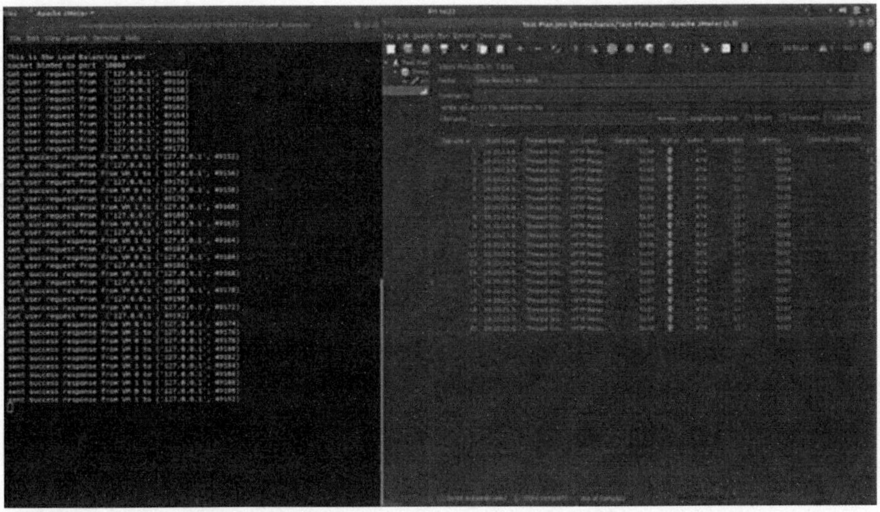

Fig. 3 Private cloud alone handling all requests

- **matlibplot.pyplot** to plot graphs
- **math.sqrt** to find root mean square error (RMSE) value after fitting curves.

A time series forecasting implementation that uses the data that we have collected from CloudSim VM traces, to predict the future load on a cloud environment. The data we are forecasting is the overall CPU and RAM loads on the public cloud.

The implemented load balancer is able to successfully balance the incoming load (the Jmeter user requests). Figures 3, 4, and 5 represent the simulation of private cloud creation and load balancing in the simulated cloud.

Figures 6 and 7 are from the implementation of the predictive load balancing algorithm using the Holt-Winter forecast method. The figures represent the behavior of a sequence of load value as x-axis over time as y-axis. The root mean square error for CPU utilization was found to be: 3.530772906523469, and the root mean square error for RAM utilization was found to be: 3.6400794728341954 in our experiment.

6 Conclusion and Future Work

Cloud computing has grown exponentially in years, and in this cloud environment, various workloads that are submitted to data centers are transferred to other data centers due to the limited resources to handle peak client demands. So balancing the load is very challenging due to heterogeneity of resources. We have proposed a time series-based load balancing model to handle load in multi-cloud environment and experimented as a prototype model. In the future, this work can be extended with actual on-premises and cloud infrastructure to create a better proof of concept.

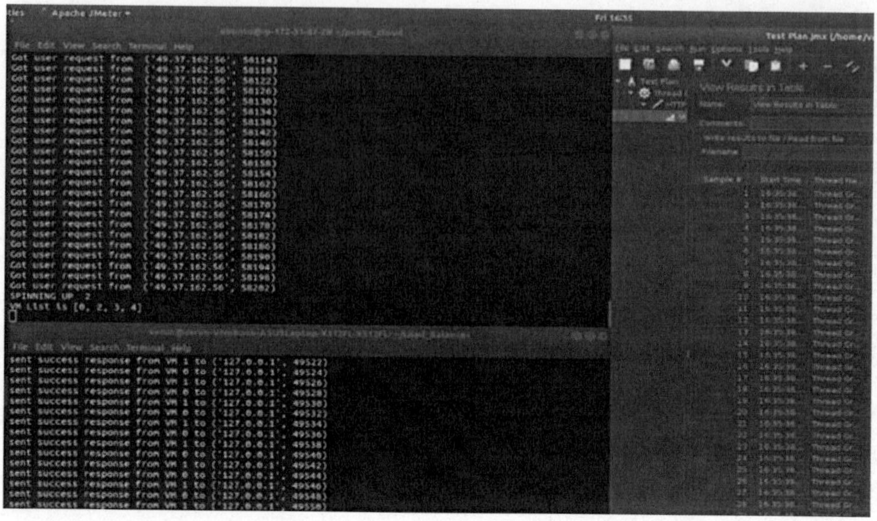

Fig. 4 Bursting into public cloud

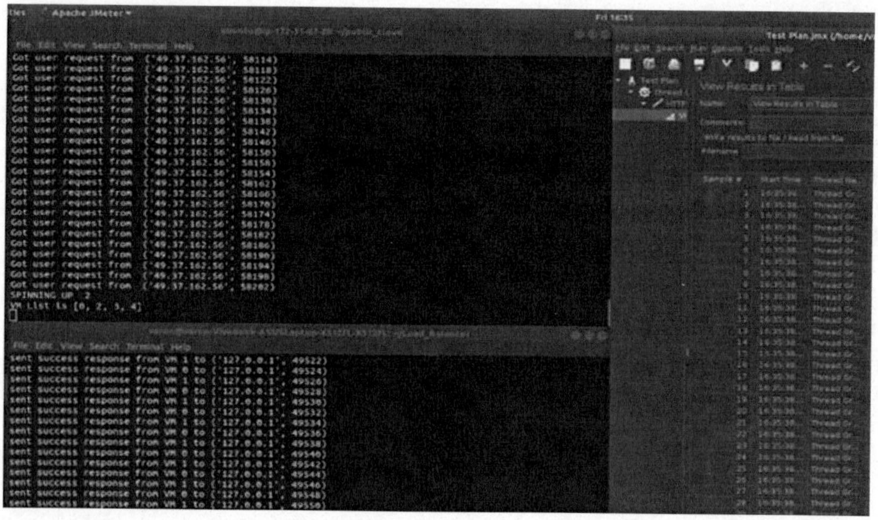

Fig. 5 Success responses from both clouds

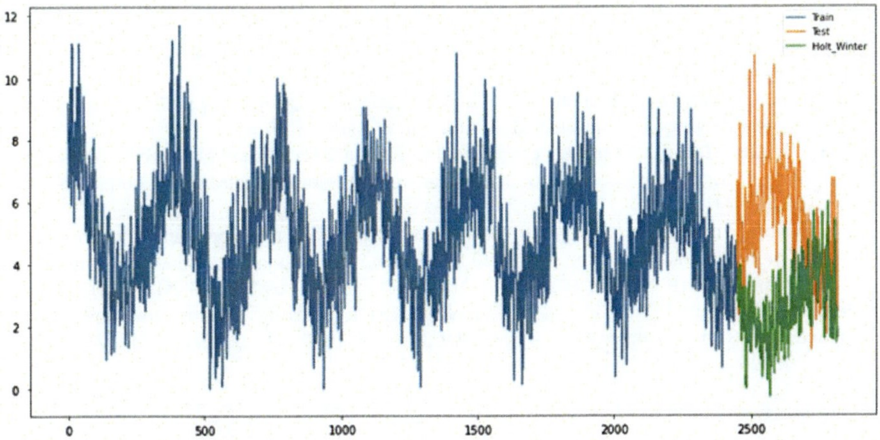

Fig. 6 Holt-winters CPU forecast

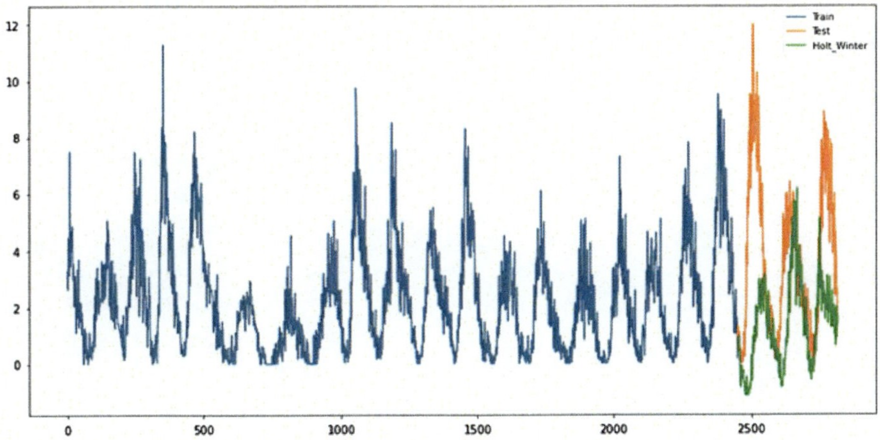

Fig. 7 Holt-winters RAM forecast

References

1. Grozev N, Buyya R (Oct 2014) "Multi-cloud provisioning and load distribution for three-tier applications". Published in ACM Transactions on Autonomous and Adaptive systems, Article No. 14
2. Panda SK, Jana PK (Sept 2014) "Efficient task scheduling algorithms for heterogeneous multi-cloud environment",. Published in 2014 International conference on advances in computing, communications and informatics (ICACCI)
3. Vhansure F, Deshmukh A, Sumathy S (2017) "Load balancing in multi cloud computing environment with genetic algorithm." published in IOP Conference Series: Materials Science and Engineering, 263(4)

4. Qu C, Calheiros RN, Buyya R (Mar 2017) "Mitigating impact of short-term overload on multi-cloud web applications through geographical load balancing," published in Concurrency and Computation Practice and Experience
5. Panda SK, Gupta I, Jana PK (Feb 2017) "Task scheduling algorithms for multi-cloud systems: allocation-aware approach", published in Information Systems Frontiers, Springer
6. Zhang B, Zeng Z, Shi X, Yang J, Veeravalli B, Li K (Jun 2021) "A novel cooperative resource provisioning strategy for Multi-Cloud load balancing". J Parallel Distributed Comput 152:98–107 (Elsevier)
7. Cui J, Chen P, Yu G (2020) "A learning-based dynamic load balancing approach for microservice systems in multi-cloud environment," 2020 IEEE 26th international conference on parallel and distributed systems (ICPADS), pp 334–341. https://doi.org/10.1109/ICPADS51040.2020.00052
8. Bermúdez J, Segura J, Vercher E (2007) Holt-winters forecasting: an alternative formulation applied to UK air passenger data. J Appl Stat 34:1075–1090. https://doi.org/10.1080/02664760701592125

ECC-Based Secure and Efficient Authentication for Edge Devices and Cloud Server with Session Key Establishment

Bhanu Chander and Kumaravelan

Abstract With the rapid progress in various technologies, Internet of things (IoT) considers as one of the hot topics in the research community. In modern days, more and more sensed data from IoT devices are store in cloud. However, the increasing number of cloud data security incidents guarantee that confidentiality of sharing data is still a significant challenge. This paper proposed a proficient and secure authentication scheme among IoT edge devices and cloud server based on ECC practice. We examine our proposed protocol design logic with the AVISPA tool. In addition, performance comparison results and security requirements show that the proposed scheme is secure against various known, unknown attacks, and it is also more resourceful than the existing schemes.

Keywords Internet of things · Authentication · Fuzzy extractor · Elliptic curve cryptography · AVISPA tool

1 Introduction

With the tremendous improvement in microelectromechanical devices, network hardware, and wireless communication technologies, now we have the Internet of things (IoT). IoT is a structure of interconnected computation devices and consents devices to connect to transfer collected information over a link without any human/computer interactions. The inclusion of IoT comfortably boosts our daily life also provides infrastructure for supplementary ideas like Internet of medical things (IoMT), Internet of energy (IoE), and Internet of vehicles (IoV). Here, merging IoT concepts with artificial intelligence (AI), machine learning (ML), and deep learning (DL) may permit several opportunities and produce excellent outcomes. However, data transfer among interrelated devices raises numerous challenges [1–3]. This means there are chances devices connected through IoT can imprisonment transmit sensitive information that effortlessly compromises users/networks confidentiality. More importantly,

B. Chander (✉) · Kumaravelan
Department of Computer Science and Engineering, Pondicherry University, Pondicherry 609605, India

A. Ullah et al. (eds.), *Proceedings of International Conference on Information Technology and Applications*, Lecture Notes in Networks and Systems 350,
https://doi.org/10.1007/978-981-16-7618-5_12

edge device contains limited resource constraints and extensively distributed, and thus, there is a chance for easily compromised or stolen by the attacker then regain entire sensitive info. So, there is a need for a robust access control mechanism to transfer data by any device in an IoT scheme [4–6]. Researchers designed various lightweight validation and key-agreement protocols for IoT networks to deal with these mentioned issues. In related work, secret keys, biometric structures, smart cards apply to design secure verification protocols for dissimilar appliances [6]. However, most of them suffer from computation complexities and are not resistant to various passive and active attacks.

Nowadays, ECC is gaining more and more interest from the research community with its high security with lesser key size than RSA and stands as one of the prolific asymmetric-key cryptanalyses. ECC-based verification is remarkably consistent then offers a valued trade-off among safety and efficiency associated with other validations. Hence, ECC-based verification is widely applied for cloud computing environments since it provides sensible security in a cloud computing environment [4–7]. Nevertheless, their safety levels are not appropriate in small-capacity IoT devices because their execution stands on the design of the device model. The remaining sections of the paper are mapped as follows. Section 2 describes the existing works of ECC-based authentications. Sections 3 and 4 discuss mathematical preliminaries and proposed protocol, respectively. Section 5, Subsection 5.1 defines security analysis, and Subsection 5.2 explains AVISPA tool verification.

2 Related Work

The authors of [8] project a systematic review of cryptography optimization, mainly describing security problems by compiler optimizations, recent compilers that enable security requirements, and various schemes designed to mitigate optimization complications. Authors [9] implemented a novel verification scheme for secure communication among IoT devices based on simple XOR and one-way hash functions. The scheme's performance evaluation in BAN logic and the AVISPA tool shows enhanced results against different attacks. Authors [10] proposed a host-identity-diet-exchange protocol (HIPDEX) to secure mutual verification among the user and fixed sensor node. The authors did a natural experiment for quantitative evaluation and found a reduction in computing latency and low energy consumption. Authors [11] implemented a novel idea where patients can get health-related tips and facilities from the intelligent surrounding of the hospital without any explicit devices. For this, the authors designed a biometric-based verification which also secures patient's data privacy. Simulation on real datasets demonstrates model resistance against numerous formal and informal attacks. Authors of [12] designed the enhanced protocol called RESEAP to overcome the security issues shown in Kumari et al. Authors prove the security enhancement of the proposed work by comparing security analysis, computational and communication costs. In Ref. [13], the authors designed an ECC-based verification scheme for cloud-assisted services—proposed technique employed among patient-doctor, healthcare-cloud, patient-healthcare, cloud-healthcare with hash functions. Simulation results show that the designed scheme satisfies various security requirements and defends known and unknown attacks. Authors [14] fabricated a trust verification procedure based on a packet cipher function for mobile nodes. In the procedure, the sink node verifies platforms and authenticates the mobile node, then pre-arranged pseudonyms and asymmetric keys utilized for realization on the anonymity of mobile nodes. At last, a block cipher executes to boost the verification functions. The proposed scheme is compared with other existing protocols and shows high accuracy and lower communication complexities. Authors [15], implemented an efficient and resource-constrained secure communication among cloud and edge devices with ECC. Both BAN logic and Scyther tools were applied to verify the proposed scheme, and results show that the scheme satisfies all the security requirements. In addition, it shows more excellent power resources and is suitable for RFID tags. Authors of [16] prepared an ECC-based verification protocol for vehicular cloud computing (VCC) containing fixed RFID tags. The authors use a random oracle model and information analysis to prove the security requirements of the proposed work. AVISPA tool verification and security analysis comparison results demonstrated that the protocol satisfies efficient communication with limited power.

3 Preliminaries

This section describes notations used in the proposed protocol, mathematical preliminaries, and ECC formation to understand the readers easily.

3.1 Theory of ECC

ECC is a special kind of public-key cryptography based on mathematically planned elliptic curves (EC), requiring smaller key sizes. Thus, it could be a suitable target for resource-constrained situations. ECC was projected by well-known researcher Neal Koblitz in 1985, respectively. An elliptic curve E_{F_p} over a finite field F_p demarcated as the set of every (x, y) ∈ Fp × Fp such that $y^2 \bmod p = \left(x^3 + ax + b\right) \bmod p \ldots$ Equation (1), here *a*, *b*, *x*, *y* ∈ *Fp* and gratifies equation $D = 4a^3 + 27b^2 \bmod p \neq 0$, and a prominent point at infinity which *O* symbolizes the additive elliptic curve group Gp fixed as $\{(x, y)$: *x*, *y* ∈ *Fp* also $(x, y) \in E_{F_p}\}U\{O\}$ here the point "*O*" is acknowledged as "*point at infinity*".

Point addition—assume two points *P*, *Q* on EC specified in Eq. (1), the straightforward-line joins two points as $P + Q = R$. Here, the straightforward line crisscrosses Eq. (1) at the point (−R), which reflects at point *R* concerning the *x*-axis. *Point-subtraction*—let consider the point $Q = (-P)$, then the line designed through

Table 1 Comparison of key sizes

ECC key size in Bits	RSA key size in Bits	Key size Ratio
163	1024	1:6
256	3072	1:12
384	7680	1:20
512	15,360	1:30

linking by P and Q as $P + Q = P + (-P) = O$. Here, the line amalgamated thru P and $(-P)$, which crisscrosses Eq. (1) at the point O termed as "point of infinity." *Point Doubling*—assume $P + P = 2P$ and $Q = 2P$, then the tangent line is drawn at point P, crisscrossing the curve of Eq. (1) at point $(-Q)$. *Scalar point multiplication*—depends on how the cyclic group GP is well-defined as $Q = x.P = P + P + P.... + P$ (x times), here $x \in Z_p^*$ where P is an originator of the cyclic group GP. However, the security strength of the ECC depends on the anxiety of solving the elliptic curve discrete logarithm problem, and it produces a comparable level of security as RSA with less key size (See Table 1). In addition, there are some strong reasons why we adopted ECC to project verification scheme for WSNs described as follows. ECC is more complex than other suitable encryption techniques and harder to break down than the factorization complications. ECC produces equal security with reduced key sizes utilizing low power and bandwidth (See Table 1). Mentioned reasons or advantages are essential when PKI implements low power networks. Therefore, conferring to the above-discussed good-looking features of ECC, we selected it to propose the projected authentication protocol.

3.2 *Biometric and Fuzzy Extractor*

In general, biometric-based confirmation consents one to authorize a specific identity. From the basis of statistical information, biometric data is unpredictable and broadly applied in scheming cryptographic explanations. Fuzzy filters are a pair of functions: *Gen*—produces the unchanging arbitrary bits from a specified input, whereas the other *Rep*—recovers the string from an input close to the unique input in a predefined threshold. Fuzzy extractor functions with mathematical background explained as follows: Gen: probabilistic functions which allow input Bio (info of the user), then in recurrence gives $\sigma \in \{0, 1\}$ as the biometric-key with the bit-length (l). Rep: takes user biometric as the input, say Bio (0) then (τ) as the public reproduction constraints, studies the Hamming distance (Bio (0), Bio) < t, where (t) is an error tolerance threshold. The production is the unique biometric key σ = Rep (Bio0, τ). With the inclusion of a fuzzy extractor, local biometric verification performs in our proposed protocol.

4 The Proposed Scheme

This section proposed an ECC-based authentication scheme, where embedded devices securely communicate with a session key with a cloud server (CS). The proposed scheme contains two phases: the registration phase and log-in and authentication phase. We accept that all entities of the designed scheme agree on all the fundamentals essential in the EC, which is the domain constraint of the scheme. See Table 2 for notation meanings.

4.1 Registration Phase

In the initial stage of communication, embedded device (D_i) must record with the server (S) via a secured communication network. It must assume that a registered device with the server has no pre-shared secret-key before the registration. It must consider a reasonable assumption that the registration phase to performed over a secured network.

Step R1: As a first step, the embedded device (D_i) creates a random number X_i and takes inputs device identity (ID_i), password (PW_i). In the meantime, the device imprints biometric data and generates (σi, τi) using Gen = (σi, τi). Then, the embedded device generates $M_{R1} = h(\text{ID}_i \oplus \text{PW}_i \oplus X_i)$. A subsequent message M_{R2} declared due to XOR, the random number X_i along with τi, as $M_{R2} = X_i \oplus$ τi. Finally, the device computes $\text{PID}_i = M_{R1} + M_{R2}$. Then (D_i) sends calculated PID_i to the cloud server and the local extracted time value T_1.

Table 2 Notations

Notation	Meaning
CS	Cloud Server
D_i	Embedded device
PW_i	Embedded device password
T_i	Timestamp value
G	Generator point of a large-order n
r_i	A random number generated by protocol party
R_i	A random number generated by server
X_i	A random number generated by device
CK_i	Cookie Information
σ_i	Secret constraint found after Gen function
τ_i	The second o/p found after Gen function
ET	Expiration time of Cookie
H ()	Hash function
$\oplus$	Bit-wise XOR function
$\|\|$	Concatenation
SK	Shared secret-key
$A_i ?= B_i$	Determine whether A is equal to B

Table 3 Registration phase

Embedded Device (D_i)	Cloud Server (CS)
Generate a random number X_i ID_i, PW_i, X_i Gen = (σ_i, τ_i) M_{R1} = h ($ID_i \oplus PW_i \oplus X_i$) $M_{R2} = (X_i \oplus \tau_i)$ $PID_i = M_{R1} + M_{R2}$	
PID_i, T_1 →	
	$T_1' - T_1 < \Delta$ T Generate Random number R_i Stores (ID_i, R_i) $T_i = (R_i \oplus$ H $(X_s) \oplus PID_i)$ CK_i = H ($R_i \oplus X_s \oplus$ ET $\oplus PID_i \oplus \sigma_i^*$) $CK_i' = CK_i$ + G A_i = H ($R_i \oplus$ H (X_s) CK_i') $A_i' = A_i$ * G Store (PID_i, ET, T_i, σ_i^*, CK_i)

Step R2: After receiving the request message from(D_i), CS checks the validity of received T_1 by comparing the extracted local timestampT_1'. If the dissimilarity is more than the threshold value known as maximum transmission delay ΔT, CS terminates the request message. If it was lesser than the threshold value, CS checks for (ID_i), whether it existed in its database or not. If it exists, the CS produces a new random number (R_i) and establishes a novel entry in the database directory then stores it as (ID_i, R_i). Then CS calculates $T_i = R_i \oplus H(X_s) \oplus \text{PID}_i$, $\text{CK}_i = H(R_i \oplus X_s \oplus \text{ET} \oplus \text{PID}_i \oplus \sigma_i^*)$, $\text{CK}_i' = \text{CK}_i + G$, $A_i = H(R_i \oplus H(\text{X}_s)||\text{CK}_i')$, $A_i' = A_i * G$. Finally, sends CK_i' to embedded device and stores (PID_i, ET, T_i, σ_i^*, CK_i) in a secure database. As soon as Di takesCK_i', stores it along with the produced PID_i securely (Table 3).

4.2 *Login and Authentication Phase*

Step LA1: To share the new session key SK with CS, embedded device (D_i) generates a random number r_1, and starts computations $P_1 = r_1 * G$, $P_2 = r_1 * \text{CK}_i'$ and $\text{PPID}_i = (\text{PID}_i * P_1)$. Then, D_i forwards the calculated computations and local extracted timestamp value (P_1, P_2, PPID_i, T_2) to CS as the required information for authentication.

Step LA2: The message(P_1, P_2, PPID_i, T_2) received by the CS verifies $T_2' - T_2 < \Delta T$. If the verification is not satisfied, the CS is aborted. Otherwise, CS discovers the data related to ID_i in its secret database from the given(P_1, PPID_i) and validates the (CK_i, ET, PPID_i) with stored T_i. CSg enerates random numberR_s, computes $R_s = T_i \oplus H(X_s) \oplus \text{PPID}_i$, and checks $P_2'? = \text{CK}_i * P_1$, if it finds a match, then it will validate D_i. In general, finding the corresponding D_i could be a time-consuming procedure, but it is complete by the server, which has a smaller amount of computational anxiety. It also has the benefit that the device will not be traceable. After successful validation of D_i, server generates a random number r_3, and compute: $P_3 = r_2 * G$, $P_4 = (\mathrm{CK}_i * \mathrm{PPID}_i * R_s) + (r_2 * r_1 * A_i^{'})$. Sever forwards required authentication information to D_i, which contains (P_3, P_4) and extracted timestamp value T_3.

Step LA3: After receiving the message from CS, D_i verifies $T_3^{'} - T_3 < \Delta T$. Once the positive timestamp confirmation is done, it validates whether $P_4^{'} ? = (\mathrm{CK}_i * \mathrm{PPID}_i * R_s) + (r_2 + r_1 * A_i^{'})$. If the validation satisfies, D_i accepts that the server is legal, then continuous the procedure by calculating $V_i = (\mathrm{PID}_i * P_3)$. And forwards the message (V_i, T_4) to CS.

Step LA4: CS verifies timestamp value and $V_i ?= (\mathrm{PPID}_i * r_2)$. If the verification is successful, the server admits the D_i as a legal device, then the shared session key $\mathrm{SK} = r_3 * P_1 = r_1 * P_3$ will be generated among CS and device D_i (Table 4).

Table 4 Login and authentication phase

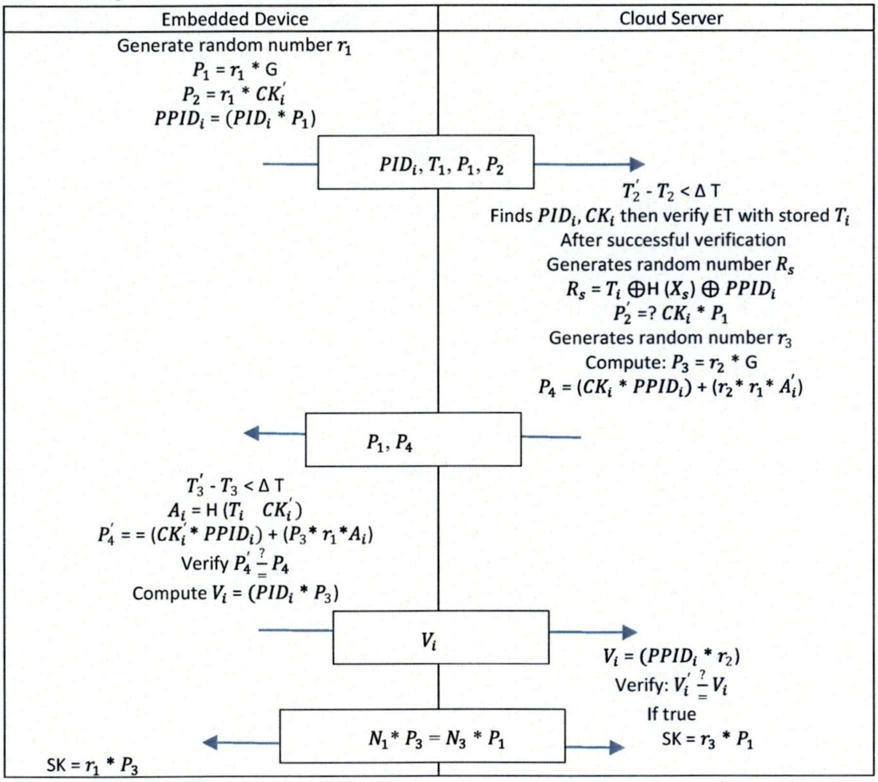

5 Security Analysis and AVISPA Verification

5.1 Security Analysis

To know the security level of any authentication protocol, there is a need to evaluate the security analysis; hence, we evaluate the security analysis of the projected verification scheme against numerous attacks.

Replay Attack: In a replay-attack, the invader observes and stores one of the broadcasted messages among CS and D_i and then tries to attach with D_i Or CS. If we check the proposed work, both D_i and CS contribute equally and randomly for each session, and any snooped session is entirely unusable for the following session. For example, consider an attacker attempts to send a long-standing message to CS in the place D_i. In this scenario, the attacker could use (PPID_i, P_1, P_2, T_2) and possibly receives ($P_3^{'}$ and $P_4^{'}$). After this, an attacker must send $V_i^{'} = \text{PID}_i * P_3^{'}$. Which is not available, and the verification of $V_i^{'} = V_i$ not happened. Hence, there is no chance for a reply attack on both D_i and CS.

Brute-Force Attack: In this attack, the invader must extract security parameters {P_1, P_2, P_3, T_4 and Timestamp values} from the broadcasted communications among D_i and CS; moreover, attackers cannot obtain random numbers r_1, r_2. So, the proposed design can oppose the brute force molest.

Impersonation Attack: In this attack, the attacker forwards (PPID_i, P_1, P_2, T_2) and V_i to the CS. After receiving the message, CS generates a random number $r_2^{'}$ then calculates $V_{i,}^{'} = r_2^{'} * \text{G}$, $P_4^{'} = (\text{CK}_i^{'} * \text{PPID}_i) + (P_3 * r_1 * A_i)$ and imagines to receive $V_i^{'} = r_2^{'} * \text{PPID}_i$ which is not accepted by the server, so the verification of V_i automatically fails. So, there is no scope for an impersonation attack on CS. For impersonate on CS, an attacker must return a valid $P_3 = r_2 * G$, $P_4 = (\text{CK}_i * \text{PPID}_i) + (r_2 * r_1 * A_i^{'})$, and hence, attacker must compute P_3, P_4 which was not possible without prior knowledge of CK_I, r_1.

Mutual Authentication: The designed scheme confirms unbeaten mutual verification between D_i and CS by evaluating ($P_2^{'}$, $V_i^{'}$). Verification of P_2 can do by a legitimated device D_i, because $P_2^{'} = ?\text{CK}_i * P_1$ here $\text{CK}_\text{i}^{'} = \text{CK} * G$. Same as invader cannot figure out V_i. Hence, the designed method accomplishes accurate mutual authentication.

Traceability Attack: The invader desires a constant value in the transported messages from CS or else device. It is not possible since in the proposed work, every session or message is randomized/transferred with new values such as T_i, and r_i.

Device Anonymity: It indicates an invader not proficient in getting the device's masked identification through broadcasted messages at both login and authentication setups. In the proposed method, CS forwards {P_1, P_2, PPID_i, T_i} to CS; here, PPID_i associated with the device's enclosed identity. In addition, P_1, P_2, P_3, P_4, PPID_i, T_i, and biometric details neither propels as any message nor store in D_i or CS. As a result, the projected idea preserves device anonymity.

Message Integrity Attack: In the proposed system, the safety of every statement/message ensure by the security level of CK_i, and PPID_i. We assumed that both (CK_i, and PPID_i) are high secured since the attacker cannot compromise (could not extract x and xG).

Man-in-the-Middle Attack: If the attacker has stolen CK_i, it does not affect the security of the proposed work since the impersonate attack on CS is not possible. The attacker could not generate $P_4 = (\text{CK}_i * \text{PPID}_i) + (r_2 * r_1 * A_i^{'})$, and not possible to verify V_i without information of PID_i.

5.2 *Formal Security Analysis by AVISPA Tool*

This section describes the security verification of the projected authentication protocol through employing automated validation of Internet-based security protocols and appliances (AVISPA) tool. It has four back-ends, those are SAT-based-Model-checker (SATMC), On-the-fly-mode-checker (OFMC), Tree-automata-based-on-automatic-approximations-for-the-analysis-of-security-protocols (TA4SP), and Constraint-Logic-based Attack-checker (CL-AtSe). These back-ends' output has a similar format and indicates whether the projected protocol procedure is SAFE or UNSAFE against numerous security attacks in the Dolev–Yao (DY) adversary model. Here we present simulation results of OFMC and Cl-AtSe. OFMC uses various numeric methods to construct an infinite tree, finds false detection, and then checks/verifies the proposed protocol built in the right manner. Cl-AtSe - The CL-AtSe checker produces a group of restrictions to effectively find attacks on the designed protocol by forming algebraic expressions. The outcome result OFMC and CL-AtSe illustrates in Fig. 1 projected that the scheme is harmless against different attacks based on the Dolve - Yalo model and achieves the goals as specified.

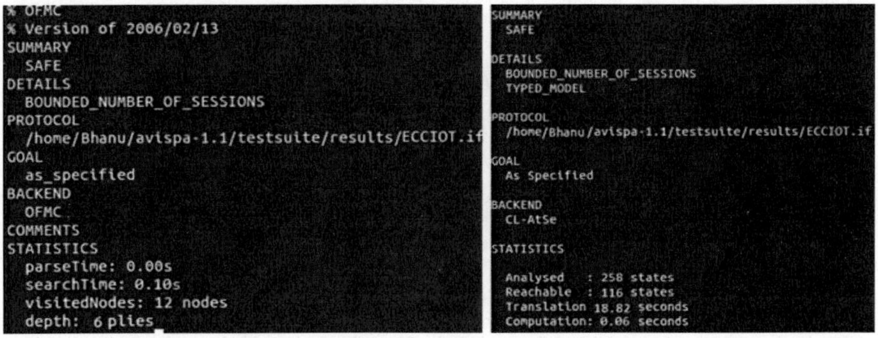

Fig. 1 OFMC and CL-AtSe result window

6 Conclusion

Security and privacy are the two critical anxieties of launching a secure verification background of the IoT in any network. This work establishes robust, flexible, and secure authentication among cloud servers and IoT edge devices based on ECC and biometric information. We mathematically analyzed our proposed work to show its resilience against various known, unknown attacks. Moreover, we formally checked our scheme with the AVISPA tool. Finally, simulation results show that our proposed work is more effective than other existing schemes and well-suited for constrained IoT devices.

References

1. Sinigaglia F, Carbone R, Costa G, Zannone N (2020) A survey on multifactor authentication for online banking in the wild. Comput Secure 95
2. Ometov A, Bezzateev S, Mäkitalo N, Andreev S, Mikkonen T, Koucheryavy Y (2018) Multi-factor authentication: a survey. Cryptography 2(1):1
3. Mahmood Z, Ning H, Ullah A, Yao X (2017) Secure authentication and prescription safety protocol for telecare health services using ubiquitous IoT. Appl Sci 7(10):1069
4. Chander B, Kumaravelan G (2020) Internet of things: foundation. In: Principles of internet of things (IoT) ecosystem: insight paradigm. Springer, Cham, pp 3–33
5. Chander B, Gopalakrishnan K (2021) Secure, efficient, lightweight authentication in wireless sensor networks. In: Machine learning, deep learning and computational intelligence for wireless communication. Springer, Singapore, pp 303–312
6. Raza S, Helgason T, Papadimitratos P, Voigt T (2017) Secure sense: End-to-end secure communication architecture for the cloud-connected internet of things. Future Gener Comput Syst 77:40–51
7. Chander GB, Kumaravelan G (2018) Simple and secure authentication in wireless sensor network using digital certification. Int J Pure Appl Math 119(16):137–143
8. Prasad A, Venkatesh S, Handadi AB, Mory M (2019) The security implications of compiler optimizations on cryptography—a review
9. Gupta A, Tripathi M, Jamil Shaikh T, Sharma A (2018) A lightweight anonymous user authentication and key establishment scheme for wearable devices. Comput Netw. https://doi.org/10.1016/j.comnet.2018.11.021
10. Bettoumi B, Bouallegue R (2016) Evaluation of authentication based elliptic curve cryptography in wireless sensor networks in IoT context. IEEE Access
11. Kumar T, Braeken A, Liyanage M, Ylianttila M (2017) Identity privacy preserving biometric based authentication scheme for naked healthcare environment. In: IEEE ICC 2017 SAC symposium e-health trac
12. Safkhani M, Bagheri N, Kumari S, Tavakoli H, Kumar S, Chen J (2020) RESEAP: An ECC-based authentication and key agreement scheme for IoT applications, vol 8
13. Kumari A, Kumar V, Yahya Abbasi M, Kumari S, Chaudhary P, Chen C-M (2020) CSEF: Cloud-based secure and efficient framework for smart medical system using ECC, vol 8
14. Wang Q, Chen W, Wang L (2019) A sink node trusted access authentication protocol for mobile wireless sensor network using block cipher algorithm based on IoT. Int J Wirel Inf Netw. https://doi.org/10.1007/s10776-019-00471-6

15. Rostampour S, Safkhani M, Bendavid Y, Bagheri N (2020) ECCbAP: A secure ECC-based authentication protocol for IoT edge devices. Pervasive Mob Comput 67:101–192
16. Kumar V, Ahmad M, Mishra D, Kumari S, Khan MK (2020) RSEAP: RFID based secure and efficient authentication protocol for vehicular cloud computing. Vehicular Commun 22

Transfer Learning Method with Deep Residual Network for COVID-19 Diagnosis Using Chest Radiographs Images

Ayesha Komal and Hassaan Malik

Abstract Nowadays, one of the most economical and conventional clinical diagnostic methods is chest radiography. Even in impoverished countries, modern digital radiography machines are easily affordable. Hence, this practice is broadly used by health experts in their medical practices for the diagnosis of COVID-19. In this research study, we applied our proposed method based on 49-Layers of convolutional neural network (CNN) with a residual network on chest X-rays images, and then we also compare the classification accuracy of our proposed model with four well-renowned transfer learning classifiers such as VGG-16, VGG-19, ResNet-50, and Inception-v3, with the aim to assists our frontline doctors in diagnosing the coronavirus at the initial stage. The novelty of this study is that our proposed classification method takes full advantage of the input data provided through chest radiographs in classifying the COVID-19 infected patients. An extensive experiment has been carried out by splitting the dataset into 70–30% ratios for the training and testing processes of the proposed model. Our proposed model has achieved the highest accuracy of 97.0%. The VGG-16, VGG-19, and ResNet-50 models show a good performance for the classification of COVID-19 with an accuracy of 92%, 83%, and 91%, respectively. The worst classification is obtained by the InceptionV3 model with an accuracy of 55% for COVID-19 cases. This research showed that deep learning models are useful in classifying COVID-19 radiography images based on the proposed model framework. The time has come to use artificial intelligence approaches in assisting physicians in the detection of COVID-19 infected patients.

Keywords Convolutional neural network · COVID-19 · Artificial intelligence · Chest radiograph

A. Komal (✉) · H. Malik
Department of Computer Science, National College of Business Administration and Economics Lahore, Sub Campus Multan, Multan, Pakistan

H. Malik
e-mail: f201928004@umt.edu.pk

H. Malik
School of Systems and Technology, University of Management and Technology, Lahore, Pakistan

A. Ullah et al. (eds.), *Proceedings of International Conference on Information Technology and Applications*, Lecture Notes in Networks and Systems 350,
https://doi.org/10.1007/978-981-16-7618-5_13

1 Introduction

At the end of 2019, the coronavirus disease appeared in the city of Wuhan Province of China. In COVID-19, patients were suffering from common symptoms of fever, fatigue, myalgia, and cough. According to World Health Organization (WHO) reports [1] coronavirus disease range is common cold to acute respiratory problems. COVID-19 roots serve lung disease symptoms, and, in such cases, patients were admitted to the intensive care unit (ICU). Initially, the RT-PCR kits were used for the diagnosis of COVID-19. The test was performed by collecting the sputum or nasopharyngeal swab of the suspected patients, but it has a low positive rate. So, chest radiography was used for the diagnosis of coronavirus disease. Most clinical experts believe that replacing nucleic acid testing with lungs radiography is one of the initial problem-solving methods for this new kind of pneumonia virus. In the medical domain, image processing techniques are widely used for medical image enhancement and organ segmentation [2, 3]. In image processing, CNN has provided the ability of nonlinear modeling for a wide range of applications [4–6]. The related studies of CNN were conducted on the analysis of pulmonary nodes [7, 8], classification of begin and malignant tumors, and pulmonary tuberculosis analysis worldwide [9, 10].

Chest radiography is considered to be one of the economical disease diagnostic methods. In this study, the proposed model and other four transfer learning classifiers were used to screen the chest radiographs images database and analyze the infection probability of COVID-19. We compare the performance of our proposed model with four transfer learning classifiers in terms of many parameters such as accuracy, sensitivity, specificity, and F1-score. The objective of this study is to design an efficient model for accurate diagnosis of coronavirus and assisting our health experts in diagnosing this deadly virus at an early stage to save human life. The novelty of this study is that our designed model takes full benefit of the input data provided through chest radiographs in classifying the COVID-19 infected patients.

The structure of this manuscript is organized as follows: Sect. 2 consists of related work. In Sect. 3, the materials and methods of the study are discussed. Section 4 contains the results and discussions. This study is concluded in Sect. 5.

2 Related Work

Recent development in the application of artificial intelligence (AI) has inspired the computer-aided diagnosis system (CAD) in many healthcare areas. AI-based deep learning model provides screening the patient's chest X-ray images with the hope to bring precision and diagnosing confirmed patients of coronavirus or other pneumonia tract diseases. It is assumed that viral pneumonia cases detecting during the COVID-19 infection context have a high probability to presume COVID-19 infection. AI methods are used for estimating infection status that occurred due to pneumonia [11–13]. The transfer learning methods and handcrafted techniques were also used to achieve remarkable performance in classifying pneumonia and COVID-19 infected X-rays. CNN model can extract features directly from the provided chest X-ray images and classify that patient into their respective classes, i.e., normal or pneumonia or COVID-19 infected. Pneumonia is an infection in the respiratory system that is caused by viruses, fungi, and bacteria that affect the human lungs [14]. The common method used to diagnose pneumonia by using chest X-ray images and computer technologies is developed on a system so that treatment and detection are possible. Many researchers used CNN-based models such as LeNet, AlexNet, VGGNet, ResNet, and GoogleNet for the diagnosis of pneumonia [15]. Deep regression framework for automatic pneumonia detection learns the multiple channel images and multi-model information to stimulate the pneumonia screening process. This framework uses visual features from multi-model channel images to provide more visual features in the single image channel and improve the ability to screen the pneumonia disease [16]. Pneumonia disease is the primary cause of death among young children and the high mortality rate worldwide. Highly trained specialists examine the chest X-rays radiograph for pneumonia detection. The computer-aided diagnosis method shows the potential for improving diagnostic accuracy. This computational approach for pneumonia regions detection is based on squeeze, extinction deep learning convolution neural network, single-shot detector, multi-task arguments, and augmentations. When the lungs are infected by a fungal infection and another bacterial disease that causes the symptoms of pneumonia disease. So, the expert doctors and radiologists analyzing chest X-ray images of infected patients. In the case of deep learning, training requires a large number of labeled datasets, which is expensive and time-consuming to obtain medical applications so, it requires the input of human experts. Using two different models as feature extractions and training on these features is an effective way to diagnose pneumonia [17]. Reverse transcription–polymerase chain reactions (RT-PCR) are used for clinical testing in the cause of pneumonia symptoms, testing may produce negative and positive results [18]. There are some drugs to treat COVID-19 patients but there is no clinical evidence available that this medicine work on patients. However, using the transfer learning approaches developing the medicine work properly on coronavirus patients. RNA of patients calculated and built a 3D model of 3C like protease and screening the drugs accordingly [19–21]. A transfer learning model-based system automatically generates the report of computed tomography for pulmonary tuberculosis (PTB) cases. Using a dataset of 501 CT imaging from 223 patients with active pulmonary tuberculosis were collected and other cases as negative sample [22]. The 3D convolutional neural model (CNN) was trained, and PTB CT–images inspection is evaluated. The transfer learning method explains the location of lesions and classifies them into the military, tubercular, infiltrative, cavity, and gaseous types similarly. Then the noisy or Bayesian function also used to generate infection and quantitative diagnostic reports generated. AI-based automated CT scan images analysis tools are also used for the detection of COVID-19 patients, and these are different from other testing tools used in clinical cases [11, 23]. Clinical studies show that most coronavirus patients are infected with a lung infection. Transfer learning is one of the most successful techniques to assist radiologists to analyze the big records of X-ray images. Applying this technique on CT scan images or X-ray detects 96.00% of COVID-19 cases and 70.65% of another viral diseases [24]. There is no automatic to active diagnostic reporol available to measure the infection volume for coronavirus patients. Transfer learning is the best method to find automatic segmentation and quantification of infection regions and also CT scans of entire lungs [25]. Deep learning based on image segmentation is now recognized as a strong tool in image segmentation. It has been widely used to distinguish the homogenous regions. So, they present a popular method that has employed deep learning for medical image segmentation. For this purpose, the CNN model is used for images segmentation [26–28]. In recent times, radiographic images have been used for screening the COVID-19.

3 Methods and Materials

3.1 X-Ray Image Dataset

For this study, we collected coronaviruses infected chest radiographs from a GitHub repository created by Cohen et al. [29]. This repository contains images collected from different hospitals and open sources. The complete metadata information is not discussed, but the average age of the COVID-19 infected patients was around 55 years (Fig. 1).

The collected radiology images from the database were of different sizes of resolution. So, we resized the images to the fixed size of resolution which is 300 × 300 pixels for our proposed model, ResNet-50, and inception v3. In addition, the collected chest X-ray images were also reshaped to a fixed size of 224 × 224 resolution by using the OpenCV library. The process of data normalization was also applied for the appropriate training of the proposed method. Hence, our datasets were prepared to be fed into the proposed and other transfer learning classifiers for the training process. All of the chest radiographs images were captured from the frontal view of

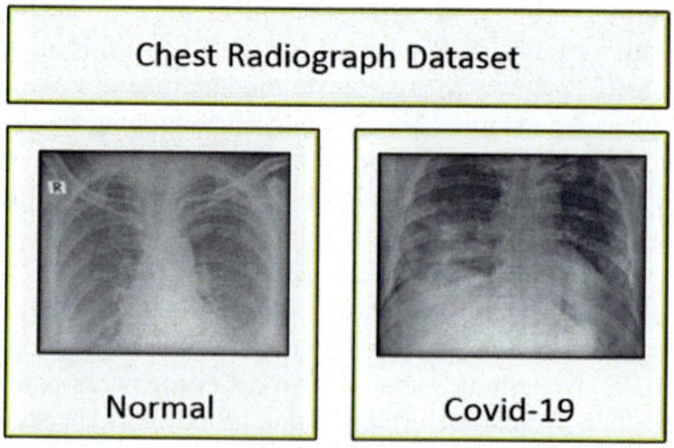

Fig. 1 Chest radiograph images of a normal and COVID-19 infected patient

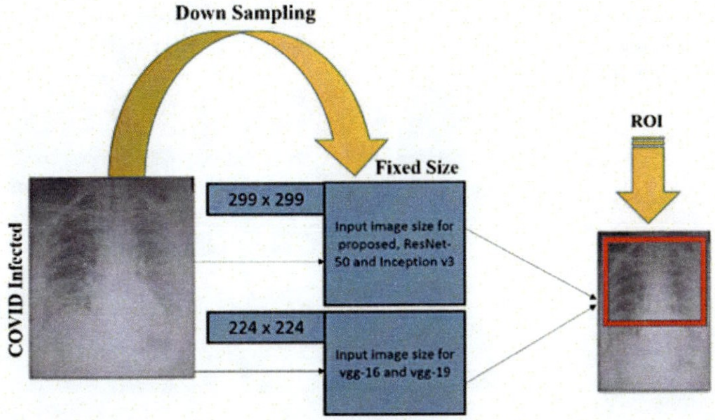

Fig. 2 Downs sampling and ROI of chest radiographs

the jpeg extension. The region of interest (ROI) was restricted to the mid part of the chest radiograph containing the infected area as shown in Fig. 2. By mentioning the ROI to include most of the COVID-19 infected portion instead of the whole image, this study evaluated the portion which plays the most distinctive role in COVID-19 classification.

There is a lot of training is required when the amount of data is scarce. Thus, we used various types of data augmentation techniques to increase the size of the dataset to avoid the model from overfitting. The parameters used for data augmentation was presented in Table 1.

After apply augmentation technique images were stored in folder augmentation. Each chest X-ray was augmented with six types of augmentation as discussed in Table 1; therefore, the size of the dataset reaches 450 chest radiographs including 264 COVID-19 and 186 normal images. Original images measured as a raw images database and have not been augmented. The detailed summary of the images after synthetically increasing the size of the dataset is shown n Fig. 3.

Table 1 Parameter used for data augmentation

Parameters	Values
Rotation	40°
Width shift	0.2
Height shift	0.2
Shear range	0.2
Horizontal flip	True
Zoom range	0.2

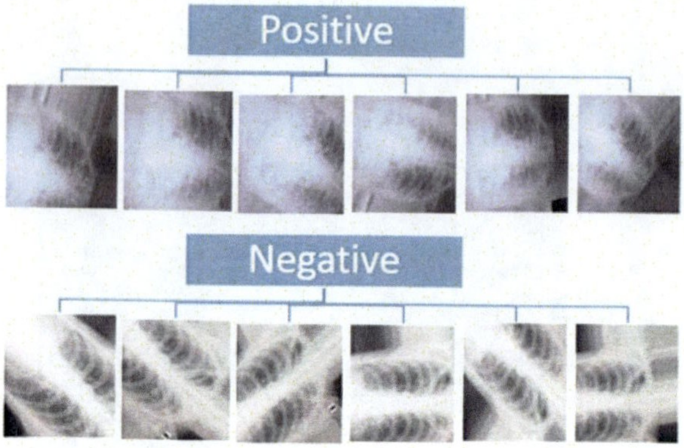

Fig. 3 Chest radiograph images after applying data augmentation techniques

3.2 *Proposed Model*

The proposed model has been designed on the CNN network to classify the coronavirus infected chest radiographs. The input of chest radiographs having a size of 300 × 300 resolutions of radiographs is fed to the proposed model for training purposes. There are RGB channels in the input image. The first layer of our model is the conv layer. This layer starts the process by using filters, which are also known as the kernel. The size of the kernel is based on two values as shown width (w) and height (h). In our work, we fixed the size of the filter to the value of 3. The three more Conv layers have been added to the model to extract the profound features from the radiographs and thus provide comprehensive patterns of the images to the model. The filters also start the conv operations with the sub-area of the radiographs. The reason for applying conv operation was to be multiplied and added the filters and the pixel values of the radiographs. The sub-area of the radiographs is also called the receptive field. Then, more Conv layers were inserted in the model step by step to increase the ability of the model for extracting the feature elements. The values of the filter are termed weights, and these weights must be learned by the model at the time of training. The filter executes the conv process from the starting of the radiographs and uninterruptedly moved across the complete radiograph. The method of performing filter conv ends when the whole image was processed. Feature maps during the radiographs yield an array of values. The value of moving the filter was confined by one more parameter known as a stride. In our study, we set the value of stride to 1 for all conv layers. By selecting the higher values of stride to reduce the spatial dimension and produces some issues for instance the receptive field can go outside of the input size. Hence, to overcome these problems, the zero-padding technique was applied which pads the zero around the border of the input radiographs to maintain the output size dimension as same as the input. Conv layers have used different numbers of filters for extracting the multiple and dominant features. The first conv layer of our model contains 16 filters. The number of filters of other Conv layers was changed, from 16 to 512. The activation function rectified linear unit (ReLU) which was applied on the conv outputs. The process of ReLU activation is to change all the negative results by zero. The purpose to use the ReLU function in our proposed CNN model is that it improves the nonlinearity and helps to increase the computational time without influencing the model accuracy. The max-pooling layer was inserted after the conv layers which helps to reduce the spatial dimension of the input radiographs. We fixed the filter of the size of 2 × 2 and also sets the value of stride to 2 for MP layers. This filter convolves all over the input volume and produced the highest value of the sub-area of the radiographs. It has been observed that MP layers consist of a particular feature's position about other features. It also reduces the computational cost by decreasing the number of weights and prevents the model from overfitting. Then, the dropout layer was added. We have applied different values of the dropout to our model such as 0.05, 0.20, and 0.25, but only the appropriate and significant threshold of 0.20 was reported in this study. The reason behind applying the different dropout values was to prevent our proposed model from overfitting. The working of this layer was to drop out the activation randomly and also confirm the model to predict the actual label. The flattened layer of our model was applied to convert the two-dimensional feature map into a one-dimensional feature vector. The data generated by the flatten layer was fed to the FCL. The FCL performed the classification process from the obtained one-dimensional feature vector. For this study, FCL contains 512 neurons. The initial dense layer of FCL passes the result of activation to the other dense layer. In the end, the model classifies the coronavirus infected radiographs and normal cases.

3.3 *Transfer Learning Classifiers*

VGG-16 trained on the ImageNet database consists of 16 convolutional layers and has a filter size of 3 × 3 but their depth is less than the VGG-19 network. Figure 4 shows the architecture of the VGG-16 model. It has a total of 5 layers and used a max-pooling layer of 2 × 2. After the max-pooling layer, 3 fully connected layers (FCL) exist. At the final step, softmax classifiers are used and then applied ReLU

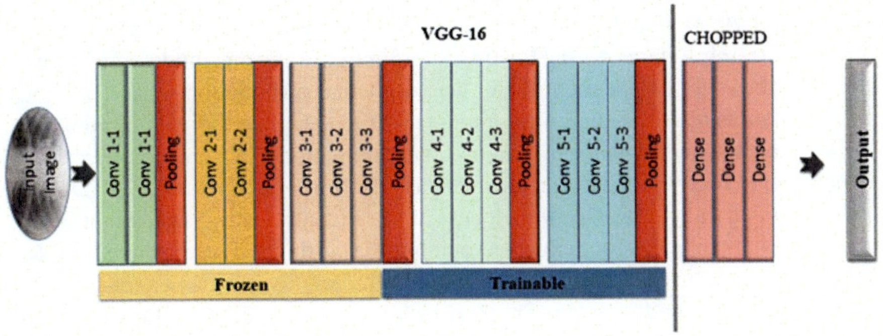

Fig. 4 VGG-16 model

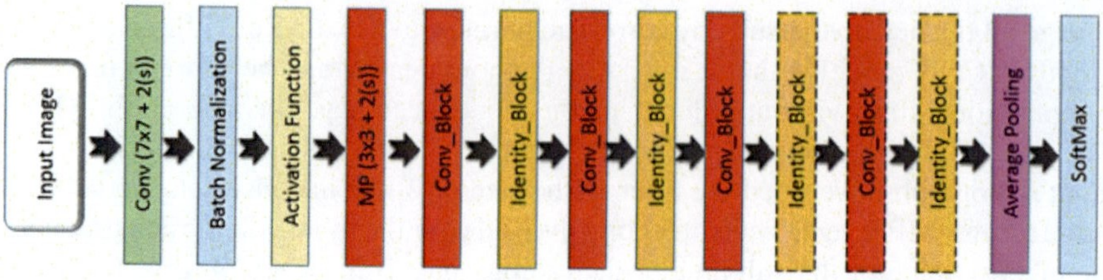

Fig. 5 ResNet50 model

activation function on all hidden layers. Inception V3 and GoogleNet won the 2014 ImageNet challenge with an accuracy of 93.3% and both are 22-layer networks [30]. Their coming version is mentioned as Inception v3 where (*n* is the number) such as inception *v*1, inception *v*2, and inception *v*3. The inception *V*3 model has many a-symmetrical and symmetrical several blocks, and each block has many branches of convolutional, max-pooling, average pooling, dropouts, and fully connected layers. Large-scale visual recognition challenge of 2014 was developed visual geometry network based on convolutional neural network (CNN) by Oxford robotics institute [31]. For image extraction functionality VGGNet used filters of 3×3. There are two types of deep neural network design VGG-19 and VGG-16 both are dissimilar in-depth and layers. VGG-19 network contains larger parameters and is more expensive than VGG-16 to train the network.

ResNet50 means residual networks having that is based on 50 layers. It is in some way similar to the VGG-16 model, but ResNet-50 additionally contains mapping functionality. This is shown in Fig. 5 ResNet-50 reduces the gradient and vanishing issue by using the alternative and shortcut way for the incline to stream through. If the current layer is not compulsory, then identity mapping used in ResNet50 permits the model to pass a CNN weighted layers. So, by avoiding these layers problem of overfitting that occurs in the dataset is solved.

The chest radiographs images of the dataset are selected for training the model and randomly divided into five folders. To estimate the model, training was performed 5-folder cross-validation, to reduce the chance of overfitting. With five-cross-validation, the database was separated into independent training, validation, and testing sets of 70, 10 and 20% split, respectively. Independent folder was applied to estimate the training status during the time of training. When one model training was completed, the other independent folder was applied as a validation set and previous validation data is also reused as a training set part to calculate the model training.

To evaluate the performance for each deep learning model, different matrices have been applied to measure the correct or misclassification of diagnosed COVID-19 in the radiograph's images. Firstly, the process of a confusion matrix is shown in Fig. 6. The confusion matrix has four expected outcomes as follows. True positive (TP) is a number for anomalies and identified with the right diagnosis. True negative (TN) is an incorrectly measured number of regular instances. False negative (FN) is a list of anomalies observed as an ordinary diagnosis. False positive (FP) is a regular instance that is classified as a regular anomaly [32].

Fig. 6 Confusion matrix

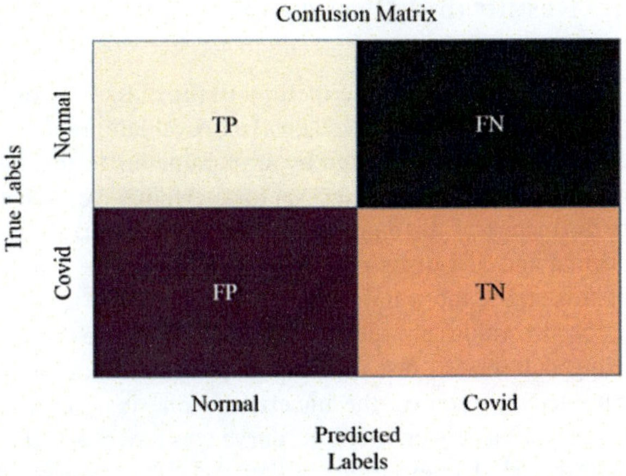

The following performance metrics were measured by using a confusion matrix.

Accuracy: The accuracy of a method determined how predicted the correct value. It is the summation of true positive and true negative divided by the total value of the confusion matrix

$$\text{Accuracy} = \frac{\text{TP} + \text{TN}}{\text{TP} + \text{TN} + \text{FP} + \text{FN}} \times 100\% \tag{1}$$

Precision: Precision gives the relationship between the true positive (TP) predicted values full predicted values.

$$\text{Precision} = \frac{\text{TP}}{\text{TP} + \text{FP}} \tag{2}$$

Recall: Recall is the ratio between the true positive values of prediction and the summation of predicted true positive values and predicted false negative values.

$$\text{Recall} = \frac{\text{TP}}{\text{TP} + \text{FN}} \tag{3}$$

F1-Score: The F1-score is the model accuracy that combines recall and precision twice.

$$\text{F1 - Score} = 2\left(\frac{\text{Precision} * \text{Recall}}{\text{Precision} + \text{Recall}}\right) \tag{4}$$

4 Results and Discussion

The determination of the method to the classification of COVID-19 patient's chest radiograph pictures is based on a train evaluation scheme. We train our model with the dataset of training, change hyper-parameters to the validation set from the network results, and then use the test set to determine overall performance. For evaluating the performance of the deep learning classifier, 450 chest radiograph images including normal and 154 diseased cases chosen from the dataset. For all CNNs architecture of this study, the training parameters are: The learning rate = 0.0001, epoch, and batch size values are set to 15 and 50, to achieve the best coverage in a few iterations using dataset. To test the response of the model in two groups, we employ four parameters based on the uncertainty matrix: accuracy, sensitivity, specificity, and F1-score. Moreover, the ROC curve was dawn, and the resulting accuracy value was determined. To determine further the classification impact of the proposed model in this article, the five most frequently used CNN 49 layers image classification architecture Reset50, VGG-19, VGG-16, and InceptionV3 have been contrasted.

4.1 Development

The experiments deep learning algorithms including the approach implies are built using Python is coded for approaches that do not have convolutional networks. The test has been conducted with the GPU NVIDIA GeForce GTX 1080Ti and 16 GB RAM.

4.2 Test Results

Here, we can see the model proposed to improve our model performance understanding. The efficiency of the proposed model is constructed with the performance of existing classification methods using chest radiographic images dataset, where the collected data is shown in Table 2.

Table 2 compares our approach to standard CNNs, such as VGG-16, VGG-19, ResNet50, and InseptionV3. To the best of our understanding, the 224 × 224 color image classification of the VGG-16 and VGG-19, while the 299 × 299 color image classification of InceptionV3. It should again be emphasized that the size of an initial input image must be set to a scale of 300 × 300 by our proposed deep learning algorithms. According to the result finds in Table. 2, the proposed approach is calculated to have the highest accuracy in the classification and F1-score in the data collection used in this analysis. Ozturk et al. [33] utilized similar steps were achieved 87% accuracy, 85% sensitivity, 92% specificity, and 87% F1-score. Loannis et al. [34] achieved 93% accuracy, 92% sensitivity, 98% specificity, and 95% F1-score. Xu

Table 2 Comparison of the proposed COVID-19 diagnostic method with other deep learning models

Reference	Performance matrix (%)				
	Method used	Accuracy	Sensitivity	Specificity	F1-Score
Ozturk et al. [33]	DarkCovidNet	87.02	85.35	92.18	87.37
Loannis et al. [34]	VGG-19	93.48	93.85	98.75	95.7
Xu et al. [35]	ResNet 18	86.7	83.3	81.3	83.9
Shayan et al. [36]	CNN	93.2	96.09	99.7	97.9
Yan et al. [37]	MSCNN	87.5	89.1	85.7	93.4
Proposed method	49-CNN	97.0	98.41	88.45	92.22

et al. [35] achieved 86% accuracy, 83% sensitivity, 81% specificity, and 83% F1-score. Shayan et al. [36] achieved 93% accuracy, 96% sensitivity, 99% specificity, and 97% F1-score. Yan et al. [37] achieved 87% accuracy, 89% sensitivity, 85% specificity, and 93% F1-score. Our proposed method achieved highest 97% accuracy, 98% sensitivity, 85% specificity, and 93% F1-score. The ROC for the proposed model and the existing classification methods are shown in Fig. 7.

The training accuracy and validation accuracy of our proposed model relating to 30 epochs are shown in Fig. 8. It has been noticed that the highest obtained training accuracy was 97.12% and validation accuracy was 94.21%. The training and validation loss of our proposed model were 0.0091 and 0.099, respectively.

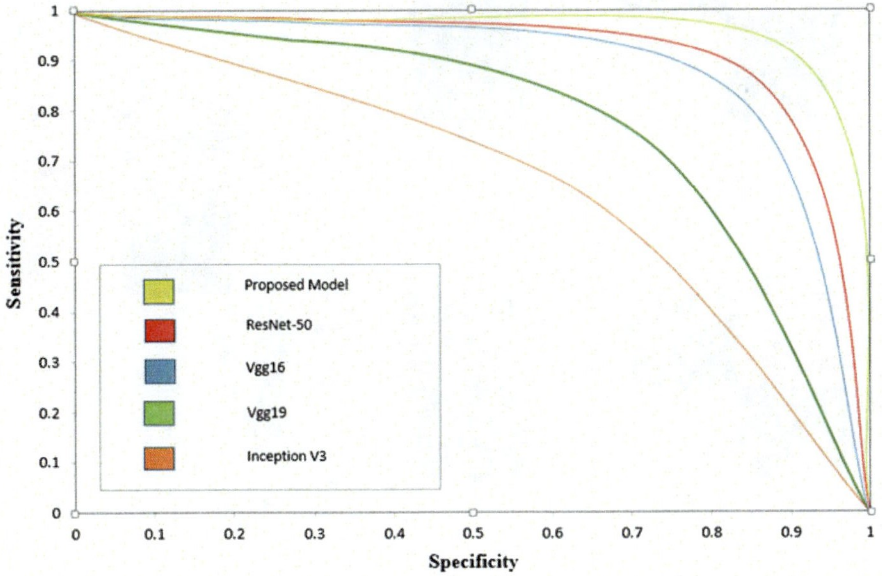

Fig. 7 ROC curves for the different methods with the radiographic chest images dataset

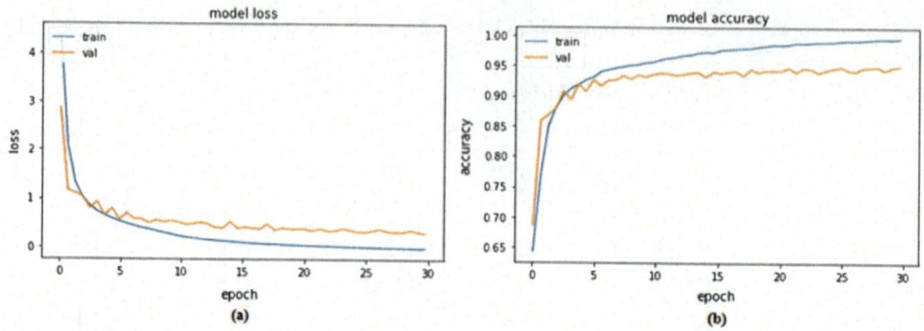

Fig. 8 Proposed model accuracy and loss

These outputs showed that our proposed model was trained well and can correctly identify the coronavirus infected radiographs versus healthy cases.

From Fig. 9, it has been observed that the CNN-49-layer model achieved 97% accuracy, precision 88%, Recall 98%, and F1-score 92%. The VGG-16 model achieved 92% accuracy, precision 96%, recall 72%, and F1-score 82%. The VGG-19 model achieved 83% accuracy, 5% precision, recall 84%, and F1-score 62%.The

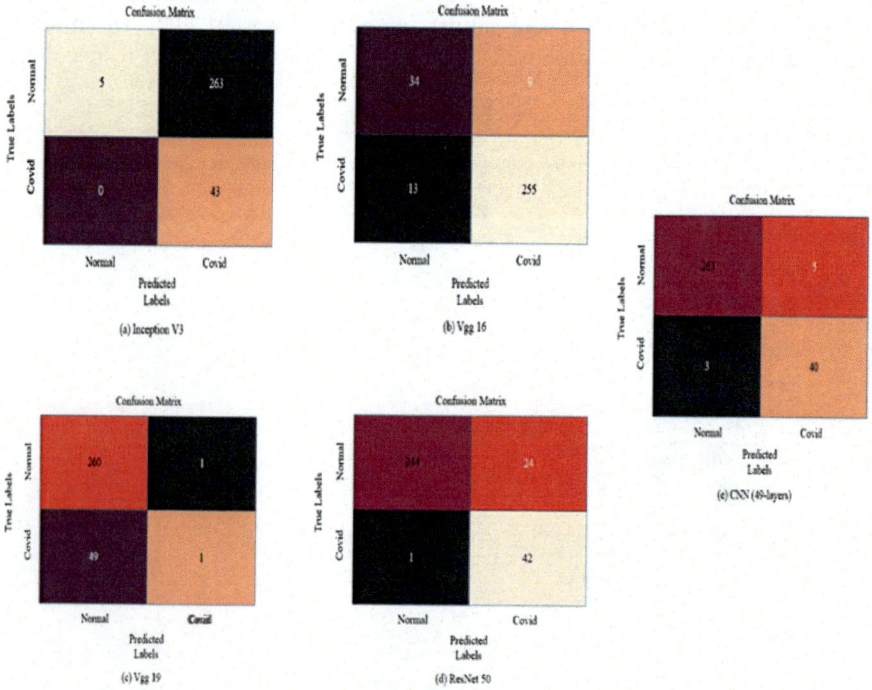

Fig. 9 Confusion matrix of all transfer learning model

ResNet50 model achieved 91% accuracy, 63% precision, 60% recall, and 67% F1-score. The Inception V3 model achieved 15% accuracy, 14% precision, 30% recall, and 40% F1-score. The performance measure is computed where the average value of this measure is 97, 98, 88 and 92% for accuracy, sensitivity, specificity, and F1-score, respectively.

5 Conclusions

In this research, we suggest an appropriate diagnostic approach in which cases of COVID-19 chest radiographic images can be detected and separated. In general, we have applied two methods: the conventional transfer learning method, a combination of pre-trained deep learning models, and the traditional classification of machine learning. The result reveals that the presented CNN 49 layers architecture is superior to its other deep learning model with higher accuracy (97.0%) and sensitivity (96.1%), respectively. For diagnosing COVID-19 patients with high efficiency, the presented CNN architecture can be used. In the next step of the report, a segmentation approach for detecting infectious areas in the human lung is presented. To accomplish this aim, we have developed a CNN 49-layer architecture to classify COVID-19 patient's radiographic images. For training data, 70% of the images are used and the remaining images are used for validation of the presented architecture. The findings indicate that 97% of the presented system can nearly locate the area of the infected lung images. This network can also be used for separating the affected area from the lung images of patients. This discovery can also be used for controlling and monitoring patients from the infected area. These approaches are almost used in a wearable monitoring system for diagnosis of disease, monitoring, and transferring to certain physicians. COVID-19 pandemic shows that science has a long-term future perspective. So, it is time to implement artificial intelligence methods in medicine and disease identification methods to help human diagnosticians.

References

1. Shan et al (2020) Lung infection quantification of COVID-19 in CT images with deep learning. arXiv.org. [Online]. Available: https://arxiv.org/abs/2003.04655
2. Zhu N, Zhang D, Wang W et al (2020) A novel coronavirus from patients with pneumonia in China, 2019. N Engl J Med. https://doi.org/10.1056/NEJMoa2001017
3. Li Q, Guan X, Wu P et al (2020) Early transmission dynamics in Wuhan, China, of novel coronavirus-infected pneumonia. N Engl J Med. https://doi.org/10.1056/NEJMoa2001316
4. Corman VM, Landt O, Kaiser M et al (2020) Detection of 2019 novel coronavirus (2019-nCoV) by real-time RT-PCR. Euro Surveill 25(3). https://doi.org/10.2807/1560-7917.ES.2020.25.3.2000045
5. Huang C, Wang Y, Li X et al (2020) Clinical features of patients infected with 2019 novel coronavirus in Wuhan, China. Lancet. PII: S0140-6736(20)30183-5. https://doi.org/10.1016/S0140-6736(20)30183-5

6. Chan JF, Yuan S, Kok KH et al A familial cluster of pneumonia associated with the 2019 novel coronavirus indicating person-to-person transmission: a study of a family cluster. Lancet. PII: S0140-6736(20)30154-9. https://doi.org/10.1016/S0140-6736(20)30154-9
7. Liu X, Guo S, Yang B et al (2018) Automatic organ segmentation for CT scans based on super-pixel and convolutional neural networks. J Digital Imaging 31(6)
8. Hesamian MH, Jia W, He X et al (2019) Deep learning techniques for medical image segmentation: achievements and challenges. J Digital Imaging 32(8)
9. Akagi M, Nakamura Y, Higaki T et al (2019) Correction to: deep learning reconstruction improves image quality of abdominal ultra-high-resolution CT. Eur Radiol 29(8)
10. Nardelli P, Jimenez-Carretero D, Bermejo-Pelaez D et al (2018) Pulmonary artery-vein classification in CT images using deep learning. IEEE Trans Med Imaging PP(99):1–1
11. Ardila D, Kiraly AP, Bharadwaj S et al (2019) End-to-end lung cancer screening with three-dimensional deep learning on low-dose chest computed tomography. Nat Med 25(6):954–961. https://doi.org/10.1038/s41591-019-0447-x
12. Li L, Huang H, Jin X (2019) AE-CNN classification of pulmonary tuberculosis based on CT images. In: 2018 9th International conference on information technology in medicine and education (ITME)
13. Huang L, Han R, Ai T, Yu P, Kang H, Tao Q, Xia L (2020) Serial quantitative chest CT assessment of COVID-19: deep-learning approach. Radiol: Cardiothoracic Imaging 2(2)
14. Apostolopoulos ID, Mpesiana TA (2020) Covid-19: automatic detection from X-ray images utilizing transfer learning with convolutional neural networks. Phys Eng Sci Med 43:635–640. https://doi.org/10.1007/s13246-020-00865-4
15. Xie et al. (2020) Comparison of different samples for 2019 novel coronavirus detection by nucleic acid amplification tests. Int J Infect Dis 93:264–267. https://doi.org/10.1016/j.ijid.2020.02.050
16. Gozes O, Frid-Adar M, Greenspan H, Browning PD, Zhang H, Ji W, Bernheim A, Siegel E (2020) Rapid AI development cycle for the coronavirus (COVID-19) pandemic: initial results for automated detection & patient monitoring using deep learning CT image analysis. arXiv.org. [Online]. Available: https://arxiv.org/abs/2003.05037
17. Zhang J et al (2020) Viral pneumonia screening on chest X-ray images using confidence-aware anomaly detection. arXiv.org. [Online]. Available: https://arxiv.org/abs/2003.12338
18. Bai et al. (2020) Performance of radiologists in differentiating COVID-19 from viral pneumonia on chest CT. Radiology 200823. https://doi.org/10.1148/radiol.2020200823
19. Wang et al (2020) Temporal changes of CT findings in 90 patients with COVID-19 pneumonia: a longitudinal study. Radiology 200843. https://doi.org/10.1148/radiol.2020200843
20. Corman et al (2020) Detection of 2019 novel coronavirus (2019-nCoV) by real-time RT-PCR. Eurosurveillance 25(3). https://doi.org/10.2807/1560-7917.es.2020.25.3.2000045
21. Hesamian, Jia W, He X, Kennedy P (2019) Deep learning techniques for medical image segmentation: achievements and challenges. J Digital Imaging 32(4):582–596. https://doi.org/10.1007/s10278-019-00227-x
22. Loey, Smarandache F, Khalifa NM (2020) Within the lack of chest COVID-19 X-ray dataset: a novel detection model based on GAN and deep transfer learning. Symmetry 12(4):651. https://doi.org/10.3390/sym12040651
23. Born J et al POCOVID-Net: automatic detection of COVID-19 from a new lung ultrasound imaging dataset (POCUS). arXiv.org. [Online]. Available: https://arxiv.org/abs/2004.12084v3
24. Wang et al (2020) A deep learning algorithm using CT images to screen for corona virus disease (COVID-19). Available: https://doi.org/10.1101/2020.02.14.20023028
25. Zhu W, Huang Y, Zeng L et al (2019) AnatomyNet: deep learning for fast and fully automated whole-volume segmentation of head and neck anatomy. Med Phys 46(2)
26. Huang P, Park S, Yan R et al (2018) Added value of computer-aided CT image features for early lung cancer diagnosis with small pulmonary nodules: a matched case-control study. Radiology 286(1):286–295. https://doi.org/10.1148/radiol.2017162725
27. Hemdan EED, Shouman MA, Karar ME (2020) COVIDX-Net: a framework of deep learning classifiers to diagnose COVID-19 in X-Ray images. arXiv preprint arXiv:2003.11055

28. Wang L, Wong A (2020) COVID-Net: a tailored deep convolutional neural network design for detection of COVID-19 cases from chest radiography images. arXiv preprint arXiv:2003.09871
29. Cohen JP (2020) COVID-19 image data collection. https://github.com/ieee8023/COVID-chestXray-dataset
30. Szegedy C, Ioffe S, Vanhoucke V, Alemi AA (2017) Inception-v4, inception-ResNet and the impact of residual connections on learning. Presented at the Proceedings of the thirty-first AAAI conference on artificial intelligence. San Francisco, California, USA
31. Simonyan K, Zisserman A (2014) Very deep convolutional networks for large-scale image recognition
32. Sokolova M, Lapalme G (2009) A systematic analysis of performance measures for classification tasks. Inf Process Manage 45(4):427–437
33. Ozturk T, Talo M, Yildirim EA, Baloglu UB, Yildirim O, Rajendra Acharya U (2020) Automated detection of COVID-19 cases using deep neural networks with X-ray images. Comput Biol Med 121:103792. https://doi.org/10.1016/j.compbiomed.2020.103792
34. Apostolopoulos ID, Mpesiana TA (2020) Covid-19: automatic detection from X-ray images utilizing transfer learning with convolutional neural networks. Phys Eng Sci Med 43(2):635–640. https://doi.org/10.1007/s13246-020-00865-4
35. Xu X, Jiang X, Ma C, Du P, Li X, Lv S, Yu L, Ni Q, Chen Y, Su J, Lang G, Li Y, Zhao H, Liu J, Xu K, Ruan L, Sheng J, Qiu Y, Wu W, Liang T, Li L (2020) A deep learning system to screen novel coronavirus disease 2019 pneumonia. Engineering 6(10):1122–1129
36. Hassantabar S, Ahmadi M, Sharifi A (2020) Diagnosis and detection of infected tissue of COVID-19 patients based on lung x-ray image using convolutional neural network approaches. Chaos, Solitons Fractals 140:110170
37. Yan T, Wong P, Ren H, Wang H, Wang J, Li Y (2020) Automatic distinction between COVID-19 and common pneumonia using multi-scale convolutional neural network on chest CT scans. Chaos, Solitons Fractals 140:11015

Deep Learning Approach for COVID-19 Diagnosis Using X-Ray Images

Muntasir Al-Asfoor and Mohammed Hamzah Abed

Abstract During the times of pandemics, faster diagnosis plays a key role in the response efforts to contain the disease as well as reducing its spread. Computer-aided detection would save time and increase the quality of diagnosis in comparison with manual human diagnosis. Artificial intelligence (AI) through deep learning is considered as a reliable method to design such systems. In this research paper, an AI-based diagnosis approach has been suggested to tackle the COVID-19 pandemic. The proposed system employs a deep learning algorithm on chest X-ray images to detect the infected subjects. An enhanced convolutional neural network (CNN) architecture has been designed with 22 layers which are then trained over a chest X-ray dataset. More after, a classification component has been introduced to classify the X-ray images into three categories. The system has been evaluated through a series of observations and experimentations. The experimental results have shown a promising performance in terms of accuracy. The system has diagnosed COVID-19 with accuracy of 0.9961% and normal subjects with accuracy of 0.96067 while it showed 0.9588 accuracy on Pneumonia.

Keywords COVID-19 · Deep learning · CNN · X-ray images · Medical system

1 Introduction

Artificial intelligent (AI) has affected health care in many ways. It radically changing the ways people are accessing health care from an online medical check, over the phone services to robot-assisted surgeries [1]. Nowadays, machine learning has shown substantial importance in the applications of medical image processing [2]. More precisely, a field of machine learning called deep learning which encompasses

M. Al-Asfoor · M. H. Abed (✉)
Computer Science Department, College of Computer Science and Information Technology, Al-Qadisiyah University, Al Diwaniyah, Iraq
e-mail: mohammed.abed@qu.edu.iq

M. Al-Asfoor
e-mail: muntasir.al-asfoor@qu.edu.iq

A. Ullah et al. (eds.), *Proceedings of International Conference on Information Technology and Applications*, Lecture Notes in Networks and Systems 350,
https://doi.org/10.1007/978-981-16-7618-5_14

more advanced and better-performed methods for training a neural network to diagnose diseases. Deep learning uses models which consist of multiple layers to learn the representations of data using many techniques. These techniques have dramatically improved the state-of-the-art in many computational applications [3]. COVID-19 a disease was first discovered in China in December 2019 and then declared a pandemic by WHO in March 2020. AI would provide a computer-assisted system to classify the huge amount of data which is pooling from around the world. Training a system to detect with precision if a person is infected can save time and lives. This paper has proposed an AI-based method to diagnose COVID-19 patients. A deep learning approach has been introduced to train a neural network about chest X-ray images of infected subjects and then use the inferred knowledge to identify future cases. The rest of this paper has been organized as follows: Sect. 2 introduces related work in the field of lung diseases diagnosis. Furthermore, Sects. 3 and 4 illustrate the concept of deep learning and materials and methodologies which have been used during this research. Section 5 demonstrates the experimental results while Sect. 6 concludes the paper and proposes suggestions for future improvements.

2 Related Work

Nowadays, medical imaging analysis combined with machine learning has emerged as a promising solution to be applied in the field of diagnosis. Deep learning as a successor of conventional machine learning is being applied in the fields of diagnosis and prediction. Xue et al. [4] have introduced a detailed review of the applications of image analysis using deep learning for the purpose of diagnosing cancer as well as survival prediction. Computerized early detection of infections would increase the probability of healing and future follow-up. Many researchers have devoted their work to design solutions that automate the process of analyzing and diagnosing lung diseases to reduce the human bias and misjudging. A lung cancer classification has been presented by Attique Khan et al. [5] with the aim of detecting lung tumor. The authors have used CT images which are then processed through number of steps to spot the infectious regions of the lung. The system has not employed deep learning which might have improved the performance in terms of time complexity. Liu et al. [6] work has focused on the detection of lung cancer using reinforcement learning. They have used deep learning models with the aim of locating lung tumor and characterizing them into different types of tumors. An advanced neural network framework has been introduced by Pasa et al. [7] to optimize the diagnosis of tuberculosis diseases. The authors have claimed that the proposed architecture would produce good training results while improving the computational time performance. They also visualized the output using the saliency map as well as the grad-CAMs. A subjective test has been done by a radiologist to determine the system efficiency in detecting the infected areas of lung. Kim et al. [8] proposed a method called "Class-selective Relevance Mapping" (CRM), which is capable of localizing and visualizing a particular region of interest (ROI) in, medical images. Their system would use image modalities to classify specific types of images to visually explain the learned behavior of a machine learning scenario.

3 Deep Learning

Deep learning is an approach of machine learning field inspired by an artificial neural network [9, 10]. It used multilayer sequencing to extract high-level features from raw images or data [11]. Depending on the problem domain and degree of complexity, the features have been selected, as well as the number of layers is increased. One of the most modern deep learning techniques is convolutional neural network (CNN) [12]. CNN is a branch of deep learning network [13], and it is majorly designed as a series of phases formed by layers. In the current research, we have proposed CNN model with an architecture that consists of 22 layers. Convolutional layer is the first layer of CNN [14, 15]. It takes a filter (kernel) and passes it through all the points in the entire image and passing at any point into a single position (output). The next layer is pooling layer which performs a down-sampling of an image, and it takes subsamples of convolutional layer output and produces a single output. There are different pooling methods such as max-pooling, mean pooling, and average pooling; basically, the flow is forwarded of extracted features from first convolution layers to the final fully connected layer and classification.

4 Materials and Methodology

This section describes the detailed design of the proposed system's architecture. The proposed model consists of two components, namely the training component and classification component. The training part has been designed based on deep learning using convolutional neural network. The purpose of it is to train a labeled chest X-ray dataset of images which are partitioned into infected with COVID-19 and normal. Figure 1 shows the design paradigm of the proposed solution.

4.1 CNN Architecture Design

The proposed CNN architecture consists of 22 layers starting from the input layer to the classification layer. Basically, the proposed model of CNN contains four phases of convolution layer for features extraction. Each one of them includes a convolutional layer, normalization layer, activation function, and pooling.

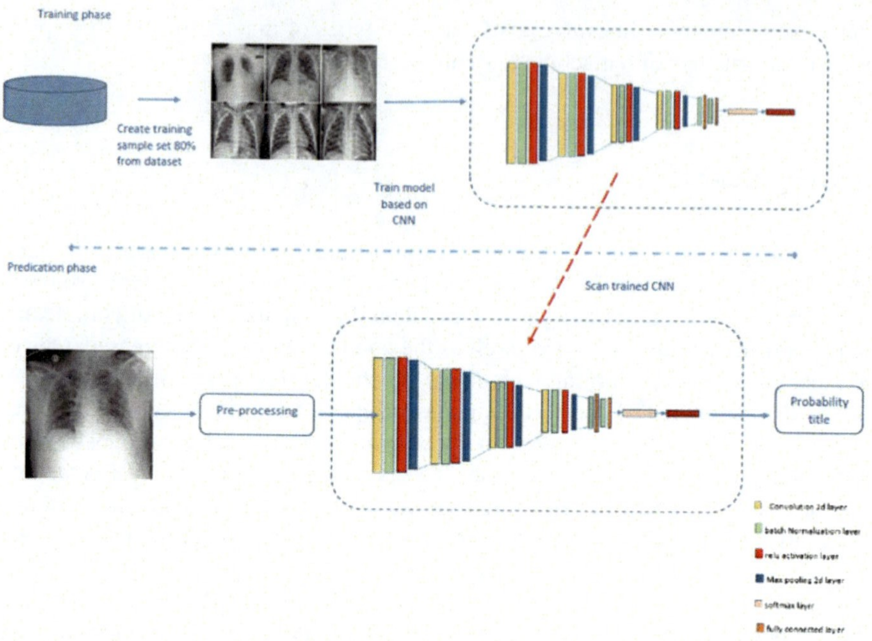

Fig. 1 Proposed model schema for COVID-19 detection

- The first layer in the proposed CNN is an input layer which is used to determine the number of channels and input images dimensions. In this model, the properties for this layer were input size 224 × 224.
- After that the convolutional layer a 2-D convolutional layer applies to slide convolutional filters to the input image. The layer scans the input matrix (image) vertically and horizontally, to compute the result of the weights and the input, and then adds a bias term.
- The third part of the first block is the normalization layer; batch normalization layer was used with epsilon 0.00001. This layer is used to normalize each input channel across a mini-batch. Then, the layer changes the input raw data by a learnable offset as well as changing the scales by a learnable scale factor. To increase the training performance of CNN and reduce the sensitivity to network initialization.
- Rectified Linear Unit ReLU Activation function is the most common activation function which is used in deep learning and CNN. ReLU is mathematically described in Eq. (1)

$$\text{ReLu } f(x) = \max(0, x) \quad (1)$$

- The next CNN layer is pooling layer, and the main function of this layer is to reduce the number of parameters. This operation applies to each features map

Table 1 Summary of COVID-19 performance

Class	Accuracy	sensitivity	precision
COVID-19	0.9961	0.9565	0.88
Normal	0.96067	0.93129	0.91044
pneumonia	0.9588	0.96679	0.97677

individually. There are many pooling functions, and in this model, max-pooling is suggested.

The above layers are repeated with different parameters for four times, to extract the high-level features from the X-ray images. Then two fully connected layers with drop-out layer in between for the classification phase. Finally, the softmax and class output layer to classify the entire image belong to which class, based on training phase. The proposed CNN model is design in such a sequential way with 22 layers, to diagnosis COVID-19 diseases. The different parameters of each layer as mentioned in Table 1 increase the accuracy of lung infection diseases by select the proper features which have the highest contribution in classification score.

5 Experimental Result and Analysis

The proposed solution has been examined and evaluated by conducting a series of experiments which has been done under different scenarios.

5.1 Experimental Setting

The experimental settings have been organized as follows.

5.1.1 Dataset Collection

The proposed CNN model has been trained and evaluated based on collected chest X-ray images datasets. To evaluate the proposed model, we trained CNN based on these images. The COVID-19 collection of images were collected from two different datasets: first one chest X-ray images (pneumonia) [16] and the second one COVID-19 images collection [17–19]. The chest X-ray dataset contains two label classes: normal and pneumonia, 1585 samples for Normal class and 4275 samples for pneumonia. The COVID-19 images collection consists of 123 samples of chest X-ray images. One of the most important challenges that we have faced during the course of this research was the limited number of chest X-ray for patients who are suffering from COVID-19. All the images were uninformed into 224 × 244, as well as enhanced by using adaptive histogram equalization.

5.1.2 Training and Implementation

The proposed model based on CNN was trained on the chest X-ray collected images. Initially, the CNN model was trained and evaluated based on sub-COVID19-collected images from the 123 images for each label class, then trained on all the dataset. The training option and parameters that used for training were as the following: momentum: 0.9000, initial learning rate = 0.0100, Max Epochs = 20 and Mini Batch Size = 64. As well as using learning rate schedule with piecewise option, updating the learning rate every certain number of epochs during the training phase. The solver of the training network was stochastic gradient descent with momentum "SGDM" optimizer. The proposed CNN was built and evaluated using the MATLAB R2019 and deep network designer tool.

5.2 *Result and Discussion*

To evaluate the performance of the proposed CNN model, some of the analysis were used to measure the result based on standard measurement, to understand the detection performance. The measurement that we calculate is an accuracy, precision, and sensitivity of a method that determines how correct class is predicted as well as confusion matrix was calculated. Figure 2 shows the confusion matrix of the chest X-ray collected dataset, as well as can see the precision and sensitivity values which's belong to the three classes.

Table 1 shows the summary of the performance of COVID-19 diagnosis model, as well as it contains value of true positive, true negative, false positive, false negative, accuracy, sensitivity, and precision for each class of COVID-19, normal and pneumonia dataset.

The accuracy ratio of COVID-19 classification was the highest between the other classes, rather than the sensitivity and precision were the highest value for pneumonia class. And its acceptable ratio for COVID-19, because of the limited number of chest X-ray images which belongs to who's infected by COVID-19 virus.

The performance of CNN proposed model in training and validation is based on chest X-ray images as shown in Fig. 3. The validation was traced on each epoch to calculate the performance based on true predication at each epoch and iterations. Table 2 shows the validation accuracy during training phase.

The success validation ratio was 92.00% during 13 iterations and the accuracy increased during training iterations, when the training terminated at maximum epoch 20, the accuracy was 95.8 overall for all three classes.

To evaluate the proposed CNN, we have to compare the accuracy result with another pre-trained CNN; in this research work, we picked AlexNet to compare with

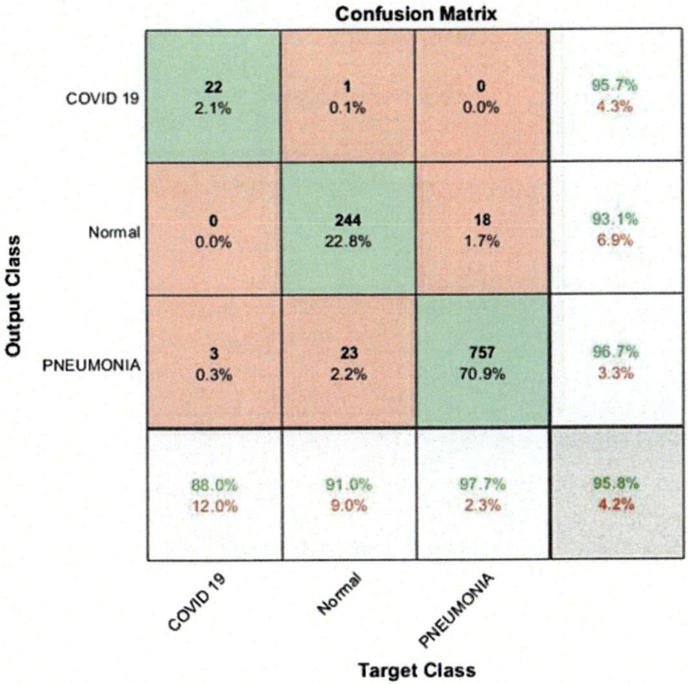

Fig. 2 Confusion matrix of proposed CNN

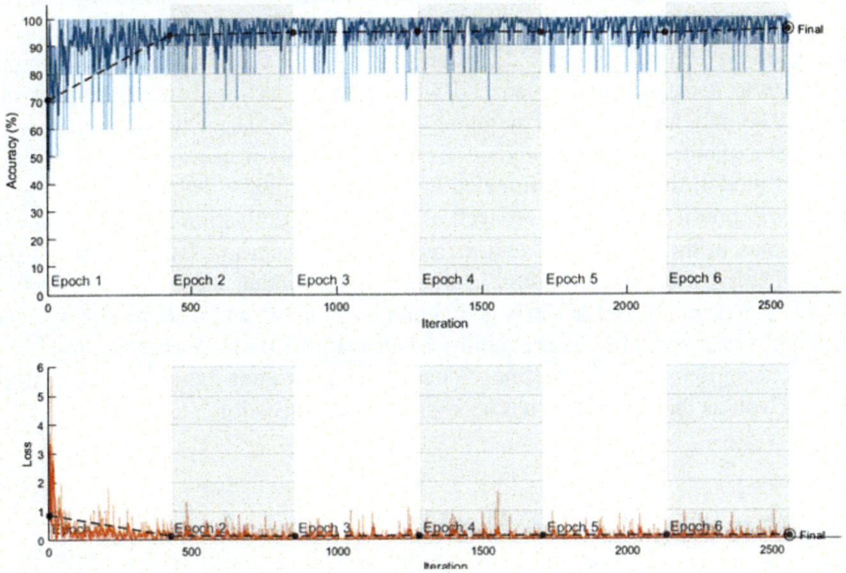

Fig. 3 Performance in the training and validation dataset

Table 2 Validation accuracy based on each epoch

Dataset	Epoch	True predication of validation (%)
Collection of chest X-ray	1	62.67
	13	92.00
	20	95.8

Table 3 Summary of accuracy comparing with different methods

Algorithm	Dataset	True Predication (%)
AlexNet	Collected chest X-ray images	95.00
Proposed CNN	Collected chest X-ray images	95.8

the proposed CNN, because the architecture design is a sequential same structure style with our model. Table 3 shows the summary of mean true predication based on total test chest X-ray images, comparison on the same dataset.

From all tables and figures, the proposed model achieved higher accuracy than pre-trained AlexNet in the case of our selected chest X-ray image "COVID19-project dataset".

5.3 *Class Activation Map*

Deep learning network is often considered as a black box that no way to figure out what the network has learned and which part of an input image responsible for predication. Class activation mapping [19] is one technique that you can use to get visual explanations of the predictions of convolutional neural networks. In the proposed model, the region of infected is more important to be detected and help the CNN in predication phase. As well as using class activation mapping helps to identify bias in the training set and increase model accuracy. By focusing on the region of interest of infected lung, in the proposed model, the class of activation model is placed at the end of CNN architecture. Figure 4a and b shows the selected sample of test image with the probability of class for COIVD 19 chest X-ray.

The red region of the CAM images has the highest contribution of predication score, as well as the other class has less score in final decision.

6 Conclusion and Future Work

This work presents a CNN model for COVID-19 diagnosis using X-ray images. The proposed CNN consists of 22 layers as well as the dataset was used for evolution of

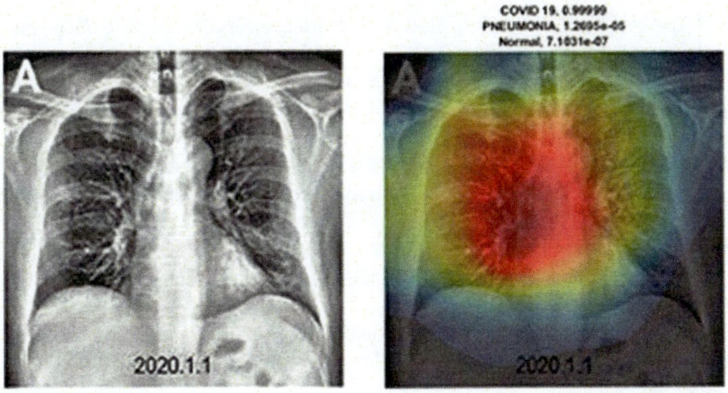

Fig. 4 COVID-19 chest X-ray image and CAM image

the model collected from two public datasets: chest X-ray images (pneumonia) and COVID image collection. The CNN makes predication based on deeper features extracted from X-ray images. The accuracy measurement was used to analyze the performance of the proposed model. The proposed model based on CNN was achieved accuracy for COVID-19: 0.9961, normal: 0.96067, and pneumonia:0.9588 as well as precision was 0.88, 0.91044, and 0.97677 for COVID-19, normal and pneumonia, respectively. And the sensitivity ratio was: 0.9565, 0.93129, and 0.96679 for COVID-19, normal, and pneumonia, respectively, as well as class activation mapping increased the diagnosis accuracy by given a visual explanation of the lung region.

References

1. Ungar L (2019) AI and healthcare: the road to modern health. Healthcare transformation
2. Wernick MN, Yang Y, Brankov JG, Yourganov G, Strother SC (2010) Machine learning in medical imaging. IEEE Signal Process Mag
3. LeCun Y, Bengio Y, Hinton G (2015) Deep learning. Nature
4. Xue Y, Chen S, Qin J, Liu Y, Huang B, Chen H (2017) Application of deep learning in automated analysis of molecular images in cancer: a survey. Contrast Media Mol Imaging 2017:10. Article ID: 9512370. https://doi.org/10.1155/2017/9512370
5. Attique Khan M, Rubab S, Kashif A, Imran Sharif M, Muhammad N, Shah JH, Zhang Y-D, Satapathy SC (2019) Lungs cancer classification from CT images: an integrated design of contrast based classical features fusion and selection. Pattern Recogn Lett
6. Liu Z, Yao C, Yu H, Wu T (2019) Deep reinforcement learning with its application for lung cancer detection in medical internet of things. Future Gener Comput Syst
7. Pasa F, Golkov V, Pfeiffer F, Cremers D, Pfeiffer D (2019) Efficient deep network architectures for fast chest X-ray tuberculosis screening and visualization. Sci Rep Nat
8. Kim I, Rajaraman S, Antani S (2019) Visual interpretation of convolutional neural network predictions in classifying medical image modalities, diagnostics
9. Tajbakhsh N, Shin JY, Gurudu SR, Hurst RT, Kendall CB, Gotway MB, Liang J (2016) Convolutional neural networks for medical image analysis: full training or fine tuning? IEEE Trans Med Imaging 35(5):1299–1312. https://doi.org/10.1109/TMI.2016.2535302

10. Kadhim MA, Abed MH (2020) Convolutional neural network for satellite image classification. Intell Stud Comput Intell. https://doi.org/10.1007/978-3-030-14132-5_13
11. Lakshmanaprabu SK, Mohanty SN, Shankar K, Arunkumar N, Ramirez G (2018) Optimal deep learning model for classification of lung cancer on CT images. Future Gener Comput Syst
12. Xu X, Jiang X, Ma C, Du P, Li X, Lv S, Yu L, Chen Y, Su J, Lang G, Li Y, Zhao H, Xu K, Ruan L, Wu W (2020) Deep learning system to screen Coronavirus disease 2019 pneumonia. https://arxiv.org/ftp/arxiv/papers/2002/2002.09334.pdf
13. Narin A, Kaya C, Pamuk Z (2020) Automatic detection of coronavirus disease (COVID-19) using X-ray images and deep convolutional neural networks. Electr Eng Syst Sci Image Video Process
14. Wang L, Wong A (2020) COVID-Net: a tailored deep convolutional neural network design for detection of COVID-19 cases from chest radiography images. arXiv:2003.09871v2. [eess.IV] 30 Mar 2020
15. Abed MH, Al-Asfoor M, Hussain ZM (2020). Architectural heritage images classification using deep learning with CNN. In: Proceedings of the 2nd International workshop on visual pattern extraction and recognition for cultural heritage understanding. Bari, Italy. http://ceur-ws.org/Vol-2602/paper1.pdf, published at http://ceur-ws.org
16. Kermany DS, Goldbaum M, Cai W, Valentim CCS, Liang H, Baxter SL, McKeown A, Ge Yang, Wu X, Yan F, Dong J, Prasadha MK, Pei J, Ting MYL, Zhu J, Li C, Hewett S, Dong J, Ziyar I, Shi A, Zhang R, Zheng L, Hou R, Shi W, Fu X, Duan Y, Huu VAN, Wen C, Zhang ED, Zhang CL, Li O, Wang X, Singer MA, Sun X, Xu J, Tafreshi A, Lewis MA, Xia H Zhang K (2018) Identifying medical diagnoses and treatable diseases by image-based deep learning. Cell 172:1122–1131. https://doi.org/10.1016/j.cell.2018.02.010
17. Cohen et al (2020) COVID-19 image data collection. https://github.com/ieee8023/covid-chestxray-dataset
18. Cohen JP, Bertin P, Chester VF (2019) A web delivered locally computed chest x-ray disease prediction system. Chester
19. Abed MH, Al-Rammahi AHI, Radif MJ (2019) Real-time color image classification based on deep learning network. J Southwest Jiaotong Univ 54(5)

A Hybrid Approach for Classifying Parkinson's Disease from Brain MRI

S. Sreelakshmi and Robert Mathew

Abstract Parkinson's disease (PD) is an incurable, neurodegenerative disease, and its early diagnosis can be done with the aid of modality magnetic resonance imaging (MRI). Biomarkers enable the easy and accurate diagnosis of this disease, and thereafter, the proper treatment can help to control the progressive loss of dopaminergic neurons. Deep learning approaches such as convolutional neural networks (CNN) are gaining much momentum recently and heavily being used in the medical domain for detecting useful patterns relevant to diseases. In this work, we are trying to develop the genetic algorithm (GA-based segmentation technique incorporated with deep CNN models for the easy and accurate diagnosis of PD. Experimental results show that the proposed approach outperforms the state-of-the-art for classifying Parkinson's disease from MRI.

Keywords Parkinson's disease · Deep learning · MRI · Medical image segmentation · Genetic algorithm

1 Introduction

Deep learning techniques are powerful and emerging machine learning techniques derived from artificial neural networks. The technological advancement over traditional computer vision problems like object recognition, classification, and detection is exceptional due to this new technology. Nowadays, the medical imaging skill has been used for the advance tracking, better diagnosis, and treatment of several diseases. Deep learning is creating exciting solutions for medical image analysis problems and is considered as a key method for future applications. Medical image segmentation, identifying the pixels of organs or lesions from background medical images, is one

S. Sreelakshmi (✉)
Indian Institute of Information Technology and Management-Kerala (IIITM-K), Thiruvananthapuram, India
e-mail: sreelakshmi.mpcs19@iiitmk.ac.in

R. Mathew
Anugraham Neurocare, Thiruvananthapuram, India

A. Ullah et al. (eds.), *Proceedings of International Conference on Information Technology and Applications*, Lecture Notes in Networks and Systems 350,
https://doi.org/10.1007/978-981-16-7618-5_15

of the most challenging and important tasks in medical image analysis that is to deliver critical information about the shapes and volumes of these organs. Some of the recent studies suggested that the nature-inspired methods such as genetic algorithms are useful for medical image analysis because they can model the inherent uncertainties of medical images caused by poor contrast, ill-defined boundaries, and subjectivity. Genetic algorithms simulate the process of biological evolution using the principle of survival of the fittest. GAs has been used for a variety of optimization problems, such as image enhancement, feature extraction from remotely sensed images, and image segmentation [1].

Parkinson's disease is a neurodegenerative disease that mostly affects old age people. Main cause of PD is the progressive loss of dopaminergic neurons in the substantia nigra (SN). SN is a human basal ganglia structure located in the midbrain that plays the critical role in reward and movement. Diagnosis of neurological diseases like PD is a growing concern and one of the most difficult challenges for modern medicine. Current diagnosis technologies produce huge quantities of data for detection, monitoring, and treatment of neurological diseases. In general, analysis of those medical big data is performed manually by experts to identify and understand the abnormalities. It is a really heavy task for a person to accumulate, manage, analyze, and assimilate such large volumes of data by visual inspection. As a result, the experts have been demanding computerized diagnosis systems, called "computer-aided diagnosis (CAD)" that can automatically detect the neurological abnormalities using the medical big data. This system improves consistency of diagnosis and increases the success of treatment, saves lives, and reduces cost and time [2].

Neuroimaging is now in the forefront of aiding in the diagnosis of Parkinson's disease. In PD, the prefrontal cortex (where the SN is located) of the human brain has been the focus of extensive study. The microscopic histological changes in the neurodegenerative diseases are inevitably associated with progressive regional and global brain atrophy, which may be assessed in vivo using MRI. Structural magnetic resonance imaging (sMRI) has been proven to be an effective tool for PD diagnosis. It reflects the typical pattern of a progressive loss of dopaminergic neurons in neuropathology. Figure 1 is an illustration of the axial view of brain MRI that shows

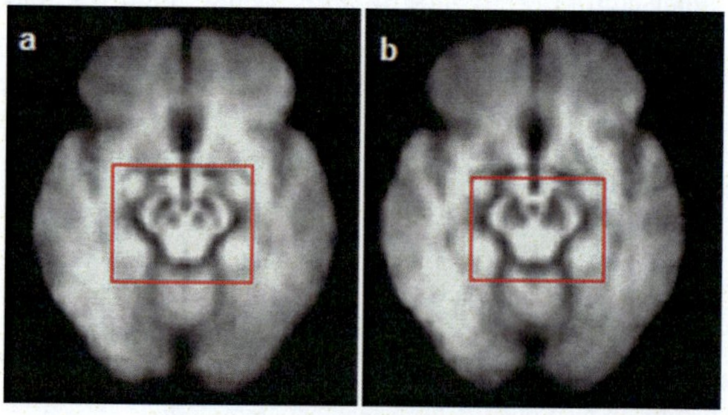

Fig. 1 An illustration of the axial view of brain MRI that shows the difference between **a** the typical brain and **b** PD affected brain

the difference between the typical brain and PD affected brain. The rectangle highlights the midbrain area that contains substantia nigra, which shows the prominent changes in case of PD.

Motivated by the above, here we propose a GA based segmentation algorithm to segment the brain regions like substantia nigra and also use the deep CNN model for the PD classification. Recently lots of segmentation techniques adorn major roles in medical image analysis. But it needs some additional techniques to analyze segmented images. The proposed model can provide precise and clear results. So it will be very useful to the doctors and also save their time in diagnosis than the existing method. The main contributions of this work may be summarized as follows:

1. Reviews some of the recently reported approaches for detecting PD from medical imaging modalities using computer vision.
2. Proposes a method for classifying PD from axial view of brain MR images using Genetic Algorithm and deep Neural Networks.

The considerations of this work are organized as follows. Some of the relevant and recent works on computer vision for detecting PD from medical images are discussed in Sect. 2. Then, propose and discuss a new methodology in Sect. 3, and Sect. 4 details the results and discussions. Section 5 consists of conclusions and future work.

2 Related Work

Some of the very prominent and recent state-of-the-art of medical image segmentation techniques and PD diagnosis techniques are reviewed here. Computer-aided diagnosis (CAD) has been used to assist doctors in diagnosing patients faster and more accurately. To use deep learning methods for medical image diagnosis, the computer must learn the features that represent the input data for the problem at hand. The most successful and popular models for image analysis currently are convolutional neural networks (CNNs), which contain many layers that transform their training images using small convolution filters into a matrix called the features matrix.

Nagasubramanian et al. [3] presented the acoustic-based deep learning techniques for detecting Parkinson disease symptoms. They analyze multivariate vocal data using multivariate speech attribute processing algorithms and using the deep architectures for design and implementation. Mohammed et al. [4] introduced a CNN-based network model for the Parkinson's disease diagnosis. And they ensure that the CNN-based network model is highly capable of discriminating PD patients based on single photon emission computed tomography (SPECT) images from normal controls. Bi et al. [5] designed a practical multimodal feature fusion method called clustering evolutionary random neural network ensemble (CERNNE) for the prediction of PD. CERNNE uses clustering evolution strategy to optimize ensemble learners to achieve better classification results. The CERNNE is also applicable for forming a multi-task analysis framework, recognizing PD patients and predicting PD-associated brain regions and genes. Solana-Lavalle et al. [6] presented the MRI based machine learning technique for predicting PD and also used the voxel-based morphometry for obtaining the regions of interest for feature extraction. They conduct separate experiments in men and women and conclude men are more prone to Parkinson's disease than women are. Fuzzy optimum path forest for automated PD identification is introduced by de Souza et al. [7], which is based on fuzzy logic and graph-based framework theory. Yasaka et al. [8] proposed a new framework to investigate whether PD patients can be differentiated from normal controls by applying a deep learning method to a parameter-weighted connectome matrix with higher performance than a number of streamlines (NOS)-based structural connectome matrix calculated from MRI obtained with a 3-T MRI unit and to find neural circuit disorders associated with PD by using gradient-weighted class activation mapping (Grad-CAM) methods. Cantürk et al. [9] used dynamic spiral test and static spiral test of PD patients who were analyzed with pretrained deep learning algorithms for early detection of PD. They used fuzzy recurrence plot technique to convert time-series signals to grayscale texture images. Geraedts et al. [10] introduced a fully automated electroencephalography (EEG)-based machine learning pipeline that was applied to deep brain stimulation (DBS) candidates with PD. And they achieved high accuracy using a compound of automatically extracted EEG biomarkers to classify PD patients according to cognition, rather than a single spectral EEG feature. A novel adaptive unsupervised feature selection approach is proposed by Huang et al. [11] by exploiting manifold learning from longitudinal multimodal data. Classification and clinical score prediction are performed jointly to facilitate early PD diagnosis. And also the proposed approach performs united embedding and sparse regression, which can determine the similarity matrices and discriminative features adaptively. Basukala et al. [12] construct a new algorithmic concept for the automatic segmentation of substantia nigra (SN) and red nucleus (RN). SN and RN are important for understanding PD. This algorithm aims to segment SN and RN from quantitative susceptibility mapping (QSM) MRI and use the results to investigate PD. A new machine learning method using conventional brain MRI for PD diagnosis introduced by Shu et al. [13]. They developed a radiomics model based on whole-brain white matter and clinical features to predict the progression of Parkinson disease (PD). And they ensure that this also supports the use of a simple radiomics signature built from whole-brain white matter features as a useful tool for the assessment and monitoring of PD progression.

Prajapati et al. [14] conduct a study to investigate the topological properties of functional brain networks within normal controls and PD patients, and also evaluate the gender difference among PD patients through graph theoretical approach. Choi et al. [15] developed a new deep learning-based cognitive signature of fludeoxyglucose brain PET adaptable for Parkinson's disease as well as Alzheimer's disease, which predicts cognitive score of the subjects and validates in patients. Pereira et al., [16] proposed a novel framework for the classification of structural magnetic resonance imaging, dopamine transporter scan data, and demographic information to differentiate PD patients, "Scans Without Evidence for Dopaminergic Deficit"

(SWEDD) patients and healthy control subjects using deep convolutional neural networks. Wingate et al. [17] introduced a deep learning model for diagnosing PD through medical images. The model consists of the analysis and use of the knowledge extracted by deep convolutional and recurrent neural networks when trained with medical images, such as magnetic resonance images and dopamine transporters scans. Mostafa et al. [18] conduct a study to examine different machine learning methods to identify PD using brain MRI. And they find out that the detection accuracy increases drastically when focusing on the gray matter and white matter regions from the MRI instead of using the whole MRI. Zhang et al. [19] introduced a new architecture named semi-supervised multi-view learning clustering (SMC) based on semi-supervised learning which also contribute to improving the effectiveness of identifying PD by previous labeled and subsequent unlabeled medical MRI data in the realistic medical environment.

Manzanera et al. [20] employed 3D CNNs in FDG-PET brain images with the purpose of discriminating patients diagnosed with PD from normal controls and also performed the hyperparameter optimization on three deep models to estimate the classification accuracy of the networks on new data. For helping doctors and people in diagnosing PD at an early stage, Johri et al. [21] combined two CNN models named VGFR spectrogram detector and voice impairment classifier. Shinde et al. [22] introduced a novel convolutional neural network-based marker for PD based on neuromelanin-sensitive magnetic resonance images. Xu et al. [23] conducted a detailed survey that aims to state an overview of existing research recently that used statistical machine learning/artificial intelligence methods to perform quantitative analysis of brain MRI data for the study of PD diagnosis. Kiryu et al. [24] conduct a detailed study to evaluate the diagnostic performance of deep learning with the deep convolutional neural networks to distinguish each representative PD using brain MRI.

3 Proposed Method

The proposed PD detection method consists of two main phases. In phase 1, the genetic algorithmic approach is implemented for segmenting the region of interest (ROI) from brain MRI, and in phase 2, the deep convolutional neural network (CNN) model is used to perform the classification of segmented images into PD and healthy control classes. Figure 2 illustrates the architecture of the proposed methodology.

A total of 300 T1 weighted axial views of live MRI data are used for this study. The brain structure, which includes the substantia nigra, is seen best in these (axial) slicing indices. After obtaining the data from the MRI scanner, the preprocessing is performed using the FreeSufer software suite. The image preprocessing aims to remove the skull and other unwanted areas (noises) from the given images. After that, the preprocessed data are implemented using GA for getting accurate segmented ROI. And finally, the segmented images are passed through the deep neural networks, which is able to classify the data.

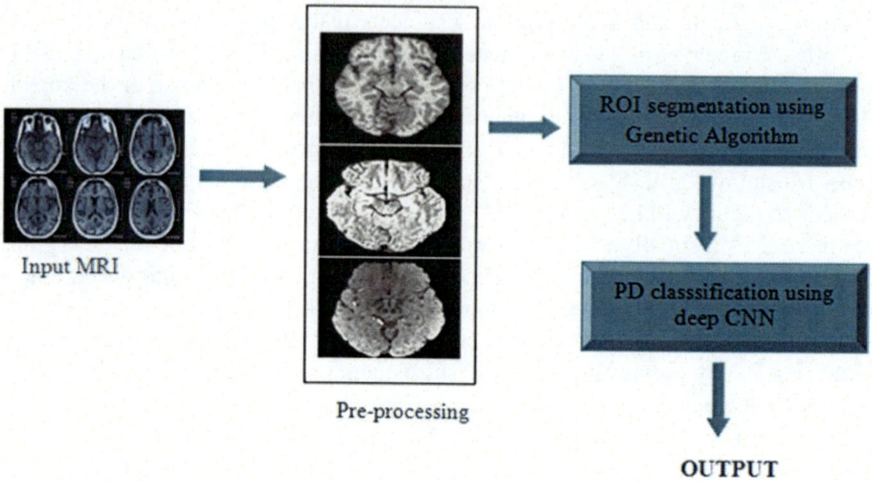

Fig. 2 Architecture of the proposed method

3.1 ROI Segmentation Using GA

Charles Darwin discovered the nature inspired algorithm based on the principle of "survival of the fittest" called genetic algorithm (GA). Many of the recent articles stated that the GA based medical image segmentation and analysis is a very promising tool for disease detection [25]. Here, we are using the parameter selection model GA for MRI segmentation. It will produce more accurate results by comparing other existing methods. The steps followed for the genetic algorithm are given in Table 1.

3.2 PD Classification Using CNN

Convolutional neural network (CNN) is the state-of-the-art for image classification tasks, and it forms a significant part of machine learning techniques. In this work, the pretrained CNN is used for automatic classification of PD, and we have to train the essential layers of CNN accordingly. The working of basic CNN for PD classification is given in Fig. 3.

4 Result and Evaluation

In this section, we study the performance of our proposed segmentation method using a live dataset. The experiments were carried out using the healthy subjects and subjects diagnosed with PD. Aim of the proposed work is segment the midbrain area

Table 1 Pseudo code for GA

```
1 START
2 Input: preprocessed MRI data
3 Computation of the image statistics
4 Generation of initial population
5 Performing segmentation using initial
parameters
6 Computation of the quality measures
7 While not(stopping conditions) Do
{
SELECT individual using REPRODUCTION
operator
GENERATE ne population using CROSSOVER and
MUTATION
SEGMENT the image using new parameters
CALCULATE the quality measures of
segmentation
}
8 Update the knowledge base
9 Output: Segmented image
10 END
```

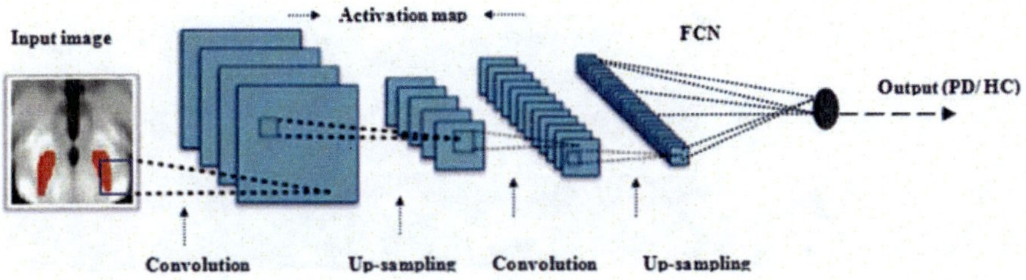

Fig. 3 Working flow of CNN for PD classification

(contain substantia nigra, which shows the significant changes in PD) from the given MRI and identify if the segment image is PD affected or not.

Among the 300 data that we took for the study consisted of 150 PD patients and 150 NC (normal control/healthy control) people with no gender discrimination. Figure 4 showcased some input sMRI data. By using the GA approach, our model accurately segmented the midbrain areas and that segmented images are gone through the basic CNN, which successfully classifies whole images into affected and healthy classes. The segmented images are shown in Fig. 5. The red circle shows the midbrain area which highlights the substantia nigra (yellow color).

To validate the effectiveness of the proposed model, adopt some metrics to evaluate the classification results. The confusion matrix of the proposed work is given in Fig. 6.

The proposed approach uses four classifiers to classify the segmented area into PD affected and healthy control. The experiments conducted with these classifiers show that CNN classifiers produce the maximum classification accuracy. The comparison

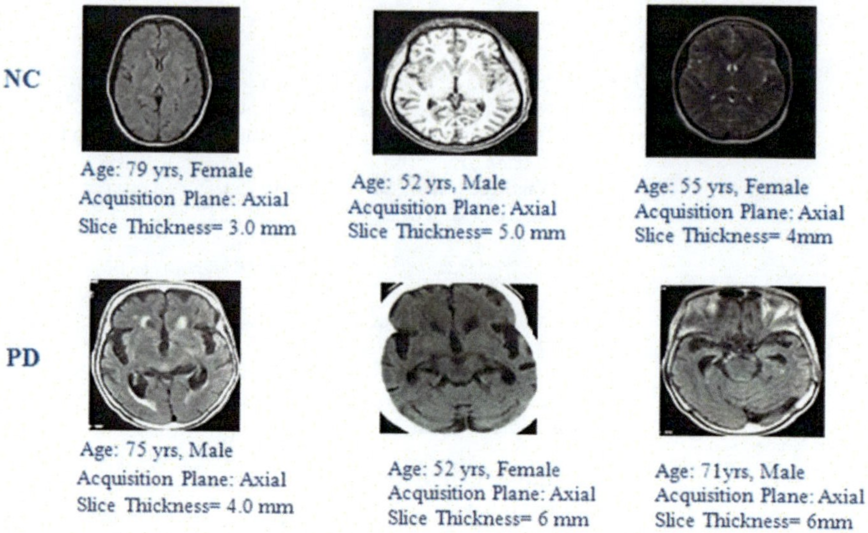

Fig. 4 Input data for both affected and healthy controls

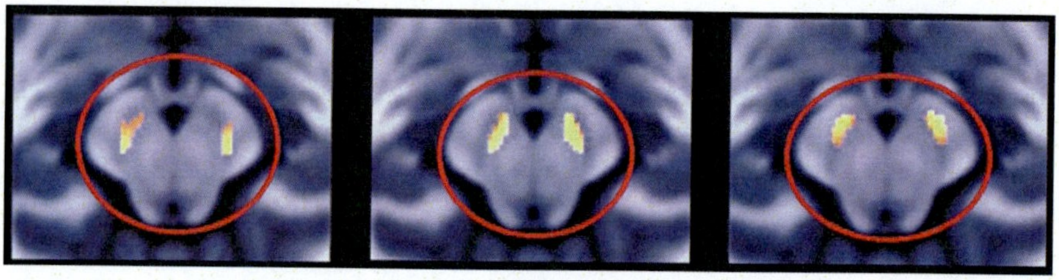

Fig. 5 Segmentation of midbrain using genetic algorithm

Fig. 6 Confusion matrix on testing data

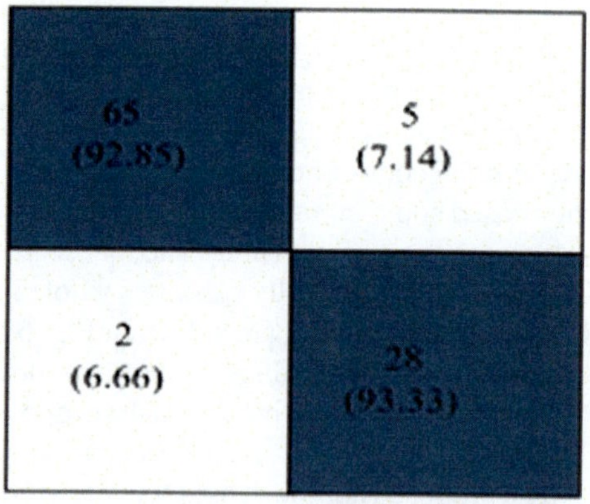

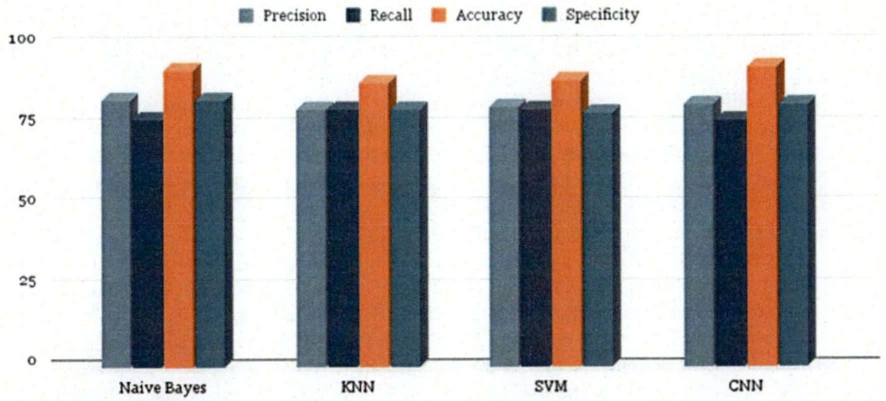

Fig. 7 Classifiers comparison chart

graph that depicts the classification accuracy obtained for Naive Bays, K-nearest neighbors, support vector machines, and CNN are shown in Fig. 7.

5 Conclusion and Future Work

This paper proposed a hybrid approach that combines genetic algorithm and deep learning approaches for classifying Parkinson's disease from brain MRI. The proposed approach uses GA for midbrain segmentation and segmentation followed by a deep learning based classification method for effectively classifying the input image into PD affected and healthy classes. The results obtained from the experiments conducted shows that the proposed method could effectively classify Parkinson's disease. As the results are promising, the others would like to explore future dimensions by training the classifier using more MR images and other segmentation techniques.

References

1. Ghosh P, Mitchell M, Tanyi JA, Hung AY (2016) Incorporating priors for medical image segmentation using a genetic algorithm. Neurocomputing 195:181–194
2. Siuly S, Zhang Y (2016) Medical big data: neurological diseases diagnosis through medical data analysis. Data Sci Eng 1:54–64. https://doi.org/10.1007/s41019-016-0011-3
3. Nagasubramanian G, Sankayya M (2021) Multi-variate vocal data analysis for detection of Parkinson disease using deep learning. Neural Comput Appl 33(10):4849–4864
4. Mohammed F, He X, Lin Y (2021) An easy-to-use deep-learning model for highly accurate diagnosis of Parkinson's disease using SPECT images. Comput Med Imaging Graph 87:101810

5. Bi XA, Hu X, Xie Y, Wu H (2021) A novel CERNNE approach for predicting Parkinson's disease-associated genes and brain regions based on multimodal imaging genetics data. Med Image Anal 67:101830
6. Solana-Lavalle G, Rosas-Romero R (2021) Classification of PPMI MRI scans with voxel-based morphometry and machine learning to assist in the diagnosis of Parkinson's disease. Comput Methods Programs Biomed 198:105793
7. de Souza RW, Silva DS, Passos LA, Roder M, Santana MC, Pinheiro PR, de Albuquerque VHC (2021) Computer-assisted Parkinson's disease diagnosis using fuzzy optimum-path forest and restricted Boltzmann machines. Comput Biol Med 131:104260
8. Yasaka K, Kamagata K, Ogawa T, Hatano T, Takeshige-Amano H, Ogaki K, Abe O (2021). Parkinson's disease: deep learning with a parameter-weighted structural connectome matrix for diagnosis and neural circuit disorder investigation. Neuroradiology 1–12
9. Cantürk İ (2021) Fuzzy recurrence plot-based analysis of dynamic and static spiral tests of Parkinson's disease patients. Neural Comput Appl 33:349–360
10. Geraedts VJ, Koch M, Contarino MF, Middelkoop HAM, Wang H, van Hilten JJ, Bäck THW, Tannemaat MR (2021) Machine learning for automated EEG-based biomarkers of cognitive impairment during deep brain stimulation screening in patients with Parkinson's disease. Clin Neurophysiol 132(5):1041–1048
11. Huang Z, Lei H, Chen G, Frangi AF, Xu Y, Elazab A, ... Lei B (2021) Parkinson's disease classification and clinical score regression via united embedding and sparse learning from longitudinal data. IEEE Trans Neural Netw Learn Syst
12. Basukala D, Mukundan R, Lim A, Hurrell MA, Keenan RJ, Dalrymple-Alford JC, ... Melzer TR (2021). Automated segmentation of substantia nigra and red nucleus using quantitative susceptibility mapping images: application to Parkinson's disease. Comput Electr Eng 91:107091
13. Shu ZY, Cui SJ, Wu X, Xu Y, Huang P, Pang PP, Zhang M (2021) Predicting the progression of Parkinson's disease using conventional MRI and machine learning: an application of radiomic biomarkers in whole-brain white matter. Magn Reson Med 85(3):1611–1624
14. Prajapati R, Emerson IA (2020) Global and regional connectivity analysis of resting-state functional MRI brain images using graph theory in Parkinson's disease. Int J Neurosci 1–11
15. Choi H, Kim YK, Yoon EJ, Lee JY, Lee DS (2020) Cognitive signature of brain FDG PET based on deep learning: domain transfer from Alzheimer's disease to Parkinson's disease. Eur J Nucl Med Mol Imaging 47(2):403–412
16. Pereira HR, Fonseca JM, Ferreira HA (2020) Combination of medical imaging and demographic data for Parkinson's disease diagnosis. In: Doctoral conference on computing, electrical and industrial systems. Springer, Cham, pp 339–346
17. Wingate J, Kollia I, Bidaut L, Kollias S (2020) Unified deep learning approach for prediction of Parkinson's disease. IET Image Proc 14(10):1980–1989
18. Mostafa TA, Cheng I (2020) Parkinson's disease detection using ensemble architecture from MR images. In: 2020 IEEE 20th International conference on bioinformatics and bioengineering (BIBE). IEEE, pp 987–992
19. Zhang X, Zhai D, Yang Y, Zhang Y, Wang C (2020) A novel semi-supervised multi-view clustering framework for screening Parkinson's disease. arXiv preprint arXiv:2003.04760
20. Manzanera OM, Meles SK, Leenders KL, Renken RJ, Pagani M, Arnaldi D, Nobili F, Obeso J, Rodriguez Oroz M, Morbelli S, Maurits NM (2019) Scaled subprofile modeling and convolutional neural networks for the identification of Parkinson's disease in 3D nuclear imaging data. Int J Neural Syst 29(09):1950010
21. Johri A, Tripathi A (2019) Parkinson disease detection using deep neural networks. In: 2019 Twelfth international conference on contemporary computing (IC3). IEEE, pp 1–4
22. Shinde S, Prasad S, Saboo Y, Kaushick R, Saini J, Pal PK, Ingalhalikar M (2019) Predictive markers for Parkinson's disease using deep neural nets on neuromelanin sensitive MRI. NeuroImage: Clin 22:101748
23. Xu J, Zhang M (2019) Use of magnetic resonance imaging and artificial intelligence in studies of diagnosis of Parkinson's disease. ACS Chem Neurosci 10(6):2658–2667

24. Kiryu S, Yasaka K, Akai H, Nakata Y, Sugomori Y, Hara S, Seo M, Abe O, Ohtomo K (2019) Deep learning to differentiate Parkinsonian disorders separately using single Midsagittal MR imaging: a proof of concept study. Eur Radiol 29(12):6891–6899
25. Ghaheri A, Shoar S, Naderan M, Hoseini SS (2015) The applications of genetic algorithms in medicine. Oman Med J 30(6):406

OntoCOVID: Ontology for Semantic Modeling of COVID19 Statistical Data

Shaukat Ali, Shah Khusro, Sajid Anwar, and Abrar Ullah

Abstract Several COVID19 statistical datasets are provided to support stakeholders for better planning and decision making in healthcare. However, the datasets are in heterogeneous proprietary formats that create data silos and compatibility issues and make data discovery and reuse difficult. Further, the data integration for analysis is difficult and is performed by the domain experts manually which is time consuming and error prone. Therefore, an explicit, flexible, and widely acceptable methodology is needed to represent, store, query, and visualize COVID19 statistical data in the datasets. In this paper, we have presented the design and development of OntoCOVID ontology for representing, organizing, sharing, and reusing COVID19 statistical data in the datasets. The OntoCOVID is a lightweight ontology providing definitions of classes, properties, and axioms to semantically represent and relate information in the COVID19 statistical datasets. The OntoCOVID is evaluated to demonstrate its completeness and information retrieval for different use-case scenarios. The results obtained are promising and advocate for the improved ontological design and applications of the OntoCOVID.

Keywords Semantic web · COVID19 · Ontology · SPARQL · OWL · Dataset

S. Ali (✉) · S. Khusro
Department of Computer Science, University of Peshawar, Peshawar, Pakistan
e-mail: shoonikhan@uop.edu.pk

S. Khusro
e-mail: khusro@uop.edu.pk

S. Anwar
Centre of Excellence in IT, Institute of Management Sciences (IMSciences), Peshawar, Pakistan
e-mail: sajid.anwar@imsciences.edu.pk

A. Ullah
School of Mathematical and Computer Science, Heriot-Wat University, Edinburgh, UK
e-mail: a.ullah@hw.ac.uk

A. Ullah et al. (eds.), *Proceedings of International Conference on Information Technology and Applications*, Lecture Notes in Networks and Systems 350,
https://doi.org/10.1007/978-981-16-7618-5_16

1 Introduction

Information is the fundamental requirement in the public healthcare domain for effective decision making regarding epidemiologic surveillance, disease outbreak, healthcare program planning and outcomes assessment, cost and expenditures, coverage and quality of services, and policy analysis [1]. Like infectious diseases, the COVID19 pandemic has posed serious threats to the general public health and has resulted into significant morbidity, mortality, and economic damages worldwide. The COVID19 statistical data (i.e., patients' cases, clinical trials, and drug efficiency) is constantly growing, and different organizations are providing datasets under the initiatives of open data on the Web (e.g., ourworldindata, worldometer, and kaggle) to support the stake holders. However, the COVID19 statistical datasets (CSDs) are in heterogeneous data formats, naming conventions, structures, encodings, data types, and resides in distributed data repositories. Mostly, the CSDs are distributed in tabular formats (i.e., CSV and Excel) with small metadata; making data integration, comparison, and reuse very difficult. In addition, the disparate CSDs are not linked despite of having related indicators and vocabularies and data formats used are inconsistent. This diversity, variety, and veracity of CSDs poses serious time-consuming and error-prone problems of data management (i.e., representation, encoding, storage, and retrieval), sharing, and integration for effective planning and decision making by the stakeholders. The explosive growth of CSDs needs to be properly explored to reveal worthy information and convert them into worthy knowledge for improved COVID19 healthcare practices. Therefore, semantically enriched and flexible data representation, analysis, querying, and visualization methods are needed for CSDs. Sharing semantically structured CSDs would enable researchers to produce advanced methods for providing high quality information to revolutionize the healthcare sector.

The Semantic Web (SW) [2] extends the Web from interlinking documents only to revolutionize global data sharing, integration, and analysis [3]. The Health Care and Life Science Interest Group (HCLSIG) is founded with the aim to develop, advocate, and support the use of SW technologies in the healthcare and life science domains [4, 5]. The healthcare domain has already adopted the SW technologies principles for enhanced semantic representation, information retrieval (IR), and visualization, which are not possible with separate data stores [6, 7]. The ontology is SW technology which provides constructs to develop structured vocabularies comprising of terms to represent entities and relationships in human and computer interpretable format [8]. In the field of informatics, ontologies provides important platform for data standardization, representation, integration, sharing, and analysis. Therefore, ontologies could provide semantic layer by annotating CSD with location, time, and description metadata from the disparate sources to enhance semantic understanding, compatibility, interoperability, and representing of connections among unmatched terms to enhance functioning of a system. Therefore, ontology is needed to give a universally agreed language and dictionary of terms for realizing the composition of information about countries, patients' cases, diagnostic tests, health facilities, population, etc., in the CSDs. The ontology will give increased expressive representations, enhanced access, domain knowledge reusability, formal analysis of information, and making knowledge in the domain unambiguous without requiring technical details understanding about data management.

The CSDs heterogeneity is an open challenge. Therefore, an explicit and flexible solution is needed to exploit potential of the CSDs for effective planning and decision making by the stake holders. This paper is aimed to design and develop OntoCOVID ontology which consists of explicit and formal conceptualization of CSD from the heterogeneous COVID19 datasets. The OntoCOVID has several properties: (1) increasing data interoperability by semantically representing and annotating of CSD; (2) combining of CSD from different sources to improve intelligent healthcare monitoring and decision making; (3) description of CSD to improve data reusability and shareability; and (5) addition of geo-spatial information to support retrieval and visualization systems. The OntoCOVID architecture is kept conservative with the open option for modifications, additions, and reusing into other COVID19 related heavy-weight ontologies. This paper provides a pragmatical idea for OntoCOVID development with no take that orthogonal, full pledge, or widely agreed CSD ontology is practicable. However, evaluation of OntoCOVID has shown promising results.

2 Related Work

The public healthcare systems and emergencies require data sharing in different domains and data systems to control and reduce the COVID19 impacts. However, this data sharing is hindered by representing and encoding relevant information in domain-specific terminologies and coding systems and resulting into loss of interpretability across data systems. The researchers have recognized the strength of using ontologies to provide solutions to the problems. The COVID19 surveillance ontology facilitates monitoring of COVID19 cases and relates respiratory conditions using data from multiple brands of computerized medical record systems [9]. The COVID19 ontology covers the role of molecular and cellular entities in virus-host-interactions, in the virus life cycle, as well as a wide spectrum of medical and epidemiological classes linked to COVID19 [10].

The COVID19 is a relatively new disease with its unique prospects. The researchers have explored ontologies to semantically organize and share COVID19 data from medical, therapeutic, and epidemiological aspects. However, none of the ontologies addresses CSD related to human cases and fatalities, planning, and hospital systems, geography and demographics of countries, etc. As discussed earlier, heterogeneous CSDs are of limited significances for effective decision-making and planning processes. Therefore, experiences from COVID19 ontology-based research can be applied to CSDs for semantically representation, organization, interlinking, sharing, reusing, and visualization.

3 OntoCOVID Development

The ontology provides a solution to develop a domain data model by formally and explicitly defining its classes and meaningful data linkages. Therefore, ontology is needed to semantically model and map CSDs for explicit, formal, and uniform representation and using in applications. The existing ontologies could be helpful in certain extent to provide matching set of definitions that are fitting the data needs but not completely. Reusing existing ontologies could increase data interoperability but could increase complexity in retrieval by not efficiently modeling the new data needs [11]. The bespoke ontology could reduce data interoperability but could allow to model data in an efficient manner [11]. Therefore, developing a bespoke ontology using 'in between' approach could be efficient solution by utilizing classes and properties from well-known existing and accepted ontologies and vocabularies. However, new data needs are to be defined in ontology explicitly.

A bespoke reference "OntoCOVID" ontology is developed for CSDs. The OntoCOVID is developed by using 'in between' approach reusing ontologies and vocabularies (i.e., Schema, Friend of a Friend, DBPedia, Dublin Core, and W3C Geospatial). However, the new classes and properties are explicitly defined in the OntoCOVID to efficiently model the new data needs. A snippet of OntoCOVID is shown in Fig. 1. The OntoCOVID is modeled using class of episodes where each episode represents CSD data of a particular country on a particular date. Each episode is associated with URIs and literal values. The values are important to numerically quantitate an episode's object, retrieving, and reasoning process. Each episode is added with geographical coordinates for useful map-based tracking and visualization as location has become basic attribute for health data [12].

3.1 *Goals and Design Principles*

The goal and design principles of the OntoCOVID are:

- The ontology should represent CSD for using in a variety of informatics systems.
- The ontology presents classes, properties, relationships, and axioms in easy to understandable way to be used by the developers easily and effectively.
- The ontology could accommodate changes and extensions to easily add new classes and relationships to increase interoperability and information sharing.
- The ontology is general by describing classes, facets, and relationships that are possibly present in a wide range of CSDs.
- The ontology supports inference methods with low resources utilizations.
- The classes and properties are describe and presented precisely to be easily accessible and provide acceptable results for quality and completeness metrics.
- The ontology to be used easily in different SW frameworks and triple stores.

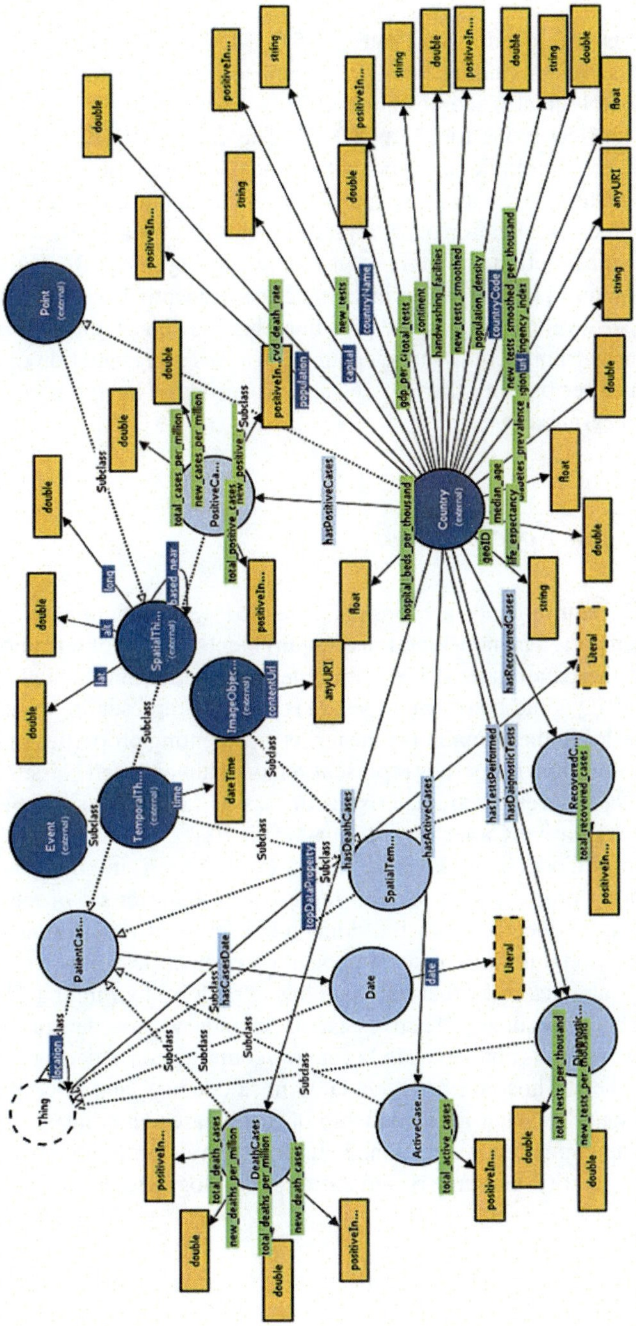

Fig. 1 Snippet of OntoCOVID classes and properties in Protégé

3.2 *Materials and Methodology*

The OntoCOVID is developed using Protégé 5.5.0. The POEM methodology [3] is used, which emphasizes on finding, reusing, reengineering, and combining of ontological and non-ontological resources and ontology design patterns, which are the basic designing considerations for OntoCOVID. The OntoCOVID emphasizes on leveraging upper-level ontologies and vocabularies to provide a baseline for defining classes and relationships in a domain-specific ontology to achieve low development time, widely agreement, realization of basic requirements, simple and quicker consolidation with other information and more robust knowledge systems [3]. The content ontology design pattern [13] is used to instantiate the conceptual contents of OntoCOVID from upper-level ontologies and give unambiguous non-logical vocabulary of the domain. Furthermore, componency pattern provides representation of classes and properties as either part of other classes and properties or having other classes or properties as integral parts.

3.3 *Requirements Specifications*

The OntoCOVID requirements are basically related with functional and non-functional requirements. The non-functional requirements consist of terminological requirements and the patterns used for naming terms. The terminological requirements of OntoCOVID could be of two types: (1) the terms representing classes and sub-classes of entities in the domain; (2) the terms representing properties of classes in the domain, where some properties represent relationship between classes (object properties) and other represent characteristics of classes (datatype properties). For example, Country, Positive Cases, Event, Diagnostic Tests, etc., are representing classes, and Country Code, Total Positive Cases, Population, Latitude, Longitude, etc., are representing properties. A lexicon representing relevant set of terminologies is identified and collected from the domain-specific and application-specific resources. The functional requirements at present consists of 43 competency questions representing intended tasks that ontology has to perform by running SPARQL queries such as What is frequency of active cases in 'a country' on a 'date or month'?; What is total number of tests conducted in 'a country/countries' on a 'date or month'?; What is total number of hospital beds available in 'a country' and how many are used on a 'date/month'?; What is relationship of death cases and diabetic patients in 'a country'? The competency questions would help in determining correctness, completeness, verifiability, and understandability of ontology requirements [3].

3.4 Classes Hierarchy

The OntoCOVID classes are constructed from classes of countries, patients' cases, spatial and temporal, diagnostic tests, medical facilities, and demographics. The identified classes are hierarchically arranged by determining their relationships that whether a class is a sub-class of another class or not. The required classes from upper-level ontologies/vocabularies (e.g., schema:Country, dc:Date, etc.) are directly reused. New classes are created either parent classes or sub-classes of the classes in the upper-level ontologies/vocabularies.

Figure 2 represents a snip of OntoCOVID taxonomic classes hierarchy. The OntoCOVID contains 19 classes, where a class is created by considering the unique requirements of the domain. The OntoCOVID:PatientCases is declared to define the COVID19 cases and OntoCOVID:ActiveCases, OntoCOVID:DeathCase, OntoCOVID:PostiveCases, and OntoCOVID: RecoveredCases are declared sub-classes of OntoCOVID:PatientCases to represent different types of COVID19 cases in a country on a particular date. The dc:Date is used to represent date information, and OntoCOVID:DiagnosticTests is declared to represent the COVID19 diagnostic tests conducted in a country. The schema:Country is declared sub-class of wgs84:Point to represent country information. The wgs84:Event is declared to represent related COVID19 event information, and schema:ImageObject represents COVID19 cases graphical information. In addition to sub-class axioms, classes are combined with other axioms to ease the instantiation of objects explicitly and semantically. For example, the disjoint axiom is described for classes of the same generation level to restrict objecs' behaviors such that OntoCOVID:ActiveCases and OntoCOVID:DeathCase are disjointed that an object can be an created from only one of them.

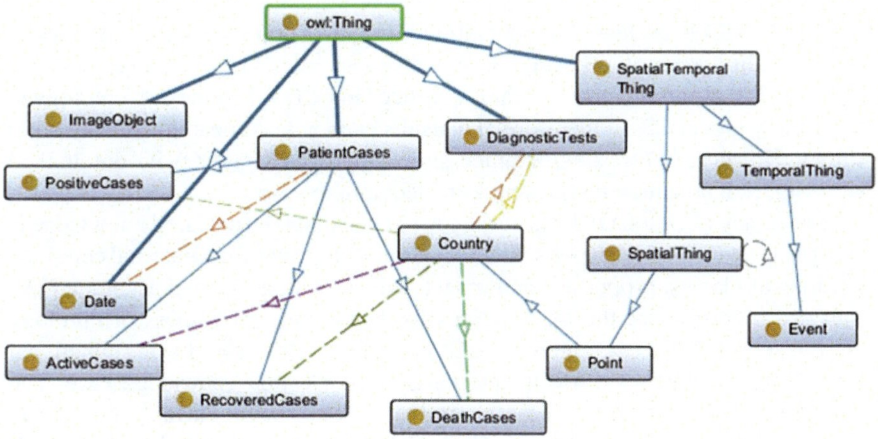

Fig. 2 Snippet of OntoCOVID classes hierarchy

3.5 *Properties and Restrictions*

The properties show relationships/links and can be object property or datatype property. The object property requires that domain and range would be objects of classes and the datatype property requires that domain would be an object of a class and range would be an object of a datatype. The object properties are either used from upper-level ontologies/vocabularies or explicitly created. The object properties available in upper-level ontologies/vocabularies are inherited and specialized for using with OntoCOVID classes. For example, foaf:based_near relates an instance of wgs84:SpatialThing with itself. In addition, several domain-specific object properties are defined to connect classes in OntoCOVID meaningfully and subtly such as OntoCOVID:hasDeathCase and OntoCOVID:hasCasesDate which relates instance of schema:Country with instances of respective classes. Table 1 shows a portion of OntoCOVID object properties. Similarly, the datatype properties in OntoCOVID are either used form upper-level ontologies/vocabularies or explicitly created. For example, wgs84:lat, dbpprop:countryCode, schema:url, OntoCOVID:new_death_cases, OntoCOVID:new_positive_cases, OntoCOVID:total_tests, dbpprop: population, etc. Table 2 shows a portion of OntoCOVID datatype properties. The classes are refined by using properties to define constraints and axioms to instantiate their objects. The property restrictions are used to describe how many occurrences and values of a property are essential to be an object of a class. For example, to be object of schema:Country requires at least one occurrence of wgs84:lat and wgs84:lat datatype properties.

4 OntoCOVID Evaluation

The OntoCOVID is evaluated using OntoQA [14] which uses different metrics to examine quality of ontology using various characteristics [15]. The overall OntoCOVID quality results are given in Table 3. Since OntoCOVID is the first ontology to ontologically model CSDs, therefore, the results cannot be compared with any other ontology. However, the results obtained show improved OntoCOVID ontological design with rich potential to represent knowledge in the domain.

The accuracy of OntoCOVID is evaluated to check that the declared knowledge is in accordance with domain experts' knowledge. Ontology with accurate and complete definitions of classes, properties, and objects will provide enhanced accuracy [15]. Recall and precision are the IR measures for examining correctness of ontology using Eqs. (1) and (2) [15]. The recall and precision are measured using functional requirements by running SPARQL queries on OntoCOVID knowledge base. The knowledge base is formed by instantiating OntoCOVID with relevant information from the CSDs using Protégé 5.5.0 built-in Cellfie plugin and MappingMasterDSL language. The knowledge base is hosted on Fuseki triplestore and SPARQL endpoint to query for retrieving results using Web-based JavaScript library Sgvizler. Figure 3

Table 1 Excerpt of OntoCOVID object properties

Domain	Object property	Range	Quantifier	Cardinality
schema:Country	OntoCOVID:hasActiveCases	OntoCOVID:ActiveCases	Existential	Exactly 1
schema:Country	OntoCOVID:hasRecoveredCases	OntoCOVID:RecoveredCases	Existential	Exactly 1
schema:Country	OntoCOVID:hasTestsPerformed	OntoCOVID:DiagnosticTests	Universal	Max 1
Wgs84:SpatialThing	Foaf:based_near	Wgs84:SpatialThing	Exis and Unil	Min 1
schema:Country	OntoCOVID:hasDataDate	dc:Date	Existential	Max 1
schema:Country	OntoCOVID:hasEvent	Wgs86:Event	Universal	Nil

Table 2 Excerpt of OntoCOVID datatype properties

Domain	Datatype property	Range	Quantifier	Cardinality
schema:Country	Wgs84:alt	Double	Existential	Exactly 1
schema:Country	dbo:countryName	String	Existential	Exactly 1
OntoCOVID:DeathCases	OntoCOVID:New_Death_Cases	Positive integer	Existential	Exactly 1
OntoCOVID:PositiveCases	OntoCOVID:New_Positive_Case	Positive integer	Existential	Exactly 1
OntoCOVID:DiagnosticTests	OntoCOVID:NewTests	Positive integer	Universal	Max 1
schema:Country	dbpprop:Population	Positive integer	Existential	Exactly 1

Table 3 Quality analysis of OntoCOVID

Ontology	Classes	Relationships	Classes richness	Relationships richness	Inheritance richness	Tree balance
OntoCOVID	19	28	87.34	73.86	2.17	1.15

```
SELECT ?countryname xsd:integer(?totalPositiveCases)
WHERE{
      ?node dbpprop:countryName ?countryname.
      ?node covid:total_positive_cases ?totalPositiveCases.
      ?node covid:continent ?continent.
      ?node dc:date ?mydate.
      FILTER regex(?mydate, "2021-01-20").
      FILTER regex(?continent, "Europe").
      FILTER (?totalPositiveCases > 300000).
}
```

Fig. 3 SPARQL query to retrieve total number of COVID19 cases of countries in Europe

shows proof-of-concept SPARQL query to retrieve the total number of COVID19 positive cases up to 20/01/2021 in the different countries of Europe where the total number of positive cases is greater than 300,000. Figure 4 shows pie chart of data retrieved by the SPARQL query in Sgvizler. The query has resulted into acceptable precision and recall values by retrieving relevant and required information. Thus, OntoCOVID effectively organizes disparate CSDs into a semantic model to answer high-level queries.

$$\text{Recall} = \left(\frac{\text{Number of relevent items retrieved}}{\text{Total number of relevant items}}\right) \tag{1}$$

$$\text{Precision} = \left(\frac{\text{Number of relevent items retrieved}}{\text{Total number of items retrieved}}\right) \tag{2}$$

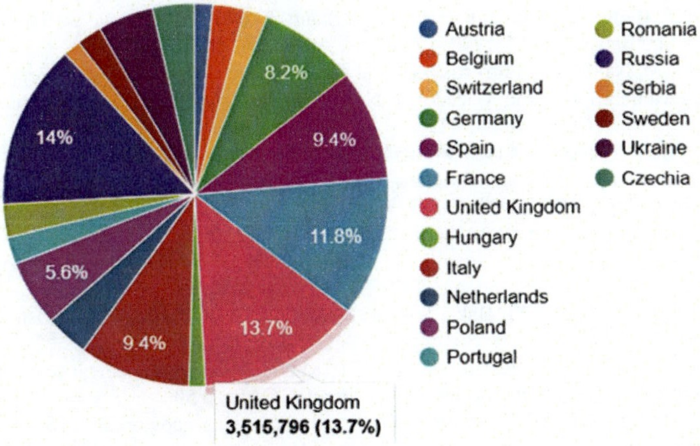

Fig. 4 Pie chart visualization of the SPARQL query

5 Conclusion

The availability of heterogeneous CSDs demands for the realization of explicit, uniform, and flexible data management techniques. The healthcare domain has already used SW technologies to enhance information management, retrieval, and visualization. In this paper, we have presented a novel approach for the design and development of OntoCOVID ontology to define an explicit, uniform, and flexible semantic data model to represent, organize, share, and reuse data in the heterogeneous CSDs. The OntoCOVID is developed and evaluated using SW technologies to show its usability and importance. The OntoCOVID includes definitions of classes, properties, relationships, and axioms. The results of the tests have shown that OntoCOVID has an enhanced extendable ontological design and is complete to answer ontology-related competency questions. The OntoCOVID can be easily extended by defining new classes and properties to represent and accommodate new data needs. The OntoCOVID can have a number of potential application areas including government and healthcare decision making and linked open data.

References

1. Tilahun B, Kauppinen T, Keßler C et al (2014) Design and development of a linked open data-based health information representation and visualization system: potentials and preliminary evaluation. JMIR Med Inform 2(2):e31
2. Berners-Lee T, Hendler J, Lassila O (2001) The semantic web. Sci Am 284(5):34–43
3. Ali S, Khusro S (2016) POEM: Practical ontology engineering model for semantic web ontologies. Cogent Eng 3(1):1193959
4. W3C Semantic Web Health Care and Life Sciences Interest Group Charter. https://www.w3.org/2011/09/HCLSIGCharter. Accessed 06 June 2020

5. Zenunia X, Raufia B, Ismailia F et al (2015) State of the art of semantic web for healthcare. Procedia Soc Behav Sci 195:1990–1998
6. Jovanovik M, Najdenov B, Strezoski G et al (2015) Linked open data for medical institutions and drug availability lists in Macedonia. New trends in database and information systems II, vol 312. Springer, Cham, pp 245–256
7. Cheung K-H, Prudhommeaux E, Wang Y et al (2009) Semantic web for health care and life sciences: a review of the state of the art. Brief Bioinform 10(2):111–113
8. He Y, Yu H, Ong E et al (2020) CIDO, a community-based ontology for coronavirus disease knowledge and data integration, sharing, and analysis. Sci Data 7(1):1–5
9. McGagh D, Liyanage H, Lusignan SD et al (2020) COVID-19 surveillance ontology. https://bioportal.bioontology.org/ontologies/COVID19. Accessed 24 Apr 2021
10. Sargsyan A, Kodamullil AT, Baksi S et al (2020) The COVID-19 Ontology. Bioinformatics 36(24):5703–5705
11. Reddy BP, Houlding B, Hederman L et al (2018) Data linkage in medical science using the resource description framework: the AVERT model. CASI, p 59
12. Andes N, Davis JE (1995) Linking public health data using geographic information system techniques: Alaskan community characteristics and infant mortality. Stat Med 14(5–7):481–490
13. Presutti V, Gangemi A (2008) Content ontology design patterns as practical building blocks for web ontologies. In: International conference on conceptual modeling, 2008. Springer, pp 128–141
14. Tartir S, Arpinar B (2007) Ontology evaluation and ranking using OntoQA. In: Proceedings of the ICSC, pp 185–192
15. Ali S, Khusro S, Ullah I et al (2017) Smartontosensor: ontology for semantic interpretation of smartphone sensors data for context-aware applications. J Sens 2017:1–26

Deep Learning Model for Thunderstorm Prediction with Class Imbalance Data

Diarmuid Healy, Zaid Mohammed, Nadia Kanwal, Mamoona Naveed Asghar, and Mohammad Samar Ansari

Abstract Weather patterns affect many sectors of human life. Extreme weather can lead to financial and human losses, with storms alone causing billions of dollars in property damage each year. Weather data are recorded as time series data, and therefore, this depends on order which needs to be exploited for learning and forecasting. Five-year hourly observations, including atmospheric pressure, wind speed, and wind direction, were used to develop and test deep learning models. However, the data showed a very large imbalance when extracted for a binary classification, i.e., thunderstorm and not-thunderstorm. The SMOTE re-sampling technique was used to study the effects on the prediction of time series meteorological data. A recurring neural network, called the LSTM deep learning model, for the thunderstorm prediction task proved to be a good predictor for thunderstorms with significant class imbalance data.

Keywords LSTM · Deep learning · Weather prediction · Weather forecasting · Thunderstorm prediction · Weather classification · Time series · Multi-layer perceptron

1 Introduction

Weather affects many aspects of human life, including both indoor and outdoor activities. Severe weather can result in disasters such as famines and wildfires, and can lead to financial losses from property damage or destruction, to loss of life. Severe

D. Healy · Z. Mohammed · N. Kanwal (✉) · M. N. Asghar
Athlone Institute of Technology, Athlone, Ireland
e-mail: nkanwal@ait.ie

N. Kanwal
Lahore College for Women University, Lahore, Pakistan

M. N. Asghar
The Islamia University of Bahawalpur, Bahawalpur, Pakistan

M. S. Ansari
Aligarh Muslim University, Aligarh, India

A. Ullah et al. (eds.), *Proceedings of International Conference on Information Technology and Applications*, Lecture Notes in Networks and Systems 350,
https://doi.org/10.1007/978-981-16-7618-5_17

weather events are estimated to have resulted in $2 trillion in economic damages and over half a million casualties during 1998–2017 [1]. Of the different types of severe weather events, in the US alone in 2019, tornadoes and thunderstorms resulted in $20.3 billion in insured losses [2]. Thunderstorms predictions both temporal-spatially and in their intensity are difficult to predict due to their relatively small spatial (1–10 km) and temporal (1–10 km) scales [3].

Due to the impacts that such storms have on human activities, there is a need to improve the accuracy of their prediction. Recent advances in machine learning and deep learning have been applied to the field of weather prediction, due to their efficiency when compared to traditional numerical weather prediction models which require vast amounts of data crunching and computing power [4] to perform predictions using physical weather models. Despite the significance of thunderstorms, their frequency is commonly low, and therefore, multi-year weather data usually contains very few instances (hours within a year) with thunderstorm. This makes the dataset highly affected by the class imbalance problem and in-turn reduces the effectiveness of some machine learning algorithm used for prediction [5, 6]. We used data augmentation to improve the data distribution between two classes using a well-known method Synthetic Minority Oversampling Technique (SMOTE).

Recurring neural networks (RNNs) [7] are one of the most popular data-based approaches used in the prediction of time series. It benefits from feedback loops which makes the output of RNN at one time step to be seen as an input in the subsequent one, and this sequence of previous information helps the model understand short-term dependencies. However, they have a long-term addictions capture issue due to the disappearance of the gradient issue [8]. To prevent this RNNs problem, another algorithm called long short-term memory (LSTM) has been proposed to capture long-term data dependencies using a memory cell and the synchronization method. Here, we used LSTM to see if it can find more inherent patterns in the data to predict thunderstorms based on strongly biased data.

This organization of the paper is as follows, firstly, existing literature is discussed on weather prediction using deep learning models. Follow-up on methodology used in this study and the data used for thunderstorm prediction are described in Sect. 3. Finally, the results are presented and discussed in Sect. 4, and the conclusion and considerations for the future work are provided in Sect. 5.

2 Literature Review

This section discusses recent related works on weather prediction using deep learning models. Hewage et al. [9] compared two LSTM models to the traditional numerical weather prediction (NWP) WRF model for the task of predicting ten different weather variables. They evaluated both a Multiple Input Single Output (MISO) and Multiple Input Multiple Output (MIMO) model. The MISO model is tested for variables individually, while the MIMO one makes all predictions together. Their findings show that both LSTM models outperform the WRF model. Additionally, they found that their MISO model outperforms the MIMO counterpart. Sadeque and Bui [4] proposed a cascaded LSTM model for the task of predicting weather data using adaptive learning. They found that the cascaded LSTM with an adaptive learning rate could yield better results than standard LSTM for weather prediction from time series data. Akram and El [10] employed a two-layer LSTM model to predict a range of different weather variables: wind speed; temperature; and humidity. It was parallelized to take advantage of GPU acceleration. They found that LSTM was able to predict these variables with a high degree of accuracy after 10 h of training. Poornima and Pushpalatha [11] proposed an intensified LSTM using sigmoid weighted linear units for the task of rainfall prediction. The weighted linear units multiply the input value(s) for the sigmoid activation function, with the aim of handling the vanishing gradient problems. Based on the RMSE metric used for evaluation, they found that LSTM outperforms the RNN and Holt-Winters models being used for comparison and their proposed intensified LSTM provided better accuracy than standard LSTM.

Chhetri et al. [12] also tackled the task of rainfall prediction, proposing a seven-layer bidirectional LSTM and GRU (BLSTM-GRU) model for this task. As compared against linear regression, multi-layer perceptron (MLP), convolutional neural network (CNN), standard LSTM, gated linear unit (GRU), and BLSTM models, their BLSTM-GRU model outperformed all of them, while standard LSTM outperformed the other control models. While also tackling the task of rainfall prediction, Xingjian et al. [13] did this using a series of radar echoes to predict future echoes using their proposed extended fully connected LSTM (FC-LSTM) model. This model uses convolutional structures in both the input-to-state and state-to-state transitions (convolutional LSTM or ConvLSTM). As compared to the state of the-art real-time optical flow by variation methods for echoes of radar (ROVER) system, their FC-LSTM model provided improved accuracy. Jin et al. [14] also tackle the task of future radar echo prediction, proposing a multi-level correlation LSTM (MLC-LSTM) model for this. The MLC-LSTM model uses three-level radar echoes combined with both a ConvLSTM fusion module and a separation module. The model was trained using a generator and discriminator pattern confrontational AI instead of a standard loss function, and the result is competitive with or outperforms ConvLSTM, trajectory GRU (TrajGRU), tracking radar echo by correlation (TREC), and optical flow models. Using an LSTM model to predict temperature and several weather parameters (precipitation, wind speed, pressure, dew point, visibility, and humidity), Fente and Singh [15] find that LSTM is highly accurate for the given task. Cong et al. [16] compare a stacked LSTM model to a deep neural network (DNN) and a random forest (RF) model for test task of temperature prediction. Their stacked LSTM model is an LSTM model stacked with either RF or DNN. The stacked models are compared to each model by itself. They found that the LSTM-only model outperforms DNN and RF models, while stacking with LSTM-DNN provided optimal results with LSTM-RF providing no difference to LSTM-only.

Karevan and Suykens [17] proposed a spatiotemporal stacked LSTM (STLSTM) model for this task. This model uses an LSTM model for each of five different locations with the hidden layers being used as input to a second LSTM layer. They found that a standard stacked LSTM model outperforms the state-of-the-art numerical prediction model and that their proposed ST-LSTM outperforms the stacked LSTM for minimum temperatures and is competitive for maximum temperatures. Peng and Ni [18] propose a hybrid support vector machine LSTM (SVM-LSTM) model for temperature prediction and compare this to traditional LSTM. Their proposed model combines the results from an SVM and LSTM model to improve the accuracy of both, using empirical model decomposition (EMD) to decompose the input. They find that standard LSTM outperforms standard SVM models, while their proposed variant outperforms both. Abayomi-Alli et al. [19] use an LSTM model to predict solar radiation as well as precipitation, wind speed, humidity, and maximum temperature, finding the LSTM model to be highly accurate for the given tasks. Qing and Niu [20] compare an LSTM model to persistence algorithms, linear least square regression, and multi-layered feedforward neural networks using back-propagation (BPNN) for the task of solar irradiance forecasting, finding that the LSTM model significantly outperforms the competitor models. Yu et al. [21] found similar results when comparing an LSTM model with auto-regressive integrated moving average (ARIMA), support vector regression (SVR), BPNN, CNN, and recurrent neural network (RNN) models.

Wang et al. [22] combine wavelet decomposition using discrete cosine transform (DCT) transformation, CNN, and LSTM for solar irradiation forecasting in inclement weather. When compared against standard LSTM, CNN-LSTM, CNN, and ANN models, they found that their DWT-CNN-LSTM model was able to outperform the other models in poor weather conditions while falling behind in sunny weather where the prediction task was more stable.

3 Methodology

The objective of this work is to design a lightweight long short-term memory (LSTM) model which can classify weather conditions and accurately predict thunderstorms. The LSTM model will be evaluated in comparison with a competitor multi-linear perceptron (MLP) model built for the same task. The choice of the LSTM is motivated by their propensity to learn patterns in long sequences (which recurrent neural networks are not good at). The MLP model is developed as a baseline against which the LSTM model performance will be compared.

3.1 Data and Data Preprocessing

Data that is used in this project is five year's observations on hourly basis including five features and a target class: temperature, humidity, air pressure, wind speed, wind direction, and the class showing weather description. This data set has approximately 45,000 observations and five features for different cities: temperature, humidity, air pressure, wind speed, and wind direction. The data set contains 45,000 observations across each of 36 cities in three countries: USA, Canada, and Israel. Data for temperature, wind speed, wind direction, pressure, humidity, and weather description are first combined based on city. Next, the target category of weather description is *binarized* to fit the goal of predicting specifically thunderstorms.

The weather descriptions now reveal an impediment which any model will have to overcome: a severe class imbalance (as can be seen from the depiction of the thunderstorm frequency by characteristics shown in Fig. 1a). Given a total of over 45,000 observations, for each location, there are fewer than 2% of the target class thunderstorm in the best case, which is Miami. As such, this city is selected for further progress as having the highest likelihood of success (shown in Fig. 1b). Looking at the temperature features for the city of Miami shows that they are on quite different scales (Table 1). As such, these numeric fields are normalized using min–max normalization.

Focusing on Miami, thunderstorms are more likely to happen in summer every year, with discrete occurrence during spring and autumn (Fig. 1b). This indicates that temporal factors could prove beneficial to predictions and will be included in the model feature set. As such, the month, day, and hour values were extracted from the datetime index and considered as independent features to be used in the classification

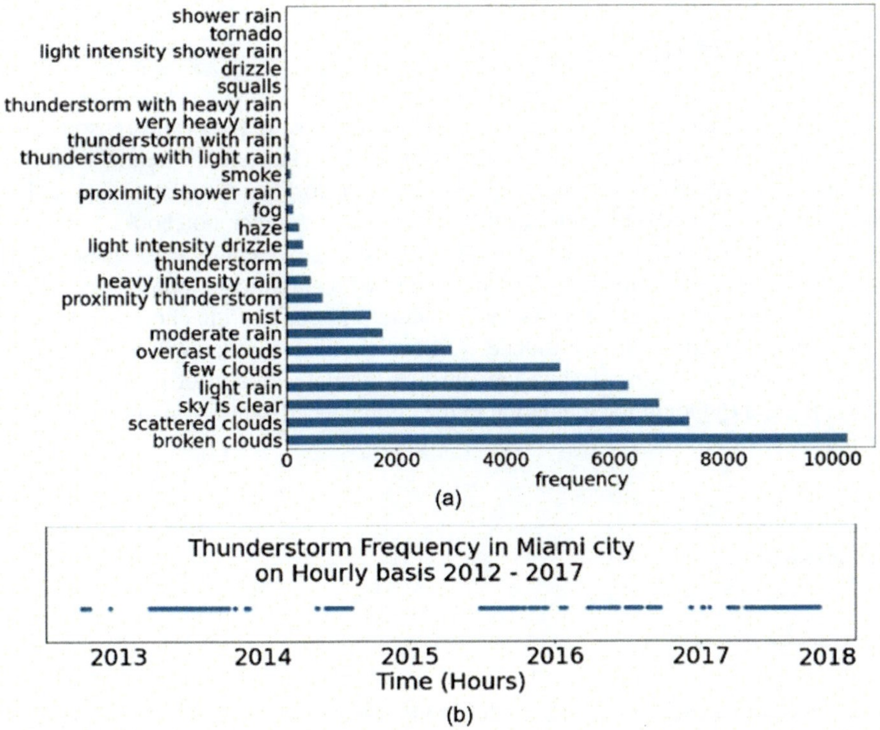

Fig.1 **a** Thunderstorm frequency by type, **b** thunderstorm frequency over time in Miami

Table 1 Data description of the weather feature set for Miami

	Humidity	Pressure	Temperature	Wind_direction	Wind_speed
Count	45,230.00	45,230.00	45,230.00	45,230.00	45,230.00
Mean	81.94	1018.91	283.93	162.19	2.46
Std	14.48	16.88	6.61	98.79	2.00
Min	12.00	810.00	245.15	0.00	0.00
25%	73.00	1012.00	279.24	81.00	1.00
50%	86.00	1017.00	283.63	148.00	2.00
75%	93.00	1023.00	288.51	243.00	4.00
Max	100.00	1100.00	307.00	360.00	25.00

process. The temperature in Miami varies between 5 °C and 35 °C, humidity is over 60% in general, pressure varies between 1010 and 1030 hPa (hectopascal), wind direction with over 300° are more frequent, and wind speed between 0 and 40 km/h is more frequent, as can be seen for Fig. 2. Finally, data is split into training and test data using a 4:1 split (i.e., 80:20).

3.2 *Tools and Technology*

The proposed deep learning models are built in Python using Keras and TensorFlow. Other libraries for data processing and evaluation include Scikit-Learn, NumPy, Pandas, and SciPy. Seaborn and Matplotlib are used for data visualization, and the majority of model development work is done using Jupyter notebook. SMOTE, or "Synthetic Minority Oversampling Technique," is a technique for generating synthetic samples of a minority class to contend with classification problems where there is a class imbalance [23]. As the thunderstorm class in the chosen dataset is heavily outweighed by the non-thunderstorm class, this technique will be applied to the data to compare results of a given model once the classes are made more balanced. SMOTE has a Python implementation in the "imbalanced dataset" module which is used.

4 Results

4.1 *MLP Model*

A multi-layer perceptron (MLP) model is first built using Scikit-Learn, using the hidden layer structure (8, 128, 4) (shown in Fig. 3a) considering the input layer is

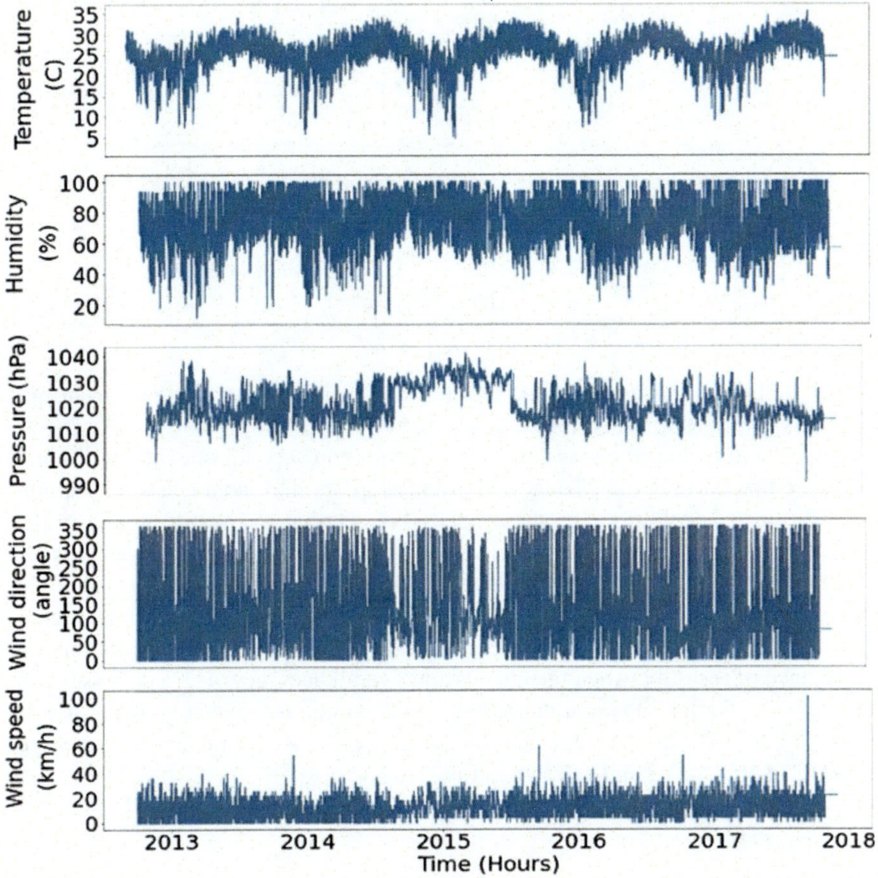

Fig.2 Different features and their variation in the Miami dataset

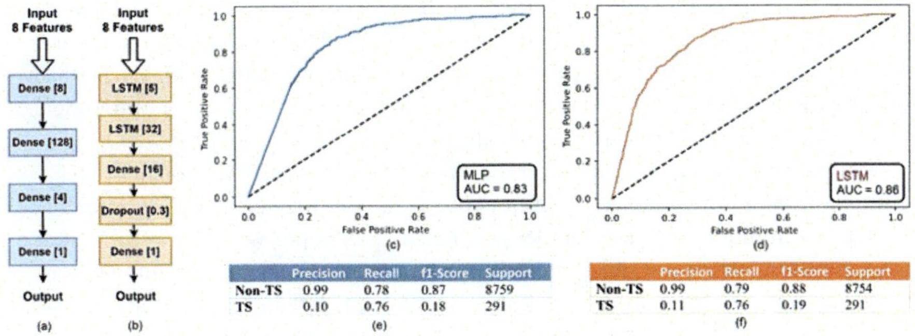

	Precision	Recall	f1-Score	Support
Non-TS	0.99	0.78	0.87	8759
TS	0.10	0.76	0.18	291

(e)

	Precision	Recall	f1-Score	Support
Non-TS	0.99	0.79	0.88	8754
TS	0.11	0.76	0.19	291

(f)

Fig. 3 **a** MLP model, **b** LSTM model, **c** AUC obtained for MLP model, **d** AUC obtained for LSTM model, **e** Loss, recall, and precision values for MLP model, and **f** Loss, recall, and precision values for LSTM model

Table 2 Recall results of each model for a range of epochs 25–1000

Epochs	Original		Na¨ıve		SMOTE 10x		SMOTE balance	
	TS	No TS	TS	No TS	TS	No TS	TS	No TS
25	0.93	0.61	0.97	0.52	1.0	0.0	0.88	0.65
50	0.76	0.79	0.9	0.6	0.88	0.64	1.0	0.0
100	0.00	1.00	0.67	0.9	1.0	0.0	0.79	0.73
200	0.00	1.00	0.85	0.73	0.7	0.78	0.67	0.78
500	0.00	1.00	0.73	0.78	0.7	0.79	0.0	1.0
1000	0.00	1.00	0.55	0.84	0.47	0.88	0.0	1.0

fed with eight features, and output layer has one neuron using rectified linear unit (ReLU) as its activation function.

Due to the imbalanced classes, the model results can vary significantly with small changes to the learning rate. To try and produce an optimal model, several learning rates were tested. Additionally, the model is susceptible to overfitting due to this imbalance. As a solution to the imbalance problem, the model used SMOTE technique by doubling the thunderstorm label ten times, and 500 iterations is used. The Adam solver is used, and the predictions for a range of learning rates can be seen in Table 2. Based on these results, the learning rate 0.001 is chosen as combining the best balance of recall between the classes and overall accuracy. The loss was 0.258, where it stopped after 334 iterations as the model has not improved for 10 continuous epochs. Values of loss, precision, and recall for this model are presented in Fig. 3e. The AUC of ROC is 83%, as shown in Fig. 3c.

4.2 LSTM Model

A long short-term memory (LSTM) model is built using Keras. This model contains two LSTM layers with 5 and 32 units, a dense layer of 16 units, dropout layer using 30% dropout rate, and finally a dense output layer (shown in Fig. 3b). Each LSTM layer and the dense layer uses the ReLU activation function, while the output dense layer uses the Softmax activation function.

Through experimentation, it was found that a batch size = 24, epochs = 50 gave the most optimal performance. Additionally, to handle the serious class imbalance between the thunderstorm/no thunderstorm classes, sample weights were included during model training to add emphasis to the observations where thunderstorms occur. Through experimentation, these weights were set as 25 for thunderstorm observations and 1 for non-thunderstorm observations. This resulted in a model with balanced recall across both classes of between 76 and 79% for the thunderstorm/no thunderstorm classes, respectively. Values of loss, precision, and recall for this model are presented in Fig. 3f. The AUC of ROC is 86%, as shown in Fig. 3d.

As the thunderstorm class in this dataset is heavily outweighed by the majority non-thunderstorm class, additional models are built to tackle this class imbalance and determine if results can be improved. Firstly, a naive duplication approach is taken: each instance of thunderstorm is copied ten times so that they occur in series of ten. Then, the sample weighting for this model is reduced from 25 to 5. Secondly, a similar approach is taken but using SMOTE to generate synthetic samples instead of simple duplication. Finally, SMOTE is used to completely balance the two classes; in this instance, there are now nearly 50% synthetic data samples. The recall results for each model are below showing results for varying numbers of epochs (presented in Table 2).

The results from handling the class imbalance problem by increasing the samples of the minority class, while competitive with the original LSTM model, do not provide major improvements over ones that were obtained using the imbalanced classes with a large sample weighting. Additionally, the original dataset provides a model that can be trained adequately within a significantly shorter time period.

The results show that even the distribution of data between all classes is crucial in time series data and cannot be addressed through conventional methods such as replication and SMOTE. Therefore, a more sophisticated approach to identifying hidden trends in the data must be identified and duplicated with the data for better data augmentation. Deep learning models such as LSTM appeared to better understand long-term dependencies in the data and make better predictions as compared to other machine learning models.

5 Conclusion

In this study, a deep learning model was built to predict thunderstorms from enormous meteorological data. In addition, a complex class imbalance problem in temporal weather data has been investigated. We used the SMOTE data augmentation methodology to improve data distribution, and machine learning and deep learning models were used to predict thunderstorms using five-year weather data. MLP and LSTM have performed well in the task of predicting thunderstorms. Between the two models, the LSTM model outperformed by a margin of 16.21% (AUC of ROC), while the LSTM model also resulted in the more balanced recall and higher recall for both classes. Further work could look at improving methods of increasing data in time series data. Consideration should also be given to further optimizing and generalizing machine learning models using more features and a larger number of geographic areas.

References

1. Wallemacq P, House R Economic losses, poverty and disasters 1998–2017. Tech. Rep. [online]. https://www.unisdr.org/we/inform/publications/61119. Accessed 27 Feb 2021
2. Insurance Information Institute. Facts + statistics: Tornadoes and thunderstorms, iii.org. [online]. https://www.iii.org/fact-statistic/facts-statistics-tornadoes-andthunderstorms. Accessed 27 Feb 2021
3. Kulikov MY, Belikovich MV, Skalyga NK, Shatalina MV, Dementyeva SO, Ryskin VG, Shvetsov AA, Krasil'nikov AA, Serov EA, Feigin AM (2020) Skills of thunderstorm prediction by convective indices over a metropolitan area: comparison of microwave and radiosonde data. Remote Sens 12(4):604
4. Al Sadeque Z, Bui FM (2020) A deep learning approach to predict weather data using cascaded lstm network. In: 2020 IEEE Canadian conference on electrical and computer engineering (CCECE). IEEE, pp 1–5
5. Amin A, Anwar S, Adnan A, Nawaz M, Alawfi K, Hussain A, Huang K (2017) Customer churn prediction in the telecommunication sector using a rough set approach. Neurocomputing 237:242–254
6. Amin A, Anwar S, Adnan A, Nawaz M, Howard N, Qadir J, Hawalah A, Hussain A (2016) Comparing oversampling techniques to handle the class imbalance problem: a customer churn prediction case study. IEEE Access 4:7940–7957
7. Elman JL (1990) Finding structure in time. Cogn Sci 14(2):179–211
8. Karevan Z, Suykens JA (2020) Transductive lstm for time-series prediction: an application to weather forecasting. Neural Netw 125:1–9
9. Hewage P, Behera A, Trovati M, Pereira E (2019) Long-short term memory for an effective short-term weather forecasting model using surface weather data. In: IFIP International conference on artificial intelligence applications and innovations. Springer, pp 382–390
10. Zaytar MA, El Amrani C (2016) Sequence to sequence weather forecasting with long short-term memory recurrent neural networks. Int J Comput Appl 143(11):7–11
11. Poornima S, Pushpalatha M (2019) Prediction of rainfall using intensified LSTM based recurrent neural network with weighted linear units. Atmosphere 10(11):668
12. Chhetri M, Kumar S, Pratim Roy P, Kim BG (2020) Deep blstm-gru model for monthly rainfall prediction: a case study of Simtokha, Bhutan. Remote Sens 12(19):3174
13. Shi X, Chen Z, Wang H, Yeung DY, Wong WK, Woo WC (2015) Convolutional LSTM network: a machine learning approach for precipitation nowcasting. arXiv preprint arXiv:1506.04214
14. Jing J, Li Q, Peng X (2019) Mlc-lstm: exploiting the spatiotemporal correlation between multi-level weather radar echoes for echo sequence extrapolation. Sensors 19(18):3988
15. Fente DN, Singh DK (2018) Weather forecasting using artificial neural network. In: 2018 Second international conference on inventive communication and computational technologies (ICICCT). IEEE, pp 1757–1761
16. Li C, Zhang Y, Zhao G (2019) Deep learning with long short-term memory networks for air temperature predictions. In: 2019 International conference on artificial intelligence and advanced manufacturing (AIAM). IEEE, pp 243–249
17. Karevan Z, Suykens JA (2018) Spatio-temporal stacked lstm for temperature prediction in weather forecasting. arXiv preprint arXiv:1811.06341
18. Peng W, Ni Q (2020) A hybrid svm-lstm temperature prediction model based on empirical mode decomposition and residual prediction. In: 2020 IEEE International conference on systems, man, and cybernetics (SMC). IEEE, pp 1616–1621
19. Abayomi-Alli A, Odusami MO, Abayomi-Alli O, Misra S, Ibeh GF (2019) Long short-term memory model for time series prediction and forecast of solar radiation and other weather parameters. In: 2019 19th International conference on computational science and its applications (ICCSA). IEEE, pp 82–92
20. Qing X, Niu Y (2018) Hourly day-ahead solar irradiance prediction using weather forecasts by LSTM. Energy 148:461–468

21. Yu Y, Cao J, Zhu J (2019) An LSTM short-term solar irradiance forecasting under complicated weather conditions. IEEE Access 7:145651–145666
22. Wang F, Yu Y, Zhang Z, Li J, Zhen Z, Li K (2018) Wavelet decomposition and convolutional LSTM networks based improved deep learning model for solar irradiance forecasting. Appl Sci 8(8):1286
23. Chawla NV, Bowyer KW, Hall LO, Kegelmeyer WP (2002) Smote: synthetic minority over-sampling technique. J Artif Intell Res 16:321–357

Multiple Parallel Activity Detection and Recognition to Avoid COVID-19 Spread-Out

Muhammad Talha Ubaid, Muhammad Zeeshan Khan, Muhammad Usman Ghani Khan, Amjad Rehman, and Noor Ayesha

Abstract COVID-19 has been affecting people around the globe. It is affecting almost every country currently, according to the World Health Organization (WHO). This virus is transmitted to another person if an asymptomatic person makes close contact with the everyday person. There is no cure for this virus, and the only solution is social distancing and avoids the people doing these activities. In this paper, we proposed a system for detecting and recognizing the activities that violate social distancing. These activities involve handshakes and hugging. We implement a system that is capable of detecting and identifying multiple parallel activities. Temporal features are extracted for similar activities in 16 frames. We use the two models for this purpose: Faster RCNN for the detection and ResNet-50 to recognize the activities. First, Faster RCNN detects the group of people and that region of interest ROI saved and passes to the ResNet-50 to recognize the activities. We also generated our dataset on the local environment in multiple parallel activities. This system achieves the accuracy of 95.03% for the detection of testing dataset and recognition of multiple parallel activities 92.88% accuracy accomplished. The system used different public datasets and generated some local datasets for handshake and hugging activities.

Keywords Object detection · Activity detection · Activity recognition · ResNet · Faster RCNN · Healthcare · Public health

M. T. Ubaid · M. Z. Khan · M. U. G. Khan
Intelligent Criminology Research Lab National Center of Artificial Intelligence, Al Khawarizmi Institute of Computer Science, University of Engineering and Technology, Lahore, Pakistan
e-mail: Talha.Ubaid@kics.edu.pk

M. Z. Khan
e-mail: Zeeshan.Khan@kics.edu.pk

M. U. G. Khan
e-mail: Usman.ghani@kics.edu.pk

A. Rehman (✉)
Artificial Intelligence & Data Analytics Lab CCIS, Prince Sultan University, Riyadh 11586, Saudi Arabia

N. Ayesha
School of Clinical Medicine, Zhengzhou University, Zhengzhou, Henan, PR China

A. Ullah et al. (eds.), *Proceedings of International Conference on Information Technology and Applications*, Lecture Notes in Networks and Systems 350,
https://doi.org/10.1007/978-981-16-7618-5_18

1 Introduction

COVID-19 is a transmissible disease. It has been affecting people around the globe. It is affecting many counties currently, according to the World Health Organization (WHO). A virus infects people with fever, dry cough, tiredness, sore throat, aches, and pain [1]. It spread through the tiny drop of saliva when an infected person sneezes, coughs, or any other way through which their saliva particles to another person. To date, there are no specific vaccines or medicines for COVID-19. This virus is transmitted to another person if an asymptomatic person contacts another person, i.e., touching, handshaking, hugging, etc. The medication for this disease is still to be discovered. So, prevention is the only answer to this.

Social distancing is the space between the two persons to avoid close contact between them. Social distancing is an effective way to get rid of this virus. By doing this, growth rate of affected people start reducing [2]. Every country tries to implement rules and regulations to avoid chaos. On the government level, authorities ask the people to do the social distancing to avoid the COVID spread. Many countries imposed a lockdown for a few weeks to avoid social gatherings. But lockdown is not the long-term solution. Now we have to live with social distancing if we want to dodge from COVID-19. In all places like offices, markets, malls, and mosques, social distancing was implemented. This is the only way to keep away from this virus. A lot of great work had been done to prevent COVID-19 spread [3].

Every mall, office, and market try to implement social distancing and restrict people's close interaction. Handshake, hugging, and activities are forbidden. Management pushes its best to limit these activities. Artificial intelligence (AI) is one of the newest fields in science and engineering and goes further, and it attempts to understand and build intelligence entities. Deep learning is a subset of machine learning, which in turn is a subset of artificial intelligence. It uses a programmable neural network that enables machines to make accurate decisions without help from humans. Like we use our brains to identify patterns and classify various types of information, deep learning algorithms can be taught to accomplish machines' same tasks. We can solve our real-world problems in the field of science, mathematics, philosophy, psychology, linguistic, and simplify and solve complex problems, such as health issues, traffic issues, marketing issues, and social distancing. We implement a system that helps detect and recognize the activities that violate social distancing, i.e., handshaking and hugging. Many people broke the rules of social distancing. It is challenging to detect and recognize that person who violated the social distancing rules. The most common thing that violates social distancing is handshaking and hugging.

Human activity detection attempts to recognize activities using information sensors such as radar, cameras, and smartphones. In the past, acquiring sensor data for activity detection was effortful, which is now quickly done by smartphones and GPS. It is challenging to relate sensory information to explicit human activity, outlining this problem as a univariate or multivariate classification task.

We proposed a system that detects and recognizes the activities that violate social distancing, such as handshakes and hugging. In this paper, two tasks have been done (1) first, we detect the region where activities performing in the input (2) these detect detected regions. Region of Interest (ROI) passed to another model for the classification of the activities. The detail of the process is explained in the methodology section. Our main contributions in this research work are:

- A novel system is capable of detecting and recognition multiple parallel activities.
- Combine the two different deep learning algorithms to get a valuable result.
- Generate the dataset of handshake and hugging activities for the local environment.

In Sect. 2 of the paper, existing related work has been discussed. In Sect. 4, we review our dataset and clarify the outcomes we accomplished through tests. Finally, in Sect. 5, we present a conclusion and recommend future work.

2 Literature Survey

For the detection of a group and human activities, different methodologies were proposed and used. In the early days, the recognition of human action was done by old techniques of computer vision. The Gaussian method was proposed in Ref. [4] to classify the group activities done by humans. It deals with the motion of the trajectories and works on the uncertainty of the object's movement. The pattern of activities was analyzed and described by the Gaussian process. Support vector machine did the rest. Azar et al. [5] implemented the convolutional relational machine (CRM) to classify individual and group human activities. It generated the activity map base on the activity which is performed in the input video. It used the convolutional neural network (CNN) after getting the activity map to recognize the input. Our problem is to detect activities that violate social distancing. But after that, the detected part should be able to recognize as well, for the detection of multiple human activities such as running and kicking.

In Patil et al. [6], histogram of oriented gradient (HOG) descriptor and support vector machine (SVM) classifier were used for the activity classification. The different feature extraction played the role of feature extractor, and afterward, SVM classifies the input video into related classes. Detection and recognition of suspicious activities in surveillance are vital tasks. They were capturing certain activities in real-time in a different environment without human intervention. To get the temporal feature, the Gaussian function and equation of OMEGA are used [7]. This activity is divided into two parts: abnormal and everyday activities. Output was extracted by using image retrieval. Ronao and Chao [8] utilized the combination of deep learning and mobile sensors to recognize human activities. Deep convolution neural network, accelerometer, and gyroscope used for the feature extraction and classification of the input activities. Working with the sensor may add overheads to the system and is not user-friendly. Most of the sensor does not work in the outer atmosphere or may affect the system's accuracy.

Mliki et al. [9] used the unmannered ariel vehicle (UAV) dataset. Ariel view videos were used for the recognition of human activities. In this process, two stages are offline and inference. The first one generates the human model for activities. In the inference part, the previously developed model is used for the detection of the activities. It classifies two times, first the frames of the video and completes the video in sequence. A framework proposed by Rahman and Laganiere [10] gets the video's temporal features using a 3D convolution neural network (3D CNN). The activity was detected on the base on the base of these temporal features. The evaluation of the system was done on the THUMOS' 14 benchmark.

Nowadays, the surveillance cameras are everywhere, especially in offices, banks, ATMs, etc. The surveillance video is of long-time duration. Nar et al. [11] detected abnormal activities from surveillance at ATM. By using the posture recognition of the human body, it detects strange activities. To get the skeleton information, 3D Kinect camera is used for the bone joint movement and the human body's position. Shi et al. [12] classifies the different human activities using spectrogram and deep convolutional network (DCNN). For the feature extraction, transfer learning base on DCNN by using the elevating training set. The pretrained DCNN was trained on a large dataset of RGB named as ImageNet.

In the research work described above, every system has some deficiency. In previous work, only recognition of the group activities was done while we have done both detection and recognition in our system, while Patel et al. [6] detected and recognized only single person activities using a surveillance camera. We proposed a system to detect and recognized the different parallel activities in single or multiple frames.

3 Methodology

Our proposed methodology comprises two different state-of-the-art algorithms: Faster RCNN and ResNet. The developed system can detect and recognize the activities. These activities violate social distancing, i.e., handshake and hugging. This trivial process is divided into two phases: detection and recognition. To detect the input activities, we implement the Faster RCNN, and recognition is done by ResNet-50. The architecture of the proposed model is shown in Fig. 1.

The motivation of work is to detect and classify the activities that break social distancing rules using deep learning techniques. The system takes the input to identify the actions. Faster RCNN [13] model is used to detect the activities. We trained our model using it. Activities that violate social distancing include persons in the form of groups. So, we first detect the two groups of people. By using the prediction of the Faster RCNN model, we get the related interested region. Faster RCNN generates the bounded box around the detected regions. Region of interest (ROI) predicted by the model is saved for further processing of the input. If we have multiple groups

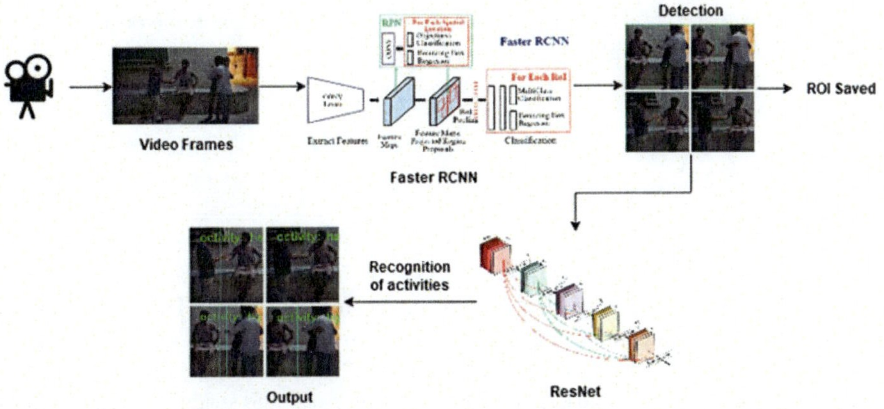

Fig. 1 System diagram

in a single frame, it also detects all parallel groups and saves the ROI. For the classification of this ROI, we implement the ResNet-50 [14]. All the ROI detected by the previous model is passed through the ResNet-50 one by one, and it recognizes the activities that are performed in these detected ROI's. The implemented model gives the results on all the ROIs detected from the input by the Faster RCNN and envisaged the label on each ROI as shown in Fig. 2. The short names of labels are used. For the handshake, used the "hs," and for a hug, "hg" is used.

From the input video, we have multiple frames, and in that, different activities were performed, and, in some cases, we have parallel activities going on a single frame. We also detect those parallel activities as we detect the group of two people

Fig. 2 Activity detection and recognition

on each frame. Afterward, that detected region goes through the ResNet-50 for the recognition of those multiple parallel activities.

3.1 Faster RCNN

Faster RCNN works by using the algorithm of selective search. It took the input and passed through the different modules of the system. In convolution layers, additional filters are used to extract the other feature maps from the input. The convolution layer, pooling layer, and fully connected layer perform the task in the convolution network. The ConvNet extracts the feature maps from the input, the size of the image is reduced, and that size depends upon the filter size and stride of the filter that extracted the feature map. The system needs to realize whether an object is available in the given picture for each purpose of the feature map and conjecture for its size.

For this reason, anchor points are put over input for each yield feature map mined by a convolutional arrangement. Such anchors characterize objects and their sizes conceivable in this situation in various measurements and different aspect ratios. These last features are sliding over the small neural network named region proposal network (RPN). Each pixel of the extracted feature maps runs on this filter to know anchors points and the bounding boxes to get the region of interest. For instance, we have a 3 × 3 convolution filter implemented with 512 units for the feature maps extraction, and we get two results, 1 × 1 convolution with 18 different units of object classification. The other is 1 × 1 with 36 units of bounded box regression. It gives the probabilities and regression unit to improve the bounded box generated around the object using anchors' coordinates—training losses calculated using Eq. (1). Here p is the probability, i is the index of the anchors, and t is the vector having direct observation of the coordinates of the bounded box. L_(cls) is the total loss of the classification layer, and L_reg represents the total regression loss. N_cls and N (reg) signify the number of training examples for classification and regression layers, respectively. For balancing the weights hyperparameter, Lambda is used.

$$L(\{p_i\}, \{t_i\}) = \frac{1}{N_{\text{cls}}} \sum_i L_{\text{cls}}\left(p_i, p_i^*\right) + \lambda \frac{1}{N_{\text{reg}}} \sum_i p_i^* L_{\text{reg}}\left(t_i, t_i^*\right) \quad (1)$$

3.2 ResNet-50

Classification of the activities like handshake and hugging are done with this model. For this purpose, we train our data on ResNet-50. ResNet residual networks are used for different computer vision tasks. ResNet-50 has five stages in its model, as shown in Fig. 3. It has an identity block in which each one has three convolution layers.

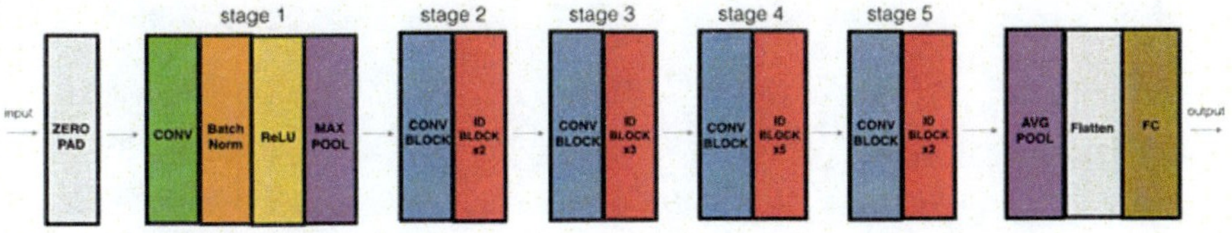

Fig. 3 ResNet-50 architecture

Skip connection is used in it. It is about adding the original input to the next block having convolution in it. It works as the vanishing gradient problem by allowing this another shortcut path for the gradient to flow through. It also helps the model learn an identity function by ensuring that the higher layer will perform at least and the lower layer. The information passed through the downsampling to the layers that are upsampled.

In this network, the input is taken, and initial convolution and max-pooling are done using the kernel of different sizes [14]. After that, in residual blocks, convolution is performed, which contains three convolution layers. Kernel size is usually 64, 64, and 128, respectively. The arrows refer to the identity connection. The convolution operation in the residual block is performed with stride 2. Hence, the input size will be reduced to half in height and width, but the channel width will be doubled. As we move from one stage to another, the channel's width is twice, and input is reduced by half. There are different variants of the residual networks. Some use the 34 layers, and some use the 121 residual layers. In this, we use the ResNet-50, and 50 layers are used. The deeper model should not produce a training error higher than its shallower counterparts.

4 Experiments and Results

4.1 Dataset

The dataset is a basic essential and integral part of the model for every deep learning system [15, 16]. The results of the model depend upon the dataset. In this, we use different datasets (i) TV Human Interaction Dataset [17], (ii) UT-Interaction dataset [18], (iii) Facial Descriptors for Human Interaction dataset [19]. We also get the dataset through Web scrapping and add it to our datasets to detect a two-person group, as shown in Fig. 4. We have 1200 frames that are annotated for the training purposes of the Faster RCNN. For the training of the ResNet-50 model, we used the TV Human Interaction Dataset, UT-Interaction dataset, and Facial Descriptors for Human Interaction dataset. These data consist of multiple and varied categories, but we extract the data from two activities: handshake and hugging. Frames are extracted from the video and cleaned for further processing. A total number of 1456 frames of each class are used for the recognition model.

Fig. 4 Dataset

4.2 Discussion and Results

We trained our two models Faster RCNN and the ResNet-50, on different datasets for the specific purpose of detection and recognition. The experiments are done on the Nvidia 1080 ti GPU of 10 GB, having 32 GB of RAM. Keras, TensorFlow, and Nvidia CUDDN libraries were used to do work. We trained the Faster RCNN on 1000 plus images and set the parameters of 1 batch size, 0.0002 learning rate, and 0.91 momenta for detection. In the beginning, the loss was 4.02323, and it decreased as the model learns, as shown in Fig. 5. A total of 200,000 epochs were run on this model that results in 0.0023 loss. Our system gave 95.03% accuracy for the detection. The evaluation is done on the testing data for accuracy. These results are further conceded to other models for recognition of the activities.

ResNet-50 model is trained on the dataset having two classes, handshake and hugging. At the start, the learning rate is $1e^{(-4)}$, and the momentum equals 0.9. The batch size is 16 trained on the 1000 plus images. We achieved an accuracy of 92.88% for recognition of the activities from the ROI's. The accuracy of the model increases as this learns more, as shown in Fig. 6. For the comparison, Vahora [20] achieved 82% accuracy using a deep convolution network for activity recognition, which is less than our model, as shown in Fig. 7.

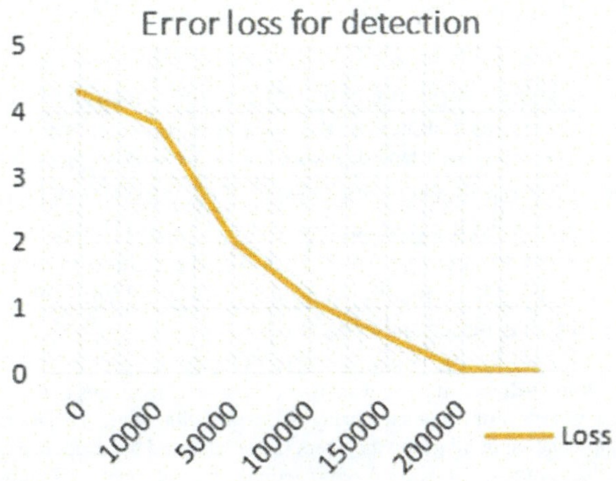

Fig. 5 Error loss for detection

Fig. 6 Error loss recognition

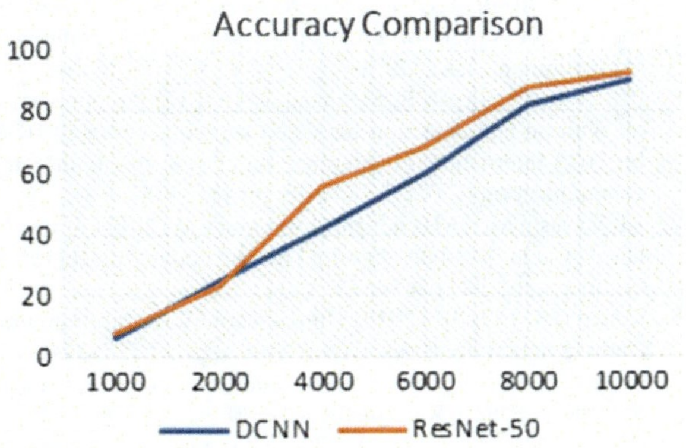

Fig. 7 Accuracy comparison

5 Conclusion

In this paper, we implement a system for detecting and recognizing the activities that violate social distancing. We use the two models for this purpose: Faster RCNN for the detection and ResNet-50 to recognize the activities. We detect and recognized multiple parallel activities. First, Faster RCNN detects the group and that region of interest ROI saved and passes to the ResNet-50 to recognize the activities. We achieved an accuracy of 95.03% for the detection and 92.88% for recognizing the activities. These results were also compared with other models and shown that our system has better and efficient.

Acknowledgements We want to offer our sincere thanks to the National Center of Artificial Intelligence for ultimately supporting our exploration work. The authors also want to acknowledge the appreciation of all group members (associates and the executives) and association (KICS) for their help, commitment, specialized meetings, and information sharing exertion for this. Without their support and encouragement, this research work could not be accomplished.

References

1. https://www.who.int/emergencies/diseases/novel-coronavirus-2019
2. Saba T, Abunadi I, Shahzad MN, Khan AR (2021) Machine learning techniques to detect and forecast the daily total COVID-19 infected and deaths cases under different lockdown types. Microsc Res Tech 84(7):1462–1474. https://doi.org/10.1002/jemt.23702
3. Khan MZ, Khan MUG, Saba T, Razzak I, Rehman A, Bahaj SA (2021) Hot-spot zone detection to tackle Covid19 spread by fusing the traditional machine learning and deep learning approaches of computer vision. IEEE Access 9:100040–100049
4. Cheng Z, Qin L, Huang Q, Jiang S, Tian Q (2010) Group activity recognition by Gaussian processes estimation. In: 2010 20th International conference on pattern recognition. IEEE, pp 3228–3231
5. Azar SM, Atigh MG, Nickabadi A, Alahi A (2019) Convolutional relational machine for group activity recognition. In: Proceedings of the IEEE conference on computer vision and pattern recognition, pp 7892–7901
6. Patil CM, Jagadeesh B, Meghana MN (2017) An approach of understanding human activity recognition and detection for video surveillance using HOG descriptor and SVM classifier. In: 2017 International conference on current trends in computer, electrical, electronics, and communication (CTCEEC). IEEE, pp 481–485
7. Al-Nawashi M, Al-Hazaimeh OM, Saraee M (2017) A novel framework for intelligent surveillance system based on abnormal human activity detection in academic environments. Neural Comput Appl 28(1):565–572
8. Ronao CA, Cho SB (2016) Human activity recognition with smartphone sensors using deep learning neural networks. Expert Syst Appl 59:235–244
9. Mliki H, Bouhlel F, Hammami M (2020) Human activity recognition from UAV-captured video sequences. Pattern Recogn 100:107140
10. Rahman MA, Laganière R (2020) Single-stage end-to-end temporal activity detection in untrimmed videos. In: 2020 17th Conference on computer and robot vision (CRV). IEEE, pp 206–213

11. Nar R, Singal A, Kumar P (2016) Abnormal activity detection for bank ATM surveillance. In: 2016 International conference on advances in computing, communications and informatics (ICACCI). IEEE, pp 2042–2046
12. Shi X, Li Y, Zhou F, Liu L (2018) Human activity recognition based on deep learning method. In: 2018 International conference on radar (RADAR). IEEE, pp 1–5
13. Ren S, He K, Girshick R, Sun J (2015) Faster r-CNN: towards real-time object detection with region proposal networks. In: Advances in neural information processing systems, pp 91–99
14. He K, Zhang X, Ren S, Sun J (2016) Deep residual learning for image recognition. In: Proceedings of the IEEE conference on computer vision and pattern recognition, pp 770–778
15. Rehman A, Khan MA, Mehmood Z, Saba T, Sardaraz M, Rashid M (2020) Microscopic melanoma detection and classification: a framework of pixel-based fusion and multilevel features reduction. Microsc Res Tech 83(4):410–423. https://doi.org/10.1002/jemt.23429
16. Mittal A, Kumar D, Mittal M, Saba T, Abunadi I, Rehman A, Roy S (2020) Detecting pneumonia using convolutions and dynamic capsule routing for chest X-ray images. Sensors 20(4):1068
17. https://www.robots.ox.ac.uk/~alonso/tv_human_interactions.html
18. https://cvrc.ece.utexas.edu/SDHA2010/Human_Interaction.html
19. https://vision.cs.hacettepe.edu.tr/interaction_images/
20. Vahora SA, Chauhan NC (2019) Deep neural network model for group activity recognition using contextual relationship. Eng Sci Technol Int J 22(1):47–54

An Inception-ResNetV2 Based Deep Learning Model for COVID-19 Detection

Tanees Riaz, Tarim Dar, Hafsa Ilyaas, and Ali Javed

Abstract COVID-19 has become the global epidemic affecting millions of people across the world. The fact that COVID-19 spreads quickly and devastating for elderly person makes this disease lethal as we witnessed a massive mortality rate in first, second, and third wave since 2020. Early diagnosis of COVID-19 is mandatory to prevent the spread and damage control, as only few nations have been able to vaccinate more than 50% of their population. The healthcare professionals commonly use the real-time polymerase chain reaction (RT-PCR) test to identify the COVID-19. Although RT-PCR test is considered more reliable among other COVID-19 detection tests; however, sensitivity of RT-PCR lies in the range of 65%-95% and took hours to diagnose the COVID-19 disease. Therefore, there exists an urgent need to develop more rapid and reliable diagnostics methods for COVID-19. In this regard, Chest X-ray and CT scan images are also being used to determine the abnormalities in the lungs of the COVID-19 patients which are found after the initial symptoms of this disease. We exploit the benefits of convolution neural network (CNN) for reliable detection of various diseases and used it for COVID-19 detection. For this purpose, we proposed a deep learning model to automatically detect the COVID-19 disease by processing the chest X-ray images. More specifically, we presented an Inception-ResNetV2 network-based deep learning model for COVID-19 detection. Performance of our model is evaluated on the publicly available COVID-19 dataset. The accuracy of 96% indicates the effectiveness of the proposed model for COVID-19 detection.

Keywords COVID-19 detection · Chest radiographs · Inception-ResNetV2

1 Introduction

The coronavirus disease, well known as COVID-19, is first identified in the Wuhan city of China during December 2019. The COVID-19 disease is caused by the virus

T. Riaz · T. Dar · H. Ilyaas · A. Javed (✉)
University of Engineering and Technology, Taxila, Pakistan
e-mail: ali.javed@uettaxila.edu.pk

A. Ullah et al. (eds.), *Proceedings of International Conference on Information Technology and Applications*, Lecture Notes in Networks and Systems 350,
https://doi.org/10.1007/978-981-16-7618-5_19

named as severe acute respiratory syndrome coronaviruse-2 (SARS-CoV-2) COVID-19 was declared as a global pandemic by the world health organization (WHO) last year due to the rapid spread of this disease around the globe in a short period of time. Till June 13, 2021, COVID-19 has affected 175,306,598 people globally and caused the deaths of 3,792,777 [1]. According to WHO, COVID-19 patients can be either symptomatic or asymptomatic. The symptomatic patients belong to the category where the infected person can have either mild or severe symptoms. In mild, the patient deals with dry cough, fever, tiredness, and in some cases diarrhea, whereas in severe case, patient can feel shortness of breath and pneumonia [2]. The patient may be asymptomatic and such patients can spread COVID-19 to many persons as they are unaware of being infected with the disease and can be more casual when in contact with other persons. Thus, asymptomatic patients become a major reason for a speedy COVID-19 escalation. Automated diagnosis of COVID-19 at earlier stage can prevent the spread and damage control. This becomes more important by considering the fact that COVID-19 vaccination drive will take much time to vaccinate majority of the people across the globe. Primarily, real-time reverse transcription polymerase chain reaction (RT-PCR) method is commonly used to identify the COVID-19 across the world. However, PCR test takes few hours to confirm the presence/absence of COVID and has low sensitivity of 65–95% [3]. Thus, these circumstances increase the urgency of developing a reliable and automated artificial intelligence (AI) tool to effectively diagnose the COVID-19.

Several studies have been presented recently where researchers detect the COVID-19 disease using radiological images (chest X-ray or CT scan). Chest X-ray is also given more importance in this regard. The studies reveal the acute lung abnormalities after the 10 days of initial indications of the disease [4]. Kong [5] found the existence of peripheral ground glass and nodular opacities in lungs. Zhao [6] declared that mostly COVID-19 patients also have vascular expansion and consolidation in the lesion along with mixed ground glass opacities (GGO). Additionally, COVID-19 detection through radiological images has more sensitivity than COVID-19 detection through RT-PCR testing technique. Kim [7] analysis reveals the high diagnostic sensitivity of the CT scan images for patients in critical conditions as compared to RT-PCR. The reason is that the swab or nasopharyngeal sample is required for RT-PCR. If the sample taken from the patient does not contain detectable amount of corona virus, then this can lead toward failure in the COVID-19 detection, thus affecting the testing accuracy of COVID-19 diagnosis. Moreover, RT-PCR technique is costly and needs more time to produce the result. Thus, automated methods to analyze the chest X-ray for diagnosis can be used to overcome the issues associated with RT-PCR test [2].

Convolution neural networks (CNNs) have been employed to solve various image processing and computer vision tasks. The significance of deep learning models for object detection and classification motivated the research community to adapt the CNNs for various medical imaging applications, i.e., lung diseases classification, brain tumor detection, tumors detection in biological membranes and cells, etc. Recently, researchers have adopted different CNN models including ResNet, AlexNet, GoogleNet, and DenseNet for COVID-19 detection. Wang [8] presented a COVID-Net deep neural network to detect COVID-19 in CXR images. This model was evaluated on COVIDx dataset. Majeed [9] proposed a CNN-x model for four types of cases (normal, COVID, non-COVID viral infected, and bacterial infected) and compared the results of different architectures for each case. This model consists of four layers having 16 filters of three different sizes in each layer. Sethy [10] proposed a method to extract the features from several pretrained CNN models and later used support vector machine (SVM) to classify between the COVID and non-COVID infections. It was concluded that Resnet-50 along with the SVM outperforms all the comparative methods for classification in COVID-19 detection. Rehman [11] classified the features collected from different CNN architectures by applying the binary and multiclass classification. For binary classification, ResNet shows 98.7% accuracy, whereas MobileNet shows 97 and 80% accuracy for three and four classes classification, respectively. Zheng [12] developed a weekly supervised system based on deep learning for automatic detection of COVID-19 by using 3D chest CT volumes. Pretrained Unet was applied to segment each patient lungs region and then fed to 3D deep neural network DeCoVNet for COVID-19 detection. This algorithm achieved an accuracy of 90.8%. Ghoshal [13] reveals that there exists a strong relationship between uncertainty of model and accuracy of prediction. To estimate uncertainty in model, Bayesian convolutional neural network was trained on COVID-19 images using the transfer learning method. Estimating uncertainty in model is necessary to avoid COVID-19 misdiagnosis and allow radiologist in identification of false predictions and unknown cases. Apostolopoulos [14] conducted a comparative analysis to investigate the performance of different CNN architectures (VGG19, MobileNet v2, Inception, Xception, Inception-ResNetv2) with transfer learning for automatic detection, reliable features extraction, and low-cost diagnosis of COVID-19 on dataset which is gathered from Github repository and online Web sites. The results suggest that the MobileNet v2 achieves the best classification accuracy of 96.78% over the rest of the CNN models. Based on the results, it was inferred that CNN models have great significance in diagnosis of COVID-19. Ismael and Şengür (2021) employed a pretrained CNN model to classify between the COVID-19 and healthy chest X-ray images. More specifically, ResNet50 was used to extract the deep features that were then used to train the SVM for classification and achieved an accuracy of 94.7%. Alam et al. [15] employed a fusion of histogram of gradient and CNN for features extraction and used them to train a VGGNet based deep learning model for classification of COVID and non-COVID chest radiographs. This method employed a fivefold cross-validation approach for testing and achieved an accuracy of 99.4%, specificity and sensitivity of 95.7% and 93.6% respectively [15].

In this paper, we proposed a deep learning model to automatically detect the COVID-19 using the chest X-ray images. The proposed model employs the Inception-ResNetV2 architecture [16] containing the hybrid of residual layers and inception blocks. The proposed deep learning model has a flexible architecture having several filters that are changeable in different layers. The major contributions of the paper are as follows:

- We present an effective deep learning model to automatically detect the COVID-19 from chest radiograph images.
- We present a computationally efficient Inception-ResNetV2 deep learning model by using inception blocks in residual-inception.

Rest of the paper is organized as follows. Section 2 presents the methodology for COVID detection. Performance evaluation is provided in Sect. 3. In Sect. 3.4, analysis and discussion is provided, and finally, we conclude our work in Sect. 4.

2 Proposed Methodology

This section provides the architectural and implementation details of the proposed COVID-19 detection method.

2.1 Architectural Details

The proposed method employs the Inception-ResNetV2 pretrained hybrid convolutional neural network for COVID-19 detection. We exploited the benefits of CNN for generating effective feature maps from the chest radiograph images over other features extraction methods. Additionally, our Inception-ResNetV2 model is computationally efficient over other CNN models. The hybrid Inception-ResNetV2 model is comprised of 164 layers which has combination of residual connections and inception architecture with the ability to classify chest X-rays. The residual model is well known for the training of deep architectures, while inception model is renowned for multi-branch architecture. The network is trained on ImageNet database and takes an input image of size 299×299. Residual connections are pooled with different sized convolutional filters avoiding the degradation caused by deep structures and reducing the computational time [17]. Shown in Fig. 1 is the architecture of Inception-ResNetV2 model.

2.2 Implementation Details

The proposed model contains the pretrained Inception-ResNetV2 model which is fine tuned to identify the COVID-19 disease in X-ray images. All layers of model except last three layers are initialized by using pretrained Inception-ResNetV2, whereas last three layers are replaced to improve the model. The fully connected layer weights are set to 10. The model is trained on dataset by using Adam optimizer. The hyper-parameters used for the training are: initial learning rate = le-4, maximum epochs = 20, minimum batch size = 16, and data shuffling in every epoch. Moreover,

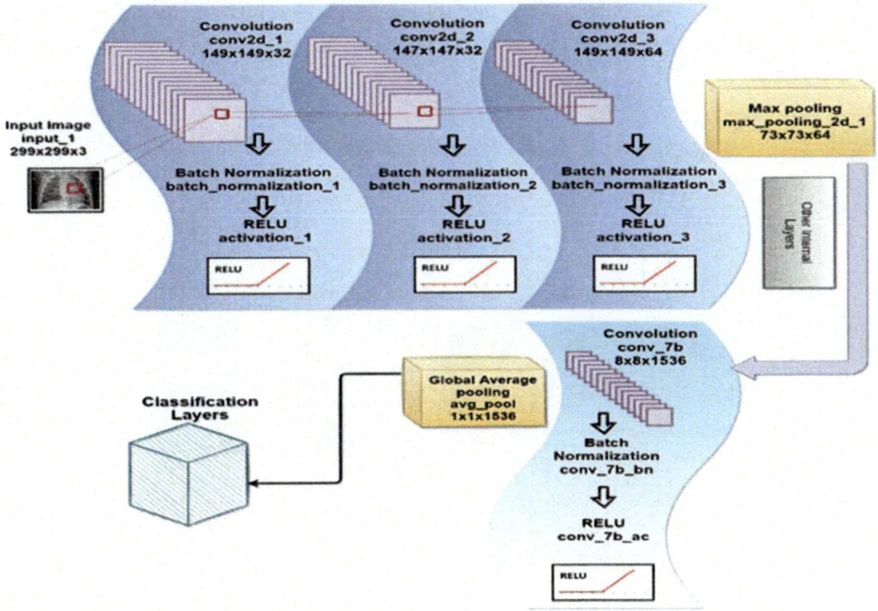

Fig. 1 Proposed Inception-ResNetV2 architecture

we employed the rotation, reflection, and shear methods for data augmentation. The input image is first resized to 299 × 299 dimension before feeding into our Inception-ResNetV2model.

Due to the limited set of X-ray images, K-fold cross-validation is used to randomly split the dataset into *k* partitions. In each fold, data is partitioned into training and testing data. We have divided the dataset into tenfold which enable us to fit the model to ten distinct training set and thus provide better performance estimation. K-fold validation is also used to reduce the biasing effect in the dataset. After dividing the data into testing and training portion, it is fed into Inception-ResNetV2 architecture that extracts the features from the chest X-rays and then fed to the classification layer. The global average pooling calculates the average value of each feature map and delivers the output for predictions to a fully connected layer. Output passes to predictions-softmax layer and then softmax assign decimal probabilities to every class and give output to classification Layer-predictions which is responsible for classification of image between the COVID-19 and non-COVID infections. Figure 2 shows the process of the classification.

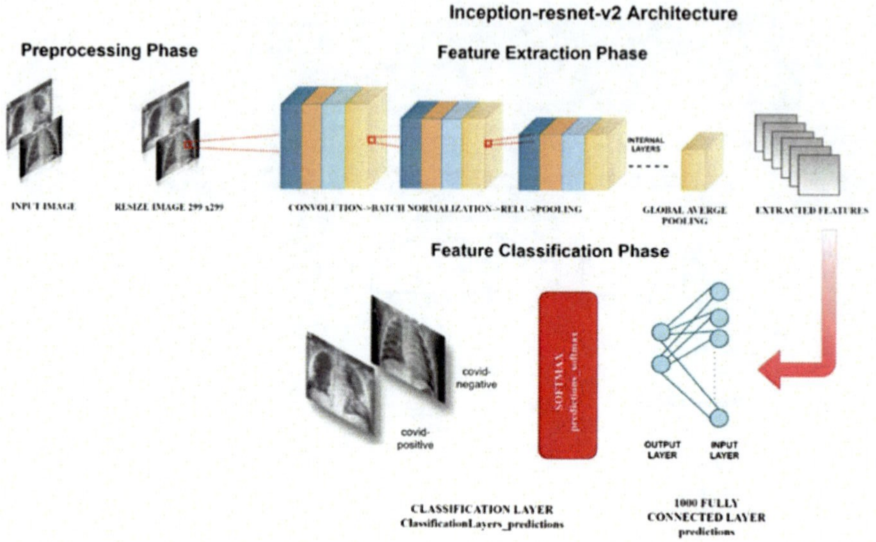

Fig. 2 Classification process

3 Experiments and Results

This section presents the results of the experiments used for performance evaluation. The details of the dataset are provided in this section. We visualize the training progress of each fold to analyze the entire process of training. We employed the accuracy, precision, recall, and F1-score as metrics for performance evaluation. The proposed model is implemented using MATLAB 2020a. We performed the experimentation on the system with the following specifications: 8 GB RAM, 1.61 GHz processor, Windows 10 operating system.

3.1 Dataset

Performance of the proposed method is evaluated on publicly available dataset of chest radiograph images [18] containing 468 COVID-19 cases. This dataset provides 468 X-ray images for COVID-19 patients. We used the 468 normal chest X-ray images from the Kaggle's chest X-ray images dataset [19] to create the final repository for experimentation. So, our dataset comprises of 936 images in total, and all images are of different dimensions. Therefore, we have resized all images to 299 × 299 dimension before feeding them into our model. Figure 3 shows the chest X-ray images of normal and COVID-19 infected patients.

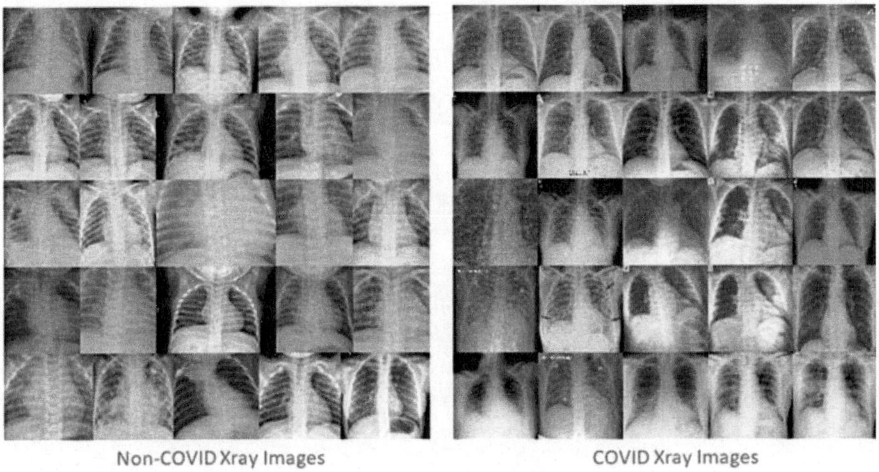

Fig. 3 X-ray images of non-COVID and COVID-19 infected patients

3.2 Performance Evaluation

We designed an experiment to measure the performance of our model for COVID detection. We measured the accuracy, precision, recall, and F1-score using the true positives (TP), false positives (FP), true negatives (TN), and false negatives (FN). In our case, TP represents the frames of COVID-19 infected images that are also detected as COVID-19 images, whereas FP represents the frames of non-COVID images that are falsely detected as COVID-19 images by our model. Similarly, TN represents the frames of non-COVID images that are also detected as non-COVID, whereas FN represents the frames of COVID-19 images that are falsely detected as non-COVID images by our model. We computed the accuracy, precision, recall, and F1-score as follows:

$$\text{Accuracy} = \frac{\text{TP} + \text{FP}}{\text{Total Positives} + \text{Total Negatives}} \tag{1}$$

$$\text{Precision} = \frac{\text{TP}}{\text{TP} + \text{FP}} \tag{2}$$

$$\text{Recall} = \frac{\text{TP}}{\text{TP} + \text{FN}} \tag{3}$$

$$\text{F1\_ score} = \frac{2 * (\text{Precision} * \text{Recall})}{\text{Precision} + \text{Recall}} \tag{4}$$

We obtained the precision of 92.6%, recall of 100%, specificity of 92%, F1-score of 96.15%, and accuracy of 96% for our method. From this experiment, we can

Table 1 Comparison with existing deep learning models

Architecture	Accuracy (%)
VGG19	90
DenseNet201	90
InceptionV3	70
ResNetV2	50
Proposed model	96

observe that our model correctly predicts all the COVID-19 cases, whereas only 4% of normal X-rays are classified as COVID-19 by our model. These results signify that the proposed method can reliably be used to detect the COVID-19 infections in the chest radiograph images.

3.3 Comparison with Existing Models

To evaluate the significance of the proposed model over existing DL models, we compared the performance of our model against the VGG19, DenseNet201, InceptionV3, and ResnetV2 DL models on the same dataset. The results in terms of accuracy are reported in Table 1. This experiment reveals that the ResNetV2 DL model performed the worst by achieving only 50% accuracy, whereas the proposed model performed the best and achieved an accuracy of 96%. It is important to mention that VGG19 and InceptionV3 also achieved better results and were joint second-best performers among all the DL models used in this experiment.

3.4 Discussion and Analysis

The results of different experiments performed in this paper revealed the significance of our Inception-ResNetV2 based model for reliable COVID-19 detection using radiological images. Moreover, it is important to mention that the inception blocks in residual-inception are computationally efficient than pure inception blocks. A filter-expansion layer is used for increasing the dimensionality reduced by each inception block. Expansion-layer is a convolutional layer without activation which expands the dimensions of the filter bank to equalize it with the depth of input.

Experimental results show that the proposed architecture has 0% false negative rate, which contributes to achieve 100% sensitivity. This level of sensitivity assures that the architecture will detect all COVID-19 patients correctly. RT-PCR testing technique is the most common technique used for the detection of COVID-19. As RT-PCR testing has a high false negative rate, thus our proposed architecture better addresses this issue.

Further analysis of results show that proposed architecture is able to successfully classify 92% of the normal X-ray images correctly while leaving only 8% of incorrect detection of normal cases. The proposed model along with current COVID testing techniques is able to help the physicians in better detection of COVID-19. Further evaluation of the proposed deep learning model for COVID-19 detection on a large-scale and more diverse dataset including cross-dataset evaluation can help to better generalize the proposed model for reliable COVID detection.

4 Conclusion

This paper has presented an effective and efficient deep learning-based Inception-ResNetV2 model for COVID-19 detection from chest radiograph images. We have presented a reliable COVID-19 detection system capable of accurately diagnosing the COVID-19 quickly. We evaluated the performance of our model on the publicly available COVID-19 X-ray images dataset. The accuracy of 96%, precision of 92.6%, and recall of 100% illustrate the effectiveness of the proposed model for COVID-19 detection. Our model will facilitate the doctors/physicians for automatic screening of COVID-19 cases using radiological images in quick time. In future, we plan to test the effectiveness of the proposed deep learning model for COVID-19 detection on a large-scale and more diverse dataset including cross-dataset evaluation.

References

1. WHO 2021 WHO. [Online] Available at: https://covid19.who.int/. Accessed 2021
2. Long C et al (2020) Diagnosis of the coronavirus disease (COVID-19): rRT-PCR or CT? Eur J Radiol 126:108961. https://doi.org/10.1016/j.ejrad.2020.108961
3. Radiologyassistant (2021) RadiologyAssistaant. [Online]. Available at: https://radiologyassistant.nl/chest/covid-19/covid19-imaging-findings. Accessed 2021
4. Ge H et al (2020) The epidemiology and clinical information about COVID-19. Nat Pub Health Emerg Collect 39:1011–1019. https://doi.org/10.1007/s10096-020-03874-z
5. Kong W, Agarwal PP (2020) Chest imaging appearance of COVID-19 infection. Radiol: Cardiothoracic Imaging 2(1). https://doi.org/10.1148/ryct.2020200028
6. Zhao W et al (2020) Relation between chest CT findings and clinical conditions of coronavirus disease (COVID-19) pneumonia: a multicenter study. Am J Roentgenol (AJR) 214(5):1072–1077. https://doi.org/10.2214/AJR.20.22976
7. Kim H et al (2020) Diagnostic performance of CT and reverse transcriptase-polymerase chain reaction for coronavirus disease 2019: a meta-analysis. Radiol Soc North Am: Radiol 296(3). https://doi.org/10.1148/radiol.2020201343
8. Wang L et al (2020) COVID-Net: a tailored deep convolutional neural network design for detection of COVID-19 cases from chest x-ray images. Sci Rep 10:19549. https://doi.org/10.1038/s41598-020-76550-z
9. Majeed T et al (2020) COVID-19 detection using CNN transfer learning from X-ray Images. medRxiv preprint. https://doi.org/10.1101/2020.05.12.20098954
10. Sethy PK, Behera SK (2020) Detection of coronavirus disease (COVID-19) based on deep features. Preprints 2020, 2020030300

11. Rehman A et al (2020) Improving coronavirus (COVID-19) diagnosis using deep transfer learning. medRxiv. https://doi.org/10.1101/2020.04.11.20054643
12. Zheng C et al (2020) Deep learning-based detection for COVID-19 from chest CT using weak label. IEEE Trans Med Imaging. https://doi.org/10.1109/TMI.2020.2995965
13. Ghoshal B, Tucker A (2020) Estimating uncertainty and interpretability in deep learning for coronavirus (COVID-19) detection. arXiv, eess. arXiv:2003.10769
14. Apostolopoulos ID, MPesiana TA (2020) COVID-19: automatic detection from x-ray images utilizing transfer learning with convolutional neural networks.. Phys Eng Sci Med. https://doi.org/10.1007/s13246-020-00865-4
15. Alam N-A-A, Ahsan M et al (2021) COVID-19 detection from chest X-ray images using feature fusion and deep learning. https://doi.org/10.3390/s21041480
16. Szegedy C et al (2017) Inception-v4, Inception-ResNet and the impact of residual connections on learning. arxiv journal arXiv:1602.07261v2
17. Elhamraoui Z (2020) Medium. [Online] Available at: https://medium.com/@zahraelhamraoui1997/inceptionresnetv2-simple-introduction-9a2000edcdb6
18. Cohen DJ (2020) github. [Online] Available at: https://github.com/ieee8023/covid-chestxray-dataset. Accessed Oct 2020
19. Mooney P (2020) Kaggle. [Online] Available at: https://www.kaggle.com/paultimothymooney/chest-xray-pneumonia. Accessed Oct 2020

Deep Learning-Based Sentiment Analysis on COVID-19 News Videos

Milan Varghese and V. S. Anoop

Abstract Coronavirus disease (COVID-19) has adversely affected all walks of human life. The whole world is confronting this deadly virus, and no country in this world remains untouched during this pandemic. There are several online news videos related to COVID-19 that are shared on various online platforms such as YouTube, DailyMotion, and Vimeo. There were several arguments on the genuineness of the contents, people watch them, share them, and most importantly express their views and opinions as comments on those platforms. Analyzing these comments can unearth the patterns hidden in them to study people's responses to videos on COVID-19. This paper proposes a deep learning-based sentiment analysis approach to people's response toward online COVID-19 video news. This work implements different deep learning approaches such as LSTM, Bi-LSTM, CNN, and GRU to classify sentiment from the comments collected from YouTube.

Keywords COVID-19 · Sentiment analysis · Computational social science · Deep learning · Long short-term memory · Bi-directional long short-term memory · Convolutional neural network · Gated recurrent unit

1 Introduction

Corona virus disease (COVID-19) has affected all walks of human life like never before. This deadly virus has its presence all over the world, and countries are utilizing their healthcare, technological, and administrative powers in its full swing to deal with it. As of June 17, 2021, a total of 176,531,710 confirmed cases and 38,26,181 deaths were reported globally as per the statistics published by the World Health Organization (WHO). From the early days, this virus causing the disease was reported in various parts of the world, several news articles and videos were published in news

M. Varghese (✉) · V. S. Anoop
Rajagiri College of Social Sciences (Autonomous), Kochi, Kerala, India
e-mail: msccs113@rajagiri.edu

V. S. Anoop
e-mail: anoop@rajagiri.edu

A. Ullah et al. (eds.), *Proceedings of International Conference on Information Technology and Applications*, Lecture Notes in Networks and Systems 350,
https://doi.org/10.1007/978-981-16-7618-5_20

and online video sharing platforms such as YouTube and Vimeo. While the genuineness of those videos published on such platforms remains a challenge [1, 2], people who watched them know about the symptoms, causes, and how the government is tackling the situation. People were not only watching these videos but also shared their views, opinions, and emotions as comments on these videos. These comments are good sources of information on which detailed analysis can be performed to identify the sentiment of people toward government policies and other initiatives [3, 4].

Sentiment analysis is an active research area in natural language processing (NLP) that attempts to analyze sentiment from large quantities of unstructured text to classify sentiment into negative, positive, and neutral [5, 6]. This area of artificial intelligence got significant attention from researchers and computational social scientists, and many approaches and techniques were reported in the literature for sentiment analysis. The people's opinions accumulated in the public platforms such as social networks are a gold mine to understand the social trends and reactions. Earlier sentiment analysis approaches were heavily dependent on rule-based mechanisms that were hand-crafted, and human-curated rules were used for classifying the sentiments [7–9]. Later, the exponential growth of data and the improved computing infrastructures paved the way to the popularity of machine learning-based approaches for sentiment analysis [10–12]. Deep learning-based sentiment analysis gained much momentum very recently [13–15]. The popularity of word embedding models such as word2vec [16] and Glove [17] fueled this growth as the features obtained from this model could work well with deep neural networks. Later, many such advanced pre-trained embedding models such as BERT got introduced that could yield better results for sentiment analysis tasks [18]. Deep neural network architectures such as long short-term memory (LSTM), bi-directional LSTM, convolutional neural networks (CNN), recurrent neural networks (RNN), gated recurrent unit (GRU) are being widely used for sentiment analysis and classification tasks [18–21].

This work proposes a sentiment analysis approach using deep neural networks for classifying sentiments expressed as comments to the COVID-19 news videos published on YouTube, which is one of the very popular online video sharing and social media platforms. Deep learning architectures such as LSTM, Bi-LSTM, CNN, and GRU are used to classify the sentiment of the comments expressed in plain text for the news videos. The classification approaches discussed in this paper will be highly useful to understand the trend, emotions, and opinions of the public toward the online news videos related to COVID-19.

The remainder of this article is organized as follows: Sect. 2 discusses some of the very recent and prominent work reported in the literature on sentiment analysis, Sect. 3 details our proposed approach, and Sect. 4 outlines the experiment conducted. In Sect. 5, the results are obtained, and a detailed discussion is presented, and in Sect. 6, the authors conclude the work and future research dimensions are presented.

2 Related Work

In the recent past, there were several approaches reported in the machine learning and natural language processing literature on sentiment analysis with varying degrees of success. This section details some of those prominent and highly related recent approaches to sentiment analysis. Ridhwan et al. [22] proposed an approach to understanding the public sentiment toward COVID-19. The authors have used VADER lexicon-based classifier which was used to classify the sentiments of the dataset, and a pre-trained RNN was used to find the correlations between real-life events and sentiment changes throughout the study [22]. Yadav et al. [23] reported a detailed review on different deep learning approaches for sentiment analysis in which different deep learning approaches such as CNN, RNN, LSTM, and GRU for sentiment analysis were discussed. Another overview paper on sentiment analysis using deep learning approaches was presented by Habimana et al. [24]. The review also highlights current issues and hypothesized possible solutions for the issue. The study suggested that we could use bi-directional encoder representations from transformers (BERT), sentiment-specific word, embedding models, cognition-based attention models, common sense knowledge, reinforcement learning, and generative adversarial networks for further developing good sentiment analysis models using deep learning [24].

Yang et al. [25] proposed an approach for sentiment analysis of e-commerce product reviews in Chinese. The authors have combined sentiment lexicon and deep learning which was used to enhance the sentiment features in the review. This approach combined deep learning models like convolutional neural networks (CNN) and attention-based bi-directional gated recurrent unit (BiGRU). This model was used for overcoming the shortcoming of existing sentiment analysis models in product reviews [25].

The proposed approach employs different deep neural architectures to analyze the sentiment of user comments and classify them into positive and negative. The details of the proposed approach are detailed in Sect. 3.

3 Proposed Approach

This section details the proposed approach that uses deep learning models for classifying the sentiment from the comments posted on YouTube for COVID-19 news videos. The overall workflow of the proposed approach is shown in Fig. 1. There are mainly five steps involved in the proposed approach. The first phase is to collect comments from YouTube, specifically the comments posted for COVID-19 related news published. As there are many videos published on COVID-19, it is nearly impossible and impractical to manually collect the comments. This work scraps the comments section of YouTube and automatically collects the comments for every video. As people can post the comments as they want, it will contain noises and

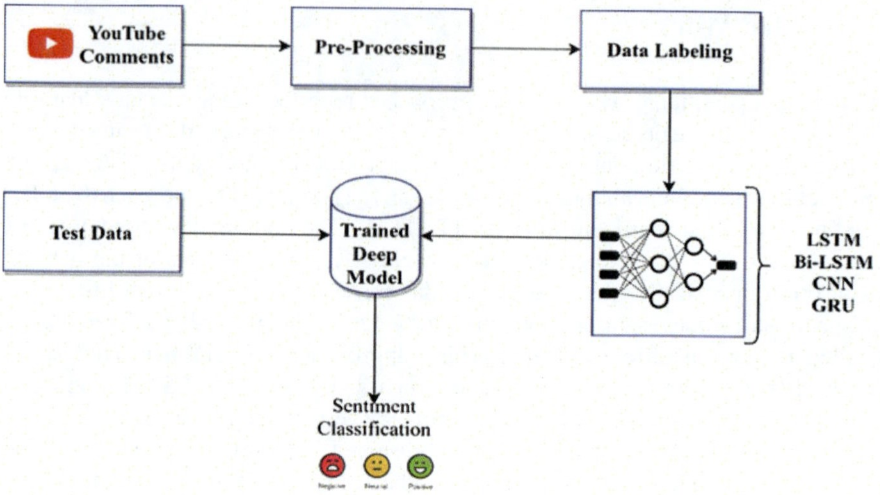

Fig. 1 Overall workflow of the proposed approach

unwanted symbols such as emojis and smileys. The comments may also contain URLs and numbers which should be removed before proceeding with the experiment. The second step of the proposed approach thus deals with the pre-processing of the collected comments. In machine learning, classification is a supervised learning algorithm that requires a large set of labeled data to work with. As the comments collected are unstructured and unlabeled, the third phase of the proposed approach is to manually label the data into positive and negative. Once the data is labeled, our experiment-ready dataset can be used for training different deep learning models used in this work such as LSTM, Bi-LSTM, CNN, and GRU. A very short description of each of these deep neural networks is as follows.

- ***Long Short-Term Memory***: Long short-term memory (LSTM) is an extension of recurrent neural networks but with a gating technique to capture dependencies which are long [26]. LSTM may be treated as a memory cell having two inputs-current input and previous hidden state. This memory cell decides which information must be persisted and which should be lost. LSTM is widely used in many areas in machine learning and natural language processing [27].
- ***Bi-Directional Long Short-Term Memory***: Bi-directional LSTM (BLSTM) network captures the structure of the sequence data in both forward and backward directions, whereas long short-term memory can do this only in the forward direction in a looped structure. Many machine learning and natural language processing applications such as sequence labeling and language modeling use this advantage of bi-directional long short-term memory.
- ***Convolutional Neural Network***: One of the popular neural network architectures that were initially proposed for image processing got significant attention and application in the text as well. Convolution can be seen as a simple application

that applies a filter to an input that in turn results in an activation [22, 23]. CNN is a type of multi-layer neural network that contains at least two hidden layers. The hidden layers are responsible for the tasks such as convolution and pooling.

- ***Gated Recurrent Unit***: Gated recurrent unit (GRU) on the other hand uses a simple gating function compared to long short-term memory and contains two types of gates for its operation—an update gate and a reset gate. The reset gate always takes input from the very previous hidden state and current input, and a sigmoid activation function is calculated.

The proposed approach implements the four deep neural network architectures outlined to classify the comments collected from COVID-19 news videos published on YouTube. The details of the experiment conducted with the proposed approach are discussed in Sect. 4.

4 Experiments

A detailed explanation of the experiments conducted using the proposed approach discussed in Sect. 3 is described here. The details on the dataset collection, pre-processing, and classifier training are discussed in detail. The proposed approach uses publicly available comments the users have generated for COVID-19 news videos on YouTube (https://www.youtube.com) platform. As the manual collection of comments is a herculean task, we have scrapped the comments using an automated process. For this, we have identified 100 YouTube videos on COVID-19 and the publicly available YouTube API and Python script for collecting the comments. We have scrapped a total of 101,000 publicly available comments from the videos. While posting the comments, the users may include emoji's, smileys, and other special characters, and removing them is crucial before training the machine learning models. This work uses a Python library called demoji, available at https://pypi.org/project/demoji/, for removing the emoji's from the plain text. In this work, we consider the comments posted in English, and the other language comments are removed from the experiment-ready copy of the dataset. This was done with the help of langdetect library available at https://pypi.org/project/langdetect/.

The other set of pre-processing such as removing the special characters and URLs is also carried out on the dataset and also expanded the contractions in the pre-processed comments to enhance the same. Contractions are words or combinations of words that are shortened by dropping letters and replacing them with an apostrophe. For example, "might've" should be replaced with "might have," "we've" should be replaced with "we have," and so on. This has been applied to all the datasets and has been done by collecting common contractions replacement details available for the English language. The punctuation in the text is also removed in the pre-processing step. After this step, the polarity value for each comment is calculated using the TextBlob library available at https://textblob.readthedocs.io/en/dev/, and the polarity category value is computed in such a way that if the polarity value is greater than zero, then the polarity category value is assigned as 1, otherwise, −1. The comments have been tokenized using Keras (https://keras.io/) tokenizer module, and after tokenizing, we got 69,989 unique tokens to work with. During the pre-processing stage, we have also removed the stop words and lemmatized the words using the natural language toolkit (NLTK) library available in Python. To create the feature vector to train the deep neural networks, this work makes use of Glove [17] pre-trained language model and glove.6B.300d.txt version of Glove is used which has 300 dimensions. Then, Keras embedding module is used for creating the word embedding layer for the deep neural network. Then, we trained all the deep neural architectures discussed in Sect. 3, such as the long short-term memory (LSTM), bi-directional long short-term memory (Bi-LSTM), convolutional neural networks (CNN), and gated recurrent unit (GRU).

5 Results and Evaluation

The results obtained from the experiment conducted as discussed in Sect. 4, using the proposed approaches outlined in Sect. 3, are presented in this section. Long short-term memory has recorded a test accuracy of 87.76% and test loss of 28.70% for 30 epochs. The graph that shows the model accuracy and the confusion matrix for LSTM is shown in Fig. 2a and b, respectively. The classification score for the LSTM model is given in Table 1. For the bi-directional long short-term memory (Bi-LSTM), the experiment has recorded a test accuracy of 87.93% and a test loss of 29.55%. The model accuracy and the confusion matrix for the Bi-LSTM model are shown in Fig. 3, and the classification scores are given in Table 2. The model accuracy and the confusion matrix for CNN is given in Fig. 4 and Table 3 respectively, and Fig. 5 and Table 4 shows the classification accuracy and confusion matrix of Gated Recurrent Unit (GRU) model.

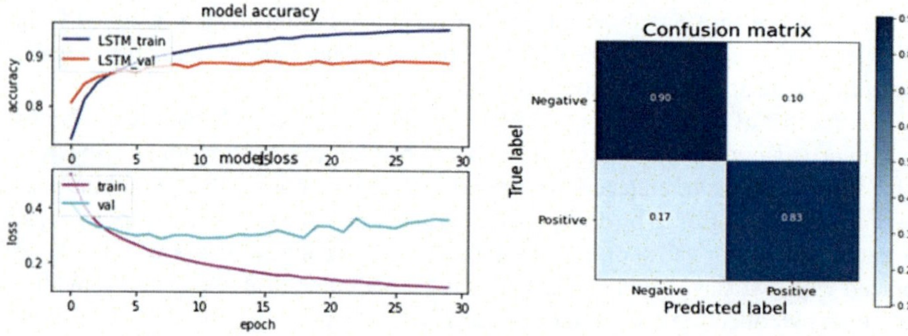

(a) (b)

Fig. 2 **a** Denotes the model accuracy and **b** denotes the confusion matrix for long short-term memory (LSTM)

Table 1 Classification scores for long short-term memory model

	Precision	Recall	F1 score	Support
Negative	0.90	0.90	0.90	15,953
Positive	0.84	0.84	0.83	9343
Accuracy	–	–	0.88	25,296
Macro average	0.87	0.87	0.87	25,296
Weighted average	0.88	0.88	0.88	25,296

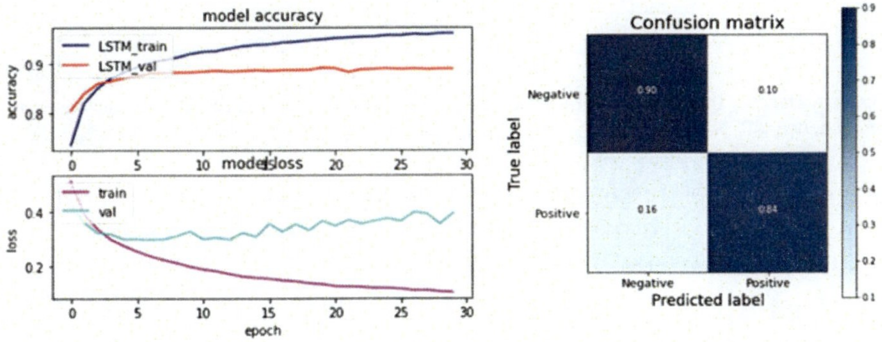

(a) (b)

Fig. 3 **a** Denotes the model accuracy and **b** denotes the confusion matrix for bi-directional long short-term memory (Bi-LSTM)

Table 2 Classification scores for bi-directional long short-term memory model

	Precision	Recall	F1 score	Support
Negative	0.91	0.90	0.90	15,953
Positive	0.83	0.84	0.84	9343
Accuracy	–	–	0.88	25,296
Macro average	0.87	0.87	0.87	25,296
Weighted avg	0.88	0.88	0.88	25,296

6 Conclusions and Future Work

This work proposed a deep learning approach for classifying the sentiment of people toward the COVID-19 videos published on YouTube. The proposed approach uses four deep neural network architectures for the classification and could report an acceptable level of accuracy. As the results are promising, the authors would like to explore further the dimensions of running the experiments with more data and also to use other pre-trained embeddings such as bi-directional encoder representations from transformers (BERT).

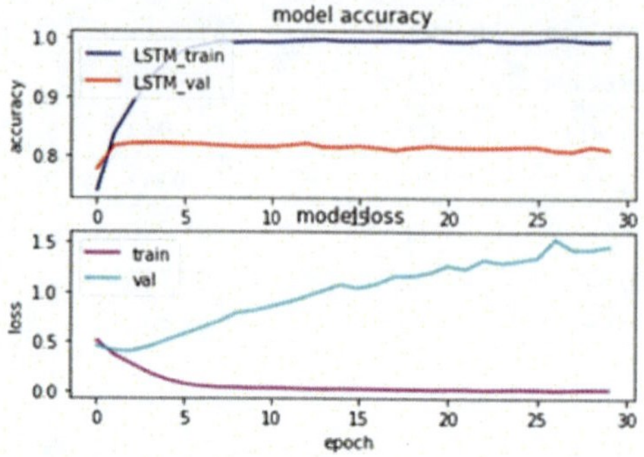

(a)

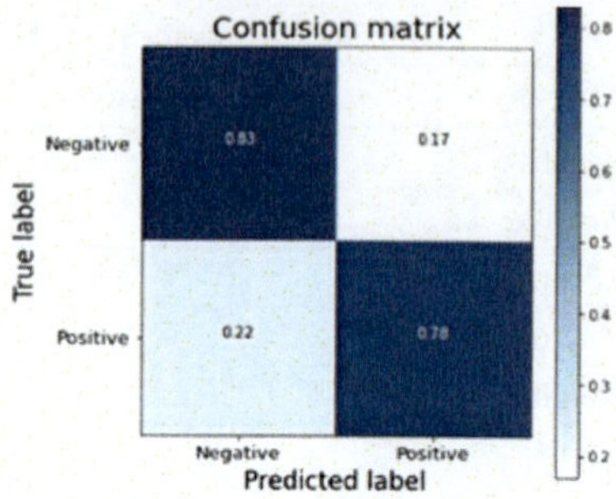

(b)

Fig. 4 **a** Denotes the model accuracy and **b** denotes the confusion matrix for the convolutional neural network (CNN) model

Table 3 Classification scores for convolutional neural network model

	Precision	Recall	F1 score	Support
Negative	0.86	0.83	0.85	15,953
Positive	0.73	0.78	0.75	9343
Accuracy	–	–	0.81	25,296
Macro average	0.80	0.80	0.80	25,296
Weighted avg	0.81	0.81	0.81	25,296

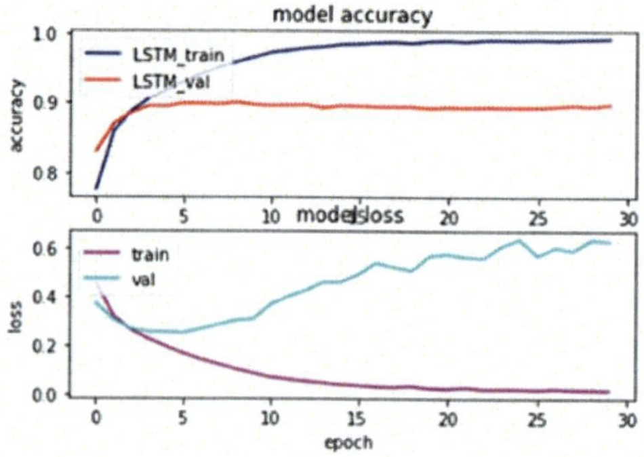

(a)

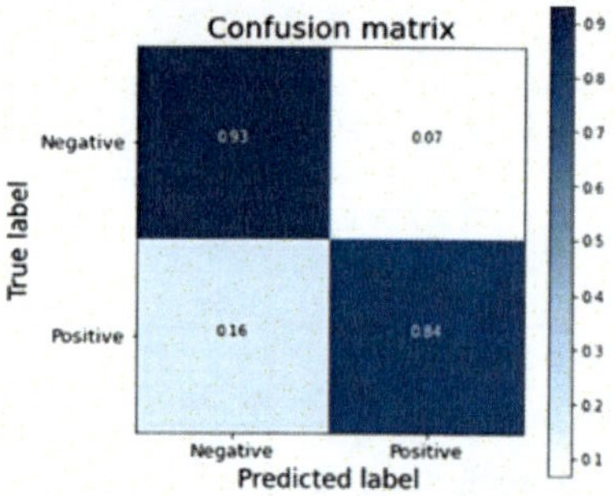

(b)

Fig. 5 **a** Denotes the model accuracy and **b** denotes the confusion matrix for the gated recurrent neural network (GRU) model

Table 4 Classification scores for gated recurrent unit model

	Precision	Recall	F1 score	Support
Negative	0.91	0.93	0.92	15,953
Positive	0.88	0.84	0.86	9343
Accuracy	–	–	0.90	25,296
Macro average	0.89	0.89	0.89	25,296
Weighted avg	0.90	0.90	0.90	25,296

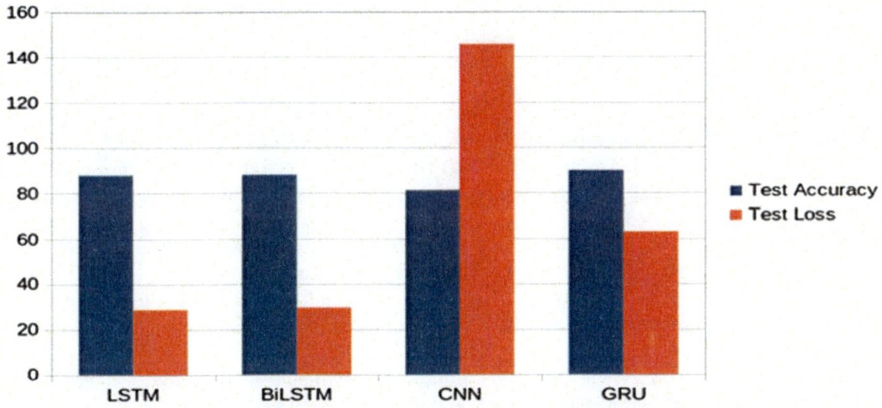

Fig. 6 Comparison of different test accuracies for deep learning models

References

1. Serrano JCM, Papakyriakopoulos O, Hegelich S (2020). NLP-based feature extraction for the detection of COVID-19 misinformation videos on Youtube. In: Proceedings of the 1st Workshop on NLP for COVID-19 at ACL 2020
2. Obadimu A, Khaund T, Mead E, Marcoux T, Agarwal N (2021) Developing a socio-computational approach to examine toxicity propagation and regulation in COVID-19 discourse on YouTube. Inf Process Manage 102660
3. Parabhoi LPL, Sahu RRSRR, Susan RSDDR, Kumar MKVDM, Seth MASMA, Parabhoi DPD (2021) YouTube as a source of information during the Covid-19 pandemic: a content analysis of YouTube videos published during January to March 2020
4. Aljameel SS, Alabbad DA, Alzahrani NA, Alqarni SM, Alamoudi FA, Babili LM, Aljaafary SK, Alshamrani FM (2021) A sentiment analysis approach to predict an individual's awareness of the precautionary procedures to prevent COVID-19 outbreaks in Saudi Arabia. Int J Environ Res Public Health 18(1):218
5. Liu B (2012) Sentiment analysis and opinion mining. Synth Lect Hum Lang Technol 5(1):1–167
6. Feldman R (2013) Techniques and applications for sentiment analysis. Commun ACM 56(4):82–89
7. Khan A, Asghar MZ, Ahmad H, Kundi FM, Ismail S (2017) A rule-based sentiment classification framework for health reviews on mobile social media. J Med Imaging Health Inform 7(6):1445–1453
8. Denecke K (2009) Are SentiWordNet scores suited for multi-domain sentiment classification? In: 2009 Fourth international conference on digital information management. IEEE, pp 1–6

9. Nakagawa T, Inui K, Kurohashi S (2010) Dependency tree-based sentiment classification using CRFs with hidden variables. In: Human language technologies: The 2010 Annual conference of the North American chapter of the association for computational linguistics, pp 786–794
10. Hasan A, Moin S, Karim A, Shamshirband S (2018) Machine learning-based sentiment analysis for twitter accounts. Math Comput Appl 23(1):11
11. Agarwal B, Mittal N (2016) Machine learning approach for sentiment analysis. In: Prominent feature extraction for sentiment analysis. Springer, Cham, pp 21–45
12. Pannala NU, Nawarathna CP, Jayakody JTK, Rupasinghe L, Krishnadeva K (2016) Supervised learning based approach to aspect based sentiment analysis. In: 2016 IEEE International conference on computer and information technology (CIT). IEEE, pp 662–666
13. Goodfellow I, Bengio Y, Courville A, Bengio Y (2016) Deep learning, vol 1, no 2. MIT Press, Cambridge
14. LeCun Y, Bengio Y, Hinton G (2015) Deep learning. Nature 521(7553):436–444
15. Li H (2017) Deep learning for natural language processing: advantages and challenges. Natl Sci Rev
16. Mikolov T, Grave E, Bojanowski P, Puhrsch C, Joulin A (2017) Advances in pre-training distributed word representations. arXiv preprint arXiv:1712.09405
17. Pennington J, Socher R, Manning CD (2014) Glove: global vectors for word representation. In: Proceedings of the 2014 conference on empirical methods in natural language processing (EMNLP), pp 1532–1543
18. Devlin J, Chang MW, Lee K, Toutanova K (2018) Bert: pre-training of deep bidirectional transformers for language understanding. arXiv preprint arXiv:1810.04805
19. Wang Y, Huang M, Zhu X, Zhao L (2016) Attention-based LSTM for aspect-level sentiment classification. In: Proceedings of the 2016 conference on empirical methods in natural language processing, pp 606–615
20. Chen T, Xu R, He Y, Wang X (2017) Improving sentiment analysis via sentence type classification using BiLSTM-CRF and CNN. Expert Syst Appl 72:221–230
21. Rhanoui M, Mikram M, Yousfi S, Barzali S (2019) A CNN-BiLSTM model for document-level sentiment analysis. Mach Learn Knowl Extr 1(3):832–847
22. Ridhwan KM, Hargreaves CA (2021) Leveraging Twitter data to understand public sentiment for the COVID-19 outbreak in Singapore. Int J Inf Manage Data Insights 100021
23. Yadav A, Vishwakarma DK (2020) Sentiment analysis using deep learning architectures: a review. Artif Intell Rev 53(6):4335–4385
24. Habimana O, Li Y, Li R, Gu X, Yu G (2020) Sentiment analysis using deep learning approaches: an overview. Sci China Inf Sci 63(1):1–36
25. Yang L, Li Y, Wang J, Sherratt RS (2020) Sentiment analysis for e-commerce product reviews in Chinese based on sentiment lexicon and deep learning. IEEE Access 8:23522–23530
26. Hochreiter S, Schmidhuber J (1997) Long short-term memory. Neural Comput 9(8):1735–1780
27. Akhil KK, Rajimol R, Anoop VS (2020) Parts-of-speech tagging for Malayalam using deep learning techniques. Int J Inf Technol 12(3):741–748

Collaborative Filtering Based Hybrid Music Recommendation System

Muhammad Umair Hassan, Numan Zafar, Haider Ali, Irfan Yaqoob, Saleh Abdel Afou Alaliyat, and Ibrahim A. Hameed

Abstract In the era of big data, valuable information is often hidden in a vast ocean of data. It is difficult for typical users to find useful information that they are looking for quickly and easily. Nowadays, fast-growing music streaming services have become more and more popular. Undoubtedly, many songs with millions of scales have brought a massive challenge to users' choices. A personalized music recommendation service can solve this problem. The method proposed in this paper is a hybrid music recommendation system that combined the collaborating filtering algorithm based on nearest neighbour users with the word embedding algorithm based on a heterogeneous text network to provide users with personalized recommendation service. The proposed approach is implemented successfully, and the project is working well. In the proposed work, the music recommendation system's rating is exponentially increased, the popularity of the items and the interest of users are enhanced. The overall recommendation ability of the music system is improved with the use of the proposed hybrid approach.

Keywords Collaborative filtering · Recommender system · Collaborative recommendation

1 Introduction

Recommender systems are generally utilized in the present systems with huge data and allow users to recommend systems through a high volume of data. These systems are being utilized progressively, more extensively, and concentrate on the zones'

M. U. Hassan · S. A. A. Alaliyat · I. A. Hameed (✉)
Department of Information and Communication Technology and Natural Sciences, Norwegian University of Science and Technology, Ålesund, Norway
e-mail: ibib@nntu.no

N. Zafar · I. Yaqoob
Department of Electrical and Computer Engineering, Clarkson University, Potsdam, USA

H. Ali
School of Information Science and Technology, Northwest University, Xian, China

A. Ullah et al. (eds.), *Proceedings of International Conference on Information Technology and Applications*, Lecture Notes in Networks and Systems 350,
https://doi.org/10.1007/978-981-16-7618-5_21

recommendations, such as music, books, news, film and Web-based business items (e-commerce). The working guideline of recommender systems is typically based on user preferences and comments and ratings on items made by users. By expanding the utilization of recommender systems, solving issues caused by algorithms utilized as a part of these systems or optimizing algorithms has been set among research questions to be worked seriously.

The most commonly used technique in music recommendation systems is collaborative recommendation. The main idea of collaborative recommendation approaches is to exploit information about the past behaviour or the opinions of an existing user community for predicting which items the current user of the system will most probably like or be interested in. The two primary approaches of collaborating filtering are user-based nearest neighbouring recommendation and item-based nearest neighbour recommendation. In the user-based approach of collaborating filtering to find similar users and the current user, the system seeks to find their likes and dislikes to predict current users' interests [1].

The embedded text-based approach for heterogeneous text is similar to the recommendation based on music content; the system can only find songs that recommend the same tastes as the user's past interests. For example, if the user recently listened to a song A, then according to the embedding of the heterogeneous text network, a song similar to the song A is found, a song that is the same as the user's past taste of interest [2]. In this paper, we try to combine the collaborating filtering algorithm based on heterogeneous text network to provide users with customized, professional and personalized music recommendation services, saving users from screening thousands of songs.

The rest of this paper is organized as follows. Section 2 presents the literature review of methods that we have followed in this study. In Sect. 3, we explained our proposed methodology and qualitative results, while Sect. 4 proposes the concluding remarks and our future plan.

2 Literature Review

The music recommendation systems are dedicated to extracting semantic and emotional information from various musical entities such as digital audio files, the main page of a band Web page, lyrics and behavioural activities concerning the user's current song. A fundamental approach in music information retrieval is to explain music through computational features [3]. This can be broadly divided into music content, music context, user attributes and user context, wherein the digital audio signal itself from the music file, i.e. the content embedded in the music file, is extracted based on the characteristics of the music content.

A few standard approaches to audio feature extraction on frame and block levels also advance similarity measures using features such as Mel frequency cepstral coefficients (MFCCs) [4]; this method is used for odelling speech to model music. For forming MFCCs, they examine two presumptions: the Mel recurrence scale's utilization to model the spectra and the utilization of the discrete cosine transform to decorate the Mel spectral vector. Moreover, block-level feature is a new and more robust feature for automatic tag prediction. A work Seyerlehner et al. [5] has compared the proposed block-level features sets to standard features sets for the automatic tag classification. In Ref. [6], pitch-class profiles are a group of pitches identified by intervals of multiple numbers of an octave treated together as an equivalence class.

The feature extraction based on the music context, Knees et al. [7] provide an overview that utilizes contextual data by emphasizing methods to explain similarity measures between artists or individual tracks. They primarily suggest the music context, which includes information on and identified with music artists and pieces. Moreover, according to Knees et al. [8], the artist's background aims to develop automatic methods to find semantic relations between musical entities by analysing text from the Web. Extraction of information identified with music artists from Web pages has been presented and implemented through music information retrieval, Web mining and information visualization by Schedl [9].

Adomavicius and Tuzhilin [10] defined collaborative filtering based on user-based systems and evaluated a user's interest for an item using the ratings for the item by other users, called neighbours, that have similar rating patterns. Schedl et al. [11] examine the user-centric retrieval and reflect on why the breakthrough in this field has not been accomplished yet.

In the rapid development of the recommendation system, there is a need to combine various recommended methods to improve its effectiveness. For example, a method based on the music content itself and a Web feature based on the music context is a hybrid music recommendation method presented in work by Schedl et al. [12]. Another hybrid music recommendation method based on content-based and collaborative filtering methods by Yoshii et al. is introduced in Ref. [13].

Wang and Wang [14] proposed a novel model for content-based recommendation model based on the probabilistic graphical model and the deep belief network, making the two stages into an automated procedure that at the same time learns features from audio content and makes personalized recommendations by maintaining the integrity of the specifications.

Yun [15] presented a hybrid method based on music genres and collaborative filtering. They calculated the geographical preference, emotional preference and style preference of music genres, weighted their integration, and then weighted integration with their collaborative filtering recommendation system to recommend mixed results to users instead of using them individually. Zhang [16] first categorized the theme of music labels and then tapped real-time user preferences from the user's music playlists and combined with traditional collaborative filtering methods. Xie [17] thoroughly considered the user's information and behaviour and combined collaborative filtering and user's demand preference.

3 Proposed Methodology

This section explains a hybrid approach of collaborating filtering algorithms based on nearest-neighbour users with the word embedding algorithm based on a heterogeneous text network.

3.1 Collaborative Filtering Algorithm Based on Nearest Neighbour

Collaborative filtering [1] is almost a household name in the recommendation community. Its superior performance has survived the precipitation of time, and it is a classic recommendation algorithm. Collaborative filtering can be roughly divided into memory-based and model-based collaborative filtering. Collaborative filtering based on the model can be divided into two categories: collaborative filtering based on singular value decomposition (SVD) and collaborative filtering based on implicit factor matrix decomposition. Its characteristic is that most of the time overhead is concentrated on the training model, i.e. the training time is long, but the forecast is fast. The memory-based collaborative filtering can also be divided into two categories: collaborative filtering based on nearest neighbours and collaborative filtering based on item nearest neighbours. It is characterized by very little time focusing on the training model, almost no, because there is no model, i.e. the training time is short, but the forecast is slow. The system will use the collaborative filtering algorithm based on the nearest neighbours' advantages and disadvantages as the primary music recommendation algorithm. For each currently active user, collaborative filtering based on nearest neighbours first finds the K-nearest neighbours closest to it, and then, based on these recent neighbours, aggregates scores for songs that the currently active user has not heard, and finally scores them. The highest previous songs are recommended for the currently active user. This is the core idea of collaborative filtering based on nearest-neighbour users, while Table 1 represents the user score matrix.

It can be seen in the user music rating's matrix table that there are ten users and ten songs, each of which corresponds to the user's rating information for each song, and rating information of 0 indicates that the user has not yet scored the song. For example, for the current user 6, the system uses a simple Euclidean distance formula to calculate the similarity:

$$S_{i,j} = \sqrt{(x^{(i)} - x^{(j)})^2} \quad (1)$$

However, its shortcomings are also apparent. For example, there are few users or even no users in the initial stage of system operation. Therefore, this system is challenging to recommend collaborative filtering according to the nearest neighbours.

Table 1 User score matrix

Users/songs	A	B	C	D	E	F	G	H	I	J
1	1	2	5	0	0	3	4	0	0	5
2	0	0	1	0	0	0	0	0	0	0
3	1	2	5	4	0	2	3	5	5	5
4	5	0	2	1	0	0	0	0	0	0
5	0	0	0	0	5	0	0	0	0	2
6	4	5	0	0	0	0	5	0	0	0
7	5	0	0	0	0	0	0	0	0	0
8	3	4	5	0	0	0	5	0	0	0
9	0	0	0	1	0	0	5	0	0	0
10	1	0	4	5	2	4	0	0	5	2

3.2 Word Embedding Algorithm Based on Heterogeneous Text

The meaningful and effective low-dimensional representation of learning text is a crucial prerequisite for many machine learning tasks such as text classification, clustering and information retrieval. The purpose of embedding is to get a low-dimensional vector representation of things. The proposed work mainly constructs a network of lyrics to lyrics and a network of lyrics to documents, as illustrated in Fig. 1. The system builds a text network between lyrics and lyrics. The lyrics symbiosis network is recorded as $G_{ww}(V, E_{ww})$, which captures the symbiosis of lyrics within a particular context where V is the vocabulary representing all words, E_{ww} represents a collection of all lyrics. The edge weights w_{ij} indicate the number of times the lyrics v_i and v_j cooperate with the context in a specific window size.

For building a textual network of lyrics to documents, remember that the lyrics document network is $G_{wd}(V \cup D, E_{wd})$, which is a typical bilateral network, where

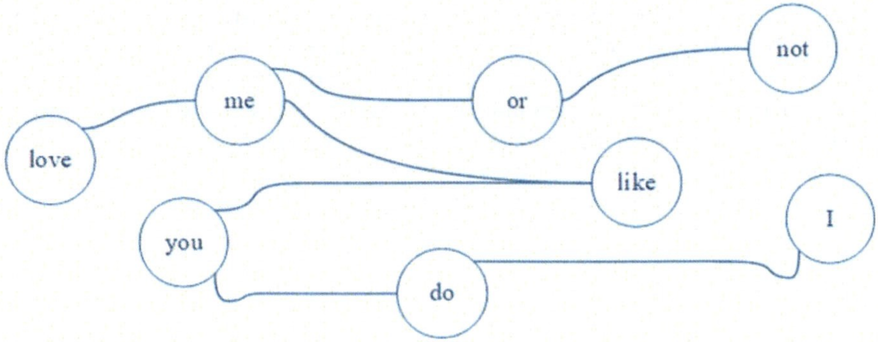

Fig. 1 Lyrics to lyrics edges

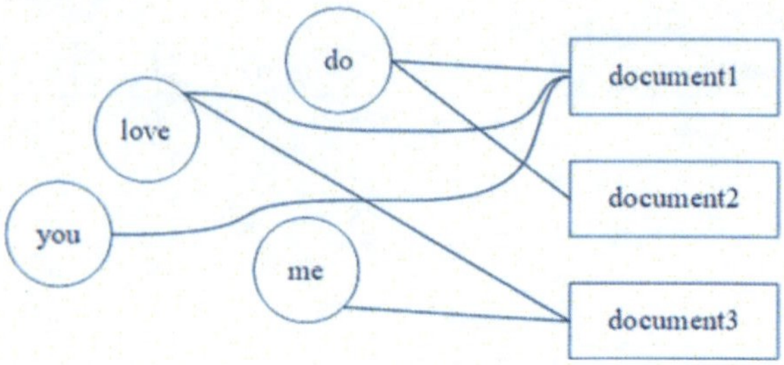

Fig. 2 Lyrics to document edge map

V is also a collection of lyrics that, as the name suggests, represents a collection of documents D. Moreover, the E_{wd} edges represent the collection of lyrics to the edge of the document. Here, the weight w_{ij} from the lyrics v_i to the document d_j is defined as the number of times the lyrics v_i appear in the document d_j. See Fig. 2.

Here, the heterogeneous text network combines the lyrics to the lyrics text network and the lyrics to the document text network. For a given heterogeneous text network, the goal of word embedding is to learn the low-dimensional vector representation of the word, and at the same time, the vector representation of the document is learned. As for how to learn? The system begins by embedding one of the following bilateral networks.

Remember the bilateral network $G = (V_A \cup V_B, E)$, where V_A and V_B are two disjoint sets of different types of nodes E are the edges between them. As mentioned in the method of predictive text prediction [2] embedded through large-scale heterogeneous text networks, the system first defines a conditional probability from a point v_i in a set V_A to a point v_j in a set V_B:

$$p(v_j|v_i) = \frac{\exp(\vec{u_i}^T \cdot \vec{u_j})}{\sum_{k \in B} \exp(\vec{u_i}^T \cdot \vec{u_k})} \tag{2}$$

Among them, $\vec{u_i}$ sets the embedded v_i nodes in V_A, and $\vec{u_j}$ is the embedded vector of v_j nodes in the set V_B. Then, for each node v_i in the set V_A, the system can obtain a conditional probability distribution $p(\cdot|v_i)$ about the set V_B to use its conditional probability distribution to represent the corresponding second-order proximity. According to Ref. [2], the system can make the conditional probability distribution as close as possible to the empirical distribution, thus preserving the second order of the network structure. Proximately, here KL divergence relative entropy is used to measure the similarity of the two distributions, as follows:

$$D(P||Q) = \sum (P(x) \log(P(x)/Q(x))) \tag{3}$$

where Q is the probability distribution, P is the true distribution and $D(P\|Q)$ is the KL divergence relative entropy representing the Q distribution to fit the Q distribution, resulting in information loss. Here, the system uses the probability distribution to fit the empirical distribution, namely:

$$O = \sum_{i \in A} \lambda_i D(\hat{p}(\cdot|v_i)||p(\cdot|v_i)) \tag{4}$$

where O is the objective function of the system and λ_i is the importance of node throughout the network, i.e. see (5).

$$\lambda_i = \sum\nolimits_{j \in B} w_{ij} \tag{5}$$

The empirical distribution can be defined as:

$$\hat{p}(v_j|v_i) = \frac{w_{ij}}{\lambda_i} \tag{6}$$

Similarly, as mentioned in (2), removing some constants, the objective function of this system can be abbreviated as:

$$O = -\sum_{(i,j) \in E} w_{ij} \log(p(v_j|v_i)) \tag{7}$$

The simplified process in which constants are removed is:

$$\begin{aligned} O &= \sum_{i \in A} \lambda_i D(\hat{p}(\cdot|v_i)||p(\cdot|v_i)) \\ &= \sum_{i \in A} \lambda_i \sum_{j \in B} \hat{p}(v_j|v_i) \log(\hat{p}(v_j|v_i)/p(v_j|v_i)) \\ &= \sum_{i \in A} \lambda_i \sum_{j \in B} \hat{p}(v_j|v_i)[\log(\hat{p}(v_j|v_i)) - \log(p(v_j|v_i))] \\ &= \sum_{i \in A} \lambda_i \sum_{j \in B} \hat{p}(v_j|v_i) \log(\hat{p}(v_j|v_i)) - \hat{p}(v_j|v_i) \log(p(v_j|v_i)) \end{aligned} \tag{8}$$

Omitting some const, then it will become as:

$$\begin{aligned} &= \sum_{i \in A} \lambda_i \sum_{j \in B} - \log(p(v_j|v_i)) \\ &= \sum_{i \in A} \lambda_i \sum_{j \in B} - \log(p(v_j|v_i)) \\ &= - \sum_{(i,j) \in E} w_{ij} \log(p(v_j|v_i)) \end{aligned} \tag{9}$$

For such an objective function, since the logarithmic function is a convex function, the system can use stochastic gradient descent to optimize training and learning. The optimization process is:

$$
\begin{aligned}
O &= -\sum_{(i,j)\in E} w_{ij} \log(p(v_j|v_i)) \\
&= -\sum_{(i,j)\in E} w_{ij} \log\left(\frac{\exp(\overrightarrow{u_i}^T \cdot \overrightarrow{u_j})}{\sum_{k\in B} \exp(\overrightarrow{u_i}^T \cdot \overrightarrow{u_k})}\right) \\
\mathrm{Cost}(i,j) &= -w_{ij} \log\left(\frac{\exp(\overrightarrow{u_i}^T \cdot \overrightarrow{u_j})}{\sum_{k\in B} \exp(\overrightarrow{u_i}^T \cdot \overrightarrow{u_k})}\right) \\
&= -w_{ij}\left[\log\left(\exp(\overrightarrow{u_i}^T \cdot \overrightarrow{u_j})\right) - \log\left(\sum\nolimits_{k\in B} \exp(\overrightarrow{u_i}^T \cdot \overrightarrow{u_k})\right)\right] \\
&= -w_{ij}\left[\overrightarrow{u_i}^T \cdot \overrightarrow{u_j} - \log\left(\sum\nolimits_{k\in B} \exp(\overrightarrow{u_i}^T \cdot \overrightarrow{u_k})\right)\right] \\
\frac{\partial \mathrm{Cost}(i,j)}{\partial \overrightarrow{u_i}} &= -w_{ij}\left[\overrightarrow{u_j} - \frac{1}{\sum_{k\in B} \exp(\overrightarrow{u_i}^T \cdot \overrightarrow{u_k})} \sum\nolimits_{k\in B} \exp(\overrightarrow{u_i}^T \cdot \overrightarrow{u_k})\overrightarrow{u_k}\right] \quad (10)
\end{aligned}
$$

3.3 *Proposed Hybrid Approach and Results*

In this system, the user's implicit feedback collectively includes the user's play record, the user's download record, and the user's favourite record of these three implicit "score" to the user based on the nearest neighbour user's collaborative filtering is recommended. Then, we use the heterogeneous text network composed of lyrics documents to learn similar songs that have been embedded for a mixed recommendation.

For the collaborative filtering based on the user's nearest neighbour, the system firstly acquires the five information lists of the user's download record list, user's play record list, user's favourite record list, user information list and song information list from the database. Then, the system builds a "score" matrix of user songs. In this process, since the system does not require the user to give an individual "score" to the song explicitly, the system constructs a user frequency matrix to approximate the user. Rating, for some systems, it is impossible to obtain the user's rating for specific items, but the system can see the user's behaviour to reflect the user's "score". For example, a user often listens to a song. Then, the system can infer that the user is very likely to like the song. In Table 2, the system uses the user information list, the song information list and the download record information list to maintain a list of song download information for each user, which stores ID information of the songs downloaded by the user, such as user songs. The download mapping is available in Table 2.

Similarly, for the user song collection mapping matrix, the system also maintains the songs information list for each user, which stores the user's favourite song ID information, such as user song collection mapping in Table 3. One of the differences is that the user's songs play the mapping matrix. This is because the system has to store additional songs on it.

Table 2 User songs download mapping

Users/songs	A	B	C	D	E	F	G	H	I	J
1	1	2	5	–	–	–	–	–	–	–
2	0	10	1	2	3	4	5	6	9	51
3	1	2	5	4	0	21	13	–	–	–
4	5	0	2	1	10	20	50	11	42	66
5	0	1	2	3	4	5	6	7	8	9
6	1	0	3	2	5	4	6	7	9	8
7	0	1	2	3	4	5	6	8	7	9
8	1	2	0	3	7	5	6	4	8	9
9	0	1	2	3	4	5	6	8	9	7
10	1	0	2	3	5	4	6	7	8	9

Table 3 User songs collection mapping

Users/songs	A	B	C	D	E	F	G	H	I	J
1	0	1	2	3	4	5	6	8	9	7
2	0	10	1	2	3	4	5	6	9	51
3	1	2	5	4	0	21	13	–	–	–
4	5	0	2	1	10	–	–	–	–	–
5	0	1	2	3	4	5	6	7	8	9
6	1	0	3	2	5	4	6	7	9	8
7	0	1	2	3	4	5	6	8	7	9
8	1	2	0	3	7	5	6	4	8	9
9	0	1	2	3	4	5	6	8	9	7
10	1	2	5	–	–	–	–	–	–	–

The system can then build a user's song "score" matrix by constructing a suitable user song download mapping information matrix, a user song collection mapping information matrix and a user song play mapping information matrix. After obtaining the "score" matrix, the system can construct the K-nearest neighbour matrix of the user according to this matrix, i.e. calculating the similarity between users according to the "score" matrix of the user song to find each user.

So, if the number of songs N in the current user's favourite list has exceeded 10, then the recent playlist is useless at this time. Similarly, if the list of favourites is less than the threshold, it is necessary to find the difference in the most recent playlist, and the system will also randomly select the difference. Simultaneously, for the problem of not discovering new tastes, collaborative filtering can also be justified by collaborative filtering since collaborative filtering can recommend the new favourite songs of the nearest neighbour user to the current user. Here, a simple splicing method is used for recommending, namely simple splicing based on the recommendation result generated by the collaborative filtering algorithm of the nearest neighbour user and the recommendation result generated by the word embedded algorithm based on the heterogeneous text network, to generate the final result for the currently activated user recommended results.

4 Conclusion and Future Work

In this paper, we combine the collaborating filtering algorithm based on nearest neighbour users with the word embedding algorithm based on a heterogeneous text network to provide users with personalized recommendation service and develop an implementation method using the software engineering development method. The music recommendation system demonstrates the "Track Stacking" algorithm. From the perspective of system developers, although the system has relatively reasonable control in terms of self-adaptation and cross-plat formed, there is a lack of stress in testing and comprehensive considerations for performance requirements such as system security and high concurrency. From the algorithm's perspective, the collaborative filtering of the system uses a memory-based approach, and the collaborative filtering algorithm based on the model is still to be further explored. For word embedding based on heterogeneous text networks, the system only considers embedding between the same language models but lacks further considerations embedded between multiple language models. The proposed approach is implemented successfully, and the project is working well. Moreover, there are still considerable challenges in the continuation of automated music playlists. Therefore, in the coming period, together with recommendation system enthusiasts and researchers worldwide, attention will be focused on the continuation of automated playlists.

References

1. Shi Y, Larson M, Hanjalic A (2014) Collaborative filtering beyond the user-item matrix: a survey of the state of the art and future challenges. ACM Comput Surv (CSUR) 47(1):1–45
2. Tang J, Qu M, Mei Q (2015) Pte: predictive text embedding through large-scale heterogeneous text networks. In: Proceedings of the 21th ACM SIGKDD international conference on knowledge discovery and data mining, pp 1165–1174
3. Knees P, Schedl M (2015) Music retrieval and recommendation: a tutorial overview. In: Proceedings of the 38th International ACM SIGIR conference on research and development in information retrieval, pp 1133–1136
4. Logan B (2000) Mel frequency cepstral coefficients for music modeling. In: ISMIR, vol 270, pp 1–11
5. Seyerlehner K, Widmer G, Schedl M, Knees P (2010) Automatic music tag classification based on block-level. In: Proceedings of sound and music computing 2010
6. Purwins H (2005) Profiles of pitch classes-circularity of relative pitch and key: experiments, models, music analysis, and perspectives
7. Knees P, Schedl M (2013) A survey of music similarity and recommendation from music context data. ACM Trans Multimedia Comput Commun Appl (TOMM) 10(1):1–21

8. Knees P, Schedl M (2011) Towards semantic music information extraction from the web using rule patterns and supervised learning. In: Workshop on music recommendation and discovery, pp 18–25
9. Schedl M (2008) Automatically extracting, analyzing, and visualizing information on music artists from the World Wide Web. Ph.D. Thesis, Johannes Kepler University Linz, Austria
10. Adomavicius G, Tuzhilin A (2011) Context-aware recommender systems. In: Recommender systems handbook. Springer, Boston, MA, pp 217–253
11. Schedl M, Flexer A, Urbano J (2013) The neglected user in music information retrieval research. J Intell Inf Syst 41(3):523–539
12. Schedl M, Schnitzer D (2013) Hybrid retrieval approaches to geospatial music recommendation. In: Proceedings of the 36th international ACM SIGIR conference on research and development in information retrieval, pp 793–796
13. Yoshii K, Goto M, Komatani K, Ogata T, Okuno HG (2006) Hybrid collaborative and content-based music recommendation using probabilistic model with latent user preferences. In: ISMIR, vol 6, pp 296–301
14. Wang X, Wang Y (2014) Improving content-based and hybrid music recommendation using deep learning. In: Proceedings of the 22nd ACM international conference on multimedia, pp 627–636
15. Yun K (2016) Research and implementation of personalized music recommendation system based on mixed mode. Computer Application Technology, Central China Normal University, Wuhan
16. Zhang J (2017) Design and implementation of label-based music recommendation system. Beijing University of Posts and Telecommunications, Computer Science and Technology, Beijing
17. Xie W (2016) Design and implementation of personalized music recommendation system based on domain ontology. Computer Science and Technology, Zhejiang University, Zhejiang

Analysis of Generative Adversarial Networks for Data-Driven Inverse Airfoil Design

Priyam Gupta, Prince Tyagi, and Raj Kumar Singh

Abstract Data-driven methods have led to new approaches in the field of aerodynamic design and have recently found success in inverse design applications. The conventional inverse design methods are analytically complex and mathematically demanding to formulate. This research attempts to perform inverse airfoil design using generative adversarial networks (GANs) with the objective to generate airfoil shapes that produce desired pressure distribution at given flow conditions. The Convolutional Neural Network-based generator extracts features from the pressure coefficient profiles and predicts the corresponding airfoil shape coordinates. These deep ConvNet structures eliminate the problems posed by shape parameterization in classical methods and extract patterns from the data with finer details. This work examines the performance of three advanced generative adversarial network architectures to obtain a model which is stable, computationally efficient and has competitive prediction accuracy. The candidate GANs include Wasserstein GAN, Boundary Seeking GAN and Bidirectional GAN. The networks are trained on a database of airfoil shapes and pressure coefficient distribution. It is shown that boundary seeking generative adversarial network produces highly accurate results and is computationally least expensive to train.

Keywords Machine learning · Generative adversarial networks · Inverse design

1 Introduction

Inverse aerodynamic design is a category of design optimization techniques wherein the objective is to design aerodynamic shapes which produce desired pressure or velocity distribution at given flow conditions. There has been a significant advancement in computational power and computational fluid dynamics (CFD) algorithms,

P. Gupta · P. Tyagi (✉) · R. K. Singh
Delhi Technological University, Shahbad Daulatpur, Main Bawana Road, New Delhi, Delhi 110042, India

R. K. Singh
e-mail: rajkumarsingh@dce.ac.in

A. Ullah et al. (eds.), *Proceedings of International Conference on Information Technology and Applications*, Lecture Notes in Networks and Systems 350,
https://doi.org/10.1007/978-981-16-7618-5_22

still, the heavy cost of high-fidelity simulation demands more efficient design optimization techniques. The forward aerodynamic shape optimization approach involves evaluating the geometry for the desired parameter using CFD simulations, followed by manipulating it until the desired parameter has been optimized [1]. This morphing of shape in every iteration makes the process computationally expensive. The inverse design methods are an efficient alternative to the time-consuming iterative process of analysing several airfoil shapes to obtain the one with desired characteristics.

The inverse aerodynamic design can be widely classified as surface flow design and flow-field design [2]. While the surface flow design needs the desired aerodynamic property specified on the surface, the flow-field design incorporates global flow-field features for shape design constraints. There are a large number of inverse methods for surface flow design and most of them are considered analytically complex and mathematically demanding to formulate [3]. This demands a computationally efficient and tractable inverse design method. A good inverse aerodynamic design model should be able to generate smooth-contoured and manufacturable designs within the enforced constraints and should avoid exploiting the weakness of the problem formulation [4].

There has been a significant advancement in the application of machine learning and data-driven algorithms in recent years due to unprecedented volumes of data and advanced computing power. The field of fluid mechanics is also incorporating these advanced algorithms for flow-field prediction, reduced-order and turbulence closure modeling [5]. Basu et al. [6, 7] conducted a comprehensive study of machine learning algorithms for pressure reconstruction framework from PIV velocity data and observed a high degree of accuracy in the pressure distribution. Generative adversarial networks (GANs) [8] have achieved impressive results in the image to image translation and have shown to be effective in capturing the essential features in training data. GANs have also been used for a one to one mapping from a parameterized supercritical airfoil to its corresponding transonic flow-field profile over the parametric space [9]. They have found effective use in supervised as well as unsupervised super-resolution reconstruction of turbulent flows [4, 10]. There is a recent surge in interest to study deep learning models for aerodynamic design applications. Chen et al. [11] implemented Bezier GANs to learn low-dimensional latent representation, encoding the variation in aerodynamic shapes. Du et al. [12] utilized a B-Spline-based generative adversarial network parameterization method along with a multi neural network-based surrogate model for predicting aerodynamic coefficients which was demonstrated on an airfoil shape optimization problem. Yilmaz et al. [4] implemented Conditional GANs and CNN based framework for surface flow inverse airfoil design conditioned on desired stall regions and downsampled desired drag polar data.

Motivated by the impressive performance of generative adversarial networks for image to image translation and observing the need for a GAN architecture that is stable to train and requires less computation, and the objective of this work is to investigate the performance of different GAN architectures for data-driven inverse airfoil design. This study examines three GAN architectures and performs a side-by-side comparison. The candidate GAN architectures are Wasserstein GAN (WGAN) [13], Bidirectional GAN (BiGAN) [14] and Boundary seeking GAN (BGAN) [15]. These are selected on the basis of their promising performance and enhanced training stability over the vanilla GAN model in the past studies.

This paper has been arranged as follows: Sect. 2 delineates the fundamental concepts and mathematical formulation of generative adversarial networks taken into consideration along with data generation methodology. Section 3 discusses the training of the candidate GAN models and compares their performance. Finally, concluding remarks are provided in Sect. 4.

2 Methodology

2.1 Generative Adversarial Network

Generative adversarial networks are a special type of neural network which consist of two elements: Discriminator and Generator. The generator takes in a random noise from a distribution L_z and tries to generate samples G_s similar to the data samples D_S. These samples G_s are then fed to a discriminator which predicts the probability of whether G_s is fake or not. The primary objective of the generator is to generate samples that discriminator is fooled to believe as identical to $D_{s,}$ while discriminator's motive is to not get fooled by the generator. During the training, the generator learns to generate such real-looking samples until the discriminator can't distinguish between real and fake samples. The architecture of a GAN is depicted in Fig. 1. The absolute expected error of the Discriminator when it is provided with "actual" and "generated" data is expressed by Eq. (1).

$$E(G, D) = \frac{1}{2}E_{x\sim p_t}[1 - D(x)] + \frac{1}{2}E_{z\sim p_z}[D(G(z))]$$

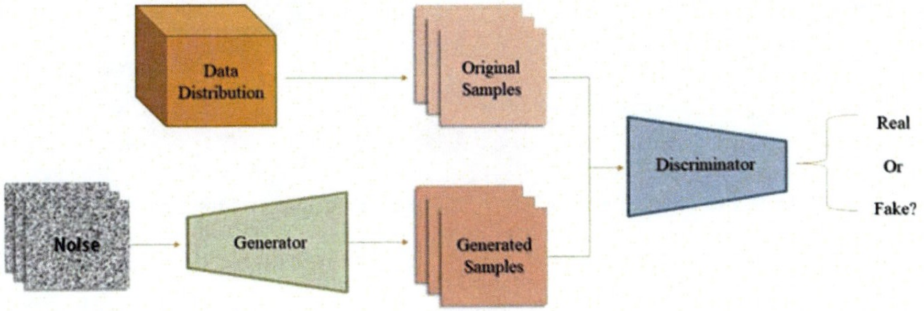

Fig. 1 Generic architecture of a generative adversarial network

$$= \frac{1}{2}(E_{x\sim p_t}[1 - D(x)] + E_{x\sim p_g}[D(x)]) \tag{1}$$

The training objective of the generator is to maximize the loss $E(G, D)$, while the discriminator attempts to minimize it.

$$\max_G(\min_D E(G, D))$$

1. *Wasserstein Generative Adversarial Networks*: The Wasserstein Generative Adversarial Network, or Wasserstein GAN, is an extended version of the generative adversarial network which enhances model's training stability and implements a loss function that correlates with the quality of generated samples. This substantially improves the approximation over the distribution of the data observed in the training set [13].
2. *Boundary Seeking Generative Adversarial Network*: GANs face a major difficulty for training discrete data when outputs of the generator are completely differentiable with respect to the generator parameters, θ. The boundary seeking GANs work well for discrete data and are equally efficient for multi-modal continuous distributions [15]. They utilize a boundary loss which ensures that the important weights have a strong connection to the decision boundary of the discriminator.
3. *Bi-Directional Generative Adversarial Network*: Bi-Directional generative adversarial networks or adversarial feature learning attempts to enhance the robustness of the generator by ensuring that the generated samples can be mapped back to input data [14]. Bi-GANs consist of an encoder in addition to the generator that is present in the standard GAN architecture which creates an inverse mapping between the generated samples and the input data. In order to ensure that the inverse mapping is done optimally, the BiGAN discriminator distinguishes not only between the generated and actual samples but also between the inverse mapped samples and the corresponding actual data. The schematic diagrams for all the algorithms are depicted in Fig. 2.

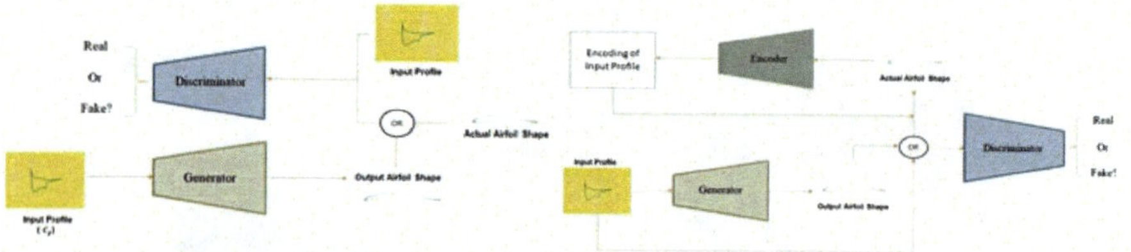

(a) WGAN and BGAN Architecture
(b) BiGAN Architecturel

Fig. 2 Schematic of generative adversarial network architectures implemented for inverse airfoil design

2.2 Data Generation

Data generation is an essential component of the proposed framework. The input data consists of the coefficient of pressure (C_p) distribution over the surface of the airfoil for an angle of attack of 3° and Reynolds Number of 100,000. The corresponding airfoil coordinates make up the output data. The C_p distribution over the airfoil is evaluated using the XFOIL simulation tool. XFOIL is considered over solving the Navier Stokes equation using CFD models because of the low-computational cost involved which facilitated the generation of a large and diverse dataset enabling the demonstration of the maximum potential of the considered models. The dataset consists of 1400 airfoil profiles obtained from the UIUC airfoil database [16]. The C_p profile of each airfoil is plotted onto an image of size 144 × 144 × 1 which is then normalized to a range of [−1, 1]. Figure 3 shows a sample of the upper and lower surface C_p plots used as the input data. The output data consists of the ordinates of the airfoil profile interpolated to a predefined set of × coordinates. This assisted in reducing the number of output parameters thereby enhancing the accuracy of the model.

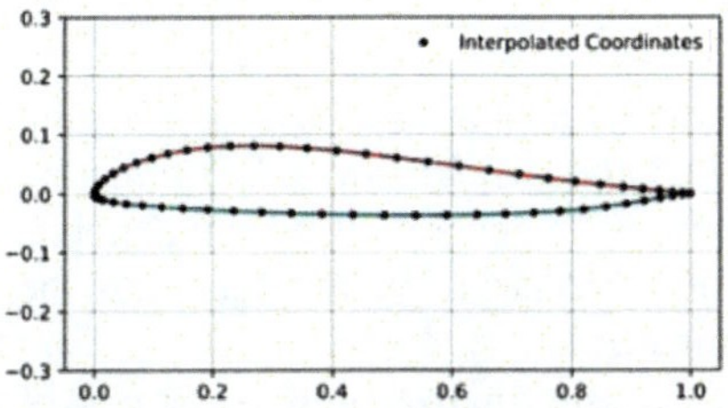

(a) Sample Airfoil from the dataset

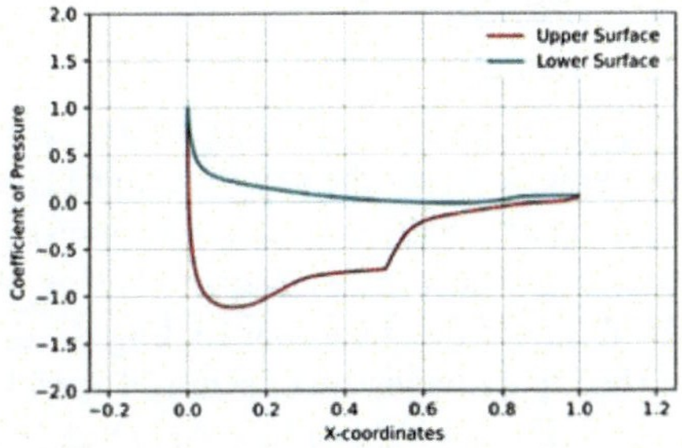

(b) C_p plot for the given sample airfoil

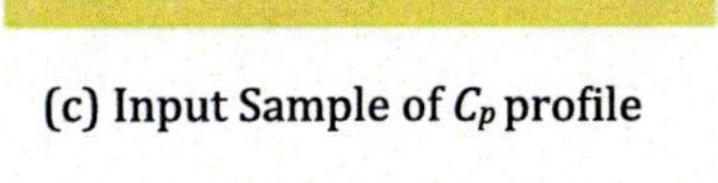

(c) Input Sample of C_p profile

Fig. 3 Training data generated using UIUC airfoil database and XFOIL

Table 1 Generator architectures with their hyperparameters

Layer type	BGAN	Bi-GAN	WGAN	Activation function
Input	$144 \times 144 \times 1$	$144 \times 144 \times 1$	$144 \times 144 \times 1$	Relu
1st convolution	$4 \times 4, 32$	$4 \times 4, 32$	$4 \times 4, 32$	Relu
2nd convolution	$4 \times 4, 64$	$4 \times 4, 64$	$4 \times 4, 64$	Relu
3rd convolution	$4 \times 4, 128$	$4 \times 4, 128$	$4 \times 4, 128$	Relu
4th convolution	$4 \times 4, 128$	$4 \times 4, 128$	$4 \times 4, 128$	Relu
Flatten	–	–	–	–
Fully connected	100	100	100	Tanh
Fully connected	100	100	100	Tanh
Output layer	70	70	70	–

The format for convolutional layer size is (convolutional filter size, number of convolutional filters). In the fully connected layer, there are 100 neurons in each layer

3 Results

3.1 Training

The dataset is split into a training and a test dataset with an 80 and 20% distribution. The network learns the weight using a *content loss* and an *adversarial loss* function. The content loss follows a simple approach wherein the generated and actual points are compared using mean squared error, while the adversarial loss is subjected to the probability of the generator fooling the discriminator.

The input images are of size $144 \times 144 \times 1$ and are normalized down to a range of $[-1, 1]$. The BiGAN and BGAN models are trained using the Adam optimizer with a learning rate of 2e−5 and a batch size of $m = 32$. Dropout layers were included after every convolution layer to avoid overfitting with a dropout probability of 0.4. The Wasserstein GAN model is trained using the RMSProp optimizer with a learning rate of 2e−5. The optimization algorithm for each model is chosen on the basis of the recommendations in their original works [13–15]. The hyperparameters for each model are tuned using the grid search method. The architecture of each GAN model is delineated in 1 and 2.

The models were written in TensorFlow and were trained on Tesla K80 GPUs. Detailed architecture of generator is depicted in Table 1 and of discriminator in Table 2.

3.2 Discussion

The training process of the GANs is analogous to a minimax game wherein two players compete with each other, i.e., the generator tries to generate real-looking

Table 2 Discriminator architectures with their hyperparameters

Layer type	BGAN	Bi-GAN	WGAN	Activation function
Input	70 × 1	70 × 1, 144 × 144 × 1	70 × 1	Leaky Relu
1st layer	512	Flatten + 512	512	Leaky Relu
2nd layer	256	256	256	Leaky Relu
3rd layer	128	128	128	Leaky Relu
4th layer	64	64	64	Leaky Relu
Output	1	1	1	Sigmoid (not in WGAN)

(In the 1st layer, 512 indicates 512 neurons in each layer.). All Layers are fully connected

samples that can't be differentiated further by the discriminator. Theoretically, GANs are said to be converged when the discriminator can no longer distinguish between fake and original data. The loss evolution of the generator and discriminator for each of the GAN model is depicted in Fig. 4. GAN loss function generally depicts vibrating and non-converging behavior, while training, therefore, a good practice to check that the model has converged is by visually observing whether the output produced is close to the training data distribution.

Initially, the comparison among the generated samples of the three proposed GANs and the vanilla GAN is done on the basis of the quality of the airfoils generated.

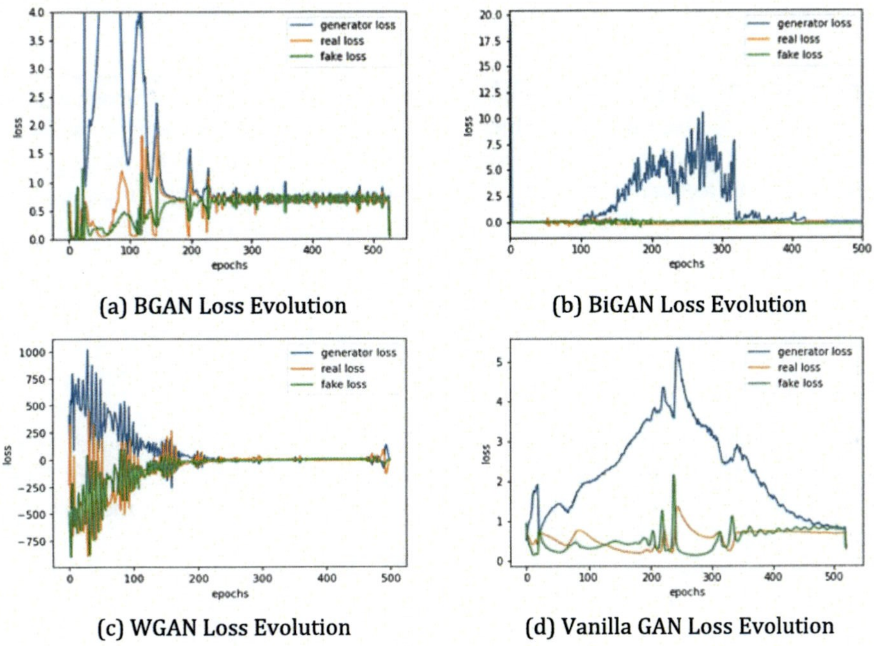

Fig.4 Loss functions during the training of the GAN models

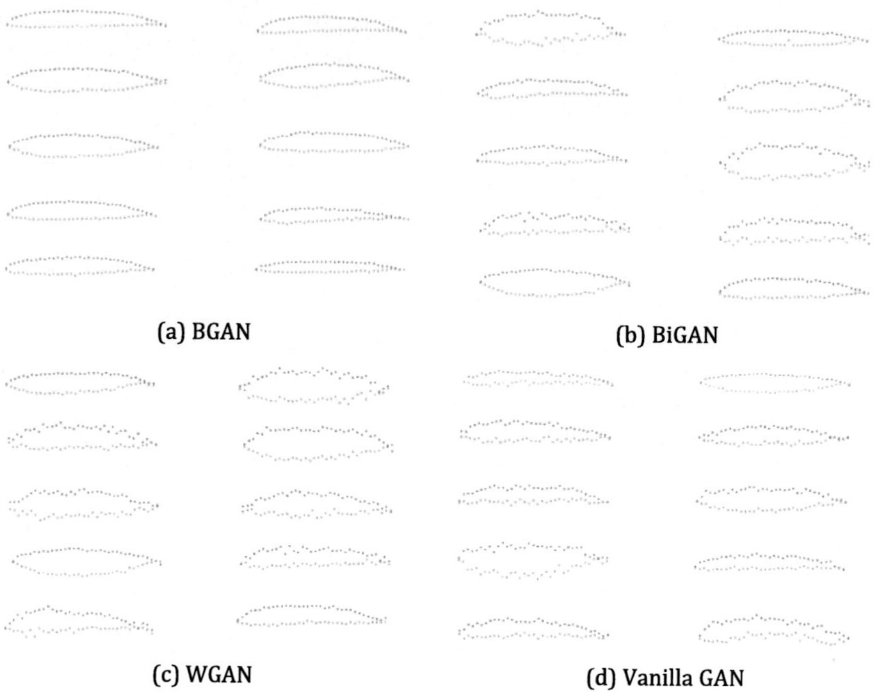

Fig. 5 Generated airfoil samples of GAN models

This can be evaluated by observing the subset of the generated samples by each model shown in Fig. 5. Since the main objective of the proposed GAN frameworks is to generate airfoil geometries that produce desired aerodynamic properties at given flow conditions, a distance metric (d_{GAN}) averaged over all the generated samples is used to evaluate the proximity of the generated airfoils to the actual samples which is given by Eq. 2.

$$d_{GAN} = \sqrt{\frac{\sum_{i=1}^{N} (y_{pi} - y_{ai})^2}{N}} \tag{2}$$

where y_{pi} is the vector containing 70 y-coordinates of each predicted sample, y_{ai} is the vector containing the actual points, and N is the total number of samples. The d_{GAN} metric for each model is shown in Fig. 7, with BGAN performing the best and vanilla GAN performing the worst.

From Fig. 5, it can be observed that the BGAN produces smooth-contoured aerodynamic shapes, while WGAN fails to do so. Though it is able to capture the features of the airfoil, it suffers from overtraining. BiGAN is able to capture the structure of the airfoil but it has quite a high-distance metric (d_{GAN}) as compared to BGAN. Vanilla GANs also showed the potential to capture the features of the airfoil, but

Fig. 6 Shape evolution over training epochs for BGAN

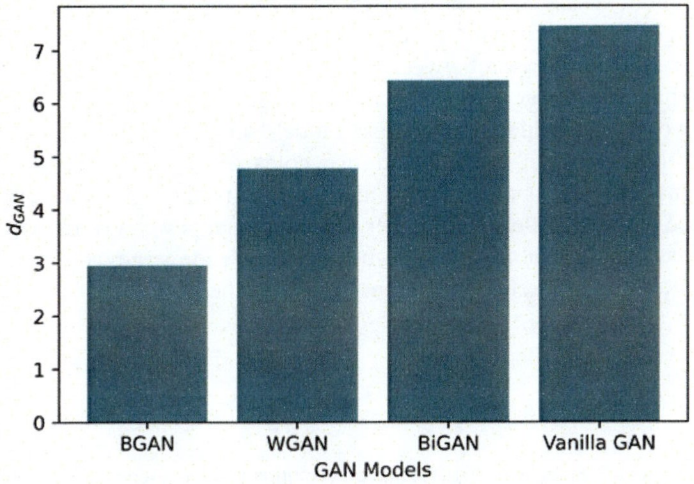

Fig. 7 Distance metric over test samples for the GAN models

they had the worst d_{GAN} which means that they were not able to generalize over all the airfoil shapes. Despite the repeated efforts to fine-tune the models, WGAN and BiGAN suffered from overtraining and mode collapse, i.e., they produced a restricted variety of output samples, while BGAN consistently produced samples close to the actual data distribution. It should be noted that all the GAN model architectures have been modified to suit a low-computational resource requirement in order to optimize the training time. Considering that the samples generated by BGAN had the lowest distance metric, its high training stability, its ability to learn the aerodynamic features of the airfoils and generalize over the airfoil data distribution, and BGAN shows the best performance among the studied GAN architectures. Figure 6 shows how airfoils are generated over the epochs during the training of BGAN and its ability to extract aerodynamic features from the data distribution (Fig. 7).

4 Conclusion

In this work, we analyzed the performance of generative adversarial networks for aerodynamic design applications based on its ability of the image to image translation and concluded upon a framework with reasonably tamed training stability and appreciable quality of generated airfoil design. We pointed out the vulnerability of GANs and worked toward a model that could minimize them.

Three modified GAN structures were tested for inverse airfoil design and were compared with the vanilla GAN based on quality of airfoils generated, proximity to the corresponding airfoil for given flow conditions and the training stability. When analyzing the application of the models, all the models were able to learn aerodynamic features of the airfoil but only BGAN produced appreciably smooth-contoured surfaces. It generated aerodynamic curves with the least distance metric which was confirmed by the comparison between the actual and generated airfoils, making it the most suitable candidate for inverse airfoil design.

Future work may include testing these models on variable flow conditions and developing more accurate datasets using Reynolds Averaged Navier Stokes based computations. The results of this work are quite motivating to test the larger scale performance of these models with higher computational power. Overall GANs have proved to have substantial potential in the application for aerodynamic design and can be employed in different applications across this field.

References

1. Kumar S, Gupta P, Singh RK (2019) A natural evolution based numerical optimisation framework to develop and enhance airfoil-slat arrangement. In: ASME international mechanical engineering congress and exposition. vol 59445. American Society of Mechanical Engineers, p V007T08A056
2. Dulikravich G, Baker D (1999) Aerodynamic shape inverse design using a fourier series method. In: 37th Aerospace sciences meeting and exhibit, p 185
3. Dulikravich GS (1992) Aerodynamic shape design and optimization-status and trends. J Aircr 29(6):1020–1026
4. Yilmaz E, German B (2020) Conditional generative adversarial network framework for airfoil inverse design. In: AIAA AVIATION 2020 FORUM. p 3185
5. Brunton SL, Noack BR, Koumoutsakos P (2020) Machine learning for fluid mechanics. Annu Rev Fluid Mech 52:477–508
6. Basu M, Kumar S, Gupta P, Singh RK (2020) A Quantitative analysis of machine learning based regressors for pressure reconstruction in PIV applications. In: ASME fluids engineering division summer meeting
7. Tyagi P, Gupta P, Singh RK (2021) Extraction of flow features for predicting pressure distribution using convolutional neural networks Tech Rep
8. Goodfellow I, Pouget-Abadie J, Mirza M, Xu B, Warde-Farley D, Ozair S, Courville A, Bengio Y (2014) Generative adversarial nets. In: Ghahramani Z, Welling M, Cortes C, Lawrence ND, Weinberger KQ (eds) Advances in neural information processing systems, vol 27. Curran Associates, Inc., pp 2672–2680. http://papers.nips.cc/paper/5423-generative-adversarial-nets.pdf
9. Wu H, Liu X, An W, Chen S, Lyu H (2020) A deep learning approach for efficiently and accurately evaluating the flow field of supercritical airfoils. Comput Fluids198:104393
10. Kim H, Kim J, Won S, Lee C (2020) Unsupervised deep learning for super-resolution reconstruction of turbulence. arXiv preprint arXiv:2007.15324
11. Chen W, Chiu K, Fuge M Aerodynamic design optimization and shape exploration using generative adversarial networks (012019). https://doi.org/10.2514/6.20192351

12. Du X, He P, Martins J A b-spline-based generative adversarial net-work model for fast interactive airfoil aerodynamic optimization (012020). https://doi.org/10.2514/6.2020-2128
13. Arjovsky M, Chintala S, Bottou L (2017) Wasserstein gan. arXiv preprint arXiv:1701.07875
14. Donahue J, Krahenbuhl P, Darrell T (2016) Adversarial feature learning. arXivpreprint arXiv: 1605.09782
15. Hjelm RD, Jacob AP, Che T, Trischler A, Cho K, Bengio Y (2017) Boundary seeking generative adversarial networks. arXiv preprint arXiv:1702.08431
16. Selig M (1996) UIUC airfoil data site. Department of Aeronautical and Astronautical Engineering University of Illinois at Urbana-Champaign

Deep Models for Analysis of Pneumonia Infection Using Chest Radiographs

Siddharth Gupta and Avnish Panwar

Abstract Pneumonia is a potentially lethal illness that is difficult to identify in its initial phases. This illness causes significant damage to one or both lungs. The findings of numerous pre-trained models and several machine learning classifiers for automated pneumonia diagnosis are presented in this study. To detect the infection, pre-trained CNN models were used to extract features from chest radiographs of pneumonia infected patients and healthy people. The collected characteristics were then fed into several machine learning classifiers. The findings showed that using the VGG16 model in conjunction with logistic regression yielded the highest accuracy of 94.4%. Doctors will be able to detect the existence of pneumonia in the lungs much more quickly if they use the technique described above.

Keywords Machine learning · Convolution neural network · VGG16 · VGG19 · Inception V3

1 Introduction

Pneumonia is inflammation in either one or both of the lungs caused by infection. This disease severely affects small children, adults, people who are suffering from any respiratory disease or patients who are on ventilators in hospital [1]. The most common people who are prone to diseases like pneumonia are from underdeveloped countries, where the population is a million and the medical facilities are almost negligible [2]. According to the documented report by the World Health Organization (WHO), every year about four million people lose their lives due to diseases like pneumonia and other respiratory diseases [3]. Also, one more report documented that 150 million people were infected with pneumonia and the majority of them were small

S. Gupta (✉)
Computer Science and Engineering Department, Graphic Era Deemed to Be University, Dehradun, Uttarakhand, India

A. Panwar
Computer Science and Engineering Department, Graphic Era Hill University, Dehradun, Uttarakhand, India

A. Ullah et al. (eds.), *Proceedings of International Conference on Information Technology and Applications*, Lecture Notes in Networks and Systems 350,
https://doi.org/10.1007/978-981-16-7618-5_23

children ranging from the age group of 1–5 years [4]. This disease can be spread by bacteria, fungus, or viruses. However, viral pneumonia causes mild infection and bacterial pneumonia causes severe infection. Fungal pneumonia mostly infects patients who are suffering from diseases like diabetes or people who have very weak immunity [5]. To overcome the dominant mortality and to stop the spread of pneumonia infection in the lungs, an early diagnosis technique is required. The advancement in the medical domain introduced a computer aided diagnostic (CAD) technique that was very helpful in early detection of pneumonia infection [6]. However, CAD based systems require handcrafted techniques for extracting features from images. To overcome this drawback, Artificial Intelligence (AI) techniques were opted [7].

The Deep Learning (DL) technique in AI was used for the feature extraction and classification of images. More precisely, the Convolution Neural Network (CNN), a form of deep neural network, has achieved very good results in classification and recognition of images. Influenced by the cortex in humans [8], CNN separates various classes of image recognition problems. The only disadvantage to using CNNs is that they require a very massive amount of data to train the model. In the medical imaging field, to train the CNN model using medical images, either the model was trained from scratch or the model was fine-tuned using dataset images [9].

Chest x-ray images were used to feed as the input images to these CNN models. Chest radiographs have become appropriate to provide as an input due to their low cost, ease of availability, and simplicity in comparison with other radiographic techniques such as computed tomography (CT), ultra-sound, etc. Chest radiography technique is the most common diagnostic method which is used in the healthcare system, even in remote areas [10]. Also, real-time chest radiographs help to diagnose the progression of disease at regular intervals. Figure 1 shows the chest radiograph of a patient suffering from a pneumonia infection and a healthy chest radiograph.

In the current work, pneumonia chest radiographs and healthy chest radiographs were collected and pre-processed. After pre-processing, these were provided as

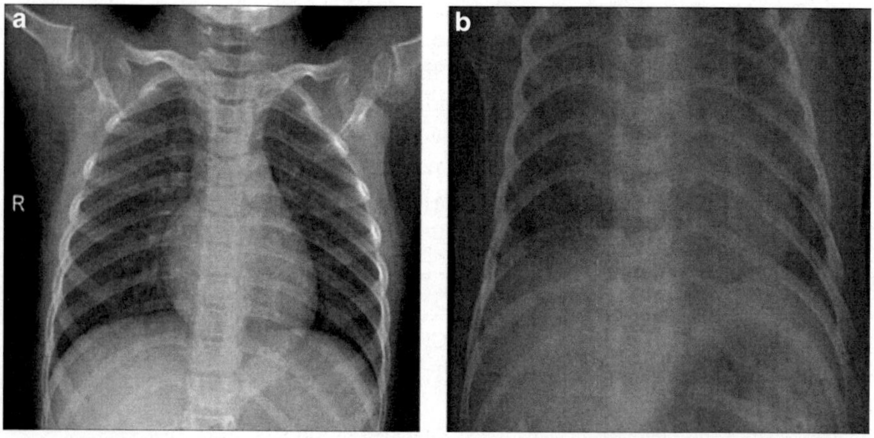

Fig. 1 Chest radiographs. **a** Healthy chest radiograph, **b** pneumonia infected chest radiograph

input to distinct CNN models for feature classification. Once the features were extracted, these images were passed to machine learning classifiers for classification of radiographs as pneumonia infected or normal chest x-ray scans.

The prime objective of this work: (1) is to present the functioning of different deep learning models for detection of pneumonia infection using chest radiographs. (2) to implement an affordable technique for the early diagnosis of pneumonia infection so that the radiologist and clinicians may take the appropriate action to stop the damage to the lungs due to pneumonia [11].

2 Methodology

2.1 Design Methodology and Pre-processing

The dataset of pneumonia x-ray images was publically available on Kaggle [12]. This dataset is adapted from the chest x-ray images (pneumonia) dataset. The total number of images was divided into three different categories: training, validation, and testing. These training observations consist of 1082 healthy chest x-ray scans and 3110 lung opacity cases. Validation comprises 267 normal chest x-ray cases and 773 lung opacity cases. Finally, testing consists of 624 images out of which 234 are of normal chest x-ray images and 390 are lung opacity cases. The total number of images was 5856 [12]. Figure 2 shows the dataset images used in the current work.

Once the chest x-ray images were extracted, they were pre-processed to make them the same shape and size. Image pre-processing techniques such as image resizing and image cropping were performed to make the images appropriate to feed into CNN models. Finally, to remove the noise from these dataset images, wiener filter was used. This filter also removes the blur shades from the images. Table 1 shows the dataset description.

2.2 Architecture of CNN Model

A CNN [13] model belongs to a class of neural network which is used to extract features from images. In general, the CNN takes an input image, trains the model based on input data, and finally classifies the features. This model consists of an input layer, an output layer and many hidden layers. Of these layers, the first layer is the convolutional layer, which is used to apply the filter to the input image. The next layer, ReLU acts as an activation function which removes all the negative values from the filtered images. These negative values were replaced by zeros. Pooling is performed to shrink the image stack to a smaller size. Finally, a fully connected layer is used for the classification. The detailed CNN architecture can be seen in Fig. 3.

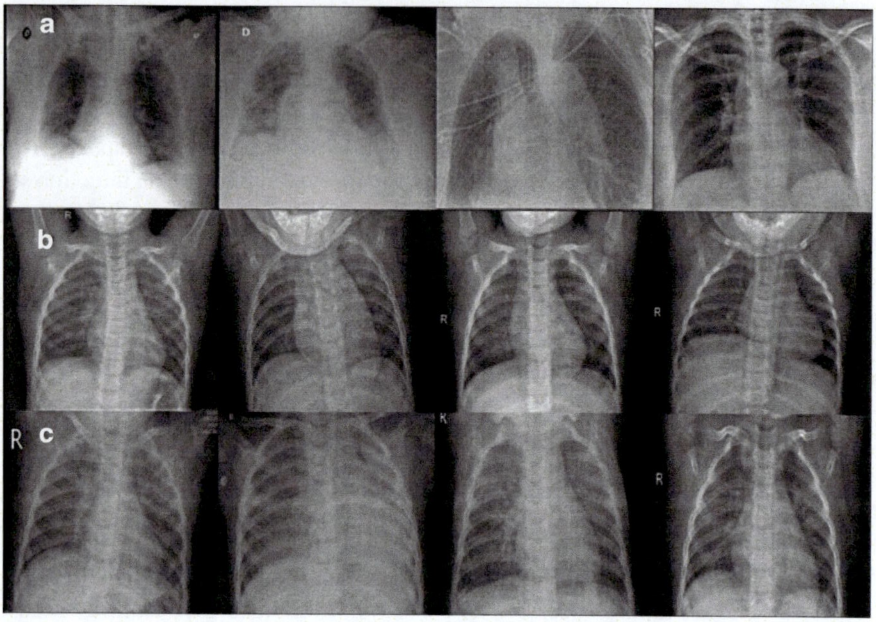

Fig. 2 Chest radiographs. **a** Pneumonia infected chest radiograph, **b**, **c** healthy chest radiograph

Table 1 Dataset description

	Training	Testing	Validation	Total
Normal	1082	234	267	
Opacity	3110	390	773	
Total	4192	624	1040	5856

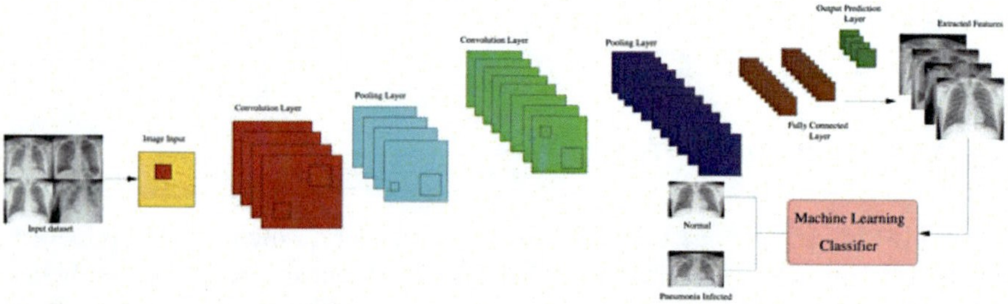

Fig. 3 Architecture of pre-trained models for analysis of pneumonia using chest radiographs

In the current work, different pre-trained CNN models were used for the purpose of feature extraction. The input dataset images were fed to the CNN model, and this model extracted the relevant features from the images. Finally, these images were passed to the machine learning classifiers.

2.3 Machine Learning Classifiers Used

After extracting the features, these images were passed to different machine learning classifiers such as kNN, tree, SVM [14], random forest [15], naïve bayes, logistic regression [16], adaboost for classifying the images as pneumonia infected or healthy images. Classifiers are used for categorizing objects into the right class. In our work, we have shown the performance of different classifiers based on several parameters, such as accuracy, precision, recall, f1-score, and AUC.

2.4 Evaluation Parameters

The given study presents the classification of pneumonia infected patients radiographs and healthy chest radiographs. All the chest radiographs were passed to pre-trained CNN models for feature extraction and finally to different ML classifiers for classification. Five different matrices were used to evaluate the performance of CNN models. These metrics were sensitivity, specificity, accuracy, precision, and f1-score [17].

$$\text{Sensitivity} = (\text{tp})/(\text{tp} + \text{fn}) \tag{1}$$

$$\text{Specificity} = (\text{tn})/(\text{tn} + \text{fp}) \tag{2}$$

$$\text{Accuracy} = (\text{tp} + \text{tn})/(\text{tp} + \text{tn} + \text{fn} + \text{fp}) \tag{3}$$

$$\text{Precision} = (\text{tp})/(\text{tp} + \text{fp}) \tag{4}$$

$$\text{F1} - \text{score} = (2 * \text{tp})/(2 * \text{tp} + \text{fp} + \text{fn}) \tag{5}$$

3 Results and Discussions

In this section, the evaluation of different CNN models and machine learning classifiers were defined. The below stated results were judged on various parameters, such as AUC, accuracy, f1-score, precision, and recall. Tables 2, 3, and 4 show the results for VGG16, VGG19, and inception v3 pre-trained models.

Table 3 shows the pre-trained VGG19 model along with the kNN, tree, SVM, RF, Naïve bayes, LR, AdaBoost classifiers.

Table 2 Evaluation result for VGG16 model

Model	AUC	Accuracy	F1 score	Precision	Recall
kNN	0.923	0.854	0.917	0.919	0.917
tree	0.712	0.769	0.829	0.830	0.829
SVM	0.931	0.942	0.942	0.942	0.942
RF	0.916	0.886	0.886	0.886	0.886
Naive	0.870	0.899	0.900	0.904	0.899
LR	0.966	**0.944**	0.944	0.944	0.944
adaBoost	0.757	0.812	0.813	0.815	0.812

Table 3 Evaluation result for VGG19 model

Model	AUC	Accuracy	F1 score	Precision	Recall
kNN	0.932	0.859	0.855	0.862	0.859
tree	0.803	0.837	0.836	0.836	0.837
SVM	0.939	0.856	0.856	0.857	0.856
RF	0.932	0.857	0.856	0.856	0.857
Naive	0.892	0.833	0.835	0.841	0.833
LR	0.960	0.897	0.898	0.899	0.897
adaBoost	0.789	0.803	0.803	0.803	0.803

Table 4 Evaluation result for Inception v3 model

Model	AUC	Accuracy	F1 score	Precision	Recall
kNN	0.951	0.888	0.889	0.894	0.888
tree	0.766	0.832	0.832	0.832	0.832
SVM	0.975	0.921	0.921	0.921	0.921
RF	0.938	0.872	0.872	0.873	0.872
Naive	0.940	0.877	0.878	0.882	0.877
LR	0.980	0.933	0.933	0.933	0.933
adaBoost	0.806	0.819	0.819	0.818	0.819

Table 4 shows the pre-trained Inception v3 model along with the kNN, tree, SVM, RF, Naïve Bayes, LR, AdaBoost classifiers.

The above tables show the performance of several pre-trained models, such as VGG16, VGG19, and Inception v3 models, along with the different classifiers. The input images were provided as an input to these pre-trained models for feature extraction. Once the features were extracted, they were passed to different machine learning classifiers for the classification of images as pneumonia infected or normal images. It may be seen from the results that the VGG16 model along with the logistic regression classifier gives a maximum accuracy of 94.4%. Hence, these values state that the dataset images that were provided to these models can be classified as pneumonia infected and healthy radiographs with an accuracy of 94.4%.

The other parameter to judge the performance of the several classifiers is the ROC curve [18]. The ROC curve is drawn between the sensitivity and the specificity. Figure 4 shows the ROC curve for the VGG16 model and different classifiers. As seen in the figure, various classifiers have variable performance when it comes to detecting and diagnosing pneumonia from chest x-ray scans.

Figure 5 shows the ROC curve for the VGG19 model and different classifiers. The pre-processed chest x-ray scans of size 224 × 224 were utilized to train the CNN model VGG19 employed in this study. The results show that an accuracy of 89.7% was achieved using the VGG19 model and the logistic regression classifier [19].

The original ImageNet dataset, which contains over 1 million training images, was used to train the Inception v3 model. Figure 6 shows the inception v3 model and different machine learning classifiers for detection of pneumonia in chest x-ray

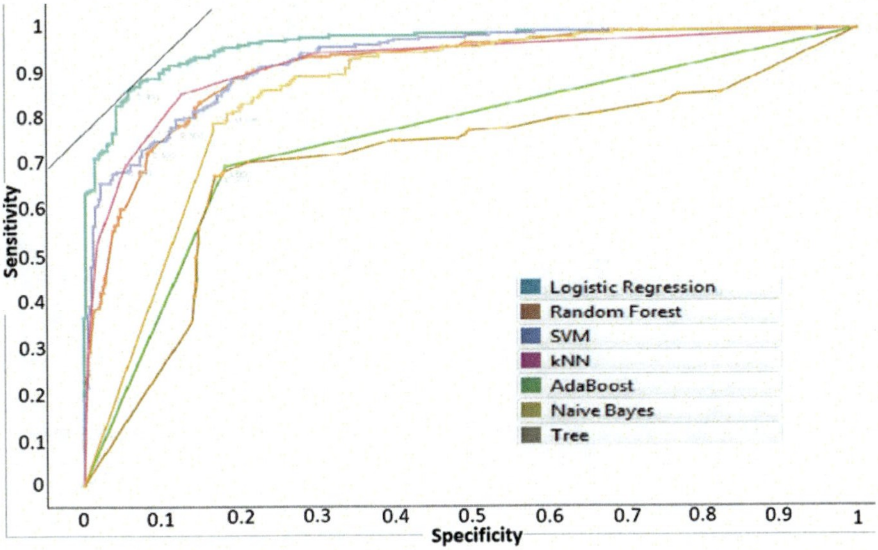

Fig. 4 ROC curve for diagnosis of pneumonia detection using VGG16 model and different classifiers

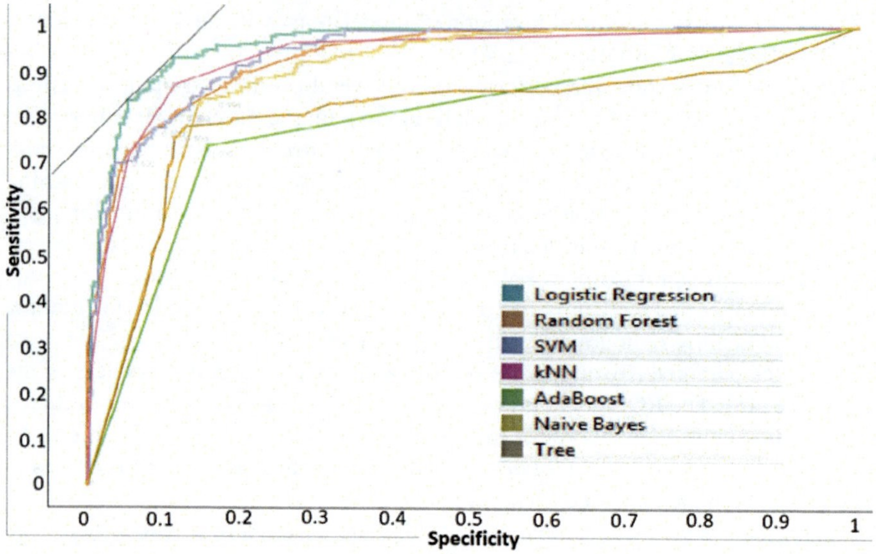

Fig. 5 ROC curve for analysis of pneumonia diagnosis using VGG19 model and different classifiers

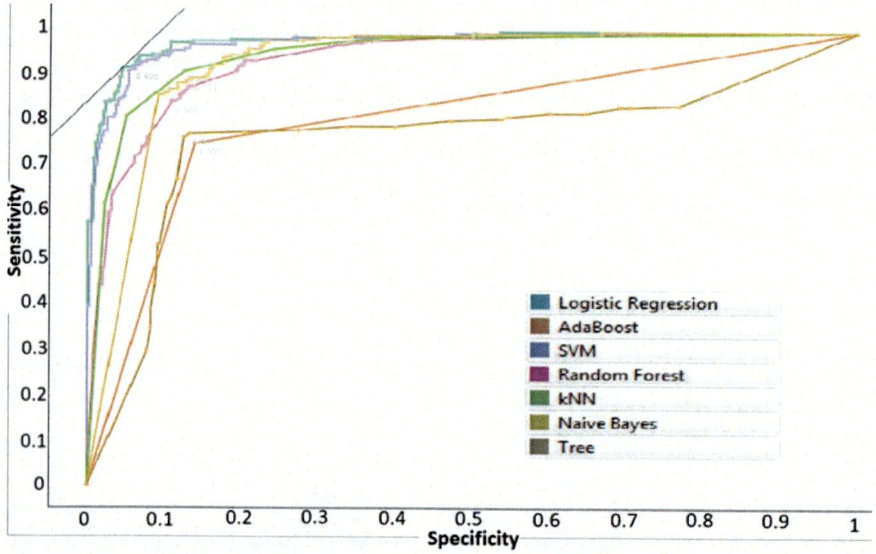

Fig. 6 ROC curve for detection of pneumonia analysis using Inception v3 model and several classifiers

scans. It may be observed that an accuracy of 93.3% was marked using this model and logistic regression classifier [19].

The above result shows the performance of different pre-trained models and different classifiers. These values indicate that the VGG16 model and Logistic Regression give the best accuracy for separating the chest radiographs of pneumonia patients from normal chest radiographs.

4 Conclusion

Pneumonia is a life-threatening disease that results in many deaths every year. This disease majorly affects small children. One of the possible solutions to overcome pneumonia is early diagnosis. In the current work, different chest radiographs were provided as an input to the different pre-trained CNN models. These CNN models were used for feature extraction. After extracting the features, these images were passed to various machine learning classifiers for the classification of chest radiographs as pneumonia infected or healthy chest radiographs. The results prove that the VGG16 model along with logistic regression gives the maximum accuracy of 94.4%. Thus, the stated results verify that whenever a new radiograph is fed to the VGG16 model and LR classifiers, and they detect pneumonia infected or healthy images with 94.4% correct prediction. In future, different models, such as AlexNet, ResNet, MobileNet, etc., will be implemented using the same image dataset or the number of images in the dataset may be increased to enhance the accuracy of the work.

References

1. Gilani Z, Kwong YD, Levine OS, Deloria-Knoll M, Scott JAG, O'Brien KL, Feikin DR (2012) A literature review and survey of childhood pneumonia etiology studies: 2000–2010. Clin Infect Dis 54(suppl_2):S102–S108
2. Siddiqi R (2019) Automated pneumonia diagnosis using a customized sequential convolutional neural network. In: Proceedings of the 2019 3rd international conference on deep learning technologies, pp 64–70
3. WHO | World Health Organization. Accessed: Mar 31, 2020. [Online]. Available: https://www.who.int/
4. World Health Organization. Accessed: Sep 8, 2020. [Online] Available: https://www.who.int/news-room/fact-sheets/detail/children-reducing-mortality
5. Mackenzie G (2016) The definition and classification of pneumonia. Pneumonia 8(1):1–5
6. Karar ME, El-Din Hemdan E, Shouman MA (2021) Cascaded deep learning classifiers for computer-aided diagnosis of COVID-19 and pneumonia diseases in X-ray scans. Complex Intell Syst 7(1):235–247
7. Ilyas M, Rehman H, Naït-Ali A (2020) Detection of covid-19 from chest x-ray images using artificial intelligence: an early review. arXiv preprint arXiv:2004.05436
8. Kim J, Sangjun O, Kim Y, Lee M (2016) Convolutional neural network with biologically inspired retinal structure. Procedia Comput Sci 88:145–154 (2016)

9. Mehra R (2018) Breast cancer histology images classification: training from scratch or transfer learning? ICT Express 4(4):247–254
10. Hira S, Bai A, Hira S (2021) An automatic approach based on CNN architecture to detect Covid-19 disease from chest X-ray images. Appl Intell 51(5):2864–2889
11. Jaiswal AK, Tiwari P, Kumar S, Gupta D, Khanna A, Rodrigues JJPC (2019) Identifying pneumonia in chest X-rays: a deep learning approach. Measurement 145:511–518
12. Dataset link: https://www.kaggle.com/pcbreviglieri/pneumonia-xray-images
13. Simonyan K, Zisserman A (2014) Very deep convolutional networks for large-scale image recognition. arXiv preprint arXiv:1409.1556
14. Maji S, Berg AC, Malik J (2008) Classification using intersection kernel support vector machines is efficient. In: 2008 IEEE conference on computer vision and pattern recognition. IEEE, pp 1–8
15. Iwendi C, Bashir AK, Peshkar A, Sujatha R, Chatterjee JM, Pasupuleti S, Mishra R, Pillai S, Jo O (2020) COVID-19 patient health prediction using boosted random forest algorithm. Front Public Health 8:357
16. Bhandari S, Shaktawat AS, Tak A, Patel B, Shukla J, Singhal S, Gupta K, Gupta J, Kakkar S, Dube A (2020) Logistic regression analysis to predict mortality risk in COVID-19 patients from routine hematologic parameters. Ibnosina J Med Biomed Sci 12(2):123
17. Gupta S, Panwar A, Goel S, Mittal A, Nijhawan R, Singh AK (2019) Classification of lesions in retinal fundus images for diabetic retinopathy using transfer learning. Int Conf Inf Technol (ICIT) 2019:342–347. https://doi.org/10.1109/ICIT48102.2019.00067
18. Dhaya R (2020) Deep net model for detection of covid-19 using radiographs based on roc analysis. J Innov Image Process (JIIP) 2(03):135–140 (2020)
19. Gupta S, Panwar A, Rawat D (2021) A comparison among distinct deep learning techniques for real-time testing of Covid-19 infected patient using chest radiography. In: 2021 2nd international conference on secure cyber computing and communications (ICSCCC), 2021, pp 394–398. https://doi.org/10.1109/ICSCCC51823.2021.9478106

An Artificially Intelligent Marine Life Surveillance and Protection System (MLSPS)

Peer Azmat Shah, Eiman Abdulla Obaid Salem Alyammahi, Meera Abdalla Mohamed Qusoom Alnaqbi, Khadija Mubarak Mohamed Awasiya Alzaabi, Asmaa Mohamed Ali Eabayed Aldhanhani, and Zeeshan Hameed

Abstract Protecting fisheries and the marine environment are a high priority area for UAE Government and regulatory monitoring is very important to detect and prevent illegal activities by the fishermen. Sustainability and developing solutions with technology are an important aspect of this domain. The technology-based solutions will raise the efficiency level and will try to meet the strategic objectives of time optimization and will be helpful to protect and preserve the fisheries and marine environment. In this work, a new Marine Life Surveillance and Protection system is proposed and developed that performs regular monitoring in the fishing area to ensure law compliance. It uses modern IoT-based drone technology and artificial intelligence-based IBM Watson cloud service. The proposed MPSPS has two parts, the cloud-based system and the IoT devices. The system is a web-based implementation using Node-RED programming tool and Watson Visual Recognition nodes in IBM Cloud. The second is the programmable Tello Drone that used the SDK and is connected with the IBM Cloud. Drone flies over the sea and captures the real-time images and videos and sends them to the IBM cloud, where data classification and analysis is performed using Artificial Intelligence techniques. The results show that the new system has improved the performance for time and detection accuracy.

Keywords Component · Marine life · Surveillance system · Cloud computing · IoT · IBM · Artificial intelligence

1 Introduction

Nowadays, technology is evolving every day and monitoring systems are used in many sectors as solutions for many security and violation problems. For our research work, we found the UAE's interest in fisheries and its sustainability, and we are

P. A. Shah (✉) · E. A. O. S. Alyammahi · M. A. M. Q. Alnaqbi · K. M. M. A. Alzaabi · A. M. A. E. Aldhanhani · Z. Hameed
Department of Computer & Information Sciences (CIS), Higher Colleges of Technology, Fujairah Women's College, Fujairah, United Arab Emirates
e-mail: pshah@hct.ac.ae

A. Ullah et al. (eds.), *Proceedings of International Conference on Information Technology and Applications*, Lecture Notes in Networks and Systems 350,
https://doi.org/10.1007/978-981-16-7618-5_24

working to develop a possible solution to support this field by implementing a system that depends on artificial intelligence and IoT.

The marine environment in the United Arab Emirates is important socially, economically and environmentally, it is an important source of desalination, fishing, and oil production [1]. Therefore, the government agencies responsible for marine areas are working on the sustainability of marine resources, protecting them from depletion and pollution, and enforcing the laws to prevent illegal actions that could affect the marine environment negatively [2]. Many popular fishes in UAE such as Hammour (orange-spotted grouper) and Kanaad (Spanish mackerel) are about to vanish from the sea, the remaining fishes from these kinds reach less than 10% in the country and about to become extinct. Those types of fishes are being fished more than their sustainable limit [3]. In 2017, about 20 dugongs (sea cows) were found dead drowned in abandoned fishing nets [4].

The idea of marine life surveillance and protection system is not new. Different researchers have designed solutions and implementations are also available. While we were searching for existing solutions, we found the Indonesian Vessel Detection System (VDS) with a similar idea to ours but they are monitoring the fishing boats using a radar located on a spaceship [5]. A similar system called automatic identification systems—AIS' is a fishery monitoring tool in Africa. Their system can monitor the beaches for far distances and the focus is not marine life protection rather its focus is surveillance. The disadvantage of it is the transparency, where the data is not encrypted and available for anyone. This system may not be useful because it is monitoring large areas, and our focus is on the fishing areas only [6].

The overcome the problem of manual surveillance and protection of marine life on the UAE shores, we have proposed a new artificial intelligence-based Marine Life Surveillance and Protection System (MLSPS) that will use drone technology to provide views of the fishing areas. The drone will fly over the sea and capture the real-time images and videos and send them to the IBM cloud, where data analysis is performed using the IBM Watson Studio using artificial intelligence techniques to identify fishermen boats, fishing tools, residues, and any illegal activity being performed at the UAE shores and the sea. The drone will be connected to a secure system that will analyze the information. The recorded videos and photos can be used as a piece of evidence. All the system data will be encrypted with strong algorithms to solve the transparency issue and increase the security of the system. Below are some of the contributions of this research work.

- Provides a precise and clear source of digital evidence to analyze digital data in the fishing areas.
- Provides a live recording of the fishing area activities and detects illegal activities, and fishing residues position.
- Detects unauthorized fishermen and boats.
- Raises the level of legal commitment.
- Can help to identify pollution problems in the fishing area.

The organization of this paper is as follows: In Sect. 2, existing approaches, their pros and cons, and a literature review is presented. Section 3 presents the proposed system design. In Sect. 4, system implementation and results have been discussed and finally, a conclusion along with future work is presented in Sect. 5.

2 Existing Solutions and Literature Review

Considering the UAE's 2021 vision, preserving fisheries is one of the essentials to ensure sustainability for future generations and enhance food security [7]. This research work seems to help in developing and serving this field in line with the UAE vision and objectives.

In the UAE, many sectors used a modern monitoring system for different purposes. We did some research intending to get information about monitoring systems, how they are used, and what are the differences between them and our project. Dubai Road and Transport Authority operates the smart monitoring system, which in turn monitors the traffic in Dubai roads using surveillance cameras and sensors. That helps to monitor traffic congestion, follow-up traffic systems to report malfunctions and maintenance operations, and control the traffic signals [8]. The system relies on fixed cameras, but the use of mobile cameras will be much easier to monitor the fishing area. Also, the traffic control system is not in line with the objectives of the fisheries protection and control system.

At IDEX 2017, a ship identification system was launched to provide permanent surveillance of marine areas to protect and increase safety. The system consists of radars, long-range cameras, tracking devices, positioning systems, and command and control systems. The system will help prevent collisions, reduce maritime traffic accidents, monitor coasts, and protect marine borders. The goal of this system is to monitor the ship's movement and borders to increase the level of security and safety [9]. However, there is no need for complex, sophisticated, and expensive components and techniques when the goal is to monitor a much smaller area, as the fishing area, and problems of less severe.

Al Maktoum International Airport uses intelligent surveillance systems with 3300 cameras distributed in lounges, gates, aircraft parking, and cargo areas. The system also sends alarms to surveillance teams to respond to suspected cases [10]. The system helps to increase the safety of the airport. However, the use of surveillance cameras in the marine area for monitoring fishing operations will not be practical because using a large number of fixed and stationary cameras will distort the landscape. And there is no need to operate cameras around the clock because fishing operations are at specific periods of the day.

In a face-to-face meeting with one of the Coast Guard of UAE, we learned that they use the drone for reconnaissance and surveillance of the marine area and the borders. The drone scanning operation is monitored directly through the control room to track the ships' movement and suspicious activities and control marine navigation and boundaries. However, the method used there lacks the feature of automatic data analysis.

Considering fisheries is one of the main pillars in enhancing food security at the national level, and there is a great effort to develop and maintain this sector. We did our search to know if there are any monitoring solutions in the fishing area, and we found that UAE still depends on observation teams for that purpose. However, our thinking at this point is how to reshape this process and make it more efficient. We started thinking to use the IoT and implement a system that helps in monitoring the fishing area and provide evidence. Rather than deploying beach observation teams to track fishermen's behavior and their commitment to fishing laws, the idea of Marine life surveillance and protection system relies on modern technology and artificial intelligence to ensure faster surveillance, better results, and less effort.

3 Proposed System Design

The Marine Life Surveillance and Protection System (MLSPS) mainly consists of three entities: the user, the drone, and the system. Each one of these entities has spe- cific roles. The starting point is authentication, where the authenticated user can access the system. This process aims to verify the identity of the user and prevent unauthorized access and usage of the system. As a result, this should improve the system's accuracy and efficiency as well as preserving the system's security.

Then, the system will be connected to the drone, in that way the system will be able to control the drone and the new mission will start. The drone will send real-time streaming video and pictures to the system, and the system will analyze and classify them. This process will help the user to have a better understanding and more detailed results. An alarm could be generated if there is a result that needs to take action toward it. The database is a critical entity that will store the data and provide backups. The database will store mission information, results, and other relevant information. Also, the system will generate a report to use as a piece of evidence later if needed. Figure 1 shows the abstract level design of the proposed system.

The functional decomposition diagram that describes the main functions in the system is shown in Fig. 2. The main functionalities the proposed system will cover are: authentication to ensure that only authenticated users can access the system, scanning which describe the process of connecting and controlling the drone, taking pictures, analysis to provide more information about the pictures, classification which will classify the pictures and generate alarms in particular situations, and finally storing function which is responsible about storing data and generating reports.

Figure 3 shows the context diagram that describes the system processes in general, and how the entities interact with them.

Data flow diagram represents how the data moves from entities to the system, and from one process to another. For our system, the main data is the videos and images that will be transferred from the drone to the system, analyzed and classified in the system processes, represented as an output to the user, then the result will be stored in the database. Figure 4 shows the data flow diagram of the proposed system.

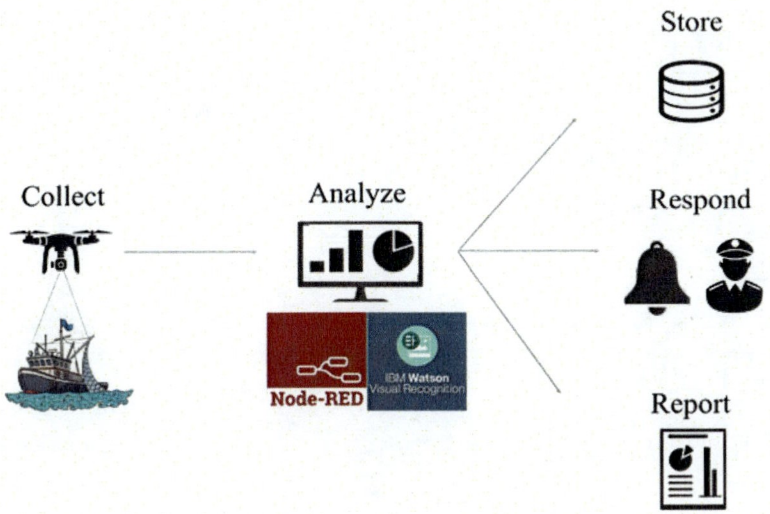

Fig. 1 Functional decomposition diagram (FDD)

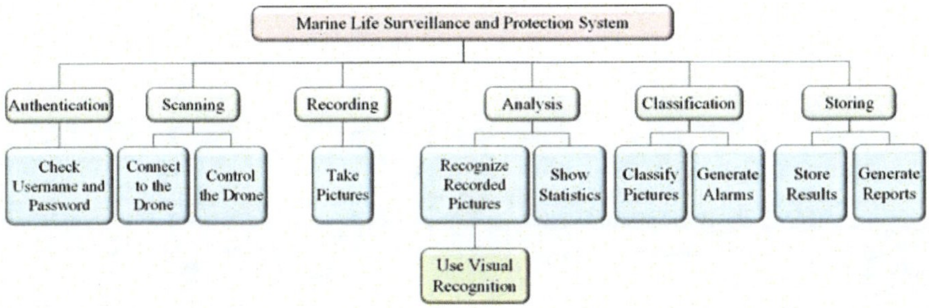

Fig. 2 Functional decomposition diagram (FDD)

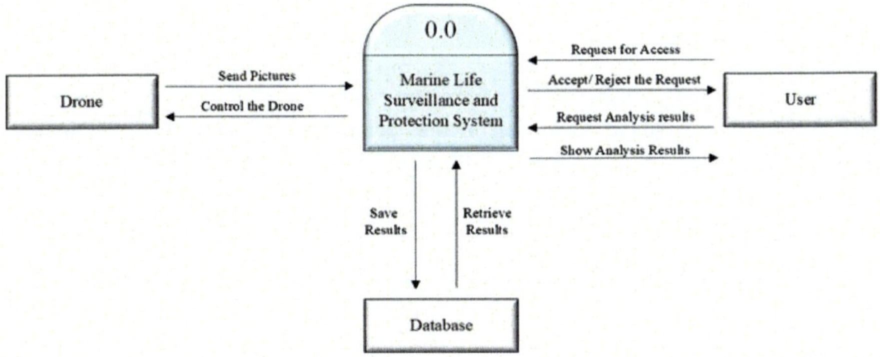

Fig. 3 Functional decomposition diagram (FDD)

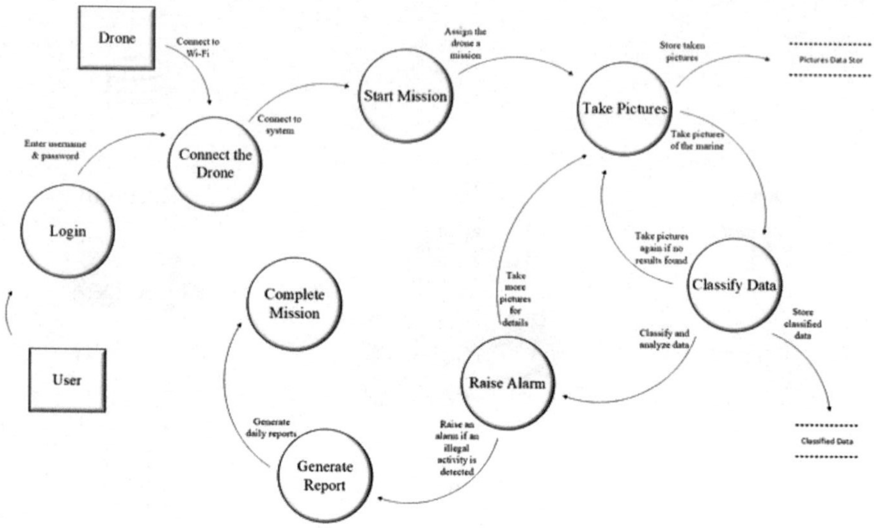

Fig. 4 Data flow diagram (DFD)

Table 1 Decision table

Process name: classification									
		1	2	3	4	5	6	7	8
Condition	Classify a picture	Y	Y	Y	Y	N	N	N	N
	Give the picture a description	Y	Y	N	N	Y	Y	N	N
	If pictures describe an illegal activity	Y	N	Y	N	Y	N	Y	N
Action	Generate alarm	X							
	Classify as normal records		X	X					
	Save results	X	X						
	No action				X	X	X	X	X

The classification was performed using the decision tree which is shown in tabular format using Table 1.

4 System Implementation and Results

The implementation was performed in the Node-Red programming tool which was installed locally on machines. An authenticated user must access the Node-Red editor with a valid username and password. Then, the drone must be connected to the laptop through a wireless connection. In our case, we used the 4G wireless Internet connectivity. The user will start a new mission by entering mission details in the system, and the mission details will be stored directly in the Cloudant database. The system will control the drone, and the system will receive data from the drones. Pictures and videos will be analyzed automatically using Watson Visual Recognition from IBM Cloud, and alarms can be generated for suspicious situations. All the results will be saved in the Cloudant database, and an HTML report will be generated and saved automatically in the local machine.

4.1 Development Environment

Instead of using the usual programming tools, we decided this time for our project to use the Node-Red programming tool. Node-Red is an open-source and free programming tool used to connect hardware devices, APIs, and other online services. Since we are building an Internet of Things system, we found it easier to use it to connect the drone to the system. It is a browser-based editor; we can easily edit and connect flows that are represented as nodes and then deploy the changes to run it with one click. Node-Red is built on Node.js, which executes JavaScript codes. The flows are stored in JSON format, which makes it easy to import and export the commands [11]. We are using the IBM Cloudant database for our project; it offers the database as a service and uses Apache CouchDB. It is only available online and uses the Erlang language that can be accessed through RESTful HTTP/JSON API. The database allows data durability to make data persistent and provide access control [12].

To connect the drone to the system, we used a 4G wireless Internet connection. We also needed to connect the system to the internet in which the Node-Red is installed on it locally, with an ethernet cable to avoid losing the connection with the drone, since it is hard to connect a PC with two different wireless Internet connections at the same time.

We used the IBM Watson Visual Recognition service to let our system understand the content of each image and video as the human does. Watson determines if the content of a picture matches a tagged picture. With Custom Object Detection Feature, Watson can identify objects in a picture, and based on this identification; it gives us feedback as a percentage of picture accuracy [13]. Figure 5 shows the control module for drones.

4.2 IBM Watson Classification

We created a classification model to classify pictures according to specific classes. Our model contains five classes: Overfishing, Pollution, Fishing Boat, Fishermen, and Fish. Figure 6 shows the Watson studio classification model.

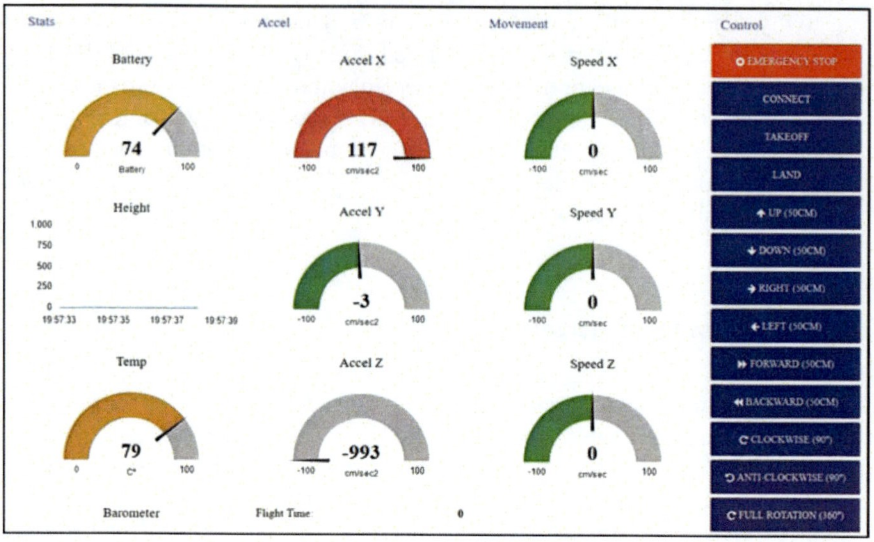

Fig. 5 Drone control system

Classes

CLASS	NUMBER OF EXAMPLES
Fish	10
Fishermen	10
Fishing Boats	10
Overfishing	10
Pollution	10

Fig. 6 Watson classification model classes

Figure 7 shows the Watson classification model results for fishing boats and pollution. It clearly shows that the confidence for pollution detection is more than 60%, and for fishing boats, detection is more than 50%.

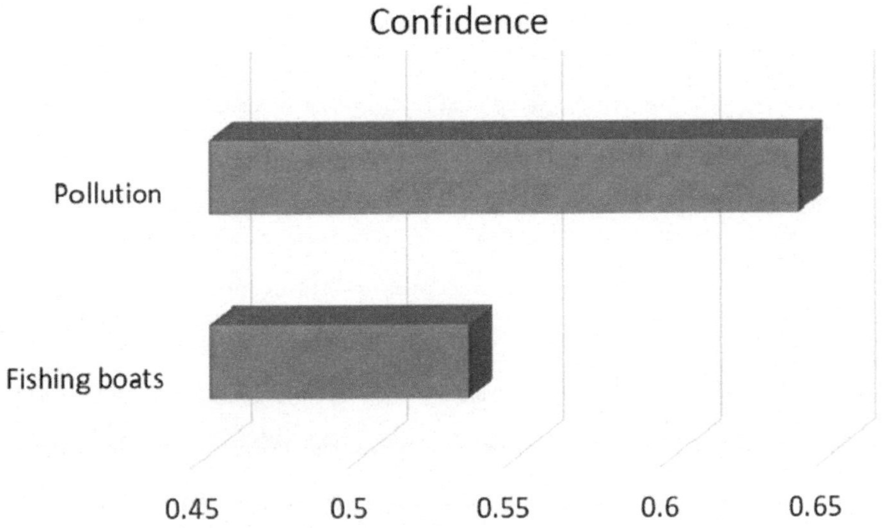

Fig. 7 Watson classification model result for fishing boats

5 Conclusion and Future Work

The Marine Life Surveillance and Protection System aims to monitor the fishing area regularly. Instead of sending people to perform this task, it will be done through artificial intelligence technology. The project contains the drone, system, and database. The drone controlled by the system records the videos and pictures from the fishing area and sends them to the cloud-based system. The system analyzes and classifies them, generates an alarm, and creates reports. As future work, the project may be implemented using the new computational intelligence techniques for improved accuracy of the analysis and classification result. This work can be used for purposes other than monitoring the fishing area, such as monitoring a wild area, natural reserves, traffic movement, and mountainous areas.

Acknowledgements The authors acknowledge the Higher Colleges of Technology, the United Arab Emirates for providing financial support to this research work.

References

1. Marine Environment & Fisheries Sustainability. (n.d.). Retrieved from Moccae.gov.ae: https://www.moccae.gov.ae/en/knowledge-and-statistics/marine-environment-andfisheries-sustainability.aspx
2. Awadh S (2019) 6 challenges facing «fisheries» and threaten their productivity at the state level. Retrieved from Alittihad: https://www.alittihad.ae/article/24543/2019/6

3. Radan S (2016) Fish vanishing from UAE waters due to overfishing. Retrieved from Khaleej Times: https://www.khaleejtimes.com/nation/general/fish-vanishing-from-uaewaters-due-to-overfishing
4. Duncan G (2018). Six dugongs found dead in Abu Dhabi as rogue fishing put species at risk. Retrieved from The National: https://www.thenational.ae/uae/environment/six-dugongs-found-dead-in-abu-dhabi-asrogue-fishing-put-species-at-risk-1.796083
5. Longépé N (2018) Completing fishing monitoring with spaceborne vessel detection system (VDS) and automatic identification system (AIS) to assess illegal fishing in Indonesia. Retrieved from Science Direct: https://www.sciencedirect.com/science/article/abs/pii/S0025326X17308330
6. The potential use of 'automatic identification systems—AIS' as a fisheries monitoring tool. (2018). Retrieved from Francisco Blaha: http://www.franciscoblaha.info/blog/2018/2/1/the-potential-use-of-automaticidentification-systems-ais-as-a-fisheries-monitoring-tool
7. Farooq S (2018) «Climate change»: throwing damaged fishing gear on beaches threatens the environment. Retrieved from Albayan: https://www.albayan.ae/across-theuae/news-and-reports/2018-11-22-1.3415865
8. Intelligent Traffic Systems (2019) Retrieved from RTA: https://www.rta.ae/wps/portal/rta/ae/home/about-rta/rta-agencies/trafficandroads/About/
9. Atlas Telecom launches coast monitoring and protection systems at IDEX 2017 (2017). Retrieved from Emirates News Agency: http://wam.ae/ar/details/1395302598767
10. Fudah M (2010) Intelligent surveillance systems to secure Al Maktoum Airport. Retrieved from Emirates Today News: https://www.emaratalyoum.com/localsection/accidents/2010-05-11-1.242624
11. Node-RED (n.d.) Retrieved from nodered: https://nodered.org/
12. System Properties Comparison Cloudant versus Couchbase versus CouchDB (n.d.). Retrieved fromDB-ENGINES: https://db-angines.com/en/system/Cloudant%3BCouchDB%3BCouchbase
13. Khan S (2019) IBM Watson Visual Recognition: Custom Object Detection Generally Available. Retrieved from Medium: https://medium.com/ibm-watson/ibm-watson-visual-recognition-custom-object-detection-generally-available-906b2ae77969

From Simulation to Deployment: Transfer Learning of a Reinforcement Learning Model for Self-balancing Robot

Sreenithi Sridharan and Talal Shaikh

Abstract Reinforcement learning algorithms are time and resource-intensive and can be influenced by the setup of the physical robot environment and its hardware capabilities. Often, in small-scale projects, it is not workable to build a physical robot equipped with good processing capabilities and to set up a fully controlled and monitored environment for reinforcement learning. During this period, it will be more cost-effective for most of the training to be conducted in a simulated environment and then transferred to a physical robot. In this project, two RL experiments were conducted on a simple two-wheeled robot model in a simulated environment. The first was to make the robot start from a completely fallen down position and learn to stand, and the second to start from a balanced position and learn to maintain the position. It was found that starting from a balanced position gave a better performance, and hence, this learned model was used as a baseline for testing on a physical robot such as the LEGO Mindstorms, but it could be seen that the LEGO hardware was not well suited for this kind of intense reinforcement learning algorithms.

Keywords Self-balancing robot · Reinforcement learning · Simulation

1 Introduction

Robots are capable of being used in a wide range of applications, such as hotel room-service and rescue missions in remote areas where humans cannot go. Due to these adversarial situations, the standard control algorithms, such as those that regulate a robot's movement on smooth terrain, may not satisfy the needs of the robot when these areas are uneven or damaged. As a result, solving problems such as self-balancing through self-learning algorithms like reinforcement learning represents one of the major focuses of research and development.

These algorithms are very resource-intensive and can be easily affected by the environment setup and hardware capabilities of the physical robot [10]. Small-scale

S. Sridharan · T. Shaikh (✉)
Heriot-Watt University, Dubai, UAE
e-mail: t.a.g.shaikh@hw.ac.uk

A. Ullah et al. (eds.), *Proceedings of International Conference on Information Technology and Applications*, Lecture Notes in Networks and Systems 350,
https://doi.org/10.1007/978-981-16-7618-5_25

projects do not always allow for the construction of an automated robot with good processing capabilities and the setup of a fully controlled and monitored environment. The most cost-effective way to train a robot in such an environment is to begin with a simulated one and then transfer it to a physical one. This project aims to perform two RL experiments on a simple 2-wheeled robot model and use this trained model as a baseline model for testing on physical robots. The first RL experiment in the simulated environment is to make the robot learn to balance itself from a completely unbalanced position; the second is to start from a balanced position and learn to maintain it. Based on the results of the experiments, the most successful model was then tested on a LEGO Mindstorms EV3 robot. In future, the project will experiment with other robot models, such as a balancing robot based on the Arduino platform.

2 Background

2.1 *Reinforcement Learning*

In reinforcement learning, the world is represented by a set of states, $\boldsymbol{S}$, and the agent can perform a set of actions, $\boldsymbol{A}$, at each state. Each point in time is considered a *time step*. At each time step $\boldsymbol{t}$, the agent looks at the current state $\boldsymbol{s_t}$ and chooses an action $\boldsymbol{a_t}$ from the set of all actions. After performing this action, the agent receives a reward $\boldsymbol{r_{t+1}}$ from the set of all possible rewards $\boldsymbol{R}$ and reaches a new state $\boldsymbol{s_{t+1}}$. The rewards may be positive or negative depending on whether the action taken by the agent leads it toward the goal or away from it. In addition, the discount factor $\boldsymbol{\gamma}$ influences the algorithm on whether it should look for later rewards or immediate rewards, and the learning rate $\boldsymbol{\alpha}$ is used to tell how much importance must be given to the previously learned knowledge and newly learned knowledge.

All the above discussed parameters can be formulated as a Q-Learning function equation as (Eq. 1) below which will be used repeatedly to get different Q-values for each state-action pair, and ultimately high Q-values for the best action at each state, defining the right action policy.

$$Q(s, a_t) \leftarrow Q(s, a_t) + \alpha(r_{t+1} + \gamma \max Q(s_{t+1}, a)) \quad (1)$$

2.2 *Self-balancing Robots*

Self-balancing robots are commonly called the inverted pendulum problem [11] due to its nature of trying to remain attached to the ground in the bottom and the unstable top body of the robot swings like a pendulum until it learns to balance on its own. In the self-balancing problem, the major task to be achieved is to keep the Center of

Mass (COM) of the robot in coordination with the wheels such that the robot can either stay balanced in a stationary position or move forward without falling down using control algorithms.

There are three approaches to developing such control algorithms on robots. The first approach is by directly developing and testing control algorithms on the physical robot [15], while the second approach is to use a completely simulated environment and robot as done by [12] and the third approach is by combining both, where the time and resource consuming training process for the control algorithm is done in a simulated environment and the trained model is then transferred and fine-tuned on the physical robot [6].

Different manual control algorithms like PID, LQR and Fuzzy controllers have been used on self-balancing robots like [9–11]. Taking this one step forward using AI control algorithms, [8] used Q-learning to solve the LQR control problem, while [12] used Q-learning independently of traditional control algorithms to make the self-balancing robot learn to balance.

This project too aims to implement a Q-learning model independent of traditional control algorithms like [12] in a completely simulated environment with a simple simulated robot model and then transferred the learnt model to a physical robot and fine-tuned for better results using the transfer learning approach like [6].

2.3 Transfer Learning

'Transfer learning' refers to learning fully or partially in one environment and transferring that knowledge to another. Recent years have seen the emergence of this method in the realms of Computer Vision and Natural Language Processing (NLP). In Convolutional Neural Networks (CNN), the first few convolutional layers of a trained model are called the 'backbone,' which is meant for performing basic image processing, while later layers focus more on solving a specific classification problem [16]. CNN models like ResNet-50 [13, 14] and VGGNet [4, 7] have been widely used through this concept of transfer learning in various industries, including medicine, civil engineering, and so on. Extending this to Robotics, tasks involving vision and control learnt using Deep Learning or Deep Q-Learning were first learnt in simulation and then transferred and fine-tuned on a robot in real-world as done by [2, 6, 17].

In this project, this idea was experimented on a simple Q-learning RL algorithm, where the Q-table learned in the simulated environment was transferred to the LEGO Mindstorms robot. The GyroBoy [5] kit for LEGO Mindstorms EV3 is known to be used as a self-balancing project. A GyroBoy's arm structure is made up of color and ultrasonic sensors as seen in Fig. 1, which are controlled by a small motor so that they can avoid obstacles. As shown in Fig. 2, these components were removed and the structure was modified accordingly. Additional extensions were added to the robot's front and back in order to prevent the robot from falling over entirely and causing damage to the hardware and connections.

Fig. 1 Official GyroBoy structure

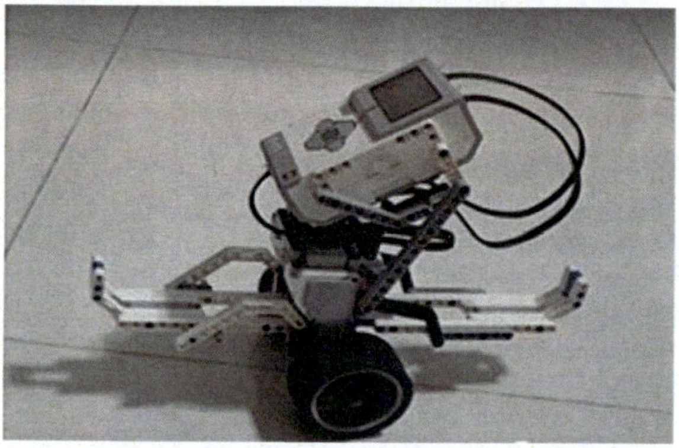

Fig. 2 Modified GyroBoy for reinforcement learning

3 Implementation

3.1 Simulation Environment and Tools

The simulation environment used was the Gazebo Simulator with the Robot Operating System (ROS), to run Python or C++ programs.

Robot Model. The "*teeterbot*" robot model[1] was used. This was a very simple structure of a block connected to 2 wheels in an empty simulation world as seen in Fig. 3.

[1] https://github.com/robustify/teeterbot.

3.2 RL Algorithm

As discussed in Sect. 2.1, a basic Q-Learning Algorithm was implemented for this project. The different components of the Q-Learning algorithm were defined as described below.

State Space. The state space was decided to be the possible angles of inclination that the robot can have. The angle of inclination at the absolute balanced position is 0°, and the inclination toward the back and front of the robot were retrieved as negative and positive angles, respectively. As the Q-table does not work with a continuous state space, the state space was discretized to integers.

There were two parts to the experiment, the first starting to learn to stand from a completely fallen down position and the second to start from a balanced position and learn to maintain the position as shown in Fig. 4a, b, respectively. Hence, 2 different state space ranges were used for the 2 experiments as below:

$$\text{Starting from a fallen position, } S = \{s| - 86^\circ \le s \le 86^\circ\} \tag{2}$$

$$\text{Starting from a balanced position, } S = \{s| - 60^\circ \le s \le 60^\circ\} \tag{3}$$

Action Space. For the simulation robot, the action space was the voltage values set to the wheel motors. The action space too was discretized, by considering only integer values. The discretized action space can be represented mathematically as below:

$$A = \{a| - 100 \le a \le 100\} \tag{4}$$

Reward Function. Taking ideas from the different approaches of [1, 3, 14], the reward function was designed corresponding to how far the robot's position is from the balanced position as described below:

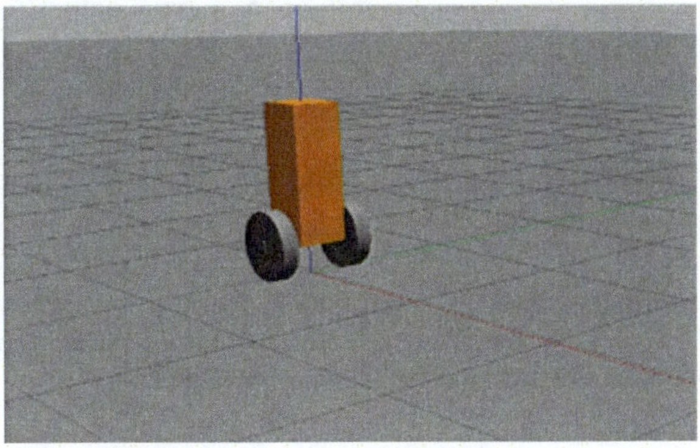

Fig. 3 Teeterbot robot 3D model by Micho Radovnikovich

$$R(s_t, a_t) = \begin{cases} 100, & \text{when } s_{t+1} \text{is the ideal state, i.e. robot is balanced at } 0^\circ \\ 5, & \text{when} - 8^\circ \leq s_{t+1} \leq -8^\circ \\ 2, & \text{when} - 40^\circ \leq s_{t+1} \leq -9^\circ \text{ and } 9^\circ \leq s_{t+1} \leq 40^\circ \\ 1, & \text{when} - 60^\circ \leq s_{t+1} \leq -41^\circ \text{ and } 41^\circ \leq s_{t+1} \leq 60^\circ \end{cases} \quad (5)$$

where (s_t, a_t) is the previous state and action, and s_{t+1} is the new state achieved when taking an action a_t from the state s_t.

In addition, to encourage the robot from getting up from an unbalanced position like s_{t+1} higher than $\pm 40^\circ$, the differences in angles were considered. So, if the new angle difference $s_{t+1} - s_t$ is positive, it implies that the robot is falling, hence it was penalized by giving a negative reward of -5 instead of 1. The higher negative reward is given so that the robot is penalized heavily and tries to avoid it and act opposite, by trying to get up from that position. Another modification to the reward function is when the robot is nearing its balanced position. Now, the robot was encouraged to maintain this balanced position, and penalized if it started falling. To do this, when the robot is at a position $\pm 8^\circ$ from the ideal 0° state, the angle differences were again considered. So if the new absolute angle difference $|s_{t+1} - s_t|$ is lesser than the previous angle difference $|s_t - s_{t-1}|$, the reward was doubled.

Learning Rate α and Discount Factor γ. A Learning rate α of value 0.7 and a high discount factor of 0.999 was used. This was done to give more importance to newly learnt knowledge, while looking for short term rewards.

Random Action Factor ε. Any RL algorithm requires a tradeoff between exploration and exploitation when selecting actions, where the robot must explore using random actions in the beginning of learning, and then go on to use the learnt Q-table as learning progresses. This was achieved using the Epsilon greedy strategy.

This difference in choosing actions is achieved by using the random action factor ε which is given a very high value initially and is reduced by some factor with each episode. The learning algorithm starts with a high value of 0.7 which gets reduced to 99% of its value through each episode. So, the epsilon value changes as follows:

Episode 1——0.7

Episode 2——0.7 $*$ 0.99

Episode 3——0.7 $*$ 0.99 $*$ 0.99

... and so on

Criteria for Early Stopping of Training Episodes. The simulation robot was trained for 1500 episodes with 1000 iterations in each episode by default. But an early stopping point for the learning process was considered to stop the learning at an earlier stage when the robot has learnt to balance well enough for that episode. The criteria was considered to be accumulated total rewards of 3000 per episode as opposed to the final state as done in [12].

Fig. 4 Simulation robot at **a** (left) fallen position (pitch angle −86°) and **b** (right) standing position (pitch angle 0°)

4 Testing and Performance Assessment

4.1 Starting from a Fallen Down Position

The RL algorithm was made to run for 1500 episodes with 1000 iterations each, starting from a completely fallen position as shown in Fig. 4a. But it was found that the robot could not get up at all, and there was no improvement in the states and rewards collected. This can be seen in Fig. 5 below by the total rewards accumulated per episode of training. Though there were some slight improvements in the rewards collected, there was not a very good difference, and the total number of rewards are still negative indicating that the robot has not gotten up at all.

Reasons of Failure. When investigating the reason of failure, it was found that the robot was not be able to lift its body from the floor just by using the motor commands as it didn't have a structure that can support it, and there was no obstacle or end point in the simulation world that the robot can hit creating a change in momentum and state of inertia that can lift the body of the robot.

4.2 Starting from a Standing Position

As a second experiment, the training was run for 1500 episodes with 1000 iterations each starting from a balanced position (pitch angle 0°) as shown in Fig. 4b.

The total rewards collected in each episode were recorded as shown in Fig. 5. This time, the average total rewards improved over the episodes and become more positive.

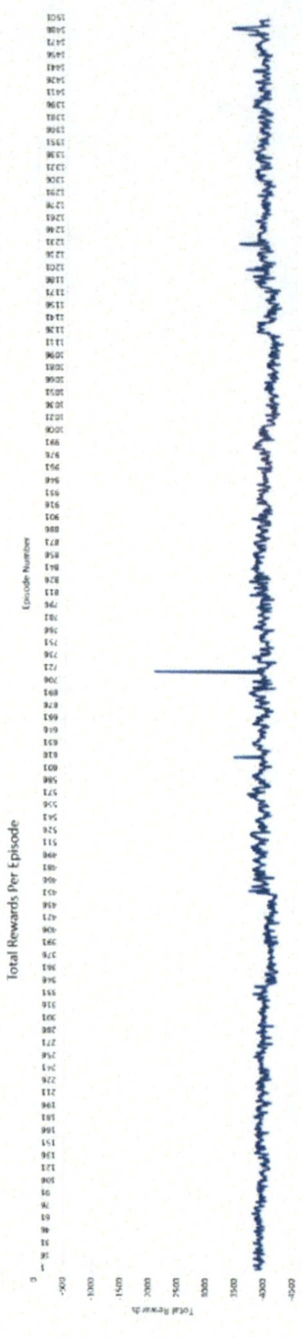

Fig. 5 Total rewards per episode for 1500 episodes starting from a fallen position

Though the overall average is positive, some epochs still had negative rewards. This is due to the fact that though the robot was allowed to start at a balanced position, the state space and reward function were still kept the same that is the robot was allowed to completely fall down during the training and attempt to get up. From the previous experiment, it could be seen that getting up from a completely fallen position was not possible for the robot, and hence would result in accumulation of higher negative rewards if the robot fell down earlier during the episode. The positive average of rewards after 1200 episodes shows that the robot has learned to maintain its balance better over the 1500 episodes.

4.3 Testing Performance of the Learnt Model

To check the maximum rewards that the robot can collect, the robot continued to train for 200 more episodes starting with the learnt model and removing the maximum rewards constraint. The results of the total rewards accumulated for each episode are shown in Fig. 6. The robot was now able to achieve higher rewards of more than 4000 and balance for a longer time, with a linear increase in reward trend.

In addition, the time taken by the robot to fall from the balanced position was recorded for 10 random episodes each at the beginning of the training and after training for 1500 episodes. The recordings can be seen in Fig. 7. The average falling time at the start of training was found to be around 4.8 s, while the average after training has significantly increased to 9.0 s. This shows that the robot has learned to balance better after training and can maintain its position for a longer time (Fig. 8).

4.4 Testing on the LEGO Mindstorms EV3 Robot

After learning for 1500 episodes in the simulated environment, the learned Q-Table was then tested on the modified GyroBoy robot built using the LEGO EV3 Kit as shown in Fig. 2. The RL algorithm was programmed in Python using a specially designed ev3dev-lang-python library. But this did not provide good results, and the robot failed to balance in its position. The robot did try to make attempts to get up from its position as can be seen in the video,[2] but could not balance consistently in its position.

The possible causes of failure were investigated to find that there were two main problems. The first is the problem of time lag in sending and receiving motor commands which was caused by the ev3-dev-lang-python library. For the balancing algorithm to work efficiently, the robot must accelerate immediately in the direction that it is falling to prevent its fall and hence prompt response of the motors is very crucial, which didn't work on the robot. Hence, the robot fell even before the motors

[2] https://youtu.be/WvAqW9NyCPc.

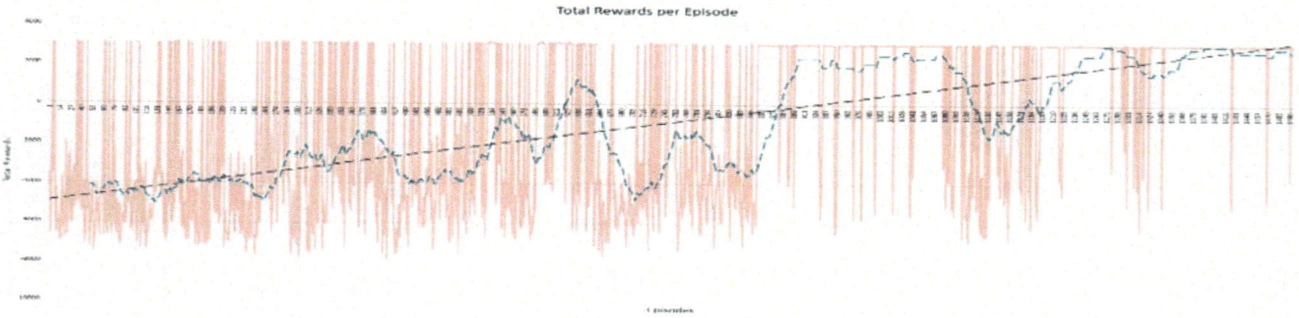

Fig. 6 Total rewards per episode for 1500 episodes starting from a standing position. The pink lines show the graph of the total rewards collected in each episode. The blue and black lines are the trend lines of the graph. The blue line represents the moving average over 50 episodes, and the black line shows the linear trend of the graph

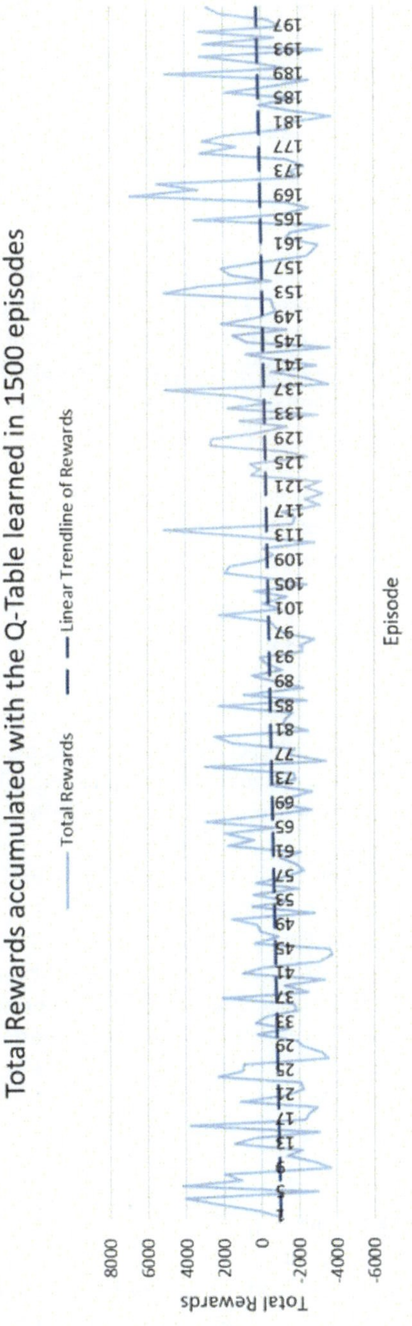

Fig. 7 Total rewards per episode for 200 episodes accumulated by the robot after learning

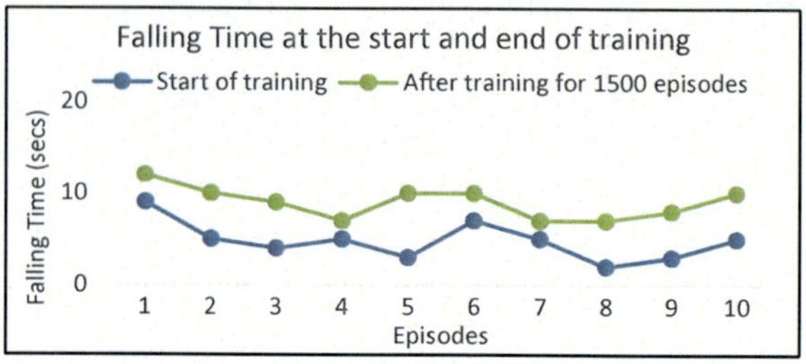

Fig. 8 Falling time when starting from the balanced position at the start and end of training

react making it unable to balance in its position. Second was the problem of Gyro Drift caused by the LEGO Gyro Sensor when the RL algorithm was made to run for a few minutes on the robot. The Gyro Drift is the phenomenon of changing gyro values with time, where the gyro sensor would read different values for the same position of the sensor. As the Gyro Sensor was the only source of input that tells the robot its current position or state, faulty values provided by the sensor meant that the RL algorithm can no longer be run on the robot.

Hence, in future, a different robot structure such as an Arduino must be used for such intense tasks and techniques such as Kalman Filters can be used to better calibrate noisy input data from the sensors.

5 Conclusion

This project explored how an RL algorithm can be used to implement a self-balancing robot structure in simulation. Two experiments were conducted, where the robot was made to learn to balance by starting from a completely fallen position and at a balanced position. It was seen that the simple inverted pendulum robot structure was not capable to get up from a fallen position independently, and hence requires a suitable design of simulation environment or a different robot structure to learn to balance, while starting from a balanced position gave better learning results. The learning results from simulation were then tested on a physical robot such as the LEGO Mindstorms, but it could be seen that the LEGO Hardware was not well suited for this kind of intense reinforcement learning algorithms.

6 Future Works

This project used a simple Q-Learning approach to solve the inverted pendulum problem. But a lot of research has gone into the field of RL and newer and efficient methods such as Deep RL or Deep Q-learning methods have been adopted. In future, this project too can be extended to try the Deep Q-Learning method in a simulation environment. In addition, the LEGO EV3 robot hardware could not handle intense

Q-learning, hence in future, robots built using alternate hardware such as an Arduino can be used to better assess the performance of the transfer learning capabilities of the learnt model.

References

1. Bonarini A, Caccia C, Lazaric A, Restelli M (2008) Batch reinforcement learning for controlling a mobile wheeled pendulum robot. In: IFIP international conference on artificial intelligence in theory and practice. Springer, Boston, MA, pp 151–160
2. Bousmalis K, Irpan A, Wohlhart P, Bai Y, Kelcey M, Kalakrishnan M, Downs L, Ibarz J, Pastor P, Konolige K, Levine S (2018) Using simulation and domain adaptation to improve efficiency of deep robotic grasping. In: 2018 IEEE international conference on robotics and automation (ICRA). IEEE, pp 4243–4250
3. Chang CL, Chang SY (2016) Using reinforcement learning to achieve two wheeled self balancing control. In: 2016 international computer symposium (ICS). IEEE, pp. 104–107
4. Cheng PM, Malhi HS (2017) Transfer learning with convolutional neural networks for classification of abdominal ultrasound images. J Digit Imaging 30(2):234–243
5. Christensen BK (2017) GyroBoy–a self-balancing robot programmed in JAVA with leJOS EV3
6. Ding Y, Gafford J, Kunio M (2012) Modeling, simulation and fabrication of a balancing robot. Harvard University, Massachusettes Institute of Technology, 151
7. Gao Y, Mosalam KM (2018) Deep transfer learning for image-based structural damage recognition. Comput-Aided Civ Infrastruct Eng 33(9):748–768
8. Guo L, Rizvi SAA, Lin Z (2021) Optimal control of a two-wheeled self-balancing robot by reinforcement learning. Int J Robust Nonlinear Control 31(6):1885–1904
9. Juang HS, Lurrr KY (2013) Design and control of a two-wheel self-balancing robot using the Arduino microcontroller board. In: 2013 10th IEEE international conference on control and automation (ICCA). IEEE, pp 634–639
10. Mahmood AR, Korenkevych D, Vasan G, Ma W, Bergstra J (2018) Benchmarking reinforcement learning algorithms on real-world robots. In: Conference on robot learning. PMLR, pp. 561–591
11. Rahman MM, Rashid SH, Hassan KR, Hossain MM (2018) Comparison of different control theories on a two wheeled self balancing robot. AIP Conf Proc 1980(1):060005
12. Rahman MM, Rashid SH, Hossain MM (2018) Implementation of Q learning and deep Q network for controlling a self balancing robot model. Robot biomimetics 5(1):1–6
13. Reddy ASB, Juliet DS (2019)Transfer learning with ResNet-50 for malaria cell-image classification. In: 2019 International conference on communication and signal processing (ICCSP). IEEE, pp. 0945–0949
14. Rezende E, Ruppert G, Carvalho T, Ramos F, De Geus P (2017) Malicious software classification using transfer learning of resnet-50 deep neural network. In: 2017 16th IEEE international conference on machine learning and applications (ICMLA). IEEE, pp 1011–1014
15. Tsai CC, Huang HC, Lin SC (2010) Adaptive neural network control of a self-balancing two-wheeled scooter. IEEE Trans Industr Electron 57(4):1420–1428
16. Visual Transfer Learning for Robotic Manipulation. https://ai.googleblog.com/2020/03/visual-transfer-learning-for-robotic.html, last accessed 2021/16/06
17. Zhang F, Leitner J, Milford M, Corke P (2016) Modular deep q networks for sim-to-real transfer of visuo-motor policies. arXiv preprint arXiv:1610.06781

Deep Oversampling Technique for 4-Level Acne Classification in Imbalanced Data

Tetiana Biloborodova, Mark Koverha, Inna Skarga-Bandurova, Yelyzaveta Yevsieieva, and Illia Skarha-Bandurov

Abstract The current technological horizon in the Internet of Things, computer vision, deep learning and healthcare systems makes it possible to monitor some pre-existing conditions as acne vulgaris, automate the assessment of the acne severity by photo and monitor the skin health by specialists. Remote analysis of the skin condition and automatic image classification has several challenges. One of the most critical problems is imbalanced data raised because the number of clinical cases for each acne grade differs. This paper proposes a deep oversampling technique for 4-level acne classification that enables to deal with imbalanced datasets. The method was validated using several criteria. The experimental results obtained for imbalanced data sets revealed that the acne classification via proposed deep oversampling outperforms benchmark approaches.

Keywords Acne classification · Remote monitoring · Deep learning · Oversampling technique · Imbalanced data

T. Biloborodova (✉)
G.E. Pukhov Institute for Modelling in Energy Engineering, 15 General Naumov Str., Kyiv 03164, Ukraine
e-mail: biloborodova@snu.edu.ua

M. Koverha
Department of Computer Science and Engineering, Volodymyr Dahl East Ukrainian National University, 43 Donetska Street, Severodonetsk 93400, Ukraine

I. Skarga-Bandurova
School of Engineering, Computing and Mathematics, Oxford Brookes University, Wheatley Campus, Oxford OX33 1HX, UK

Y. Yevsieieva
School of Medicine, V. N. Karazin, Kharkiv National University, 4 Svobody Square, Kharkiv 61002, Ukraine

I. Skarha-Bandurov
Luhansk State Medical University, 32 Budivel'nykiv Street, Rubizhne 93012, Ukraine

A. Ullah et al. (eds.), *Proceedings of International Conference on Information Technology and Applications*, Lecture Notes in Networks and Systems 350,
https://doi.org/10.1007/978-981-16-7618-5_26

1 Introduction

Social distancing and quarantine are steps applied worldwide to minimize social interaction between people and thus reduce coronavirus (COVID-19) transmission. Due to unprecedented distancing measures, frequent visits to the doctor without urgent need are undesirable. However, some physiological processes in the body require regular monitoring to assess the condition and receive further recommendations for care. Acne, also known as acne vulgaris, is one of these processes. Acne vulgaris is a long-term inflammatory skin disease that manifests itself in various skin lesions. Modern advances in the Internet of Things technologies, computer vision, deep learning and healthcare systems make it possible to monitor the skin condition, automate the assessment of the severity of acne and remotely monitor the state of the skin by specialists.

The main goal in the long-term perspective is to assess a response to acne therapy using remote skin monitoring. One of the most important components of remote skin condition monitoring is the stage of automatic recognition and assessment of the severity of acne using a photo of the patient taken at home. To provide patients with a reliable remote solution, we have analyzed tools and techniques for automated grading of acne vulgaris [1, 2].

A recent study on skin assessment for acne developed by a joint team from Microsoft and Nestlé Skin Health [3] aimed to solve the problem of using unlabeled image data and reducing noise in image data. In this study, the team conducted experiments to detect and classify acne using the convolutional neural network (CNN) image classification and regression. The RMSE of the recognition model is 0.482. The qualitative and quantitative analysis of the skin condition using different deep learning techniques allow us to identify the following problems in-home monitoring, which wield significant influence on recognition quality and, as a result, on the level of adoption of this technology:

- Image Variety: a face photo can be taken in different conditions (resolution, angle, light, distance, etc.) and other volumes that turn into data misbalancing.
- Data complexity: several criteria are used to assess the skin condition (oily skin, acne disease, skin radiance), each of which requires an individual approach.
- Preparation of marked images for training the model for specific criteria. Research in the field of image recognition shows that to this moment, high model accuracy can only be achieved using labeled images to train the model.
- Quality of recognition: Without considering the class imbalance problem, learning algorithms or constructed models can be overwhelmed by the majority class and can ignore the minority class.
- Acne grading scales variety: Medical aspect of this problem is in a wide variety of the grading scales; to this moment, more than 25 systems exist for evaluating acne severity. Moreover, acne measurers are divided into two camps, the graders and the lesion counters [4].

Grading is widely used for diagnosing acne severity. It is based on observations of the dominant lesions, evaluating the presence or absence of inflammation and estimating the extent of involvement. We analyzed several grading scales developed by US Food and Drug Administration IGA (Investigator's Global Assessment) [5], American Academy of Dermatology [6], France GEA (Global Acne Evaluation) [7], Ukraine [8], Japan [9]. As it was mentioned in [6], a unified grading system may be very helpful, but at present, there is no universal approach for grading and assessing severity of acne. Most of the scales [5–9] suggest 4-level acne severity rate. For this reason, in this study we also use a 4-type classification.

The majority of the acne cases are graded as a mild, while only about 1% have severe acne rate [10–12]. Class distribution in this case is not uniform that can negatively affect the model performance and lead to incorrect classification. Most contemporary works in class imbalance concentrate on imbalance ratios ranging from 1:4 up to 1:100 [13].

2 State-of-the-Art

There are several approaches to deal with imbalanced datasets. In [14], an algorithm for assessing the severity of acne was presented. Photos were annotated according to IGA [5], and CNN was used for classification. The models were trained separately on three different image sizes. In [14], data imbalance was noted as one of the limitations, which did not allow obtaining a classification accuracy of more than 67%. The paper [15] proposes a system for assessing the general acne severity by counting local skin lesions. At the first stage, the system extracts labels, and then their distribution is used to evaluate the overall acne severity level. All experiments were conducted on their own hand-annotated imbalanced dataset. The authors have obtained a classification accuracy equal to 84.11%. In the study [16], the authors used CNN to construct binary and multi classifiers on two balanced datasets. An increase in the data has been carried out and obtained accuracy for any class does not exceed 81%. The authors of [17] proposed a new CNN-based AcneNet that demonstrated a very high classification accuracy equal to 99.44%, given a balanced dataset to train the proposed model. CNN from GoogLeNet is used in [18]. In their study, data augmentation was applied. As a result, 325 image areas were obtained from 164 source images and used for further classification. The achieved classification accuracy is 85.6%. A higher recognition quality (up to 99.44% accuracy) is achieved for balanced datasets for all these papers. This fact determines the relevance of studies in the direction of improving classification performance by eliminating the negative impact of the imbalanced distribution of data.

The palette of the techniques used for handling imbalanced datasets includes three levels: data-layer, algorithm-layer and evaluation layer. The methods used at the data layer mostly deal with resampling and are more general because they do not rely on the chosen classifier. There are two basic techniques in this layer, and they are oversampling and undersampling [19]. Oversampling comprises from elimination of the negative impact of uneven distribution by creating new samples from the minority class while undersampling reduces negative effect of uneven distribution by discarding internal samples in the majority class. Hybrid methods utilize combinations of these two for data balancing. At first glance, the oversampling and undersampling seem functionally equivalent, since they both resize the original dataset and can provide the same proportion of balance. However, removing objects from the majority class can cause the classifier to miss important concepts related to the majority class [20]. Thus, in order to avoid the data loss, especially when dataset is not huge enough, it makes sense to solve data imbalance using oversampling.

3 Methodology

In this section, we describe our approach for automatically recognition of the severity of acne based on deep learning with series of preprocessing techniques formulated below.

Following [20], we assume imbalanced dataset S with m objects $|S| = m$, as a pair $S = \{(x_i, y_i)\}$, $i = 1, \ldots, m$, where $x_i \in X$ denotes the instance in the n-dimensional feature space $X = \{f_{1,}f_2, \ldots, f_n\}$, and $y_i \in Y = \{1, \ldots, C\}$ denotes the class identifier label associated with x_i. in particular, when $C = 2$ we deal with binary classification task. In addition, we define subsets $S_{\text{min}} \subset S$ and $S_{\text{maj}} \subset S$, where S_{min} is the set of minority objects of classes in S, and S_{maj} is the set of objects of prevailing classes in S, so that $S_{\text{min}} \cap S_{\text{maj}} = \{\Phi\}$ and $S_{\text{min}} \cup S_{\text{maj}} = \{S\}$. Based on these assumptions, the problem of acne recognition can be formulated as a 4-type classification task.

Thereafter, the multiclass classification problem is transformed into a regression task [21]. This transformation helps to reduce possible subjectivity in the assessment of the acne severity at the stage of data annotation. Since the acne levels class have a meaningful ordering, we assign the numeric values from 0 to C to each of the corresponding C classes labels. Formally, it presented as follow $y \subseteq S = \{1, \ldots, C\} \xrightarrow{transformation} y \in \mathbb{R}^n$. The values of the class variable are transformed into numbers. A numeric target is corresponding to a level of acne severity.

Thresholding is also used to transform back a regression prediction into a classification prediction $\hat{y} \in \mathbb{R}^n \xrightarrow{transformation} \hat{y} \subseteq S$, where $\hat{y}$ is a prediction.

The proposed methodology includes the several steps, as shown in Fig. 1. Firstly, we extract patches from original images of human faces using one of the two pre-trained models. Then, data augmentation, feature extraction and data oversampling are conducted. Finally, results obtained after oversampling are fed to CNN for model training and evaluation.

Step 1: Patches extraction. At the first stage, preliminary processing of images is carried out for capturing skin areas from the face image. The severity of acne does not depend on the location; it depends on the volume and severity of the lesion on the patient's face. From this prospective, the separate patches of the face image should give results that are more informative

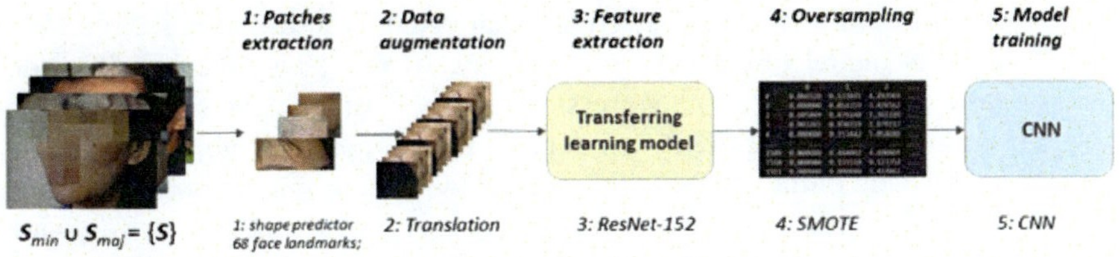

Fig. 1 Overview of the proposed approach

relatively to the entire face. The result of this step is a dataset *S* consisting of *m* patches extracted from facial images. Each patch inherits the image class label.

Step 2: Data augmentation. CNNs are spatially sensitive, which leads to insufficient recognition quality when using a limited number of images for network training. That is, in the case when the test images have new acne lesions that did not occur in the training images, the trained model is not able to generalize the training results on the test data. To overcome this issue, we use translation of an image section. Translation involves moving an image around the x and y axes in specified directions by a specified number of pixels. The sliding translation was used to enlarge these areas of the images. The result of this step is a dataset *S* consisting of an increased number of patches. Each translated patch inherits the class label of the original patch that was used for augmentation.

Step 3: Feature extraction. We utilized the transferring learning paradigm [22] and a pre-trained ResNet-152 to extract features from training set of acne images. The result of this step is a dataset $S = \{(x_i, y_i)\}$ consisting of the features extracted using the trained model, where x_i is the vector of extracted features of the patch m_i, and y_i is the class label which denotes the severity of acne associated with x_i.

Step 4: Oversampling. We use oversampling to balance the number of dataset objects for each class. Formally, oversampling can be represented as follows. Any objects generated from the dataset *S* are denoted as *E*, with disjoint subsets of $E_{\min}$ and E_{maj} representing the minority and majority of the *E* objects, respectively, whenever they are applied. The random oversampling process is implemented by adding a set *E* selected from the minority class: for a set of randomly selected minority examples in $S_{\min}$, increase the original set *S* by replicating the selected examples and adding them to *S*. Thus, the number of typical examples in $S_{\min}$ increases by $|E|$, and the balance of the class distribution of *S* is adjusted accordingly. This provides a mechanism for changing the degree of balance in the distribution of classes to any desired level. The result of this step is a dataset $S = \{(x_{\mathrm{i}}, y_{\mathrm{i}})$ consisting of extracted and generated features, where x_i is a vector of extracted and generated features of patches m_i, and y_i is the class label associated with x_i.

Step 5: Model training. The extracted and generated features are passed to train a CNN model to classify acne severity. At this stage, the classification task was changed into a regression task by defining acne severity grades as integer equivalents. It was done to reduce possible subjectivity in the assessment of the acne severity at the stage of data annotation. The inverse transformation was done using [0.5, 1.5, 2.5] as the edge list. To divide the underlying truth and predicted severity levels into categorical severity levels, we used severity label <0.5 to assign level 0, a severity label in the range 0.5–1.5 to assign level 1, labeled severity in the 1.5–2.5 range to assign level 2, and labeled >2.5 to assign level 3.

Step 6: Model evaluation. The test pipeline includes the extraction of facial image patches, feature extraction using the ResNet-152, patch classification from trained CNN, obtaining the average predicted value of the acne severity class from the face image, and assessing the quality of recognition. Model evaluation is implemented on test data. In this step, each patch of the face image was evaluated, and the average rate for all patches of the same face was assigned to the image class.

4 Results

For this study, the ACNE04 [15] the open dataset was used. The ACNE04 includes 1457 face images and expert annotations according to the Japanese rating scale. The dataset has the following acne severity annotations: level 0–Mild, level 1–Moderate, level 2–Severe, level 3–Very severe. All images were taken at an angle of approximately 70 degrees from the front of the patient and manually annotated by specialists. The distribution of objects in accordance with the class is presented in Table 1.

We used 1165 images to train the model and 292 images to test the model which corresponds to a distribution of 80% for training and 20% for testing. In accordance with a four-point scale for assessing the severity of acne, the images are labeled as follows. The images are annotated as follows: level 0 Mild assigned to class 0, level 1 Moderate-to class 1, level 2 Severe-to class 2, level 3 Very severe-to class 3. A study by Microsoft [3] was used as a benchmark for the experiment.

Table 1 Image class distribution in ACNE04 [14] used for experiments

Acne severity	Class	Dataset	Train	Test
Level 0 mild	0	513	410	103
Level 1 moderate	1	633	506	127
Level 2 severe	2	507	146	362
Level 3 very severe	3	108	103	6
Total		1457	1165	292

Table 2 Data distribution for each preprocessing step

Class	Raw data	Patched data	Augmented data	Oversampled data
0	410	1367	3556	4333
1	506	1716	4333	4333
2	146	476	1843	4333
3	103	247	1514	4333
Total	1165	3806	11,246	17,332

At the first stage, the procedure of image preprocessing was carried out. This step utilizes two pre-trained models. The first one is shape_predictor_68_face_landmarks model [23], which returns 68 characteristic points for each person given the bounding box of their face. This facial landmark model detects landmarks of frontal borders or faces slightly away from the camera while two open eyes can be observed in the image. Then, the patches of the face, cheeks and chin are extracted from the original images and stored as image files in the main directory. In case, when the face is very away from the camera and both eyes were not recognized in the image, shape_predictor_68_face_landmarks cannot reveal the boundaries of the face. When this occurs, an alternative, the One Eye model [24], was used. When the One Eye model identifies an eye in the image, areas of the skin of the face and cheeks are cropped from the original image and stored as separate image files (patches) in the main directory. In all scenarios, when the selection of patches from the photo is successful, the only selected patches with reference to the original image are used for further steps. Otherwise, if neither of these models cannot process the image, the original image was used entirely for further processing and analysis. A fragment of a set of dependencies and binding patches to the original image is given below.

After preprocessing stage, we obtained 3806 image patches. For data augmentation, sliding translation of patches was used. All images were normalized to 224 by 224 pixels. Further, the feature extraction from each patch was carried out using the ResNet-152 model [25]. Each patch is bound to the original image and thus the extracted features inherit the dependencies of the patches. At the next step, we used Synthetic Minority Oversampling Technique (SMOTE) [26]. The total number of samples for each class was fitted to the most numerous class in augmented dataset. Information about the data distribution at each stage is summarized in Table 2.

Data generated at the oversampling stage were used to train a CNN classifier. Object classes were defined as integers.

5 Discussion

The results of comparison between the proposed approach and baseline methods without oversampling are shown in Table 3.

Table 3 The results of experiments with and without oversampling

Metrics	Without oversampling	With oversampling	Delta,%
RMSE	0.422356	0.397419	5.904261
MAE	0.325285	0.297179	8.640423
MedAE	0.256841	0.223299	13.05944
EV	0.826736	0.873874	5.7017
MAPE	0.199264	0.171855	13,75,512
R^2	0.826682	0.873646	5.68102
MPD	0.102844	0.087945	14.48699
MGD	0.070191	0.061218	12.78369

Resulting RMSE, MAE, MedAE, MAPE, MPD, MGD criteria show a smaller error, while the EV, R^2 criteria show a higher value using the proposed approach, which indicates a higher quality of the model using the proposed approach. We also compared the performance of the proposed approach with other studies aimed at determining the severity of acne from a photo. The criteria RMSE, MAE, MedAE, R^2 can only be compared between models whose errors are measured in the same units, i.e., using the same data.

A set of metrics for evaluating the model quality as *Accuracy*, *error rate* (*ER*) were computed after converting the results back from continuous to discrete scale using [0.5, 1.5, 2.5] as a list of edges for separating true labels and predicted ones. The *Precision, Recall, F1-score* metrics were used for evaluating the classification performance of each class. The results are shown in Table 4.

The oversampling made it possible to achieve a higher accuracy for each class and in general for the results of classification.

Comparison of the obtained results with the benchmark models discussed above is presented in Table 5.

As can be seen, the proposed approach showed the highest accuracy and the lowest error rate in comparison with studies with imbalanced data, however, it could not surpass the results for works where input data was initially balanced, which sounds natural. At the same time, it can be argued that for current task, the proposed approach has shown a significant advantage for acne classification in conditions of imbalanced data which in turn shape expectation to use this technique for remote self-assessment.

Table 4 Evaluation of classification result

Metrics	Without oversampling (%)	With oversampling (%)
Precision	82.25	82.25
Recall	82.25	84.75
F1-score	82.5	85
Accuracy	80	85
ER	20	15

Table 5 Comparison of acne classification studies

Approach for acne classification	Balanced/imbalanced	Accuracy (%)	ER (%)
Junayed et al. [16]	Balanced	99.44	0.56
Lim et al. [13]	Imbalanced	67	33
Wu et al. [14]	Imbalanced	84.11	15.89
Ours	Imbalanced	85	15

6 Conclusion

The paper presents an approach for assessing the severity of acne from a photograph under conditions of imbalanced data. The proposed technique includes image preprocessing, data augmentation, oversampling, feature extraction, training and model evaluation. To reduce the subjectivity of assessing the severity of acne by different experts at the stage of image annotation, the classification task is transformed to a regression task by defining acne severity classes as integer equivalents. When obtaining results on test or new unlabelled data, the predicted values obtained for each sample are reduced to the values of the acne severity classes using the described scaling. To assess the quality of the model, we used different criteria that show a smaller error and a higher value comparing benchmark models, which indicates a higher quality of the model using the proposed approach. The model showed the highest accuracy for studies using imbalanced data and the lowest error rate. Thus, we can talk about the advantage of using the proposed technique for image analysis in conditions of imbalanced data and could suggest it for further investigation.

References

1. Arhubdulal/DermaVigil. Github.com [Online] https://github.com/arhubdulal/DermaVigil
2. TroveSkin. Your Smart All-In-One Skincare Tracker. Troveskin.com. [Online] https://www.troveskin.com/
3. Zhao T, Zhang H, Spoelstra J (2019) A computer vision application for assessing facial acne severity from selfie images. arXiv preprint arXiv:1907.07901
4. Witkowski JA, Parish LC (2004) The assessment of acne: an evaluation of grading and lesion counting in the measurement of acne. Clin Dermatol 22(5):394–397. https://doi.org/10.1016/j.clindermatol.2004.03.008 PMID: 15556725
5. Nast A, Dreno B, Bettoli V, Degitz K, Erdmann R, Finlay AY, Ganceviciene R, Haedersdal M, Layton A, López-Estebaranz JL, Ochsendorf F (2012) European evidence-based (S3) guidelines for the treatment of acne, 1–29. https://doi.org/10.1111/j.1468-3083.2011.04374.x
6. Zaenglein AL, Pathy AL, Schlosser BJ, Alikhan A, Baldwin HE, Berson DS, Bowe WP, Graber EM, Harper JC, Kang S, Keri JE (2016) Guidelines of care for the management of acne vulgaris. J Am Acad Dermatol 74(5):945–973
7. Dreno B, Poli F, Pawin H, Beylot C, Faure M, Chivot M, Auffret N, Moyse D, Ballanger F, Revuz J (2011) Development and evaluation of a global acne severity scale (GEA scale) suitable for France and Europe. J Eur Acad Dermatol Venereol 25(1):43–48

8. ACNE: Evidence-based clinical guideline (2017), Ukraine. [Online] https://www.dec.gov.ua/wp-content/uploads/2019/11/akn_akne.pdf
9. Hayashi N, Akamatsu H, Kawashima M, Acne Study Group (2008) Establishment of grading criteria for acne severity. J Dermatol 35(5):255–260
10. Heng AHS, Say YH, Sio YY, Ng YT, Chew FT (2021) Epidemiological risk factors associated with acne vulgaris presentation, severity, and scarring in a Singapore Chinese population: a cross-sectional study. Dermatology, 1–10
11. Tayel K, Attia M, Agamia N, Fadl N (2020) Acne vulgaris: prevalence, severity, and impact on quality of life and self-esteem among Egyptian adolescents. J Egypt Public Health Assoc 95(1):1–7
12. Durai PCT, Nair DG (2015) Acne vulgaris and quality of life among young adults in South India. Indian J Dermatol 60(1):33–40
13. Krawczyk B (2016) Learning from imbalanced data: open challenges and future directions. Prog Artif Intell 5:221–232. https://doi.org/10.1007/s13748-016-0094-0
14. Lim ZV, Akram F, Ngo CP, Winarto AA, Lee WQ, Liang K et al (2020) Automated grading of acne vulgaris by deep learning with convolutional neural networks. Skin Res Technol 26(2):187–192. https://doi.org/10.1111/srt.12794
15. Wu X, Wen N, Liang J, Lai YK, She D, Cheng MM, Yang J (2019) Joint acne image grading and counting via label distribution learning. In: Proceedings of the IEEE/CVF international conference on computer vision (ICCV), pp 10642–10651. https://doi.org/10.1109/ICCV.2019.01074
16. Shen X, Zhang J, Yan C, Zhou H (2018) An automatic diagnosis method of facial acne vulgaris based on convolutional neural network. Sci Rep 8(1):1–10
17. Junayed MS, Jeny AA, Atik ST, Neehal N, Karim A, Azam S, Shanmugam B (2019). AcneNet-A deep CNN based classification approach for acne classes. In: 2019 12th international conference on information & communication technology and system (ICTS). IEEE, pp 203–208. https://doi.org/10.1109/icts.2019.8850935
18. Alarifi JS, Goyal M, Davison AK, Dancey D, Khan R, Yap MH (2017) Facial skin classification using convolutional neural networks. In International conference image analysis and recognition. Springer, Cham, pp 479–485
19. Yijing L, Haixiang G, Xiao L, Yanan L, Jinling L (2016) Adapted ensemble classification algorithm based on multiple classifier system and feature selection for classifying multi-class imbalanced data. Knowl-Based Syst 94:88–104. https://doi.org/10.1016/j.knosys.2015.11.013
20. He H, Garcia EA (2009) Learning from imbalanced data. IEEE Trans Knowl Data Eng 21(9):1263–1284. https://doi.org/10.1109/tkde.2008.239
21. Osojnik A, Panov P, Džeroski S (2017) Multi-label classification via multi-target regression on data streams. Mach Learn 106(6):745–770. https://doi.org/10.1007/s10994-016-5613-5
22. Shorten C, Khoshgoftaar TM (2019) A survey on image data augmentation for deep learning. J Big Data 6(1). https://doi.org/10.1186/s40537-019-0197-0
23. Davisking/dlib-models. Github.com. [Online] https://github.com/davisking/dlib-models
24. OpenCV. Github.com. [Online] https://github.com/opencv/opencv/blob/master/data/haarcascades/haarcascade_eye.xml
25. The Microsoft Cognitive Toolkit. Cntk.ai. [Online] https://www.cntk.ai/Models/Caffe_Converted/ResNet152_ImageNet_Caffe.model
26. Chawla NV, Bowyer KW, Hall LO, Kegelmeyer WP (2002) SMOTE: synthetic minority over-sampling technique. J Artif Intell Res 16:321–357

Convolutional Neural Network—A Practical Case Study

João Azevedo and Filipe Portela

Abstract The convolutional neural networks had enormous success in the classification of images, and networks such as "AlexNet", "VGG", "Inception" and "ResNet" were references for this purpose. This way, it is intended to verify which networks were more successful in the "Imagenet" dataset challenge. Then, it was verified their success when classifying videos through the "Kinetics400" and "UCF101" datasets and, finally, to conclude if the success in the classification of images can also evidence a possible success in the classification of videos. To this end, the margin of error of the networks mentioned above is compared. The two networks with the lowest margin of error are selected, and these networks are studied to classify videos. Thus, if these networks are successful, they can receive input videos from sensors and accurately identify the human activity present in the video. It should be noted that the networks "ResNet" and "Inception" had very satisfactory success rates, above 70%, showing success in the approach adopted.

Keywords Convolutional neural network · Classification · Machine learning · Action recognition

1 Introduction

The research intends to conclude if the success of neural networks in the image analysis process can serve as a bridge to the success of these same networks in the video analysis process. Thus, the study focuses on an initial phase by detecting the networks whose results in image classification were more satisfactory. It can be

J. Azevedo · F. Portela (✉)
Algoritmi Research Centre, University of Minho, Braga, Portugal
e-mail: cfp@dsi.uminho.pt

J. Azevedo
e-mail: a84182@alunos.uminho.pt

F. Portela
IOTECH—Innovation on Technology, Trofa, Portugal

A. Ullah et al. (eds.), *Proceedings of International Conference on Information Technology and Applications*, Lecture Notes in Networks and Systems 350,
https://doi.org/10.1007/978-981-16-7618-5_27

concluded whether their success in video classification is also satisfactory, and then use the most effective networks to process the videos received by the sensors.

This article provides an overview of the convolutional neural networks and studies the relationship between them, successfully used in classifying images and their results in videos.

So, the error margin of images classification was compared, and the two networks with a smaller margin of error were selected to classify the videos. Finally, their effectiveness is obtained through these networks, and the success/unsuccess of the investigation is concluded. For the case under study, it is considered as failure values effectiveness lower than 70% and success values equal to or greater than 70%.

Therefore, the article has a Background section to conclude on the "State of The Art" of convolutional networks, with a Material and Methods topic where the collected information is presented and a section that presents the Case Studies and the final results.

2 Background

A convolutional neural network is a learning algorithm with connected nodes that function like the human brain's neurons [1]. These power hidden pattern recognition, correlation in raw data, as well as clustering and classification. CNNs constitute a continuous learning process. They learn from the data passed to them and constantly improve [2]. In the architecture of a simple CNN (Fig. 1), five different layers are identified, input layer, convolution layer, clustering layer, fully connected layer and output layer, each of which has a specific function [3, 4] (Fig. 2).

The input layer receives the image to be processed and runs through the entire network. In turn, high-level attributes that may be relevant in the classification are detected in the convolution layer. Between the convolution and clustering layers,

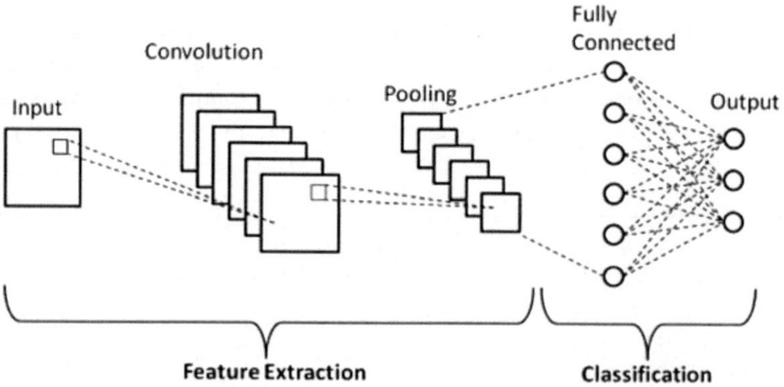

Fig. 1 Basic CNN architecture—from [4]

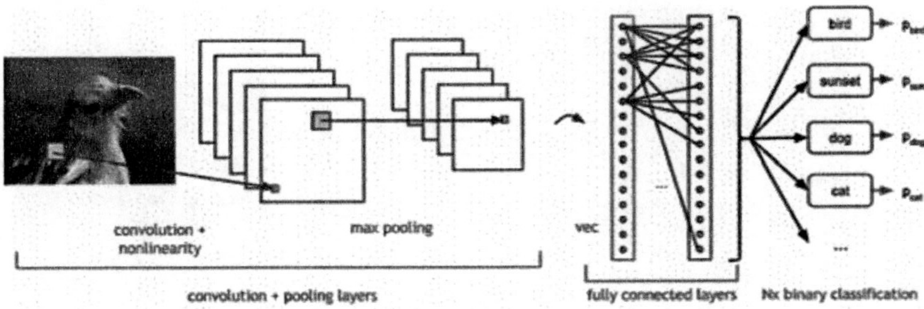

Fig. 2 Concrete example of a CNN—from [5]

there is an activation layer. The ReLU layer is governed by the function $f(x) = \max(0, x)$ and allows the network training to be faster with the same performance since it attenuates the gradient dissipation effect that decreases when passing between layers and would reduce the network training speed. The clustering layer clusters the data since the attribute will be found in the convoluted image no matter its location. The fully connected layer relates the attributes obtained with the object classes. For example, if an image of a bird is inserted, it would have a high correspondence with attributes such as "beak" and "wings" [5]. Next, a concrete example of how a CNN is processed is presented from the input to the output layer.

2.1 AlexNet

AlexNet was the first convolutional network to win an Imagenet challenge [6]. Its architecture, depicted in Fig. 3, consists of 5 convolutional layers, three max-pooling layers, two normalisation layers, two fully connected layers and one softmax layer.

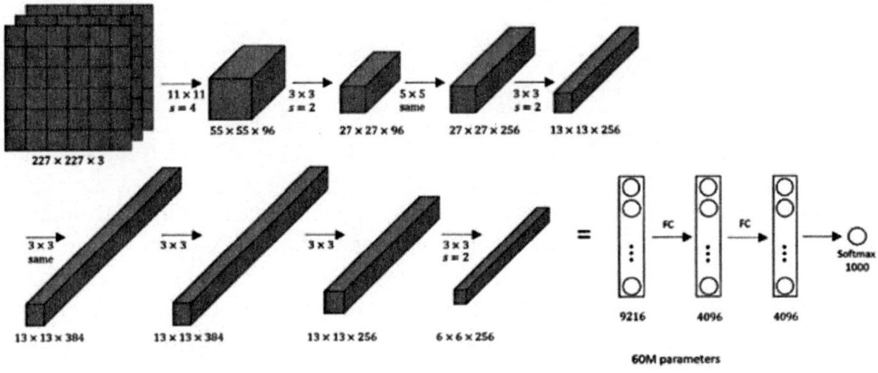

Fig. 3 AlexNet architecture—from [7]

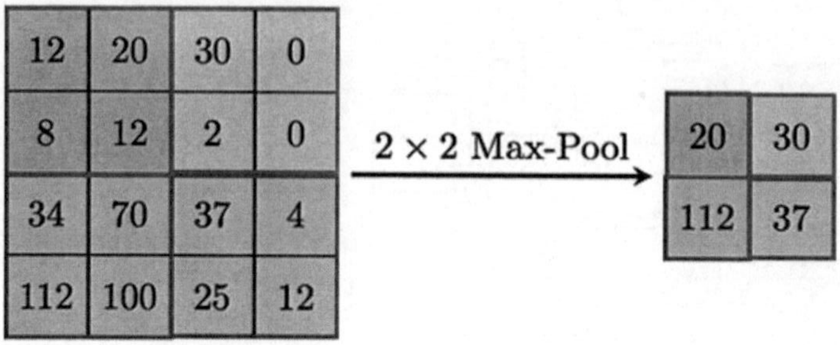

Fig. 4 Max-pooling—from [8]

Each convolutional layer consists of convolutional filters and a nonlinear activation function (ReLU). The input size is fixed due to the existence of fully connected layers, so this will be 224 × 224 × 3 without padding and 227 × 227 × 3 when padding is checked [7].

In the architecture in question, overlapping was used in max-pooling; to reduce the image size, the pixel whose value is larger is stored, and assumptions are made about the pixels not represented. Figure 4 illustrates a max-pooling operation [8].

2.2 *Vgg-16*

VGG-16 is a convolutional network proposed by K. Simonyan and A. Zisserman of Oxford University for image recognition and was distinguished as the most successful network in the Imagenet competition with an error margin of less than 8%. Its architecture, shown in Fig. 5, is an update of the AlexNet architecture replacing the large kernel filters with several 3 × 3 kernel filters in a row [9]. Its simplicity characterises it. It uses only 3 × 3 convolutional layers, uses max-pooling to reduce image size, and has two fully connected layers with 4096 nodes each, followed by a softmax classifier [10].

2.3 *GoogleNet/Inception*

Inception is a convolutional network proposed by collaborators from the Universities of Michigan, North Carolina, and Google in a paper entitled "Going Deeper with Convolutions". Its architecture is based on LeNet. Still, it has implemented a new concept: "the inception module", which is based on small convolutions that aim to reduce the number of parameters under evaluation. The training of these networks is greatly facilitated [11].

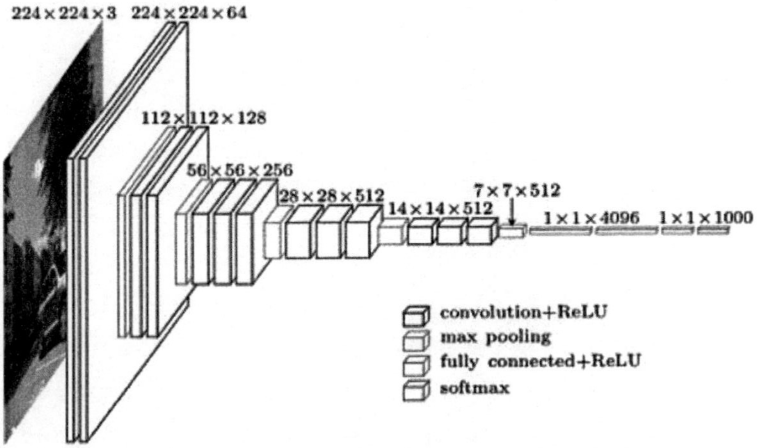

Fig. 5 VGG-16 architecture—from [9]

Thus, the rationale is to extend the network and reduce its depth to reduce overfitting and detect objects even when they undergo larger variations. Its constitution makes it possible to identify the initialisation layers in the network stem, auxiliary classifiers and the nine inception sequential modules. Still, authors can identify average pooling at the exit of the last inception module [12]. The representation of the architecture is in Figs. 6 and 7.

2.4 *ResNet*

The main idea of ResNet was the introduction of an identity shortcut connection that allows jumping through one or more layers, as illustrated in Fig. 8, constituting a residual learning algorithm. This solution indicates that a model deeper in the network should not contain a larger error than a less deep model. The following equation presents the residual block,

$$H(x) = f(x) + x,$$

unlike architectures such as Inception in which $H(x) = f(x)$. Thus, ResNet indicates that residual networks show better optimization results with changing the input space of the network.

Kaiming He is the creator of Resnet. He stated that the key to training such a deep network lies in the ability to jump between connections. ResNet won the ImageNet challenge in 2015, with a profound network consisting of 152 layers, where the signal that starts one layer was added to the output of a later layer [13].

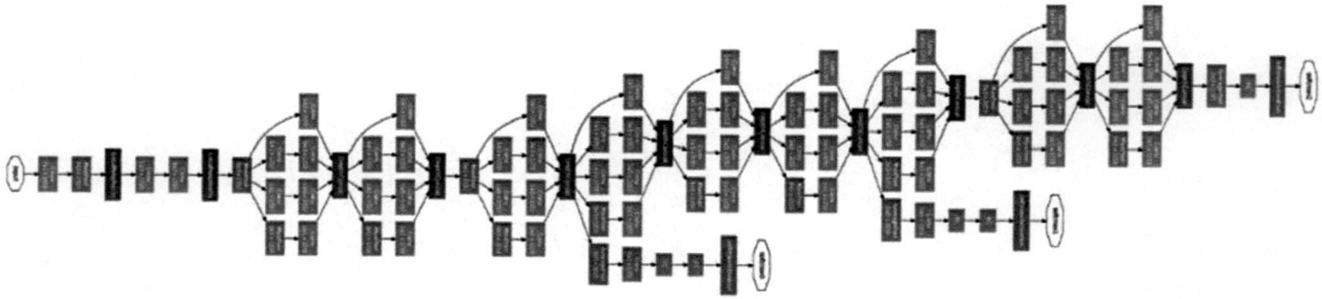

Fig. 6 Inception architecture [11]

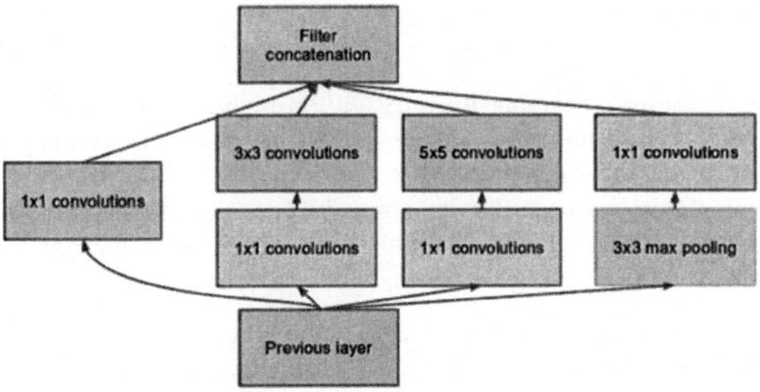

Fig. 7 Inception modules—from [11]

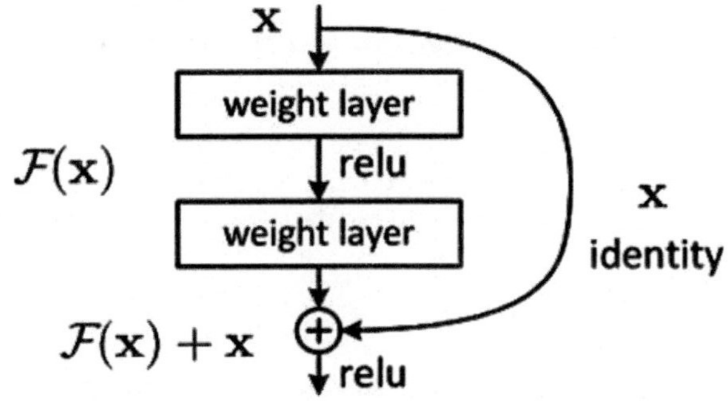

Fig. 8 Residual block—from [6]

Resnet with inter-layer hopping allows the activation layer to be an input from a layer deeper in the network. So, instead of waiting for gradients with higher magnitude to propagate layer by layer, it allows them to hop intermediate layers by performing a gradient update during the "backpropagation" process inherent to inter-layer hopping [14].

3 Material and Methods

For a better understanding, the "CRISP-DM" model was used in the study carried out. The project started by exploring the competition of ImageNet and the results obtained in it, going through the understanding and preparation of the data. Next, the modelling and evaluation stage is based on the effectiveness of the networks in image classification and compares this effectiveness with that obtained in video classification. In a later stage, deployment, sensors may provide real-time information and help companies obtain relevant information to respond to identified problems.

3.1 What Is Imagenet, and How Can It Help?

Imagenet is a dataset containing over 15 million high-quality "labelled", that is classified, images. It is the most popular dataset for researchers aiming to test a model for image classification [15].

Based on research on image classification with convolutional networks using the ImageNet dataset [16], I decided to choose four years in which the winners obtained an error rate below 20% to perform the study of this article.

- AlexNet, Imagenet challenge winner 2012-Error Rate 15.3%.
- VGG-16, Imagenet challenge winner 2013-Error Rate 7.3%.
- Inception, Imagenet challenge winner 2014-Error Rate 6.67%.
- ResNet, Imagenet challenge winner 2015-Error Rate 3.57%.

3.2 Comparison of the Margin of Error in the Different CNNs

By analysing the previous data, it is possible to conclude that the temporal evolution coincided with the evolution of the networks. From 2012 to 2015, the network margin reduced by around 12%.

3.3 Selection of the Two Networks with the Smallest Margin of Error

Through the analysis carried out previously, it can be seen that the networks whose margin of error is smaller are Inception and ResNet; therefore, these networks will be studied and analysed in the margin of error of video classification.

4 Case Study

To proceed with the training of neural networks in video classification, the datasets Kinetics400 and UCF101 were selected, which have several videos of daily human actions in their constitution.

The dataset Kinetics400 has about 300,000 videos, divided into 400 classes of human actions, and is currently the most used dataset to train convolutional networks in recognition of activities in videos [17]. The UCF101 dataset, on the other hand, has 17,000 videos inserted in 101 classes [18] and served as the basis for kinetics 400.

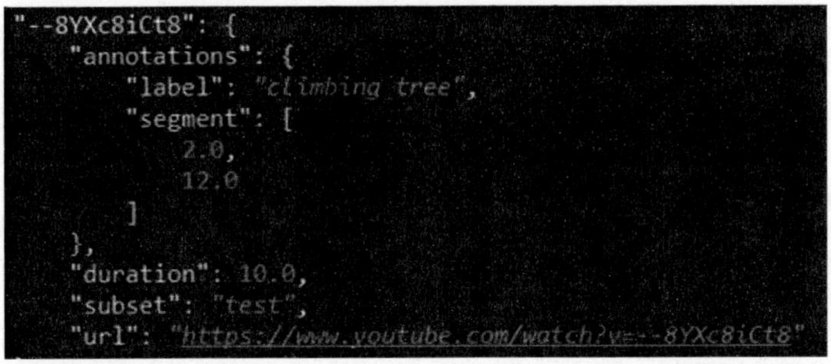

```
"--8YXc8iCt8": {
    "annotations": {
        "label": "climbing tree",
        "segment": [
            2.0,
            12.0
        ]
    },
    "duration": 10.0,
    "subset": "test",
    "url": "https://www.youtube.com/watch?v=--8YXc8iCt8"
```


Fig. 9 Json file

As far as datasets are concerned, these are made up of a rather particular set of files. This way, for each category (test, train and validate), there are two files, one CSV and one json. The excel files present five attributes, being them "label" that intends to identify the human action present in the video, "Youtube_id" that identifies the video, "time_start" and "time_end" that indicate the time interval to be evaluated and "split" that may have the value of the three categories (test, train, validate).

In turn, the json file (Fig. 9) is identical in its constitution to the excel file; however, it presents the attributes with different names and has the insertion of the "duration" and the "URL", which presents the link to the video.

4.1 Practical Case

With the goal to check the study's integrity, a ResNet network was configured, pre-trained on the ImageNet dataset and then trained on the Kinetics400 dataset, as indicated in the study mentioned above.

Through the analysis of the images in Fig. 10, it is possible to see that these networks are very effective in video classification, being able to detect what human activity is being performed. It is noteworthy that in both tests were inserted as "input" videos of a person making a pizza and a person washing his hands, respectively,

Fig. 10 Practical examples—from [19]

activities that the network detected and placed as the label "making pizza" and "washing hands", revealing the success of the investigation/study performed.

5 Results

After selecting the networks and datasets, they were pre-trained on the ImageNet dataset and tested on the chosen datasets, Kinetics 400 and UCF101, obtaining the results shown in Fig. 11.

Thus, the authors can conclude that for the dataset Kinetics400 the network best suited for video classification is the Resnet. In turn, in the dataset UCF101, the network that presents greater effectiveness is the Inception.

After the study, it was possible to elaborate on Table 1, which presents the performance of the networks presented before the different cases under study and which configurations were used in the networks.

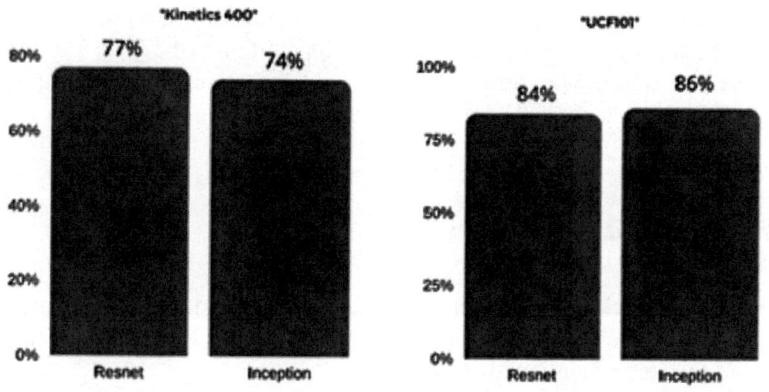

Fig. 11 Effectiveness of networks on datasets

Table 1 Network comparison

Type of CNN	Dataset to pre-train the network	Dataset to train the network	Network configuration	Error margin in the network (%)	Effectiveness of the network (%)
Inception	ImageNet	UCF101	1 CNN with 144 crops	6.57	86
Inception	ImageNet	Kinetics400			74
ResNet	ImageNet	UCF101	ResNet with 152 layers	3.57	84
ResNet	ImageNet	Kinetics400			77

6 Conclusion

It can be concluded that the stipulated objectives were met and that the successful networks in image classification are also effective when classifying actions present in videos. Thus, the research carried out allows concluding that the Resnet network, with a margin of error lower than 4% in image classification, and the Inception network, with a margin of error lower than 7% in image classification, are effective and successful in detecting human activities present in videos.

Thus, the results obtained were higher than expected, and this classification method showed robustness, classifying human activities with high efficiency, with minimum values of 74% and maximum values of 86%. It shows that CNNs pre-trained on an image dataset can be successful applied on video classification datasets. In the digital age we live in, as well as in the middle of a pandemic; these models could in the future be adapted and trained to detect the presence/absence or improper use of masks in people present in spaces where their use is mandatory. Thus, cameras and sensors would serve as real-time input for the model that would alert in cases of misuse or even the non-existence of masks.

Acknowledgements This work has been supported by FCT—Fundação para a Ciência e Tecnologia within the R&D Units Project Scope: UIDB/00319/2020 and supervised by IOTECH.

References

1. Yamashita R, Nishio M, Do R, Togashi K (2018) Convolutional neural networks: an overview and application in radiology. Retrieved from https://insightsimaging.springeropen.com/articles/10.1007/s13244-018-0639-9
2. SAS (2019) Neural networks what they are & why they matter. Retrieved from https://www.sas.com/en_us/insights/analytics/neural-networks.html
3. Ebermam E, Krohling R (2018) Uma Introdução Compreensiva às Redes Neurais Convolucionais: Um Estudo de Caso para Reconhecimento de Caracteres Alfabéticos. Retrieved from http://www.fsma.edu.br/si/edicao21/FSMA_SI_2018_1_Principal_08.pdf
4. Balaji S (2020) Binary image classifier CNN using TensorFlow. https://medium.com/techiepedia/binary-image-classifier-cnn-using-tensorflow-a3f5d6746697
5. Florindo J (2018) Redes Neurais Convolucionais—deep learning. Retrieved from https://www.ime.unicamp.br/~jbflorindo/Teaching/2018/MT530/T10.pdf
6. Amor E (2020) Important CNN architectures. Retrieved April 16, 2021, from https://www.topbots.com/important-cnn-architectures/
7. Hacker D (2018). B CNN AlexNet. Retrieved from http://datahacker.rs/deep-learning-alexnet-architecture/
8. Team GL (2020) AlexNet: the first CNN to win image net. Retrieved from https://www.mygreatlearning.com/blog/alexnet-the-first-cnn-to-win-image-net/
9. Neurohive (2018) VGG16—convolutional network for classification and detection. Retrieved from https://neurohive.io/en/popular-networks/vgg16/
10. Simonyan K, Zisserman A (2015) Very deep convolutional networks for large-scale image recognition. Retrieved from https://arxiv.org/pdf/1409.1556.pdf
11. Sermanet P (Google I., Liu W (University of N. C., Reed S (University of M. (2014) Going deeper with convolutions. Retrieved from https://arxiv.org/pdf/1409.4842.pdf

12. Karim R (2019) 10 CNN Architectures. Towards data science. Retrieved from https://towardsdatascience.com/illustrated-10-cnn-architectures-95d78ace614d#6872
13. Feng V (2017) An overview of ResNet and its variants. Towards Data Science. https://towardsdatascience.com/an-overview-of-resnet-and-its-variants-5281e2f56035
14. He K, Zhang X, Ren S, Sun J (2015). Deep residual learning for image recognition. Retrieved from https://arxiv.org/pdf/1512.03385.pdf
15. Sharma P (2020) 7 popular image classification models in ImageNet challenge (ILSVRC) competition history. Retrieved from https://machinelearningknowledge.ai/popular-image-classification-models-in-imagenet-challenge-ilsvrc-competition-history/
16. Krizhevsky A, Sutskever I, Hinton G (2016) ImageNet classification with deep convolutional neural networks. Retrieved from https://www.cs.toronto.edu/~kriz/imagenet_classification_with_deep_convolutional.pdf
17. Google (2017) The kinetics human action video dataset. Retrieved from https://arxiv.org/pdf/1705.06950.pdf
18. Soomro K, Zamir A, Shah M (2012) UCF101: a dataset of 101 human actions classes from videos in the wild. Retrieved from https://arxiv.org/pdf/1212.0402.pdf
19. Rosebrock A (2019) Human activity recognition with OpenCV and deep learning. Retrieved from https://www.pyimagesearch.com/2019/11/25/human-activity-recognition-with-opencv-and-deep-learning/

Software Engineering

Deep-Learning Approach for Sentiment Analysis in Software Engineering Domain

Aneesah Abdul Kadhar and Smitha S. Kumar

Abstract Sentiment analysis (SA) is an automated process that is increasingly being applied on software engineering (SE) text to study developers' sentiments and to improve productivity. Prior studies have reported varying performance for software engineering datasets depending on the techniques employed. In this paper, we propose deep-learning approach to perform sentiment analysis to identify the sentiment polarities of Jira [20] dataset. We developed and compared the performance of stacked CNN, stacked LSTM and stacked BiLSTM classifiers and found that the stacked BiLSTM classifier outperformed the other classifiers developed on the above-mentioned dataset. Additionally, this study investigated the impact of domain-customization on the performance of the classifiers by employing SE-specific word embeddings learnt from StackOverflow posts and compared with the pre-trained open-domain word embeddings learnt from google news posts. The classifiers employing Google News (GN) embeddings outperformed the SE-customized based classifiers. The better performance of GN embedding classifiers is due to the large generic training corpus of the GN word embedding model which identifies the sentiments expressed accurately when compared to SE-specific word embedding.

Keywords Sentiment analysis · Software engineering · Deep learning · CNN · LSTM · BiLSTM · Word embedding · Word2Vec

1 Introduction

Sentiment analysis (SA) is the study of text to determine the sentiment or emotional tone (positive, negative, neutral) expressed in the words [4]. SA to analyze developers' emotions is making a steady emergence in recent years. Study conducted by Guzman et al. in [11] indicates that the emotions expressed by the developers

A. A. Kadhar · S. S. Kumar (✉)
Heriot-Watt, Dubai, UAE
e-mail: smitha.kumar@hw.ac.uk

A. A. Kadhar
e-mail: aa2045@hw.ac.uk

A. Ullah et al. (eds.), *Proceedings of International Conference on Information Technology and Applications*, Lecture Notes in Networks and Systems 350,
https://doi.org/10.1007/978-981-16-7618-5_28

play an important role and have a large impact on the overall productivity and task quality at the early stage of software development. Emotions of developers are often expressed in crowdsourcing platforms, commit comments of various open-source platforms and issue tracking tools that handle their collaborative work. The sentiments expressed can then be used to improve various software libraries, Software Engineering tools, code suggestion tools [11, 18]. Given the wide variety of uses of SA in the SE domain, prior studies involved researchers using off-shelf SA tools for SE domain datasets achieving poor results [14, 15, 18]. This prompted the need for a customized SE-specific sentiment analysis tool, as SE text differs from articles, books, spoken language and consists of technical jargon, code snippets, and URLs [18]. However, recent techniques for SE domain sentiment analysis uses SE-specific lexicon methods [14], machine learning algorithm [1, 4] have reported an improvement, but are unreliable and provide varying results when applied on datasets other than the original datasets they have been trained on [22]. More recent studies conducted have shown that deep-learning algorithms when used in sentiment analysis outperform the machine learning based approaches in open-domain datasets [13]. This provides the motivation to apply deep-learning algorithms to identify sentiment polarity in SE text. This study aims to classify the sentiments of manually annotated dataset to positive, negative, and neutral sentiment polarity. Deep-learning algorithms are used to identify sentiment polarity and reduce misclassifications. Additionally, the study aims to address the following two research questions.

RQ1: What are the best deep-learning techniques to achieve better classification results on a software engineering dataset?

RQ2: What is the impact on the performance of the classifier when generic and software specific word embeddings are used?

2 Related Work

This section discusses the various work related to SA in SE domain employing multiple approaches such as lexicon, machine learning (ML) and deep-learning (DL) techniques. The studies conducted in [9, 11] provided the evidence that software developers express their sentiments and emotions on crowdsourcing platform, code repositories, issue tracking systems. Garcia et al. in [9] and Guzman et al. in [11] studied the relation between the sentiments expressed and their overall contribution to the project. Guzman et al. in [11] used SentiStrength, a lexicon analyzer to classify the sentiments. However, Jongeling et al. in [15] compared four domain independent sentiment analyzers (SentiStrength [24], Alchemy,[1] NLTK [2], StandfordCoreNLP [23]) found that existing sentiment classifiers disagreed among themselves and the labeled sentiments. Hence, substantiating the claim that off-the-shelf sentiment analyzers yield low accuracy when used in SE domain. Hence, customization of SA tools specific to that domain is necessary. To address the problem of

[1] http://www.alchemyapi.com/products/alchemylanguage/sentiment-analysis/.

domain dependency, Islam and Zibran in [14] proposed a SE-specific sentiment analysis built on top of SentiStrength called SentiStrength-SE. SE domain-specific dictionary was developed and built upon the SentiStrength dictionary. Evaluation conducted reported that SentiStrength-SE outperformed the original SentiStrength [24] sentiment analyzer. Thus, proving that incorporating domain-specific knowledge in the SA tool can substantially improve the classification result. However, the accuracy of the analyzer is restricted by the quality and the fixed size of the sentiment dictionary. Alternatively, machine learning and deep-learning techniques do not require a domain-specific dictionary but need a labeled dataset to train and identify patterns present to classify the sentiment polarity of the given text [1, 4].

Novielli et al. in [4] developed a ML classifier to identify positive, negative and neutral sentiments of SE text. The study aimed to reduce the misclassifications of neutral sentiment texts as negative polarity which was achieved by extracting lexicon, keyword, and semantic features from the text. The classifier employed SVM algorithm and obtained f-score of 87% when trained on StackOverflow dataset. Therefore, presenting strong evidence that extracting lexicon, keyword and semantic features boosts the performance of the classifier, by successfully reducing misclassification of sentiment polarity for the input text. Furthermore, Novielli et al. [4] and Ahmed et al. [1] compared the accuracy of the developed classifiers with SentiStrength and substantiated the claim that ML algorithms outperform off-the-shelf Lexicon Analysers. However, a comparative study on SE domain SA tools (SentiStrength [24], Senti4SD [4], Emotxt [5]), conducted by Islam and Zibran in [13] on three datasets from StackOverflow posts, Jira issue comments, and code review comments found the accuracy of the above-mentioned tools varying across the datasets. It was observed that the tools performed best on their originally tested dataset and the overall accuracy decreased when tested on other SE domain datasets. Hence, there is a need for a reliable Sentiment Analysis tool.

Deep-learning (DL) algorithms extract all the features present in the dataset automatically, without the need for explicit feature extraction techniques. The comparison study conducted by Kansara and Sawant in [16] compared traditional ML approaches with DL approaches for SA. ML models such as Naives Bayes, Logistic Regression and Random forests were compared with DL models namely, LSTM, CNN and a hybrid model of CNN + LSTM. These models were trained and evaluated on, the IMDB movie reviews, tweets, and hotel reviews datasets. The study concluded with the finding that the DL models significantly outperformed the machine learning models. Additionally, DL classifiers built in [17] employed single layer of CNN, RNN LSTM, and BiLSTM algorithms and yielded high accuracies when trained on open-domain datasets. However, the word2vec features are highly complex and often require more than one layer to extract the features [25]. The study conducted by Zhou et al. in [25], observed that stacking the BiLSTM layers decreased the prediction loss and increased the prediction accuracy and f-measure. The author attributes the better performance to the structure of stacked model that makes the model deep, thereby extracting the complex features word2vec more effectively. Similarly, studies conducted in [7, 21] employed stacked layers of CNN, LSTM and BILSTM algorithms in the sentiment classifiers and achieved f-measure results significantly higher than the classifiers employing single layers of CNN, LSTM and BiLSTM algorithms for the open domain.

The above studies direct us to build a sentiment classifier for SE domain by stacking layers of the DL algorithms.

3 Data Preparation

To train and evaluate the developed deep-learning classifiers, SE-related dataset from Jira is used. The Jira dataset developed by Ortu et al. in [19] is released online and are annotated with sentiments for the purpose of SE-sentiment classification.

3.1 Jira Dataset

Following the approach adopted by Chen et al. in [6], we excluded the issue comments of Group-1. The comments labeled as love, joy are assigned a positive sentiment label, whereas the comments labeled as sadness and fear are awarded a negative polarity and the comments labeled as neutral are given a neutral sentiment polarity. Furthermore, issue comments containing surprise and conflicting emotion labels are excluded. The final dataset comprises 2571 issue comments from Group-2 and Group-3. Text data is the most unstructured form of data; hence, preprocessing is considered as a vital task in SA to remove absurd information from raw text and has a direct impact on the performance of the deep-learning classifiers.

3.2 Data Processing

The preprocessor was written using the Regex library of python in which contractions were expanded, URLs and code snippets were removed. Additionally, we adopted a conservative stop-word removal approach. In this approach, stop words that frequently appeared and did not carry any sentiment information were removed. Moreover, we adopted a conservative approach to punctuation removal and removed all punctuation except '.','!', '?' as they are often used as delimiters and also express sentiments [8]. Furthermore, normalization of the text is done by converting it all to lowercase.

4 Word Embedding

This study compared the effect on the performance of the classifiers when SE-specific and generic word embeddings are used in the embeddings layer of the classifiers. The generic embeddings include the pre-trained Google News word embeddings trained on 100 billion google news headlines. The model was developed by following a configuration of 300 vector dimension, resulting in a vocabulary of size 3 million keywords. To create software-specific embeddings, 2 million StackOverflow posts with answers were mined. The posts were preprocessed and the tokenized sentences were input to the Word2Vec tool in Continuous Bag of Words architecture. The model was then trained for 3 epochs with a vector dimension of 100 and a window size of 5 (words that occurred less than 5 times were ignored). The vector dimension of 100 was chosen, since the dataset is small and contains limited vocabulary. Hence lower dimension is adequate for capturing the important features. The resulting Word2Vec model comprises of a vocabulary of 220,232 keywords.

5 Deep-Learning Classifiers

The deep-learning classifier models-stacked LSTM, stacked BiLSTM and stacked CNN are trained on the Jira dataset. The dataset was shuffled, and Stratified K-Fold cross validation (where $K = 5$) technique was employed to ensure that each fold contained equal proportion of classes present in the dataset.

5.1 Stacked CNN Classifier

CNN work by recognizing patterns, thereby identifying the correlated words present in the text. The CNN model consists of an embedding layer that learns the embeddings for all the words present in the dataset, followed by 2-layers of one-dimensional CNN layer of 16 filters and a kernel size 3, max-pooling layer, flatten layer and a dense layer of 3 neurons to output the sentiment. Additionally, L2-regularization of 0.01 was and dropouts of 0.4 and 0.2 are incorporated in the CNN layers to prevent overfitting. Categorical cross entropy is used as the loss function to find the optimal weights of the neural network [10, 12]. Furthermore, the model employs SGD optimizer to manipulate the weights and learning rate to minimize the losses.

5.2 *Stacked LSTM Classifier*

The stacked LSTM model comprises of an embedding layer, 2-layers of LSTM, followed by dense layers to output the classified sentiment label. In this configuration, each LSTM layer was built comprising of 16 neurons and was found that the model performed well without overfitting and underfitting. Additionally, overfitting of the neural network model was prevented by incorporating a dropout, recurrent dropout of 0.3 and L2-regularization of 0.001 was added in LSTM layers of the model. Categorical cross entropy and RMSprop were used as the loss function and optimizer, respectively.

5.3 *Stacked BiLSTM Classifier*

Similar to the above models, the stacked BiLSTM model comprises of an embedding layer, 2-layers of BiLSTM algorithm of 16 neurons and dense layers to output the sentiment. The model incorporated a dropout and a recurrent dropout of 0.3 and L2-regularizer of 0.001 in the BiLSTM layers to prevent overfitting. Furthermore, RMSprop and categorical cross entropy was used as the optimizer and loss function, respectively.

6 Baseline Methods

Our study employs the two existing sentiment analysis methods for comparisons, including Senti4SD and SentiStrength-SE. These classifiers are described briefly below.

6.1 *Senti4SD*

Senti4SD developed by Novielli et al. in [4] is a machine learning based sentiment analyzer that leverages lexicons features based on the sentiment word list of, keyword features such as unigrams and bi-grams and semantic features. The classifier employs SVM algorithm to classify the sentences.

6.2 SentiStrength-SE

SentiStrength-SE [14] is an extension of SentiStrength adapted to the SE domain. The lexicon dictionary of SentiStrength-SE was built by running SentiStrength on the Group-1 subset of the Jira dataset to analyze the reasons for the misclassification and thereby customized the lexicon dictionary accordingly.

7 Results

The performance of various models are reported in terms of precision, recall and F-measure as shown in Tables 1, 2 for the Jira dataset. The overall performance of the model is calculated by employing macro-averaging as the aggregated technique.

RQ-1 What are the best deep-learning techniques to achieve better classification results on a software engineering dataset?

Based on the experiments and an analysis of the results present in Tables 1, 2, it is observed that the stacked BiLSTM model performs the best for the Jira dataset, when SO and Google News embeddings are used. The superiority of the stacked BiLSTM when compared to CNN and LSTM models can be attributed due to the structure of BiLSTM, since they are able to learn dependencies that help in differentiating the strong and the weak sentiments present in the sentences.

RQ-2 What is the impact on the accuracy of the classifier when generic and software specific word embeddings are used?

Table 1 Performance of SE-embedding based deep-learning classifiers

Dataset	Word-embedding	Class	Metric	LSTM	BILSTM	CNN
Jira	SO	Positive	Precision	0.92	0.93	0.90
			Recall	0.87	0.89	0.83
			F-measure	0.89	0.91	0.86
		Negative	Precision	0.73	0.82	0.77
			Recall	0.72	0.80	0.75
			F-measure	0.73	0.81	0.76
		Neutral	Precision	0.68	0.76	0.76
			Recall	0.75	0.81	0.71
			F-measure	0.71	0.78	0.73
		Overall	Precision	0.78	0.84	0.81
			Recall	0.78	0.84	0.76
			F-measure	0.78	0.84	0.78

Table 2 Performance of generic embedding based deep-learning classifiers

Dataset	Word-Embedding	Class	Metric	LSTM	BILSTM	CNN
Jira	GN	Positive	Precision	0.90	0.93	0.90
			Recall	0.91	0.91	0.89
			F-measure	0.90	0.92	0.90
		Negative	Precision	0.81	0.86	0.81
			Recall	0.79	0.85	0.83
			F-measure	0.80	0.85	0.82
		Neutral	Precision	0.74	0.79	0.75
			Recall	0.75	0.83	0.75
			F-measure	0.74	0.81	0.75
		Overall	Precision	0.82	0.86	0.82
			Recall	0.82	0.86	0.82
			F-measure	0.82	0.86	0.82

As observed from Tables 1, and 2, although the performance of the classifiers employing SO word embedding is similar to that of the classifiers incorporating Google News embeddings, the performance of the former does not surpass that of the latter. This better performance could be attributed to the differing Word2Vec model parameters that were used in the development of the SO word embeddings and Google news embeddings. The Word2Vec model of Google news embeddings were developed by training the model on one hundred billion posts and so comprises a significantly larger vocabulary of size 3 million words [3]. Furthermore, the Google News posts are written by professionals, and so the Google News model is less prone to spelling errors, in contrast to the developed SO model, thus yielding better results [3].

8 Comparisons to Baseline Methods

From Table 3, it is observed that the overall performance of stacked BiLSTM classifiers employing Google News word embeddings surpasses that of SentiStrength-SE, Senti4SD in the Jira dataset, hence proving that deep-learning classifiers outperform lexicon and ML classifiers.

9 Conclusions

The focus of this study was to discover which of the deep-learning models perform better in the SE-specific dataset and found that stacked BiLSTM models tend to

Table 3 Comparison of stacked BiLSTM-GN classifier with baseline methods

Dataset	Class	Metric	BILSTM-GN	SentiStrength-SE	Senti4SD
Jira	Positive	Precision	0.93	0.93	0.88
		Recall	0.91	0.91	0.92
		F-measure	0.92	0.91	0.90
	Negative	Precision	0.86	0.87	0.83
		Recall	0.85	0.73	0.79
		F-measure	0.85	0.79	0.81
	Neutral	Precision	0.79	0.71	0.74
		Recall	0.83	0.84	0.73
		F-measure	0.81	0.84	0.73
	Overall	Precision	0.86	0.84	0.81
		Recall	0.86	0.83	0.82
		F-measure	0.86	0.82	0.81

perform better across the Jira dataset. Additionally, we compared our results with the state-of-art models-Senti4SD, SentiStrength-SE and found that the stacked BiLSTM classifier outperformed both the classifiers across the Jira dataset. Additionally, we studied the impact on the performance of classifiers when generic and software specific word embedding were used and found that the developed classifiers incorporating the general Google News word embedding model performed better in identifying sentiments when compared to SO-embedding based classifiers. By incorporating Google News word embedding in the stacked BiLSTM model, we found the above classifier achieve improved results.

References

1. Ahmed T, Bosu A, Iqbal A, Rahimi S (2017) Senticr: a customized sentiment analysis tool for code review interactions. In: 2017 32nd IEEE/ACM international conference on automated software engineering (ASE), pp 106–111. https://doi.org/10.1109/ASE.2017.8115623
2. Bird S, Klein E, Loper E (2009) Natural language processing with Python, 1st edn. O'Reilly Media, Inc.
3. Biswas E, Vijay-Shanker K, Pollock L (2019) Exploring word embedding techniques to improve sentiment analysis of software engineering texts. In: 2019 IEEE/ACM 16th International conference on mining software repositories (MSR), pp 68–78. https://doi.org/10.1109/MSR.2019.00020
4. Calefato F, Lanubile F, Maiorano F, Novielli N (2018) Sentiment polarity detection for software development, pp 128–128. https://doi.org/10.1145/3180155.3182519
5. Calefato F, Lanubile F, Novielli N (2017) Emotxt: a toolkit for emotion recognition from text, pp 79–80. https://doi.org/10.1109/ACIIW.2017.8272591
6. Chen Z, Cao Y, Lu X, Mei Q, Liu X (2019) Sentimoji: an emoji-powered learning approach for sentiment analysis in software engineering. In: Proceedings of the 2019 ACM Joint European

software engineering conference and symposium on the foundations of software engineering, ESEC/FSE'19
7. Dashtipour K, Gogate M, Adeel A, Larijani H, Hussain A (2021) Sentiment analysis of Persian movie reviews using deep learning. Entropy 23. https://doi.org/10.3390/e23050596
8. Ding J, Sun H, Wang X, Liu X (2018) Entity-level sentiment analysis of issue comments. https://doi.org/10.1145/3194932.3194935
9. Garcia D, Zanetti MS, Schweitzer F (2013) The role of emotions in contributors activity: a case study on the gentoo community. In: 2013 International conference on cloud and green computing, pp 410–417. https://doi.org/10.1109/CGC.2013.71
10. Goodfellow I, Bengio Y, Courville A (2016) Deep learning. MIT Press (2016), http://www.deeplearningbook.org
11. Guzman E, Azócar D, Li Y (2014) Sentiment analysis of commit comments in github: an empirical study. 11th Working conference on mining software repositories, MSR 2014—proceedings. https://doi.org/10.1145/2597073.2597118
12. Hameed Z, Zapirain B (2020) Sentiment classification using a single-layered bilstm model. IEEE Access 8:73992–74001. https://doi.org/10.1109/ACCESS.2020.2988550
13. Islam MR, Zibran MF (2018) A comparison of software engineering domain specific sentiment analysis tools. In: 2018 IEEE 25th international conference on software analysis, evolution and reengineering (SANER), pp 487–491. https://doi.org/10.1109/SANER.2018.8330245
14. Islam MR, Zibran MF (2018) Sentistrength-se: exploiting domain specificity for improved sentiment analysis in software engineering text. J Syst Softw 145:125–146. https://doi.org/10.1016/j.jss.2018.08.030, https://www.sciencedirect.com/science/article/pii/S0164121218301675
15. Jongeling R, Datta S, Serebrenik A (2015) Choosing your weapons: on sentiment analysis tools for software engineering research. In: 2015 IEEE international conference on software maintenance and evolution (ICSME), pp 531–535. https://doi.org/10.1109/ICSM.2015.7332508
16. Kansara D, Sawant V (2020) Comparison of traditional machine learning and deep learning approaches for sentiment analysis, pp 365–377. https://doi.org/10.1007/978-981-15-3242-935
17. Kim Y (2014) Convolutional neural networks for sentence classification. In: Proceedings of the 2014 conference on empirical methods in natural language processing. https://doi.org/10.3115/v1/D14-1181
18. Lin B, Zampetti F, Bavota G, Di Penta M, Lanza M, Oliveto R (2018) Sentiment analysis for software engineering: how far can we go? In: 2018 IEEE/ACM 40th international conference on software engineering (ICSE), pp 94–104. https://doi.org/10.1145/3180155.3180195
19. Murgia A, Ortu M, Tourani P, Adams B, Demeyer S (2018) An exploratory qualitative and quantitative analysis of emotions in issue report comments of open source systems. Empirical Softw Eng 44:1. https://doi.org/10.1007/s10664-017-9526-0
20. Ortu M, Murgia A, Destefanis G, Tourani P, Tonelli R, Marchesi M, Adams B (2016) The emotional side of software developers in jira. https://doi.org/10.1145/2901739.2903505
21. Ouyang X, Zhou P, Li CH, Liu L (2015) Sentiment analysis using convolutional neural network. In: 2015 IEEE international conference on computer and information technology; ubiquitous computing and communications; dependable, autonomic and secure computing; pervasive intelligence and computing, pp 2359–2364. https://doi.org/10.1109/CIT/IUCC/DASC/PICOM.2015.349
22. Shen J, Baysal O, Shafiq MO (2019) Evaluating the performance of machine learning sentiment analysis algorithms in software engineering, pp 1023–1030. https://doi.org/10.1109/DASC/PiCom/CBDCom/CyberSciTech.2019.00185
23. Socher R, Perelygin A, Wu J, Chuang J, Manning C, Ng A, Potts C (2013) Recursive deep models for semantic compositionality over a sentiment treebank. EMNLP 1631:1631–1642
24. Thelwall M, Buckley K, Paltoglou G (2012) Sentiment strength detection for the social web. J Am Soc Inf Sci Technol 63:163–173. https://doi.org/10.1002/asi.21662
25. Zhou J, Lu Y, Dai HN, Wang H, Xiao H (2019) Sentiment analysis of Chinese microblog based on stacked bidirectional LSTM. IEEE Access 7:38856–38866. https://doi.org/10.1109/ACCESS.2019.290504

Managerial Conflict Among the Software Development Team

Madnia Ashraf, Abdallah Tubaishat, Feras Al-Obeidat, and Ali Raza

Abstract Requirement engineering is one of the most essential and first steps of software development. This phase plays a vital role in the success of software. In this paper, we focus on managerial conflict, one of the types of software conflicts in a particular stage of software development that is a development phase. As there are many stakeholders and massive requirements, conflict comes across. This research proposed a framework to identify the cause of managerial conflicts and respond to these conflicts among the development team to solve them. The proposed framework helps to enlighten conflict management skills in the development team to increase productivity. When the development team knows how to work effectively in the face of managerial conflicts that will undoubtedly come their way, they can more easily make decisions quickly. Those decisions will help to prevent software from delayed deployment and failure of software.

Keywords Requirement engineering · Managerial conflicts conflict development team software development

1 Introduction

The first step of software development is requirement engineering, as software development is not the name of programming. In a complex system, there are a hundred stakeholders and thousands of requirements that will be the functionality of the

M. Ashraf
Punjab Information Technology Board, Lahore, Pakistan

A. Tubaishat · F. Al-Obeidat
College of Technological Innovation, Zayed University, Abu Dhabi, United Arab Emirates
e-mail: abdallah.tubaishat@zu.ac.ae

F. Al-Obeidat
e-mail: feras.al-obeidat@zu.ac.ae

M. Ashraf · A. Raza (✉)
Department of Computer Science & IT, University of Sargodha, Sargodha, Pakistan

A. Ullah et al. (eds.), *Proceedings of International Conference on Information Technology and Applications*, Lecture Notes in Networks and Systems 350,
https://doi.org/10.1007/978-981-16-7618-5_29

product [1, 2]. Every stakeholder wants each described requirement implementation. However, there are many other factors like budget, time, resources, conflict between stakeholder and software development team that makes the all-requirement implementation not feasible [3]. There are many defined techniques to prioritize the requirement in the requirement engineering phase. For any software development, requirement engineering is one of the initial steps, and the final product quality and success rate are most affected by this phase [3, 4]. Its principle is to appliance the most core requirements of stakeholders concerning budget, quality, available resources, and delivery time [5]. However, beyond these requirement challenges, there are many conflicts that a software development team faces. To respond to requirement conflicts, most researchers think the most crucial technique is requirement negotiation. Mainly in requirement negotiation, they focus on win–win technique, easy win–win technique, and quality assurance techniques. In recent ICSE (International Conference of Software Engineering), presenters talk about the requirement negotiation technique to respond to conflicts as it is more important how we negotiated than how were requirement specified. However, the major problem of requirement negotiation is how to identify the conflicting requirement, how to resolve the managerial conflict of requirements from the different stakeholders [6], particularly among the development team. In this research work, we study the type of conflict and the technique to solve these conflicts. Mainly focus on managerial conflict among the software development team [7]. To respond to requirement conflicts, most researchers think the most crucial technique is requirement negotiation, from which the most prevalent techniques are distributive, team, integrative, and multiparty negotiation [8]. In most organization managers who implement must implement strategy to response conflicts typically impose their thoughts or opinions onto development team and the conflict often ends with undesirable results. Such managers their viable interests, which may result in others moving away from the achievement of the project goal [9]. What are the present methods used for conflict resolution in prioritizing requirements among the development team? How requirement prioritization technique resolves conflict in different areas and especially in the development team? Which paper resolves the conflict for the process model? This research paper is having seven sections: Sect. 2 has motivation to lead this research work, Sect. 3 has research methodology to address the existing technique challenges, Sect. 4 has literature review, Sect. 5 has proposed framework, Sect. 6 has the validation of the proposed framework, and Sect. 7 has conclusion of research work.

2 Motivation

Effective development of software systems needs a set of complete, consistent, and clear requirements. In software development, hundreds and thousands of different stakeholders have different backgrounds partake in the requirement engineering process. Consequently, it is not easy to satisfy every stakeholder's requirement, so the probability of conflicts increases, mainly leading the software to delay and fail. The requirement engineers trade-off the needs of stakeholders with the project resources and the project constraints [9, 10].

3 Research Methodology

To answer the research question, I accomplished a systematic literature review according to the research guidance provided by B Kitchenham [5]. By adopting the defined research methodology and guidelines, I searched only quality research papers [11] on software requirement prioritization technique, software conflict type, and software conflict resolution technique. There are several research papers available online on the topic of requirement prioritization technique and conflict resolution [12]. This SLR aims to comprehend and summarize the conflicts in requirement prioritization, how to handle requirements conflict, and find fields for more research to accompany the accomplishments of existing conflicts techniques [13].

3.1 Search Strategy

The detailed description of the search strategies developed in this research consisted of search terms, literature resources, and the search process.

3.2 Search String

Following five steps are considered while building the search string (1) derived effective term from the defined research question (2) identify five most suitable synonym is for each significant term (3) identify the keyword in the relevant article, books, and research paper (4) Use of the Boolean "OR" to combine substitute spellings and synonyms (5) Use the Boolean operation "AND" to link the significant terms. General String: Managerial Conflict resolution in requirement prioritization. Parent Search String: (("Conflict" OR "clash" OR "incompatible" OR "disagreement") AND ("resolution" OR "intent" OR "Resolve" OR "design" OR "object" OR "Plan" OR "motion" OR "revise") AND ("Requirement" OR "need" OR "wish" OR "demand" OR "want" OR "essential") AND ("prioritization" OR "listening" OR "arrange" OR "order")).

3.3 Database Selection

We used five electronic database resources to extract data for synchronizations in this research primarily. These include Science Direct, Springer, Google Scholar, IEEE Xplore, and ACM SLR and have to do a broad search of all relevant sources about the topic of discussion [14, 15]. A thorough search was launched on the five electronic database sources, and the returned results (papers) were assembled assets of Digital Library. Select the research paper by checking title, abstract, and index terms; we also search for published journal papers, conference proceedings, workshops, symposiums, book chapters, and IEEE bulletins.

Prospective papers. The reference lists of all relevant papers were perused to detect additional relevant papers and then combine them with the ones in stage 1 (Fig. 1; Table 1).

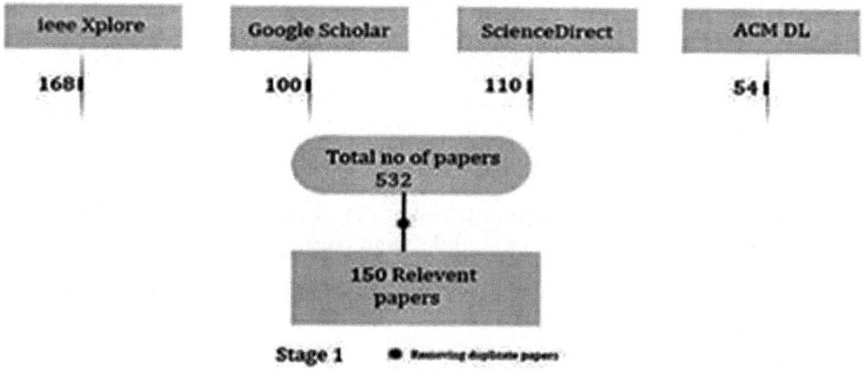

Fig. 1 Database finding

Table 1 Inclusion and exclusion criteria

S. N.	Inclusion criteria	Exclusion criteria
1	Papers mainly focus in requirement conflicts	Papers having not focus on research question
2	Relevant papers from 2000 to 2020	Gray papers
3	Papers that are at least focusing at least one question answer	Duplicate papers

3.4 *Quality Assessment*

Selected research quality was accessed by the quality assessment of these articles. That was implemented after the final selection of research article that must be included in the final literature to answer the research question (Table 2).

QA1. Is that study focusing on the causes of managerial conflicts? Rendering to well-defined research question, quality assessment of search studies is achieved, calculated with the help of quality scores.

4 Literature Review

We do a literature review by selecting the same paper that targets the same problem and gives a solution or proposed framework. The first category paper targets the requirement prioritization technique, second paper targets the conflict that face while prioritizing requirements, how they handle the conflict, and which type of conflict to solve by adopting these studies. In contrast, the last and third category try to include the literature about managerial conflict among the development team by narrowing down and studying these studies' existing solutions and future work. According to [16–19], when prioritizing the requirement the parameter keep in mind risk, cost, and its importance in our all-project development due to fix these parameter conflict comes from different development team. Due to defined time and budget, resolving the requirement conflict is essential. Comparison of existing ten requirement prioritizing techniques based on accuracy, time, usability, and ease of use. This study does not conclude which technique is effective and efficient for RP (requirement prioritization) and does not target the conflict resolution among them. It depends on the condition under which it is used—prioritizing the requirement based on cost, effort, and threat associated with the corresponding requirement. According to [20–22], the software development team may face conflicts without it being unavoidably dangerous to throughput. Nevertheless, mostly we think that conflicts may reason to damage or negativity among the development team. The leading reason why conflicts can be damage according to Folberg and Taylor "conflicts are not essentially mean something bad or negative, our organization often outlooks conflicts as negatively

Table 2 Total number of studies stated by any cause

Question and answer rule	"No. of causes identify"	Score containing	"Assessed quality result"
Rule1	If less than or equal to 2	0.005	Ignored
Rule2	If less than and equal to 3	1.1	Accepted to ok
Rule3	If ¿3 and ¡ = 5	1.8	Accepted

because it associated with both win/lose situation and mostly observed by the contributors as a disaster [22–25]. A disaster way of thinking injects itself to damaging processes because people will frequently scurry to use everything, they think will mitigate the conflict." However, it is not an effective approach to avoid harm, to respond the conflicts one or some of the members reply to conflicts creatively. Team members who responded to the conflicts and the impact of conflicts have a direct relation. No conflicts have a simple impact even it seems simple, conflict mostly conflicts managerial have complex to handle [25–28] (Table 3).

Beneath every single conflict, there is a presenter of possible opposing along with attuned motives of the conflict. Challenges that cause the conflict (1) limited resources (2) time management (3) Lack of trust (4) Goals versus requirements [16, 30, 31].

Table 3 Findings of SLR

References	Conflict analysis techniques or approach	Conflict identification method	Scope of these approach	Evaluation
[24]	Heuristic algorithm	Manual	Identify	No
[25]	Root requirements analysis	Manual	Identify and analyze	No
[26]	Non-mathematical or non structured techniques	Manual	Identify, analyze (3 type) resolve	No
[27]	Tractability approach	Automatic	Identify	Yes
[12]	Method of graphical representation using meta model	Automatic	Identify	Yes
[29]	Graphical method using NDT meta model	Automatic	analyze	Yes
[30]	Rules of validation	Automatic	Identify analyze (7 types) resolving for each type	Yes
[34]	Interaction level three detection framework	General Frame-Work	Identify	Yes
[31]	Research of investigation on non-Functional Requirements and catalogue building for NFRs conflicts	General frame-work	Analyze the absolute, relative, no conflict in development	No
[32]	Graphical method using problem diagram	Manual	Identify	Yes

4.1 Conflict Major Cause

1. Massive requirement
2. Managerial factor (cost, time, stakeholder significance)
3. Creeping requirement
4. Complex domain system.

4.2 Problem Statement

What are existing support system and models for the identification of conflicts and determining the managerial conflicts among a massive number of software requirements that are resulting from different interpretations of numerous participants that have to benefit from accomplishing a deal, and it would be helpful to reduce the risks so that neither revelry feels hard-pressed to give up too ample? Additionally, [16, 31] Humphrey argues, there is no conflict among software team members can be settled that cause the conflict among development team?

5 Proposed Framework

A very significant and first rule of settling the managerial conflict at any level in the development phase, interactive to transnational, is to maintain the lineup of communication open, even if ferry diplomacy makes use of a tertiary party is essential. [16] According to Humphrey, "Below these situations in front of intrusion is normally needed."

5.1 Execution of Framework

The execution for proposed model is as follows:

- Start ()
- Get Input from Excel
- Calculate total no conflict
- Separate the managerial conflicted requirements
- Initial population is built randomly by generating values for attributes within domain.
- Run Fitness (Conflict on each solution)
- Apply Crossover between managerial conflicted requirements
- Apply Mutation between managerial conflicted requirements

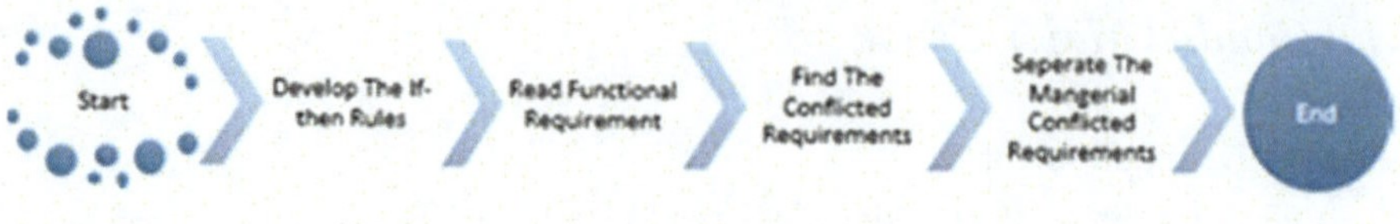

Fig. 2 Abstract execution of proposed framework

- Repeat Fitness to Mutation until Stopping Criteria
- Stop () (Fig. 2).

6 Validation of Proposed Framework

The validation of the proposed framework has been done by the expert judges who work in a particular organization for a specific software development time. Criteria for the organization's selection to validate the proposed framework is based on two variable times for software development and software process model for software development. An organization in which V and RAD model is used for the software development and minimum time to complete software in six months and maximum in two-year time required to complete a software development (primarily software related to FinTech) is our targeting organization for the validation of proposed framework.

6.1 Validation Criteria

We validate the proposed framework from seven different organizations that meet our validation criteria. X1, X2, X3, X4, X5, X6, X7 (in percentage) are the organization and response from these organization is in Table 4. As a result, poor planning to respond to conflicts and communication gaps between the developer and requirement engineer and unnotified change in the development team mainly cause managerial conflicts. Furthermore, these causes lead to failure or delay in software development.

7 Conclusion

The proposed framework is not for global software development and is not useful for the organization that used Agile software development for software development. In the future, we implement Global Software Development by overcoming the communication gaps in local organization implementation. One of the major reasons is also related to the communication gap between the requirement engineer and development team. Managers who worked on the requirement conflicts technique have no clear

Table 4 Organization response

S. N.	Cause of managerial conflicts	Response	X1	X2	X3	X4	X5	X6	X7	Avg
1	Not identified conflicts	Conflicts ignore mostly	80	45	56	78	80	59	46	63
2	Frequent change in requirement	Follow proper template to requirement change	80	45	56	78	80	59	46	63
3	Poor planning to response conflicts	Consider minor conflicts cause	67	70	80	98	87	94	65	80
4	Communication gap between developer and requirement engineer	Schedule the proper meeting	78	56	75	78	67	78	78	72
5	Unnotified change in development team	Change by proper planning and observing its need	65	64	45	56	89	65	88	67
6	Poor categorization of conflicts	Categories conflicts in a proper category	66	45	67	54	63	78	59	61

idea about the conflict concerning different causes and challenges. This research helps to identify the cause of managerial conflicts among the development team. To timely respond to managerial conflicts that prevent software from failure and overcome delays in software deployment. In the future, managerial conflicts cause by global software development may target and solve by expanding the proposed framework.

References

1. Adnan SM, Ilyas M, Razzaq S, Maqbool F, Wakeel M, Adnan SM (2020) Code smell detection and refactoring using AST visitor. Tech J 25(01):59–65
2. Paielli RA, Erzberger H (2019) Trajectory specification for terminal airspace: conflict detection and resolution. J Air Transp 27(2):51–60
3. Nuseibeh B, Easterbrook S (2000) Requirements engineering: a roadmap. In: Proceedings of the conference on the future of software engineering, pp 35–46. Author F, Author S (2016) Title of a proceedings paper. In: Editor F, Editor S (eds) CONFERENCE 2016, LNCS, vol 9999, pp 1–13. Springer, Heidelberg
4. Macaulay LA (2012) Requirements engineering. Springer Science Business Media
5. Van Lamsweerde A (2000) Requirements engineering in the year 00: a research perspective. In: Proceedings of the 22nd international conference on software engineering, pp 5–19
6. Kotonya G, Sommerville I (1996) Requirements engineering with viewpoints. Softw Eng J 11(1):5–18
7. Neill CJ, Laplante PA (2003) Requirements engineering: the state of the practice. IEEE Softw 20(6):40–45
8. Eric SK, Giorgini P, Maiden N, Mylopoulos J (eds) (2011) Social modeling for requirements engineering. MIT Press
9. Borchert A, Heisel M (2021) Conflict identification and resolution for TrustRelated requirements elicitation a goal modeling approach. J Wirel Mob Netw Ubiquitous Comput Dependable Appl 12(1):111–131
10. Sivrikova N, Ptashko T, Chernikova E, Perebeynos A, Kharlanov V (2021) Assessment of conflict resolution training at university. In: SHS web of conferences, vol 90. EDP Sciences, p 02010
11. Adnan A, Anwar S, Zia T, Razzaq S, Maqbool F, Rehman ZU (2018) Beyond Beall's Blacklist: Automatic Detection of Open Access Predatory Research Journals. In 2018 IEEE 20th international conference on high performance computing and communications; IEEE 16th international conference on smart city; IEEE 4th international conference on data science and systems (HPCC/SmartCity/DSS). IEEE, pp 1692–1697
12. Achimugu P, Selamat A, Ibrahim R, Mahrin MNR (2014) A systematic literature review of software requirements prioritization research. Inf Softw Technol 56(6):568–585
13. Van Lamsweerde A (2001) Goal-oriented requirements engineering: a guided tour. In Proceedings fifth IEEE international symposium on requirements engineering. IEEE, pp 249–262
14. Sheng Y, Hassan A, Spear M (2021). Semantic conflict detection for transactional data structure libraries. In: Proceedings of the 33rd ACM symposium on parallelism in algorithms and architectures, pp 446–448
15. Chaki D, Bouguettaya A (2020) Fine-grained conflict detection of IoT services. In: 2020 IEEE international conference on services computing (SCC). IEEE, pp 321–328
16. Babar MI, Ramzan M, Ghayyur SA (2011) Challenges and future trends in software requirements prioritization. In: International conference on computer networks and information technology. IEEE, pp 319–324

17. Aasem M, Ramzan M, Jaffar A (2010) Analysis and optimization of software requirements prioritization techniques. In: 2010 international conference on information and emerging technologies. IEEE, pp 1–6
18. Berander P, Andrews A (2005) Requirements prioritization. In: Engineering and managing software requirements. Springer, Berlin, Heidelberg, pp 69–94
19. Khan K (2006). A systematic review of software requirements prioritization
20. Azar J, Smith RK, Cordes D (2007) Value-oriented requirements prioritization in a small development organization. IEEE Softw 24(1):32–37
21. Berander P (2004).Using students as subjects in requirements prioritization. In Proceedings. 2004 international symposium on empirical software engineering, 2004. ISESE'04. IEEE, pp 167–176
22. Lehtola L, Kauppinen M, Kujala S (2004) Requirements prioritization challenges in practice. In: International conference on product focused software process improvement. Springer, Berlin, Heidelberg, pp 497–508
23. Pergher M, Rossi B (2013) Requirements prioritization in software engineering: a systematic mapping study. In: 2013 3rd international workshop on empirical requirements engineering (EmpiRE). IEEE, pp 40–44
24. Perini A, Susi A, Avesani P (2012) A machine learning approach to software requirements prioritization. IEEE Trans Software Eng 39(4):445–461
25. Lima DC, Freitas F, Campos G, Souza J (2011) A fuzzy approach to requirements prioritization. In: International symposium on search based software engineering. Springer, Berlin, Heidelberg, pp 64–69
26. Sadiq M, Jain SK (2013) A fuzzy based approach for requirements prioritization in goal oriented requirements elicitation process. In: International conference of software engineering and knowledge engineering (SEKE), pp 54–58
27. Ahmad KS, Ahmad N, Tahir H, Khan S (2017) Fuzzy MoSCoW: a fuzzy based MoSCoW method for the prioritization of software requirements. In: 2017 International conference on intelligent computing, instrumentation and control technologies (ICICICT). IEEE, pp 433–437
28. Tabassi AA, Abdullah A, Bryde DJ (2019) Conflict management, team coordination, and performance within multicultural temporary projects: evidence from the construction industry. Proj Manag J 50(1):101–114
29. Elsood MAA, Hefny HA, Nasr ES (2014) A goal-based technique for requirements prioritization. In: 2014 9th international conference on informatics and systems. IEEE, pp SW-18
30. Bajaj P, Arora V (2013) Multi-person decision-making for requirements prioritization using fuzzy AHP. ACM SIGSOFT Softw Eng Notes 38(5):1–6
31. Vestola M (2010) A comparison of nine basic techniques for requirements prioritization. Helsinki University of Technology, pp 1–8
32. Tufail H, Qasim I, Masood MF, Tanvir S, Butt WH (2019) Towards the selection of optimum requirements prioritization technique: a comparative analysis. In: 2019 5th International conference on information management (ICIM). IEEE, pp 227–231
33. Zave P (1997) Classification of research efforts in requirements engineering. ACM Comput Surv (CSUR) 29(4):315–321
34. Cheng BH, Atlee JM (2007) Research directions in requirements engineering. In Future of software engineering (FOSE'07). IEEE, pp 285–303
35. Khan JA, Rehman IU, Khan YH, Khan IJ, Rashid S (2015) Comparison of requirement prioritization techniques to find best prioritization technique. Int J Mod Educ Comput Sci 7(11)
36. Hassanian ZM, Sadeghi A, Farhadi M, Rahrovan A (2019) The relationship between leadership style and conflict management among nurses in educational and health centers. Acta Facultatis Med Naissensis 36(4):326–333

Decentralisation of FinTech Business Models

Fátima Leal, Maria Emília Teixeira, and Fernando Moreira

Abstract Digital transformation could change our daily routines, from currency to the purchases. Specifically, financial sector aims to efficiently supply distinct services using an infrastructure by relying on a centralised authority to manage and coordinate transactions at specific prices. In turn, final customers constantly seek to lower costs, quality, and reliability. Moreover, costumers desire less documentation and bureaucracy which hinder the access to financial services and credits. Financial system needs to create a digital strategy at the organisational level to address technology challenges and satisfy user's needs. In this scenario, financial and technology domains have been integrated originating FinTech business models with significant impact. Therefore, this paper contributes not only with a technology background which allow to explore new FinTech-based business models but also describing existing business models employed in the market. Moreover, we present a comparison among traditional and FinTech-based models allowing to analyse the financial progress. The paper identifies, still, future research opportunities including regulation of decentralised financial business models.

Keywords Blockchain · Decentralisation · Business models · FinTech · Smart contracts · RegTech · InsurTech

F. Leal · F. Moreira (✉)
REMIT, Universidade Portucalense, Porto, Portugal
e-mail: fmoreira@upt.pt

F. Leal
e-mail: fatimal@upt.pt

M. E. Teixeira · F. Moreira
IJP, Universidade Portucalense, Porto, Portugal
e-mail: emiliat@upt.pt

F. Moreira
IEETA, Universidade de Aveiro, Aveiro, Portugal

A. Ullah et al. (eds.), *Proceedings of International Conference on Information Technology and Applications*, Lecture Notes in Networks and Systems 350,
https://doi.org/10.1007/978-981-16-7618-5_30

1 Introduction

The exponential growth in digitalisation has transformed all sectors, including financial services. The digital economy has impacted the standard financial business models which normally operate upon centralised systems controlled by a single authority. Such models not only incorporate gaps which boost opportunities for fraud and crime but also tend to increase costs employing extra fees. In addition, its centralised nature is vulnerable failures and attacks.

As the society is moving into digital economy, financial technology (FinTech) has started to substitute some services provided by financial institutions. Despite reducing transaction costs and expand its scope, FinTech has not removed the power of central authorities or intermediaries [1], such as banks and technology companies. In this scenario, technologies such as blockchain have raised as a promising solution for decentralised finance (DeFi). Specifically, DeFi provides the same financial services without any traditional central authority or intermediaries since it is built on blockchain frameworks [2].

Digital transformation has become imperative for any organisation which wants to stay competitive and follow the society needs. This transformation become obvious for financial sector, mainly with blockchain and its decentralisation [3]. Decentralised financial models came to stay revolutionising the financial sector.

Scant research has been addressed in terms of comparing FinTech-based models analysing its vantages and disadvantages and indicating new future directions. In this context, this paper aims to describe the business impact of FinTech, comparing the traditional and FinTech-based models. Additionally, we surveyed Fintech-based models giving the decentralised vision which has been explored in financial domain.

As results, the paper presents an analysis which allows to verify where decentralised models can mitigate standard problems of current FinTech business models.

The remaining paper is organised as follows. Section 2 describes the business impact of FinTech comparing the traditional with FinTech-based models. Section 3 details the current FinTech models. Section 4 provides a decentralised vision supported by blockchain and smart contracts. Section 5 contains the final remarks of the work.

2 Business Impact of Financial Technology

FinTech put together both finance and technology domains which brought a new reality for financial world and corresponding services [4]. FinTech is arguably characterised by digital transformation based on the proliferation of robotisation, artificial intelligence (AI), blockchain, and other innovative and disruptive technologies.

Towards this scenario, FinTech is changing the traditional financial models employing technological innovations and forcing the standard business system to

Table 1 Comparison of financial business models before and after FinTech

Traditional models	FinTech-based models
Hands-based processes	Digital information
High service costs	Cost reduction
Physical infrastructures	Virtual platforms
Human resources dependency	Technology dependency
Centralised processes	Decentralised processes
Limited models	Ubiquitous models
Manual, slow, and high regulated processes	Digital, automatic, smart, and rapid processes
Limited progress margin	New competitiveness landscape
Customers' acceptance	Highly unknown concept by customers

quickly adapt their processes. Since costumer needs have changed, financial business models require significant transformations. FinTech has had a strong impact in financial sector which is extremely regulated and inaccessible to common users. Table 1 enumerates multiple progresses of FinTech causing significant society impacts in terms of employment, regulations, financial intermediaries, etc.

The differences among financial domain before and after technology are enormous.

Traditional financial institutions rely on physical infrastructures and heavy logistics processes with high costs including: (*i*) human resources; (*ii*) maintenance; (*iii*) costumer communication, and (*iv*) documents. FinTech involves emerging technologies such as blockchain, big data analytics, or robots. Such technologies automate and decentralise the business models obsoleting several of traditional models and the corresponding stakeholders. FinTech is not forced to be in a single territory allowing ubiquitous models. Those features forced new business models. However, traditional entities are excessively bureaucratic hindering FinTech-based services.

3 Fintech Business Models

European Commission has elaborated a plan for a more competitive and innovative financial sector [5] integrating solutions as digital identification, mobile applications, cloud computing, big data analytics, artificial intelligence, blockchain, and distributed ledger technologies. There are conditions to insert FinTech-based business models in the market being conducted by the same rules as traditional financial models to protect both consumer and service providers. Undoubtedly, FinTech raised

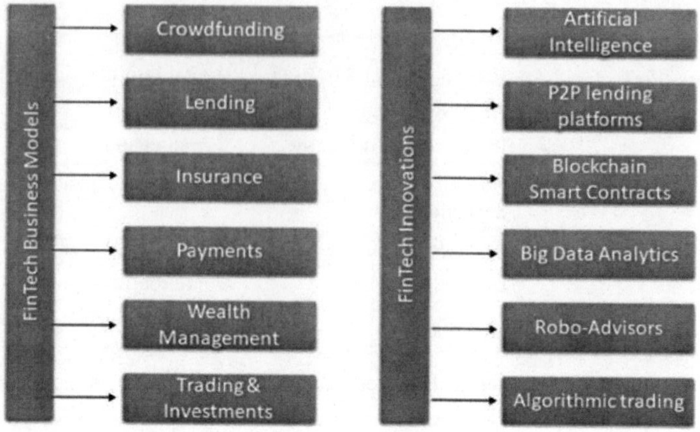

Fig. 1 FinTech business models and corresponding innovations

more flexibility and versatility to financial domain which was centralised and paralysed without margin to progress. Therefore, FinTech has been competing with traditional financial services, offering customer-centric services using technology. Nevertheless, FinTech will not replace the traditional financial model but complement it adding technological methods.

Fintech-based business models have included multiple innovations embedding technologies as illustrated in Fig. 1 [6, 7]. Those innovations have enabled to automate the existing financial services or create new products. The expansion of FinTech has revolutionised the financial domain, although, raising new risks. This section details the FinTech models and innovations as well as its gains and risks.

3.1 Models and Services

Employing multiple innovations, FinTech has been distinguished in multiple models [8]: (*i*) crowdfunding; (*ii*) peer-to-peer (P2P) lending (*iii*) insurance; (*iv*) payments; and (*iv*) wealth management. FinTech ecosystem (i.e. FinTech start-ups, government, technology developers, financial customers, and traditional financial institutions) has started the innovation, stimulating the economy with new financial business models [9].

- **Crowdfunding** refers mainly to online crowdfunding platforms where it is possible to obtain funding for projects or business ideas instead of recurring to traditional financial institutions. This FinTech model omits a detailed credit analysis being fast, simple, and without documentation. Kickstarter,[1] Indiegogo,[2] and GoFundMe[3] use this model to obtain and return the financed value to the investor.

- **Peer-to-peer Lending** involves digital platforms enabling individuals and businesses to lend and borrow money. This model matches the needs of both investor and investee concerning credit contracts. Since the standard credit processes are slow, difficult, and bureaucratic, lending-based models offer an efficient structure, and an improved lending process for stakeholders. Lending platforms increased the number of individual entities able to grant loans benefiting the access to the credit. As benefits, P2P lending offers rapid processes, flexibility with reduced regulation, and accessibility. However, since it involves unknown entities, the model represents a higher risk. Lending platforms mitigate the risk with smaller loans amounts and shorter terms [10]. Lending Club,[4] Prosper,[5] SoFi,[6] Zopa,[7] and RateSetter[8] are examples of P2P lending platforms.
- **Insurance** embraces emergent technologies to renovate their services. This originated the term InsurTech [11], i.e. insurance technology. InsurTech enables to perform a faster and effective risk analysis. It involves data analytics which personalise insurance products to meet costumer's needs [12] in the multiple insurable sectors. CoverFox,[9] The Zebra,[10] Sureify Labs,[11] and Ladder[12] are InsurTech platforms. Savings and insurance personalisation are the main vantages. However, there is still a lack of: (*i*) comparison between InsurTech service providers; and (*ii*) legal protection for InsurTech companies. InsurTech model integrates, still, robo-advisers who provide a digital support avoiding face-to-face interactions.
- **Payments** models integrate services like online and mobile payments. It includes: (*i*) pre-payed credit cards; (*ii*) systems to process payments; (*iii*) money transfer applications among bank accounts, and (*iv*) decentralised capital markets. FinTech payments services include mobile wallets, P2P mobile payments, foreign

[1] www.kickstarter.com.
[2] www.indiegogo.com.
[3] www.gofundme.com.
[4] www.lendingclub.com.
[5] www.prosper.com.
[6] www.sofi.com.
[7] www.zopa.com.
[8] www.ratesetter.com.
[9] www.coverfox.com.
[10] www.thezebra.com.
[11] www.sureify.com.
[12] www.ladderlife.com.

Table 2 Summary of FinTech products and services of each business model

Model	Products
Crowdfunding	Donation-, reward-, and equity-based crowdfunding platforms
Lending	P2P lending platforms, lending marketplaces
Insurance	Robo-advise, risk analysis, personalised packages
Payments	Digital wallets, cryptocurrencies, P2P transfers, foreign exchange
Wealth management	E-trading, social trading platforms, robo-advise, personalised investments

exchange, real-time payments, and cryptocurrencies [13]. PayPal,[13] Bigcommerce,[14] Affinipay,[15] Circle,[16] etc., implement FinTech payments model. As benefits, this model is simple, processed at lower cost, and fast. Nevertheless, it influences reckless spending.

- **Wealth Management** aims to handle financial assets or investments in its entire chain since banking, budgeting, tax services, etc. In this context, wealth management includes not only asset-based services but also financial planning and advising services. With FinTech, wealth management has introduced text analytics, natural language processing, robo-advisory, risk analysis, and algorithmic trading in their services [14]. Namely, robo-advisory has disrupted this business model assisting investors without human intervention automating wealth management services. Wealth management business model incorporates trading and investments. While trading involves short periods of time and higher risks, investments require larger timeframes but lower risks. FinTech has enriched trading and investments models with: (*i*) automated management and advice [15] (e.g., FutureAdvisor,[17] Wealthfront[18]), (*ii*) social trading [16] (e.g., eToro,[19] Covestor,[20] StockTwits[21]), and (*iii*) algorithmic trading [17] (e.g., QuantConnect[22]).

FinTech business models together with the corresponding innovations enable to create personalised products using big data analytics (e.g., machine learning, natural language processing, data visualisation). It is important to highlight the role of robo-advisors in insurance and wealth management models providing an automated investment advice. Table 2 summarises the most used FinTech products.

---

[13] www.paypal.com.

[14] www.bigcommerce.co.uk.

[15] www.affinipay.com.

[16] www.circle.com.

[17] App.futureadvisor.com.

[18] www.wealthfront.com.

[19] www.etoro.com.

[20] www.interactiveadvisors.com.

[21] www.stocktwits.com.

[22] www.quantconnect.com.

3.2 Advantages and Disadvantages

FinTech business models came to stay benefiting both consumers and economy. Moreover, the financial markets became more accessible for ordinary consumers. Most of the services reduced their costs or suffered a fee optimisation. The ubiquity of the services increases the flexibility, enables fast decisions, and a real-time monitoring. Despite such advantages, there is still scepticism concerning technology in financial services. Indeed, FinTech business models remains at its infancy requiring further research into its characteristics and attributes [8, 18]. In this context, Table 3 enumerates advantages and disadvantages raising several sensitive issues: (*i*) security, (*ii*) privacy, (*iii*) threats, and (*iv*) cyber-attacks. This paper aims to explore solutions of current FinTech business models where decentralisation emerges as the next step of FinTech.

Table 3 Advantages and disadvantages of FinTech business models

FinTech business models	Advantages	Disadvantages
Crowdfunding	Fast funding	Reputation damaged risk
	Market validation	Idea plagiarism
	Visibility	Platform fees
Lending	Rapid processes	Risk
	Flexibility	Smaller loans
	Accessibility	Frequently repayments
Insurance	Personalised products	Lack of comparison
	Paper reduction	Transparency
	Personalisation	Scepticism
Payments	Lower costs	Security concerns
	Time saving	Reckless spending
	Simple to use	Limited merchants
Wealth management	Permanent market analysis	Technology failure
	Full-time diligence	Fluctuate profits
	Personalised wealth planning	Frauds

4 Decentralisation of FinTech Models

Decentralisation has been seen as the future in several domains. It relies on distributed ledger technology (DLT) which aims to interact directly with the final client eliminating intermediary entities such as agents, brokers, or bankers.

DLTs enable the application of decentralised business models where authorised network members have access to the distributed ledger without relying on a single or centralised authority. Specifically, the distributed ledger uses a peer-to-peer environment which is continuously updated and synchronised. Malicious behaviours are easily detected and banned since blockchain holds security, authenticity, immutability, and transparency as main characteristics [19]. Standard financial business models will be revolutionised if decentralisation continue to build *momentum*, being the blockchain-based DeFi the next step to the revolution.

DeFi employs blockchain technology [20] to build a financial infrastructure without central authorities. Originally, blockchain-based applications were designed for cryptocurrencies. Then, the introduction of smart contracts automated the negotiation of a contract [21]. At moment, it is emerging non-cryptocurrency-related distributed ledger applications [22] which may be seen as the start of the revolution.

The decentralisation of conventional FinTech business models using DLTs is a novel business topic. In this context, we gather blockchain features which may address several of the issues analysed in Table 3.

- *Crowdfunding* platforms based on blockchain eliminates the platform fees. Additionally, fraudulent behaviours are easily detected and punished. Blockchain-based crowdfunding platforms use smart contracts among stakeholders to reach an optimal solution [23–25].
- *Lending* with distributed networks can minimise attacks and reduce frauds and thefts. Here, cryptocurrencies have been explored to mitigate those problems. In this context, blockchain-based lending has been emerged where members can receive loans through digital assets or using cryptocurrencies as collateral.
- *Insurance* has been reinvented by several start-ups, which are creating decentralised insurance products leveraging blockchain and smart contracts. Blockchain-based insurance models introduce cryptographically secured forms of shared information [26] to save time, reduce costs, and improve transparency. Decentralised insurance with blockchain has been explored to keep track of accidents, material, and moral losses during the insurance period preventing insurance frauds [27].
- *Payments*, with bitcoin, was the first decentralised mechanism explored by FinTech models. Blockchain provides a secure way of sending payments eliminating central authorities.
- *Wealth Management*, despite being in very early stages, has started to explore smart contracts together with decentralised applications mainly for asset and investment management. Blockchain raises solutions for: (*i*) managing the asset life cycle providing an end-to-end traceability and preventing frauds as well as data falsification and (*ii*) safely exchange assets on decentralised networks.

This scenario indicates that blockchain may revolutionise the financial domain. However, it is raising new realities not covered by the actual legislation. In fact, the legal authorities are not prepared for the financial revolution and the corresponding impacts. This unknown world hinders to understand which will be the illicit behaviours. Preparing legislation to protect both consumers and service providers will be a hard task. The legislation must cover new schema frauds for decentralised models. This context originated the RegTech concept. It is the management of regulatory processes for FinTech including regulatory monitoring, reporting, and compliance. Therefore, the decentralisation of FinTech highly depends on legislation to become a reliable path.

5 Conclusions

FinTech has innovated several financial services assisting both businesses and costumers to manage financial operations. Financial products and services were automated via emerging technologies seeking to mitigate costumer needs. In this context, this paper describes the business impact of FinTech analysing the evolution of the conventional financial models. At moment, most of the conventional financial models, i.e. crowdfunding, lending, payments, insurance, and wealth management, are technology dependants presenting several benefits and drawbacks. The main concerns analysed focus the risk of frauds, transparency, and failures. The decentralisation of FinTech models has been mentioned as the next step for its business models since it can solve several problems.

Decentralisation relies on blockchain technologies which offers authenticity, consensus, immutability, and faster/cheaper transactions. Therefore, this paper also provides a decentralised vision for the FinTech business models.

The analysis indicates the benefits of decentralisation. However, the main limitation for the direct application of blockchain technologies in production is the lack of regulations. In the financial domain, this is a strong barrier, due the tight regulation of this type of sector. However, blockchain is becoming one of the key technologies for enabling digitalisation. Following these conclusions, as future work, we intend to explore: (*i*) decentralised business models and (*ii*) regulatory solutions for decentralised FinTech.

References

1. Chen Y, Bellavitis C (2020) Blockchain disruption and decentralized finance: the rise of decentralized business models. J Bus Ventur Insights 13. https://doi.org/10.1016/j.jbvi.2019.e00151
2. Salami I (2020) Decentralised finance: the case for a holistic approach to regulating the crypto industry. Butterworths J Int Bank Finan Law 35(7):496–499

3. Nguyen QK (2016) Blockchain—a financial technology for future sustainable development. https://doi.org/10.1109/GTSD.2016.22
4. Leong K, Sung A (2018) FinTech (financial technology): what is it and how to use technologies to create business value in Fintech Way? Int J Innov, Manag Technol 9(2). https://doi.org/10.18178/ijimt.2018.9.2.791
5. European Commission (2018) Fintech action plan: for a more competitive and innovative European financial sector
6. Mehrban S et al (2020) Towards secure fintech: a survey, taxonomy, and open research challenges. IEEE Access 8. https://doi.org/10.1109/ACCESS.2020.2970430
7. Legowo MB, Subanidja S, Sorongan FA (2021) FinTech and bank: past, present, and future. Jurnal Teknik Komputer AMIK BSI 7(1). https://doi.org/10.31294/jtk.v4i2
8. Laidroo L, Koroleva E, Kliber A, Rupeika-Apoga R, Grigaliuniene Z (2021) Business models of FinTechs—difference in similarity? Electron Commer Res Appl 46. https://doi.org/10.1016/j.elerap.2021.101034
9. Lee I, Shin YJ (2018) Fintech: ecosystem, business models, investment decisions, and challenges. Bus Horiz 61(1). https://doi.org/10.1016/j.bushor.2017.09.003
10. Ferretti F (2021) Peer-to-Peer lending and EU credit laws: a creditworthiness assessment, credit-risk analysis or ... neither of the two? Ger Law J 22(1):102–121. https://doi.org/10.1017/glj.2020.100
11. Chuang L-M, Liu C-C, Kao H-K (2016) The adoption of Fintech service: TAM perspective. Int J Manag Adm Sci (IJMAS) 3(07)
12. Nicoletti B (2017) A business model for Insurtech initiatives. Fut FinTech. https://doi.org/10.1007/978-3-319-51415-4_8
13. Milian EZ, de M Spinola M, de Carvalho MM (2019) Fintechs: a literature review and research agenda. Electron Commer Res Appl 34. https://doi.org/10.1016/j.elerap.2019.100833
14. Phoon K, Koh F (2018) Robo-advisors and wealth management. J Altern Investments 20(3). https://doi.org/10.3905/jai.2018.20.3.079
15. Creamer G, Freund Y (2010) Automated trading with boosting and expert weighting. Quant Finance 10(4). https://doi.org/10.1080/14697680903104113
16. Oehler A, Horn M, Wendt S (2016) Benefits from social trading? Empirical evidence for certificates on wikifolios. Int Rev Finan Anal 46. https://doi.org/10.1016/j.irfa.2016.05.007
17. Nuti G, Mirghaemi M, Treleaven P, Yingsaeree C (2011) Algorithmic trading. Computer 44(11). https://doi.org/10.1109/MC.2011.31
18. Suryono RR, Budi I, Purwandari B (2020) Challenges and trends of financial technology (Fintech): a systematic literature review. Inf (Switz) 11(12). https://doi.org/10.3390/info11120590
19. Leal F, Chis AE, González-Vélez H (2021) Multi-service model for blockchain networks. Inf Process Manag 58(3), 2021. https://doi.org/10.1016/j.ipm.2021.102525
20. Li Z, Zhong RY, Tian ZG, Dai HN, Barenji AV, Huang GQ (2021) Industrial blockchain: a state-of-the-art survey. Robot Comput-Integr Manuf 70. https://doi.org/10.1016/j.rcim.2021.102124
21. Aggarwal S, Kumar N (2021) Blockchain 2.0: Smart contracts☆☆Working model. Adv Comput 121. https://doi.org/10.1016/bs.adcom.2020.08.015
22. di Francesco Maesa D, Mori P (2020) Blockchain 3.0 applications survey. J Parall Distrib Comput 138. https://doi.org/10.1016/j.jpdc.2019.12.019
23. Kumari S, Parmar K (2021) Secure and decentralized crowdfunding mechanism based on blockchain technology. https://doi.org/10.1007/978-981-15-7533-4_7
24. Hassija V, Chamola V, Zeadally S, BitFund: a blockchain-based crowd funding platform for future smart and connected nation. Sustain Cities Soc 60. https://doi.org/10.1016/j.scs.2020.102145
25. Jacynycz V, Calvo A, Hassan S, Sánchez-Ruiz AA (2016) Betfunding: a distributed bounty-based crowdfunding platform over Ethereum. Adv Intell Syst Comput 474. https://doi.org/10.1007/978-3-319-40162-1_44

26. Tasca P (2019) Insurance under the blockchain paradigm. In: Business transformation through blockchain, 2019. https://doi.org/10.1007/978-3-319-98911-2_9
27. Yıldırım İ (2020) Blockchain in the insurance industry: use cases and applications. In: Cross-industry use of blockchain technology and opportunities for the future, 2020

A Road Map Toward Crowdsourcing Actors, Platforms and Applications, a Review-Based Study

Abdullah Khan and Shah Nazir

Abstract The crowdsourcing term was first defined as the outsourcing of tasks or work to a group of an unknown workers in an open call for the purpose of utilizing human intelligence. Crowdsourcing nowadays gained popularity in industries and academia around the globe. Various organization and communities takes interest in it because of its parallel and cost-efficient nature. While crowdsourcing, it is necessary to have a complete knowledge of crowdsourcing process. Crowdsourcing activity consist of three main actors. Current study briefly discussed the analyzed crowdsourcing actors crowd, client and platforms. Features of the crowd and role of the client have been discussed. Various crowdsourcing platforms have been analyzed and categorized in order to get deep knowledge of the tasks performed on it. The authors concludes their study by identifying various crowdsourcing applications in a variety of disciplines.

Keywords Crowdsourcing actors · Crowdsourcing platforms · Crowdsourcing application

1 Introduction

The word crowdsourcing was first introduced by Jeff Howe in 2006; he defined it as the outsourcing of various task or work to a diverse crowd (group of workers) in an open call in order to utilize the workers intelligence [1]. Nowadays, the interest of crowdsourcing in communities and organization is increasing day by day because of performing parallel processing of tasks in cost- and time-efficient way due to online labor markets. In the field of computer science, the concept of crowdsourcing have been used by various crowdsourcing platforms such as AMT and CrowdFlower. The process of crowdsourcing is an efficient and fast way of finding information by social networks [2]. In the current study, all the crowdsourcing actors and application will

A. Khan · S. Nazir (✉)
Department of Computer Science, University of Swabi, Swabi, Pakistan
e-mail: shahnazir@uoswabi.edu.pk

A. Ullah et al. (eds.), *Proceedings of International Conference on Information Technology and Applications*, Lecture Notes in Networks and Systems 350,
https://doi.org/10.1007/978-981-16-7618-5_31

be discussed briefly. Crowdsourcing platforms will be discussed briefly along with their tasks to be performed on it. The following are the contribution of the study:

First of all, the crowd and its features will be analyzed in context of crowdsourcing, and then, the clients associated with crowdsourcing will be analyzed. Furthermore, the authors will analyze several famous crowdsourcing platforms in order to highlight the relevancy of the platforms with different type of tasks. The authors will also highlight several crowdsourcing applications and then will categorize these applications.

2 Related Work

The existing literature consist of various crowdsourcing platforms used for the solution of different problems. Among all these crowdsourcing platforms, Amazon Mechanical Turk and TopCoder are the most famous platforms used for transcription, classification and software development tasks, respectively [3, 4]. For microtasks, various crowdsourcing platforms have been analyzed during literature study which include Recaptcha, TestBirds, Clickchores, Quora, Fiverr, Hollywood, Stock Exchange, Galaxy Zoo, TripAdvisor, Microtask, Procter & Gamble, Solve, YourEncore, buzzwords, KicKstarter, Yet2, Avery Dennison, AppStori, ShortTask and MicroWorkers [1, 4–9]. Various crowdsourcing platforms which are related to software engineering and development are also analyzed which consist of TopCoder, GetACoder, Guru, TestFlight, Mob4Hire, TaskCity,uTest, BugCrowd, Kaggle, StackOverflow, oDesk, eLance and vWorker [1, 5, 7, 10, 11]. Similarly, a variety of crowdsourcing applications have been analyzed during literature which include software requirement, designing, coding, testing and maintenance along with aggregation, database design, data processing, problem solving, multimedia, software development, design-based applications and creative applications. As the crowdsourcing actors consist of client, crowd and platforms, therefore, are these three actors are analyzed and discussed briefly during literature study. Various important crowd features have also been analyzed during the study which include skill, autonomous, appropriate, qualified/educated, problem solving, fast, expertise, competent, experience, energetic, identifiable, professionals, smart, right, knowledge, reliable/efficient, loyal, flexible, eager, creative, trustworthy, cooperative, coordinative, collaborative and adaptable [12].

3 Methodology

Performing the process of crowdsourcing in an efficient way is a challenge for everyone. Therefore, it is very important to know about crowdsourcing activity before crowdsourcing. All crowdsourcing activities have been discussed briefly in the current paper. The process of crowdsourcing consist of three main actors which include crowd, client and platforms. The crowdsourcing activity cannot be completed by avoiding any of these actors. The role of each of these actors have also been discussed. These crowdsourcing actors are discussed in subsections. Section 3.1 discusses crowd, Sect. 3.2 discusses client, and Sect. 3.3 discusses crowdsourcing platforms. In Sect. 3.4, crowdsourcing applications have been discussed. Section 4 consists of conclusion.

3.1 Crowd

Crowd also known as workers are the individuals which performs the tasks or work posted by the clients. When an individual want to be a crowd or worker he must register himself with a platforms. Before performing a task, every individual register himself and give full information during registration process which is then required for performing the work and also help in matching task and worker relevancy. As various crowd selection techniques use crowd profile history for matching tasks and workers, therefore, an individual should properly fill all the required information at the time of registration [1]. Crowd features [12] have been listed in Table 1.

3.2 Clients

Clients are all those individuals who posted the job or work to be performed by the crowd. Clients must post a task which is properly labelled so that everyone can easily recognize it. The posted task by the clients is also known as human intelligence tasks (HIT's) [5].

3.3 Platforms

Platforms are one of the important crowdsourcing actors which act as a broker or an intermediator between crowd and clients. The clients posted the tasks through platforms and then the crowd takes it through platform; thus, platforms play a very important role during crowdsourcing activity [1, 5]. Various crowdsourcing platforms have been analyzed during literature are shown in Table 2.

These are famous crowdsourcing platforms which have been highlighted during literature. AMT and TopCoder are most famous platforms among these crowdsourcing platforms.

TopCoder is the world largest crowdsourcing platforms use for various software engineering tasks. TopCoder platform have 750,000 software engineers in March 2015 and have been awarded dollar 67,000,000 as a monetary rewards for performing various software engineering tasks [1, 5].

Table 1 Crowd features

S. No.	Features
1	Trustworthy
2	Skill
3	Professionals
4	Smart
5	Right
6	Knowledge
7	Identifiable
8	Reliable/efficient (workers)
9	Qualified/educated
10	Problem solving
11	Loyal
12	Flexible
13	Fast
14	Expertise
15	Experience
16	Energetic
17	Eager beavers
18	Creative
19	Coordinative
20	Cooperative
21	Autonomous
22	Appropriate
23	Competent
24	Collaborative
25	Adaptable

Various software engineering tasks including software development is performed on GetACoder and TopCoder platform. Some of these platform are specific such as BugCrowd and uTest which is used for software testing and security analysis. AMT, Freelancer and oDesk are general purpose crowdsourcing platforms which is used for every type of tasks [6, 7, 25, 37]. Table 3 shows tasks performed on these platforms.

3.4 Crowdsourcing Applications

Crowdsourcing application is increasing day by day in the industries as well as in academia. Variety of tasks have been performed in an online labor market.

Table 2 Crowdsourcing platforms

S. No.	Platforms	Citation	Tasks performed
1	Amazon Mechanical Turk (AMT)	[3, 6, 13–24]	Image clustering, document labelling, creative designing, transcription, classification or categorization tasks
2	TopCoder	[1, 4, 5, 7, 10, 11, 14, 18, 20, 25, 26]	Software development tasks
3	CloudCrowd	[5, 6, 13]	Editing, searching, writing, categorization, data entry and translation tasks
4	Freelancer	[1, 2, 6, 27, 28]	Service marketplaces
5	Recaptcha	[8]	Microtasks
6	Galaxy Zoo	[8, 23]	Microtasks
7	TripAdvisor	[8]	Microtasks
8	Hollywood Stock Exchange	[8]	Microtasks
9	InnoCentive	[6–9, 16, 29–31]	Problem solving, innovation projects, business, scientific problems. Chemistry, computer science, engineering, statistics, design and mathematics
10	99 Designs	[6, 20, 31]	Competition
11	Delicious	[6]	Content-rating
12	IdeaScale	[6]	Idea
13	Upwork	[28, 31, 32]	Macrotasks
14	ClickWorker	[33]	For managing e-commerce data
15	TestFlight	[1]	Software engineering tasks
16	NineSigma	[5]	Microtasks
17	Mob4Hire	[1]	Software engineering tasks
18	GetACoder	[1, 25, 34]	Software development tasks
19	BugCrowd	[1]	Software testing, security analysis
20	TestBirds	[34]	Microtasks
21	Clickchores	[1]	Microtasks

(continued)

Table 2 (continued)

S. No.	Platforms	Citation	Tasks performed
22	Microtask	[1]	Microtasks
23	Quora	[24, 31]	Microtasks
24	Kaggle	[10]	Software development tasks
25	SurveyMonkey	[31]	Microtasks
26	MyGov	[31]	Contributions to policy making (India)
27	Fiverr	[27]	Microtasks
28	KicKstarter	[4]	Microtasks
29	StackOverflow	[1]	(Question/answer website) provides crowdsourced programming knowledge for software developers
30	Bountify	[1]	Micro programming tasks
31	Yet2	[5]	Microtasks
32	YourEncore	[5]	Microtasks
33	CrowdFlower or SmartSheet	[1, 3, 6, 15, 19, 21, 22, 32, 33, 35, 36]	Cleaning, labelling, designing/managing projects, classification or categorization tasks, tasks, data collection,
34	CastingWords or SpeechInk	[37]	Transcription services include transcription from audio to text
35	oDesk	[1, 5, 37]	Web development and Web programming, transcription, data-cleansing
36	Taskcn	[9, 16]	Advertisement posting, business proposal writing, website development,
37	eLance	[11]	Software project development
38	vWorker	[11]	Software project development
39	Guru,	[11]	Software project development
40	TaskCity	[11]	Software project development

(continued)

Table 2 (continued)

S. No.	Platforms	Citation	Tasks performed
41	buzzwords	[8]	Microtasks
42	Procter & Gamble	[9]	Microtasks
43	Solve	[9]	Microtasks
44	Avery Dennison	[9]	Microtasks
45	AppStori	[1, 7]	Microtasks
46	uTest	[1, 7, 34]	Software testing, security analysis,
47	iStockPhoto	[6, 38]	Sells photographs, video clips and animation, content markets,
48	Threadless	[6, 23]	Design process of t-shirts
49	MicroWorkers	[4, 19, 33]	Microtasks
50	ShortTask	[6]	Microtasks
51	Crowdcast	[6]	Prediction
52	XPrize	[6]	Innovation prizes

Table 3 Platforms categorization

S. No.	Platforms	Offered task
1	TopCoder	Software development tasks
	GetACoder	
2	BugCrowd,	Security analysis, software testing
	uTest	Software testing, security analysis
3	Amazon Mechanical Turk,	General purpose tasks
	Freelancer,	Document labelling, creative designing, classification tasks Service marketplaces
	oDesk	Web development and Web programming, Transcription, Data-cleansing
4	TestFlight,	Software engineering tasks
	TopCoder,	
	AppStori	
	uTest,	
	Mob4Hire	

Crowdsourcing applications in software engineering consist of software requirements, software designing, coding of softwares, testing and verification of softwares, evolution and maintenance of softwares. Several crowdsourcing platforms analyses during literature are shown in table 4.

Table 4 Crowdsourcing applications

S.NO	Field	Applications	Citations
1	Software engineering	Software requirements, software designing, software coding, software testing and verification, software evolution and maintenance	[1]
2	IT	Labeling large data sets, creative applications, design-based applications, text translation to different languages	[39]
3	IT	Recommendation systems, software development, multimedia, database design, and research engine evaluation	[39]
4	Crowdsourcing tasks	Aggregation, problem solving, generation of the contents, and processing of the data	[8]

4 Conclusion

The interest of the communities and organization is increasing day by day in the field of crowdsourcing. Crowdsourcing plays very important role in industries as well as academia and thus gain popularity throughout the world. During crowdsourcing activity, it is very important to know about crowdsourcing process. In the current study, the authors discusses crowdsourcing actors such as crowd, client and platforms. Crowd performs the selected tasks while client posts a task to be performed by the crowd, and the platform acts as intermediator between client and crowd. The authors also analyses several features of the crowd. The client role has also been highlighted. Then, various crowdsourcing platforms have been highlighted, and the tasks to be performed on these platforms have been analyzed. The authors also highlighted various crowdsourcing application in a variety of disciplines.

References

1. K. Mao, L. Capra, M. Harman, and Y. Jia, "A survey of the use of crowdsourcing in software engineering," *Journal of Systems and Software,* vol. 126, pp. 57–84, 2017/04/01/ 2017.
2. Y. Baba, H. Kashima, K. Kinoshita, G. Yamaguchi, and Y. Akiyoshi, "Leveraging non-expert crowdsourcing workers for improper task detection in crowdsourcing marketplaces," *Expert Systems with Applications,* vol. 41, pp. 2678–2687, 2014/05/01/ 2014.
3. M.-C. Yuen, I. King, and K.-S. Leung, "Task recommendation in crowdsourcing systems," presented at the Proceedings of the First International Workshop on Crowdsourcing and Data Mining, Beijing, China, 2012.
4. P. Wu, E. W. T. Ngai, and Y. Wu, "Toward a real-time and budget-aware task package allocation in spatial crowdsourcing," *Decision Support Systems,* vol. 110, pp. 107–117, 2018/06/01/ 2018.
5. A. Sarı, A. Tosun, and G. I. Alptekin, "A systematic literature review on crowdsourcing in software engineering," *Journal of Systems and Software,* vol. 153, pp. 200–219, 2019/07/01/ 2019.
6. N. Luz, N. Silva, and P. Novais, "A survey of task-oriented crowdsourcing," *Artificial Intelligence Review,* vol. 44, pp. 187–213, 2015/08/01 2015.

7. K.-J. Stol and B. Fitzgerald, "Two's company, three's a crowd: a case study of crowdsourcing software development," presented at the Proceedings of the 36th International Conference on Software Engineering, Hyderabad, India, 2014.
8. D. Geiger and M. Schader, "Personalized task recommendation in crowdsourcing information systems — Current state of the art," *Decision Support Systems,* vol. 65, pp. 3–16, 2014/09/01/ 2014.
9. H. Zheng, D. Li, and W. Hou, "Task Design, Motivation, and Participation in Crowdsourcing Contests," *International Journal of Electronic Commerce,* vol. 15, pp. 57–88, 2011/07/01 2011.
10. Z. Wang, H. Sun, Y. Fu, and L. Ye, "Recommending crowdsourced software developers in consideration of skill improvement," presented at the Proceedings of the 32nd IEEE/ACM International Conference on Automated Software Engineering, Urbana-Champaign, IL, USA, 2017.
11. K. Mao, Y. Yang, M. Li, and M. Harman, "Pricing crowdsourcing-based software development tasks," presented at the Proceedings of the 2013 International Conference on Software Engineering, San Francisco, CA, USA, 2013.
12. G. Wang, F. Ali, J. Yang, S. Nazir, T. Yang, A. Khan, *et al.*, "Multicriteria-Based Crowd Selection Using Ant Colony Optimization," *Complexity,* vol. 2021, p. 6622231, 2021/01/23 2021.
13. G. Xintong, W. Hongzhi, Y. Song, and G. Hong, "Brief survey of crowdsourcing for data mining," *Expert Systems with Applications,* vol. 41, pp. 7987–7994, 2014/12/01/ 2014.
14. H. Javadi Khasraghi and A. Aghaie, "Crowdsourcing contests: understanding the effect of competitors' participation history on their performance," *Behaviour & Information Technology,* vol. 33, pp. 1383–1395, 2014/12/02 2014.
15. A. Tarasov, S. J. Delany, and B. Mac Namee, "Dynamic estimation of worker reliability in crowdsourcing for regression tasks: Making it work," *Expert Systems with Applications,* vol. 41, pp. 6190–6210, 2014/10/15/ 2014.
16. H. Ye and A. Kankanhalli, "Investigating the antecedents of organizational task crowdsourcing," *Information & Management,* vol. 52, pp. 98–110, 2015/01/01/ 2015.
17. Zhang X, Yang Z, Liu Y, Tang S (2019) On Reliable Task Assignment for Spatial Crowdsourcing. IEEE Trans Emerg Top Comput 7:174–186
18. T. Alelyani and Y. Yang, "Software crowdsourcing reliability: an empirical study on developers behavior," presented at the Proceedings of the 2nd International Workshop on Software Analytics, Seattle, WA, USA, 2016.
19. R. M. Borromeo, T. Laurent, and M. Toyama, "The Influence of Crowd Type and Task Complexity on Crowdsourced Work Quality," presented at the Proceedings of the 20th International Database Engineering & Applications Symposium, Montreal, QC, Canada, 2016.
20. E. R. Q. Weidema, C. López, S. Nayebaziz, F. Spanghero, and A. v. d. Hoek, "Toward microtask crowdsourcing software design work," presented at the Proceedings of the 3rd International Workshop on CrowdSourcing in Software Engineering, Austin, Texas, 2016.
21. A. Sales Fonteles, S. Bouveret, and J. Gensel, "Trajectory recommendation for task accomplishment in crowdsourcing – a model to favour different actors," *Journal of Location Based Services,* vol. 10, pp. 125–141, 2016/04/02 2016.
22. G. Li, Y. Zheng, J. Fan, J. Wang, and R. Cheng, "Crowdsourced Data Management: Overview and Challenges," presented at the Proceedings of the 2017 ACM International Conference on Management of Data, Chicago, Illinois, USA, 2017.
23. B. Morschheuser, J. Hamari, J. Koivisto, and A. Maedche, "Gamified crowdsourcing: Conceptualization, literature review, and future agenda," *International Journal of Human-Computer Studies,* vol. 106, pp. 26–43, 2017/10/01/ 2017.
24. Yang P, Zhang N, Zhang S, Yang K, Yu L, Shen X (2017) Identifying the Most Valuable Workers in Fog-Assisted Spatial Crowdsourcing. IEEE Internet Things J 4:1193–1203
25. Yu D, Zhou Z, Wang Y (2019) Crowdsourcing Software Task Assignment Method for Collaborative Development. IEEE Access 7:35743–35754
26. Yu D, Wang Y, Zhou Z (2019) Software Crowdsourcing Task Allocation Algorithm Based on Dynamic Utility. IEEE Access 7:33094–33106

27. L. Nassar and F. Karray, "Overview of the crowdsourcing process," *Knowledge and Information Systems,* vol. 60, pp. 1–24, 2019/07/01 2019.
28. Jiang J, An B, Jiang Y, Shi P, Bu Z, Cao J (2019) Batch Allocation for Tasks with Overlapping Skill Requirements in Crowdsourcing. IEEE Trans Parallel Distrib Syst 30:1722–1737
29. Ali-Hassan H, Allam H (2016) Comparing crowdsourcing initiatives: Toward a typology development. Canadian Journal of Administrative Sciences / Revue Canadienne des Sciences de l'Administration 33:318–331
30. Ghezzi A, Gabelloni D, Martini A, Natalicchio A (2018) Crowdsourcing: A Review and Suggestions for Future Research. Int J Manag Rev 20:343–363
31. L. M. S and Y. Reddy, "Crowdsourcing security: opportunities and challenges," presented at the Proceedings of the 11th International Workshop on Cooperative and Human Aspects of Software Engineering, Gothenburg, Sweden, 2018.
32. G. Li, J. Wang, Y. Zheng, and M. Franklin, "Crowdsourced Data Management: A Survey," in *2017 IEEE 33rd International Conference on Data Engineering (ICDE)*, 2017, pp. 39–40.
33. R. M. Borromeo and M. Toyama, "An investigation of unpaid crowdsourcing," *Human-centric Computing and Information Sciences,* vol. 6, p. 11, 2016/08/05 2016.
34. L. Vaz, I. Steinmacher, and S. Marczak, "An Empirical Study on Task Documentation in Software Crowdsourcing on TopCoder," in *2019 ACM/IEEE 14th International Conference on Global Software Engineering (ICGSE)*, 2019, pp. 48–57.
35. C. Qiu, A. C. Squicciarini, B. Carminati, J. Caverlee, and D. R. Khare, "CrowdSelect: Increasing Accuracy of Crowdsourcing Tasks through Behavior Prediction and User Selection," presented at the Proceedings of the 25th ACM International on Conference on Information and Knowledge Management, Indianapolis, Indiana, USA, 2016.
36. T. Awwad, N. Bennani, K. Ziegler, V. Sonigo, L. Brunie, and H. Kosch, "Efficient Worker Selection Through History-Based Learning in Crowdsourcing," in *2017 IEEE 41st Annual Computer Software and Applications Conference (COMPSAC)*, 2017, pp. 923–928.
37. D. Schall, B. Satzger, and H. Psaier, "Crowdsourcing tasks to social networks in BPEL4People," *World Wide Web,* vol. 17, pp. 1–32, 2014/01/01 2014.
38. A. Moayedikia, W. Yeoh, K.-L. Ong, and Y. L. Boo, "Improving accuracy and lowering cost in crowdsourcing through an unsupervised expertise estimation approach," *Decision Support Systems,* vol. 122, p. 113065, 2019/07/01/ 2019.
39. M. Hosseini, A. Shahri, K. Phalp, J. Taylor, and R. Ali, "Crowdsourcing: A taxonomy and systematic mapping study," *Computer Science Review,* vol. 17, pp. 43–69, 2015/08/01/ 2015.

Dynamic Offloading in Fog Computing: A Survey

Mariam Orabi, Raghad Al Barghash, and Sohail Abbas

Abstract Recently, a variety of heavy applications emerged, such as augmented reality, virtual reality, online gaming and data analysis, which demand resources higher than the capabilities of the end devices and require high response. To provide feasible distributed computing scheme, fog computing networks are employed. However, fog nodes and edge devices have limited resources, which may cause violations in quality of service (QoS) requirements. Offloading provides a solution for this issue by outsourcing tasks to other nodes. An important decision to implement offloading is how to pick the destination node, which can be solved by dynamic offloading that leverage the current information about the resources in the network to address this challenge. This work reviews literature that provides solutions for dynamic offloading. A categorization of the literature and comparisons between different approaches is proposed. As a result, trends in the literature are analyzed and future directions are suggested.

Keywords Fog computing · Resource management · Offloading · Dynamic offloading

1 Introduction

With the emergence of diverse applications requiring high computations and storage capabilities, migration of computations, data and processes became essential for most corporations and individual. First, the traditional solutions proposed for such scenarios are centralized cloud-based servers, which represent high performance computational and storage machines and provide pay-per-usage services. Central

M. Orabi (✉) · R. Al Barghash · S. Abbas
College of Computing and Informatics, University of Sharjah, Sharjah, UAE
e-mail: morabi@sharjah.ac.ae

R. Al Barghash
e-mail: U19105793@sharjah.ac.ae

S. Abbas
e-mail: sabbas@sharjah.ac.ae

A. Ullah et al. (eds.), *Proceedings of International Conference on Information Technology and Applications*, Lecture Notes in Networks and Systems 350,
https://doi.org/10.1007/978-981-16-7618-5_32

clouds are reliable solutions, highly used by enterprise users, as reported in the Right Scale survey conducted in 2019 by Flexera, about 94% of enterprise Internet users depend on cloud-based services [9].

However, in the cloud-based solutions, central cloud servers are expensive machines that are located far from end users. In other words, for real-time applications, such solutions may fail because of the delay resulting in transferring the tasks to and from the cloud. Moreover, many medium and lightweight applications do not require the high capabilities of a cloud server; a machine with ordinary capabilities can do the job with lower cost. Fog-like computing networks address these flaws by constructing an intermediate layer between the end users devices and the central cloud servers. This intermediate layer consists of distributed machines that provides resources and services for end users called fog nodes.

Figure 1 shows an example of a fog computing network architecture. End users can be Internet of things (IoT) nodes or users that request services from fog or cloud nodes. The communication between end user devices and the fog nodes is usually handled by gateways, which transmits the transformed tasks to the fog manager. A fog manager decides the destination node that will handle the task, which can be a fog or cloud node, and schedules the tasks among fog nodes [1, 10]. The dashed line in the figure represents a traditional end user to access the cloud.

Outsourcing tasks is considered to be one of the main resource management techniques in the field of fog computing [5], also known as offloading. An end device may want to offload tasks to fog or cloud nodes as a type of computational migration. Additionally, a fog node may request offloading some tasks. That is, a fog node can get overloaded and may cause reduction in Quality of service (QoS). A solution for this scenario can be migrating some tasks from the overloaded node to another (comparatively less overloaded) node that can handle them.

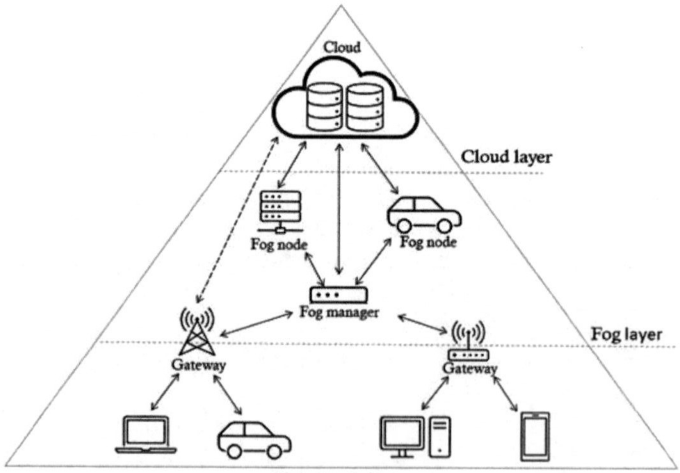

Fig. 1 Fog computing network architecture

Offloading tasks demands many challenging decisions. Specifically because fog computing has several dynamic aspects, namely the dynamic nature of the networks, the dynamic behavior of nodes in the fog network and the dynamic availability of resources in fog nodes, etc., [28]. These decisions require answers for some critical questions, such as how to pick tasks to be offloaded and how to select the destination node for the offloading. For the latter question, two approaches exist to address such questions, namely static and dynamic offloading.

In static offloading, the destination node is predefined regardless of the current status of the nodes and the network. Even though static offloading provides simple settlement to these questions, yet QoS and real-time applications require better offloading techniques that can fully and efficiently leverage resources in fog networks. On the other hand, dynamic offloading can provide promising solutions, where the decision is made based on the current status of the network and the needs of the migrating task.

Existing surveys about fog offloading are few and some provide review of IoT applications only [1]. Moreover, this field is developing quickly, so a recent review is encouraged. This work focuses on dynamic offloading aiming to direct and help researchers by reviewing state-of-the-art solutions that address the following questions related to offloading destination: (1) What is the type of the offloading node; a cloud, fog, or edge node? (2) To which node the task will be assigned? The contributions of this work can be summarized as follows:

- State-of-the-art literature review about dynamic fog offloading solutions for any application.
- A clear taxonomy for dynamic fog offloading is proposed.
- Trends in the literature are analyzed and summarized; future directions are also recommended.

In this paper, recent publications of high quality are chosen, analyzed and summarized. A categorization of the chosen works is proposed. Comparisons between them are provided to the reader in tabular format. Then, trends in research based on the temporal order of publications are identified. Finally, gaps in research and future directions are provided.

The rest of this work is organized as follows: Sect. 2 provides a review of related surveys and identifies the gap that this review addresses, Sect. 3 introduces the taxonomy and brief summary of the literature, Sect. 4 provides comparison of the literature, a discussion of trends noticed in the literature and possible future directions. Finally, Sect. 5 concludes the findings, limitations, and recommendations of this work.

2 Related Work

Many authors reviewed research related to fog computing, but more work was done on edge offloading than fog offloading as shown in Fig. 2. An early survey was proposed

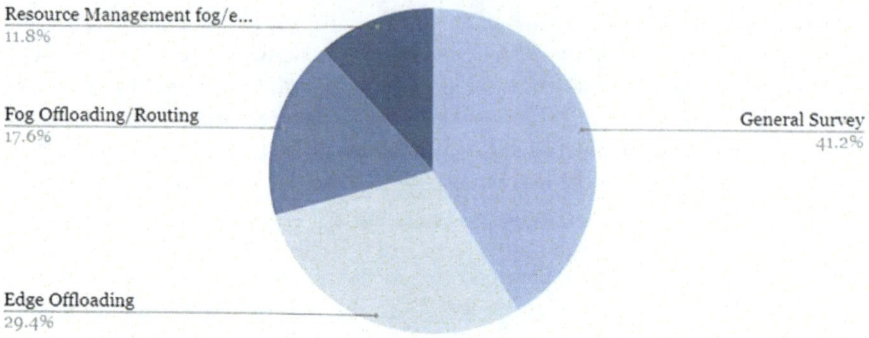

Fig. 2 Count of different related survey types in the years 2015 to 2020

by **Yi et al.** [28], which provides a review of the definition, application scenarios, and potential issues in fog computing. Several surveys followed, an overview of fog computing and comparisons to edge and cloud computing was presented by **Hu et al.** in a comprehensive survey [8], in which key technologies, applications and possible challenges are demonstrated. **Mahmud et al.** published a book chapter, in which research focusing about fog computing is reviewed and mapped to a taxonomy [16]. Similarly, **Mukherjee et al.** extensively reviewed research works related to fog computing, and provided a summary about the fundamentals of fog computing, resource allocation solutions, applications and challenges of fog computing [20]. **Mouradian et al.** enriched the research area with a survey focusing on the architecture and algorithms that are used in fog computing [19]. Moreover, **Martinez et al.** reviewed literature that provide solutions for fog computing design and resource management focusing on the technical aspects in research, such as installations and evaluations [17]. A recent comprehensive review was proposed by **Yousefpour et al.** providing a taxonomy of the literature [29].

The aforementioned works provide a general overview of the arising fog computing models. They provide strong background information necessary to understand the design, architecture, applications and challenges of fog computing in general. Nonetheless, a specialized reader interested in resource allocation and offloading problem and proposed solutions in the research area will need more in-depth surveys rather than general surveys. **Hong and Varghese** reviewed resource management in fog/edge computing architecture [6], where offloading is one the data flow architectures reviewed. Comparably, **Rejiba et al.** focused only on service migration in fog/edge or similar computing architectures [23]. Even though several works focus offloading in edge computing architecture [13–15, 24, 25], yet only few reviews are constructed for fog-based architecture. In a slightly different approach, **Okay and Ozdemir** reviewed routing schemes for IoT applications in fog computing, the authors consider offloading as one of the services that require routing to find an optimal target for the computational load to migrate to [21].

Task placement and offloading as well as load distribution are resource management approaches in the field of fog computing as **Ghobaei-Arani et al.** provided in their review for recourse management in fog computing [5]. However, none of the works mentioned earlier provide a specialized review for tasks offloading specifically. A relevant review was conducted by **Aazam et al.** that reviewed offloading scenarios for IoT applications in fog and cloud architectures [1]. The authors propose a taxonomy for the reviewed literature and suggest possible future directions for research. Additionally, a short survey with a similar focus was proposed by **Jha et al.** [12]. These two works are related to the main objective of this work and provide a comprehensive overview of literature. However, both works focus on IoT applications. While IoT is one of the main fields in which fog computing is used, yet there exist other application scenarios that are not included in the review. Additionally, this area is rapidly evolving and new techniques are arising. Therefore, a recent survey that reviews variety of scenarios and applications of task offloading in fog computing is required. Thus, up to the best of our knowledge, this is the first work that addresses this gap.

3 Literature Review

This survey focuses on dynamic offloading schemes in fog computing-enabled systems. Research in this area can be categorized into two classes, as shown in Fig. 3. First, works that develop solutions for static fog networks where all fogs are stationary with fixed locations. Second, works that develop solutions for dynamic fogs networks where fog nodes can be mobile.

These classes can be further categorized to centralized and decentralized offloading approaches. In centralized approaches, the offloading destination node is selected by a centralized unit. That is, all required information about the network, resources and application requirements are gathered in a central unit, where algorithms to select the offloading node are executed. This approach can provide global optimized decisions, but may suffer from high cost and single point of failure issues.

In contrast, in decentralized offloading approaches, selecting offloading node can be decided in two scenarios:

- Several distributed units gather information from local nodes with local optimized decision.

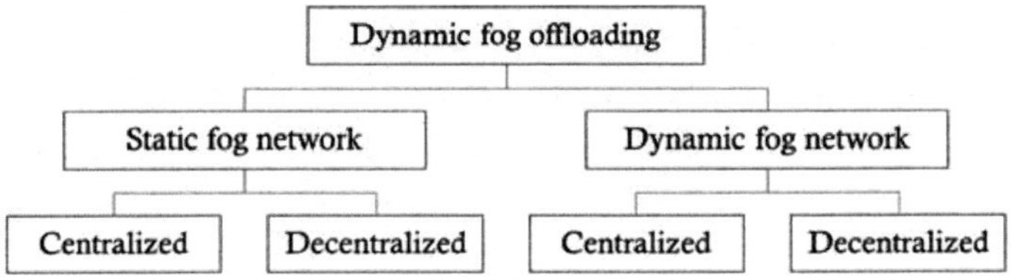

Fig. 3 Dynamic offloading in fog-enabled systems taxonomy

- The node that generated the offloading request is responsible for selecting the offloading node.

  Even though this approach minimizes failures impact, yet this approach is dependent on the capacity of the node generating the request, which may reduce the effectiveness of such approach. Additionally, security poses a challenge in such approaches because the required information must be shared among the nodes in the system.

The rest of this section provides review of works following the categorization given in Fig. 3.

3.1 Static Fog Networks

Centralized offloading Misra and Saha [18] introduced Detour: a task offloading paradigm from IoT devices to fog nodes based on software-defined networking (SDN). In other words, detour has a centralized SDN controller that collects network information and manages the network path based on the current requirements of the system. The authors take into account delay, energy and network path when modeling the problem as an integer linear program and propose an approximate solution based on a greedy heuristic approach, which showed acceptable performance when the number of offloaded tasks is relatively large. Simulations of Detour were compared with two other approaches, namely a delay-aware approach with greedy path, and a random offloading scheme with random path. Results show that Detour is capable of reducing delay and optimize energy consumption better than the aforementioned approaches. However, in Detour, an SDN controller manages the network paths from IoT devices to fog nodes. To apply this paradigm, it requires implementing many switches in the network, which is expensive and unrealistic.

To overcome this drawback, **Phan et al.** [22] utilized offloading among fog nodes leveraging SDN to insure a programmable network structure that is capable of enhancing QoS. The system is composed of three layers: fog nodes, controller and a services layer. The controller, which is the central layer, contains an SDN controller and a fog orchestrator, which periodically collects information about the network and fog nodes. In other words, the controller provides the information needed by the higher layer, services layer. If a fog node is overloaded, the offloading service assigns the offloading destination node by filtering the nodes capable of handling the task, which are ranked based on the current resources available and the network status of each node. OpenFlow switches are used to guarantee bandwidth in the network. Compared to a distributed offloading architecture that does not employ SDN, the proposed system showed lower latency. However, due to the centralized architecture and application-dependent network management, SDN-based systems are more prone to suffer from security and protection problems.

Another solution proposed by **Zhang et al.** [31] is FTO: fair task offloading scheme. The structure of the proposed scheme is represented as cluster of passive and active fog nodes (passive and active) distributed around a terminal node, where passive nodes have are powered by battery unlike active nodes that have unlimited access to circuit power. New tasks are handled by the terminal node, which studies nearby nodes status of workload, computation power, priority rate and computation rate. Then, it selects fog nodes to process sub-tasks and send the results back to the terminal node to combine them and make another decision for next task. The proposed FTO model was compared with three other models: (1) selects nodes based on processing rate, (2) assigns equal offloading with no priority, and (3) no offloading is applied. FTO model achieved better fairness, less delay and energy consumption results than all other three models.

Decentralized offloading Adhikari et al. [2] proposed DPTO: a deadline-aware framework for IoT devices to fog or cloud nodes offloading. DPTO performs task scheduling as well as selecting the suitable offloading destination node. Based on the tasks deadline, it specifies priority to the tasks and classifies based on the specified priority. Then, tasks are allocated to the suitable multi-level feedback queue in the IoT gateway to be properly scheduled. IoT gateway collects nodes information and assign each task to the appropriate nodes, using a rule-based algorithm, where nodes that minimize download/upload, have enough available resources and can meet the deadline. The authors suggest preemptive task scheduling at fog nodes as a future direction. Experiments compared this approach to three different approaches, (1) random offloading approach, (2) offloading to the node with the best downloading and uploading time approach, (3) and offloading to the node that has enough available resources. The proposed solution performed better in reducing waiting and offloading time, meeting tasks' deadlines and increase throughput.

A time-sensitive IoT application architecture was developed by **Al-Khafajiy et al.** [3]. The proposed architecture consists of three layers: (1) Things layer; collects data from IoT devices that are connected with communication protocol to transmit data to next layer. (2) Fog layer responsible for processing and computation of data transmitted to reduce overload on cloud layer and provide faster request-response for IoT devices. (3) Cloud layer with full access to the shared resources and higher processing power. Also, requests are split in a queue to light and heavy requests, where the latter is processed in nodes with less load and higher processing power to avoid starvation. The system starts scheduling processes by computing requests type and queue size of the fog nodes. Then, it flags up requests that cannot be processed at fog nodes that have congestion to be offloaded to other nodes. Offloading is requested based on time cost function, nodes locations and neighbor nodes load. The proposed architecture was evaluated on a healthcare monitoring system-based which showed high efficiency, good response and reduced delay time.

3.2 Dynamic Fog Networks

Centralized Offloading **Jazayeri et al.** [11] proposed a task offloading model from IoT devices to fog or cloud nodes, by implementing GASDEO; a Greedy Auto-Scaling DEep reinforcement learning-based offloading. In this paper, the authors started with a smart gateway that uses MAPE loop; monitor–analyze–plan–execute for automatic filtering of operations to be performed locally or offloaded based on the input workload. This step depends on monitoring data collected from mobile devices (MD) sensors and the needed resources. Then, this data is analyzed to prepare for planning stage that uses deep learning models; this stage to check if the MD can process the input workloads locally or not. The implemented DRL algorithm along with a hierarchical greedy method used to automatically find the best destination of the offloaded tasks. The model performed better in reducing energy consumption, total execution cost, total network usage, and application delay than three baseline models with 1, 4, 2 and 0.07%, respectively.

Decentralized Offloading In a different approach, **Wang et al.** [27] proposed FORT, an offloading scheme from Internet of vehicles (IoV) devices to fog nodes or cloudets. The objective of FORT is to manage traffic information received from vehicles in real-time manner to optimize response time. FORT assumes that traffic information has local value only, so the regions of a geographical area are locally monitored by cloudets. Therefore, it provides locally distributed traffic management and consists of three layers: global clouds, local cloudets and vehicular fog nodes. Local Road Side Units (RSU) are distributed over the regions and are responsible for scheduling tasks from passing vehicles either to a cloudet or to a vehicular fog node. In FORT, vehicles close to RSU are picked as fog nodes, which can be stationary (parked) or dynamic (moving). The authors modeled this offloading to dynamic fogs problem in queuing theory and showed that it can be modeled as an $M/M/1$ queue, so an approximate solution addresses the offloading problem by breaking it into two sub-problems. Simulations showed that FORT can significantly reduce response time and the number of required servers on cloudets compared to a randomized approach.

Alam et al. [4] proposed a multi-agent distributed reinforcement learning algorithm and a solution architecture for mobile fog computing. The authors used agents to learn from the environment of fog nodes including storage, processing and networking bandwidth capability, then finds the best way to offload basic blocks from MDs to mobile fog nodes. Basic blocks can be offloaded to the same fog or to any adjacent fog node, and can be executed in parallel if they are independent, which reduces execution latency. Evaluations considered energy consumption and response time in mobile phones, clouds and mobile fog nodes. It was shown that mobile phones and clouds takes more time to respond than mobile fogs, as fog nodes are close to users and have high computational powers and because parallelism was utilized.

Tan et al. [26] tackled offloading from end user devices to fog nodes in dynamic fog structures. The authors deal with the problem as a Multi-Armed Bandit problem and extend the Upper Confidence Bound (UCB) algorithm to model it. In other words, reinforcement learning techniques are implemented to pick the offloading destination node, where delay time is used as the regret factor in the proposed algorithms. Three different algorithms are introduced; one in case of stationary fog nodes, and two in case of moving fog nodes. The algorithms are compared to the UCB algorithm and simulations show that they learns better and are capable to reduce delay more. However, it is assumed that users' devices are responsible to generate offloading requests periodically and to run the proposed algorithms in order to select the destination fog node. Even though this assumption can make learning from experience easier, but this will overburden users' devices that have limited computational and storage capacities.

Another offloading framework developed by **Yousefpour et al.** [30] focused on reducing the delay of IoT services by distributing the load and implementing fog-to-fog, IoT-to-cloud and fog-to-cloud communications. Workload balancing depends on three main things: queue length, types of service requests and required processing time. The fog layer can be implemented in a distributed or centralized manner depending on the applications it is processing. The policy uses a limit for each fog node to offload tasks to reduce delay and balance workload. When a request reaches a fog node, it studies the waiting time as well as offload limit to decide either to process or offload the request to the best neighboring node or to the cloud. Each offloaded request will wait until it enters the queue of a fog node, otherwise it will be offloaded again. All types of requests, heavy and light, can be offloaded to the fog nodes and processed. The authors evaluated the framework with two other models: light fog processing, processes only light requests in fog nodes, and no fog processing, does not process any request in the fog node. The proposed model outperforms the other models with significantly reduced delay.

4 Trends and Future Directions

After reading and analyzing the chosen works, comparisons between the reviewed papers, as well as, trends and gaps are proposed in this section. A comparison of the reviewed research is shown in Table 1 and Fig. 4. Some trends have been noticed in the temporal order of the publications. In general, earlier approaches mainly focused on delay and response time optimization [27], while later approaches consider energy consumption as one of the main optimization factors [11, 18]. Additionally, SDN-based fog networking is a powerful tool that is believed to provide improvements to the network structure and bandwidth. It was noticed that recent works provide schemes for SDN-enabled fog networks [18, 22]. However, these solutions initially were assumed to handle offloading from end-user to fog or cloud nodes [18]. But, due to the expensive implementation of the required hardware to make such schemes applicable, recent research assumed that the SDN can only handle fog-to-fog offloading instead [22].

Table 1 Comparisons between reviewed literatures

Refs.	Offloading pattern	Performance metric	Utilized techniques	Advantages	Drawbacks
[31]	Fog to fog	Delay and energy consumption	Probability of a scheduling metric and fog-based architecture	Considers power consumption priority scheduling	One terminal node is handling all tasks, possibility of single point of failure
[18]	IoT devices to fog	Delay and energy consumption	Integer linear program and a greedy heuristic algorithm with SDN	Reduced delay and energy consumption	Many switches are needed, which is expensive
[22]	Fog to fog	Response time and throughput	SDN-based solution with cost function	Light cost function. Increase throughput and response	Single point of failure for SDN controller and the fog orchestrate. Security challenges
[3]	IoT devices to fog	Delay and response time	IoT-Fog-based architectures and scheduling based on time cost function	Reduces delay and response time Supports real-time monitoring for patients	Energy consumption was not evaluated. Processing cost was not considered
[2]	IoT devices to fog or cloud	Waiting time, offloading time and throughput	Priority scheduling, multi-level feedback queuing and rule-based algorithms	Deadline-aware. Avoid starvation of low-priority tasks. Increase throughput and reduce waiting and offloading time	Local optimized decisions. Network traffic and energy consumption are not considered
[11]	End user to fog or cloud	Energy consumption, time delay, learning rate, and time interval	Deep reinforcement learning (DRL)	Scalability. Reduces delay	Renewable energy usage was not considered. Processing cost was not considered

(continued)

Table 1 (continued)

Refs.	Offloading pattern	Performance metric	Utilized techniques	Advantages	Drawbacks
[4]	User to fog	Energy consumption, execution time, and latency	Reinforcement learning based	Parallel and sequential execution of computations based on dependencies. Distributed reinforcement learning	Did not consider cost of processing Scalability was not assured
[30]	IoT devices to fog or cloud	Delay	Offloading and scheduling policy	No restricted architectural design of fog-cloud layers. Delay minimization	Energy consumption was not evaluated. Processing cost was not considered
[27]	IoV devices to fog or cloudets	Response time	Queuing theory and an approximate algorithm	Real time management of traffic in IoV. Parked and moving cars are utilized as fog nodes. High response	Energy consumption was not evaluated. Processing cost was not considered
[26]	User to fog	Delay	MAB reinforcement learning problem with a solution extending UCB algorithm	Models offloading as a MAB problem. Propose algorithms for stationary and moving fog nodes. Reduce delay	Energy was not evaluated. Overburdens users' devices

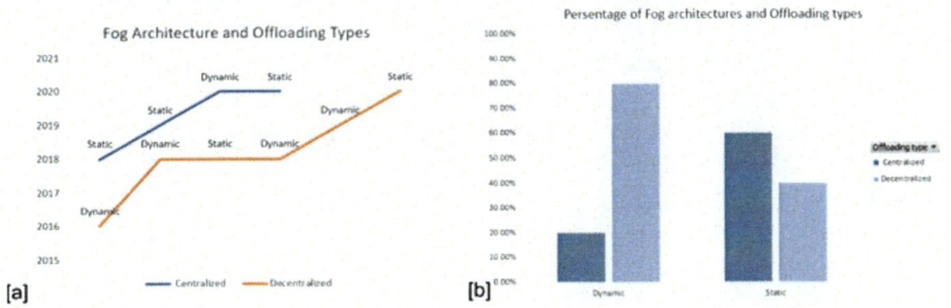

Fig. 4 **a** Fog Architecture and Offloading types selected and sorted in this survey, **b** comparison between work done on static and dynamic centralized and decentralized fog nodes in the literature

It is noticed that research is oriented to extend the solutions from static fog networks to dynamic fog networks. The reason is that dynamic fog networks, such as vehicular fog nodes, are flexible and can be changed based on the requirements of the system [27]. Nonetheless, it is found that there are new works that provide solutions for static fog networks because such networks are still the most implemented frameworks in real life [7]. Moreover, machine learning (ML), reinforcement learning (RL) and deep learning (DL) techniques started to emerge in research concerned in offloading, especially in dynamic fog networks [4, 11, 26]. It is highly encouraged that research explore this area more, as it provides promising results in multi-aspect optimization in fog offloading.

As a result of this survey, multiple gaps in research can be summarized as follows:

- Extending SDN-based solutions for dynamic fog networks is not explored yet. This area is highly needed due to the great benefits of dynamic fogs and the network guarantees resulting from leveraging SDN.
- Security solutions for single point of failure issue arising in the centralized SDN-based solutions is highly demanded. Such issue was not addressed in the reviewed research.
- More exploration of fog-to-fog offloading is required. It can be seen in Table 1 that only few works covered this issue, while many considered the offloading from end user to fog or cloud nodes. This area can provide increased QoS for the end user while reducing the offloading overhead from the end user devices.

It is worth noting that this survey is a short survey and the results need to be extensively investigated by future surveys. It is encouraged to conduct recent surveys covering this area to support findings of this survey and provide a robust theoretical background for research community.

5 Conclusions

Fog computing provides feasible solutions for real-time applications requiring services. In fog computing, edge devices and fog nodes have limited resources. Therefore, possible violation of QoS requirements can occur if a node is having shortage of a resource. Offloading tasks from such node to other nodes can provide solutions to maintain QoS as these resources will be available and accessible all the time. This review, focuses on dynamic offloading approaches, which provide promising solutions by taking advantage of the current status of the resources and the network. Literature is summarized, analyzed and gaps are identified. Comparisons between the approaches are provided, as well as, trends and future directions. As a result, research is highly encouraged to extend solutions from static fog networks to dynamic fog networks, such as SDN-based solutions. Nonetheless, security solutions for SDN-based approaches are demanded. Additionally, ML and RL techniques show good performance in offloading to dynamic fog networks, and more exploration of these techniques in fog offloading is required. It is important to mention that extensive survey of this area is needed by covering more approaches in this area. Moreover, reviewing research that focus on how to pick tasks to be offloaded is required.

References

1. Aazam M, Zeadally S, Harras KA (2018) Offloading in fog computing for IoT: review, enabling technologies, and research opportunities. Future Gener Compu Syst 87:278–289. https://doi.org/10.1016/j.future.2018.04.057, https://www.sciencedirect.com/science/article/pii/S0167739X18301973
2. Adhikari M, Mukherjee M, Srirama SN (2020) DPTO: A deadline and priority-aware task offloading in fog computing framework leveraging multilevel feedback queueing. IEEE Internet Things J 7(7):5773–5782. https://doi.org/10.1109/JIOT.2019.2946426
3. Al-Khafajiy M, Baker T, Waraich A, Al-Jumeily D, Hussain A (2018) IoT-fog optimal workload via fog offloading. In: 2018 IEEE/ACM international conference on utility and cloud computing companion (UCC Companion), IEEE, pp 359–364
4. Alam MGR, Tun YK, Hong CS (2016) Multi-agent and reinforcement learning based code offloading in mobile fog. In: 2016 International conference on information networking (ICOIN). IEEE, pp 285–290
5. Ghobaei-Arani M, Souri A, Rahmanian AA (2019) Resource management approaches in fog computing: a comprehensive review. J Grid Comput 1–42
6. Hong CH, Varghese B (2019) Resource management in fog/edge computing: a survey on architectures, infrastructure, and algorithms. ACM Comput Surv 52(5). https://doi.org/10.1145/3326066, https://doi.org/10.1145/3326066
7. Hosseinioun P, Kheirabadi M, Kamel Tabbakh SR, Ghaemi R (2020) aTask scheduling approaches in fog computing: a survey. Trans Emerging Telecommun Technol e3792
8. Hu P, Dhelim S, Ning H, Qiu T (2017) Survey on fog computing: architecture, key technologies, applications and open issues. J Netw Comput Appl 98:27–42 (2017). https://doi.org/10.1016/j.jnca.2017.09.002. https://www.sciencedirect.com/science/article/pii/S1084804517302953
9. IL I (2019) Rightscale 2019 state of the cloud report from flexera. https://www.flexera.com/2019-cloud-report
10. Islam MSU, Kumar A, Hu YC (2021) Context-aware scheduling in fog computing: A survey, taxonomy, challenges and future directions. J Netw Comput Appl 180:103008. https://doi.org/10.1016/j.jnca.2021.103008
11. Jazayeri F, Shahidinejad A, Ghobaei-Arani M (2020) Autonomous computation offloading and auto-scaling the in the mobile fog computing: a deep reinforcement learning-based approach. J Ambient Intell Hum Comput 1–20
12. Jha AK, Patel M, Pawar T (2019) Fog offloading: review, research opportunity and challenges. In: 2019 International conference on smart systems and inventive technology (ICSSIT), pp 1224–1227. https://doi.org/10.1109/ICSSIT46314.2019.8987905
13. Jiang C, Cheng X, Gao H, Zhou X, Wan J (2019) Toward computation offloading in edge computing: a survey. IEEE Access 7:131543–131558. https://doi.org/10.1109/ACCESS.2019.2938660
14. Lin H, Zeadally S, Chen Z, Labiod H, Wang L (2020) A survey on computation offloading modeling for edge computing. J Netw Comput Appl 169:102781. https://doi.org/10.1016/j.jnca.2020.102781. https://www.sciencedirect.com/science/article/pii/S1084804520302551
15. Mach P, Becvar Z (2017) Mobile edge computing: a survey on architecture and computation offloading. IEEE Commun Surveys Tutor 19(3):1628–1656. https://doi.org/10.1109/COMST.2017.2682318
16. Mahmud R, Kotagiri R, Buyya R (2018) Fog computing: a taxonomy, survey and future directions. Springer, Singapore, pp 103–130

17. Martinez I, Hafid AS, Jarray A (2021) Design, resource management, and evaluation of fog computing systems: a survey. IEEE Internet Things J 8(4):2494–2516. https://doi.org/10.1109/JIOT.2020.3022699
18. Misra S, Saha N (2019) Detour: dynamic task offloading in software-defined fog for IoT applications. IEEE J Sel Areas Commun 37(5):1159–1166. https://doi.org/10.1109/JSAC.2019.2906793
19. Mouradian C, Naboulsi D, Yangui S, Glitho RH, Morrow MJ, Polakos PA (2018) A comprehensive survey on fog computing: state-of-the-art and research challenges. IEEE Commun Surveys Tutor 20(1):416–464. https://doi.org/10.1109/COMST.2017.2771153
20. Mukherjee M, Shu L, Wang D (2018) Survey of fog computing: fundamental, network applications, and research challenges. IEEE Commun Surveys Tutor 20(3):1826–1857. https://doi.org/10.1109/COMST.2018.2814571
21. Okay FY, Ozdemir S (2018) Routing in fog-enabled IoT platforms: a survey and an SDN-based solution. IEEE Internet Things J 5(6):4871–4889. https://doi.org/10.1109/JIOT.2018.2882781
22. Phan LA, Nguyen DT, Lee M, Park DH, Kim T (2021) Dynamic fog-to-fog offloading in SDN-based fog computing systems. Future Gener Comput Syst 117:486–497. https://doi.org/10.1016/j.future.2020.12.021, https://www.sciencedirect.com/science/article/pii/S0167739X20330831
23. Rejiba Z, Masip-Bruin X, Marín-Tordera E (2019) A survey on mobility-induced service migration in the fog, edge, and related computing paradigms. ACM Comput. Surv 52(5). https://doi.org/10.1145/3326540, https://doi.org/10.1145/3326540
24. Shakarami A, Ghobaei-Arani M, Masdari M, Hosseinzadeh M (2020) A survey on the computation offloading approaches in mobile edge/cloud computing environment: a stochastic-based perspective. J Grid Comput 1–33
25. Shakarami A, Ghobaei-Arani M, Shahidinejad A (2020) A survey on the computation offloading approaches in mobile edge computing: a machine learning-based perspective. Comput Netw 182:107496 https://doi.org/10.1016/j.comnet.2020.107496, https://www.sciencedirect.com/science/article/pii/S1389128620311634
26. Tan Y, Wang K, Yang Y, Zhou MT (2019) Delay-optimal task offloading for dynamic fog networks. In: ICC 2019—2019 IEEE international conference on communications (ICC), pp 1–6 (2019). https://doi.org/10.1109/ICC.2019.8761113
27. Wang X, Ning Z, Wang L (2018) Offloading in internet of vehicles: a fog-enabled real-time traffic management system. IEEE Trans Ind Inf 14(10):4568–4578. https://doi.org/10.1109/TII.2018.2816590
28. Yi S, Li C, Li Q (2015) A survey of fog computing: concepts, applications and issues. In: Proceedings of the 2015 workshop on mobile big data, Mobidata'15. Association for Computing Machinery, New York, NY, USA, pp 37–42. https://doi.org/10.1145/2757384.2757397, https://doi.org/10.1145/2757384.2757397
29. Yousefpour A, Fung C, Nguyen T, Kadiyala K, Jalali F, Niakanlahiji A, Kong J, Jue JP (2019) All one needs to know about fog computing and related edge computing paradigms: a complete survey. J Syst Architect 98:289–330. https://doi.org/10.1016/j.sysarc.2019.02.009. https://www.sciencedirect.com/science/article/pii/S1383762118306349
30. Yousefpour A, Ishigaki G, Gour R, Jue JP (2018) On reducing IoT service delay via fog offloading. IEEE Internet Things J 5(2):998–1010
31. Zhang G, Shen F, Yang Y, Qian H, Yao W (2018) Fair task offloading among fog nodes in fog computing networks. In: 2018 IEEE international conference on communications (ICC). IEEE, pp 1–6

Public Administration and New Information: Increasing Utility

João Rodrigues dos Santos and Maria José Sousa

Abstract The utility (in the economic perspective of the utilitarianism) generated for the populations from the available public resources is likely to be increased due to the systematic incorporation of new information in the decision-making and working processes of the public administration. This paper proposes to highlight, theoretically, the relationship described above, addressing the importance of Information Systems/Information and Communications Technologies (ICT) in guaranteeing a real impact (increasing efficiency) on the quality of public work provided to citizens.

Keywords Utilitarianism · Public administration · Information and communications technologies · Efficiency

1 The Nature of Public Goods and Services

Public administration has an underlying structural objective: to provide goods and services of a public nature. Public goods and services have some characteristics that distinguish them from goods and services of a private nature.

Public goods and services are indivisible, without "rivalry" associated with their consumption, and the exclusion of consumption of a public good or service is practically impossible. It is for these reasons, and above all because of the impossibility of excluding consumption, that private markets are not interested in the commercialization of these goods and services.

Public Administration and New Information: Increasing Utility

---

J. R. dos Santos (✉)
Economics and Management Department, IADE/Universidade Europeia, Lisbon, Portugal
e-mail: joao-rodrigues.santos@universidadeeuropeia.pt

M. J. Sousa
Political Science and Public Policy Department, ISCTE—Instituto Universitário de Lisboa, Lisbon, Portugal
e-mail: maria.jose.sousa@iscte-iul.pt

A. Ullah et al. (eds.), *Proceedings of International Conference on Information Technology and Applications*, Lecture Notes in Networks and Systems 350,
https://doi.org/10.1007/978-981-16-7618-5_33

"The standard account of public goods found in economics texts fixates on two attributes of such goods: their 'non-rivalrousness' and their 'non-excludability'" [6, p. 109].

> By non-rivalrous, it is meant that one person's usage of a good is not diminished by another person's use. By non-excludable, it is meant that one person's usage of a good does not exclude other people from using the same good in the same time or space [3, p. 3].

Private goods and services, on the other hand, are divisible and can be sold divided into units small enough to be purchased by individual consumers. A private good or service is a "rival" in consumption, because when consumed by one individual, it cannot be consumed by another. On the other hand, a private good or service is subject to the exclusion principle, that is, individuals who cannot or do not want to pay are excluded from the utility/satisfaction generated during the act of consumption.

In this way, the economic difference between public goods and services and private goods and services lies exclusively in technical and objective considerations, which are very evident in the following assumption: It is not possible to exclude "non-payers" from access to public goods and services (without "rivalry" in consumption) such as, for example, public security, national defense or the cleaning and hygiene of public space.

2 Utilitarianism

Following the previous paragraphs, it is possible to understand, relatively easily, that goods and services of a public nature are likely to generate much more total utility (in the sense of utilitarian doctrine) per monetary unit invested than private goods and services.

In this regard, the doctrine of utilitarianism is of particular interest. Utilitarianism is a branch of moral philosophy that "focuses on the moral concerns resulting from inequalities in income distribution" [14, p. 64].

Bowden [2] states that utilitarianism provides an "effective method of moral decision-making" (p. 14). "As an ethical theory, it sets down what individuals should do to improve their own situation, and what should be done by every individual and by the collectivity to improve collective welfare" [1, p. 2].

The law of diminishing marginal utility, which derives from utilitarianism, states that the human being accumulates marginal/satisfaction/happiness utility to total utility up to a certain level of consumption and at a diminishing rate. In this regard, Greene and Baron [4] have concluded that "subjects made no distinction between goods and their utilities, and they showed declining marginal utility for both" (p. 16).

Even so, "preferences are far too complex to represent any and all diminishing marginal utility by a single curvature parameter" [8, p. 30].

2.1 *The Law of Diminishing Marginal Utility and the Equimarginal Principle*

Still, what is extremely relevant for this paper is the conceptual nature of the diminishing marginal utility. Given the theoretical and absolutely unequivocal foundations of this principle, it seems logical to conclude that the utility generated from the scarce resources available reaches its potential level (maximum efficiency) as long as it is ensured that all citizens of a given territory consume exactly the same levels of consumption (primary), assuming that everything else remains constant. This conception works, at least, in theory.

From an operational point of view, the rational expenditure rule, or equimarginal principle fully attests to this reasoning. The equimarginal principle underlies the law of decreasing marginal utility. This principle explains the rationality of individual consumer decisions and is based on the assumption that a consumer, with a fixed income, faced with prices of variable markets, will achieve maximum satisfaction when the marginal utility of the last monetary unit spent on each good or service is exactly the same as the marginal utility of the last monetary unit spent on any other good.

In the theoretical framework, it is this conception that, above all, sets the basis that guides governments in the redistribution process, which is embodied through the respective tax systems.

3 The Laffer Curven

However, this coin has two sides. For example, a state redistributes for universal health and education, social care and public goods and services. It is from the tax revenue, and underlying taxes, that the states finance these guarantees and their functions.

However, the balance between tax revenue and the level of public expenditure is much more complex than it may appear at first. According to Laffer [9], "not only will we reduce tax revenues [...] if we raise tax rates [...], but we will hurt the economy" (pp. 9–10).

In this regard, the Laffer Curve describes the relationship between the level of taxes and the total tax revenue collected. If taxes are too high along the curve, there will be a disincentive to productivity, with the most taxed activities slowing production and, globally, the productivity of the economy falling more or less sharply. This means, paradoxically, that after a certain limit, more taxes mean less productivity and, therefore, less tax revenue collected by States.

Certifying this idea Hájek et al. [5], citing Moszer [11] report that, after a certain level, "any increase in the tax rate cannot lead to an increase in tax revenue, as with a higher tax burden, the taxpayer responds to this fact, and the work itself becomes insufficiently profitable due to the substitution effect" (p. 4).

Thus, the complex exercise of governments involves calibrating, at each juncture, the right level of taxes (for the purpose of redistribution) that guarantees the maximization of the total utility (efficiency) reached from the resources available without discouraging the productivity of economies.

4 Applying the Essential of the Equimarginal Principle to Understand the Increasing Efficiency of the Public Sector

At the same time, when the logic of the equimarginal principle (rational expenditure rule) is partially applied, it is easy to conclude that the investment in a public good or service, due to the specific characteristics of this type of good or service (see point 1), will allow it to generate more or less total utility, depending on the number of people who benefit from its consumption. For example, for purely academic purposes, let us consider the example in the following paragraph that corresponds conceptually to the essentials of decisions taken in the sphere of public administration.

If, following a public investment of 1.000 euros, there is a benefit of 100 utility (*ut*) units for each individual in the community that benefits from that investment, then we can conclude that the utility per euro invested, in this case, will be: equal to 0.1 if the investment covers 1 individual (100 ut/1.000 euros); equal to 1.0 if it covers 10 individuals (1.000 ut/1.000 euros); equal to 10.0 if it covers 100 individuals (10.000 ut/1.000 euros); and so on.

It is about efficiency in the management of public resources that we are dealing with. In this sense, according to Sethuramalingam [15], "the economic growth of a country is significantly impacted when exposed to indicators of bad governance" (p. 7).

Following the reasoning of the previous paragraph, the public utility generated from the resources available to the public administration may increase with a decrease in the equation denominator (investment measured in monetary unit) and/or an increase in the numerator (value in utility generated with a level of investment).

In practice, it is possible to increase the numerator of the equation by adjusting the public budgets available to the real needs of the community. Thus, if more adjusted investment decisions are made, the real utility generated from each monetary unit of public investment will increase. Therefore, the efficiency in the management of public funds will improve.

And it is in this context that information, access to it and its correct management and application are of paramount importance.

5 Information and Communications Technologies and Increasing Utility Generated from Public Sector Spending

Due to the nature and speed of ICT transformation, the need to understand its effects at every moment increasingly stimulates study and research.

We live in the so-called information society. Never before has information taken such a preponderance for individual and collective development. This paradigm, which is consolidated and reinforced every day, is a reality that results from the fact that the growth of knowledge is sequential and incremental. As the industry develops and specializes due to the production of equipment and computer applications that store and process information, the greater the propensity and capacity of society to generate new information. It is an iterative process.

In the last two decades, the technological transformation has been colossal, and it is not possible to safely predict how the world will be in next two.

"Revolutionary changes in information and communication systems have shaped cyberspace, which allows not only to expand the range of information and communication services, but also to increase the growth of public (state and municipal services)" (Alhawawsha 2020, p. 42).

The use of ICT in public sector and "the explosion of digital information throughout society, offers the possibility of a more efficient, transparent, and effective government" [7, p. 633].

In this sense, in order to guarantee decision-making processes that result in the maximization of utility – only generated from the most correct choices – it is crucially important to ensure that public administration entities are technologically modernized. Only in this way will the public sphere be able to process the knowledge currently available and translate this increased capacity into more efficiency in the management of public funds.

6 Conclusions and Implications

In conclusion, it is expected that the theoretical approach to utilitarianism (in the economic perspective) in this paper introduces an additional level of reflection, which may contribute, in the future, to the emergence of methodologies that allow to measure more objectively the impact that application of public resources has on people's quality of life—knowing in advance that this impact will be greater the greater the ability to integrate new information into the functioning processes of public organizations.

Implications for Public Administration

The development of information and communication technologies in recent decades has contributed exponentially to the radical change in habits and ways of acting, both in public organizations and their employees, with implications at different levels for all stakeholders active in these changes: public servants, other official institutions and citizens.

In this way, governments and official institutions must educate and sensitize companies to the benefits and potential of new technologies:

Cost Reduction—overhead and space costs, as digital can minimize costs in buildings, and in other resources.

Increased Productivity—productivity increases between due to greater time optimization.

Organizational Flexibility—digital implies to process of restructuring organizations in a more virtual conception, and more flexible in responding to the challenges of technology.

In addition, public administration can reduce asymmetries through the implementation of digital technologies, which allow a better communication and also optimization of time and resources.

Implications for Policy

National Public Policies can promote initiatives to raise awareness about the opportunities to digital work, reducing costs and implementing work–life balance practices.

In the framework of the EU, peripheral regions have the possibility to fit into development projects related to IT infrastructures, and in a transversal way developing digital services to citizens, optimizing the resources, and contributing for a more quality relationship with the society.

Governments must define strategies that consider the digital challenges and prepare the changes that will occur in societies.

References

1. Baujard A (2013) Utilitarianism and anti-utilitarianism. In: GATE working paper No. 1332. https://doi.org/10.2139/ssrn.2357441
2. Bowden P (2009) In defense of utilitarianism annual conference: Australian association for professional and applied ethics. https://doi.org/10.2139/ssrn.1534305
3. Chohan UW, D'Souza A (2020) Club theory: a contemporary economic review. https://doi.org/10.2139/ssrn.3512557
4. Greene JD, Baron J (2000) Intuitions about declining marginal utility. J Behav Decis Mak 14:243–255. https://doi.org/10.2139/ssrn.231183
5. Hájek J, Šafr K, Rotschedl J, Cadil J (2020) The Laffer curve decomposed. Ekonomický časopis. Retrieved from https://ssrn.com/abstract=3749646
6. Hockett RC, Omarova ST (2018) Private wealth and public goods: a case for a national investment authority. In: Cornell legal studies research paper No. 17–22, https://doi.org/10.2139/ssrn.2939309
7. Ramon Gil-Garcia J, Sharon SD, Theresa AP (2018) Digital government and public management research: finding the crossroads. Public Manage Rev 20:5, 633–646. https://doi.org/10.1080/14719037.2017.1327181

8. Kimball MS, Ohtake F, Reck D, Tsutsui Y, Zhang F (2015) Diminishing marginal utility revisited. Univ Michigan. https://doi.org/10.2139/ssrn.2592935
9. Laffer A (2012) The Laffer curve and the failure of stimulus spending. In: IEA current controversies Paper No. 38. https://doi.org/10.2139/ssrn.2182070
10. Lawsky SB (2011) On the edge: declining marginal utility and tax policy. Minnesota Law Rev 95(3). https://ssrn.com/abstract=1760050 (UC Irvine School of Law Research Paper)
11. Moszer M (1981) A comment on the Laffer model. Cato J Cato J Cato Inst 1(1):23–44. http://www.cato.org/sites/cato.org/files/serials/files/cato-journal/1981/5/cj1n1-2.pdf (Spring)
12. Mohammad A (2020). Development of the concept of electronic government construction in the conditions of synergetic threats. Technol Audit Prod Reserves 3(2(53)). https://doi.org/10.15587/2706-5448.2020.207066
13. Moszer M (1981) A comment on the Laffer model. Cato J 1(1):23–44. http://www.cato.org/sites/cato.org/files/serials/files/cato-journal/1981/5/cj1n1-2.pdf (Spring)
14. Santos JR (2020) Economia Indispensável. Lisbon International Press, Lisboa
15. Sethuramalingam D (2020). Digital governance: theory, policy and practice. https://doi.org/10.2139/ssrn.3679639

Case Base Maintenance: Clustering Informative, Representative and Divers Cases (C IRD)

Asma Chebli, Akila Djebbar, and Hayet Farida Merouani

Abstract Maintenance is generally defined in the field of software and knowledge engineering as an activity that takes place after the development of the system is complete, and the application has already been deployed in operation. The success factor for case-based reasoning (CBR) systems is the quality of their case bases, in order to guarantee this quality, a maintenance process must be planned, and this is how the field of case base maintenance (CBM) emerged. The goal of this paper is to propose a case base maintenance strategy that delivers a small case base size, removes irrelevant cases from the case base, and targets only valuable cases to be retained to increase classification accuracy. We propose a case base maintenance approach C_IRD that focuses on balancing the efficiency of case retrieval and the competence of a case base by employing a soft clustering technique: FCM. The method delivers interesting abilities and is able to maintain case bases with satisfactory accuracy by reducing its size, which leads to a reduction in retrieval time.

Keywords Case-based reasoning · Case base maintenance · Soft clustering · Fuzzy C-means

1 Introduction

CBR is a branch of artificial intelligence (AI) that combines machine learning and reasoning techniques to solve new problems by adapting solutions that have worked for similar past experiences, with these previous experiences being stored as cases, to form a case base (CB) [1]. Compared to other AI problem-solving approaches, CBR mimics human thinking, it is based on memory, making it very similar to the human reasoning process. CBR has been used to develop several applications in different fields, such as manufacturing, medicine, law, quality control. Case-based reasoning systems are implemented to operate over a long period of time, supporting active learning of new cases through the retention phase of the classical CBR life cycle. A

A. Chebli (✉) · A. Djebbar · H. F. Merouani
Computer Science Department, LRI Laboratory, University of Badji Mokhtar Annaba, Annaba, Algeria

A. Ullah et al. (eds.), *Proceedings of International Conference on Information Technology and Applications*, Lecture Notes in Networks and Systems 350,
https://doi.org/10.1007/978-981-16-7618-5_34

long-term CBR application ultimately leads to the recognition of the importance of maintaining the case-based reasoner.

Retention of cases in the knowledge container at the end of each CBR lifecycle leads to very rapid growth of the knowledge container, which can negatively affect the quality of the CBR system and results in slow query execution at the retrieve phase. This performance degradation is due to memory flooding or exposure to harmful experiences, both of which affect the overall utility of a CBR system.

To avoid performance degradation, CBR systems must be maintained, and several works have been proposed to address the mentioned challenges. All of them have the same objective, namely to ensure and improve the efficiency of the CBR process.

Case base maintenance has been described as the process of improving the performance of CBR systems: case base maintenance implements policies to revise the organization or content of the case base to facilitate future reasoning for a particular set of performance objectives [10].

Aiming to maintain or even improve the quality of the case base which might have degraded after several rounds of CBR reasoning, we turned to a particular branch of research. This branch focuses on the partitioning of the case base which allows to build a refined CB structure and to maintain it. In this paper, we propose a maintenance approach that addresses the drawbacks that follow an operational CBR system and the blind retention of cases.

The proposed approach focuses on balancing the efficiency of case retrieval and the competence of a case base by employing a flexible clustering technique **FCM**. The method could be able to maintain the case bases giving satisfactory accuracy by reducing its size, which leads to a reduction in search time.

2 Case Base Maintenance: Related Works

Different categorization of CBM strategies have been proposed in literature, among other, we mention: "Optimization and partitioning of the CB" [15], "Volume, Variety, Velocity and Value" [9], and "Direct, Hybrid and Case property models" [11].

Regardless of their categorization, all maintenance strategies seek to achieve the same purpose: build and restructuring the CB with enhanced quality and thus better performance compared to its initial state, using different criteria. The success of maintenance is correlated not only to the policy itself, but also to the way it is integrated into the CBR process [5].

A significant number of researchers interpret CBM as a case reduction problem and propose algorithms that target noisy and redundant cases [9], The proposed case reduction methods attempt to increase classification accuracy and improve retrieval efficiency. Researchers then started to look at the machine learning literature and use it as a source to achieve their goals. In the following subsection, we discuss a selection of different strategies presented in the literature:

- CNNENN: Several approaches have been proposed to reduce case base memory; these strategies are based on the Edited Nearest Neighbors (NN) rules. One of the first attempts at CBM is Hart's Condensed Nearest Neighbors (CNN) [8], and Wilson's Edited Nearest Neighbors (ENN) [19], CNN is considered a skill-preserving algorithm while ENN is more of a skill-enhancing algorithm [11] and both have inspired other algorithms dedicated to CBM: CNN is an incremental algorithm, which starts with an empty CB, to which are added the cases misclassified by the other cases currently in the CB.

  However, ENN is a decremental algorithm where the cases misclassified by their K nearest neighbors are removed from the CB, as they are considered as noise. The experimental results demonstrate the effectiveness of ENN in increasing the average accuracy of CBR techniques. Subsequently, based on the ENN algorithm, a repeated ENN (RENN) method was proposed by Tomek [18], The ENN algorithm is repeated, and cases are removed from the CB until no more noise is detected.
- Competence model: Thereafter, researchers focused on the competence of the case-based reasoner to solve problems, as case deletion can reduce the later, Smyth and Keane [16] proposed a competence model used for the evaluation of individual contribution of cases. Cases are then categorized according to their competence characteristics which helps guide the selection of case deletion. This competence-guided case deletion is considered as a safe way to remove cases from a growing CB, as it saves the systems from the adverse effects of the utility problem [14] while coping with reduction in competence. As an alternative to case deletion, Smyth and McKenna [17] used the competence model for a guided case addition algorithm, based on the notions of coverage and reachability to construct a compact competent CB, using a relative coverage (RC) measure to estimate the unique competence contribution of an individual case. The proposed algorithm adds cases to the CB using the RC metric in combination with CNN to prioritize cases expected to make the largest competency contribution, given their RC value.
- Relative Performance (RP): In another paper based on the competence model of Smyth and McKenna [17], Leake and Wilson [10] argued for a more direct integration of performance consideration into case addition, Cases are added based on the performance advantage (PB) they provide through their retention to assess this contribution, To guide the maintenance, a Relative Performance (RP) metric is used in conjunction with CNN algorithm, whose input are ordered by the RP value, assessing the contribution of a case to the adaptation performance of the system,
- Competence preserving: Later on, a competence preserving method was presented by Zhu and Yang [20], the authors use two theories, which aim to find the similarity metrics and adaptation cost to determine the best coverage value to build a CB with near optimal property,
- Blame-Based Noise Reduction (BBRN): Following the same concept as Smyth and McKenna[17], Leake and Wilson(Leak and Wilson,1998), the Blame- Based Noise Reduction (BBRN) algorithm was proposed by Delany and Cunningham

[6], where for each case a property is measured called Liability, The proposed property for a case C is defined as the set of all the cases misclassified because of C, Liability is an information integrated into the CB, in the same way as Coverage and Reachability, to determine the performance of each case under certain tasks or conditions,

- Generalized Condensed Nearest Neighbors (GCNN): CNN algorithm inspired so many other works, GCNN is another example [4], Iteratively samples are selected and others ignored, the ones ignored are the samples that can be absorbed or represented by those selected, The use of a stronger absorption criterion helped to strengthen GCNN, by the addition of a threshold for the difference in distance of the closest class members and the nearest cases to another class.

3 Proposed Approach: Clustering Informative, Representative and Divers Cases (C IRD)

The main objective of CBM is to reduce the size of the case base while maintaining or even improving the quality of the CBR system as much as possible. Maintaining or improving the quality of the system depends on the quality of its case base, a knowledge container that needs to be given full attention. When we focus on the quality of the system, we are required to carefully select cases for deletion, without depreciating the efficiency of CBR and the competence of the case base [15].

Building a high-quality case base requires a CBM strategy that delivers a small CB size, removes irrelevant cases from the case base and targets only valuable cases to be retained to increase classification accuracy. We use a clustering algorithm for our proposed approach to cluster the initial CB into smaller clusters, to enable the identification of valuable cases using sampling criteria, traditionally used in active learning. We determined three types of valuable cases.

3.1 Soft Clustering to Target Valuable Cases to Retain

In an attempt to target only the valuable cases to be maintained in the case base, we use a soft clustering algorithm to divide the original case base into smaller clusters, in order to facilitate the identification of cases deemed valuable [3].

FCM algorithm attempts to partition an ensemble of elements $X = \{x_1, x_2, \ldots, x_N\}$ into an ensemble of fuzzy clusters. FCM clustering algorithm can be outlined with the following equation:

Given a finite ensemble of data, FCM returns a list of c centroids $\{C = c_1, c_2, \ldots, c_c\}$ and a partition matrix.

$W = w_{i,j} \in [0, 1]$, $i = 1, \ldots, n$, $j = 1, \ldots, c$, $w_{i,j}$ tells the degree of membership to which an element x_i belongs to cluster c_j.

The FCM algorithm aims to minimize an objective function [12]:

$$\text{argmin}_c = \sum_{i=1}^{n}\sum_{j=1}^{c} w_{i,j}^{m}||x_i - c_k^2||, \tag{1}$$

where

$$w_{i,j} = \frac{1}{\sum_{k=1}^{c}\left(\frac{x_i - c_j}{x_i - c_k}\right)^{\frac{2}{m-1}}} \tag{2}$$

3.2 Which Cases Should Be Retained and Why?

In order to have a good CB quality, after clustering the case base using FCM algorithm, we should retain valuable cases whose deletions directly reduce the competence of the system, namely (see Fig. 1):

1. *Informative cases:* these cases add rich information to the CB, and we target them using two methods:

   (a) an informative case is a case with no other cases within the CB similar to it. This type is referred to as outliers and also known as isolated cases, which should be retained, as their removal directly diminishes the competence of the system.

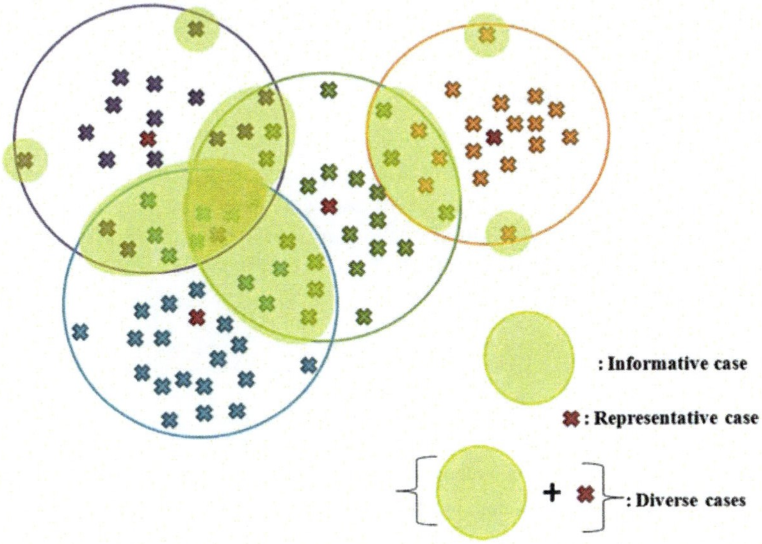

Fig. 1 C_IRD valuable cases to retain in the CB

(b) In informative cases, it is also selected using a degree of membership, two versions of FCM are used for the selection of informative cases using the membership value:

- The first version is referred to as $C\ IRD_1$ where a case that belongs to more than one cluster is considered as an informative case,
- The second version is referred to as $C\ IRD_2$, using the membership value, we try to remove redundant cases; we further refine the selected cases by checking if there are any redundant cases with the same membership value, and then select only one of these cases to retain.

2. *Representative cases:* a representative case is a case among a group of similar cases, the case with a high representation is selected, which means that it has a high density, so it represents more neighboring.
3. *Diverse cases:* diversity aims at keeping cases that scatter the entire input space, instead of focusing on one small region of it. Diversity cases are a direct consequence of the retention of informative and representative cases. By keeping representative and informative cases in the case base, we directly create diversity by assuming that cases from different clusters can be considered diverse.

After we partition the original case base using FCM, an unsupervised soft clustering algorithm and select the valuable cases (informative, representative and diverse), for each cluster, the cases that were not selected are removed. The removed cases simply consume space within the case base and increase the response time of the CBR system. Accordingly, as an attempt to maintain or improve the quality of the original case base, by applying our C IRD method, we build a new reduced case base that includes the cases that do not decrease the competence of the system.

4 Results and Discussion

For the experimental evaluation of our C IRD approach, it is tested on five case bases from the U.C.I. machine learning repository datasets [7], considering only numerical data. These datasets have contributed to the evaluation of many studies, a detailed description of which is presented in Table 1.

Table 1 Case bases description

Case base	#instances	#attributes
Iris	150	4
Breast cancer Wisconsin (Diagnostic)	569	19
Blood transfusion service center	749	5
Mammographic mass	962	6
Yeast	1483	8

The goal of our C_IRD approach is to reduce the case base size while maintaining as much as possible the competence of the system. Thus, the following main criteria will be considered:

1. **Storage size (S%)**: average storage percentage: we measure the rate of decrease in the size of the case base. The main objective of our ASSM method is to reduce the case base. The final percentage refers to the percentage of cases that were included in the original CB and are in the new reduced case base.
2. **Percentage of correct classification (PCC%)**: the main applications in CBR use for the retrieve phase the nearest neighbor algorithm [13] technique. It is a simple approach whose goal is to calculate the similarity between the cases stored in the CB and the new input case.

   Therefore, we choose to select the 1-Nearest-Neighbor (1-NN) method to calculate the percentage of correct classification: The training set contains 80% of the case base and the test set contains the remaining 20% of the cases. For each dataset, we apply the 1-NN algorithm with a five-fold cross-validation task to compute the percentage of correct classification,
3. **Time**: the retrieval time in seconds: the goal of C IRD method is to reduce the case base and thus the search time. Therefore, we choose the criterion "Time" to highlight the performance of our method in the reduction of retrieval time [15].

As we have already mentioned, the objective of our C IRD approach is to maintain the case base by preserving or even improving their performance (retrieval time) and competence (storage size and accuracy) during problem-solving. Thus, for the first experiment, we propose to evaluate the effectiveness of C IRD compared to the results obtained with the initial non-maintained case bases, which we will refer to as (CBR) (Table 2).

It is worth discussing the results obtained with the two versions of FCM used at this point, we can observe that in terms of size reduction, C IRD2 reduces the size of the case base more than C_IRD1. This is due to the fact that the latter removes cases that have the same degrees of membership, in the view that these are redundant cases, but this simply reduces the classification rates. It is necessary to emphasize that the cases belonging to several classifiers should not be concerned by the deletion; these

Table 2 Comparing CBR to the two versions of C_IRD

Datasets	CBR			C IRD1			C IRD2		
	S (%)	PCC (%)	Time (s)	S (%)	PCC (%)	Time (s)	S (%)	PCC (%)	Time (s)
IRIS	100	93.06	0.454	55.4	91.35	0.052	12.13	64.07	0.034
Breast-W	100	93.38	0.623	67.51	94.58	0.238	67.25	94.05	0.236
Blood-T	100	71.60	0.354	4.34	79.90	0.11	2.97	78.84	0.090
Mammographic	100	75.94	0.593	3.96	83.41	0.204	3.62	53.17	0.081
Yeast	100	51.37	0.515	8.51	47.42	0.212	6.42	34.07	0.105

are cases rich in information (informative) and even if the degrees of membership are the same for some cases, it does not necessarily mean that we are dealing with a redundancy. The size reduction difference between C IRD1 and C IRD1 varies between 0.26 and 43.12%. This reduction in size results in a reduction of accuracy rates (between 0.13 and 30%), which supports our hypothesis that the cases belonging to several clusters are very significant for learning and should not be deleted.

We are conscious that our approach has shortcomings, but there are positive points that can be explored to improve C_IRD1. Therefore, we have decided to compare C IRD to other well-known reduction techniques in the literature (Tables 3, 4 and 5): CNN [8], RENN [18], ENN [19] and instance-based learning IBL schemes [2] on the datasets presented above.

Table 3 shows that our proposed method did not yield the best reduction rate on all case bases, namely IRIS and Breast-W where it retains respectively 44.6% and 32.49% of the initial case base, compared to other methods, for example, IB2 and IB3 which retain only 24.00% of IRIS CNN which retain only 16.3% for Breast-W.

Table 3 Comparing storage size S (%)

Datasets	C IRD1	CNN	RNN	ENN	IB2	IB3
IRIS	55.4	27.63	93.33	95.33	24.00	24.00
Breast-W	67.51	16.3	16.87	81.69	35.48	30.46
Blood-T	4.34	37.3	38.72	32.1	26.09	26.00
Mammographic	3.96	64.21	54.26	82.52	53.48	53.93
Yeast	8.51	12.34	14.02	69.59	69.79	69.98

Table 4 Comparing classification accuracy PCC (%)

Datasets	C IRD1	CNN	RNN	ENN	IB2	IB3
IRIS	91.35	73.00	94.23	91.60	91.67	91.67
Breast-W	94.58	68.18	67.05	94.66	69.69	70.56
Blood-T	79.90	67.94	66.65	71.63	74.69	74.21
Mammographic	83.41	70.82	78.60	77.04	66.28	66.40
Yeast	47.42	83.56	83.92	88.08	73.82	73.30

Table 5 Comparing retrieval time in seconds

Datasets	C IRD1	CNN	RNN	ENN	IB2	IB3
IRIS	0.052	0.011	0.010	0.013	0.002	0.002
Breast-W	0.238	0.430	0.350	0.734	0.244	0.227
Blood-T	0.11	0.098	0.183	0.194	0.203	0.197
Mammographic	0.204	0.208	0.199	0.815	0.339	0.032
Yeast	0.212	0.640	0.604	0.134	0.595	0.581

This is due to the fact that these two bases are too small, comparing to other case bases.

For the other case bases obtained, the sizes are generally reduced by more than 95%, compared to the initial sizes of the CBRs that contain all the instances (100% of the cases). For example, C IRD1 removes more than 96.04% of the cases for the "Mammographic" case base.

The same remarks are valid for the accuracy, where the accuracy provided by our C IRD1 shows slightly better values (see Table 4). It is indeed better than that of CBR which retains all instances (Table 1), especially for the "Mammography" case base. It reaches 83.41% PCC, which is a significant difference compared to the one given by CBR: 75.94% PCC.

This proves that our method not only maintains the quality of the case base but also can improve it. Moreover, we observe that the PCC obtained by our C IRD method is better than those obtained by the other well-known reduction techniques. For example, for the "Breast-W" case base, the PCC provided by our approach is higher than 94%, while it is lower than 71% for the other methods.

Regarding the response time, as shown in Table 4, the results presented by C IRD1 are better than those given by CBR. Our maintenance method reduces the search time since the case bases have been reduced. For example, for the "Mammography" case base, since C IRD1 retains only 3.96% of the cases, the retrieval time is about twice as better as CBR. Similar observations are made regarding the comparison with CNN, RNN, ENN and IBL in which C IRD1 has, in general, the shortest retrieval time.

5 Conclusion

This paper presents an approach that seeks to maintain or even improve the quality of the case base built during the development phase that might have degraded after several rounds of CBR reasoning. We have presented a case base maintenance approach called C_IRD, our approach uses a soft clustering algorithm to target the valuable cases to be saved in the case base while removing other cases, leading to better results in terms of classification accuracy hence CBR quality.

The experimentation demonstrate that our method is an interesting case base maintenance approach able to be efficient in terms of reducing the size of the case base and the retrieval time and to obtain a satisfactory classification accuracy.

We are conscious that our approach has shortcomings, but there are some positives that can be explored to improve the strategy in future work by further exploring the track of membership values of cases belonging to more than one cluster. Furthermore, to show the scalability of our C IRD approach, we intend to evaluate it on and real case bases.

References

1. Aamodt A, Plaza E (1994) Case-based reasoning: Foundational issues, methodological variations, and system approaches. AI Commun 7(1):39–59
2. Aha DW, Kibler D, Albert MK (1991) Instance-based learning algorithms. Mach Learn 6(1):37–66
3. Chebli A, Djebbar A, Marouani HF, Lounis H (2021) Case-base maintenance: an approach based on active semi-supervised learning. Int J Pattern Recogn Artif Intell S0218001421510113. https://doi.org/10.1142/S0218001421510113
4. Chou CH, Kuo BH, Chang F (2006) The generalized condensed nearest neighbor rule as a data reduction method. In: 18th International conference on pattern recognition (ICPR'06), vol 2. IEEE, pp 556–559
5. De Mantaras RL, McSherry D, Bridge D, Leake D, Smyth B, Craw S, Faltings B, Maher ML, Cox M, Forbus K et al (2005) Retrieval, reuse, revision and retention in case-based reasoning. Knowl Eng Rev 20(3):215–240
6. Delany SJ, Cunningham P (2004) An analysis of case-base editing in a spam filtering system. In: European conference on case-based reasoning. Springer, pp 128–141
7. Dua D, Graff C (2017) UCI machine learning repository. http://archive.ics.uci.edu/ml
8. Hart P (1968) The condensed nearest neighbor rule (corresp.). IEEE Trans Inf Theory 14(3):515–516
9. Juarez JM, Craw S, Lopez-Delgado JR, Campos M (2018) Maintenance of case bases: current algorithms after fifty years. IJCAI
10. Leake DB, Wilson DC (1998) Categorizing case-base maintenance: dimensions and directions. In: European workshop on advances in case-based reasoning, Springer, pp 196–207
11. Nakhjiri N, Salamó M, S'anchez-Marré M (2020) Reputation-based maintenance in case-based reasoning. Knowl-Based Syst 193:105283
12. Nayak J, Naik B, Behera H (2015) Fuzzy c-means (fcm) clustering algorithm: a decade review from 2000 to 2014. Comput Intell Data Mining 2:133–149
13. Sabharwal Y, Sharma N, Sen S (2006) Nearest neighbors search using point location in balls with applications to approximate Voronoi decompositions. J Comput Syst Sci 72(6):955–977
14. Salamó M, Lopez-Sánchez, M.: Adaptive case-based reasoning using retention and forgetting strategies. Knowl-Based Syst 24(2):230–247 (2011)
15. Smiti A, Elouedi Z (2014) WCOID-DG: An approach for case base maintenance based on weighting, clustering, outliers, internal detection and Dbsan-Gmeans. J Comput Syst Sci 80(1):27–38
16. Smyth B, Keane MT (1995) Remembering to forget. In: Proceedings of 14th IJCAI. Citeseer, pp 377–382
17. Smyth B, McKenna E (1998) Modelling the competence of case-bases. In: European workshop on advances in case-based reasoning. Springer, pp 208–220.
18. Tomek I et al (1976) An experiment with the edited nearest-nieghbor rule
19. Wilson DL (1972) Asymptotic properties of nearest neighbor rules using edited data. IEEE Trans Syst Man Cybern 3:408–421
20. Zhu J, Yang Q (1999) Remembering to add: competence-preserving case-addition policies for case-base maintenance. IJCAI 99:234–241

Smart Control System for User Confirmation Based on IoT

Asfandyar Khan, Mukhtiar Ahmad, Javed Iqbal Bangash, Abdullah Khan, and Muhammad Ishaq

Abstract The smart systems play a vital role to facilitate human beings in the era of modern technologies. Nowadays, the control systems used for users' confirmation play a notable role that returns the improved performance and achievement contributed to any organization. Traditional users' confirmation system needs a lot of time and considered as inefficient. Several automatic users' confirmation systems are available for the confirmation of the users' attendance such as biometric, barcode. All these systems are time-consuming as the users have to stand in the queue to scan cards for marking his/her attendance. The proposed research work focuses on one of the effective algorithms of face recognition along with RFID technology that marks users' attendance automatically. The proposed framework is evaluated for the students' attendance in any academic institute. Furthermore, the system provides information service by using a web-based interface to the student required in the academic domain. The proposed framework authenticates and marks the student attendance in four phases that are face detection, face recognition, marking attendance, and providing information services. Based on results, the proposed IoT-based smart system for users' confirmation is time-effective and eliminates the efforts of documentation as well.

Keywords Smart system · Users' confirmation · IoT · RFID · Face detection · Face recognition · Arduino

A. Khan · M. Ahmad · J. I. Bangash (✉) · A. Khan · M. Ishaq
Institute of Computer Sciences and IT, The University of Agriculture, Peshawar, Pakistan
e-mail: Javed.bangash@aup.edu.pk

A. Khan
e-mail: asfandyar@aup.edu.pk

M. Ishaq
e-mail: drmishaq@aup.edu.pk

A. Ullah et al. (eds.), *Proceedings of International Conference on Information Technology and Applications*, Lecture Notes in Networks and Systems 350,
https://doi.org/10.1007/978-981-16-7618-5_35

1 Introduction

Smart systems have made the human's life easier and comfortable. They combine different types of technology such as sensors and/or actuators along with the control to define and examine a situation. Moreover, they are capable of making decisions based on the available data. RFID is an acronym for "radio frequency identification" which refers to a technology that has the ability whereby digital data encoded in RFID tags attached to an object are captured by a reader from a distance of inches to meter without direct line of sight [1]. Standardized identification technologies are optical strip, voice recognition, fingerprint, optic character recognition (OCR). RFID technology utilizes radio waves to transmit data from RFID tag attached to an object through the reader to accomplish tracking and monitoring of the object by which efficiency of the system has increased [2]. In the recent era, the use of RFID technology has been increased marvelously in many fields such as: patient monitoring, item-level inventory tracking, IT assets management, and real-time location system [3].

A RFID system consists of three basic core components: (1) a RFID tag: also called a transponder, which consists of a semiconductor silicon microchip and antenna [4]. RFID tag contains a unique identification code that can be transmitting wirelessly toward readers. Tags are of two types: active tags (battery power) and passive tag (without battery) [5]. (2) RFID reader: The RFID reader is also called an interrogator that uses radio wave to transmit and receive data from RFID tag via antennas. A RFID reader contains multiple antennas responsible for sending and receiving radio waves. There are four types of the RFID reader: read, read/write, static, and mobile [5]. (3) RFID controller: The controller (interrogator) is the electronic device that receives the data from the antenna, or transmits data to the antenna, and usually communicates this data to a host computer or PLC. The controller manages the information between the RFID products and the host system. Some controllers are equipped with programmable I/O points that can be used to make local decisions.

In the development of the academic sector, information technology played a notable rule. However, the academic sector is still facing the issue of student's attendance in all public education institutes [6]. The conventional method used in academic sectors for student attendance in the class is "calling student names" or taking student signature which is more time-consuming and laborious job. These methods did not show the arrival time of the student to the class. Additionally, as the participation sheet is passed around the class, a few students may unintentionally or intentionally sign another student name. In this case, the student misses out their name and leads to an incorrect attendance record.

The second problem of having the participation record in a printed copy and there is a chance that a lecturer may lose the participation sheet [7]. As a result, it is difficult for a lecturer to trace out the particular semester attendance record of an individual or overall class. Aside from that, a lecturer has constrained access to single hard copy, for example, just at the workplace. In terms of attendance analysis, the instructor likewise needs to perform a manual calculation to get the student participation percentage in the class which regularly devours a great deal of time. Having the above limitation in the traditional attendance system, there is a need to automate the attendance system for the academic sector based on RFID technology [8]. The primary thought behind the framework is to record student attendance in a semi-automated way where the student needs to streak their RFID embedded tags (student cards) at the RFID reader while entering the class. Along these, the student ID is automatically captured by the rea der, and information is sent to the online server for validation and record purpose [9]. While having the ability to upload student attendance data instantly to the online server without any data loss leads to efficient student attendance. This system will provide access to the authorized personnel as long as they are associated with the Internet that will have a check on the attendance process [10]. The system will be capable to automatically calculate the individual attendance percentage of the student.

In Sect. 2, we review the previous work in RFID-based attendance system. Sect. 3 discusses the methodology and the materials used in the proposed system. Section 4 presents the result and discussion, while in Sect. 5, we conclude the paper.

2 Related Work

In [11], it stated that traditional method of student attendance is time-consuming and inefficient. Student attendance has been identifying as one of the difficult tasks for achieving the performance that contributes to any academic sector. In his study, the author proposed attendance and information service for the academic domain by utilizing radio frequency identification and detection technology, programmable logic circuit, and web-based application. The presented framework was used to keep the record of student attendance and have the ability to monitor the student absentee as well. In addition, the supporting information system provides student marks, daily timetable, and time of classes, number of classroom, and other information related to the student. Based on comparison with other systems, the proposed system was easy and less time-consuming. In [12], the author presented a system that helped to solve the issue faced in traditional attendance system by utilizing RFID technology. The proposed system helped the managing authority to manage the attendance record in an efficient and timely manner. The proposed system has been implemented in a prototype system, and the result showed that the system had eased the attendance system by ma king it semi-automated with the help of RFID technology. The proposed system was easy to use and inexpensive, so it could be easily implemented in the academic sector to save time and reduce human interventions in attendance management. In [13], author reported that attendance is key element for academic achievements, so it needs to be efficient and error-free. The traditional methods of attendance are time-consuming and laborious. A RFID and web-based attendance system was innovated that marks student attendance, generates report of attendance in real time, and accessible via LMS. The results show that the system has been provided reliable, error-free and empowered the attention to maximize the utilization of RFID cards.

In [14], an automatic system for attendance with GSM notification to guardian was presented for professor and student. The result shows that in comparison with other technologies like bar code, biometric, RFID technology is more secure and easy to use with low overhead.

In [15], the author stated that radio frequency identification is a matured technology that has been implemented in different fields to automate the system. A RFID system was implemented to automate the attendance system. The proposed system contained main parts: the software and hardware parts. The hardware parts contain RFID reader which operates on 125 kHz (low frequency). The reader is attached to a host computer to a serial. A GUI-based attendance system is implemented in VB.net which facilitates the user by displaying, registering ID, record attendance system, like ID tag transaction, and other minor functions. The GUI-based attendance system was installed on the host computer. The result shows that presented makes the process of attendance system easy and well organized which will further decrease the time of analysis. To support faculty staff and save time, online attendance system that used GSM-based RFID system is an ideal solution. For this, in [16], RFID-based system was presented that uses GSM for sending reports and web interface for monitoring student record. The proposed framework offers a semi-automated mechanism for attendance, i.e., only flash RFID card at the reader. Result shows that the proposed system offers a numerous advantages, i.e., provides a central repository of student record, allows authorized person to update record and view record online. Home security is the basic need of every individual. In [17], they implement an easily accessible and cost-effective home security system that operates on remote embedded control system (RECS), PCA, and GSM technology for authentication. The presented system uses automatic electromagnetic lock and ZigBee module to interact with security system and offers to monitor real-time activity via web interface. The proposed system uses ARM1176JZF's microcontroller, and from the result, it was revealed that this system provides an efficient security to home. In [18], they conclude that reliable security and privacy are needs of each individual. Traditional method of authentication uses ID, password, or key to enter a protected zone which is more vulnerable and prone to attacks. In this research, a face recognition-based door access control system was developed in OpenCV that includes image capturing, face detection, face recognition, and notification to authority. Furthermore, GSM module sends image of the person to the management authority, and the door lock status can be changed remotely by authorized person for safety purpose. Finally, it is revealed that irregularity of student effects academic performance. A number of institutions still follow traditional time-consuming method. In [19], the author implemented a RFID and GSM-based attendance system to manage the student attendance. Each student has a tag that contains a unique identification number of that student. The student will swipe the RFID tag to the RFID reader when they enter to class. The reader is further attached to a microcontroller. The RFID reader will authenticate the student information at backend database; if the information is correct, the door of the class will be unlocked and attendance of that student will be marked in the database. In addition, the GSM module attached to the microcontroller will be activated and will send a message to the guardian about their children.

3 System Implementation

In this section, the proposed system will be described and presented. Here, all the tools, i.e., hardware and software requirements needed for designing and developing the circuit and steps required for managing the student attendance will be discussed. To implement the proposed system, software and hardware component needed to establish the implementation process is dependent on three measurements, i.e., cost, accessibility, and simple programming. As an open circuit system, the Arduino Uno microcontroller is connected to the IP cam and RFID reader through a pin. The Arduino Uno microcontroller is further connected to the Ethernet shield. When a student arrives at the doorstep of the class, the MFRC522 reader will read his/her RFID tag, and the webcam will instantly capture his/her picture and will be sent to the Arduino Uno. The produced signals will then be transferred to the server by means of Ethernet link. To record student attendance, MySQL, Wamp server, and PHP will be used. Before storing the attendance record on the server, the PHP will compare the image of the student that has been captured while reading the student ID card with the picture which is already stored in the database. Besides this, the presented system facilitates the students by displaying information such as daily class schedule, number of classroom, grading marks, upcoming tests/assignments, and all concerned information through the LCD display. The proposed system and its required information are shown through a block diagram in Fig. 1.

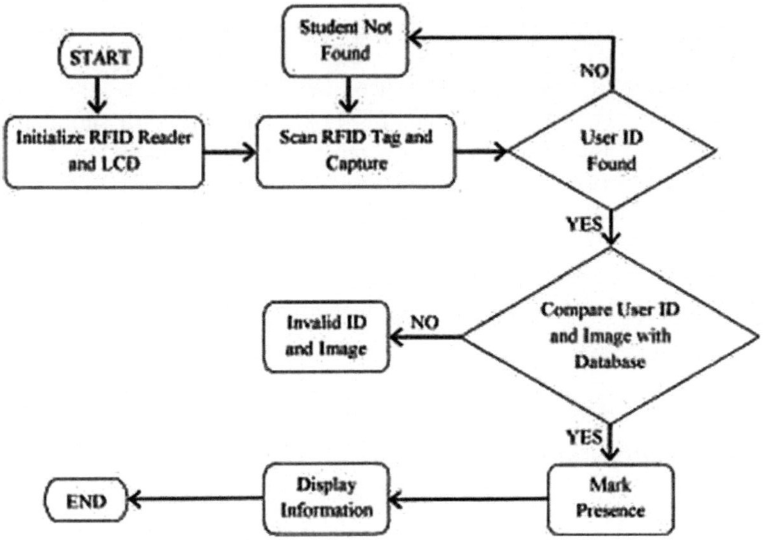

Fig. 1 Block diagram for the proposed system architecture

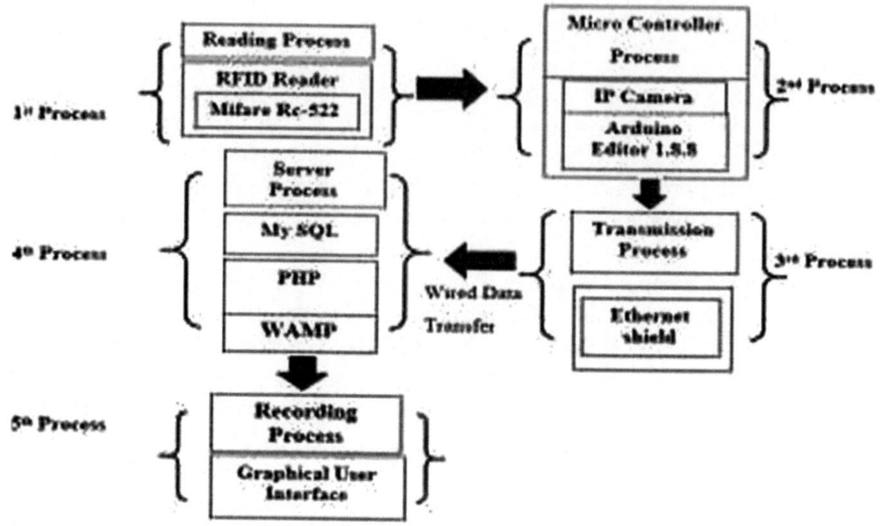

Fig. 2 Stepwise approach for student's attendance and record collection phase

3.1 *Authentication Phase*

This segment provides complete information about all the procedures of the presented system. Every step and procedure required for the management of the student attendance of the proposed system is depicted and exhibited in Fig. 2. The students will flash RFID tag in front of RFID reader where RFID reader will read the unique ID of the student and the webcam attached to the Arduino Uno will simultaneously capture the image of the student. The ca ptured image and unique ID of the student will then be transmitted from the Arduino board to Wamp server (MySQL and PHP) via Ethernet shield to collect, manipulate; furthermore, SQL query will be used to fetch the student record and retrieve the image already stored in the database. After that, PHP will be used to compare the images and store attendance information of the student and display the attendance of the student by an application which is web-based.

3.2 *Face Recognition*

Face recognition is the biometric application of the computer that is used to verify a person by analyzing and comparing pattern based on facial contours. This technology is widely used in security applications. In this study, the following step shows the methodology used for face recognition as shown in Fig. 3.

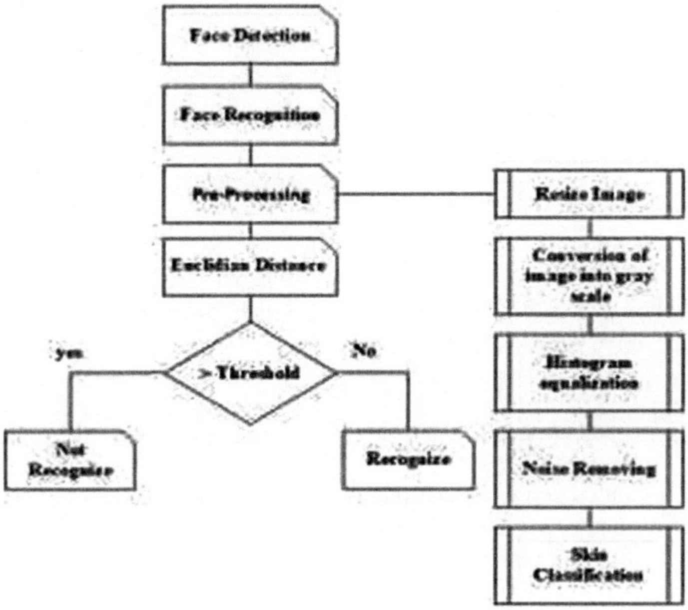

Fig. 3 Flow diagram of face recognition

3.2.1 Viola–Jones Algorithm for Face Detection

Viola–Jones algorithm was designed by Paul Viola and Michael Jones in 2001. This algorithm was the first algorithm based on object detection that offers a high detection rate. For a facial detection, Viola–Jones algorithm uses three techniques.

1. The Haar-like features for the feature extraction are of a rectangular type that is determined by an integral image.
2. AdaBoost is a machine learning method for detecting the face. The term "boosted" determines the classifiers that are complex in itself at each stage, which is built of basic classifiers using any one of the four boosting techniques.
3. Cascade classifier is used to combine many of the features efficiently. The term "cascade" in a classifier determines the several filters on a resultant classifier.

Preprocessing

Preprocessing phase of the image is an important phase. Preprocessing improves data of an image that enhances feature of an image or reduces undesired distortion for analysis task and processing. In this research study, the following preprocessed steps were taken for image enhancement.

Resize Image

Image resize is the step in which the image is shrinked to a particular fixed size that helps in quick processing. In this study, the image is resized into 320 * 240 pixel.

Image Conversion to Grayscale

Sometimes, the image acquired from a camera may have extra brightness for appropriate result that is needed to be eliminated. To get the precise result, first, the acquired image is converted into grayscale image.

Histogram Normalization

Histogram normalization is a technique used for contrast enhancement. To enhance contrast in an image, histogram equalization technique is used. By removing the contrast, the image is equalized, and equalized image's histogram is generated.

Noise Removal

An image acquired from the camera device usually contains some noise which is needed to be filtered. The efficient method of removing noise from a picture is median filtering applied here.

Skin Classification

In this phase, skin classification is performed, in which all pixels that are not related to skin are made black except the pixels that are associated to skin. After this classification, the accuracy of face recognition is improved.

Face Recognition

Euclidean distance is commonly used to measure the distance. Euclidean distance examines the root of square between the coordinates of two objects commonly known as Pythagorean Theorem. In our case to calculate, minimum distance between images captured from the camera and an image already stored in database. We use Euclidean di stance classifier, which calculates the Euclidean distance between the known projection and image projection. If the value of Euclidean is less than defined threshold, we can say that the images are similar. We can get Euclidean distance d2 from the following formula.

3.3 Information Retrieval Phase

Here, in this stage, the RFID transponder will peruse student's tag; Arduino UNO will be utilized to exchange the student data to the web server via a link through ESP8266 shield. The student will flash their RFID tag to RFID reader, which will peruse the student information, embedded in the student ID card. The IP camera attached to the Arduino UNO will take a picture of the student simultaneously. The Arduino Uno will send this information to the web server (PHP and MySQL). On the web server (PHP and MySQL), it is utilized to track the student's record by comparing the unique identity and the image already stored in the database. The information from the database will be exhibited on the monitor as shown in Fig. 4. The provided information will have the data related to the student, for example, timetable, student's name, and classroom no., student's grading, and instructor's name, as well as this system will show all guidelines which are advised by the director to the concerned student.

4 Results and Discussions

The presented system achieves two main goals; the first aim is to register, manage, and authenticate student using RFID tag and face recognition, while the second objective is to provide information related to student class number, timetable, and lecture time displayed on LCD or screen. The traditional method consumes workforce r requirements, imposes time-consuming, duplication of effort, and inefficiency because the

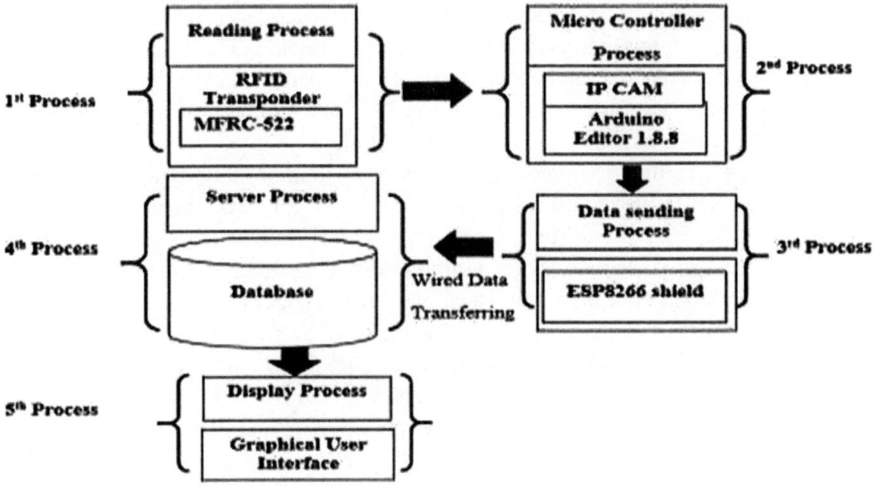

Fig. 4 Step-by-step approach of student information phase

student attendance report is normally recorded on paper and displayed on notice board. Based on different parameters, Table 1 presents a comparison between the proposed system and traditional system of attendance system [19–24].

Table 1 illustrates the comparison between proposed system and traditional system regarding various parameters. To start with result of first parameter which is human interference which is quite clear from results that traditional system is completely dependent on human effort. Whereas in proposed system, there is no dependency on human interference. Secondly, the traditional system was highly time-consuming; in contrast, the proposed system as a result shows that the proposed system required less than two minute for authentication which was half of the time consumed by traditional system. Coming toward, the processing speed in traditional system was very slow going process, whereas the proposed system processing provides a quick result. Followed by the results of system security, the traditional system is more vulnerable as it is easily accessible; in addition, the proposed system provides access to authenticated person only. The upcoming parameter resource document indicated in traditional system utilizes much resource in terms of paper work, registers, stationary, etc., while the proposed system needs only RFID card. Also, traditional authentication system shows low data accuracy as compared to the proposed authentication system. Furthermore, the traditional authentication system consumes more time for registration of a user as multiple entries were needed to register a user at multi-location, while in proposed system, the user needs to register at a central location (database), and each department is able to access it. Finally, the traditional system does not provide a user-friendly interface to the user, while the proposed system provides a user-friendly interface to the user.

Secondly, a number of automated attendance frameworks such as biometrics, RFID attendance system, magnetic stripe are appropriate for various needs and prerequisites. Traditional attendance system, for example, magnetic stripe, barcode, QR code are error-prone and required a long time for registration and also need artificial identification, manual management for record keeping. These systems are not eco-friendly as it needs documentation and paper attendance. A comparison between the common standard automatic attendance system and the proposed system based on different parameter is discussed and described in Table 2.

Table 2 shows results of comparison of common standard automatic authentication system and RFID-based authentication system regarding various parameters. To begin with the first parameter, purchasing cost which is clear from the result that biometric system is more costly as compared to the rest of automatic authentication technologies, i.e., barcode, magnetic stripe, and RFID. Secondly, the value of data density parameter for biometric is high, whereas it shows a low value for barcode and magnetic stripe; however, the data density of a RFID based system is very high. Coming toward the functionality parameter, the result shows that automatic authentication system such as biometric, barcode, magnetic strip, and proposed system had a wide functionality. Followed by the result of speed and security, from the result, it is quite clear that automatic authentication technology has a very high speed and security as compared to traditional manual authentication system. The upcoming

Table 1 Comparison between proposed system and traditional system

Parameters	Human interference	Time-consuming	Efforts Spend	Speed	System security	Resources (Documents)	Data accuracy	Registration time	User-friendly
Traditional system	Yes	More than four minutes	Yes	Slow	More vulnerable	More paper work	Low	More than eight minute	No
Proposed system	No	Less than two minutes	No	High	Authenticated person only	Only one RFID card	High	1–3 min	Yes

Table 2 Comparison of common standard automatic authentication system and RFID-based system

Parameters	Biometric	Barcode	Magnetic stripe	RFID	Face recognition
Purchasing cost	High	Low	Low	Low	High
Resource accuracy	High	High	High	Very high	High
Data density	High	Low	Low	Very high	Low
Functionality	Wide	Wide	Wide	Wide	Low
Speed and security	High	High	High	High	High
Power consumption	Moderate	High	High	Moderate	Moderate
Operating cost	High	Low	Low	Low	Low
Distance of Reader and Data Carrier	Direct contact	0–50 cm direct LOS	Direct contact	0–6 m Depend on the frequencies Used	50–150 cm direct LOS
Influence direction of not applicable as Reader and Data Carrier	Not applicable as Direct contact is required	Failure if no LOS communication	Failure if no slot of communication	No influence as data are transferred via radio wave	Failure if no LOS communication

parameter power consumption indicates from the result that automatic authentication technology, i.e., biometric and RFID need less power as compared to the rest of automatic authentication technology (magnetic stripe and barcode). Followed by the result of operating cost, it reveals that the operating cost of automatic authentication technology biometric was high, whereas the operating cost of magnetic strip and barcode is moderate; on the other hand, it can be seen from the result that the operating cost of RFID-based authentication system was low. After this, the result of parameter distance of reader and data carrier illustrates that biometric and magnetic stripe technology need a direct contact with reader; besides this, barcode authentication technology needs a direct line of sight from a distance of 0–50 cm; in the same way, RFID technology has the ability to read a card up to 6 meter depending upon the frequency utilization. Finally, it was concluded from the results that influence direction of reader and data carrier was not applicable in biometric and barcode technology as it needs direct LOS; on the other hand, magnetic stripe technology completely fails if there is no slot of communication; however, the RFID technology utilizes radio frequency for data transfer, so there is no influence direction of reader and data carrier.

The presented system facilitates both staff and students by minimizing time to take attendance as well as provides a database (DB) that contains the information

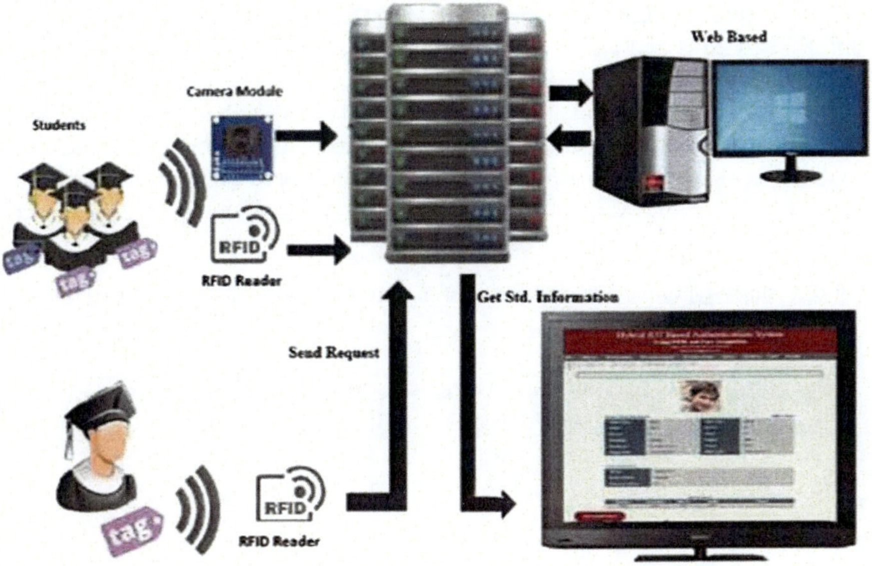

Fig. 5 Flowchart of the IoT-based authentication system

of all students. Furthermore, there is no need of paper work and archiving shelf. The presented system provides facility of new student registration, displaying warnings and reports, and it also displays result of the tests and assignment and other important notification, i.e., lecture canceling and staff appointment. From the faculty or staff point of view, it would be a lot of easier to follow the student record, and it does not require to review and check a number of paper to gather information of a student. Simply tapping the required student name to edit/add warnings and notification or specific information. From student perspective, the presented system provides a student attendance mechanism and also informs the student about their absence report and warning via the screen. In addition, the proposed system presents a weekly statistical report of the student attendance automatically. Figure 5 shows the flowchart of the IoT-based authentication system.

4.1 Comparison of Previous RFID-Based Authentication System Based on Critical Review

This section explains a comparison between the proposed system and past authors that implements RFID-based authentication system in academic sector in detail and highlights their main findings and functionality. The feature of the proposed framework incorporates a few characteristics, for example, monitoring record, maintenance of the record, sending reports, authentication of the student using face recognition, and lastly displaying information to the student. These works had been accompanied from 2013 to 2018 and were meant to govern and eradicate the flaws of conventional methods by using the RFID-based frameworks. Table 3 reviews the functionality measurement of the past framework in practice. We utilize two unique kinds of marking to demonstrate the system functionality. This does not have the functionality. We evaluate every framework functionality based on the functionality of our proposed framework, i.e., authentication via face recognition, sending reports,

Table 3 Comparison of the proposed system using RFID and previous studies for RFID-based authentication systems

Authors	System functionality					
	Tracking students	Data management	Maintenance record	Sending reports	Face Recognition	Information services
Yadav and Nainan [20]	✔	✔	✔	✔	✘	✘
Arbain et al. [21]	✘	✔	✘	✘	✘	✔
Kurniali [22]	✘	✔	✘	✘	✘	✘
Tewari et al. [23]	✔	✔	✔	✘	✘	✔
Farpat and Chavan [24]	✘	✔	✔	✘	✘	✘
Paranali et al. [25]	✔	✔	✘	✔	✘	✘
Kumar and Kumar [26]	✘	✔	✘	✔	✘	✘
Kong et al. [27]	✘	✔	✘	✘	✘	✘
Kuriakose and Vermaak [28]	✘	✔	✘	✘	✘	✘
Srinidhi and Roy [29]	✔	✔	✔	✔	✘	✔
Rjeib et al. [11]	✔	✔	✔	✘	✘	✔
Proposed system	✔	✔	✔	✘	✔	✔

record maintenance, and lastly information services to student. As we have evaluated in Table 3, several outcomes are found by comparing authentication management systems.

Table 3 illustrates the comparison of the proposed system and RFID-based authentication systems. Yadav and Nainan [20] proposed an attendance system to record and manage student attendance automatically in the laboratory by using RFID-ARDUINO approach in web-based laboratories settings. The system contains the following functionality: (1) tracking student, (2) data management, (3) sending reports to the parents; in contrast, the system does not provide the functionality, i.e., : (1) record maintenance, (2) face recognition, (3) information services. Secondly, Arbain et al. [21] present an attendance system to manage and record attendance of the student automatically by uti-lizing Arduino and RFID and web-based interface. The result shows that system includes the function, i.e.,: data management and information services; in addition, the system does not have any student tracking, sending reports to parents, face recognition, and record maintenance functionality. After this Kurniali [22] present an attendance management framework for student by utilizing a web-based approach to handle the attendance record of student in Higher Education Institute of Indonesian. The presented system only utilizes data management functionality; moreover, the system does not have any student tracking, record maintenance sending reports to the parents, face recognition, and information service facility.

Following this Tewari et al. [23] developed a student attendance system based on GPRS which can be accessed easily by professor through a web interface to monitor and check student record, as well as the system encompasses the functionality, i.e., tracking student, data management, and record maintenance and information services; on the other hand, the presented system did not utilize the functionality to send reports to parent and face recognition for authentication. Later on, Farpat and Chavan [24] present an automatic attendance framework for classroom by utilizing RFID technology. The presented system provides data management and record maintenance functionality; in contrast, the system has a deficiency of student tracking functionality, sending reports to parents, face recognition, and information services.

Following, Paranali et al. [25] proposed a system based on Mifare technology for student monitoring and attendance to locate the position of student in a campus environment. The proposed framework provides functionality, i.e.,: tracking student, data management, sending reports to the parents, whereas the system does not have record maintenance, face recognition, information services functionality. Another research study, Kumar and Kumar [26] developed a framework based on IoT and RFID that keep record in cloud transmit SMS to a number of phones, and information of the student can be accessed from anywhere. The presented framework includes the data management and sending reports to guardians; in contrast, the system does not have any student tracking, record maintenance, face recognition, and information service facility. Furthermore, Kong et al. [27] designed RFID and Arduino-based automatic attendance system connected to a real-time database to maintain the attendance of workers in a firm. The designed framework contains data management functionality, but it did not provide the functionality, i.e., student tracking, data maintenance, report sending, face recognition, and information services. After that, Kuriakose and Vermaak [28] conducted an automatic registration and attendance system by using RFID technology in Java-based environment to monitor and manage student attendance in South Africa, Central University of Technology. The developed system provides facility of data management; besides this, the system does not have functionality of student tracking, data maintenance, report sending, face recognition, and information services.

Following this (Srinidhi and Roy [29]) author developed an automatic monitoring and attendance management system with secure and safe system advantage based on web interface that utilizes biometric and RFID technology for academic domain. The developed system contains the functionality of student tracking, data management, record maintenance, report sending, and information services, whereas in the presented system, it did not contain face recognition technology to eliminate proxy authentication. After this, Rjeib et al. [11] present an attendance management system that evaluates the student attendance performance and provides information services for faculty staff and students. The presented system provides the functionality, i.e., student tracking, data management, record maintenance, report sending, and information services; besides this, the system does not contain face recognition functionality to authenticate user and eliminate proxy.

Finally, the proposed study focuses on one of the effective algorithms that marks the student attendance automatically by using face recognition and RFID technology. Furthermore, the system provides information service by using a web-based interface to the student required in academic domain. The presented framework authenticates the user in four phases that are face detection, face recognition, marking attendance, and providing information services; the proposed system includes all the functionality, i.e., tracking user, data management, record maintenance, face recognition, and information services.

4.2 *Evaluation of Proposed Study with Previous Studies for Face Recognition-Based Authentication System*

The following section provides a comparison between proposed system and previous face recognition-based authentication system that highlights their functionality and main findings. Table 4 reviews the functionality measurement of the past framework in practice. We utilized two unique kinds of markings to demonstrate the system functionality. The sign "✔" represents that the framework has one of the measurements, and the sign "✘" represents that the framework does not have the functionality. We evaluated every framework functionality based on the functionality of our proposed framework, i.e., authentication via face recognition, sending reports, record maintenance, and lastly providing information services to student. As we have evaluated in

Table 4 Evaluation of proposed study with previous studies for face recognition-based authentication systems

Authors	System functionality					
	Tracking students	Data management	Maintenance record	Sending reports	Face Recognition	Information services
Selvi et al. [30]	✘	✔	✔	✘	✔	✘
Kar et al. [31]	✘	✔	✘	✘	✔	✘
Wagh et al. [32]	✔	✔	✔	✘	✔	✘
Proposed system	✔	✔	✔	✘	✔	✔

Table 4, several outcomes have been found by comparing authentication management systems.

Table 4 illustrates the comparison of the proposed system and face recognition-based authentication system. Selvi et al. [30] proposed an attendance system to record and manage employee attendance automatically by using face recognition approach in collage environment. The system contained the following functionality: (1) data management, (2) maintenance record, (3) face recognition; in contrast, the system did not provide the functionality, i.e., (1) tracking student, (2) sending reports, (3) information services. Secondly, Kar et al. [31] presented a face recognition attendance system to manage and record attendance of the student automatically by utilizing PCA face recognition algorithm. The results showed that system includes the function, i.e., data management and face recognition. In addition, the system did not have any student tracking, record maintenance, sending reports to parents, and information services functionality. After that, Wagh et al. [32] developed a student attendance system based on eigen face recognition algorithm which can be accessed easily by professor through a web interface to monitor to check student record, as well as the system encompassed the functionality, i.e., tracking student, data management, record maintenance, and face recognition. On the other hand, the presented system does not include the functionality to send reports to parent and information services for authentication.

Finally, the proposed study focuses on one of the effective algorithms that marks the student attendance automatically by using face recognition and RFID technology. Furthermore, the system provides information service by using a web-based interface to the student required in academic domain. The presented framework authenticates the user in four phases that are face detection, face recognition, marking attendance, and providing information services. The proposed system includes all the functionality, i.e., tracking user, data management, record maintenance, face recognition, and information services.

5 Conclusion

The traditional method used in academic domain for attendance is "taking student signature" or "calling name of the student" which is a laborious and time-consuming job. The above method is not capable to show the student arrival time to the class. In some cases, the attendance sheet is circulate around the class and there is a chance of proxy or a few student may intentionally or unintentionally sign other participant name that leads to an incorrect attendance record. In other cases, the attendance record is in a hard copy, so there is a chance that a lecturer may lose the attendance sheet. Additionally, for a lecturer, it is difficult to trace out the attendance record of an overall class or an individual student. Aside from that, a lecturer needs to perform manual calculation to get the percentage of student attendance in class which is a difficult and time-consuming task. Having the above limitation in the traditional attendance system, there is a need to automate the attendance system for the academic sector based on RFID technology. The primary goal is to record attendance of the student in an automated way where student flash their RFID cards at the reader while entering the class. The student ID is automatically sensed by the reader, and the camera attached to the Arduino UNO will capture the picture of the student. The retrieved information is sent to the online server for validation and record purpose. The system will be capable to recognize the student by utilizing face recognition and RFID tag. The goal of the proposed research work is to design and develop a hybrid IoT-based authentication system using RFID and face recognition.

A hybrid IoT-based authentication system using RFID and face recognition has been presented to maintain student different information, i.e., authentication of the student by using face recognition and RFID for attendance of the students, academic records, and providing information about student accounts, timetable, room number, and other related information. Furthermore, the presented system offers easiness to the faculty staff as no extra paper works and extra lockers are needed for keeping the data safe. The achieved result of hybrid IoT-based authentication system using RFID and face recognition proved reliable source of information to the attendance system for academic sector. It can be measured as a fruitful framework. We notice from the result in Sect.4 that two of the presented attendance management systems of Sirnidhi and Roy [29] and Rjeib et al. [11], respectively, have the same functionality and criteria except face recognition of the proposed system.

References

1. Wu Y, Ranasinghe D, Sheng QZ, Zeadlly S, Yu J (2011) RFID enable traceability network. In: A survey distributed parallel database, vol 29 pp 397–443
2. Akbari B, Ajami S (2015) Radio frequency identification and rescue victims in earthquake. Int J Health Syst Disaster Manage 3(5):1–5
3. Domdouzis K, Kumar B, Anumba CJ (2007) Radio-frequency identification (RFID) applications: a brief introduction. Adv Eng Inform 21(4):350–355

4. Ahsan K, Shah H, Kingston P (2015) RFID applications: an introductory and exploratory study. IJCSI Int J Comput Sci Issues 7(1):3
5. Zudor EI, Kemeny Z, Egeri P, Monostri L. RFID technology and its current application. In: The modern information technology in the innovation processes of the industrial enterprises—(MITIP), pp 29–36. ISBN 963
6. Sayanekar P, Rajiwate A, Qazi L, Kulkarni A (2016) Customized NFC enabled ID card for attendance and transaction using face recognition. Int Res J Eng Technol (IRJET) 3(09):1366–1368
7. Mahyidin MF (2008) Student attendance using RFID system. Available from http://umpir.ump.edu.my
8. Yuru Z, Delong C, Liping T (2013) The research and application of college student attendance system based on RFID technology. Int J Control Autom 6(2)
9. Tokiwa Y, Nonobe K, Iwatsuki M (2009) Web-based tools to sustain the motivation of students in distance education. In: Frontiers in education conference, FIE'09. 39th, vol 20
10. Zonggang H, Junsheng Z (2009) Design and implementation of student attendance management system based on MVC. In: International conference of IEEE, pp 1–4
11. Rjeib HD, Ali NS, Farawn AA, Al-Sadawi B, Alsharqi H (2018) Attendance and information system using RFID and web-based application for academic sector. Int J Adv Comput Sci App. (IJACSA) 9(1):266–274
12. Zaman H, Hossain J, Anika T, Choudhury D. RFID based attendance system. In: 8th International conference on computing, communication and networking technologies
13. Kurniali S, Mayliana (2014) The development of a web-based attendance system with RFID for higher education institution in Binus University. In: EPJ web of conferences, vol 68
14. Yadav R, Nainan S (2014) Design of RFID based student attendance system with notification to parents using GSM. Int J Eng Res Technol (IJERT) 3(2):1406–1410
15. Shukla S, Shah S, Save P (2013) RFID based attendance management system. Int J Electr Comput Eng (IJECE) 3(6):784–790
16. Tiwari AS, Tiwari AS, Ade NM, Sheikh SGK, Patel NR, Khan AR (2014) Optimized design of student attendance system using RFID. In: International conference on machine learning, electrical and mechanical engineering. pp 134–139
17. Sahani M, Subudhi S, Mohanty MN (2016) Design of face recognition based embedded home security system. KSII Trans Internet Inf Syst 10(4):1751–1767
18. Prasad V et al. Using a light up feature of a mobile device to trigger door access. U.S. Patent Application No. 16/155,521
19. Kakade VM, Deshmukh P (2017) Attendance monitoring system using RFID technology. Int J Innov Emerging Res Eng 4(3):1–5
20. Yadav R, Nainan S (2014) Design of RFID based student attendance system with notification to parents using GSM. In: Int J Eng Res Technol (IJERT) 3(2):1406–1410
21. Arbain N, Nordin NF, Isa NM, Saaidin S. LAS: web-based laboratory attendance system by integrating RFID-ARDUINO technology. In: 2nd International conference on electrical, electronics and system engineering (ICEESE), pp 89–94
22. Kurniali S (2014) The development of a web-based attendance system with RFID for higher education institution in Binus University. In: EPJ web of conferences, vol 68, no 38
23. Tiwari AS, Tiwari AS, Sheikh NMS, Patel NR, Khan AR (2014) Optimized design of student attendance system using RFID. In: International conference on machine learning, electrical and mechanical engineering, pp 134–139
24. Farpat S, Vyas D, Chavan S (2015) Monitoring of attendance using RFID and GSM technology. 3(2):1–8
25. Pranali S, Mayuri S, Manisha T, Varsha J (2015) An intruder tracking and attendance monitoring system using Mifare technology. Int J Emerging Technol Adv Eng 5(4):49–53
26. Praveen Kumar M, Mani Kumar B (2015) RFID based Attendance monitoring system Using IOT with TI CC3200 Launchpad. Int J Mag Eng Technol Manage Res 2(7):1465–1467
27. Kong S, Zhao J, Shi G, Wu C, Zhao W, Liu T (2015) The design and implementation of the attendance management system based on radio frequency identification technology. In: International conference on electronic science and automation control

28. Kuriakose RB, Vermaak HJ (2015) Developing a Java based RFID application to automate student attendance monitoring. In: Pattern recognition association of South Africa and robotics and mechatronics international Conference (PRASA-RobMech), IEEE, pp 48–53
29. Srinidhi MB, Roy R (2015) A web enabled secured system for attendance monitoring and real time location tracking using Biometric and Radio Frequency Identification (RFID) technology. In: 2015 International Conference on IEEE computer communication and informatics (ICCCI), pp 1–5
30. Selvi, KS, Chitrakala P, Jenitha AA (2014) Face recognition based attendance marking system. Int J Comput Sci Mobile Comput 3(2):337–342
31. Kar N, Debbarma MK, Saha A, Pal DR (2014) Study of implementing automated attendance system using face recognition technique. Int J Comput Commun Eng 1(2):100–103
32. Wagh P, Chaudhari J, Thakare R, Patil S (2015) Attendance system based on face recognition using Eigen face and PCA algorithms. In: International conference on green computing and internet of things (ICGCIoT)

Exploiting Modularity Maximisation in Signed Network Communities for Link Prediction

Faima Abbasi and Muhammad Muzammal

Abstract An important aspect of social networks is to enable people to express their bias. Modelling this behaviour in the form of sentiment links results in a network topology commonly referred to as a signed network. Contemporary work in the signed network domain suggests that negative links can play a key role in signed link analysis. However, signed networks suffer with the data sparseness problem as there is little proportion of signed links in the complete network. The complexity of the problem increases as negative links are much more infrequent than positive ones, indicating that majority of users are more prone towards positive temperament instead of negative one. In this research, we focus on the challenges associated with signed link analysis and address the negative link prediction problem. We introduce a set of ensembles which exploits the hidden interactions of communities and scales up the negative link prediction task in a signed network. Using a scalable embedding framework, network representation is learned, which is given to a set of ensembles for the prediction of negative links. Robustness and efficiency of the approach are tested through experiments conducted on real-life signed network datasets.

Keywords Signed network · Ensembles · Negative link prediction

1 Introduction

Growing popularity of social media platforms has allowed users to share their views and communicate with one another. This kind of user behaviour provides insights into user preferences and social needs. However, some network platforms only capture relationship in the form of presence or absence of links. In contrast, there are signed networks, which model many real-world relationships using both polarity and magnitude. Variety of social relationships that exist in signed networks impart dominant

F. Abbasi (✉) · M. Muzammal (✉)
Department of Computer Science, Bahria University Islamabad, Islamabad, Pakistan
e-mail: faima.buic@bahria.edu.pk

M. Muzammal
e-mail: Mmuzammal.buic@bahria.edu.pk

A. Ullah et al. (eds.), *Proceedings of International Conference on Information Technology and Applications*, Lecture Notes in Networks and Systems 350,
https://doi.org/10.1007/978-981-16-7618-5_36

characteristics of the network in form of positive and negative relations between users [8]. The polarity associated with links can be expressed as trust, distrust or animosity, just like a voter in election network can decide to give a positive vote, a negative vote, or no vote. Similarly, in case of YouTube, a viewer may like or dislike a video or may choose not to respond at all [18]. Positive sentiment links are valuable in supporting users to extract online information and provide benefits to applications like recommendation and information filtering. Negative links provide support to decision makers for reducing vulnerability in the content and uncertainty associated with decisions. Link analysis in signed network closely focuses on the formation of links and is used in a variety of tasks, i.e. community detection [19] and node classification [17]. While addressing link prediction problem in signed network, two challenges are emphasised, (1) among unconnected node pairs which pairs would likely to form connections and (2) what would be the sentiment signs of these links.

Sparseness in signed network makes research-oriented task difficult as negative links are sparse and missing, which gives rise to a lot of noisy data. Lesser availability of interactive data having negative sentiments shows that users are more inclined to give positive sentiment whether it is a network interaction or product rating site. With all aforementioned challenges, in this work, we intent to predict negative links preserving network topology of signed network. Every single node is represented into low dimensional space through *DeepWalk* embedding approach, which mitigates disruption caused by sparsity. Edge embedding is computed interpolating the topology of network, followed by the prediction of negative links through the deployment of a set of ensembles. Significance of our approach lies in exploiting hidden interactions available in signed network communities.

1.1 Our Contributions

Summarised below are contributions of our work:

1. A modularity maximisation approach is proposed to generate multiple communities from training data to address the data sparsity issue.
2. A scalable hidden network embedding learning approach is adopted to transform positive and negative network structure into a low dimensional space, and a set of ensembles is developed for the link prediction task.
3. The proposed solution is evaluated on real-world signed network datasets in terms of scalability and robustness.

This rest of this work is organised as follows: Sect. 2 presents literature review. Section 3 formally illustrates the negative link prediction problem with important labels and notations. We present extensive solution design in Sect. 4. Experiments are discussed in Sect. 5, while Sect. 6 states the conclusion.

2 Related Work

In majority of social media sites, users are not able to distinguish between the people they trust and distrust. Identification of friend and acquaintance is obscure and vague in social media sites. Hence, the traditional approaches used for unsigned networks are not appropriate for exploring and predicting links in signed social network [7]. Unique characteristics are imposed on signed network due to availability of positive and negative links. Positive links have contrasting properties than negative links, i.e. small diameters, high transitivity, and notable local clustering parameters. Sociological theories such as status and balance theory provide deeper insight into patterns of positive and negative links which can result into various kind of relationship. Balance theory works on the notion of *friend of my friend is my friend* and *friend of my enemy is my enemy* [6]. However, status theory is based on the status difference which exists in our society. These theories can be helpful in achieving sign inference task, which involves inferring unknown signs of already existing links [2].

Tasks such as community detection [19], clustering [10], node classification [3], link prediction [9 , 11], and sign prediction [16] are studied for signed network. Community detection involves identification of node pairs which are densely and deeply associated to one another other than remaining network given in [4, 5]. Extracting features of trust and distrust relations is important phase of prediction, which is done through network embedding approach. This approach takes network graph as input and gives low dimensional latent factors of node pairs and is applied to different tasks including sign prediction [13].

Link and sign prediction tasks [9, 12] involve existence of link and identification of sign associated with unlabelled link. The negative link prediction task involves identification of existence of links as well as the negative sign assignment of the links. The focus is on negative links as they are sparser than positive one in network data and make negative link prediction more challenging [1].

3 Preliminaries and Problem Statement

Preliminaries Some useful notations are as follows: $G(U, R, \delta)$—signed graph; U or V—vertex set; R—mapping function; R^+ and R^-—positive and negative links; E and E_x—actual and expected edges; E_m—ensemble model; **Z**—low dimensional space; A^m or A—adjacency matrix; D^A—average degree; W and L—random walk and length of walk; $M_{\max}$—modularity maximisation parameter.

Problem Statement Let directed weighted graph be $G(U, R, \delta)$, where U is the vertex set, $R \subseteq U \times U$ represents relation between vertices U, and $\delta \rightarrow R\ (-1, 0, +1)$ is mapping function that labels the edges. If there exists a distrust relation between vertices U_i and U_j, -1 is assigned; if trust relation exists between U_i and U_j, $+1$ is assigned; otherwise, 0 is assigned. Negative link prediction problem is formally stated as: *Given* G *denoted by a sparse adjacency matrix* A^m, *we develop an ensemble* *model E_m, to predict subsequent missing negative links, observing already existing relations (R^+, R^-) and structural features of signed network.*

4 Proposed Framework

In this work, we propose an ensemble model E_{m} which is able to map adjacency matrix A^{m} to a low dimensional space $\mathbf{Z}$ for the prediction of subsequent missing negative links, preserving topological properties of the signed network.

The problem of negative link prediction is premeditated as a classification problem; however, deep network embedding and prediction of links are significant part of the experiments. We split our network into training (60%), validation (20%), and test (20%) graph for this experimental study.

4.1 Data Preprocessing

A hierarchical method is used to generate communities from training graph proposed in [14], which uses concept of edge betweeness measuring total number of shortest paths the edge is part of and removes the vertex having the highest betweeness value. Constant time is taken to compute edge betweeness and find candidate edge for removal. For generating communities, we consider the following steps:

Given a directed weighted graph, breadth-first search (BFS) is performed, for a randomly chosen vertex U_x. Shortest path from U_x is computed for same BFS level nodes. Edge with the highest betweeness is removed to generate communities. Edge removal is stopped by evaluating the modularity of each community formed at different levels and is informally stated as difference of E_x from E within communities. Edge removal process is repeated until modularity of undivided community is larger than total modularity score of the partitioned communities. The modularity score of each community is computed by Eq. 1 for a network with K communities.

$$M_{\max} = \sum_{n=1}^{K} \sum_{x,y \in C_n} A_{x,y} - \frac{D_x, D_y}{\varepsilon} \tag{1}$$

The notation states that modularity is a measure of community strength taking degree distribution in context. For a network of edges, E, we can compute expected edges, i.e. E_x between two nodes U_x and U_y having degrees D_x and D_y given in Eq. 1, while $A_{x,y}$ depicts the existence of an edge between two nodes. If there is an edge $A_{x,y} = (+1, -1)$, otherwise $A_{x,y} = 0$.

Community as a structural element of the network is stated by modularity maximisation concept. It has two assumptions, (1) user residing in same community is expected to share an edge either positive or negative, leading to higher value of M_{max} and (2) user not residing in same community is least expected to share an edge, leading to lower value of M_{max}.

Further, the communities are split for maintaining ratio of both classes in the training network. The splitting of communities is done according to sub-sampling criterion stated below:

Given, a set of minority class instances be N^- *and majority class instances be* P^+ *such that* $N^- < P^+$. *Let sub-communities be* K_S, *randomly select sub-community* P^S *from* P^+ *such that* $P^S < P^+$ *and* N^-. *Learning model* H_m *is trained using all sub-communities* P^1, P^2, ..., P^C, *and* N^- *such that* $Y = 1$ *to* K_S.

The appropriate choice of community generation method is significant as most of the class instances are neglected by majority of under-sampling methods.

4.2 Structure Preserving Feature Extraction

In this section, we seek to learn social representation of sub-communities. The intuition is to preserve the similarity of sub-community and represent each node pair by a vector in a low dimensional space. These vectors capture the structure of community by the position of vectors in the embedding space. The position of vertex specifies the learning representation of vector of a vertex. The context of node can be constructed by the interactions in which a specific vertex is involved in.

A deep learning unsupervised feature learning approach namely *DeepWalk* was proposed by Perozzi et al. [15]. We consider the same approach for learning features associated with vertices of each sub-community.

A number of random walks ρ are conducted for each vertex using the approach; i.e. *randomly choose a node* U_i *and initiate an arbitrary walk having length L. Sequence of nodes in an arbitrary paths is generated having* U_i *as source node and* U_n *as nth node in the path.*

Random walks are generated for each vertex U_i, which are updated by stochastic gradient descent. Skip-gram model was used to compute vector embedding using the objective function illustrated in Eq. 2. Skip-gram model has a characteristics of iterating over all the possible representations in random walk, appearing in window of predefined size and mapping each vertex to its current representation having vector $\varphi(U_i)$. *Deepwalk* algorithm details for directed weighted graph are stated in Algorithm 1.

Algorithm 1: An outline of negative link prediction ensemble framework

Input: Network G(U, R, δ) (signed)

Output: Prediction graph $T^0{}_G$

1: Divide G into training graph T^G, validation graph V^G, and test graph T^0_G

2: Append missing links generating adjacency matrix A for T^G

(continued)

(continued)

Algorithm 1: An outline of negative link prediction ensemble framework

Divide T^G into K communities using modularity maximisation approach
Divide each K community into K_S sub-community using sub-sampling method
for $i = 0$ to K_S **do**
    RandomWalk(K_S, U_i, L*)*
    SkipGram(Φ,W_v,*w)*
    Minimise objective function in Eq. 2
    Compute edge embedding using Eq. 3
    Train classifiers using node and edge embedding
end for
K_S communities are validated using V^G
Final prediction is made using T_G^0 graph 14: Results are combined through majority voting scheme
return

$$\text{Min}_{\varphi} - \log P(U_i \ldots U_n)|\varphi(U_i) \tag{2}$$

Node embedding algorithm generates semi-supervised method for learning rich feature representation for each user in directed signed network. A Pair of nodes is required for inferring edge embedding, describing affinity structure of underlying network. Equation 3 gives insight of edge embedding operator for given two nodes, U_i and U_j. This was used to generate representation $g(U_i, U_j)$ $g \rightarrow R_n$ where n is the representation size of vertex pairs (U_i, U_j) and is shown in Eq. 3.

$$f(U_i)f(U_j) = f(U_i) * f(U_j) \tag{3}$$

Using *Hadamard* approach, we computed edge embedding associated with each pair of users in signed social network. These node and edge embedding were used to train machine learning classifier in order to predict negative links as well as missing links in signed graphs.

4.3 Negative Link Prediction

Prediction of sparser links is the most sententious task of experimental study. We consider vector representation of sub-communities to train the classifiers such as decision tree, logistic regression, and random forest. Being most powerful tool of classification, decision tree constructs flowchart like structure. Attribute test is depicted by internal node, test outcome is represented by branch, and class labels are illustrated by terminal nodes. Learning phase involves splitting of source set into multiple subsets using attribute value criteria having similar values.

Logistic regression predicts the probabilities as an output; on the basis of those probabilities, we select the class label by setting up a threshold. It finds out the probabilities by $Y = \frac{1}{1+e^{-Y}}$, having Y as probability and $\mathbf{Y}$ as relationship our model has developed between the class label and features.

An extension of bagging method in ensembles is random forest classifier. *For given training examples $E_i \ldots E_n$, we pick a sample randomly to create a decision tree. Size is maximised using CART approach, where m selected values were searched in repetitive manner at each node. Learner's prediction was aggregated, and final estimation is made through majority voting scheme.*

Algorithm 1 and Fig. 1 give outline of the learning model $E_{\mathrm{m},}$ illustrating phases of study for negative link prediction task on signed network. Using modularity-based approach, training graph was split into K communities and K_{S} sub-communities. Further sub-community representation was learned using structure preserving feature learning approach. Lastly, final prediction was made on test graph using set of ensembles and majority voting scheme.

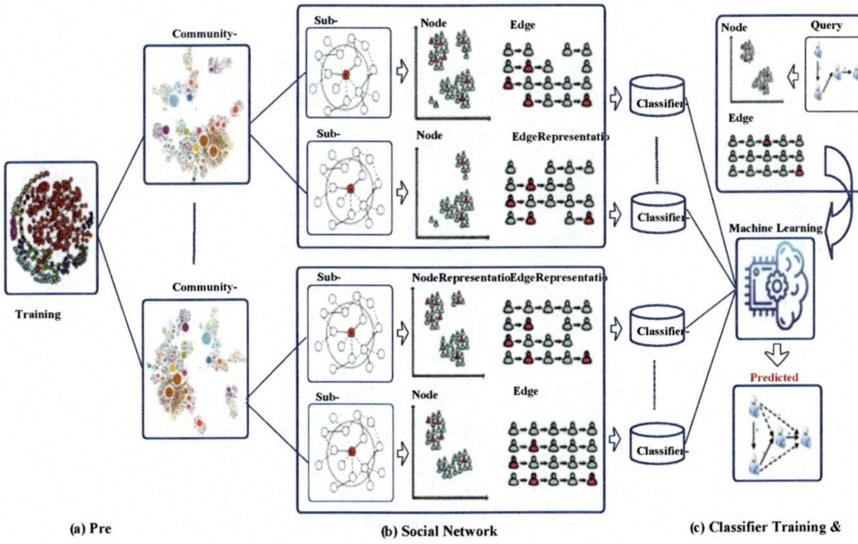

Fig. 1 Negative link prediction ensemble framework

Table 1 Network details

Components	Epinions	Slashdot-Zoo
U	131,828	82,081
R^{+}	143,630	85,146
R^{-}	24,645	24,681
D^{A}	12.7	13.08

5 Experimental Evaluation

In this section, we provide details of the experiments performed on two datasets. Further, we illustrate the results of these experiments. Some evaluation measures are also discussed which were used to evaluate the performance of the model.

5.1 *Data Analysis*

Signed social network datasets were collected from Koblenz network collection,[1] where users are able to express opinions to other users creating signed links. Statistics are given in Table 1, illustrating that Epinions consists of 13%, while Slashdot Zoo consists of 22% of negative links. Individual post review data in Slashdot and product rating data availability in Epinions can help user to approximate emotions towards one another; however, the personality information is not available in these networks.

5.2 *Performance Results*

Performance evaluation of model is very much essential task of experimental study. Precision $\left(\frac{\text{Tp}}{\text{Tp+Fp}}\right)$, recall $\left(\frac{\text{Tp}}{\text{Tp+Fn}}\right)$, and $F1$-score $\left(\frac{2\times\text{precision}\times\text{recall}}{\text{precision}+\text{recall}}\right)$ are used as performance measures due to sparseness of signed network. These measures were able to illustrate the performance of the model on test signed graph depicted in Fig. 2. Three classifiers were trained on signed network to predict negative and missing links. Cross-validation treating a signed sub-graph as a test graph T_{g} was used, where $A_{ij} = 0 \in T_{\text{g}}$, indicating the scenario where information concerning the link between two nodes is missing. As an input to predictor representation, A_{ij} was fed.

Observing the values in Fig. 2, we can infer that if negative and missing links prevalence in training graph increases, then the precision in the test graph increases. In Epinions, the percentage of negative links is 13% where decision tree and random forest classifiers give precision of 0.16 and 0.15. In case of Slashdot Zoo, the

[1] http://konect.uni-koblenz.de/

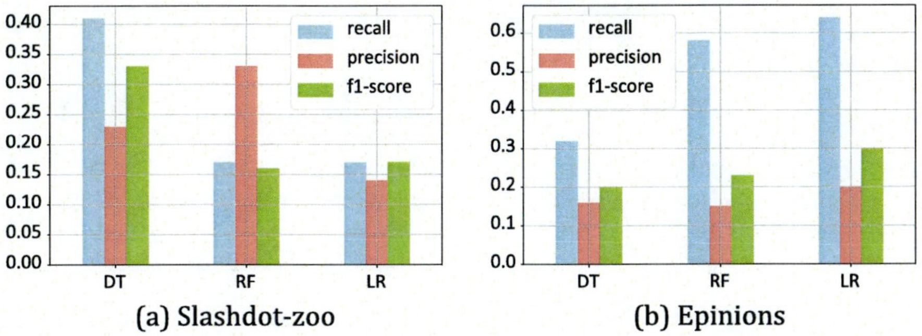

Fig. 2 Negative link prediction performance on T_g

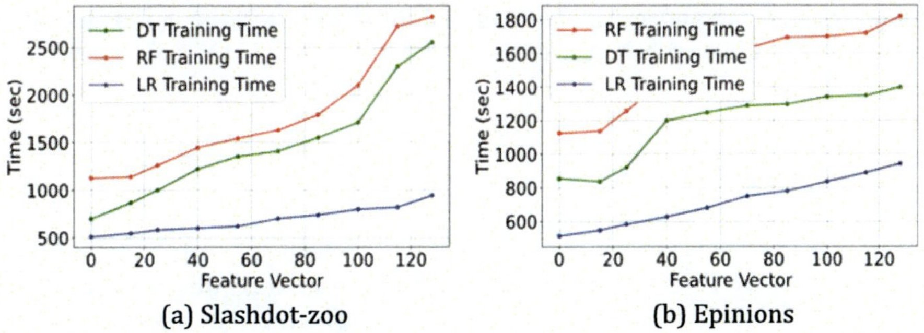

Fig. 3 Associated time (s) against feature vector

percentage increases to 24% of negative links, and decision tree and random forest classifiers give precision of 0.28 and 0.30 illustrated in Fig. 2.

We measure the training time against each feature vector for both signed network datasets for evaluating the scalability of our model on certain parameters. Figure 3 illustrates that training on Epinions, the model is computationally cheap, when compared with Slashdot Zoo, against each feature vector. This can be due to the different learning biases associated with each classifier.

Avoiding uncertainty and evaluating classifier skill are two important phases of study, which aim to identify the threshold giving notable precision and recall for negative and missing links. Reliability diagrams in Fig. 4 are used to visualise the calibrated probabilities. Isotonic calibration is used to calibrate classifier in this experimental study. Formulations used for isotonic calibration involve model prediction and true labels, illustrated as $G_T = m(Q_P) + \varepsilon$, where m is non-increasing function, G_T is true labels, and Q_P is model predictions. Figure 4 shows calibrated and un-calibrated histograms using logistic regression and random forest on signed network for model evaluation. Observing Fig. 4a we can summarise that random forest histogram shows peak at 0.9, while probabilities close to 0 and 1 are very rare for Slashdot Zoo, while

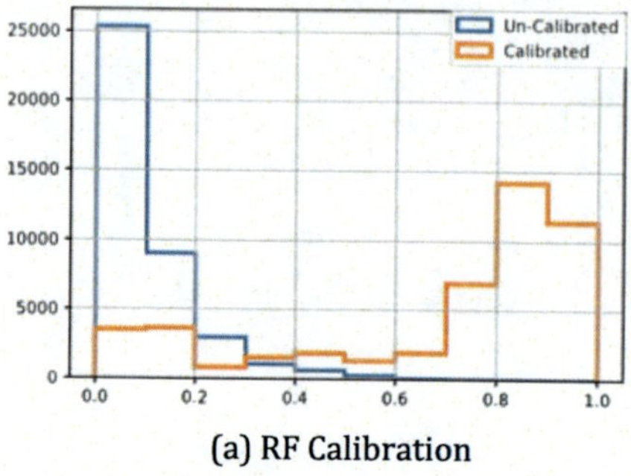

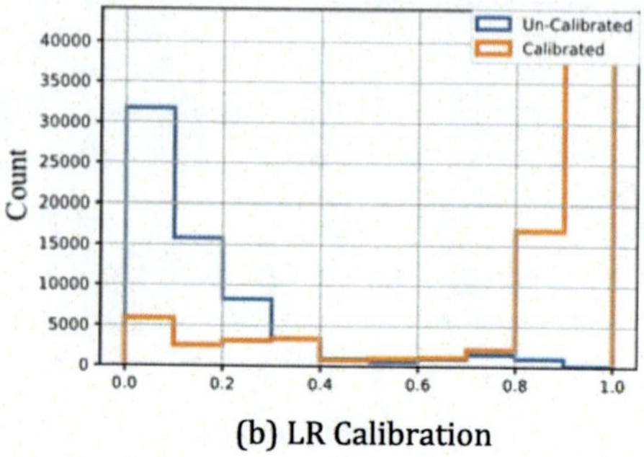

Fig. 4 Classifier calibration on T_g **a** Slashdot Zoo, and **b** Epinions

from Fig. 4b, we can infer that in case of logistic regression, histogram shows peak at 0.9 and 1.0, while probabilities close to 0.1 are very limited for Epinions.

6 Conclusions

A negative link prediction approach in directed weighted graphs is presented that relies on the phases, which preserve structure of signed network. We used modularity maximisation approach to split the training graph in multiple communities. *Deepwalk* and *Hadamard embedding* were used to learn structure preserving features. These features were used by learning classifier for prediction of missing and negative links in the test graph. Issues like sparseness were addressed followed by the observation that integrating missing links makes the network even more sparse and imbalanced. Validation of experiments on two real-life datasets indicates robustness and efficiency of the proposed approach.

This approach can further be used for node classification task in signed network, which is helpful in classifying nodes in different categories. Subsequently, similar recommendations can be thrown to users belonging to same interest based on prevalence of negative links in the network.

References

1. Beigi G, Tang J, Liu H (2020) Social science-guided feature engineering: a novel approach to signed link analysis. ACM Trans Intell Syst Technol (TIST) 11(1):1–27
2. Bongiorno C, Challet D (2020) Nonparametric sign prediction of high-dimensional correlation matrix coefficients. arXiv:2001.11214
3. Bosch J, Mercado P, Stoll M (2018) Node classification for signed networks using diffuse interface methods. arXiv:1809.06432
4. Che S, Yang W, Wang W (2020) A memetic algorithm for community detection in signed networks. IEEE Access

5. Chen J, Liu D, Hao F, Wang H (2020) Community detection in dynamic signed network: an intimacy evolutionary clustering algorithm. J Ambient Intell Hum Comput 11(2):891–900
6. Derr T (2020) Network analysis with negative links. In: Proceedings of the 13th international conference on web search and data mining. pp 917–918
7. Ghorbanzadeh H, Sheikhahmadi A, Jalili M, Sulaimany S (2021) A hybrid method of link prediction in directed graphs. Expert Syst Appl 165:113896
8. Girdhar N, Bharadwaj K (2016) Signed social networks: a survey. In: International conference on advances in computing and data sciences. Springer, pp 326–335
9. Gu S, Chen L, Li B, Liu W, Chen B (2019) Link prediction on signed social networks based on latent space mapping. Appl Intell 49(2):703–722
10. Hua J, Yu J, Yang MS (2020) Fast clustering for signed graphs based on random walk gap. Soc Netw 60:113–128
11. Jing Y, Wang H, Shao K, Huo X (2021) Relation representation learning via signed graph mutual information maximization for trust prediction. Symmetry 13(1):115
12. Muzammal M, Abbasi F, Qu Q, Talat R, Fan J (2020) A decentralised approach for link inference in large signed graphs. Future Gener Comput Syst 102:827–837
13. Naderi PT, Taghiyareh F (2020) Looklike: Similarity-based trust prediction in weighted sign networks. In: 2020 6th international conference on web research (ICWR). IEEE, pp 294–298
14. Newman ME (2004) Detecting community structure in networks. Euro Phys J B 38(2):321–330
15. Perozzi B, Al-Rfou R, Skiena S (2014) Deepwalk: Online learning of social representations. In: Proceedings of the 20th ACM SIGKDD international conference on Knowledge discovery and data mining, pp 701–710
16. Singh R, Adhikari B (2017) Measuring the balance of signed networks and its application to sign prediction. J Stat Mech: Theory Exp 2017(6):063302
17. Tang J, Aggarwal C, Liu H (2016) Node classification in signed social networks. In: Proceedings of the 2016 SIAM international conference on data mining. SIAM, pp 54–62
18. Tang J, Chang Y, Aggarwal C, Liu H (2016) A survey of signed network mining insocial media. ACM Comput Surv (CSUR) 49(3):1–37
19. Yan C, Chang Z (2020) Modularized convex nonnegative matrix factorization for community detection in signed and unsigned networks. Phys A: Stat Mech Appl 539:122904

Comparison of Camera-Based and LiDAR-Based Object Detection for Agricultural Robots

Sercan Sari

Abstract Due to the increasing rate of population in the world, the demand for food has also risen a lot day by day. There are several solutions to deal with this issue, and one of them is that maximizing the yield for agricultural products. It can be achieved by designing such robots that can analyze the field for yield estimation. While there are a lot of robotics solutions that leverage the camera or light detection and ranging (LiDAR) sensor data, almost none of them compared these two methods in the scope of agricultural robotics. In this study, we have simulated different kinds of object detection methods and we try to find which one is more suitable for specific tasks such as color detection and size distinction in the agrarian industry. The main contribution is that the advantages of two sensors are utilized, with the goal of agricultural robotics. Results show that while camera-based object detection is better for color-oriented operations, LiDAR-based object detection is preferable for the operations where size distinction is important. We can achieve over 95% accuracy rate on average for both object detection methods.

Keywords Robotics · Vision · LiDAR · Object detection · Agriculture · Yield estimation

1 Introduction

The increasing rate of global demand for food affects agricultural productivity in recent years. Not only is that one of the concerns increasing the rate of global demand for food but also rising global temperatures around the world is another significant issue [1]. In order to increase crop production, there are mostly two solutions which are clearing more land for agriculture or increasing the crop yields. Due to the development in microprocessors, computer vision, and sensors, robotic agriculture studies start to work on agricultural productivity.

S. Sari (✉)
Department of Computer Engineering, Yeditepe University, Istanbul, Turkey
e-mail: ssari@cse.yeditepe.edu.tr

A. Ullah et al. (eds.), *Proceedings of International Conference on Information Technology and Applications*, Lecture Notes in Networks and Systems 350,
https://doi.org/10.1007/978-981-16-7618-5_37

In the scope of this study, we are going to analyze the robotic solutions seeking answers for challenges in agriculture. While some solutions use computer vision to detect or harvest the crops in the field [2–7], other solutions leverages the LiDAR-based sensor data to interact with the crops [8–14]. There are also hybrid studies that utilize both camera and LiDAR sensors to collect data [15].

In [3], the authors try to estimate the yield estimation using camera-based object detection in a vineyard. In [6], the authors study on strawberry harvesting robots and real-time performance of localization algorithms. Another study [7] focuses on maize tassels detection by leveraging computer vision techniques. The authors in [10] conduct a study which is LiDAR-based in-field phenotyping and navigation. They use the Gazebo simulator and ROS-based framework. In [5], the authors create a simulation of robotic harvesting of fruit and vegetables using V-REP, ROS, and MATLAB.

As we have reviewed the literature, there are so many studies conducted experiments in a real world or simulation environment. Although almost all solutions manage to leverage the camera-based or LiDAR-based sensor data, almost none of them compared the camera-based and LiDAR-based detection for an agricultural robot. While there are some studies which compare these two methods in the scope of self-driving cars [16], they do not consider the agricultural robotics. In this study, we compare these two techniques and conduct an experimental setup to analyze the differences between camera-based and LiDAR-based detection for an agricultural robot using the Webots simulation environment.

This paper is organized as follows: Sect. 2 briefly gives details about the theoretical background and the related work; Sect. 3 describes the methodology; Sect. 4 mentions the experimental setup; Sect. 5 presents the detailed tests and results; Sect. 6 concludes the paper.

2 Background and Related Work

Agricultural robotics has an important role in the future farming. By using robots, so many different kinds of operations such as harvesting can be accomplished in the field. In order to make the robot do something in the field, a camera and sensors are utilized to collect the data. In the scope of this study, we divide the object detection mechanisms into two groups which are camera-based and LiDAR-based object detection. We are going to briefly explain them and give example studies about these two methods.

2.1 Camera-Based Object Detection

In the literature, there are so many studies that leverage data coming from the camera sensor on a robot [2–7]. While some of them use the data to detect and calculate the yield of the products, some others utilize the data to harvest the products from the field. In [6], the authors achieve to build the strawberry harvesting robot using R-YOLO. There are some other target detection models such as YOLO [17], R-CNN [18], faster R-CNN [19], and SSD [20]. Target detection is the first phase for automatic harvesting. Image processing, machine learning, and deep learning methods are used to detect the object in the field. In [4], support vector machine (SVM)-based algorithms are used to detect the aubergines in the field.

One of the drawbacks of using camera-based detection mechanisms is that the effects of lights in the agricultural field [21]. However, if the LiDAR-based object detection mechanisms are used in the field, this type of problem can be eliminated easily. On the other hand, LiDAR-based detection methods may have other drawbacks.

2.2 LiDAR-Based Object Detection

Due to the aforementioned drawbacks of camera-based object detection mechanisms, some studies leverage the LiDAR sensor data to perform some tasks such as object detection, plant modeling, and yield estimation in the field [8–14]. In [12], the authors develop a method using a 2D LiDAR sensor to simplify the detection of crops. Another study [14] uses FX6 3D LiDAR sensor to detect maize in the field. Koenig et al. [11] extract features from the LiDAR data and apply a decision tree classifier algorithm to detect the object is whether harvest residue or bare soil. LiDAR-based sensors have been also used to conduct in-field phenotyping and navigation [10].

3 Methodology

In this study, we use the Webots simulator to build and analyze a yield estimator robot and to compare two different object detection mechanisms which are camera-based and LiDAR-based.

To build the yield estimator robot, we need sensors, rotational and linear motors, a camera, and a LiDAR sensor in the simulation environment. In Fig. 1, it can be seen the main paths in order to make the yield estimator robot.

In the first phase of the study, we have built the main parts of the robot. The main parts of the robot are the same for both camera-based and LiDAR-based robots. After we have built the main parts of the robot, in order to make it move, we have written a controller to interact with the robot. Camera-based detector robot can be seen in Fig. 2.

After the movement of the robot, we have designed the controller so that our robot can get rid of the obstacles. To do that, we have added a distance sensor in front of the robot and configure the necessary setting in the controller. Since our main purpose is

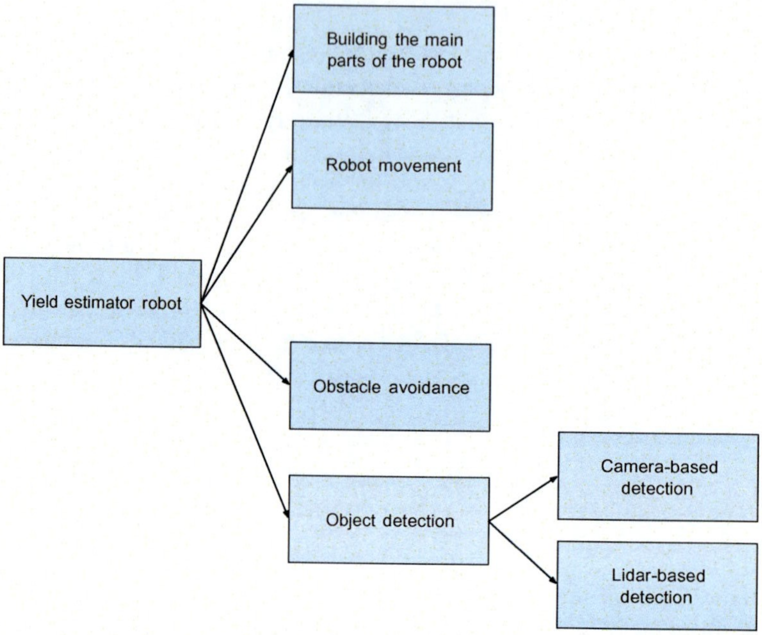

Fig. 1 System design of the yield estimator robot

Fig. 2 Camera-based detector robot

that to detect some objects in the field, we have added a stick on the robot and then added a camera on this stick. Figure 3 depicts the camera node used in this study.

After that, we have designed our controller for the camera settings. By doing these operations, we make our robot able to detect some objects in the angle of the camera.

In order to compare camera-based object detection with LiDAR-based object detection, we have also built a new world and changed our robot and its controller

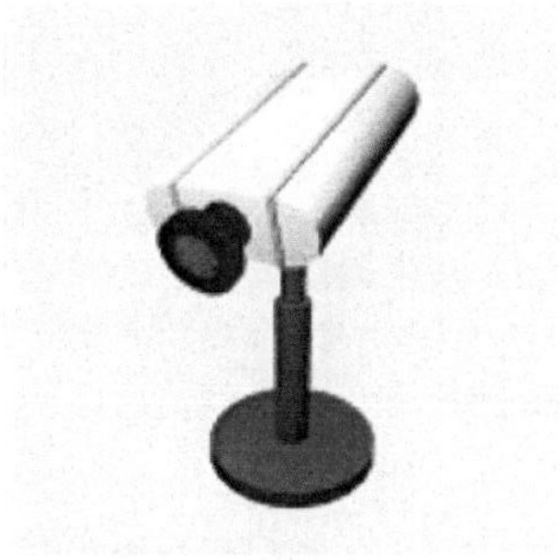

Fig. 3 Camera node

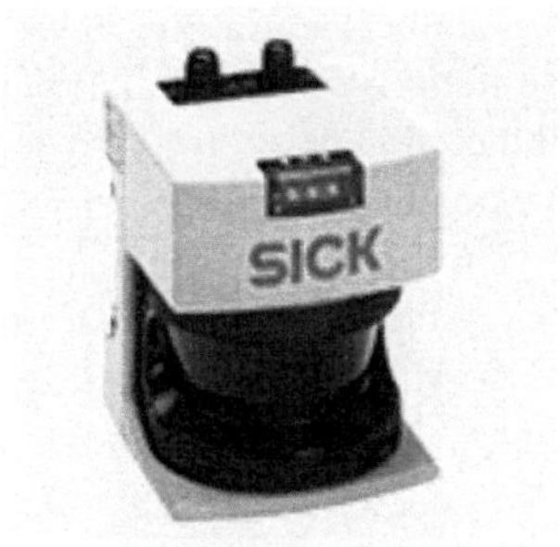

Fig. 4 Model of Sick LMS291 (LiDAR device)

for LiDAR-based object detection. We use the model of Sick LMS291 (LiDAR device) as can be seen in Fig. 4.

4 Experimental Setup

We have considered maize and tomato fields in order to evaluate and compare camera-based and LiDAR-based object detection for the agricultural robot. In Table 1, you can see the details of the seed spacing chart for the maize and tomato.

Table 1 Seed spacing chart [22]

Distance	Maize (cm)	Tomato (cm)
On the row	25	50
Between rows	70	100

Fig. 5 View from the camera-based detection

4.1 Color-Oriented Experimental Setup

As can be seen in Fig. 5, we build the world with specific seed spacing. Due to the limitations of the simulation environment, we assume the cylinders as a tomato crop and configure our robot to detect them properly. To validate our detection algorithm, we conduct several experiments.

Our algorithm can also distinguish the red tomato and the green tomato which can be considered as a non-growing tomato. With the ability of the camera node used in the study, we can also detect the height and the width of the object recognized; however, in our algorithm, we did not use this feature.

4.2 Size-Oriented Experimental Setup

The maize field can be seen in Fig. 6. Each green cylinder represents the maize crop. Our algorithm can recognize the maize which has a height of 30 cm or more. However, in LiDAR-based object detection, our sensor has no ability to detect the color of the object.

Fig. 6 View from the LiDAR-based detection

5 Tests and Results

As we have mentioned in previous section, we investigate the seed planting and spacing information to conduct various test cases appropriately. As can be seen in Table 1, maize and tomato have different seed spacing. This proves our method is working for different kinds of crops. Table 2 exhibits the features of two different methods.

We have conducted different types of tests which are color-oriented and height-oriented for camera-based and LiDAR-based detection methods, respectively. Results can be seen in Tables 3 and 4.

5.1 Color-Oriented Test Results

For the camera-based object detection method, we assume that our robot is in the tomato field. We consider the scenario that some tomatoes are red and some tomatoes are green. Results show that our method has 100% accuracy for camera-based object recognition. Even though the camera node has able to detect the height and width of the object, we do not use this feature. We have conducted only color-oriented tests for camera-based object detection. We vary the number of tomato crops from 12 to 24 in the field. While changing the total number of tomato crops, we have also changed the number of red and green tomatoes in the field so that we can examine the accuracy of our method according to color-oriented tests.

Table 2 Features of the camera and LiDAR sensors

	Camera sensor	LiDAR sensor
Color detection	YES	✖
Size detection	YES	YES

Table 3 Test cases and results for camera-based object detection

# of tomatoes	Red	Predicted red	Green	Predicted green
12	9	9	3	3
24	18	18	6	6

Table 4 Test cases and results for LiDAR-based object detection

# of maize	Height < 30 cm	Height > 30 cm	Total predicted
12	1	11	12
24	3	21	21

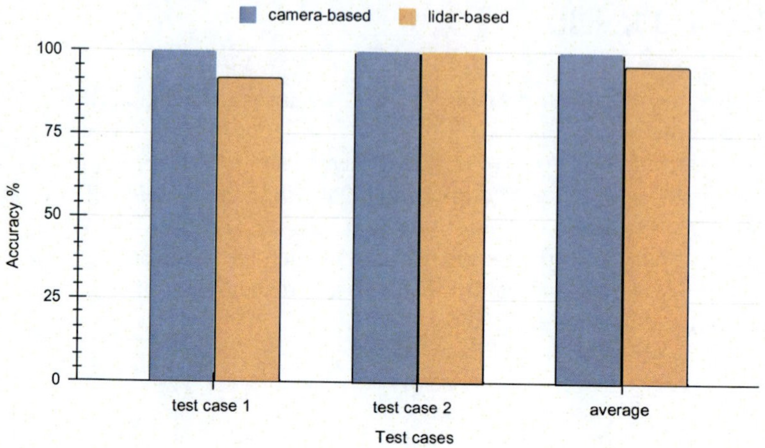

Fig. 7 Accuracy rates of test cases for the camera-based and LiDAR-based object detection

5.2 *Size-Oriented Test Results*

Since the LiDAR sensor used does not have the ability to recognize the color of an object, we have applied the height-oriented tests for the LiDAR-based object detection method. We vary the scenarios according to the number of maize in the field and the height of the maize in the field. Results show that we can achieve up to 91.6% accuracy rate for 12 maize. When we increase the number of maize up to 24, our method gives more accurate results.

5.3 *Overall Results*

When we look at the overall results in Fig. 7, we can say that both camera- based and LiDAR-based object detection for an agricultural robot work well. On average, both methods give over 95% accuracy. However, we can say that if we have specific tasks where we need information about the color of objects, LiDAR-based object detection may not be an option.

6 Conclusion

To sum up, increasing the rate of global demand for food and the rising global temperatures around the world are significant issues. In order to increase crop production, there are mostly two solutions which are clearing more land for agriculture or increasing the crop yields. Due to the development in microprocessors and sensors, robotic agriculture studies start to work on agricultural productivity.

As we have reviewed the literature, there are so many studies conducted experiments in a real world or simulation environment. Although almost all solutions manage to leverage the camera-based or LiDAR-based sensor data, al- most none of them compared the camera-based and LiDAR-based detection for an agricultural robot.

In this study, color-based and LiDAR-based object detection methods are implemented for an agricultural robot. Necessary test environments are conducted to analyze the differences between the two methods. Both methods give over 95% accuracy on average. However, we can say that if we have specific tasks where we need information about the color of objects, LiDAR-based object detection may not be a good option.

References

1. Challinor AJ, Watson J, Lobell DB, Howden S, Smith D, Chhetri N (2014) A meta-analysis of crop yield under climate change and adaptation. Nat Clim Chang 4(4):287–291
2. Azevedo F, Shinde P, Santos L, Mendes J, Santos FN, Mendonça H (2019) Parallelization of a vine trunk detection algorithm for a real time robot localization system. In: International conference on autonomous robot systems and competitions (ICARSC). IEEE, pp 1–6
3. Lopes CM, Gra¸ca J, Sastre J, Reyes M, Guzm´an R, Braga R, Monteiro A, Pinto PA (2016) Vineyard yield estimation by vinbot robot-preliminary results with the white variety Viosinho. In: Jones G, Doran, N (eds) Proceedings 11th international terroir congress. Southern Oregon University, Ashland, USA, pp 458–463
4. SepúLveda D, Fernández R, Navas E, Armada M, González-De-Santos P (2020) Robotic aubergine harvesting using dual-arm manipulation. IEEE Access 8:121889–121904
5. Shamshiri RR, Hameed IA, Karkee M, Weltzien C (2018) Robotic harvesting of fruiting vegetables: a simulation approach in V-REP, ROS and MATLAB. In: Proceedings in automation in agriculture-securing food supplies for future generations
6. Yu Y, Zhang K, Liu H, Yang L, Zhang D (2020) Real-time visual localization of the picking points for a ridge-planting strawberry harvesting robot. IEEE Access 8:116556–116568
7. Zou H, Lu H, Li Y, Liu L, Cao Z (2020) Maize tassels detection: a benchmark of the state of the art. Plant Methods 16(1):1–15
8. Bietresato M, Carabin G, Vidoni R, Gasparetto A, Mazzetto F (2016) Evaluation of a LiDAR-based 3D-stereoscopic vision system for crop-monitoring applications. Comput Electron Agric 124:1–13
9. Höle B (2013) Radiometric correction of terrestrial LiDAR point cloud data for individual maize plant detection. IEEE Geosci Remote Sens Lett 11(1):94–98
10. Iqbal J, Xu R, Sun S, Li C (2020) Simulation of an autonomous mobile robot for LiDAR-based in-field phenotyping and navigation. Robotics 9(2):46
11. Koenig K, Höfle B, Hämmerle M, Jarmer T, Siegmann B, Lilienthal H (2015) Comparative classification analysis of post-harvest growth detection from terrestrial LiDAR point clouds in precision agriculture. ISPRS J Photogram Remote Sens 104:112–125
12. Shen B, Sarow T, Hironaka K, Zhang SH (2008) Development of laser crop row sensor for automatic guidance system of tractor-mounted implements. J Jpn Soc Agric Mach 70(6):90–96
13. Weiss U, Biber P (2011) Plant detection and mapping for agricultural robots using a 3D LiDAR sensor. Robot Auton Syst 59(5):265–273

14. Weiss U, Biber P, Laible S, Bohlmann K, Zell A (2010) Plant species classification using a 3D LiDAR sensor and machine learning. In: International conference on machine learning and applications. IEEE, pp 339–345
15. Chebrolu N, Lottes P, Schaefer A, Winterhalter W, Burgard W, Stachniss C (2017) Agricultural robot dataset for plant classification, localization and mapping on sugar beet fields. Int J Robot Res 36(10):1045–1052
16. Zhang F, Clarke D, Knoll A (2014) Vehicle detection based on LiDAR and camera fusion. In: 17th International conference on intelligent transportation systems (ITSC). IEEE, pp 1620–1625
17. Redmon J, Divvala S, Girshick R, Farhadi A (2016) You only look once: Unified, real-time object detection. In: Proceedings of the IEEE conference on computer vision and pattern recognition, pp 779–788
18. Uijlings JR, Van De Sande KE, Gevers T, Smeulders AW (2013) Selective search for object recognition. Int J Comput Vis 104(2):154–171
19. Ren S, He K, Girshick R, Sun J (2015) Faster R-CNN: towards real-time object detection with region proposal networks. arXiv:1506.01497
20. Liu W, Anguelov D, Erhan D, Szegedy C, Reed S, Fu CY, Berg AC (2016) SSD: single shot multibox detector. In: European conference on computer vision. Springer, pp 21–37
21. Edan Y, Han S, Kondo N (2009) Automation in agriculture. Springer handbook of automation, pp 1095–1128
22. Seed planting and spacing. https://www.seedsnow.com/blogs/news/cheat-sheet. Accessed 10 Aug 2021

The Challenges and Case for Urdu DBpedia

Shanza Rasham, Anam Naz, Zunaira Afzal, Waleed Ahmed, Qandeel Abbas, M. Hammad Anwar, Muhammad Ejaz, and Muhammad Ilyas

Abstract DBpedia is a broad open-source knowledge base continually upgraded and broadened by a large community since 2007. DBpedia's main goal is to retrieve structured content from the articles of Wikipedia and make it available on the World Wide Web for semantic web technologies. As DBpedia services are available in 127 languages, very few contributions are from the Urdu language. This article objectively presents and discusses the concepts and challenges involved in initiating the Urdu chapter of DBpedia, the contributions required, and conclusions about the issues that need to be addressed effectively. First, a brief overview of the DBpedia concepts is presented. Second, the approach to develop Urdu DBpedia, and the issues encountered during its development are discussed. Third, the limitations and challenges are stated. Lastly, it concluded about the creation of Urdu DBpedia chapter and the complexities that can be faced during its creation.

Keywords DBpedia · Internationalization · Ontology · Linked data · Mappings · RDF · Semantics

1 Introduction

DBpedia was raised as a community project to organize the information included in Wikipedia articles. The structured content is presented on the web and made accessible in the format of a knowledge graph once information is extracted from Wikipedia. The DBpedia knowledge base automatically updates as changes occur in any information on Wikipedia. It covers a wide range of topics, including individuals, places, creative and literary works, various organizations and institutions. DBpedia was initially limited to the English Wikipedia edition, but it gradually expanded to different linguistic editions via cross-language linkages.

S. Rasham (✉) · A. Naz · Z. Afzal · W. Ahmed · Q. Abbas · M. H. Anwar · M. Ejaz · M. Ilyas
Department of Computer Science and Information Technology, University of Sargodha, Sargodha, Pakistan

M. Ilyas
e-mail: m.ilyas@uos.edu.pk

A. Ullah et al. (eds.), *Proceedings of International Conference on Information Technology and Applications*, Lecture Notes in Networks and Systems 350,
https://doi.org/10.1007/978-981-16-7618-5_38

Wikipedia maintains both structured and unstructured data. However, it has limited search capabilities and does not enable complex query searches. For example, a question concerning Soccer players that are members of particular clubs and belong to the countries with populations over five million. Wikipedia is edited by a diverse group of individuals who speak different languages. This causes a slew of problems other than textual search such as a Frenchman might spell the same word differently than an Englishman which can result in an ambiguous query search. For example, Wikipedia requires exactly matched keywords for a query search, and multiple Wikipedia pages may have redundant material. In contrast to Wikipedia, DBpedia enables detailed and sophisticated query search using the SPARQL query language [1].

DBpedia being an open knowledge graph (OKG) which is built upon the mappings of Wikipedia infoboxes, it supports creating manual and automatic mappings from Wikipedia to the DBpedia Ontology. It stores data in a machine-readable format and provides the public with a means to store, share, and organize data to be easily accessible in different languages [2]. The attributes from Wikipedia's infoboxes, abstracts, and other data are mapped with the Ontology classes. Linked data formats, or RDF triples, are then used to make these mappings available on the web. Also, it facilitates live web queries.[1]

Regarding the internationalization of the DBpedia project, non-English DBpedia chapters have been created to integrate resources into English Dbpedia through inter-language links. These DBpedia language chapters are a step toward improving the extraction of data from localized editions of Wikipedia. Currently, the existing DBpedia chapters are in 21 languages including English DBpedia [3]. The very first edition which initialized the DBpedia non-English language chapters was Dutch DBpedia [4].

DBpedia also maintains its localized version in 127 different languages. Within these local versions, there are 38.3 million instances in Things class. Out of these, 28.8 million instances are confined representations of things that additionally exist in the English form of DBpedia [5]. The complete DBpedia knowledge base embodies 28 million title tags and abstract information in 127 distinct dialects, 25.2 million connects to pictures, and 29.8 million redirect links to other articles. DBpedia database is also associated with other linked datasets[2] to support semantic web queries in the form of RDF triples. Nearly, 50 million RDF links are used to connect DBpedia with other linked datasets.

The DBpedia Ontology consists of classes, object properties, and data properties of those entities are used in DBpedia and are primarily based upon the attributes found in the Infoboxes of Wikipedia articles. The DBpedia Ontology is developed manually on basis of the data and information retrieved from Wikipedia. Over the years, DBpedia has released various versions to incorporate enhancements in its ontology.

---

[1] http://live.dbpedia.org/ontology. For More info:
http://mappings.dbpedia.org/index.php/OntologyClass:AcademicConference,
http://live.dbpedia.org/ontology/AcademicConference.

[2] A dataset is a collection of numbers or values that correspond to a particular topic.
For more info: "https://en.wikipedia.org/wiki/DBpedia#Dataset".

Each new version of the ontology refers to a new edition of DBpedia dataset that contains data taken from Wikipedia's multiple language editions. In the initial version 3.2, DBpedia provided rules and guidelines for manually generating mappings of the data which was extracted from Wikipedia infoboxes in accordance with the classes present in Dbpedia Ontology. It also addressed various issues related to the structure of Wikipedia infoboxes such as multiple infoboxes associated with a similar class name are having the same name for different properties and different names for the same kind of properties, and finally, no common template and datatype structure is followed in infoboxes [6]. Because of these stated issues, DBpedia supported the hand-operated extraction of data from infoboxes and then mapping those values with their respective classes manually. Thus, the data in DBpedia became more structured and cleaner than in Wikipedia.

With the release of DBpedia 3.5, it allowed the public to contribute by providing mapping guidelines and enabling them to generate the mapping of infobox with ontology and modify the existing mappings and classes in ontology. It resulted in an increase in mapping statistics with a greater number of added classes and properties by the external contributors [6].

As of now after DBpedia version 3.7, the classes in DBpedia Ontology are arranged in a hierarchical order of parent classes with subclasses. For example: As a musician, "Michael Jackson" refers to the class "Musician," which is a subclass of "Artist," which is itself a subclass of "Person." The class "Person" is a child of "Agent," while the "Thing" class is the parent of all subclasses. Any instance which does not belong or map to any class is automatically mapped with the "Thing" class as stated in *schema.org*.

DBpedia Ontology defines the set of classes formulated manually in a hierarchy. The classes are categorized based on the information extracted from Wikipedia's infoboxes. There are currently 786 classes and about 3000 distinct properties in the ontology. Extracted data from Wikipedia's infobox are mapped to DBpedia Ontology Classes and Properties. The ontology now has a continuously growing number of overall instances of different classes which are roundabout to 4,828,418 instances [7].

The most recent release of DBpedia contains almost three billion RDF triple data, of which 580 million belong to the Wikipedia in English, and the remaining 2.46 billion were retrieved and mapped from all the other linguistic versions [8].

The Urdu chapter in DBpedia does not exist currently due to the lack of mappings from Urdu Wikipedia with ontology classes, missing attributes, and missing of Urdu dependencies in framework extraction files. There are very few contributions made regarding the construction of Urdu DBpedia. By its creation, we can generate and integrate the structured data from unstructured and raw form in Urdu. There is lack of uniform template structure in Urdu Wikipedia which makes it challenging to automatically perform extractions and mappings of infobox properties and Ontology classes. Additionally, Urdu Wikipedia is not as rich and organized in information as the English version; specifically, infoboxes of both languages are quite different and sometimes share different templates as well.

There are almost no contributions regarding Urdu DBpedia. The main reason might be that Urdu Wikipedia is still inconsistent. Since DBpedia extracts all data and information from Wikipedia, it can be the case for less attention toward Urdu DBpedia due to the weaknesses of Urdu Wikipedia. But, we ignored this myth and start working without having proper infoboxes and their properties in Urdu Wikipedia by manually mapping attributes and classes.

The Urdu Wikipedia maintains highly unstructured data. For mapping, it is required to first extract the infobox table values either from Wikipedia dumps or live extraction. Due to different template structure, missing attributes, different and incomplete information in Urdu Wikipedia infoboxes comparative to its English version of infoboxes, the extraction framework fails to properly extract and map properties with Ontology classes. One can also adopt manual mapping approach first until the problems of Urdu Wikipedia infrastructure are resolved.

In this article, we are going to highlight the challenges faced by Urdu contributors and possible solutions to overcome them. In Sect. 2, related work is presented. Why the need raised for separate Urdu DBpedia although Dbpedia exists and the challenges of Urdu DBpedia are explained in Sect. 3. In Sect. 3.1, we discuss the challenges and limitations that our team has faced. In Sect. 4, we describe the use cases and applications of DBpedia. In Sect. 5, we wrap up our research with a conclusion, and in Sect. 6, we discuss our plans for Urdu Dbpedia future.

2 Related Work

DBpedia deals with information from domains like terrestrial information, individuals, businesses, movies, music, genetic factors, medicines, books, and scientific journals. The DBpedia knowledge base spontaneously grows as Wikipedia is modified and is multilingual. By the live extraction of Wikipedia updates, the knowledge base of DBpedia depicts the actual state of Wikipedia. Infobox templates from Wikipedia are mapped to DBpedia ontology which increased the quality of information. The creation of RDF linkages from DBpedia to several other data sources aided editors in linking their data with DBpedia. The extraction framework further performs dump-based and live extraction of information from infoboxes. Different people use different methods to define identical things or use different labels for similar attributes, e.g., birthplace and placeofbirth. DBpedia handles such state using two different extraction techniques in parallel, i.e., generic infobox extraction providing complete coverage of infoboxes and their attributes and mapping-based infobox extraction approach providing high data quality [9].

DBpedia applications also facilitate the community with access to its various open datasets and interfaces such as linked data, RDF Dumps, SPARQL query protocol, and many more. Many web editors can also incorporate DBpedia knowledge through its datasets into their Web sites [10]. This approach also connects DBpedia with other databases, and in this way, DBpedia plays the role of a hub for emerging open data web.

As DBpedia improves the integration and display of information with other semantic and linked data technologies, it also collaborates with various databases and other projects resources such as YAGO [11], MediaWiki [12], Freebase [13], and SPARQL endpoint [14].

Daily, a vast amount of short textual document collections is generated as a result of the growth of micro-blogging and social media networks, requiring the use of effective organizational and classification methods. We may use the DBpedia Spotlight framework to enhance short text documents with information obtained from the DBpedia Ontology, [15]. Despite the availability of numerous medical standard vocabularies, effectively identifying domain ideas of medical records in electronic format remains a challenge. The annotation methodologies and knowledge graphs utilized may have caused differences in the coverage and abstraction of these texts' annotations, resulting in considerably different results in the automatically processed annotations. The [16] proposed a semi-supervised method based on DBpedia for extracting medical topics from EMRs,[3] also for evaluating the impact of adding these subjects to the characteristics used to describe EMRs in the job of predicting hospitalizations.

Moreover, information from DBpedia is also being integrated in applications like chatbots [17] which provide a controlled interface to simulate and generate intelligent conversations with humans.

3 Urdu DBpedia

The Urdu language is a member of the Indo-Aryan linguistic group. Urdu is spoken by nearly 170 million people in the world. Out of which, seventy million people communicate in Urdu as their primary language in Pakistan, while the other hundred million people know Urdu as their secondary language in India [18]. It is Pakistan's official language and is also officially recognized in the constitution of India. Some other people who belong to Bangladesh, Afghanistan, the UAE, the UK, and the USA also use the Urdu language for communication.

There is a need for Urdu DBpedia to integrate with DBpedia like other multilingual editions and further compare their attributes to identify ambiguities in extraction and obsolete information. Similarly, quality assurance is also required to detect issues and inadequacies relevant to the quality of extraction, mappings of the data to ontology and provide feedback accordingly.

[3] Electronic medical records (EMRs) offer essential informative records about a health of the patient, and their examination should allow for the prevention of the future diseases that could affect the patient.

3.1 Challenges

Regarding existing mappings of DBpedia in Urdu, there are missing labels including classes, properties, and datatypes. The latest count shows that there are 786 classes, 2886 properties, and 382 datatypes missing from Urdu Dbpedia Ontology mappings [19]. Overcoming these aspects, Urdu DBpedia poses a difficult challenge.

Mapping is usually done through the DBpedia framework for extraction. For the extraction framework, some languages have their dependencies in extraction files, but Urdu dependencies are missing in those files. All dependencies in mapping UI and MongoDB are present but fail to establish the connection.

Another challenge faced while working on Urdu DBpedia is identifying Urdu Wikipedia templates, properties, and infoboxes, finding the most occurring instances, and ensuring if any of them were already mapped.

Another challenge is to fix ontology mapping issues and ambiguities related to mapping classes of ontology in English with Urdu. For instance, DBpedia reports an error when we map references to an ontology label (class or property) that does not exist. But, the challenge we face is that when we generate a mapping, even if a class or property exist, we cannot know for sure if the ontology label and the mappings against it are correct. Like, if the template or property we are mapping exists or does not exist in Urdu Wikipedia, we will not know if the class onto which it is mapped is the right one or not. Some other scenarios like this one also causing issues in ontology mapping.

The next challenge was to remove issues and inconsistencies of the Urdu DBpedia Ontology. Again, if the list of classes or properties is empty and not presented in the ontology, then the mapping will report an error. But the classes that do exist, the mappings of Urdu Wikipedia either could be difficult or wrong in the DBpedia Ontology. Because of language differences, when we map a template of Urdu Wikipedia to a class of DBpedia Ontology, but the interconnected links of the classes with other linguistic properties could be wrong due to language differences and we cannot confirm if they are correct semantically. Due to the uncertainty in mapped properties, it is also difficult to ensure the validity of the domains and ranges found in the class's properties.

The existing ontology, their classes, and properties are named based upon UK (British) English vocabulary, while other countries, including Pakistan, use USA-based English vocabulary which poses a challenge for us to face when it comes to generate mappings using name-based translations. Because the spellings could differ in USA and UK English. For example, while mapping a property/class of Urdu with English, our habit of using USA spellings like "color" could pose a risk because they could be reported missing or unavailable in the ontology because of ontology's British vocabulary basis which accepts "colour" as the correct mapping.

Due to the few existing mappings of infoboxes in Urdu, all the manual mappings were based upon the ontology classes and mappings in English. Because of this, the ontology properties and classes being used are based on assumptions and translation of English words into Urdu. So, there might conflict between infobox mappings and their properties. That is why, even working on the existing language ontology of DBpedia is also a risk.

Moreover, the mapping statistics of Urdu are also not as available on the web as they were a year ago [20]. Due to the absence of current statistics, it is not possible to track the mapping progress and the number of mapped properties, values, and templates. As per the DBpedia community, these statistics can be made available through modifications in the DIEF web tool.

If we critically analyze, the resources available for Urdu are not as rich as available resources for languages like English, Greek, or German. Likewise, there is also no dataset or WordNet for Urdu available on the Web which can facilitate the world of semantics. Moreover, the Urdu Wikipedia has many issues like missing infoboxes, attributes, and articles as well. The increase in the coverage of Urdu Wikipedia still requires a lot of community effort.

4 Use Cases and Applications

Urdu DBpedia can be used to develop semantic integrated machine learning applications including but not limited to data mining [21, 22], plagiarism detection [23], and information retrieval [24].

DBpedia Spotlight is a DBpedia application that allows you to integrate text to the LOD cloud [25]. The fragments and URIs for sources mentioned in the text can be provided by the user using the DBpedia Spotlight annotation. Urdu DBpedia may serve as a link between Urdu web material and the linked data cloud. One way to accomplish this is to publish and link more Urdu datasets as linked data, so creating a bigger Urdu knowledge base (i.e., Urdu cloud). NLP technologies might then be utilized to annotate and connect Urdu materials to the English DBpedia and ultimately the complete LOD cloud.

In addition, the Urdu DBpedia could be used to support their data with micro-data and semantic tags following a specified schema. It is used for information extraction and query construction, which enables users to search the Urdu DBpedia and get complicated semantic data. We can use specialized browsers for data visualization using Urdu DBpedia. It has the capability of navigating and searching Urdu Web sites (Urdu Point,[4] Geo News Urdu,[5] Samaa Urdu TV,[6] etc.). These Web sites make use of Urdu DBpedia's micro-data and tags to structure their content and make it machine readable for data extraction.

A tensor factorization [26] or recommendation system like [27, 28] for recommending music and movies in Urdu can also be formulated over DBpedia. We may also discover the similarities among the films through characteristics, such as some categories, using SPARQL searches (e.g., action films).

---

[4] https://www.urdupoint.com/.

[5] https://urdu.geo.tv/.

[6] https://www.samaa.tv/urdu/.

There are abundant applications that can work on semantics with the support of Urdu DBpedia. One such application is to review and identify the relations and links in the Police Reports in Pakistan as work has been done in [29]. This will help the Police Department in Pakistan to identify the crime rate, nature of the crimes, evaluate the similarities in crimes of the same nature, and analyze the cases semantically.

We can implement a similar interactive chatbot like [30] that allows for real-time discussion with end-users and can answer difficult questions using a simple graphical user interface in Urdu. The chatbot will accept questions in the Urdu language, allowing users to explore the accessible data in real time. The query will be automatically converted to a SPARQL query, and later, it will be forwarded to the DBpedia SPARQL endpoint to respond to the end-user in Urdu.

Another potential application for Urdu DBpedia could be a news filter based on semantic linking between words and sentences that repeat across different newspapers Web sites [31]. The headlines and their short descriptions from native newspapers can be extracted to draw entity relationship links among the aired news in different Urdu newspapers and can further create a clustered knowledge graph like created in [32].

5 Conclusion

The presented work explores the insights and challenges encountered in the construction of Urdu DBpedia and the generation of mappings from Urdu Wikipedia with DBpedia Ontology classes. As with the help of Urdu DBpedia, we can achieve the structured integration of information from Urdu web pages with DBpedia and its semantic interconnection. Moreover, the query search will also be enhanced in the result of it. Along with the Urdu chapter, the need for the quality of automatic extraction of infoboxes, their accurate mapping with Ontology classes, the problems encountered due to the template structure of Urdu Wikipedia, missing labels and attributes, and many other challenges and limitations are highlighted through this paper.

6 Future Work

With DBpedia Urdu chapter, we aim to provide a complete search engine of structured data for Pakistanis and other Urdu-spoken people. In the future, we will work on the extraction framework and add Urdu dependencies in its file. Moreover, the availability of Urdu live statistics is also required through changes in the DIEF web tool.

References

1. O'Connor MJ, Das AK (2009) SQWRL: a query language for OWL. In: OWLED, vol. 529, No. 2009
2. Auer S, Bizer C, Kobilarov G, Lehmann J, Cyganiak R, Ives Z (2--7) DBpedia: a nucleus for a web of open data. In: The semantic web. Springer, Berlin, Heidelberg, pp 722–735
3. Frey J, Hofer M, Obraczka D, Lehmann J, Hellmann S (2019) DBpedia FlexiFusion the best of Wikipedia> Wikidata> your data. In: International semantic web conference. Springer, Cham, pp 96–112
4. DBpedia overview. https://www.dbpedia.org/members/chapter-overview/. Last accessed 12 June 2021
5. Liu Z, Lin Y, Sun M (2020) World knowledge representation. In: Representation learning for natural language processing. Springer, Singapore, pp 163–216
6. Dbpedia ontology. https://www.dbpedia.org/resources/ontology/. Last accessed 12 June 2021
7. Mendes PN, Jakob M, Bizer C (2012) DBpedia: a multilingual cross-domain knowledge base. In: LREC. Citeseer, pp 1813–1817
8. Guo, X, Gao H, Zou Z (2020) WISE: workload-aware partitioning for RDF systems. Big Data Res 22:100161
9. Bizer C, Lehmann J, Kobilarov G, Auer S, Becker C, Cyganiak R, Hellmann S (2009) DBpedia—A crystallization point for the web of data. J Web Semantics 7(3):154–165
10. Auer S, Bizer C, Kobilarov G, Lehmann J, Cyganiak R, Ives Z (2007) DBpedia: A nucleus for a web of open data. In: The semantic web. Springer, Berlin, Heidelberg, pp 722–735
11. Tanon TP, Weikum G, Suchanek F (2020) YAGO 4: a reason-able knowledge base. In: European semantic web conference. Springer, Cham, pp 583–596
12. Rogushina JV, Grishanova IJ (2020) Ontological methods and tools for semantic extension of the media WIKI technology. Prob Program 2–3:61–73
13. Färber M, Bartscherer F, Menne C, Rettinger A (2018) Linked data quality of DBpedia, Freebase, Opencyc, Wikidata, and YAGO. Semantic Web 9(1):77–129
14. Bonifati A, Martens W, Timm T (2020) An analytical study of large SPARQL query logs. VLDB J 29(2):655–679
15. Flisar J, Podgorelec V (2020) Improving short text classification using information from DBpedia ontology. Fund Inform 172(3):261–297
16. Gazzotti R, Faron-Zucker C, Gandon F, Lacroix-Hugues V, Darmon D (2020) Injection of automatically selected DBpedia subjects in electronic medical records to boost hospitalization prediction. In: Proceedings of the 35th annual ACM symposium on applied computing, pp 2013–2020
17. Følstad A, Araujo T, Papadopoulos S, Law ELC, Granmo OC, Luger E, Brandtzaeg PB (2020) Chatbot research and design. Springer
18. Khan MM, Shahzad K, Malik MK (2021) Hate speech detection in Roman Urdu. ACM Trans Asian Low-Resour Lang Inf Process (TALLIP) 20(1):1–19
19. Missing labels for language ur, http://mappings.dbpedia.org/server/ontology/labels/missing/ur/. Last accessed 11 June 2021
20. Dbpedia mapping statistics. http://dief.tools.dbpedia.org/server/statistics/ur/. Last accessed 15 Aug 2021
21. Razzaq S, Maqbool F, Hussain A (2016) Modified cat swarm optimization for clustering. In: International conference on brain inspired cognitive systems, pp 161–170. Springer, Cham
22. Razzaq S, Maqbool F, Ahmed Farid FS, Irfan K, Anwar MA (2008) An optimized knowledge associated, storage and retrieval of digital X-rays databases. In: Proceedings of the world congress on engineering, vol 1
23. Ilyas M, Malik N, Bilal A, Razzaq S, Maqbool F, Abbas Q (2021) Plagiarism detection using natural language processing techniques. Tech J 26(01):90–101

24. Adnan A, Anwar S, Zia T, Razzaq S, Maqbool F, Rehman ZU (2018) Beyond Beall's blacklist: automatic detection of open access predatory research journals. In: 2018 IEEE 20th international conference on high performance computing and communications; IEEE 16th international conference on smart city; IEEE 4th international conference on data science and systems (HPCC/SmartCity/DSS). IEEE, pp 1692–1697
25. Mendes PN, Jakob M, García-Silva A, Bizer C (2011) DBpedia spotlight: shedding light on the web of documents. In: Proceedings of the 7th international conference on semantic systems, pp 1–8
26. Al-Obeidat F, Rocha Á, Khan MS, Maqbool F, Razzaq S (2021) Parallel tensor factorization for relational learning. Neural Comput Appl 1–10
27. Passant, A.: dbrec—music recommendations using DBpedia. In: International semantic web conference. Springer, Berlin, Heidelberg, pp 209–224
28. Mirizzi R, Di Noia T, Ragone A, Ostuni VC, Di Sciascio E (2012) Movie recommendation with DBpedia. In: IIR, pp 101–112
29. Carnaz G, Quaresma P, Nogueira VB, Antunes M, Ferreira NNF (2019) A review on relations extraction in police reports. In: World conference on information systems and technologies. Springer, Cham, pp 494–503
30. Kondylakis H, Tsirigotakis D, Fragkiadakis G, Panteri E, Papadakis A, Fragkakis A, Papadakis N (2020) R2D2: A DBpedia chatbot using triple-pattern like queries. Algorithms 13(9):217
31. Roznama Express. https://www.express.com.pk/. Last accessed 15 Aug 2021
32. Sakor A, Singh K, Vidal ME (2019) FALCON: an entity and relation linking framework over DBpedia. In: CEUR workshop proceedings, vol 2456. RWTH, Aachen, pp 265–268

A Vietnamese Festival Preservation Application

Ngan-Khanh Chau, Truong-Thanh Ma, Zied Bouraoui, and Thanh-Nghi Do

Abstract Preservation and development of intangible cultural heritage are one of the burning problems in rich cultural countries. They are significant characteristics to express the substantial identity of a nation. At present, leading artificial intelligence (AI) into the cultural development strategy is a promising approach to introduce remarkable features to the world community, especially in tourism. However, detecting and discovering the festival knowledge at a local place in a country is not only an exciting problem for travelers but also a challenge for researchers and managers. We, therefore, propose a festival information preservation framework that applied AI techniques to seek and exploit the festival knowledge in Vietnam. We provide a Vietnamese festival lightweight ontology. We investigate and implement a practical application for identifying and mining the Vietnamese festival knowledge in Mekong Delta for the first work. Our initial approach concentrates on providing a Web application that classifies the snapshot and automatically answers questions.

Keywords Application · Vietnamese festival · Cultural heritage · Machine learning · Ontology · Chatbot · BERT

N.-K. Chau (✉)
An Giang University, VNU-HCM, Ho Chi Minh City, Vietnam
e-mail: cnkhanh@agu.edu.vn

T.-T. Ma · Z. Bouraoui (✉)
CRIL, Université d'Artois and CNRS, Arras, France
e-mail: bouraoui@ril.fr

T.-T. Ma
e-mail: ma@ril.fr

T.-N. Do (✉)
Can Tho University, Can Tho, Vietnam
e-mail: cnkhanh@agu.edu.vn

A. Ullah et al. (eds.), *Proceedings of International Conference on Information Technology and Applications*, Lecture Notes in Networks and Systems 350,
https://doi.org/10.1007/978-981-16-7618-5_39

1 Introduction

Intangible cultural heritage (ICH) is one of the priceless assets of a country and an important condition to promote tourism. Therefore, taking advantages of the development of artificial intelligence (AI), several applications and researches are developed in the area, including machine learning (ML) [12], natural language processing (NLP) [26], knowledge representation and reasoning (i.e., ontology) [21] and others. For instance, the authors of [22] provided a tool to extract the Vietnamese dance concepts. They used an ontology to store the information and machine learning algorithms to classify the dance movements. Their output deeply concentrated on investigating a dance. However, applying one of these AI approaches will not be sufficient to advertise the tourism aspects. It will be lost the essential and desired information that tourists are interested in. One situation assumes that a tourist would see an ongoing festival (no information in the current standing place). And then, the primary data will be an actual image from their medium-resolution smartphone camera. A desired outcome is several pieces of the festival information. In this case, if the machine learning technique is used, its outcome will only be a specific classification (i.e., the festival name, the ethnic group). To deal with this problem, we will combine several different AI techniques.

Since ontology has the hierarchy structure and the reasoning ability, it will be a reasonable and reliable foundation for storing the dataset. Moreover, based on the strengthening of query-answering ability, it will give more flexible results. In this paper, we, therefore, use ontology to store the knowledge and detect/extract the essential information. In another approach, a neural network is one of several machine learning algorithms that can help solve classification problems. Its unique strength is creating complex predictions dynamically. Several studies of classifying images are developed rapidly over the past decades based on deep learning, such as [16, 31]. We also take advantage of this technique to implement our approach.

In addition, if the application only applies both of the above approaches (*ontology and machine learning*), the attraction in providing the desired information to tourists will naturally be limited and reduced. Hence, a question-answering system (of NLP) will be one suitable selection for this work. Several studies are related to NLP for query answering as [6, 28]. Therein, BERT, provided by Devlin et al. [10], is one of the emerging and successful approaches. From this point on, we suggest utilizing a pre-trained language model (ALBERT [19]) in our framework. Our primary goal is to answer the tourists' questions, i.e., "which festival are you seeing? What are pieces of information related to this festival?" For a first work, the scope of our approach is the ICH in Vietnam [11]. Namely, we crucially concentrate on the Mekong Delta region festivals.

The main contributions of this paper are as follows: (1) a Vietnamese festival preservation framework, (2) implementing an AI application (including festival classification and an automatic chatbot), and (3) a Vietnamese festival ontology and an image dataset. Note that, our application is developed on the Web platform.

2 Related Work

Several studies and applications have been developed and deployed for the cultural heritage aspect in practice, including [3, 8]. For instance, a group of authors [25] proposed a framework to evaluate the Thai dance movements. The system took advantage of machine learning to estimate the dancer's score via the actual images from using the Microsoft Kinect sensor. This system will support teaching and learning in education.

Furthermore, many AI approaches are related to the tourism aspect, such as [4, 33]. Some papers of the tourism research used ontology for storing, including [24, 27]. However, the authors only concentrated on storing famous landmarks and remarkable places. They ignored the cultural perspective. For instance, a group of authors [18] proposed a system for the tourism industry of Indian. The proposed approach utilized ontology derived from the Indian tourism domain for efficient information retrieval. Another one, the authors in [9] developed an ontology-based framework for tourism recommendation services. A suite of tourism ontologies was developed and engaged to enable a prototypical e-tourism system with various knowledge-based recommendation.

Regarding using machine learning for tourism, several AI scientists focused on this aspect to exploit, i.e., [1, 32]. Several applications related to local hotels and restaurant also developed, i.e., [2, 29]. Moreover, the approach's image classification issues have been attracting many attentions, including [5, 17]. We take advantage of deep learning to classify the festival images. The relevant works closed to our approach are [12, 15].

3 Background

In this section, we briefly introduce the basic information of Vietnamese festivals and provide the techniques which will be applied in our proposed approach.

Vietnamese Festivals Vietnam is a multi-ethnic country with its own unique cultures. These cultural values have been passed by tradition from one generation to another. One of these values is in traditional festivals. It is known as a form of cultural activities, spiritual products of many ethnic communities formed and developed in the historical process. They reflect **the cultural life of the ethnic groups** as well as **of the regions**. Vietnamese festivals keep an important role in the community, and they also significantly contribute to the development of tourism and other professions. We rely on these features to structure the festival ontology. In this paper, we focused on selecting, classification, and querying the information related to popular festivals in the Mekong Delta region to preserve and promote cultural heritage through traditional festivals. An overview of our initial experimental festivals is shown in Fig. 3.

Convolutional Neural Networks (CNNs) A CNN (or ConvNet) is a specific type of deep neural network (DNN) [13] and is excellent at analyzing visual imagery. It is a supervised learning algorithm that uses machine learning units called perceptrons to analyze data. The building blocks of CNN are filters or kernels. Kernels extract relevant and correct features from the inputs using convolution operations. Since deep CNN is very flexible and efficient in processing image data, we use this architecture for image classification in the study.

Bidirectional Encoder Representations from Transformers (BERTs) BERT is a pre-trained model, was developed by Google LLC, and was introduced in 2018. Until October 2019, BERT officially appeared in the Google search engine and applied to query information using the English language, and so far, the BERT algorithm has been applied to up to 70 languages, including Vietnamese. In this work, we utilize BERT's question-answering strengthen to build an automatic Q&A chatbot system. Namely, we use ALBERT, introduced by Lan et al. [19], for our implementation. This novel version has solved the problems about lower memory consumption and increased the original BERT's training speed.

4 A Vietnamese Festival Preservation Framework

This paper proposes a framework for classifying Vietnamese festivals and querying the relevant information as Fig. 1. It includes three main steps: (1) festival image classification; (2) festival identification; (3) query and answering festival information. The first two steps are for identification, and the last is for the query.

For the beginning, the framework's input will be three pieces of information, including a snapshot from the user (or a visitor), the location, and the date. This step is to identify the festival name (classification). Here, we collect three results (top accuracies) from the image classification model and then combine the information about the location and date to determine the festival name. The reason why we select the three results is that the festivals' characteristics are quite similar (i.e., temples, rivers, costumes); hence, the classification accuracy between the two images will be close to each other (i.e., 57 and 43%). Indeed, if we consider top three results and some pieces of the additional information, the accuracy will be increased. For a specific description, the two steps in detail are as follows (1) the trained CNN-DNN model provides the three festivals that keep the highest probability during the classifying process. Next, (2) these results, the user's current location, and date will be transmitted to the Vietnamese festival ontology, which takes them as inputs for performing query-answering statements. The outputs are the identified festival name and its description. At the following step, (3) we use the ALBERT model with the input considering festival description. The automatic chatbot system will answer the user's question directly.

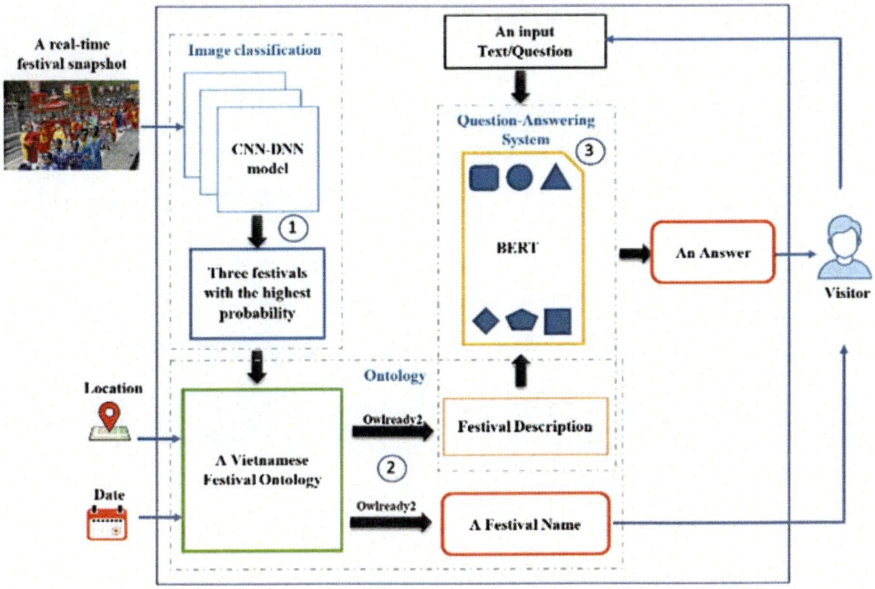

Fig. 1 Framework for the Vietnamese festival preservation

Note that, our main contribution is proposing a new method for preserving traditional ethnic festivals, thereby contributing to the widespread dissemination of Vietnam's cultures. Next, we will present how to build a festival ontology.

5 A Vietnamese Festival Ontology

With an aim of storing the information of festivals, we build a Vietnamese festival ontology aiming at query answering for the tourist's questions.

Regarding the festival perspective, each distinct area/region has different festivals with distinguishable ethnic groups. From this point, we build a lightweight ontology in which the festivals are corresponding to each city/province and according to those areas' ethnic groups. Furthermore, each province and city is belonging to a particular region. Therefore, we use this feature to classify and structure the ontology; i.e., the Mekong delta belongs to the south region of Vietnam.

Our festival ontology is shown in Fig. 2 and consists of two primary parts, including ethnic groups (Dan Toc) and festivals (Le Hoi). For (1), we construct this part based on the clusters of each ethnic group. The reason why we mention this part is that the festival departs from ethnic groups. Regarding (2), we organize the ontology hierarchy based on the geography of Vietnam. The correspondence of each area is ethnic groups and relevant festivals. Therein, we provide each festival's essential information, such as the time, place, name, descriptions, and others. Note that,

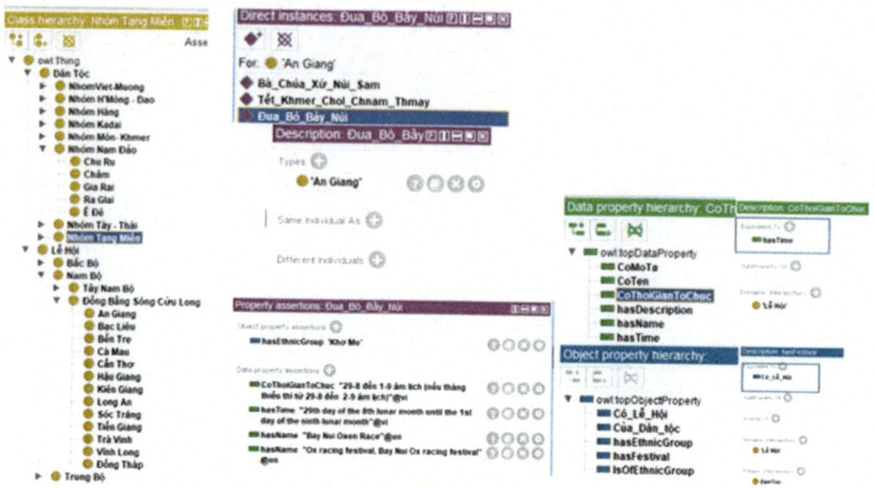

Fig. 2 Cultural festival ontology for the Vietnamese tourism

there are festivals with the same name, but they are different by each ethnic group (i.e., ThanhMinh festival). Another situation is the same name and ethnic group, but the event's timing is different (i.e., Nghinh-Ong festival). Hence, the significant features (characteristics) include the time, place (position), and ethnic groups.

Another primary task of ontology is to classify the festival name based on the reasoning and querying ability. It is also known as the second step in the framework. We build the Vietnamese festival ontology in two languages, such as English and Vietnamese. Therefore, object properties and data properties also construct with the

Id	Vietnamese Name	Numbers of Collected Images	English Name	Province	Date	Ethnic Group
F_1	Lễ hội Bà Chúa Xứ núi Sam	167	Ba Chua Xu temple festival	An Giang	23rd to the 27th, April	Kinh
F_2	Lễ hội đua bò bảy núi	295	Bay Nui Oxen Race	An Giang	29th, August to 1st, September	Khmer
F_3	Lễ hội OkOmBok	217	Ooc Om Bok festival	Bac Lieu Soc Trang Tra Vinh	15th, October	Khmer
F_4	Lễ hội Nghinh Ông	488	Nghinh Ong festival	Bac Lieu	9th - 10th, March	Kinh
				Ben Tre	15th, June	
				Ca Mau	14th - 16th, February	
				Kien Giang	15th - 16th, October	
				Soc Trang	21st - 23rd, March	
				Tien Giang	9th - 10th,March	
				Tra Vinh	11th - 12th, May	
F_5	Lễ hội Chol-Chnam-Thmay	163	Chol Chnam Thmay festival	An Giang Can Tho Soc Trang Vinh Long	around middle of April (15th, April	Khmer
F_6	Lễ đấu đèn Chùa Ông	93	Lantern fighting festival at Ong Pagoda	Can Tho	26th, June	Hoa
F_7	Lễ hội Dừa	80	Coconut festival	Ben Tre	14th to 20th, November	Kinh
F_8	Lễ hội vía Bà Thiên Hậu	35	Ba Thien Hau Workship festival	Ca Mau	23rd, March	Hoa
F_9	Lễ giỗ Đức Khai Trấn Mạc Cửu	30	Festivals Germany anniversary declaration Mac Cuu district	Kien Giang	1st to 30th, May	Kinh
F_{10}	Lễ hội Quan Thánh Đế Quân	20	Quan Thanh De Quan festival	Hau Giang Tien Giang	10th - 13rd, January	Hoa
Total		1588				

Fig. 3 Vietnamese festivals' collected database

two languages. This implementation will make it easier to retrieve information and make use of the ontology's inference capabilities at the same time. Notably, all the festival's information will be stored in the properties "hasDescription." It directly supports the third step of our framework. Next, we introduce our Web application and experiments.

6 An Application Description and Experiments

6.1 Experiment and Results

From the proposed framework, we will provide the main four parts in this sub-section. They include collecting and preprocessing festival images, classifying the festival names, building ontology, and querying the festival information. All our dataset and implementations are published on GitHub.[1]

Collecting and Preprocessing Dataset We gather an image database including ten Vietnamese festivals known in the Mekong Delta provinces (provided in Column 3 of Fig. 3). For ease of presentation, let F be a festival and $\mathcal{F} = \{F_i\}$, $i \in [1, 10]$. Regarding the first step, the images are collected using Google's image database (Google Images) on the Internet. We used a short script to download images from the Internet applying the festivals' names as the query information. Next, a preprocessing step is performed. Several screening steps are carried out to eliminate errors in the collected data, including (1) removing images that do not belong to a specific festival, (2) eliminating images with content error, (3) standardizing images in general format.

And then, we collect 1588 photographs of different scenes (backgrounds) for the festivals. Notice that, regarding each festival, we collect the images in the entire three steps of the festival organization process as presented in the above background section. This collection ensures that the dataset can fully cover the festival process.

Festival Classification Regarding image classification, this assignment may be influenced by many factors such as image quality, background, brightness, and others. The state-of-the-art algorithms for this problem are support vector machine (SVM), random forest, and artificial neural networks [20].

Here, we compare the state-of-the-art algorithms showed in Fig. 4. Namely, we train the dataset with three CNN architectures, including InceptionV3 [30] (2048 features), MobileNet [14], (1024 features), LeNet, EfficientNetB0, and the machine learning algorithms. All our experiments are conducted on a machine Linux Ubuntu 18.04, Intel(R) Core i7-7700 CPU, 3.6 GHz, 8 cores, and 32 GB main memory and the GeForce RTX 280 Ti GPU. We split the collected dataset, including regarding each festival, 75% for training (1191 images in total) and 25% for testing (397

[1] https://github.com/researchteamkt/VietnameseFestivalApp.

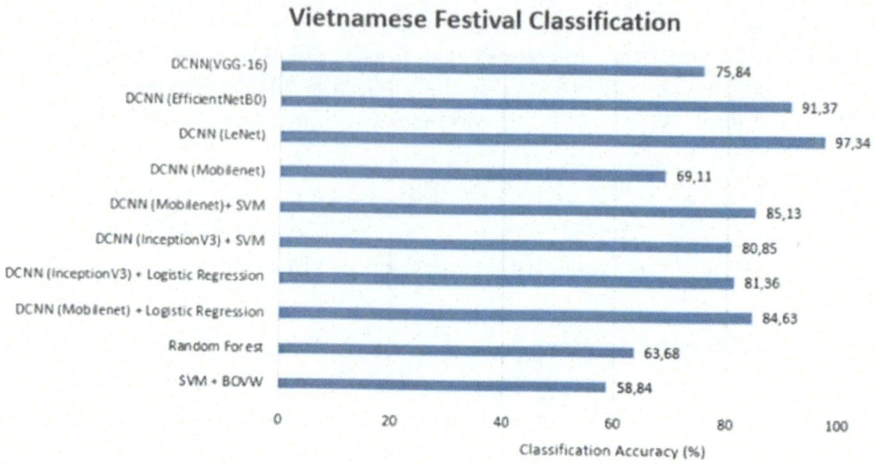

Fig. 4 Accuracy of machine learning algorithms

images in total). The algorithm's parameters of training are as follows: SVM (kernel = linear), logistic regression (random state = 9), random forest (*n*-estimators = 100). Moreover, we also reshape/resize the input image size as 32 × 32 such that the classification time is fast and efficient in real time.

From Fig. 4's result, this paper uses a deep convolutional neural network (deep, CNN) for solving the mentioned problem. An efficient and powerful library for deep learning models, Keras (written in Python), developed by Chollet [7], is utilized for training and classifying images of Vietnamese festivals under the deep convolutional neural network architecture. From the collected dataset, our DCNN model has reached an accuracy of 97.34% (Fig. 4) with the LeNet architecture. We tested the DCNN with the four epochs, including 25, 50, 125, and 150. Naturally, the accuracy will be increased when the epoch rises. We selected 150 epochs to build the classification model for the collected dataset.

Building an Ontology and Query the Information Next, we use a Protege tool[2] to construct the festival ontology. Moreover, in order to support the reasoning and querying the information, we implemented our approach by the owlready2 library[3] with the HermiT reasoner underlying the Python language. The query processing is carried out in two parts: (1) to classify the festival name and (2) to provide the festival descriptions for the chatbot. Regarding (1), after using the DCNN to classify the festival name, we select the top three accuracies for the result showed at (2) in Fig. 5a. And then, we rely on additional information, including the current time and location [(3) in Fig. 5] to classify the festival name. Since we take advantage of querying and answering the ontology to find the final result, the accuracy will increase. Indeed, we tested the result of this approach 20 times, and then, the accuracy

2 https://protege.stanford.edu/.

3 https://owlready2.readthedocs.io/en/latest/.

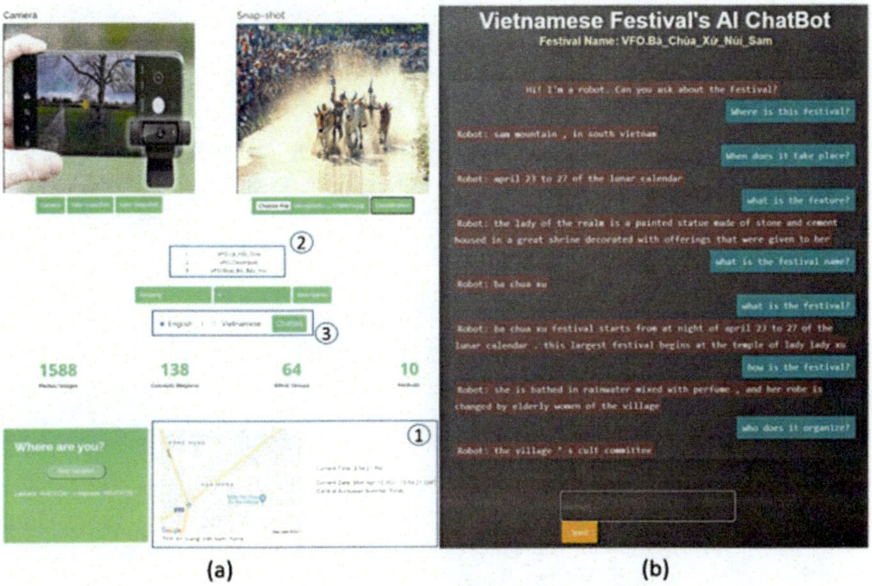

Fig. 5 Vietnamese festival application. **a** Classification and **b** Chatbot

is almost perfect (achieved 100%). For (2), when a festival name was determined, it will have a corresponding festival description. Here, our application supports two options, including reading the description and using the chatbot. For the first work, our festival ontology is constructed with 138 concepts and 655 axioms. It stores the festival information in two languages (Vietnamese and English).

A Web Application and Chatbot For the first work, we implement the application on the Web platform (using Flask[4]). The Web application is shown in Fig. 5.

Next, we also implemented an automatic chatbot application to allow the tourists to ask many relevant questions presented in Fig. 5b. We implemented the "transformers" library[5] to query in real time. We used the ALBERT model and the festival description to support answering the questions. Namely, we use the function in the "BertForQuestionAnswering" class. The query function's inputs are a festival description and a question of the user. The output will be the answers extracted from the description. The ALBERT model to work effectively requires that the description to be complete, concise, and cover most of the festival's information.

Note that, the chatbot supports two languages (English and Vietnamese). Hence, we implemented two options for this chatbot function showed at (3) in Fig. 5.

---

4 https://flask.palletsprojects.com/en/1.1.x/.

5 https://huggingface.co/transformers/.

Table 1 The festival common question

Id	Question
Q_1	Which place does the festival take place?
Q_2	When is the festival held?
Q_3	What is the feature of the festival
Q_4	What name is this festival?
Q_5	Which people community is the festival of?
Q_6	Who is the festival worshiping?
Q_7	How is the festival?
Q_8	What is the festival?

6.2 Evaluation

The result of the classification process is effective with high accuracy and a suitable time. Moreover, the accuracy is increased when utilizing the additional information and querying information in the ontology. The execution time average is 0.458 s for each classification (ten times). To evaluate our approach's efficiency, we create eight common questions Q_i, $i \in [1, 8]$ in English to discover more information. In other words, we evaluate information coverage of the ontology and the ALBERT's suitability for the system. Hence, we tested the accuracy of each festival with these questions. These questions and results of testing are shown in Tables 1 and 2 (1: correct, 0: fail).

Regarding the case of question Q_6, two results are failed because they do not have any information from the ontology's description to answer this. For Q_3 of the Festival F_6, the application still provides a result, but it does not cover the festival's information fully. The reason is that the description of this festival is not comprehensive and

Table 2 The chatbot system evaluation

Id	Q_1	Q_2	Q_3	Q_4	Q_5	Q_6	Q_7	Q_8	$\sum$
F_1	1	1	1	1	1	1	1	1	8
F_2	1	1	1	1	1	1	1	1	8
F_3	1	1	1	1	1	0	1	1	7
F_4	1	1	1	1	1	1	1	1	8
F_5	1	1	1	1	1	1	1	1	8
F_6	1	1	1	1	1	1	1	1	8
F_7	1	1	0	1	1	1	1	1	7
F_8	1	1	1	1	1	0	1	1	7
F_9	1	1	1	1	1	1	1	1	8
F_{10}	1	1	1	1	1	1	1	1	8
F_{10}	10	10	9	10	10	8	10	10	77

ambiguous overlapping yet. In general, the description information in the ontology is relatively complete. It can answer the most common questions. Namely, the result achieved 77 correct answers, i.e., 77 of 80 in total. The accuracy reaches 96:25%. Note that, we will not mention the evaluation of Vietnamese questions in detail because the ALBERT model is working well in English. The evaluation result in Vietnamese is 76:25% $(61 = 80)$. We also plan to improve our Vietnamese approach by a PhoBERT model trained in Vietnamese provided by Nguyen and Nguyen [23].

7 Conclusion

Our main contribution in this paper is a framework to preserve the intangible cultural festival in Vietnam. It directly puts attention to the tourism aspect to support visitors. Notably, we provided a Vietnamese festival ontology investigating in the demonstration at the Mekong Delta of Vietnam's south. We also collected a set of festival images and built a deep learning model to classify the images efficiently. Moreover, a tool of chatbot is provided to aid in query answering the tourist's questions. For this work, we implemented our approach in Python language on the Web platform.

Our next plan is to extend the investigation scope in all provinces/regions of Vietnam. We will develop an entire system with more information. We also plan to implement a multi-language system.

References

1. Afsahhosseini F, Al-Mulla Y (2020) Machine learning in tourism. In: ICMLMI'20
2. Alotaibi E (2020) Application of machine learning in the hotel industry: a critical review. J AAUTH
3. Barile S, Saviano M (2015) From the management of cultural heritage to the governance of the cultural heritage system. Springer, Cham, pp 71–103
4. Bulchand-Gidumal J (2020) Impact of artificial intelligence in travel, tourism, and hospitality. Springer, Cham, pp 1–20
5. Chang M, Yuan Y, Yue Q, Mincheol H (2020) A CNN image classification analysis for clean-coast detector as tourism service distribution. SST 18:15–26
6. Chen Y, Li H, Hua Y, Qi G (2020) Formal query building with query structure prediction for complex question answering over knowledge base. In: IJCAI-20, pp 3751–3758
7. Chollet F (2017) Deep learning with Python. Manning Publications Co., USA
8. Colace F, De Santo M, Greco L, Chianese A, Moscato V, Picariello A (2013) CHIS: cultural heritage information system. IJKSR 4:18–26
9. Daramola O, Adigun M, Ayo C (2009) Building an ontology-based framework for tourism recommendation services. In: ICA'09, pp 135–147
10. Devlin J, Chang MW, Lee K, Toutanova K (2019) BERT: pre-training of deep bidirectional transformers for language understanding. In: ACL'19, pp 4171–4186
11. Do PQ (2007) Traditional festivals in Viet Nam. World Publisher, p 253
12. Do TN, Pham TP, Pham NK, Huu Hoa N, Tabia K, Benferhat S (2020) Stacking of SVMs for classifying intangible cultural heritage images, pp 186–196
13. Heaton J (2021) Applications of deep neural networks

14. Howard AG, Zhu M, Chen B, Kalenichenko D, Wang W, Weyand T, Andreetto M, Adam H (2017) MobileNets: efficient CNNs for mobile vision applications
15. Jankovic R (2019) Machine learning models for cultural heritage image classification: comparison based on attribute selection. information (Switzerland) 11
16. Jin R, Dou Y, Wang Y, Niu X (2017) Confusion graph: detecting confusion communities in large scale image classification. In: IJCAI-17, pp 1980–1986
17. Kim D, Kang Y, Park Y, Kim N, Lee J (2019) Understanding tourists' urban images with geotagged photos using convolutional neural networks. In: SIR 28
18. Laddha S (2018) Indian tourism information retrieval system: an onto-semantic approach. Procedia Comput Sci 132C:1363–1374
19. Lan Z, Chen M, Goodman S, Gimpel K, Sharma P, Soricut R (2020) ALBERT: a Lite BERT for self-supervised learning of language representations. In: NCLR-20
20. Lu D, Weng Q (2007) A survey of image classification methods and techniques for improving classification performance. Int J Remote Sens 28(5):823–870
21. Ma TT, Benferhat S, Bouraoui Z, Tabia K, Do TN, Nguyen H (2018) An ontology-based modelling of Vietnamese traditional dances. In: SEKE
22. Ma TT, Benferhat S, Bouraoui Z, Tabia K, Do TN, Pham NK (2019) An automatic extraction tool for ethnic Vietnamese Thai dances concepts. In: ICMLA'19
23. Nguyen DQ, Nguyen AT (2020) PhoBERT: pre-trained language models for Vietnamese
24. Prantner K, Ding Y, Luger M, Yan Z, Herzog C (2007) Tourism ontology and semantic management system: state-of-the-arts analysis. IADIS'07
25. Sribunthankul P, Sureephong P, Tabia K, Ma TT (2019) Developing the evaluation system of the Thai dance training tool. ECTI DAMT-NCON'19, pp 163–167
26. Tanasijevic I, Pavlovic-Lazetic G (2020) HerCulB: content-based information extraction and retrieval for cultural heritage of the Balkans. The Electronic Library
27. Virmani C, Sinha S, Khatri SK (2017) Unified ontology for data integration for tourism sector. In: ICTUS'17. pp 152–156
28. Wang P, Wu Q, Shen C, Dick A, van den Hengel A (2017) Explicit knowledge-based reasoning for visual question answering. In: IJCAI-17, pp 1290–1296
29. Wilcock G (2018) Using a deep learning dialogue research toolkit in a multilingual multidomain practical application. In: IJCAI-18, pp 5880–5882
30. Xia X, Xu C, Nan B (2017) Inception-v3 for flower classification. In: ICIVC'17, pp 783–787
31. Xue W, Wang W (2020) One-shot image classification by learning to restore prototypes. AAAI'20, vol 34, pp 6558–6565
32. Zhang H, Wu H, Sun W, Zheng B (2018) DeepTravel: a neural network based travel time estimation model with auxiliary supervision. In: IJCAI-18, pp 3655–3661
33. Zhang L, Sun Z (2019) The application of artificial intelligence technology in the tourism industry of Jinan. J Phys: Conf Ser 1302:032005

Object-Oriented Software Testing: A Review

Ali Raza, Babar Shah, Madnia Ashraf, and Muhammad Ilyas

Abstract Object-oriented (OO) software systems present specific challenges to the testing teams. As the object-oriented software contains the OO methodology and its different components, it is hard for the testing teams to test the software with arbitrary software components and the chance of errors could be increased. So different techniques, models, and methods researchers identified to tackle these challenges. In this paper, we are going to analyze and study the OO software testing. For handling challenges in OO software testing, different techniques and methods are proposed like UML diagrams, evolutionary testing, genetic algorithms, black-box testing, and white-box testing. The methodology used for research is literature review (LR) of the recent decay.

Keywords Object-oriented testing · SDLC · Quality · UML · Design component

1 Introduction

The quality of software can be measured and determined by testing. The software testing may help deliver the top quality product and tell how effective its results are either it is good and effective for future use or not. By applying software testing techniques and methods, the drawbacks of the product should be removed and handled. For the faster testing phase, automated tools are used, which helps in proper generation and reporting the result which is going to be tested [1, 2]. In order to determine

A. Raza (✉) · M. Ashraf · M. Ilyas
Department of Computer Science and IT, University of Sargodha, Sargodha, Pakistan

M. Ilyas
e-mail: m.ilyas@uos.edu.pk

B. Shah
College of Technological Innovation, Zayed University, Abu Dhabi, UAE
e-mail: babar.shah@zu.ac.ae

M. Ashraf
Punjab Information Technology Board, Lahore, Pakistan

A. Ullah et al. (eds.), *Proceedings of International Conference on Information Technology and Applications*, Lecture Notes in Networks and Systems 350,
https://doi.org/10.1007/978-981-16-7618-5_40

the efficiency and effectiveness of end product working by oracle, testing is an important process. Object-oriented testing is usually carried out for the software which is developed using object-oriented methodologies. OO testing produces different issues because of prominent features of OO [3] paradigm. The time for testing an OO software was found to be increased and compared with testing the procedural software [4]. The OO concepts aim to make people more conscious of objects than operations or methods. The OO analysis and designs are faster for developing software [5]. In OO software, different mechanisms are considered and classified according to their uses. The inheritance mechanism should allow a class to be again used as a base for some other class, and then classes create different hierarchies [6]. Different testing techniques are considered to review an OO software testing, like class testing, boundary-value testing, white-box testing, black-box testing, stress testing, and user interface testing [7]. These techniques should play an essential role in the life cycle of OO software testing [3]. In OO software, different inheritance characteristics, polymorphous, and dynamic binding may affect the complexity of its testing [3]. The OO technology is much considered in the software development environment nowadays. More systems and languages are introduced and implied in the organizations, and many practices and work are done [4, 8]. The rest of this research paper consists of object-oriented software testing introduction, challenges and issue of testing, technique, method, cost challenges, and conclusion.

2 Object-Oriented Software Testing

In object-oriented software systems, most of their developments are based on analysis, design, and the coding phases [9]. The testing in any phase of development is essential, but in object-oriented, the testing could not speed up the progress. Many factors were involved which affected the software testing [10, 11]. Today, OO software has become more attractive in which the test methods differ from conventional software [12]. Testing the object-oriented software requires new theories and methods that are different from the already present procedures. The authors also suggest different tools and methods for OO testing [13, 14]. The metrics-based methods and studies are also helping and effectively used to determine the critical and changing-prone parts in OO software [15, 16].

2.1 *Challenges and Issues of Testing in Object-Oriented Software*

For testing the OO software, its requirements should be different from conventional programs as OO may enhance the reusability, reliability, and interoperability [17]. Traditional testing methods are not applicable in OO programs as they do not contain

OO concepts [18]. These methods can make it a challenge to test an OO software or program and can lead to issues like the encapsulation of class methods may cause barriers in testing as methods are invoked through the object of any class and state of an object at the time of invocation. Also, the traditional method has not found problems that occurred in inheritance and polymorphism [19]. In object-oriented software testing, it is mostly tough to detect faults after integrating classes because of polymorphism, inheritance, and dynamic binding [20].

Testing of object-oriented software should be done with different aspects and should be viewed in detail like the problems caused due to security, association, polymorphism, inheritance, and various problems [21, 22]. The OO model has a set of testing and maintenance problems. Also, the inheritance, aggregation, and association relationships between object classes made OO programs difficult for testing [22]. The main problem observed for object-oriented software testing is that the standard testing methodologies may not be helpful and practical as the programs are not executed sequentially and combined in an arbitrary order, so their test case definition should become more complex [23, 24].

2.2 Existing OO Software Testing Techniques and Approaches

The literature shows several surveyed testing techniques, but the testing techniques cover specific subsets for OO programs. At different levels of software, different testing techniques can be applied: unit, subsystem, class, integration, and system testing [25]. Most of the research on OO paradigm has focused on analysis, design, and programming fundamentals [12, 25]. Traditional techniques must be evaluated to determine if they are helpful concerning OO systems, and new techniques must be developed [23, 26]. Tonella [18] proposed a method for classes' evolutionary testing (ET) in the latest research. It is used for optimal test parameter search that having the combinations which will satisfy predefined test criteria. This criterion is defined using a cost function that measures how each of the automatically generated optimization parameters satisfies the given criteria. ET uses a kind of meta-heuristic search technique [27, 28]. ET is a subcategory of the search-based testing approach. It formulated the task of test generation as problem search and applied evolutionary algorithms (EA) like genetic algorithm (GA) or genetic programming (GP) [6, 29]. Similarly, a number of machine learning [30] and data mining techniques [31, 32] may have potential application in software testing.

2.3 *Object-Oriented Testing Methods*

Three basic methods are used for performing OO Testing: state-based testing, fault-based testing, and scenario-based testing. State-based testing is used to verify that class methods are interacting properly with one another [20]. FSM is used to represent the states of objects. Fault-based is used to determine a set of feasible faults. In comparison, scenario-based is used to detect errors caused by incorrect specifications [21, 33].

2.4 *Testing Based on Modeling*

Tools for automated testing are based on definite models of programs and algorithms [1, 34]. People must involve in the design and implementation of software with the increasing complexity of software programs [35]. The UML gives us a standard way to create a system's draft and all the processes and plans written in a specific programming language, schema, and components [22, 36]. The diagrams used in UML are use case, object, sequence, class, collaboration, state machine, activity, component, and deployment diagrams [37, 38].

2.5 *The Components of OO Testing Tools*

The following are the features for automated testing tools of OO programs: 1. import file feature, 2. export file feature, 3. GUI, 4. maintenance tool, 5. logging menu results, 6. path generation, 7. changing identifiers of classes, 8. diagrams displayer, 9. test tools for test order generation and test case generation. The quality of test cases produced by the random way is less than the quality of test cases produced by genetic algorithms because the algorithm directly addresses test cases to the desirable range fast [39, 40]. While in a computer system, the first interaction of a user with an operating system and secondly interact with the application software such as a word processor either using a command-line interface or graphical user interface [20].

2.6 *Cost Changes in Testing OO Software*

The testing cost may vary with changes or not be done correctly [41]. The goal test coverage is crisscrossing of each arm in a self-control graph, which requires many test cases, metrics, and paths [22, 28]. Nevertheless, formal elements are usually prepared to solve one problem so they would be able to get a restricted number of inputs and outputs. So, if somebody intends to unravel problems of the same type,

Table 1 Findings of literature

S. No.	Dissection of OO software testing	Categories of testing	Quality factor	Challenges	Techniques to perform	References
1	Unit testing	Functional, structure, and intuitive testing	Correctness, security	Time taking, good design review	Manual, automated	[9, 42, 41]
2	Subsystem testing	Regression, thread based, and use-based testing	Usability, portability	Test planning and scheduling issues, not enough automation, lack of right test tool, not handle enough to creep of requirement	manual	[9, 4, 17]
3	System testing	Alpha beta, acceptance	Reusability, maintainability	Communication issue	Automated, manual	[9, 5, 14, 41]

so many programs are required to write for the problem and it takes much time and the cost may increase according to it [19– 21] (Table 1).

3 Conclusion

This paper presents a review about object-oriented software testing and to handle the challenges with the proposed techniques, methods, and different modeling approaches. OO testing is complex as it contains arbitrary components, and it is harder for testing teams than traditional models containing sequential components. Different UML diagrams representation may help test with testing tools efficiently to test OO software components. Different features in automated tools may help the OO programs testing.

References

1. Fredericks EM, Hariri RH (2016) Extending search-based software testing techniques to big data applications. In 2016 IEEE/ACM 9th international workshop on search-based software testing (SBST). IEEE, pp 41–42
2. Adnan SM, Ilyas M, Razzaq S, Maqbool F, Wakeel M, Adnan SM (2020) Code smell detection and refactoring using AST visitor. Tech J 25(01):59–65

3. Jain A, Patwa S (2016) Effect of analysis and design phase factors on testing of Object oriented software. In: 2016 3rd International conference on computing for sustainable global development (INDIACom). IEEE, pp 3695–3700
4. He W, Zhao R, Zhu Q (2015) Integrating evolutionary testing with reinforcement learning for automated test generation of object-oriented software. Chin J Electron 24(1):38–45
5. Wu CS, Huang CH, Lee YT (2013) The test path generation from state-based polymorphic interaction graph for object-oriented software. In 2013 10th International conference on information technology: new generations. IEEE, pp 323–330
6. Singh NP, Mishra R, Debbarma MK, Sachan S (2011) The review: lifecycle of object-oriented software testing. In: 2011 3rd international conference on electronics computer technology, vol 3. IEEE, pp 52–56
7. Labiche Y (2011) Integration testing object-oriented software systems: An experiment-driven research approach. In: 2011 24th Canadian conference on electrical and computer engineering (CCECE). IEEE, pp 000652–000655
8. Augsornsri P, Suwannasart T (2013). Design of a tool for checking integration testing coverage of object-oriented software. In: 2013 International conference on information science and applications (ICISA). IEEE pp 1–4
9. Chen HY, Tse TH (2013) Equality to equals and unequals: a revisit of the equivalence and nonequivalence criteria in class-level testing of object-oriented software. IEEE Trans Software Eng 39(11):1549–1563
10. Panigrahi CR, Mall R (2014) A heuristic-based regression test case prioritization approach for object-oriented programs. Innov Syst Softw Eng 10(3):155–163
11. Eski S, Buzluca F (2011). An empirical study on object-oriented metrics and software evolution in order to reduce testing costs by predicting change prone classes. In 2011 IEEE fourth international conference on software testing, verification and validation workshops. IEEE, pp 566–571
12. Suri PR, Singhani H (2015) Object Oriented Software Testability (OOSTe) Metrics Analysis. Int J Comput Appl Technol Res 4(5):359–367
13. Lee D, Lee J, Choi W, Lee BS, Han C (1997) A new integrated software development environment based on SDL, MSC, and CHILL for large-scale switching systems. ETRI J 18(4):265–286
14. Dubey SK, Rana A (2011) Assessment of maintainability metrics for object oriented software system. ACM SIGSOFT Softw Eng Notes 36(5):1–7
15. Sneed HM, Verhoef C, Sneed SH (2013) Reusing existing object-oriented code as web services in a SOA. In: 2013 IEEE 7th international symposium on the maintenance and evolution of service-oriented and cloud-based systems. IEEE, pp 31–39
16. Chen HY, Tse TH, Chen TY (2001) TACCLE: a methodology for object-oriented software testing at the class and cluster levels. ACM Trans Softw Eng Methodol (TOSEM) 10(1):56–109
17. Kung D, Gao J, Hsia P, Toyoshima Y, Chen C, Kim YS, Song YK (1995) Developing an object-oriented software testing and maintenance environment. Commun ACM 38(10):75–87
18. Binder RV (1996) Testing object-oriented software: a survey. Softw Testing Verification Reliab 6(3–4):125–252
19. Barbey S, Strohmeier A (1970) The problematics of testing object-oriented software. WIT Trans Inf Commun Technol 9
20. Whittaker JA (2000) What is software testing? And why is it so hard? IEEE Softw 17(1):70–79
21. McGregor JD, Sykes DA (2001) A practical guide to testing object-oriented software. Addison-Wesley Professional
22. Runeson P (2006) A survey of unit testing practices. IEEE Softw 23(4):22–29
23. Onita C, Dhaliwal J (2011) Alignment within the corporate IT unit: an analysis of software testing and development. Eur J Inf Syst 20(1):48–68
24. Naik K, Tripathy P (2011) Software testing and quality assurance: theory and practice. Wiley
25. Wappler S, Lammermann F (2005) Using evolutionary algorithms for the unit testing of object-oriented software. In Proceedings of the 7th annual conference on Genetic and evolutionary computation, pp 1053–1060

26. Kuhn DR, Wallace DR, Gallo AM (2004) Software fault interactions and implications for software testing. IEEE Trans Softw Eng 30(6):418–421
27. Jorgensen PC (2013) Software testing: a craftsman's approach. Auerbach Publications
28. Bertolino A (2007) Software testing research: Achievements, challenges, dreams. In: Future of software engineering (FOSE'07). IEEE, pp 85–103
29. Gelperin D, Hetzel B (1988) The growth of software testing. Commun ACM 31(6):687–695
30. Al-Obeidat F, Rocha A, Khan MS, Maqbool F, Razzaq S (2021) Parallel' tensor factorization for relational learning. Neural Comput Appl 1–10
31. Razzaq S, Maqbool F, Hussain A (2016) Modified cat swarm optimization for clustering. In: International conference on brain inspired cognitive systems. Springer, Cham, pp 161–170
32. Maqbool F, Bashir S, Baig AR (2006) E-MAP: efficiently mining asynchronous periodic patterns. Int J Comput Sci Netw Secur 6(8A):174–179
33. Zhang X, Yu L, Hou X (2016) A method of metamorphic relations constructing for object-oriented software testing. In 2016 17th IEEE/ACIS international conference on software engineering, artificial intelligence, networking and parallel/distributed computing (SNPD). IEEE, pp 399–406
34. Drave I, Hillemacher S, Greifenberg T, Kriebel S, Kusmenko E, Markthaler M, Orth P, Salman KS, Richenhagen J, Rumpe B, Schulze C, Wortmann A (2019) SMArDT modeling for automotive software testing. Softw: Pract Exp 49(2):301–328
35. Gao K (2021) Simulated software testing process and its optimization considering heterogeneous debuggers and release time. IEEE Access 9:38649–38659
36. Böhme M (2019) Assurances in software testing: a roadmap. In: 2019 IEEE/ACM 41st international conference on software engineering: new ideas and emerging results (ICSE-NIER). IEEE, pp 5–8
37. Alferidah SK, Ahmed S (2020). Automated software testing tools. In: 2020 international conference on computing and information technology (ICCIT-1441). IEEE, pp 1–4
38. Alyahya S, Alsayyari M (2020) Towards better crowd sourced software testing process. Int J Cooper Inform Syst 29(01–02):2040009
39. Dadkhah M, Araban S, Paydar S (2020) A systematic literature review on semantic web enabled software testing. J Syst Softw 162:110485
40. Durelli VH, Durelli RS, Borges SS, Endo AT, Eler MM, Dias DR, Guimarães, MP (2019) Machine learning applied to software testing: A systematic mapping study. IEEE Trans Reliab 68(3):1189–1212
41. Ciupa I, Leitner A, Oriol M, Meyer B (2007) Experimental assessment of random testing for object-oriented software. In Proceedings of the 2007 international symposium on Software testing and analysis, pp 84–94
42. Wappler S, Wegener J (2006) Evolutionary unit testing of object oriented software using strongly-typed genetic programming. In Proceedings of the 8th annual conference on Genetic and evolutionary computation, pp 1925–1932

Requirements Engineering Education: A Systematic Literature Review

Shahzeb Javed, Khubaib Amjad Alam, Sahar Ajmal, and Umer Iqbal

Abstract Requirements engineering is an essential activity of software development lifecycle (SDLC) and its importance has enormously grown in recent years due to exponential growth of software industry. In this context, requirements engineering education has become largely indispensable for computing graduates and trainees in software industry. Several studies have explored different dimensions of requirements engineering education and relevant pedagogical techniques over the past few years. This systematic review aims at identifying and highlighting the extant research on REE that is available and to select useful approaches that can help in expediting the growth in this field. The systematic literature review is performed to classify existing literature related to REE into four categories: method/techniques, tools, comparison studies, and frameworks used in REE. After a rigorous evaluation process, 32 primary studies were selected and classified according to these categories. Results of this SLR indicate a significant shift toward this dimension. We have highlighted several research trends and gaps that need to be addressed by the RE community.

Keywords RE · Requirements engineering education (REE) · SLR · SDLC · Software process

S. Javed · S. Ajmal · U. Iqbal
Riphah International University, Faisalabad Campus, Rawalpindi, Pakistan
e-mail: shahzeb.javed@riphahfsd.edu.pk

S. Ajmal
e-mail: sahar.ajmal@riphahfsd.edu.pk

U. Iqbal
e-mail: umer.pervaiz@riphahfsd.edu.pk

K. A. Alam (✉)
Department of Computer Science, National University of Computer and Emerging Science (NUCES-FAST), Islamabad, Pakistan
e-mail: khubaib.amjad@nu.edu.pk

A. Ullah et al. (eds.), *Proceedings of International Conference on Information Technology and Applications*, Lecture Notes in Networks and Systems 350,
https://doi.org/10.1007/978-981-16-7618-5_41

1 Introduction

Requirements engineering (RE) is one of the most prominent software process activities that identifies, analyzes, specifies, and validates the required software system functionality (functional RE) and constraints (non-functional RE, e.g., constraints on quality and costs) on software systems [1, 2]. RE is concerned with the relationships that these factors have detailed descriptions of software behavior and their progression over time along with software families. It is a multidisciplinary movement that sets up a variety of tools and techniques at different levels of development along with different type of application domains according to their interests. Different researchers and people identify the failures of software and applications are embedded in requirement activities. Moreover, it is identified that the main cause of the failure is a lack of skill and knowledge of people or a team involved in RE activities [3]. Therefore, the correct knowledge can improve the software industry and educational institutes if properly taught at universities. Thus, it became a challenge to provide proper knowledge about RE to employees and students in firms along with educational institutes, respectively. Education in RE should be ideally be provided as an integral part of skill development and software engineering technical skills before entering into the market. It is important to integrate RE into the curriculum of universities for undergraduate and postgraduate students [4]. There are large number of studies carried out in the field of REE. In this paper, complete systematic literature review is provided. In addition, a formal protocol is applied to perform SLR. This process initially includes 338 papers which were further narrow down to 11 papers, thus hereby using proper filtration process. The main contribution of this research work is to identify the methods, tools, and techniques used to educate students in the field of RE and significance of RE education for successful completion of a software project.

2 Related Work

The primary focus of SLR is to interpret, assess, and identify the existing researches related to the explicit field. A systematic literature review ought to complete with an appropriate pursuit plan which needs to be unbiased. This method must guarantee the outcome of the search for the assessment process. There is not a proper way to deal with correctness and reliability of existing research. To extract quality research in REE, a proper assessment is performed. This SLR is to fill the gap by conducting a survey by following the guidelines defined by Kitchenham and Charters [5]. Our SLR process comprises of various steps. A review protocol is developed initially followed by the conduction of the systematic review. This is done by the extraction of impartial outcomes. The procedure is completed by announcing the outcomes and examining the results.

Table 1 Electronic databases

Identifier	Database	URL
ED-1	ACM	http://dl.acm.org/
ED-2	ScienceDirect	http://sciencedirect.com
ED-3	IEEE Xplore	http://ieeexplore.ieee.org
ED-4	Springer	http://link.springer.com
ED-5	Google Scholar	http://scholar.google.com

Table 2 Criteria for studies inclusion

Identifier	Inclusion criteria
IC-1	Articles that contain threats to validity
IC-2	Articles from peer-reviewed publication venues
IC-3	Articles focused on requirement engineering education
IC-4	The inclusion of articles from 2013 to 2018

2.1 Research Questions

The primary research question of this systematic literature review is "How to educate requirements engineering and which tools and techniques are proposed?" We divide this primary question into three research questions. These three main questions have been discussed in this paper.

RQ1 What are the approaches that were reported in REE?
RQ2 What is the research contribution in the field of REE?
RQ3 What is the overall productivity in this research domain?

2.2 Electronic Sources

Five electronic databases have been used for searching process of articles. The details of these electronic databases are mentioned in Table 1.

The inclusion and exclusion criteria based on which the studies are kept or excluded are briefly given in Tables 2 and 3, respectively.

2.3 Search Terms

To search for articles in each electronic database, the search string is used. The search string comprises of various search terms because of many correlated words which can be utilized in articles that are that can be used to find our primary studies. The query

Table 3 Criteria for studies exclusion

Identifier	Exclusion criteria
EC-1	Articles that are not in the English language
EC-2	Articles without any validation
EC-3	Articles providing general focus on software engineering
EC-4	Articles not focusing on education
EC-5	Editorial, short papers, posters, extended abstracts, blogs information, or Wikipedia data will not be included

used to find articles from electronic databases is *(Requirement* AND Engineering) AND (education OR teach* OR pedagog*).*

2.4 Selection Process

This study selection process of this SLR consists of three phases, each phase of this study selection was observed properly to ensure complete unbiased and to make sure that proper steps are followed in each phase of selection process. Hence, this guarantees that our primary studies are not biased. While getting results from google scholar, various similar articles may be skipped because maximum matching of relevant keywords cannot be provided. To handle this issue, we had used five databases as defined in Table 1. Therefore, a proper flow had been followed to include results in the studies.

2.5 Inclusion and Exclusion Criteria

The criteria (inclusion and exclusion) which had been characterized for the determination of essential investigations are referenced in Tables 2 and 3. The motivation behind these criteria is to omit such articles that have no relevance to the topic under consideration. Through these criteria, the important studies will be identified. These criteria were connected upon the second and third part of our study choice process. In second phase, the studies are extracted on the basis of title, further it is filtered on the basis of abstract. Lastly, it is filtered on the basis of inclusion and exclusion criteria along with quality assessment criteria. These two criteria are connected parallel on the full content of every one of the 28 articles. After the utilization of the two criteria, a sum of 32 articles was chosen for the last rundown of study.

The studies that confirm the inclusion/exclusion criteria are analyzed against the quality assessment criteria that are defined in Table 4.

Table 4 Quality assessment criteria

Identifier	Quality assessment criteria
QC-1	Primary studies must contain validation of the tool/techniques
QC-2	Goals and objectives of primary studies must be clearly defined
QC-3	Research in the paper must assist the aims of RE education
QC-4	Primary studies should define how to use approaches for the education of requirement engineering
QC-5	Limitations of primary studies must be clearly defined

Table 5 Quality level of studies

Quality level	Number of studies	Percentage (%)
Very high (score = 4.0–5.0)	16	47.00
High (score = 3.5–4.0)	17	49.87
Medium (score = 3)	2	3.125

2.6 *Quality Assessment Criteria*

The process of value evaluation is performed as the last step of our study determination process. To remove partialities, each of the 43 articles was evaluated by the given quality assessment criteria. This model contains five points so size of one to three had been characterized to demonstrate to satisfy the quality evaluation criteria. The quality levels of all the studies are described in Table 5 below according to the quality criteria mentioned above. Each quality criteria are divided into three levels with corresponding scores (fully = 1, partially = 0.5, and none = 0.0).

The sum of the scores of each study is calculated and mentioned above in Table 5. Most of the studies belong to very high and high-quality level, and there is only one study falls in medium group and as mentioned in inclusion criteria only medium and high-level studies are included.

After quality assessment, 35 articles were chosen which were further selected on the basis of inclusion and exclusion criteria. The resultant excluded articles are eight in number. The complete process of filtration is shown in Fig. 1.

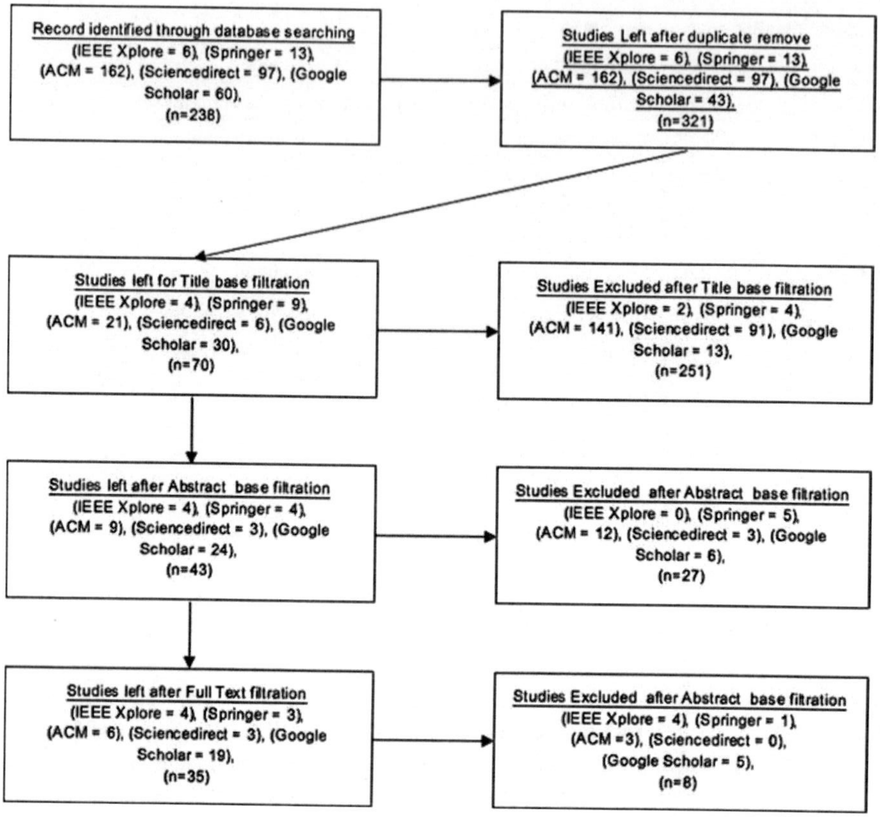

Fig. 1 Study selection procedure

3 Discussion and Results

In this section, we will discuss and answer the research questions that have been defined.

RQ1: What Are the Approaches that Were Reported in REE?

The main goal of this research question was to identify the most commonly used research approaches used in the context of REE. Table 6 gives a brief overview of the approaches with the number of studies in which the method was used and the referenced. The most used approach is role playing which is used by two studies [6, 7]. Then, there are multiple approaches and models like motivational modeling [8, 9] and requirement modeling tool [10, 11]. Competence-oriented didactics (COD) [12, 13] that were used by two studies, others are only used once by different articles.

Table 6 Approaches used in REE

Approaches	Number of studies	References
Role playing	2	[6, 7]
Motivational modeling	2	[8, 9]
Requirement modeling tool	2	[10, 11]
Competence-oriented didactics (COD)	2	[12, 13]
Rebok	2	[14, 15]
Scoring rubrics	1	[16]
Workshop videos	1	[17]
Agile	1	[18]
IOT aware model	1	[19]
Direction framework	1	[20]
KOAS model	1	[6]
Soft system methodology SSM	1	[21]
Pedagogical strategy	1	[22]
Formal methods	1	[23]
Sound didactical	1	[24]
Schematic framework for KMS	1	[25]
Multilevel assignment	1	[26]
Others	3	[26–28]

RQ2: What Are the Research Contributions in the Field of REE?
This research question was designed to find out the contribution type of studies that were finalized. The trend that can be visualized in Fig. 2 clearly shows that most of the papers provided technique/method that covers 45% of the whole pie chart. After that, 21% of the papers gave comparisons of different techniques. Papers that contribute to providing a model is 16% and tools are given by 10% studies. A few considering 5% articles are proposed frameworks.

Another finding against this research question is to identify the top publication venues which can be seen in Table 7. Maximum number of studies are published in REET, followed by requirement engineering and JTECE.

RQ3: What is the Overall Productivity in This Research Domain?
The overall research in the field of REE can be overviewed in this question. Figure 3 shows the publication channels, and it can be seen that most studies are published in conferences and a few in journals [16, 19].

Figure 3 shows there are about 28 articles are published in conferences, while seven articles are in journal.

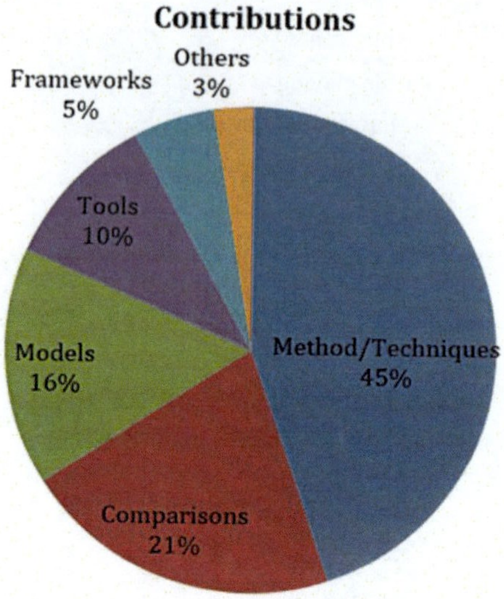

Fig. 2 Research contribution

Table 7 Top three publication venues

Publication venues	Number of studies	References
International workshop on requirements Engineering education and training (EET)	4	[8, 11, 18, 23]
Requirements engineering	2	[7, 17]
Journal of telecommunication, electronic, and computer engineering	2	[16, 19]

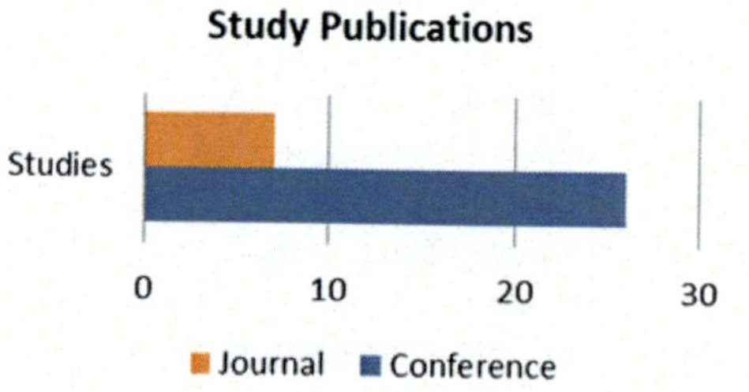

Fig. 3 Publication channel

The number of studies published in different years can be viewed in Fig. 4. Most papers published in the year 2017. However, we can see a decreasing trend in the number of publications from the year 2014–2016 and then a sudden increase in 2017 after that fall in 2018–2021.

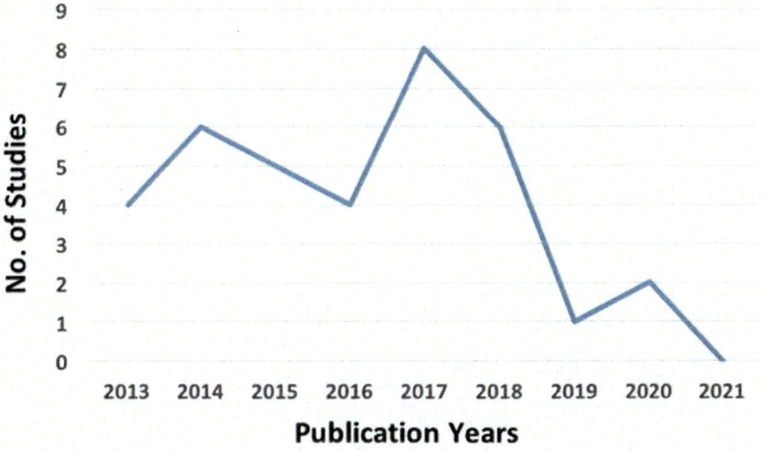

Fig. 4 Paper publication from 2013 to 2021

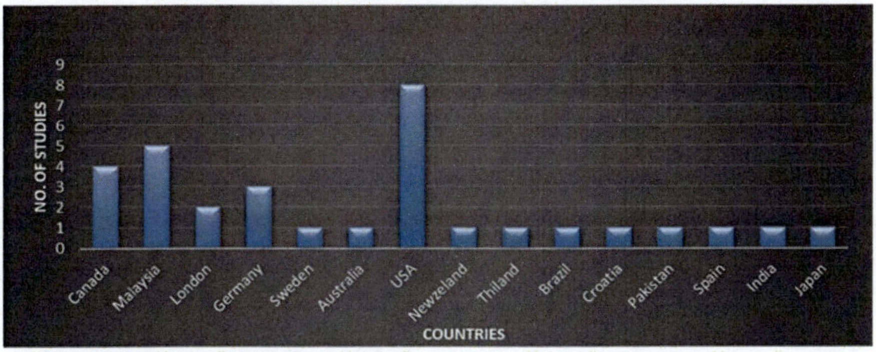

Fig. 5 Geographical distribution of papers

Figure 5 shows the most active countries in REE research, and the numbers represent that the USA is at the top with most papers in this field. Then comes Malaysia followed by Canada. Germany also showed a reasonable number of papers in REE.

4 Threats to Validity

This section highlights the limitation of the research work that we had performed. As this is a systematic literature review and it had done in a procedural way with proper proof and authentications but the databases used for the extraction of results are limited. We had targeted only five databases (IEEE, ACM, ScienceDirect, Springer, and Google Scholar) but many other reputed repositories such as Web of Science

(WoS) had not been included in our SLR due to non-availability of facility in our region. Although these repositories may contain very relevant material because of including limited databases, our results may have some restrictions.

5 Future Work

As we had already highlighted the issue of using limited databases in the previous section so to resolve this issue of biasness from the results, we will try to include some other digital libraries and electronic databases such as Web of Science (WoS), Scopus, Wiley, and Taylor and Francis.

6 Conclusion

Requirements engineering education has become an indispensable and essential course for computing students in higher education institutions as it is considered as the most crucial part of software development. This paper presents the research trends along with recent advancement in REE and active countries in the field of REE. Research techniques and models are described and presented, but a limited paper is published on these models. Requirement engineering tools and methods are described and their drawbacks are also discussed. Furthermore, the impacts of REE are shown. Research trends in recent years are described and also from recent two years, there are significant number of studies are published, while in 2019, there is no studies are published yet.

References

1. Ouhbi S, Idri A, Fernández-Alemán JL, Toval A (2015) Requirements engineering education: a systematic mapping study. Requir Eng 20(2):119–138
2. Nuseibeh B, Easterbrook S (2000) Requirements engineering: a roadmap. In: Proceedings of the conference on the future of software engineering, pp 35–46
3. Memon RN, Ahmad R, Salim SS (2010) Problems in requirements engineering education: a survey. In: Proceedings of the 8th international conference on frontiers of information technology, pp 1–6
4. Regev G, Gause DC, Wegmann A (2009) Experiential learning approach for requirements engineering education. Requir Eng 14(4):269
5. Kitchenham B, Charters S (2007) Guidelines for performing systematic literature reviews in software engineering. EBSE technical report EBSE
6. Nakamura T, Kai U, Tachikawa Y (2014) Requirements engineering education using expert system and role-play training. In: 2014 IEEE international conference on teaching, assessment and learning for engineering (TALE), pp 375–382

7. Sedelmaier Y, Landes D (2018) Systematic evolution of a learning setting for requirements engineering education based on competence-oriented didactics. In: 2018 IEEE global engineering education conference (EDUCON), pp 1062–1070
8. Lorca AL, Burrows R, Sterling L (2018) Teaching motivational models in agile requirements engineering. In: 2018 IEEE 8th international workshop on requirements engineering education and training (REET), pp 30–39
9. Garg K, Varma V (2015) Systemic requirements of a software engineering learning environment. In: Proceedings of the 8th India software engineering conference, pp 147–155
10. Subhiyakto ER, Utomo DW (2017) RMTool; Sebuah Aplikasi Pemodelan Persyaratan Perangkat Lunak menggunakan UML. J Nas. Tek. Elektro dan Teknol. Inf. 6(3):268–274
11. Morales-Ramirez I, Alva-Martinez LH (2018) Requirements analysis skills: how to train practitioners? In: 2018 IEEE 8th international workshop on requirements engineering education and training (REET), pp 24–29
12. Svensson RB, Regnell B (2017) Is role playing in requirements engineering education increasing learning outcome? Requir Eng 22(4):475–489
13. Schlingensiepen J (2014) Competence driven methodology for curriculum development based on requirement engineering. Procedia Soc Behav Sci 141:1203–1207
14. Kakeshita T, Yamashita S (2015) A requirement management education support tool for requirement elicitation process of REBOK. In: 2015 3rd International conference on applied computing and information technology/2nd International conference on computational science and intelligence, pp 40–45
15. Penzenstadler B, Fernández DM, Richardson D, Callele D, Wnuk K (2013) The requirements engineering body of knowledge (rebok). In: 2013 21st IEEE international requirements engineering conference (RE), pp 377–379
16. Mkpojiogu EOC, Hussain A (2017) Can scoring rubrics be used in assessing the performance of students in software requirements engineering education? J Telecommun Electron Comput Eng 9(2–11):115–119
17. Fricker SA, Schneider K, Fotrousi F, Thuemmler C (2016) Workshop videos for requirements communication. Requir Eng 21(4):521–552
18. Horkoff J (2018) The influence of agile methods on requirements engineering courses. In: 2018 IEEE 8th international workshop on requirements engineering education and training (REET), pp 11–19
19. Lim T-Y, Chan G-Y (2017) Teaching and learning software requirements engineering: our experience, reflection and improvement. J Telecommun Electron Comput Eng 9(3–4):51–55
20. Memon RN, Ahmad R, Salim SS (2013) A Direction framework to address problems in requirements engineering education. Malaysian J Comput Sci 26(4):294–311
21. Memon RN, Nizamani SZ, Memon F, Memon I, Kumar P (2016) A problem analysis method based on soft system methodology in requirements engineering process. Quaid-e-Awam Univ Res J Eng 15(1)
22. Portugal RLQ, Engiel P, Pivatelli J, do Prado Leite JCS (2016) Facing the challenges of teaching requirements engineering. In: 2016 IEEE/ACM 38th international conference on software engineering companion (ICSE-C), pp 461–470
23. Westphal B (2018) An undergraduate requirements engineering curriculum with formal methods. In: 2018 IEEE 8th international workshop on requirements engineering education and training (REET), pp 1–10
24. Sedelmaier Y, Landes D (2017) Experiences in teaching and learning requirements engineering on a sound didactical basis. In: Proceedings of the 2017 ACM conference on innovation and technology in computer science education, pp 116–121
25. Hassan HC (2013) A framework for user requirement assessment in technical education facility planning: a knowledge engineering approach. Procedia-Soc Behav Sci 107:104–111
26. Köppe C, Pruijt L (2014) Improving students' learning in software engineering education through multi-level assignments. In: Proceedings of the computer science education research conference, pp 57–62

27. Garcia I, Pacheco C, Leon A, Calvo-Manzano JA (2020) A serious game for teaching the fundamentals of ISO/IEC/IEEE 29148 systems and software engineering—Lifecycle processes—Requirements engineering at undergraduate level. Comput Stand Interfaces 67:103377
28. Epifânio JC, Miranda É, Trindade G, Lucena M, Silva L (2019). A qualitative study of teaching requirements engineering in universities. In: Proceedings of the XXXIII Brazilian symposium on software engineering, pp 161–165
29. Daun M, Tenbergen B (2020) Teaching requirements engineering with industry case examples. In SEUH, pp 49–50

Knowledge Models for Software Testing in Robotics

Cristina Nicoleta Turcanu

Abstract Robots are very complex systems requiring intensive testing before being released as final products on the market. Moreover, due to the fast pace of robotics technological advancements, it would be difficult to keep track of all the knowledge behind without the implementation of a proper knowledge management system. This paper presents a methodology for the usability of knowledge models in robot's software testing. A knowledge model was implemented and utilized using Celonis tool, and some conclusions were drawn from the experimentation based on a robot's simulation within Webots.

Keywords Knowledge management (KM) · Knowledge models · Software quality assurance (QA) · Mutation testing · Robots · Webots · iRobot · YAML · Celonis

1 Introduction

Knowledge management has gained popularity in various industries in the past two decades, enabling business processes to add value and to be competitive. The current meaning of knowledge is information effective in action, information focused on results [1], reflected in technologies advancements, and converted into a discipline. In the traditional approach, a knowledge management system is usually considered as a container to be filled with knowledge extracted from experts, wherein the knowledge modeling technology allows the decision support system to help an operator in detecting and diagnosing problems in a dynamic system [2]. Knowledge engineering methods and tools provide disciplined approaches for storing and accessing the knowledge, and a significant number of companies use this valuable entity to carry out tasks and create new knowledge.

Knowledge management applied in software testing supports all the needed functions of a QA work, from knowledge creation about how the system should perform in each case, to knowledge sharing and updating among the team. In other words, there is a need to adopt KM to the software testing core processes and to obtain the

C. N. Turcanu (✉)
University of Pitesti, Pitești, Romania

A. Ullah et al. (eds.), *Proceedings of International Conference on Information Technology and Applications*, Lecture Notes in Networks and Systems 350,
https://doi.org/10.1007/978-981-16-7618-5_42

benefits that it provides in terms of cost, quality, and efficiency [3]. Testing is a very crucial phase in any software development, and mutation has been ranked as one of the most effective testing techniques. Mutation testing makes use of the purposeful fault injection techniques. In other words, as explained in [4], if the mutant generates a different result than expected, this suggests that the program contains a syntactic error that needs to be corrected. In this case, the test suite is said to be efficient and the mutant is referred to as a killed mutant. Otherwise, if the test suite has not been able to identify the presence of an error, then it is inadequate and the mutant is said to be alive. [5]. The novelty of the current paper, comparing to other studies, is that the focus is not on the mutation coverage criteria themselves, but on the role of knowledge models in mutation testing.

Before applying the proposed methodology on real autonomous systems, it is very useful to use a simulation environment, as simulators are faster, less expensive, and more convenient to use. One of the most popular simulators among universities and research centers worldwide since 1998 is Webots [6]. Developed by Cyberbotics Ltd, Webots is an open-source and multi-platform desktop application [7] that were utilized for this paper purpose to generate the robot's mutation-based event logs. On the other side, the knowledge models were developed using another powerful commercial tool, Celonis execution management system (EMS), which also provides a free software academic license [8]. For the implementation of the knowledge model in the experimental work, the YAML data serialization standard language [9] was used being already integrated with Celonis.

The work is concentrated in several phases: (i) Modify the Webots robot's code to generate event logs from the simulated performance, (ii) generate the data model of this event log using Celonis tool, (iii) create in the Celonis EMS studio, the knowledge model of the robot, (iv) add some KPIs definitions in the knowledge model, and (v) use the obtained knowledge model for the mutation testing.

In this section, some aspects related to simulation in robotics (subsection A), knowledge management in software testing (subsection B), and knowledge models in Celonis studio (subsection C) are presented. Also, some research papers in the direction of knowledge management applied in robotics are discussed in the related work (subsection D).

The rest of the paper is organized as follows: Section 2 contains an overview of the methodology, Sect. 3 describes the experimental work, and Sect. 4 concludes the paper and discusses some future work.

A. Simulation in Robotics

Different fields of robotics need simulation. Simulators are written in numerous programming languages like Python, C, C++, Java, C#, and other languages that decide their platform compatibility [10], however independently of the platform, there is of high importance to test the robot's behavior considering every possible environment change. Simulators like Cyberbotics's Webots, Energid's actin, and Laminar Research's X-plane are few examples of widespread commercial tools. For this paper's purpose, Webots was chosen due to being a popular platform whose robots' controllers can be transferred to real robots like Aibo, Lego Mindstorms, Khepera, Koala, Hermission, Boe-Bot, e-puck, and others [10].

B. Knowledge Management in Software Testing

As concluded in [11], the trend is toward an increasing interest in addressing knowledge management in software testing. Most of the papers mentioned in this study highlights aspects related to providing automated support for managing software testing knowledge by means of a knowledge management system.

As mutation testing is considered, one of the most effective software testing techniques, this paper's aim is to apply it on robots, comes from the acknowledgment that this kind of testing provides valuable guidance toward improving the test suites of a safety–critical industrial software system [12].

C. Knowledge Models in Celonis Studio

Important business knowledge entities such as records, key performance indicators (KPIs), variables, and filters can be described using some knowledge representation language or data structure that enables the knowledge to be interpreted by both humans and software. Knowledge models in Celonis [8] are configured in YAML language and can be thought as a dictionary of important knowledge management entities. Knowledge models provide a way to use stored data models in a standardized way. A data model is created in Celonis based on the footprints of the historical or real-time analyzed process. A data model can be linked to several knowledge models for enabling knowledge definition, clarity, and consistency and to allow app analysts, engineers, and managers to tap into a wide range of existing content developed from working with their peers. KPIs value definitions are some of the most important entities that can be defined in the knowledge models using Celonis tool. A KPI is specified by an appropriate process query language (PQL) statement in the knowledge model. Besides its calculation, a knowledge model can also store particular information on targets (decreasing or increasing) that are applicable to a certain KPI.

D. Related Work—Knowledge Management in Robotics Industry

This paper [13] presents a practical solution to the knowledge management and communication problems stemming from the robotics issues, describing the implementation of a knowledge architecture of iRobot Roomba cleaning robots. The solution is inspired by semantic Web principles, spanning over several layers: RFID tags on objects, process interaction in a single robot via a main memory datastore and a rule system, central database for a swarm.

Another study, part of a project aiming to implement a knowledge management for a robot capable of being adaptable to any process technology used in a company in the automotive industry [14]. The aim is to use a programmable robot, able of performing repetitive tasks. The conclusion given is related to how enterprises are willing to cope with the customer requirements, for which purpose they need to make continuous improvements on their products, and this is possible only using an adequate knowledge management.

A relevant study regarding ontology-based knowledge management in robotics is [15] where the development of a system with specific modules to manage robots' knowledge and reasoning, command analysis, decision-making, and talking interaction is presented.

2 Methodology

This section describes the methodology of applying knowledge models for mutation testing of a robot's simulated behavior. This process starts by generating the event logs containing the footprints of each of the robot's movements within a simulator.

The following block diagram describes the work organization (Fig. 1).

The methodology starts from defining states for each robot's movement type (e.g., forward, backwards, etc.). Each movement itself represents an activity and an arbitrary number of activities are part of a case. One case should include the following information: (i) case key, (ii) datetime, (iii) state, and (iv) sorting (i.e., activity number). When generating an arbitrary number of cases from the robot's simulation, these are stored into event logs which are connected with Celonis for creating the data model. Mutation testing faults (or mutants) have been introduced into the robot's code, and consequently, the event logs have been generated again. Subsequently, the original traces and the mutant's traces are compared and tested using the knowledge model for determining the mutant's occurrence. The methodology is demonstrated using the iRobot's simulation within Webots. Details of the knowledge model's implementation and employment in the mutation testing will be presented in the next section.

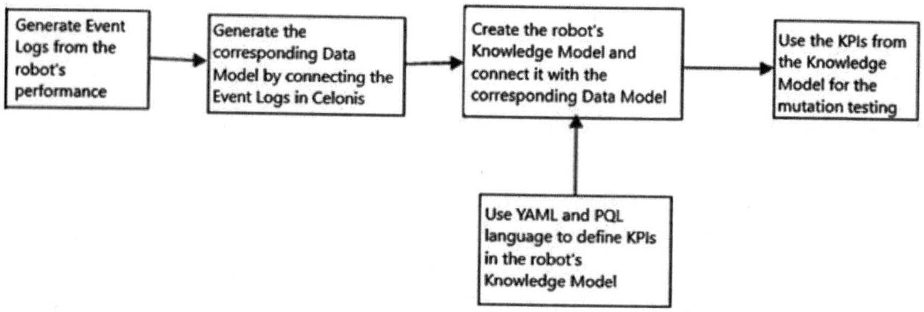

Fig. 1 Steps of the methodology

3 Experimentation

Webots simulator platform is used as basis for providing an environment to model, alter, simulate robot's behavior, and obtain the traces of the robot's performance collected into the corresponding event logs. The iRobot customizable framework based on the famous Roomba vacuum cleaning platform produced by iRobot [7] was adapted for this paper purpose. For simplicity, only two states of iRobot are considered in the knowledge model: (1) The state "turn angle from left obstacle," defined as always returning a positive angle value in a correct performance, and (2) The state "turn angle from right obstacle," defined as always returning a negative value when functioning properly.

The main body of the knowledge model in Celonis is displayed in Fig. 2. Celonis knowledge models could be utilized as a central storage for the KPIs and any other business-relevant definitions. By making use of these capabilities, the enterprise no longer has to maintain and manage the valuable knowledge in several places.

Next time, it is needed to adjust a definition (Fig. 2, lines 6–42), it is enough to do so once and this will be copied automatically everywhere it is being used [8].

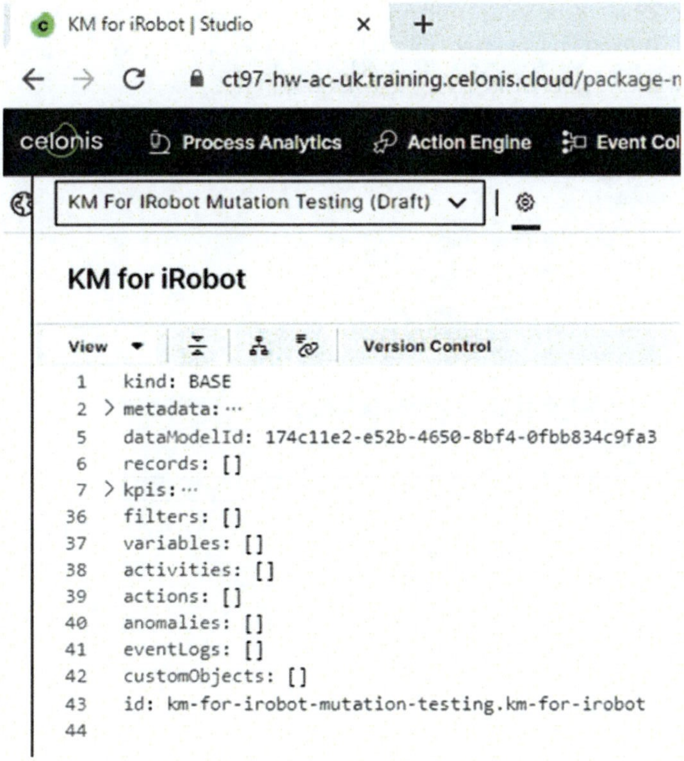

Fig. 2 The main body of KM

The knowledge model is implemented using YAML, but only the KPIs are relevant for the purpose of this paper. Two KPIs corresponding to the two states previously mentioned have been defined by filling the following characteristics (Fig. 3): (i) id: It helps to keep consistency around the entire knowledge model, (ii) description: It mentions important details regarding the KPI correct definition, (iii) PQL: It translates the business definition into an executable query, (iv) format: It states the indicator numerical value type and decimal places, (v) desired direction: It can indicate the correct direction of the KPI, i.e., whether it should decrease or increase.

The use of KPI ANGLE FROM RIGHT OBSTACLE (lines 7–16 from Fig. 3) will be explained in Section A for a correct case and Section B for the mutation testing. Subsequently, the use of KPI ANGLE FROM LEFT OBSTACLE (lines 18–27 from Fig. 3) will be exemplified in Section C and Section D, respectively.

A. *KPI ANGLE FROM RIGHT OBSTACLE results for correct case*

The corresponding KPI description states that its value should be always negative (Fig. 3, line 9). Therefore, when performing correctly, the PQL's implementation: MAX(CASE WHEN "KM_iRobot_xlsx_KM." "STATE" = 'Turn angle from right obstacle' THEN "KM_iRobot_xlsx_KM." "VALUE" ELSE 0.0 END) should always return the value 0 (Fig. 4).

It can be seen that the value 0 is correct when comparing it with the maximum value of the state "turn angle from right obstacle" from Webots (i.e., with value −3.088670

← KM for iRobot Documentation Preview

```
1   kind: BASE
2   metadata:
3     key: km-for-irobot
4     displayName: KM for iRobot
5   dataModelId: ${{km}}
6   kpis:
7     - id: KPI_ANGLE_FROM_RIGHT_OBSTACLE
8       displayName: KPI ANGLE FROM RIGHT OBSTACLE
9       description: THE VALUE OF THE ANGLE SHOULD BE ALWAYS NEGATIVE
10      pql: MAX(CASE WHEN "KM_iRobot_xlsx_KM"."STATE" = ' Turn angle from right
11        obstacle' THEN "KM_iRobot_xlsx_KM"."VALUE" ELSE 0.0 END)
12      filterIds: []
13      unit: ""
14      format: ',.6f'
15      parameters: []
16      desiredDirection: DECREASE
17
18    - id: KPI_ANGLE_FROM_LEFT_OBSTACLE
19      displayName: KPI ANGLE FROM LEFT OBSTACLE
20      description: THE VALUE OF THE ANGLE SHOULD BE ALWAYS POSITIVE
21      pql: MIN(CASE WHEN "KM_iRobot_xlsx_KM"."STATE" = ' Turn angle from left
22        obstacle' THEN "KM_iRobot_xlsx_KM"."VALUE" ELSE 0.0 END)
23      filterIds: []
24      unit: ""
25      format: ',.6f'
26      parameters: []
27      desiredDirection: INCREASE
```

Fig. 3 KPIs implementation from the KM

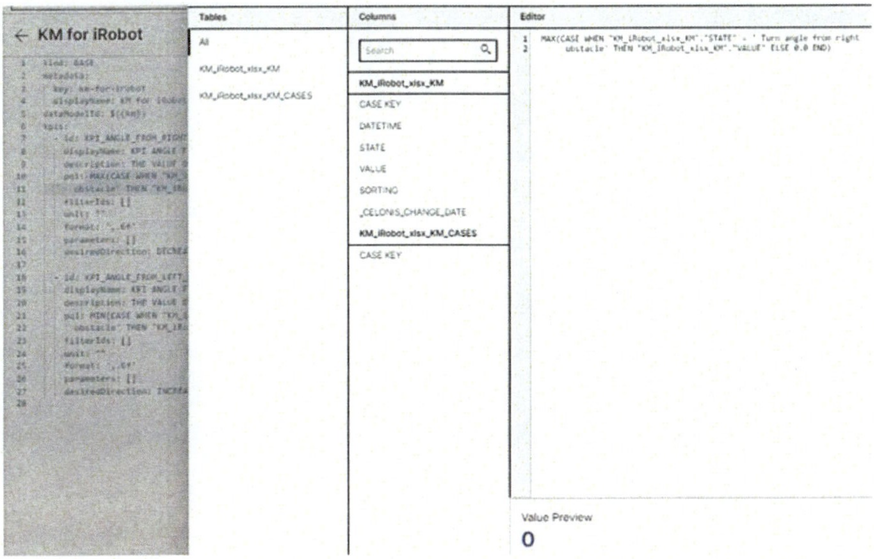

Fig. 4 KPI ANGLE FROM RIGHT OBSTACLE returned value in correct case

in case key 21, activity number 924 printed in the bottom of Fig. 5). This indicates that no mutant has been identified in the iRobot's Webots code corresponding to this KPI.

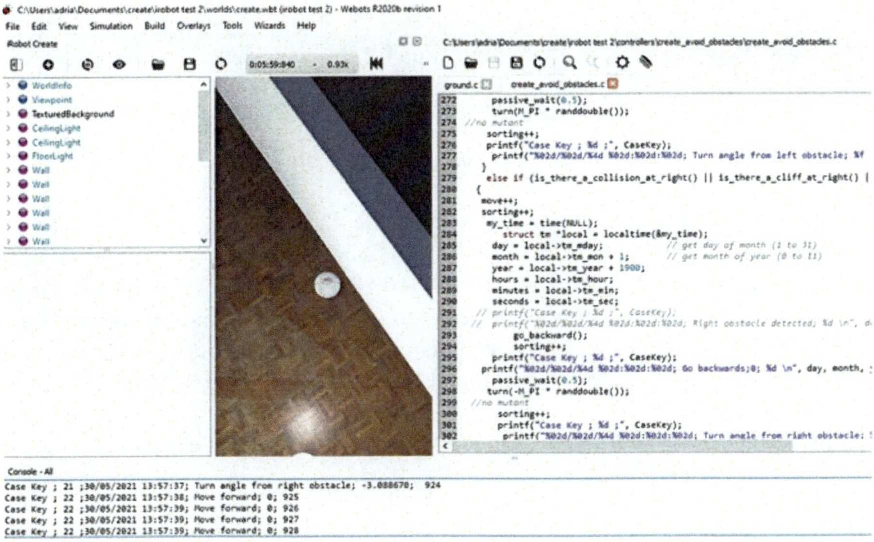

Fig. 5 Webots iRobot in correct case turn angle from right obstacle always negative

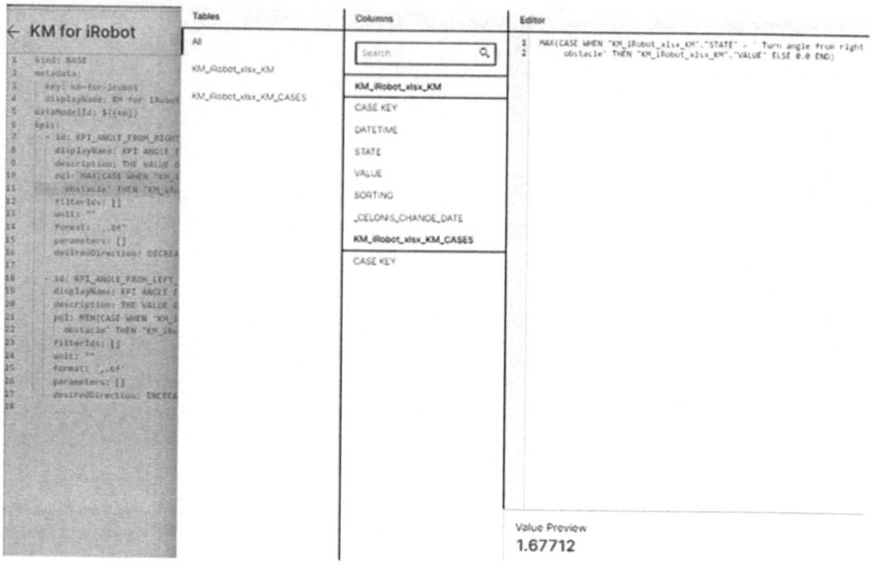

Fig. 6 KPI ANGLE FROM RIGHT OBSTACLE returned value for a mutant

B. *KPI ANGLE FROM RIGHT OBSTACLE mutant testing results*

The mutant was obtained by altering in the iRobot's simulator, the rotation angle's sign: turn(-M_PI * randdouble()) became turn(M_PI * randdouble()) (Fig. 7, line 298). Tested using the KPI, the PQL returns a positive value (i.e., 1.67712, Fig. 6). According to this KPI description, it indicates the mutant's presence. Looking in Webots can be seen the same maximum value for this state (Fig. 7, activity 1578).

C. *KPI ANGLE FROM LEFT OBSTACLE testing results for correct case*

This KPI description states that its value should be always positive (line 20 from Fig. 3).

Therefore, when there is no mutant in the code, the PQL's: MIN(CASE WHEN "KM_iRobot_xlsx_KM." "STATE" = 'Turn angle from left obstacle' THEN "KM_iRobot_xlsx_KM." "VALUE" ELSE 0.0 END) returned value should be 0. This KPI's PQL takes the minimum between 0 and the minimum value of "turn angle from left obstacle" from the Webots (i.e., value 2.988386 of case key 24, activity number 1062, Fig. 9), so no mutant is detected when the result is 0.

D. KPI ANGLE FROM LEFT OBSTACLE testing results for mutation

The mutant was obtained by altering the rotation angle's sign: turn(M_PI * randdouble()) became turn(-M_PI * randdouble()) (Fig. 11, line 273). The PQL value in the knowledge model (Fig. 10) is negative (−3.108612), contradicting this KPI description (Fig. 3, line 20) and indicating the mutant's presence.

Furthermore, this KPI PQL's returned value (Fig. 10) is confirmed in Fig. 11.

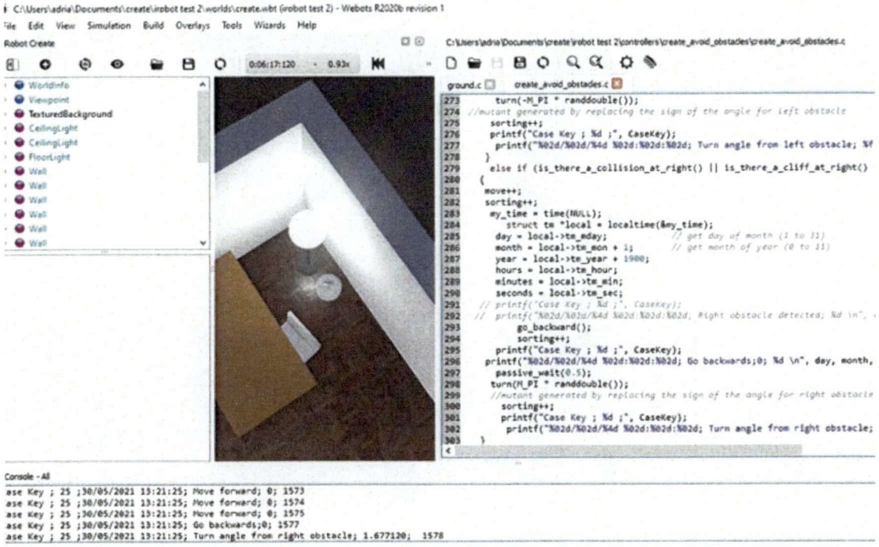

Fig. 7 Webots iRobot performance with mutant in state turn angle from right obstacle

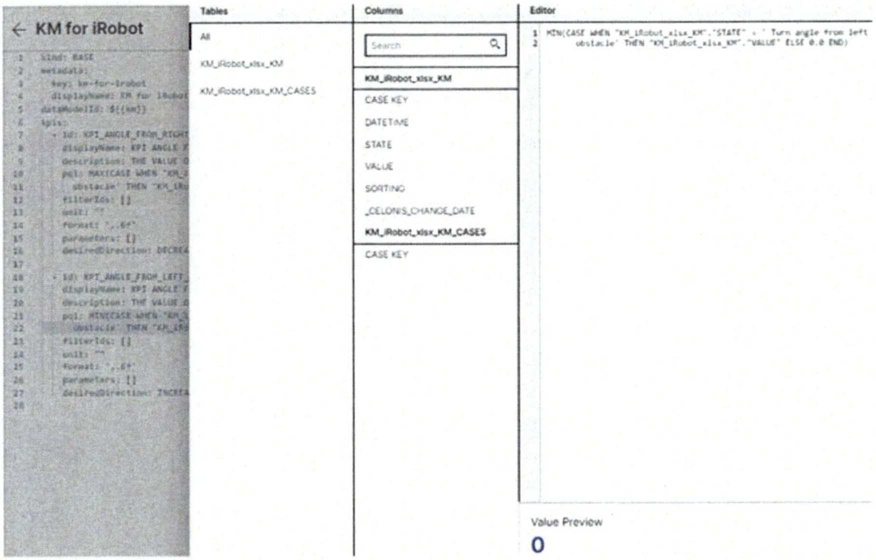

Fig. 8 KPI ANGLE FROM LEFT OBSTACLE returned value in correct case

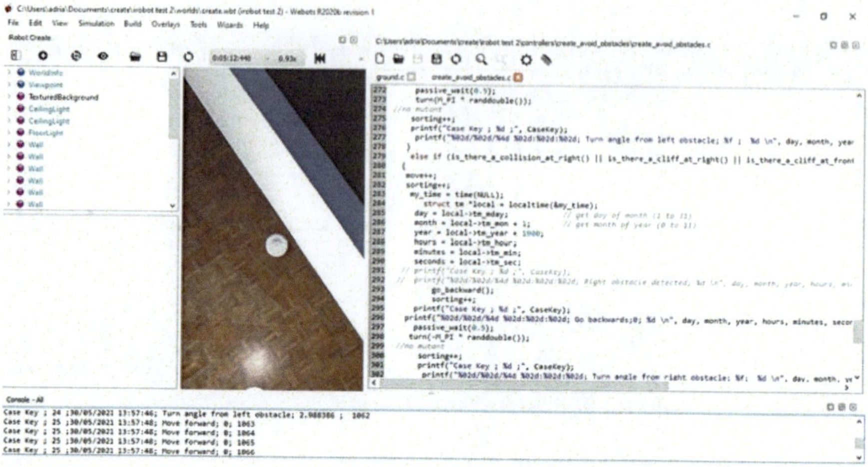

Fig. 9 Webots iRobot in correct case turn angle from left obstacle always positive

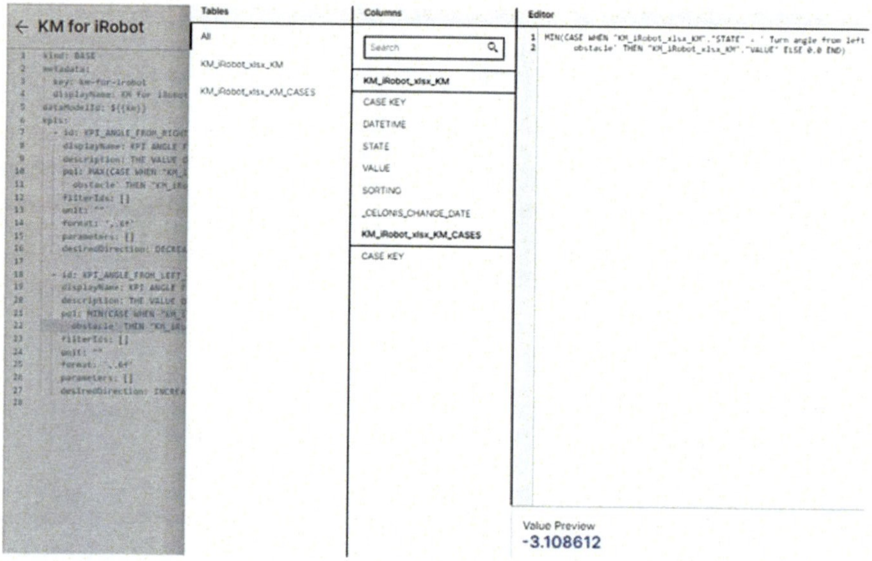

Fig. 10 KPI ANGLE FROM LEFT OBSTACLE returned value for a mutant

4 Conclusions and Future Work

This paper aims to demonstrate how knowledge models can facilitate the quality assurance in robotics. This approach is using business knowledge entities, such as KPIs, to evaluate the success of the robot's particular states in which it engages.

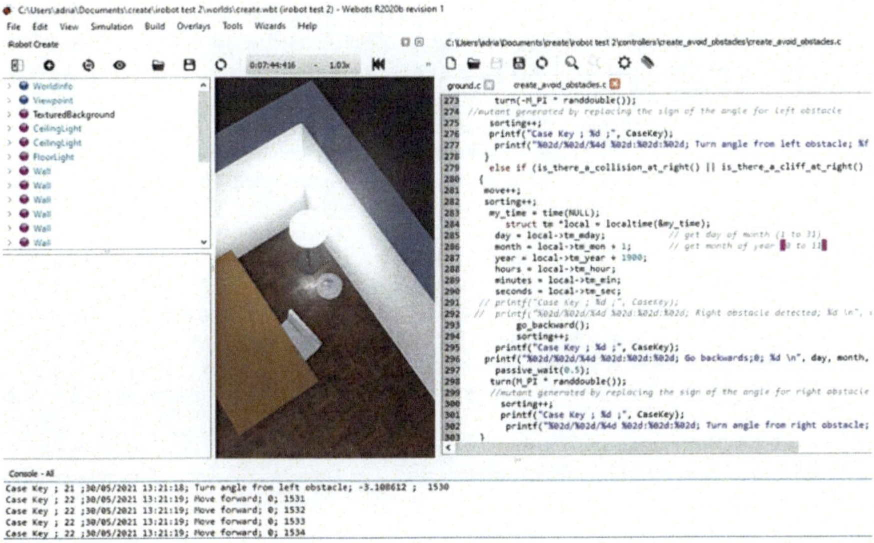

Fig. 11 Webots iRobot performance with mutant in state turn angle from left obstacle

The knowledge model was based on Webots iRobot's simulation and was implemented using Celonis EMS platform, as Webots allows the exploration of robots' characteristics and the code customization, and on the other side, Celonis knowledge models can be constantly monitored and improved, supporting decision-making, and preserving domain expertise. The future work will focus on developing a complete knowledge model with interactive views and action skills.

Acknowledgements The author would like to thank Celonis Academic Alliance and Cyberbotics Ltd.

References

1. Drucker P (1985) Innovation and entrepreneurship: practice and principles. Harper & Row, New York
2. Cuena J, Molina M (2000) The role of knowledge modelling techniques in software development: a general approach based on a knowledge management tool. Int J Hum Comput Stud 52:385–421
3. Wnuk K, Garrepalli T (2018) Knowledge management in software testing: a systematic snowball literature review. e-Informatica Softw Eng J 12(1):51–78. Politechnika Wroclawska
4. Falah B, Bouriat S, Achahbar O (2015) Reducing mutation testing cost using random selective mutation technique. Malaysian J CS 28(4)
5. Hamimoune S, Falah B (2016) Mutation testing techniques: a comparative study. In: 2016 international conference on engineering & MIS (ICEMIS)
6. Michel O (2004) Cyberbotics Ltd. WebotsTM: professional mobile robot simulation. Int J Adv Robot Syst

7. Webots (2021) http://www.cyberbotics.com. Accessed 30 May 2021
8. Celonis (2021) https://www.celonis.com/. Accessed 31 May 2021
9. YAML (2021) https://yaml.org/. Accessed 31 May 2021
10. Kumar K, Reel PS (2011) Analysis of contemporary robotics simulators. In: 2011 international conference on emerging trends in electrical and computer technology, pp 661–665
11. de Souza ÉF, de Almeida Falbo R, Vijaykumar NL (2015) Knowledge management initiatives in software testing: a mapping study. Inform Softw Technol 57:378–339
12. Roman A, Mnich M (2021) Test-driven development with mutation testing—an experimental study. Softw Qual J 29:1–38
13. Tammet T, Reilent E, Puju M, Puusepp A, Kuusik A (2010) Knowledge centric architecture for a robot swarm. In: IFAC proceedings volumes
14. Mușat F, Mihu F (2018) Collaborative robots and knowledge management—a short review. Acta Universitatis Cibiniensis, Technical Series
15. Gómez LV, Miura J (2021) Ontology-based knowledge management with verbal interaction for command interpretation and execution by home service robots. Robot Auton Syst 140. ISSN 0921-8890

FuseIT: Development of a MarTech Simulation Platform

Célio Gonçalo Marques, Giedrius Romeika, Renata Danielienė, and Hélder Pestana

Abstract This article presents a working prototype of a digital simulation platform dedicated to training skills and competences in marketing technologies. This research covers the description of the use of simulation methods in marketing training, as well as the development and functionalities of the simulation platform. This research was conducted within the scope of the FuseIT project activities. This project aims to analyze, design, develop, and implement an updated curriculum in the field of marketing in Europe with e-learning materials, blended learning environments, skills self-assessment, and knowledge assessment systems.

Keywords Marketing simulations · Marketing technologies · MarTech · Technology-based training methods

1 Introduction

In the *proposal for a council recommendation on key competences for life-long learning* [1], the European Commission states that formal education and training should equip everyone with a broad range of skills, which opens doors to personal fulfillment and development, social inclusion, active citizenship, and employment . These include literacy, numeracy, science, and foreign languages,

C. G. Marques (✉) · H. Pestana
TECHN&ART | LIED.IPT, Polytechnic Institute of Tomar, Tomar, Portugal
e-mail: celiomarques@ipt.pt

G. Romeika · R. Danielienė
Vilnius University, Vilnius, Lithuania
e-mail: giedrius.romeika@knf.vu.lt

R. Danielienė
e-mail: renata.danieliene@knf.vu.lt

R. Danielienė
Information Technologies Institute, Vilnius, Lithuania

A. Ullah et al. (eds.), *Proceedings of International Conference on Information Technology and Applications*, Lecture Notes in Networks and Systems 350,
https://doi.org/10.1007/978-981-16-7618-5_43

as well as transversal skills and key competences such as digital competences, entrepreneurship, critical thinking, problem solving or learning to learn, and financial literacy.

These skills fully meet the nature of the marketing phenomena, because of its open-endedness as a discipline, which allows students to develop skills in the context of topics such as consumer behavior, microeconomics, copywriting, big data analysis, Web developing, and among others. The versatility of skills gained while studying marketing determines a wide range of professional career paths, from industry to commerce, and facilitates personal lifelong learning activities [2].

What type of teaching methods should be used to teach marketing skills has been widely discussed. Research shows that, in Europe, traditional methods tend to be favored, while researchers often insist on using more technology-based training methods, including simulation games [3]. Many researchers recognize simulation games as an effective tool to develop skills and build competencies [4].

Given this, the FuseIT project (future competences pathways for marketing and ICT education) was created aiming to address the education and labor market requirements for professional skills and knowledge for students and university graduates (including lifelong learning activities). FuseIT involves researchers from educational institutions in Lithuania, Latvia, Portugal, and Romania and is funded with the support of the European Union.

The project aims to analyze, design, develop, and implement an up-to-date curriculum, a set of e-learning materials, a blended learning environment, as well as a simulation system for marketing ICT knowledge/skills self-evaluation and certification. This paper presents the most visible outcome of this project: A working prototype simulation platform dedicated to marketing technologies skills and competences training.

FuseIT covers not only theoretical topics, like the fundamentals of marketing, customer loyalty, strategic creativity, but also topics in digital marketing, like mobile marketing, e-mail marketing, social media marketing, spreadsheets, etc. This poses several challenges in developing the digital simulation platform.

In this research, document analysis, questionnaires, and interviews were used to determine the contents. Statistical data processing and correlation were used to analyze the collected data. An agile software development methodology was used to create the platform.

In this paper, we start by analyzing the usage of simulations in field of marketing. After, this follows a description of how the study was conducted and the presentation of the simulation platform. At the end, some considerations are made and future work is specified.

2 Background

Simulations-based situations are widely used in various fields, such as health care [5], military [6], aviation [7], communication networks [8], and in business and marketing [9, 10], where real life situations are provided, and human decisions are required to take the right actions or make decisions invoking greater effectiveness.

Because of these consequences that follow the marketing practice in the real world, simulations provide a possibly more suitable learning environment. By using simulations-based learning, students do not affect real-life situations. The consequences with feedback are provided in text or graphical format. Students could execute simulations several times, and therefore, it is expected that they will learn more and gain additional skills, like problem solving, will involve them more in the course materials and encourage their motivation.

Most learning management systems, like Moodle, do not have integrated simulations development tools, including gamification elements and the possibility to develop tree structure scenarios with text and graphical feedback. There are plenty of questionnaire form tools, like Google Forms, Microsoft Forms, which could be used for tree structure type scenarios, but they do not have gamification elements either. Unfortunately, the majority simulation specialized tools, like BranchTrack [11] or iSpring Suite Max [12], are not charge-free, which does not make them accessible to everyone.

Based on this, it was decided that the project would develop its own digital simulation platform, including additional features like multimedia usage, tree structure, gamification elements, and feedback responses after each decision, situations randomization, and multi-language environment for all partners' languages (English, Lithuanian, Latvian, Portuguese, and Romanian). Several simulation projects were analyzed, namely the research and exploration on the practice teaching mode of economics and management, improving students' comprehensive ability based on multilevel virtual simulation training by Heyuan Polytechnic (China) [13], and the learning environment for Web marketing boot camps created by Simbound, Grenoble Alpes University (France), University of Milano-Bicocca (Italy) and Dunarea de Jos University of Galati (Romania) [14].

3 This Study

This study started in 2019 and is currently in its final procedures. The first phase of study was dedicated to the analysis of the current state of the field. Firstly, an analysis of existing programs and curricula at projects targeting digital marketing competences (DMC), available at the project partners' organizations, was conducted. This was then compared with existing digital competence frameworks such as the digital competence framework for consumers and the digital competence framework for citizens 2.1. The comparison was conducted by similarity of analyzed study programs and the key learning outcomes and course information with competences presented in digital competence frameworks of the curricula. Four study programs and 26 study, modules were included in the comparison.

Secondly, a survey dedicated at clarifying future DMC requirements was carried out. The survey was performed in all partner countries: Lithuania Latvia, Portugal, and Romania, and it consisted out of five components: (a) General information about respondents; (b) required level of competences necessary for digital marketing specialist; (c) professional skills necessary for a digital marketing specialist; (d) professional positions for which digital marketing competences and professional skills are necessary; (e) business entities/types of organization for which digital marketing competences and professional skills are necessary.

In the survey, 355 respondents participated. All the multiple choices listed on survey were selected according to information provided in digital competence frameworks (the digital competence framework for consumers and the digital competence framework for citizens 2.1) and business reports (such as the 2020 workplace learning trends report: The skills of the future by Udemy for business and the top five marketing jobs in 2020 by 10digital.co). They were also supplemented by considering the best practice samples from project partners. By analyzing the results, it is possible to state that the most necessary competencies for a digital marketing specialist are those related with strategic and foundational aspects of the marketing discipline, as well as with digital marketing (Web experience management; business intelligence; targeting and optimization; usability/design). The most necessary skills are related with ICT tools directly designed for marketing purposes (e.g., social media, video, e-mail marketing; digital analytics; customer relationship management; search engine marketing; search engine optimization), with the only exception being Microsoft Excel skills.

The second phase of study was dedicated to the core curriculum, based on the results of context analysis development. It consisted of detailed knowledge index as well as general learning/teaching ideas, learning content specifications, and other methodological guidelines. A detailed syllabus of blended learning course was designed.

Thirdly, it was content creation regarding the detailed syllabus with 21 topics being created. Each topic consisted of material dedicated to students and material dedicated to teachers. The volume of material is adjusted to contact and individual working hours.

The fourth phase of the project was dedicated to the design of simulation platform. Once the contents were established, the general requirements of the platform were defined. Among them: The need for a customized and attractive graphical interface for the Web platform that would promote a good user experience; the possibility of having unlimited scenarios, with random questions and answers in different levels, translated into the different partners' native languages; the provision of visual feedback to the user expressing emotions; free registration, to allow users to keep track of their evolution; the use of gamification to enhance learning through competition; an administration area, to allow each partner the possibility of managing and importing new scenarios.

After an analysis of the available platforms for the scenario implementation, it was decided that a customized tool that corresponded to all objectives proposed for this project was to be developed.

We begin by determining the functional requirements, which express the functionalities or services that the system is expected to produce (input/output), and non-functional requirements, which state the quality, performance, safety, ease of use of the system (usability), effectiveness, efficiency, and satisfaction expected.

For the general requirements of the scenarios, it was stipulated that

- Each simulation would have a "question tree" with three levels.
- In each level, the user would have at least three options: a correct one (two points), a semi-correct one (one point), and an incorrect one (zero points).
- After choosing the correct one, the user would advance further (positive feedback would also appear).
- If they chose the semi-correct answer, an explanation of why it was semi-correct would show up (feedback), and the user would progress further.
- If they chose the incorrect answer, an explanation that the chosen solution was incorrect would show up (feedback), explaining why it was incorrect and a request for the user to "try again."
- If the student answered incorrectly, they would also be invited to read more about the subject (the specific parts where the subject matter is addressed would be indicated).
- A virtual agent would express emotions according to the student's answers.

In the development of the platform, heuristic evaluation and evaluation with future users were used. Several improvements were made based on the results of this assessment.

Preliminary performance tests, usability, and search engine prototype optimization were carried out using Webmaster Tools (e.g., Google Tools), which returned scores within the recommended values. This indicates that the tool complies with best practices.

So far, 80 scenarios have been produced, covering all the topics developed in this project, of which 40 are local topics. The scenarios, transversal to all partners, are available in the English, Latvian, Lithuanian, Portuguese, and Romanian (Table 1).

4 Results and Discussion

For the development of the simulation platform, the rapid application development (RAD) methodology was used. This methodology was chosen due to its simplicity and prototype generation possibilities and due to the small size of the development team [15].

The data structure was carried out according to the requirement analysis to obtain the model for the information system database. In this sense, the entities (and attributes) relevant to the system were identified, namely "module," "scenario," "question," "answers," "level," "user," "answer," "simulation," "language," "country," and "administrator."

The relationships between the entities and their degree of multiplicity was also identified, with the purpose of creating the physical model of the system database.

Table 1 Number of scenarios per course content

Course content: topic breakdown	Number of scenarios
Basics of marketing	1
Introduction to market research	1
Customer loyalty, satisfaction, and engagement	1
Strategic creativity	1
Customer experience management	2
CRM analytics	2
Mobile marketing	3
Design thinking	1
Digital analytics	4
Optimization of advertisement in Web	4
Video marketing	1
Web experience management (WEM)	2
E-mail marketing	3
Decision making and business intelligence	2
Digital marketing	4
Social media	4
Excel	4
Local topics	40

Some of the entities initially identified were transformed to attributes such as "level" or "administrator," the vast majority remained.

Once the analysis and model phase were completed, the prototype of the simulation system was developed. Regarding the technologies for the simulation platform, the choices were necessarily on the server side, since the client side is limited to standard Web languages. An open-source solution: Linux, Apache, Mysql, and PHP (LAMP) was adopted.

For the provision of the system to the general public, the domain http://www.fuseit.eu was acquired. Figure 1 shows the platform's home page, where several modules (thematics) appear. By selecting a module, the associated simulations appear.

Through the main menu, it is also possible to switch the language of the user interface, see the rankings, login to the account and register in the platform.

Figure 2 illustrates the page of a specific module, where the user can choose one of the available scenarios. If the user is registered on the platform, information about the scenarios, they have already responded to and their respective score is shown.

To start a simulation, the user must be registered on the platform. After this, the process starts with a general description of the functioning of the simulation and with a contextualization of the scenario (goals; prehistory; where the action takes place; what actors are involved, who they are brief description of the situation). After the user

Fig. 1 Main page of the FuseIT simulation system

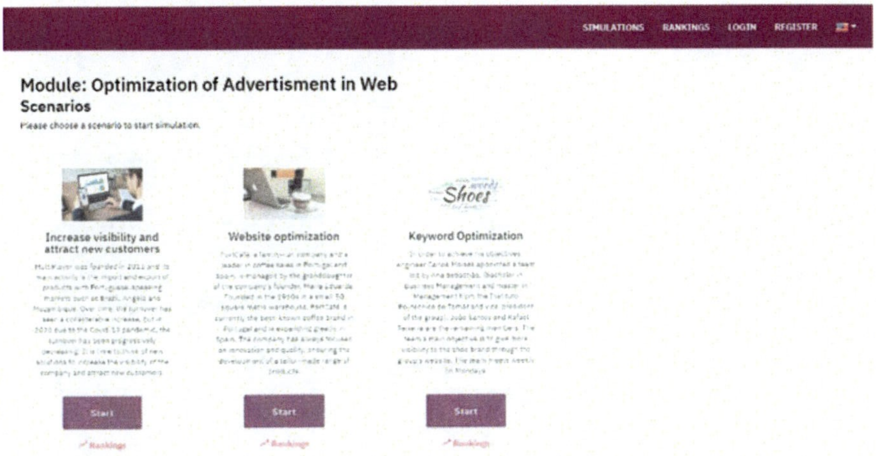

Fig. 2 Scenario selection

clicks on the button to initiate the simulation, a "root question" appears. Answers are made available to the user randomly and the user can only select one answer (Fig. 3).

Scenarios include images, videos, and links that will help the user to get a more realistic perspective of the situation. By choosing a correct answer, the user accumulates two points, the semi-correct answer accumulates one point and the wrong answer accumulates zero points. When choosing any of the answers, visual feedback is always displayed to the user (image, supplementary text, and bibliographic references) (see Fig. 4).

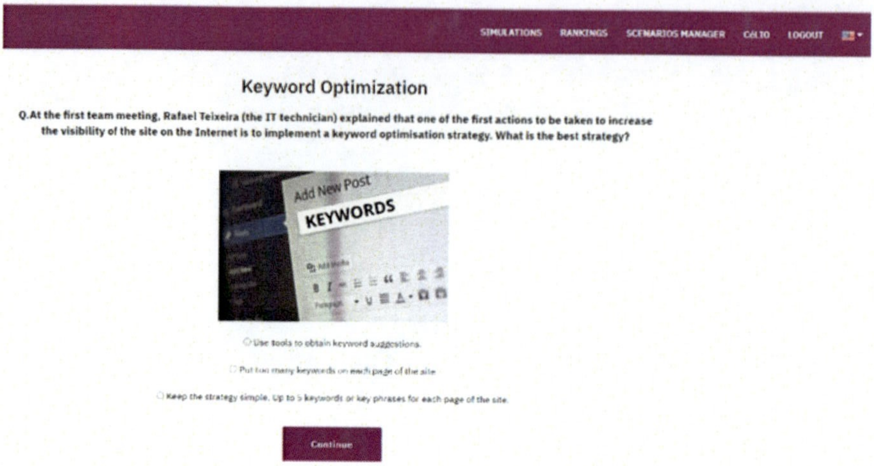

Fig. 3 Root question

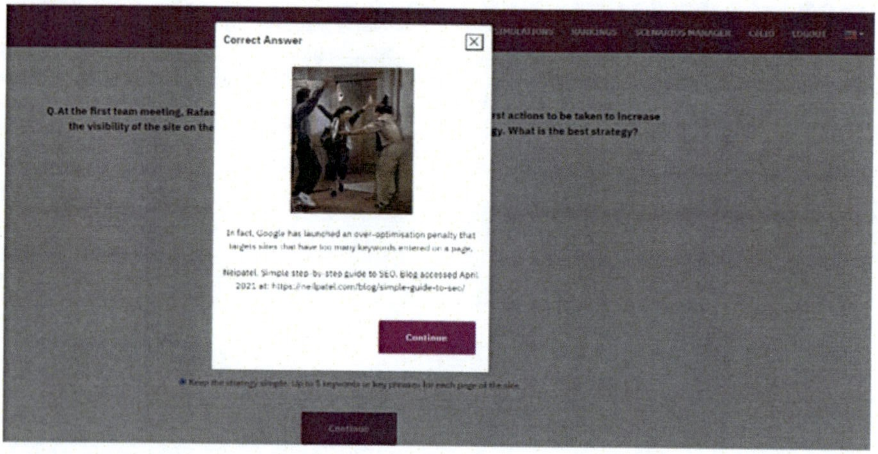

Fig. 4 Visual feedback after selecting one answer

The simulation proceeds to the next level question until it reaches the highest level. At the end, the final score is presented to the user and they can view their score in the global ranking (by module or in all modules), (see Fig. 5).

For scenario configuration, the platform has a management section reserved only to users with administration privileges, where each scenario and its question-and-answer branches (and respective feedback) can be set up (see Fig. 6).

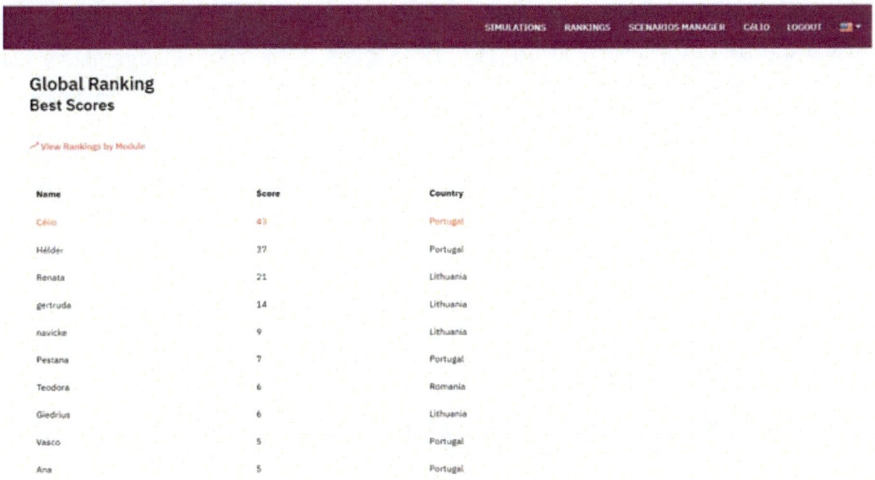

Fig. 5 Global ranking

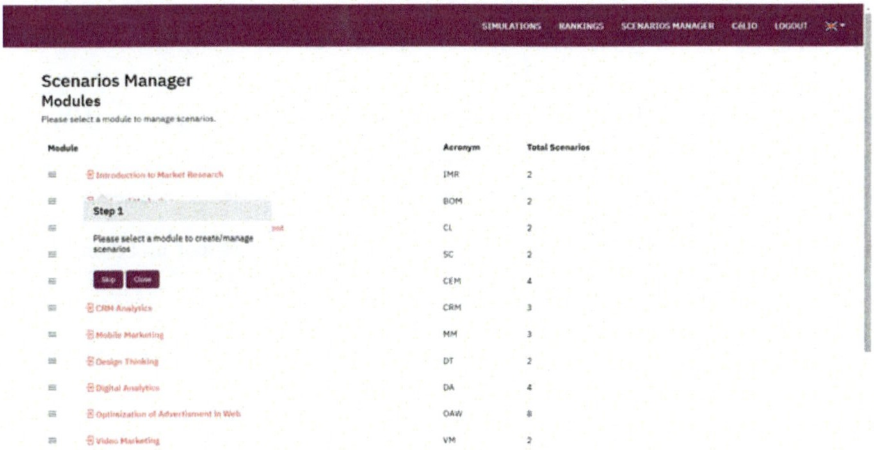

Fig. 6 Management section

5 Conclusions

The purpose of the development of a MarTech simulation platform is to allow users to consolidate their knowledge about marketing and ICT, using scenarios that allow them to simulate contexts where they would have to apply practical skills in order to solve the proposed challenges. The users will be stimulated to demonstrate critical thinking and logical reasoning and prove solid knowledge to reach the end of the process. A great advantage, apart from the playful and challenging aspect that gamification can bring, is that it also allows learning from mistakes, since in each semi-correct or wrong answer, the user is led to understand why they should have chosen a different answer and how to bridge gaps in their knowledge. They can also repeat the process until they understand the most appropriate solutions at each moment.

By placing users in realistic situations and challenging them to experience the responsibilities that marketing, and ICT professionals face daily, content consolidation is fostered and helps to prepare them to enter the labor market.

Future research activities related to this simulation tool will be dedicated to creating solutions for testing in real-time conditions. This platform will be presented to potential end-user target group's representatives. Within the research activities potential end-user feedback collection will be organized and regarding collected data tools, adjustments will be provided.

Acknowledgements This project has been funded with support from the European Commission. This publication reflects the views only of the authors, and the Commission cannot be held responsible for any use which may be made of the information contained therein.

References

1. Commission E (2018) Proposal for a council recommendation on key competences for lifelong learning. Commission Staff Working Document, Brussels
2. Danieliene R, Romeika G, Marques CG (2020) Perspectives of Martech competencies development. In: Proceedings of ALTA'20 advanced learning technologies and applications. short learning programmes. Lithuania, pp 82–89
3. Gupta A, Singh K, Verma R (2010) Simulation: an effective marketing tool. Int J Comput Appl 11(4):8–12
4. Ranchhod A, Gurau C, Loukis E, Trivedi R (2014) Evaluating the educational effectiveness of simulation games: a value generation model. Inf Sci 264:75–90
5. Seila AF, Brailsford S (2009) Opportunities and challenges in health care simulation. international series in operations research & management science. In: Alexopoulos C, Goldsman D, Wilson JR (eds), Advancing the frontiers of simulation. Springer, pp 195–229
6. Bruzzone AG, Massei M (2017) Simulation-based military training. In: Guide to simulation-based disciplines. Springer International Publishing, Cham, pp 315–361. https://doi.org/10.1007/978-3-319-61264-5_14
7. George Brown College: how simulation tools are transforming education and training, https://www.etcourse.com/simulation-tools-transform-education-and-training.html, last accessed 2021/06/14
8. Zarrad A, Ismad I (2017) Evaluating network test scenarios for network simulators systems. Int J Distrib Sens Netw. https://doi.org/10.1177/1550147717738216
9. Zulfiqar S, Sarwar B, Aziz S, Chandia KE, Khan MK (2018) An analysis of influence of business simulation games on business school students' attitude and intention toward entrepreneurial activities. J Educ Comput Res 57(1):106–130. https://doi.org/10.1177/0735633117746746
10. Stummer C, Kiesling E (2021) An agent-based market simulation for enriching innovation management education. CEJOR 29:143–161. https://doi.org/10.1007/s10100-020-00716-3
11. Branch Track, www.branchtrack.com, last accessed 2021/08/14
12. iSpring suite, https://www.ispringsolutions.com/ispring-suite, last accessed 2021/06/14

13. Deng W (2018) Research on the construction of multilevel economic management virtual simulation experimental teaching center. In: Proceedings of the 2018 5th international conference on education, management, arts, economics and social science (ICEMAESS 2018). Advances in social science, education and humanities research, 332–335
14. Capatina A, Bleoju G, Rancati E, Hoareau E (2018) Tracking precursors of learning analytics over serious game team performance ranking. Behav Inf Technol 37(10–11):1008–1020. https://doi.org/10.1080/0144929X.2018.1474949
15. Martin J (1991) Rapid application development. Macmillan, New York

Using the Rodin Platform as a Programming Tool

Adrian Turcanu and Florentin Ipate

Abstract Being the result of several research projects that involved at the same time academic and industrial partners, the Rodin platform is designed to implement and analyse mathematical models of transitional systems based on Event-B language. Several plugins were integrated in the platform including ProB, a model checker that can be used to verify LTL properties or animate the model. In this paper, we introduce a methodology for using the Rodin platform and ProB as a powerful, integrated, programming and analysis environment. Our approach is based on assigning an FSM to an algorithm, then implementing the corresponding Event-B model in Rodin, and finally verifying its correctness and finiteness using the facilities of the platform.

Keywords Mathematical modelling · Event-B language · Model checking · Algorithms · Rodin platform · FSM · Formal verification

1 Introduction

This is developed by J. R. Abrial as an extension of the B language, Event-B is a rigorous mathematical modelling language used to implement discrete systems [1]. Both B and Event-B are rooted in arithmetic, predicated logic and set theory and are proving support for modelling various data structures as relations, functions and sequences.

Rodin [2], an Eclipse-based platform dedicated to Event-B models, is the result of two European Union Projects: RODIN (2004–2007) and DEPLOY (2008–2012) [3]. Apart providing support for the implementation of Even-B models, the platform has been extended with plugins such as ProB [4], Camille [5] or iUML-B state-machines [6].

A. Turcanu (✉)
School of Mathematical and Computer Sciences, Heriot-Watt University, Dubai Campus, Dubai, UAE
e-mail: A.Turcanu@hw.ac.uk

F. Ipate
Department of Computer Science, University of Bucharest, Bucharest, Romania
e-mail: florentin.ipate@ifsoft.ro

A. Ullah et al. (eds.), *Proceedings of International Conference on Information Technology and Applications*, Lecture Notes in Networks and Systems 350,
https://doi.org/10.1007/978-981-16-7618-5_44

One of the most interesting Rodin features is the possibility to develop the models gradually. The process is called refinement and consists in starting the modelling with a core version of the system under analysis, and then adding additional details in the future developments.

Another aspect that makes Rodin a reliable modelling and verification platform is the automatic checking of models' correctness that is ensured through generating and verifying proof obligations.

This is introduced by Clark et al. [7], model checking is an automatic verification technique of finite-state systems. The techniques are applied using a tool called model checker that is capable to verify if a certain model satisfies a given property and to identify a counterexample when this is violated. Such counterexamples are usually very useful for the developer in identifying the source of the error and correcting the model [8]. However, since all the possible states of the system are explored in a brute-force manner, the technique has an associated common main issue called the state explosion problem [9].

Offering support for the verification of B, Event-B, CSP-M, TLA+ or Z models, ProB is a model checker that was integrated in the Rodin platform as a plugin. It is being used by various companies like Siemens, Alstom or Thales for the modelling and formal verification of safety critical systems [4]. Once a model is implemented, it can be animated and analysed using ProB. This includes automated consistency checking, random execution of a given number of enabled events, verification of properties specified using the LTL or CTL formalisms or visualising the state space.

An Event-B model is a collection of contexts and machines. Contexts contain the static structure of the system: sets, constants and axioms defining the main properties of sets and constants. Machines contain the dynamic structure of the system: variables, invariants and events. Invariants are representing properties of variables that have to hold in each state of the system. Events define the dynamic of the transition between various states of the system. Each event is enabled if and only if all its guards are true, and the execution of an event involves a set of simultaneous actions.

Event-B and the Rodin platform have been used in the past few years in various domains like membrane computing [10, 11], ontologies [12], control systems [13, 14] or robotics [15].

This paper aims to introduce a methodology for developing Event-B models of classical algorithms, implement them using the Rodin platform and verify their correctness using ProB, the associated model checker. This paper is organised as follows: the next section is dedicated to a detailed description of the three stages of the proposed methodology, this is further applied on some case studies in Sect. 3, and finally, Sect. 4 is used to draw some conclusions and propose some future work.

2 Methodology

In this section, we describe a methodology for constructing the Event-B model of an algorithm and then using the Rodin platform and ProB to implement and analyse such a model. The main three stages of this are described in the following subsections, and the methodology is represented in the following simplified diagram:

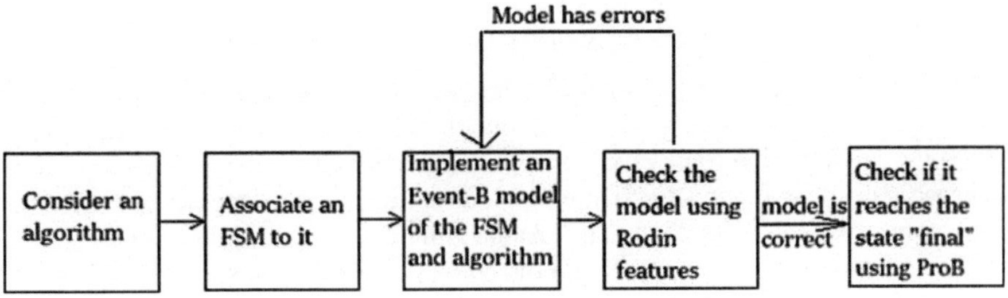

2.1 *The FSM Associated to an Algorithm*

The first step in our approach is to associate a simple finite-state machine (FSM) to an algorithm. In general, this contains three types of states: initial, intermediate and final. Whilst the intermediate and final state are unique, multiple intermediate states are used when an instruction is included in the body of another instruction.

At the beginning of the computation, all variables are assigned with their initial values and the state is set to 'initial'. Once the execution of the algorithm starts, the state is changed to 'intermediate', and in the case of imbricated instructions, additional intermediate states are used to indicate that the inner instruction is executed. Once this is done, the state of the system is changing again to the 'main' intermediate value. Finally, when the computation is coming to an end, the model is reaching the final state.

2.2 *The Event-B Model of a Programme*

In this section, we are describing how to obtain the Event-B model of a programme and the FSM associated to it according to the methodology previously introduced.

The Event-B model contains one context that is used to set up the set of all possible states and one machine that 'sees' the context.

The machine contains.

- several variables and the corresponding invariants,
- an initialisation event that is used to set up the initial values of all variables, including a variable corresponding to the state which has the 'initial' value assigned,
- one or more events corresponding to the transitions between the initial state and an intermediate state or between the intermediate states when the model has multiple intermediate states,
- one event corresponding to the transitions to the final state.

Since the execution of an event implies, the simultaneous execution of all its actions, in some cases additional events may be required to handle the situation in which the value of a certain variable is to be updated and it depends on values of other variables that are updated in the previous actions of the same event. When this is the case, an addition Boolean flag variable will give the control to an event whose purpose is to perform this update, and once this is done the same flag variable will be used to exit that event. The second case study considered in Sect. 3 involves such a situation.

If the Rodin facilities dedicated to the formal verification of the models are identifying any error in the Event-B model, then the model is analysed and corrected, otherwise it can be proceeded to the next step dedicated to verifying the algorithm's finiteness.

2.3 Using ProB to Verify the Algorithms Finiteness

One key aspect of our approach is that we treat all the algorithms in a similar way in the sense that we are checking if the computation is coming to an end by using the model checked ProB to check the LTL property: $G\{\text{state} /= \text{final}\}$. Here, '$G$' stands for the LTL operator 'globally', therefore, we are checking if the variable corresponding to the state is reaching the value 'final', which means that the computation is coming to an end. Obviously, if the model is correct, ProB will return a counterexample and more than that it will identify a sequence of events that are leading to this end.

The idea of testing with model checkers was explored by different authors and in different contexts [8], but the novelty of our approach consists in using it on Event-B models of classical algorithms with the purpose to ensure their correctness and finiteness. Some other advantages of our approach are discussed in Sect. 4.

3 Case Studies: Implementing Some Classic Algorithms in Rodin

This section is dedicated to three case studies representing three well-known algorithms in classical programming: Calculating the GCD and the LCM of two given number, generating a term of the Fibonacci sequence and sorting a vector using bubble sort. In each case, the methodology, we introduced in Sect. 3, is applied and some results of simulating and verifying the corresponding Event-B models are discussed. All these case studies and other examples of Event-B models of classical algorithms built based on our methodology are available at [16].

We consider that the complexity of the considered case studies is appropriate for demonstrating how the introduced methodology works and the method can be easily applied to other algorithms.

3.1 The GCD and the LCM of Two Given Numbers

In this section, an Event-B model dedicated to the computation of the GCD and LCM of two given numbers is discussed. The computation is based on Nicomachus's algorithm for GCD:

whilst $(a\,/= b)$.

if $(a > b)$ then $a = a - b$.

else $b = b - a$.

We recall that at the end of the computation both values of a and b are equal with the value of the GCD, and the value of the LCM can be simply calculated by dividing the product of the two given numbers to the GCD.

In Fig. 1, the results of checking the LTL property $G\{\text{state}/ = \text{final}\}$ against the GCD Event-B model are shown using the numbers 240240 and 150,150 as initial values for the two input variables a and b. As can be seen, the correctness of the

Fig. 1 Calculating the GCD and LCM of two numbers using Rodin

model is confirmed by the validity of all invariants, and once the computation has come to an end none of the events 1, 2 and 3 is enabled. Moreover, a sequence of events generating the counterexample is shown in the 'history' window.

3.2 *Fibonacci Sequence*

In this section, we will describe the Event-B model of the Fibonacci sequence based on the well-known algorithm with three variables: *a*, *b* and *c* corresponding to three consecutive terms in the Fibonacci sequence.

Our objective variable will be *c*, that, when the final state is reached will store the value of the required term.

The Event-B model contains four events: initialisation, the event corresponding to the transition from the initial to the intermediate state, an auxiliary event in which the values of variables *a* and *b* are updated such that they contain the previous two terms of the sequence, and the event dedicated to the end of the computation that enables the final state. The reasoning behind the auxiliary event is that since all actions of an event are executed in the same, we cannot update the values of variables *a* and *b* in the same event in which we are updating the value of variable *c*, since b should store the updated value of *c*.

As can be seen in Fig. 2, by checking the same LTL property as before, but now against the Event-B model of the Fibonacci sequence algorithm, the model checker found a counterexample. Therefore, when the final state is reached, the value of the variable *c* is the 1000th term of the Fibonacci sequence. The 'history' window contains a sequence of events which are leading to the final state, but, due to obvious space constraints, not all the events are shown in Fig. 2. Again, the correctness of the model is confirmed by the validity of the invariants and the fact that none of the

Fig. 2 Calculating the 1000th term of Fibonacci sequence using Rodin

Fig. 3 Sorting a vector using Rodin

events can be further applied. Moreover, the computation of such a large value is not possible with most of the commonly used programming languages.

3.3 Sorting a Vector

In this section, we are describing the Event-B implementation of the bubble sort algorithm. In this case, an additional intermediate state is used to distinguish the situation in which two elements of the matrix, say a_i and a_j, are and are not in the correct order, respectively. In the first situation, the machine remains in the 'intermediate' state and the value of the variable j is increased with one unit, and in the second case, the state of the machine is changing to 'intermediate 1', a_i and a_j are swapped and again j is increased with one unit.

Figure 3 is showing the results of verifying the same LTL property on the Event-B implementation of the bubble sort method. It can be seen that the elements of the vector a are in ascending order, all the invariants are true, the computation stopped and a sequence of events leading to this end was also identified.

4 Conclusions and Future Work

In this paper, a methodology to develop Event-B models based on classical algorithms and an underlying FSM has been introduced. Some case studies have been considered, and in each case, the computation was proved to come to an end by using the model checked ProB to verify the same LTL property against each model. The correctness of each model was also confirmed by the validity of the invariants and by the sequence of events that can be executed to reach the final state identified by ProB.

The main advantages of implementing the algorithms in a rigorous mathematical modelling language integrated within a platform that provides tools for automatic model verification are.

- the correctness of the models is ensured by the fulfilment of the proof obligations,
- the finiteness of the algorithm is ensured by checking a simple property,
- using ProB for model checking, we avoid common issues in the code as deadlocks or invariant violations,
- the power of computation is higher than that of a large number of commonly used programming languages.

Some disadvantages are

- the length of some models might overcome the one of the implementations of the same algorithms in a common programming language,
- the values of the variables are restricted to Boolean and integers.

Our future work will concentrate on implementing more complicated algorithms, comparing the computational power of our models with the computational power of the same algorithms implemented in a classical programming environment, extending our approach to NP-complete problems, building a tool for automatic translation of an algorithm given in pseudocode to the corresponding Event-B model and integrating this tool as a Rodin plugin.

References

1. Abrial JR (2010) Modeling in event-B. Cambridge University Press
2. RODIN website: http://rodin.cs.ncl.ac.uk
3. DEPLOY project website: http://www.deploy-project.eu/index.html
4. ProB website: https://prob.hhu.de/
5. Bendisposto J, Fritz F, Jastram M, Leuschel M, Weigelt I (2011) Developing Camille, a text editor for Rodin. In: Software practice and experience, 41, pp 189–198
6. UML-B website: https://www.uml-b.org/
7. Clarke EM, Grumberg O, Peled DA (1999) Model checking. The MIT Press
8. Faser G, Wotawa F, Ammann P (2009) Testing with model checkers: a survey. Softw Testing, Verification Reliab 19(3):215–261
9. Clarke EM, Klieber W, Nováček M, Zuliani P (2011) Model checking and the state explosion problem. LNCS, volume 7682
10. Lefticaru R et al (2012) Towards an integrated approach for model simulation, property extraction and verification of P systems. In: Proceedings of the tenth brainstorming week on membrane computing. Sevilla, 291–318
11. Ţurcanu A et al (2014) Modelling and analysis of *E. coli* respiratory chain. In: Frisco P, Gheorghe M, Pérez-Jiménez M (eds) Applications of membrane computing in systems and synthetic biology. Emergence, complexity and computation, vol 7. Springer
12. Ait Ameur Y, Ait Sadoune I, Hacid K, Mohand Oussaid L (2018) Formal modelling of ontologies: an Event-B based approach using the Rodin platform. In: Electronic proceedings in theoretical computer science, 271: 24–33

13. Hussain S, Farid S, Alam M, Iqbal S, Ahmad S (2018) Modeling of access control system in event-B. Nucleus 55(2)
14. Predut SN, Ipate F, Gheorghe M, Campean F (2018) Formal modelling of cruise control system using event-B and Rodin platform. IHPCC/SmartCity/DSS, 1541–1546
15. Turcanu A, Shaikh T, Mazilu CN (2020) On model checking of a robotic mechanism. J Robot Autom 4(1)
16. Turcanu A, Ipate F (2021) Using the Rodin platform as a programming tool, GitHub repository, 2021. https://github.com/aTurcanu85/Event-B-models-of-classic-algorithms.git

Network Security

P2PRC—A Peer-To-Peer Network Designed for Computation

Akilan Selvacoumar, Ahmad Ryad Soobhany, and Benjamin Jacob Reji

Abstract This paper focuses on developing the peer-to-peer rendering and computation (P2PRC) framework, which is a distributed framework for executing computationally demanding tasks that a personal machine with limited processing power will struggle to run such as graphically demanding video games, rendering 3D animations, and protein folding simulations. A custom peer-to-peer network was implemented to decentralize the execution of tasks either on central processing unit (CPU) or graphical processing unit (GPU), in order to increase the bandwidth for running tasks. To prevent the tasks in the peer-to-peer network from corrupting the operating system (OS) of the server, they will be executed in a virtual environment in the server. The user acting as the client is provided full flexibility on how to batch the tasks, and the user acting as the server has complete flexibility on tracking the container's usage and killing the containers at any time. The effectiveness of the network and the performance of the distributed task execution of the distributed framework were evaluated using Horovod and TensorFlow benchmarks. Preliminary results are very promising with 86 and 97% improvements for CPU and GPU distribution, respectively.

Keywords P2P networks · Distributed computation · Virtualization · P2PRC

1 Introduction

Scientific simulations and video games with 3D rendering or imaging applications are computationally demanding for the CPU or GPU of one computer. Most domestic users or small labs do not run computationally intensive tasks on a daily basis. Thus, buying a powerful computer every few years to run a bunch of heavy tasks, which are not executed frequently to reap the benefits of the powerful CPU or GPU, can be costly and inefficient utilization of hardware. Many users rely on PCs/laptops or servers

A. Selvacoumar (✉) · A. R. Soobhany · B. J. Reji
Heriot-Watt University, Dubai, UAE
e-mail: me@akilan.io

A. R. Soobhany
e-mail: r.soobhany@hw.ac.uk

A. Ullah et al. (eds.), *Proceedings of International Conference on Information Technology and Applications*, Lecture Notes in Networks and Systems 350,
https://doi.org/10.1007/978-981-16-7618-5_45

that belong to a server farm to run computationally intensive tasks. Renting servers to run the computationally heavy tasks can be really useful. Ethically speaking, this is leading to a monopolization of computing power similar to what is happening in the Web server area. By using peer-to-peer principles, it is possible to remove the monopolization factor and increase the bandwidth between the client and server [1]. Based on Moore's law, there would need better distribution and communication between computers for certain heavy tasks [2]. The manufacturer Nvidia changed the license conditions for its GeForce and Titan graphics cards at the end of 2017 and prohibits the use of inexpensive cards in data centers. Instead, Nvidia offers the professional graphics card Tesla (starting at approx. 8000 USD) for data centers [3]. This means that it is very expensive to rent dedicated GPUs from data centers.

This paper introduces a custom a peer-to-peer (P2P) network framework peer-to-peer rendering and computation (P2PRC), where a user can use the P2Pnetwork to act as a client (i.e., sending tasks) or the server (i.e., executing the tasks). An open-source implementation has been developed by the authors of this paper, which comes bundled with a P2Pmodule and it is possible to execute Docker containers or virtual environments across selected nodes. The financial incentive part of distributed rendering and task execution is not within the scope of this paper.

Research looking into computation for distributed or peer-to-peer networks has been performed for several decades. There have been a lot of recent developments in this area. Some of these developments have led to the implementation of different applications and frameworks such as Folding@Home and Golem network among others.

Folding@home is a distributed computing project used to help scientists simulate protein folding, which uses a client and server architecture. The clients (run by volunteers) request the server for tasks to process. The Folding@Home architecture is really flexible and can be used to support other types of projects [4]. The Golem network claims to be the first truly decentralized supercomputer. The main goal of Golem is to create a global market of computing power. Golem connects nodes/computers in a decentralized network (i.e., peer-to-peer network). The user can both request and lend resources. Golem has similar objectives to this research, which is to provide an alternative to cloud rendering providers like AWS and Google Cloud. Golem network uses libp2p [5] and IPFS [6] to run its peer-to-peer network. Golem uses ethereum smart contracts to run as its financial incentive layer [7].

BitWrk's main focus is to provide peer-to-peer rendering for Blender. BitWrk can be executed in any local network with other nodes to reduce bandwidth cost. BitWrk uses bitcoin as the mode of payment for using computing power. The whole implementation of the application is written in the Go programming language. The client enables control of trade currently in progress [8].

DrQueue is an open-source render farm system, similar to Farmer-Joe-Render. DrQueue is used in the visual effects, science, and finance industries. The objective of DrQueue is to batch tasks and runs as a task management tool for multiple nodes. It is compatible with Windows, macOS, Linux, Irix, and FreeBSD. DrQueue can be used with any renderer that supports a command line interface (CLI) [9].

Horovod is a distributed deep learning framework for TensorFlow, Keras, PyTorch, and Apache MXNet. The objective of Horovod is to take a single GPU training script and make it scale across multiple GPUs in parallel. Horovod has developed benchmarks which prove that it has a 90% scaling efficiency in certain runs. The Horovod project is open source under the Apache license [10]. A Docker container is a standard unit of software that packages code and all its dependencies so that the application executes quickly and reliably from one computing environment to another. A Docker container image is a lightweight, standalone, and executable package of software that include everything needed to run an application: code, runtime, system tools, system libraries, and settings. Docker is great to create containerized environments [11].

Based on the analysis of the tools discussed, the closest network similar to this project is the Golem network, which lacks creating virtual environments for the specific task (i.e., all the virtual environments have to be preset). The novelty of the work presented in this paper is to implement the features that are not implemented in the Golem network and to develop a simpler algorithm to run custom tasks on the peer-to-peer network. The initial plan was to ensure that Horovod is compatible to run on this project. This paper is organized by providing a detailed description of the design architecture of the proposed framework in Sect. 2. The implementation process of the framework is explained in Sect. 3, the experimental setup and results, with a discussion of the results, are described in Sect. 4. Finally, the paper is concluded in Sect. 5.

2 P2PRC Design Architecture

The objective would be to have a good understanding on the purpose of each module and how they interact with each other. The design architecture was inspired and based on the Linux kernel design [12]. The project is segmented into various modules. Each module is responsible for certain tasks in the project. The modules are highly dependent on each other, hence the entire codebase can be considered as a huge monolithic chunk that acts as its own library.

2.1 Client Module

The client module interacts with the P2P module 2.3 and Server module. It is responsible for interacting with the server module and appropriately updating the IP table on the client side. It connects to the server using the server's REST APIs. It is also the primary decision maker on how the IP table is updated is on the client side. This is because each user can have requirements like how many number of hops they would want to do to update their IP table. Hops are the number of times the client is going to download the IP table from different servers, once it gets the IP tables from the previous servers (Fig. 1).

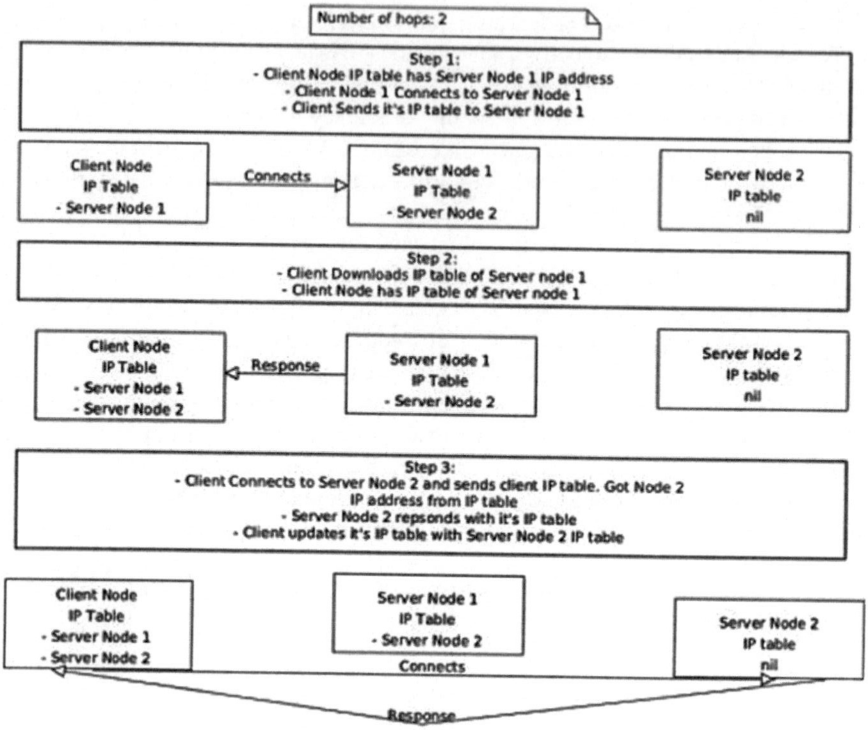

Fig. 1 Refers to how new server addresses are detected by hopping through the network

2.2 *Server Module*

The server module takes care of setting and removing the virtualization environment (i.e., containers) for accessing and doing the appropriate computation. It also interacts with the P2Pmodule to update the IP table on the server side. The server module accesses information regarding CPU and GPU specifications of the machine running the server module. To do speed tests, the server has routes that allow it to upload and download a 50 MB file.

2.3 *Peer-To-Peer Module*

The P2P module (i.e., peer-to-peer module) is responsible for storing the IP table and interacting with the IP table. In the following implementation of the P2P module, the IP table stores information about servers available in the network. In the other functionality, the P2P module takes care of is doing the appropriate speed tests to the servers in the IP table. This is for informing the users about nodes that are close by and nodes that have quicker uploads and downloads speeds. The module is responsible to ensure that there are no duplicate server IPs in the IP table and to remove all server IPs which are not pingable.

3 Implementation

The programming language used for this project was Golang [13]. The reason Golang was chosen because it is a compiled language. The entire codebase is just a single binary file. When distributing to other Linux distributions, the only requirement would be the binary file to run the code. It is easy to write independent modules and be monolithic at the same time using Go [14] (Fig. 2).

3.1 Command Line Interface Module

The CLI (i.e., command line interface) is the only one in which the user can directly interact with the modules in the project. The objective when building the CLI was to have the least amount of commands possible. The CLI was built using the library called urfave CLI v2. They were two major files created named as flags.go and actions.go.

The flags.go file is responsible to create the appropriate flags for the CLI. There are two types of flags called Boolean and string. Each of the flags outputs is assigned to a variable to be handled. The flags can also detect environment variables set.

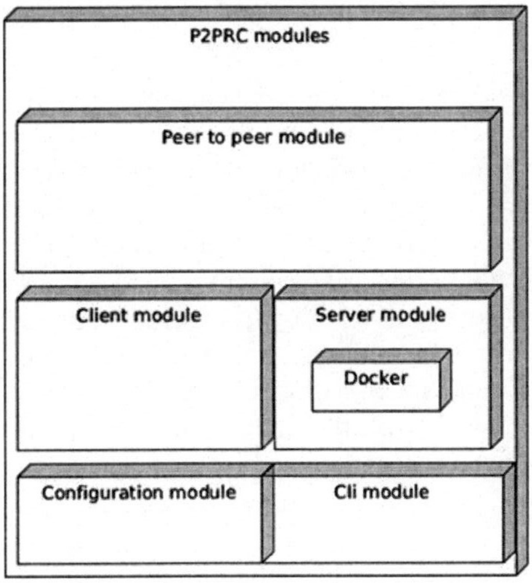

Fig. 2 Refers to all the modules that packed in the entire project and generated binary

The actions.go file is implemented to call the appropriate functions when the flags are provided. It interacts directly with the modules in the project. Action.go checks if variables are not an empty string or the Boolean value is true.

3.2 Server Module Implementation

This section focuses on an in-depth understanding of the server module implementation. The server module can be split into various sections, where each section will provide information on how a certain feature works.

The Web framework used for the server module is called Gin, which was chosen due to its wide use and strong documentation available on the official GitHub repository. The default port used is 8088. For version 1.0 of the project, the server needs to keep port 8088 open to ensure that other clients and servers can detect it. The possible requests available are GET and POST for this implementation. The possible responses are either a string or JSON response or a file. In the majority of routes, a string response refers to an error when calling the following routes. The route implemented for the Web framework is explained in details on the GitHub repository [15].

The Docker API section provides information on how the server module interacts with the Docker containers. The server calls two routes that either create or remove the Docker container. Docker has a huge advantage because it takes less than 20 s to spin up a new container once it is built and executed at least once. For Docker operations, a separate module/package has been created.

3.3 Peer-To-Peer Module Implementation

The peer-to-peer implementation was built from scratch. This is because other peer-to-peer libraries were on the implementation of the distributed hash table. At the current moment, all those heavy features are not needed because the objective is to search and list all possible servers available. The limitation being that to be a part of the network the user has to know at least one server and has to have DMZ enabled from the router if the user wants to act as a server out of the users local network. The advantage of building from scratch makes the module super light and possibility for custom functions and struct. The sub topics below will mention the implementations of each functionality in depth.

The IP table file is a JSON as the format with a list of servers IP addresses, latencies, downloads, and uploads speeds. The functions implemented include read file, write file, and remove duplicate IP addresses. The remove duplicate IP address function exists because sometimes servers IP tables can have the same IP addresses as what the client has. The path of the IP table JSON file is received from the configuration module (Fig. 3).

Fig. 3 Sample structure of the IP table file

```
{
"ip_address": [
  {
    "ipv4": "<ipv4 address>",
    "latency": "<latency>",
    "download": "<download>",
    "upload": "<upload>"
  }
 ]
}
```

The speed test functions populate the fields which are latency, download, and upload speed. Before, the speed test begins for each server IP address. The P2Pmodule ensures that each server IP address is pingable. If the server IP address is not pingable, then it removes that IP address from the struct.

3.4 Client Module Implementation

The client module is in charge of communicating with different servers based on the IP addresses provided to the user. The IP addresses are derived from P2Pmodules. The objective is to show how the client module interacts with the P2Pmodule and server module.

For updating the IP table, the client module initially calls the P2Pmodule to get the local IP table. Based on the servers IP addresses available, it calls the speed test function from the peer-to-peer module to update IP addresses with information such as latencies, download, and upload speeds. Once this is done, the client module does a REST API call to the server to download its IP table. Once the hops are done, it writes the appropriate results to the local IP table. Once this is done, it prints out the results. For reading server specifications, the client module calls the route /server specs and reads the JSON response. If the JSON response was successful, then it just calls the pretty print function which just prints the JSON output in the terminal. The client module uses the servers REST APIs to create and delete containers.

4 Experimental Setup and Results

This section evaluates the implementation by doing certain tests and benchmarks. The first test was to test the effectiveness of the network, and the second one was a benchmark to prove the performance boost when running code distributed on the following project.

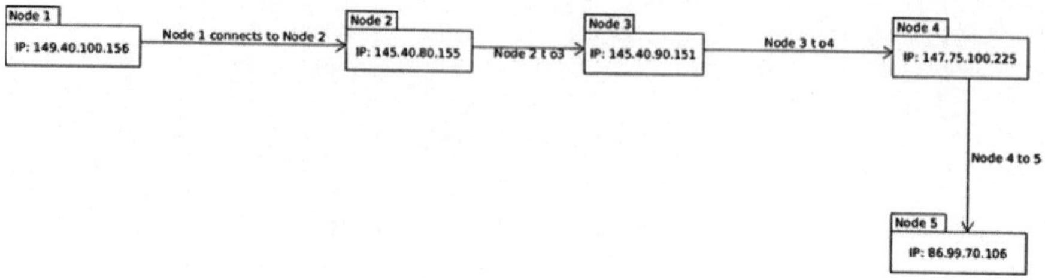

Fig. 4 Visual representation of test network scenario

4.1 Testing Network Scenario

The objective is to test the P2Pnetwork and the effectiveness of updating the IPtables. The client and the server are given the impression of a zero configuration setting. For testing, there will be a test network set. In the testing scenario, all nodes can be client and server because the IP table does not store clients IP addresses. At current number of hops would be three as default. The test network consists of five nodes acting as a client and server. The objective would be to have the entire IP table updated in each node with interacting with only one node once. Each node has knowledge of one node only (Fig. 4).

All nodes except node one were able to have information of IP addresses in the test net. This was due to the three hops set as default. Node one had in its IP table IP addresses of Node two, Node three, Node four. Once the number of hops was set to four, the objective of the test was achieved.

4.2 P2P Network Computation Performance

The P2PRC was tested using Horovod, a distributed deep learning training framework. The TensorFlow2 [16] synthetic benchmark.py available in the GitHub repository of Horovod was used for benchmarking performance. They were five different scenarios to test the capability and performance of P2PRC. A tenfold stratified cross-validation was performed, where the benchmark was run ten times on each scenario with each image set changed at each iteration. The average value was recorded. The results depict how many images were processed in each run and how much variations were present between iterations. Each benchmark consists of ten warm-up iterations, followed by ten iterations with ten image batches being loaded (each fold). The model used for testing was ResNet50. All machines used for testing shared identical resources containing a singular CPU and GPU.

A discussion of the results presented in Fig. 5 is performed below for each scenario:

- Single Machine: This was a baseline case to compare other results to. The benchmark was run on a machine without P2PRC and directly using Horovod.

	CPU Horovod	GPU Horovod
Single Machine	6.8 +-0.8	71.4 +-0.5
P2PRC - Case1	3.2 +-0.3	30.3 +-0.7
P2PRC - Case2	12.1 +-0.6	138.9 +- 0.4
P2PRC - Case3	17.7 +-0.4	204.2 +- 0.9
P2PRC - Case4	23.5 +-0.3	277.1 +-0.2

Fig. 5 Shows the average image processed per second in each scenario

- P2PRC Case 1: This scenario depicts that a single machine with two nodes running uses shared resources, and hence, the images processed is effectively halved since two benchmarks are being run simultaneously. A slight reduction in performance can be observed when multiple nodes share common host resources due to the added overhead of executing the tasks and sharing processes with the OS.
- P2PRC Case 2: This scenario resembles the most common usage for the P2PRC tool. Here, two different machines each with a single node running are used to run the benchmark and we can see the improvement in the score. The score is nearly doubly for both CPU and GPU benchmarks. This is because both machines with their discrete devices can be used to solve a singular task.
- P2PRC Case 3: This is an extension of case 2 with three discrete machines to run the benchmark, and the performance is roughly thrice the baseline performance, thanks to the added processing power available to compute the benchmark.
- P2PRC Case 4: This is an extension of case2 with four discrete machines to run the benchmark, and the performance is roughly quadrupled compared to the baseline performance. This gives us a rough idea of having more nodes on discrete hardware allows us to parallelize, and thus reduces the computation time needed.

The performance improvement by using four machines results in the number of images processed increase by 86.4 and 97% for the CPU and GPU, respectively. The results show that P2PRC can be used to improve the distributed computation of computationally intensive tasks.

5 Conclusion

This paper focused on developing a framework to decentralize the computation of batch tasks to nodes, instead of using centralized servers, to increase bandwidth. The results showed that P2PRC can be used to improve the distributed computation of computationally intensive tasks. Future work will include the client module to be able to read the number of hops needed from the configuration file. The client must be able to send gRPC calls instead of REST API calls. The P2Pmodule should provide for UPNP and suggest users to not use DMZ. IP tables should store the port number the different servers are running on.

References

1. Kilcioglu C, Rao J (2016) Competition on price and quality in cloud computing. In: Proceedings of the 25th international conference on world wide web, pp 1123–1132. https://doi.org/10.1145/2872427.2883043
2. Furber S (2008) The future of computer technology and its implications for the computer industry. Comput J 51:735–740
3. Comment B, Why computing power needs to change. , https://www.datacenterdynamics.com/en/opinions/why-computing-power-needs-change/
4. Beberg A, Ensign D, Jayachandran G, Khaliq S, Pande V (2009) Folding@home: Lessons from eight years of volunteer distributed computing. IEEE international symposium on parallel and distributed processing, pp 1–8
5. Libp2p docs, https://docs.libp2p.io/
6. Benet J (2014) IPFS—content addressed, versioned, P2P file system
7. Golem.Network wiki, https://docs.golem.network
8. Eschenburg J. indyjo/bitwrk (2021) https://github.com/indyjo/bitwrk, original-date: 2013--31T21:47:43Z
9. Seppa¨la¨ H, Suomalainen N, Integrating open source distributed rendering solutions in public and closed networking environments
10. Sergeev A, Del Balso M (2018) Horovod: fast and easy distributed deep learning in tensorflow. ArXiv:1802.05799 *[cs, Stat]*. http://arxiv.org/abs/1802.05799
11. Cito J, Ferme V, Gall H (2016) Using Docker containers to improve reproducibility in software and web engineering research
12. Mauerer W (2008) Professional Linux Kernel architecture. Wrox Press Ltd.
13. Westrup E, Pettersson F (2014) Using the go programming language in practice
14. Selvacoumar A, Akilan1999/p2p-rendering-computation (2021) https://github.com/Akilan1999/p2p-rendering-computation, original-date: 2020-10-31T18:54:18Z
15. Selvacoumar, A. Akilan1999/p2p-rendering-computation (2021) https://github.com/Akilan1999/p2p-rendering-computation/blob/4f10936664da7ddd6758a1301403e2eb180b739d/Docs/ServerImplementation.md, original-date: 2020-10-31T18:54:18Z
16. Abadi M, Barham P, Chen J, Chen Z, Davis A, Dean J, Devin M, Ghemawat S, Irving G, Isard M, Kudlur M, Levenberg J, Monga R, Moore S, Murray D, Steiner B, Tucker P, Vasudevan V, Warden P, Zhang X (2016) TensorFlow: a system for large-scale machine learning

Touchless Biometric User Authentication Using ESP32 WiFi Module

Rikesh Makwana and Talal Shaikh

Abstract Due to the ubiquitous nature of WiFi, the use of WiFi signals for Biometric User Authentication (BUA) is ongoing research which has previously focused on using multi-antenna commercial off-the-shelf (COTS) devices such as Intel 5300 or Atheros 9390. However, due to high cost and limited availability, COTS devices are restricted to small scale deployment. To overcome this issue, researchers propose using Espressif ESP32, an inexpensive single antenna microcontroller equipped with WiFi and Bluetooth modules capable of capturing detailed WiFi Channel State Information (CSI). This paper explores and extends the application of ESP32 by proposing a model for device-less and touch-less BUA systems using a simple client–server architecture. The system identifies users as they perform day-to-day activities by recognizing behavioural and physiological characteristics using LSTM—a deep learning approach. Furthermore, the paper describes the Python tool developed for parsing and filtering WiFi CSI data.

Keywords Channel state information · Biometric user authentication · Ubiquitous computing · ESP32 microcontroller

1 Introduction

Increasing interest in biometric technology stems from the need to provide users with more secure authentication options. The widespread use of biometrics has numerous applications, ranging from unlocking devices with a fingerprint to providing sophisticated law enforcement security systems using fingerprint, iris, facial, and voice biometrics.

Biometric authentication is a better alternative to traditional user authentication due to the following reasons:

1. Ease of use and accessibility.

R. Makwana (✉) · T. Shaikh
Heriot-Watt University Dubai, Dubai Knowledge Park, Dubai, UAE

T. Shaikh
e-mail: t.a.g.shaikh@hw.ac.uk

A. Ullah et al. (eds.), *Proceedings of International Conference on Information Technology and Applications*, Lecture Notes in Networks and Systems 350,
https://doi.org/10.1007/978-981-16-7618-5_46

2. Biological biometrics (fingerprints, iris, and gait) have a higher level of authenticity than passwords and authentication tokens.

However, existing systems use dedicated sensors to capture biometric data such as cameras, scanners and wearable devices, exposing the system to vulnerabilities. To eliminate the need for these additional sensors, a number of recent studies suggest using WiFi signals to capture behavioural and physiological characteristics. As the future of technology is envisioned to be device-less and touch-less, there is a growing interest in WiFi signals and their applications.

In recent years, there have been more and more applications developed using WiFi signals. Studies conducted by [9] and [10] suggest using gait patterns to authenticate a person. The system, however, is not suitable for practical use because it requires walking through predetermined paths to capture the gait. In an effort to overcome these limitations, other researchers [5, 8] have proposed recognizing both stationary and walking activity for a BUA system that does not require any active participation. According to [1] and [4], current deployments rely on COTS devices, such as the Intel 5300 or the Atheros 9390, which are only suitable for small deployments. They suggest using the inexpensive and widely available Espressif ESP32 as a microcontroller equipped with Bluetooth and WiFi modules capable of capturing detailed WiFi CSI.

This paper analyses the existing WiFi CSI systems and proposes a BUA system that uses ESP32. Furthermore, a Python tool is developed for parsing and filtering WiFi CSI, along with the ESP32 CSI Toolkit [4].

2 Related Works

Adhering to the vision of device-less and touch-less technology, several researchers propose the use of WiFi CSI sensing [6, 8–10]. This measure is an amalgamation of signals reflected by the human body and the environment. A CSI system can be deployed and operated regardless of the lighting conditions, making it suitable for multiple applications.

In contrast to traditional methods that rely on cameras and additional sensors to determine the user's gait pattern, the study [9] introduces *WiFiU*, a gait recognition system using COTS WiFi devices. Figure 1 shows how the system captures WiFi CSI data on a laptop using an Intel 5300 WiFi NIC. The system detects the user's gait patterns based on the deviation between CSI signals caused by the user walking in the designated path.

The extraction of unique gait patterns takes place as follows,

1. Conversion of amplitude from CSI to a spectrogram in the time–frequency domain.
2. Spectrogram enhancement and Principal Component Analysis (PCA) using 180 Principal Components (PCs) to denoise the data. Data is denoised by ignoring the first PC, then using the rest for feature extraction.

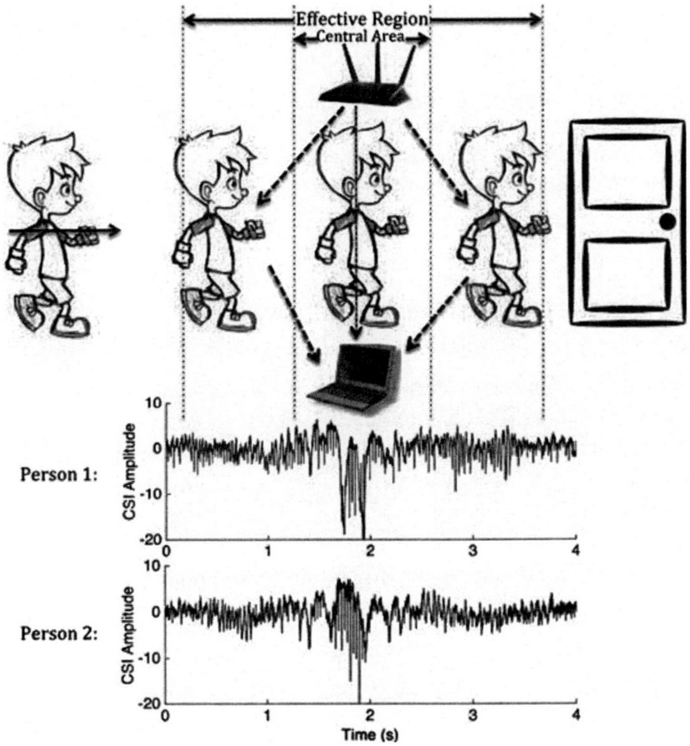

Fig. 1 Discrepancy in CSI signals caused by the user walking [10]

3. Furthermore, Spectrogram Signatures can also be determined using gait parameters such as speed and step length.

Data gathered by performing tests show the accuracy of WifiU to be 80–90%. However, certain limitations, such as the subject walking a predetermined path, are placed on the experimental environment to achieve this high accuracy.

To address these limitations, [8] introduced a robust BUA tool by collecting WiFi CSI data from the user's daily activities, including walking and stationary activities such as entering a room, using personal devices, and writing.

The extraction of unique characteristics takes place as follows,

1. Stable and reliable subcarriers that represent activity characteristics are selected using the developed subcarrier selection algorithm.
2. The data is denoised while maintaining the features using a Bandpass Butterworth Filter.
3. Human activity is detected by applying Short Time Energy (STE) to the CSI amplitude. STE is sensitive to subtle body movements in WiFi signals, making it possible to detect walking and stationary activities.

4. Further, a three-layered Deep Neural Network (DNN) based on AutoEncoder is used to get a high-level abstraction of the features.

According to the results of this experiment, the amplitude and phase of CSI can both be used to achieve over 90% accuracy for activity recognition and user identification. This high accuracy was recorded even with a limited training set of 4 subjects.

3 System Design and Implementation

Linux 802.11n CSI Tool and Atheros CSI Tool are two well-known tools for collecting CSI data using COTS Intel and Qualcomm (Atheros) WiFi NIC, respectively. The CSI collection process involves using two devices, transmitter (Tx) and receiver (Rx). Existing COTS systems use Personal Computer (PC) equipped with Intel and Atheros NIC as receiving devices and COTS routers as transmitting devices.

Recent papers [1, 4] indicate the practical limitations of these devices and propose the use of ESP32 to be advantageous due to the following reasons:

- Intel 5300 and Atheros NIC consist of 30 and 56 subcarrier groups, respectively, operating at 20 MHz. However, ESP32 provides rich CSI measures from 64 (and above) subcarriers [4] working at 20 and 40 MHz, making ESP32 suitable for tasks such as breathing, heart rate, activity monitoring.
- While Intel 5300 and Atheros require additional hardware for both Tx and Rx, ESP32 operates in both modes without the need for additional devices.
- ESP32 enables large scale implementation and deployment capability. Due to the ability to operate as a standalone device, ESP32 has a flexible architecture compared to Intel 5300 and Atheros.

This paper explores the application of ESP32 in BUA by recognizing human behavioural and physiological characteristics while performing day-to-day activities such as walking and writing. The proposed system uses two single antennae ESP32 devices in a simple client-server configuration. Figure 2 illustrates the system overview.

3.1 *Experimental Setup*

Hardware Setup: Two ESP32 WiFi modules are used in the proposed experimental system to create a simple client-server architecture. The ESP32 can be configured in a variety of ways depending on the application. The specific modules used in this project is the WROOM32-U development kit (dev kit) with a Haysenser 10dBi high gain antenna connected via a U.Fl connector and IPEX pigtail cable (Fig. 3).

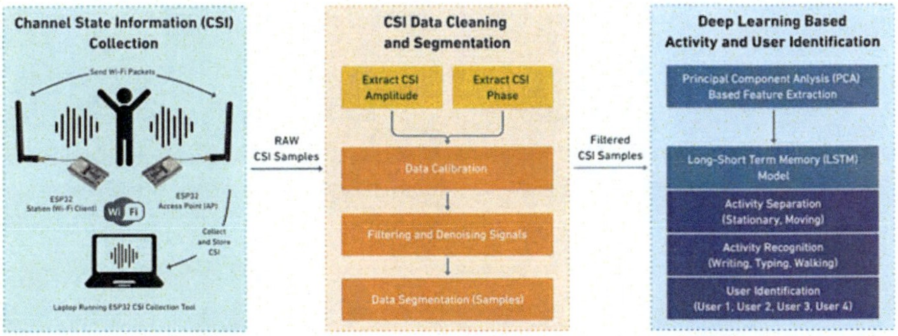

Fig. 2 Proposed ESP32 biometric user authentication system overview

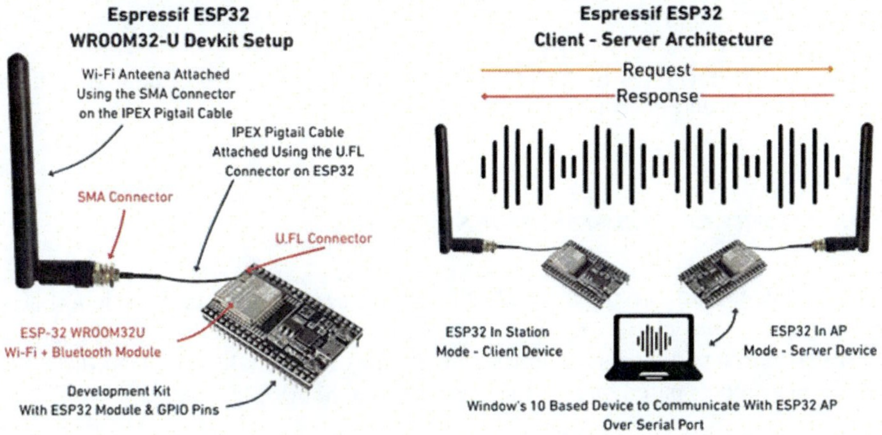

Fig. 3 Espressif ESP32 experimental hardware setup for a simple client—server architecture

As shown in Fig. 3, the two ESP32s are configured as a Station (client) and an Access Point (AP) (server) respectively, with a PC to communicate with the AP.

Software Setup: The ESP32 CSI Toolkit introduced by Hernandez and Bulut[1] [4] is an open-source toolkit that allows direct access to the CSI from the ESP32 device. The code repository contains sub-programmes for the various configuration modes (station and AP). Besides being a new toolkit, it is well documented and includes instructions for configuring and collecting CSI data, which is its primary purpose.

The AP here is configured to actively collect CSI data, so the station device(s) can connect and request packet information. ESP32 defaults to a transmitting rate of 100 packets/second. When collecting high-grained CSI measures, the packet rate is set to 1000 packets/seconds for the AP device. As noted in [4], the packet rate of devices can vary depending on their environment and distance from each other. The

[1] https://github.com/StevenMHernandez/ESP32-CSI-Tool

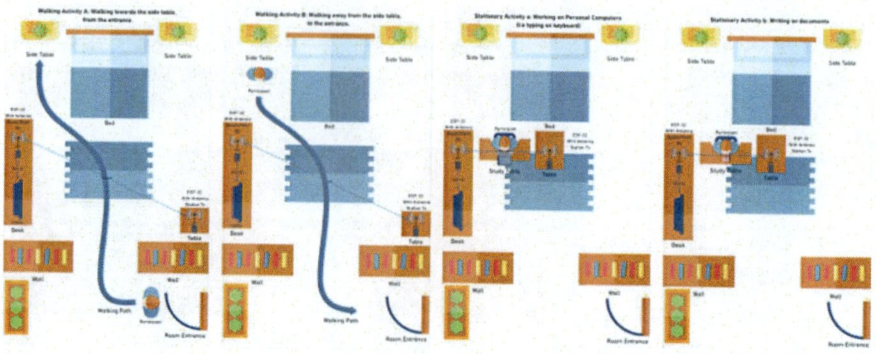

Fig. 4 Walking and stationary activities in a home setting. **a** Entering the room. **b** Exiting the room **a** Working on PC and **b** Writing on documents

original work by [8] and [4] was extended by taking advantage of both amplitude and phase information provided by ESP32.

3.2 *Activity and Environment Setup*

The experiment aims to uniquely identify users based on human behavioural and physiological characteristics while performing day-to-day activities. The study consisted of four participants (3 females and 1 male). Figure 4 illustrates two different ESP32 placements based on activities performed at home.

1. **Walking Activity—Public Space Proxemics:** The ESP32s are posi tioned across the room at a distance of 3.1 m from the user.
2. **Stationary Activity—Personal Space Proxemics:** The ESP32s are positioned across the study table at a distance of 0.8 m from the user.

3.3 *Experimental Protocol*

Each activity is completed by one user at a time. The participants are required to complete a set of walking tasks and stationary tasks.

Walking Activity: The participant is initially instructed to stay outside the room for five seconds before beginning the activity. Once the observer starts the timer,

- The participant walks across the room from the entrance (estimated time 10 s).
- Before walking back and leaving the room, the participant stands still for 5 s.
- The participant must repeat the above steps for a total of 10 rounds.

Stationary Activity: The participant is initially instructed to sit still at the study table for five seconds before beginning the activity. Once the observer starts the timer,

- The participant writes from the provided text sample(s) for the next 20 s.
- At the end of the writing activity, the participant sits still for 5 s.
- The participant must repeat the above steps for a total of 10 rounds.

3.4 Collecting and Understanding CSI Data

Recording and Labelling Data: The ESP32 CSI Toolkit outputs the collected data in a simple CSV file, due to its ease of use with any programming language (Python in our case). As a convenience measure, the CSI data is collected for a fixed duration of time and the dataset is named according to a predefined convention. The Python tool developed (Sect. 3.5) facilitates extraction of amplitude and phase information from the CSI data, as well as its filtration.

Understanding CSI Data: The CSI data collected from ESP32 is a sequence of two bytes of signed (positive/negative) characters [3] that represent the Channel Frequency Response (CFR) between the Tx and Rx subcarriers. These values determine the signal's amplitude and phase.

According to the Espressif ESP32 specifications [3], there are three CFR types based on the type of packet received. After filtering the signal with the Python tool, the data collected is of the following form:

1. **Signal Mode**: 40 MHz transmits packets at high bandwidth, allowing more fine-grained data to be transmitted (high throughput).
2. **Spacetime Blocked Code (STBC) Information**: All signals are Non- STBC due to the presence of a single antenna on each ESP32 node. ESP32 does not support multi-antenna operation but can receive data from any device with multiple antennas.
3. **CFR Subcarriers**: At 40 MHz and Non-STBC mode, the subcarrier has a length of 384 bytes or 192 subcarriers consisting of Long Training Field (LTF) (128 bytes = 64 channels) and High-Throughput LTF (HT-LTF) (256 bytes = 128 channels). Figure 5 illustrates all 192 subcarriers received. On examination of the graph, it is clear that signals are composed of Guard or Pilot subcarriers that contain no data and can be removed, resulting in 166 data subcarriers. Additional code and images available on GitHub[2].

3.5 ESP32 CSI-Python-Parser Tool

A Python tool has been developed for parsing, processing (filtering), and plotting ESP32 CSI data. The lack of a Python open-source tool and the relatively new state of ESP32 CSI inspired the creation of this tool. MATLAB tools for Intel 5300 and Atheros are available, but no similar tools exist for Python. As a result, the

[2] https://github.com/RikeshMMM/ESP32-CSI-Python-Parser/tree/main/examples

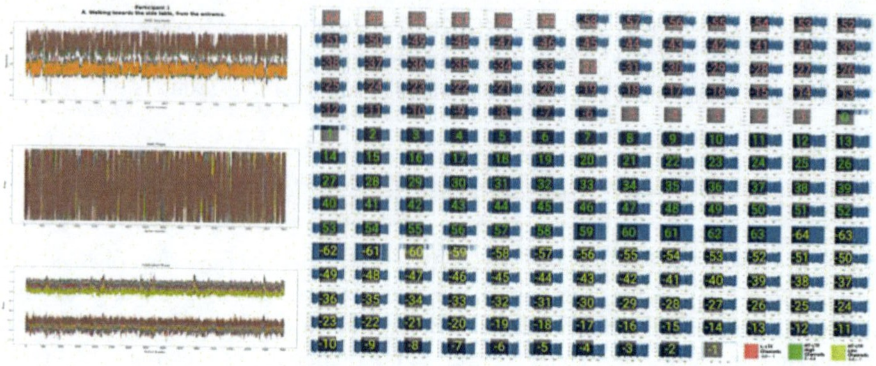

Fig. 5 LLTF and HT-LTF subcarriers for 40 MHz signal

project contributes to the open-source community. The repository can be accessed at GitHub.[3]

3.6 *Deep Learning-Based Activity and User Identification*

Model Training: In previous studies, LSTMs have been shown to be an effective model for high accuracy device-free WiFi sensing tasks [2, 7]. This experiment uses supervised learning to train the LSTM model to identify users based on the activity CSI data. A total of 160 CSI samples are collected, of which 70% are used for training and 30% for testing the model. The following are the steps for training and testing the model:

- LSTM requires that the training data be formatted in a specific three- dimensional way (sample, timestep, features). The 'samples' identify the rounds of activity, the 'timestep' is the duration of activity (number of packets) and the 'features' specify the number of subcarriers available. As an example, the walking activity is split into 10 samples, each lasting 1000 packets with 166 subcarriers. Finally, the training input shape obtained is (10, 1000, 166).
- The number of epochs and batch size are set to 100 and 64, respectively, with each LSTM layer containing 100 memory units. Data from a pilot study has been used to establish these values.
- For multi-class classification (type of activity and the user performing it), the final layer of the network must be a dense layer with softmax activation.
- Finally, the model is compiled using categorical cross-entropy where the probability of the data in each class, is identified.

[3] https://github.com/RikeshMMM/ESP32-CSI-Python-Parser

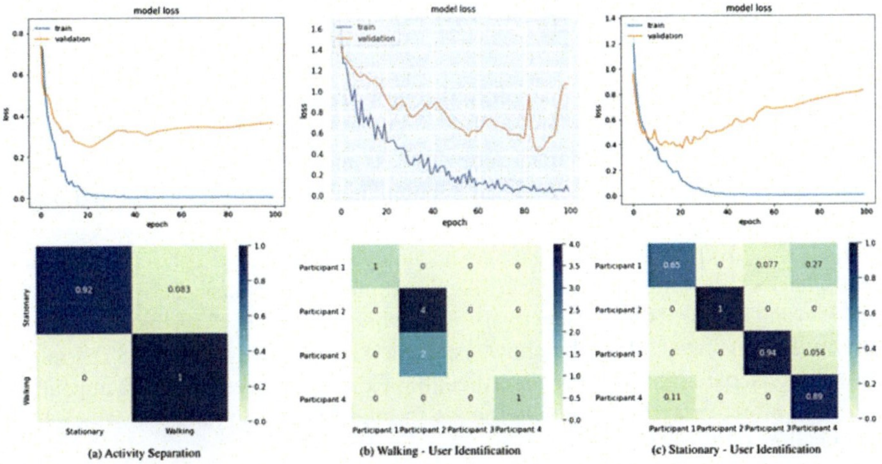

Fig. 6 LSTM model loss and confusion matrix

Model Evaluation

Activity Separation: The first step was to determine if the activity is 'Walking' or 'Stationary'. The activity separation LSTM model was initially trained for 100 epochs and it resulted in an accuracy of 90%. Figure 6 shows the results of this evaluation. It was apparent from the confusion matrix that the model could classify moving activities easily but not stationary activities. In the loss graph, notice that increasing the epoch to 100 allowed the model to learn detailed features (subcarrier details), but overfit the model.

Walking—User Identification: The model performed with an accuracy of 64% (±11%). The model was evaluated by considering the average accuracy over 10 runs. The training vs validation loss graph indicated that the model overfits. The confusion matrix showed that the model was not able to identify participant no. 3. During the experiment participant no. 3 was found to have a greater walking pace when compared with the rest of the participants. This implies that the model requires more data to learn characteristics of users with varying paces.

Stationary—User Identification: The model performed with an accuracy of 85% (±10%). The model was evaluated by considering the average accuracy over 10 runs. The training vs validation loss graph indicated that the model overfits. The confusion matrix showed that the model is not able to identify a pattern(s) in the typing sample of participant no. 1. In the experiment participant no. 1 was observed to write with their left hand (away from the Rx), as opposed to the other participants who wrote with their right hand (closer to the Rx).

3.7 Conclusion and Future Works

The purpose of this paper is to explore the applications of the ESP32, which is an inexpensive single antenna microcontroller with WiFi and Bluetooth modules that can capture detailed WiFi CSI data. A device-less and touch-less BUA sys- tem were proposed to uniquely identify users by recognizing human behavioural and physiological characteristics while performing day-to-day such as walking and working on their PC. Overall the model gave an accuracy of about 90% for activity separation. The accuracy of user identification in walking and stationary activity was 64% (±11%) and 85% (±10%), respectively. This work contributes to the field of Device-Free WiFi Sensing (DFWS) by showcasing the high accuracy of the LSTM model for activity separation and user identification. Comprehensively, ESP32 appears to be a cost-effective and scalable alternative to Intel 53000 and Atheros, which have applications in diverse scenarios.

Due to the limited number of participants, the project has its limitations. When given enough training data, ESP32 can perform closer to Intel 53000 and Atheros. The LSTM model can thus learn and train on a diverse set of day-to-day human activities. This experiment sought to study a simple client-server architecture for ESP32 CSI DFWS. However, the coverage area of CSI data collection is limited by the simplicity of the architecture. Expressif[4] proposes several possible architectures that might be explored in the future for various applications.

References

1. Atif M, Muralidharan S, Ko H, Yoo B (2020) Wi-ESP—a tool for CSI-based device-free wi-fi sensing (DFWS). J Comput Des Eng 7(5):644–656. https://doi.org/10.1093/jcde/qwaa048
2. Dargan S, Kumar M (2020) A comprehensive survey on the bio- metric recognition systems based on physiological and behav ioral modalities. Expert Syst Appl 143:113114. https://doi.org/10.1016/j.eswa.2019.113114, http://www.sciencedirect.com/science/article/pii/S0957417419308310
3. Espressif: Esp32 (2020) https://www.espressif.com/en/products/socs/esp32/overview
4. Hernandez SM, Bulut E (2020) Lightweight and standalone IoT based WiFi sensing for active repositioning and mobility. In: 21st international symposium on "a world of wireless, mobile and multimedia networks" (WoWMoM) (WoWMoM 2020). Cork, Ireland
5. Liu J, Dong Y, Chen Y, Wang Y, Zhao T (2018) Leveraging breath ing for continuous user authentication. In: Proceedings of the 24th annual international conference on mobile computing and networking. MobiCom '18, Association for Computing Machinery, New York, pp 786–788. https://doi.org/10.1145/3241539.3267743
6. Liu J, Wang Y, Chen Y, Yang J, Chen X, Cheng J (2015) Tracking vital signs during sleep leveraging off-the-shelf wifi. In: Proceedings of the 16th ACM international symposium on mobile ad hoc networking and computing. MobiHoc '15, Association for Computing Machinery, New York, pp 267–276. https://doi.org/10.1145/2746285.2746303
7. Ma Y, Zhou G, WangS (2019) Wifi sensing with channel state information: a survey. ACM Comput Surv 52(3). https://doi.org/10.1145/3310194

[4] https://github.com/espressif/esp-csi

8. Shi C, Liu J, Liu H, Chen Y (2017) Smart user authentication through actuation of daily activities leveraging wifi-enabled iot. In: Proceedings of the 18th ACM international symposium on mobile ad hoc networking and computing. Mobihoc '17, Association for Computing Machinery, New York. https://doi.org/10.1145/3084041.3084061
9. Wang W, Liu AX, Shahzad M (2016) Gait recognition using wifi signals. In: Proceedings of the 2016 ACM international joint conference on pervasive and ubiquitous computing. UbiComp '16, Association for Computing Ma chinery, New York, pp 363–373. https://doi.org/10.1145/2971648.2971670
10. Zhang J, Wei B, Hu W, Kanhere SS (2016) Wifi-id: Human identification using wifi signal. In: 2016 international conference on distributed computing in sensor systems (DCOSS), pp 75–82. https://doi.org/10.1109/DCOSS.2016.30

The Substructure for Estimation of Miscellaneous Data Failures Using Distributed Clustering Techniques

Abdul Ahad, Sreenath Kashyap, and Marlene Grace Verghese

Abstract Miscellaneous data prediction provides an analytical mechanism for explaining the failure experience, which can include the method based on distributed clustering technique. The principle drawback of the fashions is that the error estimation is completely formulated through using a supervised control mechanism where the failure rate is estimated. Although, at the same time as stretching any advanced software, the past failure records or the previous failure statistics may not be available, and in such conditions, it is much tough to become aware of the failure modules with the aid of using the above-mentioned miscellaneous data process techniques. Hence, it is very important to increase the reliability that could take care of those situations efficiently. This failure statistics are used to examine the accuracy of Uniqueness, Sensitiveness, and Frequency rate. The unsupervised cluster and its type procedure uses the distributed clustering technique to become aware of the malfunctioning and flawless factors. Initially, it identified the average and the values much smaller than the average taken into consideration of flawless factors. Finally, they are considered to be flawed factors. Using this method, the whole dataset is converted to dual form and a distributed clustering technique is applied.

Keywords Miscellaneous data · Distributed clustering · Supervised mechanism · K-means clustering · Flawless factors · Data failures

1 Introduction

To find the failing and imperfect modules, we have at first analyzed the middle and the qualities considerably less than the middle considered as non-breaking down

A. Ahad (✉)
Department of AI, Anurag University, Hyderabad, Telangana, India

S. Kashyap · M. G. Verghese
Department of CSE, Vidya Jyothi Institute of Technology, Hyderabad, Telangana, India
e-mail: sreenathkashyap@kpritech.ac.in

M. G. Verghese
e-mail: csehod@kpritech.ac.in

A. Ullah et al. (eds.), *Proceedings of International Conference on Information Technology and Applications*, Lecture Notes in Networks and Systems 350,
https://doi.org/10.1007/978-981-16-7618-5_47

modules and the end are to be mulled over as inadequate modules. Utilizing this strategy, the entire dataset is changed into double data and the appropriated grouping approach [1] is applied. Unsupervised gadget knowledge method is applied to analyze the first-class of the software within the absence of earlier fault knowledge or fault proneness labels. Clustering helps to institution the unlabeled data values derived from the software metrics into fault susceptible and non-fault inclined modules.

The fault susceptible modules can be the impact of noise in the facts. Hence, powerful observation of every cluster allows the software analyst to decide whether the fault is a virtually generated fault or because of noisy. Generally, the noisy records inside the software termed as faults can be much less in variety and can be effortlessly tested and removed with the aid of the software program professionals. In order to distinguish the fault and non-malfunctioning modules, which might be used the K-mean mechanism [2]. K-means is an important clustering set of rules or centroid primarily based on process. It can be used for clustering the graph based on a data set using hierarchical strategies. However, the techniques, which might be if not known as similarity primarily, based techniques groups the facts primarily based on proximity measure.

2 Related Work

The Miscellaneous data process technique shows the failures encountered up to a specific amount of time. It considers the fixing procedure as a counting model by using the Multi Value Feature (MVF). The parameters of the version will be projected by using both least rectangular estimation and Maximum Likelihood Estimation (MLE) model based on certain assumptions [3, 4]. Software reliability assessment has emerged as an obligatory activity and trying out any advanced software product. Usually, the debugging and testing methods reduces the range of errors, but will increase the improvement value. Later, there is a need to find out forestall and release the software into the marketplace. Many fashions were targeted within the literature to determine the forestall time of checking out any software program. Most of this version is on miscellaneous data forecast, due to the fact they offer a framework for software program failure phenomena in the checking out segment.

Proposed work is expressed in different sections for better understanding: The general introduction about miscellaneous data process described in Sect. 1 and the study of K-Mean clustering is described in Sect. 2. The investigation of disseminated bunching procedure is examined in Sect. 3. The Sect. 4 arrangements with likelihood thickness work estimations of every one of bunches acquired utilizing the double grouping strategy, and the likelihood thickness work esteems are determined in a trial result [5]. The Sect. 5 is the end lastly references are in Sect. 6.

3 Problem Statement

The techniques, which might be brought on with this technique, consist of the techniques primarily based on Software Reliability Growth Model. However, the main downside of the use of those fashions is that the error estimation [6, 7] is formulated with the aid of the use of a supervised getting to know mechanism wherein the failure price is estimated. At the same time, as with developing any advanced software, the earlier expertise of the failure records might not be had, and in such conditions, it is far difficult to pick out the failure modules by the use of the above-mentioned distributed clustering techniques. Hence, it is far necessary to increase efficient fashions, which can deal with these conditions effectively. Utilizing double K-means bunching, the arrangement of rules performs two commitments all the while. These commitments are records decrease and trademark recognizable proof. This interaction keeps until every one of the records of items inside the huge measurements set is grouped [8]. There might be a couple of exceptions, which probably will not have a place, any class. Those measurements can be analyzed and may be dealt with as anomalies.

4 Methodology for Distributed Clustering Technique

Clustering can be treated as the way toward recognizing the information having similitude in any possibility and gathering that information in a class. Grouping the double information is a commonplace interaction due to recognizing the dissemination examples and connections among the information things, particularly in enormous datasets. For enhancing the double grouping measure, an elective least squared technique is utilized iteratively. This double bunching K-means [9, 10] calculation performs two undertakings at the same time like information decrease and highlight recognizable proof. This interaction proceeds until all the information things in the enormous informational index are bunched. This dual K-means clustering model gives the best result when compared with other clustering models.

Algorithm: The dual K-means clustering

- Take i as input
- E is the quantity of beginning mistakes
- Cluster (M, E); where M is the all-out number of modules
- Let C[i] .... C[E] to be irregular segment of M into E parts
- Repeat
-     For i = 1 to E.
-     X[i] = centroid of C[i].
-     C[i] = empty.
- For j = 1 to N
- X[q] = nearest to M[j] of X[i] to X[E]
- Add S[j] to C[q]

- Until the change to C is adequately little.

This Distributed Clustering Technique [11] is utilized to foresee the product in an unaided way. The fundamental benefit of this appropriation is that it serves to strategy the information when the state of the recurrence bend is symmetric and ringer molded, the ordinary guess may not fit precisely. Hence, in order to technique such distribution, advanced symmetry distribution is used. Using this technique, the fault values of the data are considered [12]. The probability density function is given by the Eq. (1).

$$f(Z,\mu,\sigma^2) = \frac{\left(2+\left(\frac{z-\mu}{\sigma}\right)^2\right)e^{\frac{-1}{2}\left(\frac{z-\mu}{\sigma}\right)^2}}{3\sigma\sqrt{2\pi}}, \quad -\infty < Z < \infty, -\infty < \mu < \infty, \sigma > 0 \tag{1}$$

5 Experiment Result

To obtain the probability density function (PDF) value of each cluster by using the dual clustering method [13] and probability density function values are calculated. To work out the Maximum Likelihood Estimate cost by using the probability density function values. The most extreme probability gage cost one can discover the loss information well ahead of time. Finally, we estimated the accuracy of each cluster with respect to Uniqueness, Sensitiveness, and F-measure [14, 15] by using a distributed clustering technique.

Table 1 addresses the likelihood thickness work esteems against every one of the audit esteems got. The last mistakes inform after the essential testing stage is expressed as contribution to the likelihood thickness work on the consolidated procedure given in Eq. (1) [16, 17]. In Table 1, the estimation of record 14 is thought of and it is expected to be the Median. The genuine disappointments and bogus disappointments are recognized in Table 2.

Table 2 represents the expected number of failures. The x, y, and $z(t)$ are determined.

The value of $x = 78.29$, $y = 22.72$, and $z(t) = 78.2698$.

The accuracy of the model is tried against the measurements of Uniqueness, Sensitiveness, and F-Score and computations [18, 19] are introduced in Table 3 (Figs. 1 and 2).

Table 1 PDF values using DCT

Time for text (in weeks)	Failures after estimation	Values of probability density function
001	77	0.03186
002	83	0.03124
003	88	0.04264
004	92	0.04787
005	95	0.05882
006	98	0.06267
007	100	0.06989
008	101	0.07176
009	102	0.07878
010	102	0.07776
011	102	0.08251
012	117	0.08682
013	122	0.08987
014	**125**	**0.08876**
015	132	0.09962
016	137	0.018182
017	141	0.05156
018	145	0.06687
019	152	0.10247
020	157	0.10848
021	162	0.108415
022	165	0.10967
023	172	0.020892
024	178	0.020898
025	181	0.045678
026	184	0.045682

6 Conclusion

The proposed model offers with the identity of factual liabilities and flawed modules based on distributed clustering technique. The K-mean clustering is used to identify the first estimation. The primary failures recognized and estimated using probability density functions are considered as inputs and the possibility densities are estimated from these estimated values. Primarily based on the maximum likelihood concepts, the factual liabilities and the flawed facts are diagnosed. The advanced method is examined on facts set as taken into consideration. The final results obtained are as compared with that of the proposed technique primarily based on distributed clustering technique.

Table 2 Genuine failures and Bogus failures by SDCT

Genuine failures	Bogus failures
125	77
132	83
137	88
141	92
145	95
152	98
157	100
162	101
165	102
172	102
178	102
181	117
184	122

Table 3 Precision of SDCT

GP	BP	Uniqueness	Sensitiveness	F-score
0.724	0.192	0.647	0.654	0.680
0.625	0.104	0.623	0.665	0.634

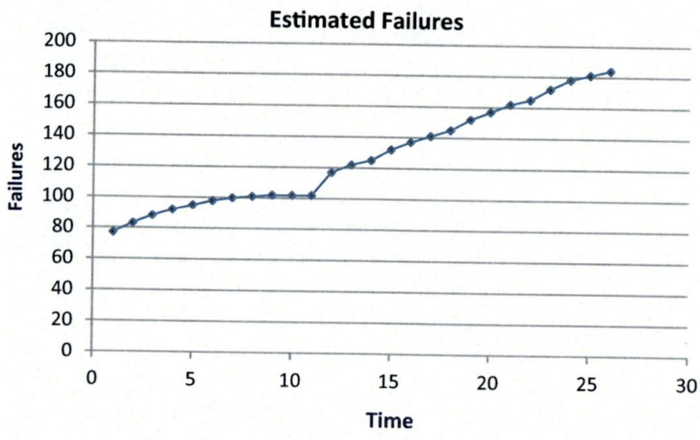

Fig. 1 Estimated failures

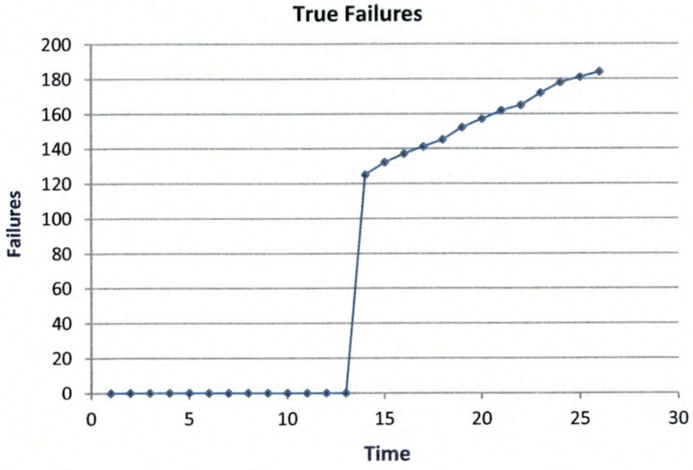

Fig. 2 True failures

References

1. Kantipudi MVV, Prasad et al (2021) Time series data analysis using machine learning-(ML) approach. Libr Philos Pract (e-journal)
2. Bendechache M, Tari AK (2019) Parallel and distributed clustering framework for big spatial data mining. Int J Parallel Emergent Distrib Syst 34
3. Kumari V (2018) Software cost estimation using soft computing techniques. https://doi.org/10.13140/RG.2.2.29166.05449
4. Kantipudi MVV, Prasad, Suresh HN (2018) Simulation and performance analysis for coefficient estimation for sinusodial signal using LMS, RLS and proposed method. Int J Eng Technol 7(1.2)
5. Ahad A et al (2018) Multi-level tweets classification and mining using machine learning approach. J Eng Sci, 1818–7803
6. Kantipudi MVV, Prasad, Suresh HN (2018) An Efficient parametric model-based framework for recursive frequency/spectrum estimation of nonstationary signal. Int J Eng Technol 7(4.6)
7. Venkataiah V, Mohanty RK (2017) Review on intelligent and soft computing techniques to predict software cost estimation. Int J Appl Eng Res 12(22):12665–12681
8. Ahad A et al (2016) A new approach for integrating social data into groups of interest. Springer Series, 978-81-322-2755-7
9. Kantipudi MVV, Prasad, Suresh HN (2016) Resolving the issues of capon and APES approach for projecting enhanced spectral estimation. Int J Electr Comput Eng (IJECE) 6(2):725–734
10. Améndola, Carlos (2015) Moment varieties of Gaussian mixtures. J Algebraic Stat. https://doi.org/10.18409/jas.v7i1.42
11. Kantipudi MVV, Prasad, Suresh HN (2015) A survey of spectral estimation using improved recursive least square (RLS) algorithm. Book chapter ERCICA, Volume 1. Springer India, pp 363–376
12. Rodrigues PP, Gama J (2014) Distributed clustering of ubiquitous data streams. WIREs Data Min Knowl Discov 4(1):38–54
13. Toto-Zarasoa V et al (2011) Maximum likelihood BSC parameter estimation for the Slepian-Wolf problem. IEEE Commun Lett 15(2):232–234
14. Jain AK (2010) Data clustering: 50 years beyond K-means. Pattern Recogn Lett 31(8):651–666
15. Colinet E, Juillard J (2010) A weighted least-squares approach to parameter estimation problems based on dual measurements. IEEE Trans Autom Control 55(1):148–152

16. Daleyand DJ, Vere-Jones D (2007) An introduction to the theory of point processes, Volume II: general theory and structure. Springer Science & Business Media. pp 166–167. ISBN: 978-0-387-21337-8
17. Arthur D (2007) K-means++: the advantages of careful seeding. In: Proceedings of the eighteenth annual ACM-SIAM symposium on discrete algorithms. Society for industrial and applied mathematics, pp 1027–1035
18. Li T (2005) A general technique for clustering dual data. In: Proceedings of the eleventh ACM SIGKDD international conference on knowledge discovery in data mining, pp 188–197
19. Sahoo R, Squillante MS (2004) Failure data analysis of a large-scale heterogeneous server environment. https://doi.org/10.1109/DSN. 2004.1311948

Performance Enhancement of SAC-OCDMA System Using an Identity Row Shifting Matrix Code

Mohanad Alayedi, Abdelhamid Cherifi, Abdelhak Ferhat Hamida, Boubakar Seddik Bouazza, and C. B. M. Rashidi

Abstract In order to face the shortcomings and problems, in spectral amplitude coding-optical code-division multiple access (SAC-OCDMA) systems, presented by multiple access interference (MAI) and its accompaniment phase-induced intensity noise (PIIN), limited capacity, etc., that prevent them to function effectively as well make performances poor. For these reasons, this paper offers a solution by proposing a novel encoding technique namely identity row shift matrix (IRSM) code based on an identity matrix and shifting property with target of beating the challenges aforementioned. Our proposed code is featured by zero cross-correlation (ZCC) property restricting the MAI effect as well neglecting PIIN which in turn positively reflects on system performance. Mathematical results appear the ability of IRSM code to improve the performance of SAC-OCDMA system as well outperform reported codes such as: diagonal permutation shift, (DPS), modified double weight (MDW), and random diagonal (RD) codes. For example, in term of system

M. Alayedi (✉)
Scientific Instrumentation Laboratory (LIS), Department of Electronics, Faculty of Technology, Ferhat Abbas University of Setif 1, 19000 Setif, Algeria

A. Cherifi · B. S. Bouazza
Technology of Communication Laboratory (LTC), Electronics department, Faculty of Technology, Dr. Tahar Moulay University of Saida, 20000 Saida, Algeria

A. Ferhat Hamida
Laboratory of Optoelectronics and Components (LOC), Electronics department, Faculty of Technology, Ferhat Abbas University of Setif 1, 19000 Setif, Algeria

C. B. M. Rashidi
Advanced Communication Engineering, Center of Excellence School of Computer and Communication Engineering (ACE-Co-SCCE), Universiti Malaysia Perlis, (UniMAP), Perlis, Malaysia
e-mail: rashidibeson@unimap.edu.my

A. Ullah et al. (eds.), *Proceedings of International Conference on Information Technology and Applications*, Lecture Notes in Networks and Systems 350,
https://doi.org/10.1007/978-981-16-7618-5_48

capacity, it amounts to 25, 48, 58, and 88, respectively, based on DPS, MDW, RD, and IRSM codes referring a remarkable enhancement. Additionally, mimic result demonstrates that IRSM code is suitable for attaining optical communication requirements through producing BER and Q-factor reach 10^{-14} ($\leq 10^{-9}$) and 7.59 dB ($\geq$6dB), consecutively.

Keyword Optical CDMA · SAC-OCDMA · IRSM code · RD code · Q-factor

1 Introduction

In recent two decades, optical networks have been attracted the operator's interests even they have been become the preferred option for them because of their services presented by improve the system performance due to improvement of service quality, users' data can be transmitted at high speeds, etc. As a result, these benefits, mentioned previously, make them convenient overall for systems with large capacities [1, 2].

With the passage of time, many technologies have been inserted into optical networks in order to increase their efficiency such as optical code-division multiple access (OCDMA) technology that has been added more optimizations such as possibility of synchronous and asynchronous accesses for several users, transmission of users' data at higher security level against eavesdropping, effective cost and ease of network expansion by adding new users [3].

Besides that, it is possible to use OCDMA technology with target of performing switching, multiplexing, and append or drop of signals of various channels through the main or single network for multiple access techniques rather than wavelength division multiple access (WDMA) and time division multiple access (TDMA) [4]. Notwithstanding, there is a main drawback imposes OCDMA systems to be limited performance called multiple access interference (MAI) where MAI restricts the system capacity as well generates an asymptotic threshold to the bit error rate (BER) or quality factor [5].

In this context, there are lots of codes which are reported for SAC-OCDMA system, including diagonal permutation shift (DPS) code [6], random diagonal (RD) code [7], and modified double weight (MDW) code [8] in literature investigation. In fact, none of these codes are able to restrict PIIN effect totally due to constant in-phase cross-correlation (IPCC) value estimated by one exactly, as first drawback. Secondly, much long code length which in turn requires provide a great number of filters to extract the desired wavelengths. Thirdly and finally, regarding to design steps of them, it is observed that described by complexity. In order to avoid the existing restrictions above, a novel SAC-OCDMA code, i.e., identity row shifting matrix (IRSM) code has been designed with zero cross-correlation (ZCC) and resilience traits to select the desired code weight and number of users. This paper is organized as follows:

Section 2 depicts the steps of designing our proposed code, Sect. 3 analyzes the system performance. Concerning to Sect. 4, it contains the simulation results using Matlab and Optisystem software, and at last, this paper is terminated by a conclusion in Sect. 5.

2 IRSM Code Construction

Before starting in code design steps depiction, let at first define the auto- and cross-correlation functions by taking two various code sequences "*A*" and "*B*" which equal to $A_j = \{A_1, A_2, A_3, \ldots\ldots, A_{N_u}\}$ and$B_j = \{B_1, B_2, B_3, \ldots\ldots, B_{N_u}\}$, respectively. They can be respectively, defined as follows [9]:

$$\begin{cases} \lambda_a = \sum_{i=1}^{K} A_j \times A_{j+t} & (1) \\ \lambda_c = \sum_{i=1}^{K} A_j \times B_{j+t} & (2) \end{cases}$$

The IRSM code is distinct by these parameters: (*L*, *w*, *C*) where each one of them denotes to code length, code weight, and system cardinality, respectively. Even be the new code suitable for SAC-OCDMA systems, it has to construct it with high auto-correlation and null cross-correlation values. It can depict the process construction of IRSM code by the next steps [10]:

A. Step 1

Originate an identity matrix (C_i) with order (i). In math science, identity matrix is considered a square matrix where all elements are zeros other than the elements of main diagonal in which equal to one. Let's take an example which ($i = 3$), it can be written according to the following form:

$$C_3 = \begin{bmatrix} 1\,0\,0 \\ 0\,1\,0 \\ 0\,0\,1 \end{bmatrix}_{3\times3} \tag{3}$$

B. Step 2

Relying on shifting of rows feature and for an identity matrix that has (i) of order, C_3 requires two times of shifting in accordance with ($i - 1$) rule. Likewise, the generated matrix is also shifted by two times. Consequently, this permits us to procure these matrices:

$$C_3^{'} = \begin{bmatrix} 0\,1\,0 \\ 0\,0\,1 \\ 1\,0\,0 \end{bmatrix} \text{ and } C_3^{''} = \begin{bmatrix} 0\,0\,1 \\ 1\,0\,0 \\ 0\,1\,0 \end{bmatrix} \tag{4}$$

C. Step 3

Write C_3 matrix according to the form below:

$$C_3 = \begin{bmatrix} L_1(C_3) \\ L_2(C_3) \\ L_3(C_3) \end{bmatrix} \quad (5)$$

where $L_1(C_3)$, $L_2(C_3)$, and $L_3(C_3)$: references to respectively to 1st row, 2nd row, and 3rd row of C_3 matrix. In a similar way, $C_3^{'}$ and $C_3^{''}$ matrices can be also written as:

$$C_3^{'} = \begin{bmatrix} L_1(C_3^{'}) \\ L_2(C_3^{'}) \\ L_3(C_3^{'}) \end{bmatrix} \text{ and } C_3^{''} = \begin{bmatrix} L_1(C_3^{''}) \\ L_2(C_3^{''}) \\ L_3(C_3^{''}) \end{bmatrix} \quad (6)$$

D. Step 4

Reformate each of matrices C_3, $C_3^{'}$ ', and $C_3^{''}$ to row vector as shown below in Eq. (7).

$$\begin{cases} Re(C_3) = \left[L_1(C_3)\ ,\ L_2(C_3)\ ,\ L_3(C_3) \right] \\ Re(C_3^{'}) = \left[L_1(C_3^{'})\ ,\ L_2(C_3^{'})\ ,\ L_2(C_3^{'}) \right] \\ Re(C_3^{''}) = \left[L_1(C_3^{''})\ ,\ L_2(C_3^{''})\ ,\ L_3(C_3^{''}) \right] \end{cases} \xrightarrow{\text{leads}} \begin{cases} Re(C_3) = \left[1\,0\,0\,0\,1\,0\,0\,0\,1 \right]_{1\times 9} \\ Re(C_3^{'}) = \left[0\,1\,0\,0\,0\,1\,1\,0\,0 \right]_{1\times 9} \\ Re(C_3^{''}) = \left[0\,0\,1\,1\,0\,0\,0\,1\,0 \right]_{1\times 9} \end{cases} \quad (7)$$

E. Step 5

Lastly, by merging between these reformatted matrices or in other sense between row vectors in Eq. (7), this enables us to procure the novel ZCC code as:

$$\text{IRSM} = \begin{bmatrix} I_3 \\ I_3^{'} \\ I_3^{''} \end{bmatrix} \rightarrow^{\text{leads}} \text{IRSM} = \begin{bmatrix} 1\,0\,0\,0\,1\,0\,0\,0\,1 \\ 0\,1\,0\,0\,0\,1\,1\,0\,0 \\ 0\,0\,1\,1\,0\,0\,0\,1\,0 \end{bmatrix}_{3\times 9} \quad (8)$$

By virtue of the new ZCC code matrix above in which namely IRSM code and had dimension: three rows and nine columns. Respectively, they express number of users (N_u) and code length (L). Consequently, it can deduce the relation between them based on the follows:

$$\begin{cases} N_u = i \\ L = i^2 \end{cases} \rightarrow^{\text{implies}} L = w^2 = N_u^2 \quad (9)$$

Besides that, number of users and the code weight have an equal value. For that, Eq. (9) will be written in another formula as below:

$$L_{\text{IRSM}} = w.N_u \tag{10}$$

In accordance with the IRSM code matrix, the code-word for each user would be:

$$\text{codewords} = \begin{cases} \text{1st User} \rightarrow \lambda_1, \lambda_5, \lambda_9 \\ \text{2nd User} \rightarrow \lambda_2, \lambda_6, \lambda_7 \\ \text{3rd User} \rightarrow \lambda_3, \lambda_4, \lambda_8 \end{cases} \tag{11}$$

3 System Performance and Analysis

Firstly, to analyze the SAC-OCDMA system readily without any complexity, many assumptions are recommended to take them a consideration which are [11, 12]:

1-/ Light source is ideally non-polarized as well its spectrum is flat over $[f_0 - \nabla f,\ f_0 + \nabla f]$ of bandwidth. Both $f0$ and Δf are respectively the optical central frequency and the optical bandwidth. Further, they are estimated by Hertz unit.

2-/ Each transmitter is provided the same power comparing to others.

3-/ All spectrum components of various users are congruent.

4-/ Each bit flow from each user is coincided.

Secondly, in order to evaluate the SAC-OCDMA system based on IRSM code, two of main noises sources are only taken a consideration which are shot noise I_{sh}^2 and thermal noise I_{th}^2. Meanwhile, PIIN I_{PIIN}^2 is ignored since our proposed code has ZCC feature. For that, the dark current noise $I_{\text{Total - noise}}^2$ can be given through Eq. (12) below [13, 14].

$$\begin{aligned} \left\langle I_{\text{Total - noise}}^2 \right\rangle &= \left\langle I_{\text{th}}^2 \right\rangle + \left\langle I_{\text{sh}}^2 \right\rangle + \left\langle \underbrace{I_{\text{PIIN}}^2}_{=0} \right\rangle \\ &= \frac{4K_b B_e T}{R_l} + 2eB_e I_{SDD} + 0 \end{aligned} \tag{12}$$

where K_b, B_e, T, R_l, e, and I_{SDD} indicate to Boltzmann's constant, electrical bandwidth, absolute temperature, load resistor, electron charge, and output of direct current, respectively. The output current of spectral direct detection (SDD) can be given by the following relation:

$$I_{\text{SDD}} = \int_0^\infty P(f)df$$

$$= R\int_0^{\infty}\frac{P_r}{\nabla f}\sum_{k=1}^{N_u} d_k \sum_{j=1}^{L} C_m(j) \times C_n(j) \bigvee(f, j)df \tag{12}$$

where $P(f)$ indicates to the power spectral density (PSD), R indicates to photodiode (PD), responsively, P_r indicates to the amount of power at receiver level, d_k indicates the data bit which is likely "0" or "1". The $\bigvee(f, j)$ function can be defined as:

$$\bigvee(f, j) = U\left\{f - f_0 - \frac{\nabla f}{2L}(-L + 2j - 2)\right\}$$
$$- U\left\{f - f_0 - \frac{\nabla f}{2L}(-L + 2j)\right\} = U\left\{\frac{\nabla f}{L}\right\}$$

where $U(f)$ presents the unit step function defined as:

$$U(f) = \begin{cases} 1 & f \geq 0 \\ 0 & \text{else} \end{cases} \tag{13}$$

Let $C_m(j)$ presents the ith element of N_u^{th} user and in accordance with IRSM code, the CC of SDD can be written as:

$$\sum_{j=1}^{L} C_m(i) \times C_n(i) = \begin{cases} w \text{ for} & m = n \\ 0 \text{ else} & \end{cases} \tag{14}$$

where the upper term refers to auto-correlation, meanwhile the down term refers to cross-correlation. By virtue of Eqs. (13) and (15), (12) of output photocurrent will be written as:

$$I_{\text{SDD}} = R\frac{P_r.w}{L} \tag{15}$$

Replacing Eq. (16) in Eq. (11), in addition to deem that both of probability of (0) or (1) bit transmission is the same and estimated by (0.5). As a result, the noise current expression will become:

$$I^2_{\text{Total-noise}} = \frac{4K_b B_e T}{R_l} + eB_e R\frac{P_{sr}w}{L} \tag{16}$$

The average signal-to-noise ratio (SNR) can be given as:

$$\text{SNR} = \frac{I^2_{\text{SDD}}}{I^2_{\text{Total - noise}}} = \frac{\left[R\frac{P_{sr}w}{L}\right]^2}{\frac{4K_b B_e T}{R_l} + eB_e R\frac{P_{sr}w}{L}} \tag{17}$$

Counting on the Gaussian approximation, the bit error rate (BER) can be expressed with assist of the next relation [15–17]:

$$\text{BER} = 0.5 \times \text{erfc}\sqrt{\text{SNR}/8} \tag{18}$$

4 Results and Debate

Counting on Eq. (19), the BER of IRSM code is computationally tested comparing with other reporting SAC-OCDMA codes: DPS, MDW, and RD codes for fixed parameter which is code weight at 4 value. The values of parameters existent in Eq. (18) are located Tables 1 and 2.

In Fig. 1, it is evident that the proposed code, in term of BER in front of number of concurrent users, outperforms DPS, MDW, and RD codes. For instance, the optical demand feature can be attained with 88, 58,49,, and 25 users for IRSM, RD, MDW, and DPS codes, consecutively. According to the above, the proposed code improves overall 3.52, 1.79, and 1.52 times the SAC-OCDMA system performance depended on DPS, RD, and MDW codes, respectively. These improvements aforementioned come back ZCC feature existing in IRSM code characteristics, which in turn has been worked cancel PIIN effect quietly. Moreover, it can compute the increased percent thanks to our proposed code as follows:

Table 1 Comparison between DPS, DCS, RD codes, and IRSM code

Code	Code weight	Code size	Code length	Cross-correlation
DPS code [6]	4	9	12	≥ 1
MDW code [8]	4	4	18	≥ 1
RD code [7]	4	4	9	Variable
IRSM code	4	4	16	0

Table 2 Used parameters in numeral analysis

Parameter	Value	Parameter	Value
PD responsivity (R)	0.75 A/w	Boltzmann's constant (K_b)	1.38×10^{-23} J.s^{-1}
Receiver load resistor (R_l)	1030 Ω	Number of active users (N_u)	60
Code weight (w)	4	Effective source power (P_r)	−10 dBm
Receiver noise temperature (T)	300 K	Spectral width (∇f)	5 THz
Electrical bandwidth (B_e)	0.311 GHz	Data bit rate (R_b)	0.622 Gb/s

$$\frac{\text{Proposed code} - \text{Compared code}}{\text{Compared code}} \times 100$$

For that, we have:

$$\begin{cases} \frac{88-25}{25} = 252\% \\ \frac{88-49}{49} = 79\% \\ \frac{88-58}{58} = 52\% \end{cases}$$

By transiting to Fig. 2, it interests to study the BER variation in front of data rate for the limited active users number at 60. It is noted that the proposed SAC-OCDMA system depended on IRSM code can be employed in order to rapid more transmission of data. Particularly, the optical transmission is assured with 1.33 Gb/s, that is 11.08, 2.77, and 2.29 times that of DPS, MDW, and RD codes, respectively, since they support up to 0.12, 0.48, and 0.58 Gbps of data rate, consecutively.

At the last, Fig. 3 interests in studying the spectral efficiency (SE) variation in front of the code weight with limiting data rate at 622 Mbps as well received power at −10 dBm. The SE can be defined by dividing the aggregate information rate on total spectral bandwidth where it can be given as [18]:

$$\text{SE} = \frac{N_{uBER=10^{-9}}}{\Delta\vartheta \times w} \tag{19}$$

where $N_{uBER=10^{-9}}$ refers to the maximum system cardinality, and $\Delta\vartheta$ refers to the bandwidth of each optical wavelength which is estimated by 90 nm. It is plainly that

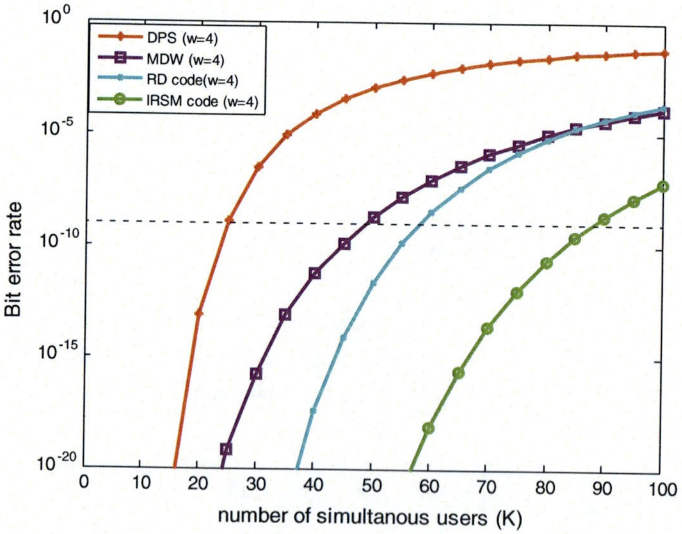

Fig. 1 BER against number of concurrent users for ($w = 4$)

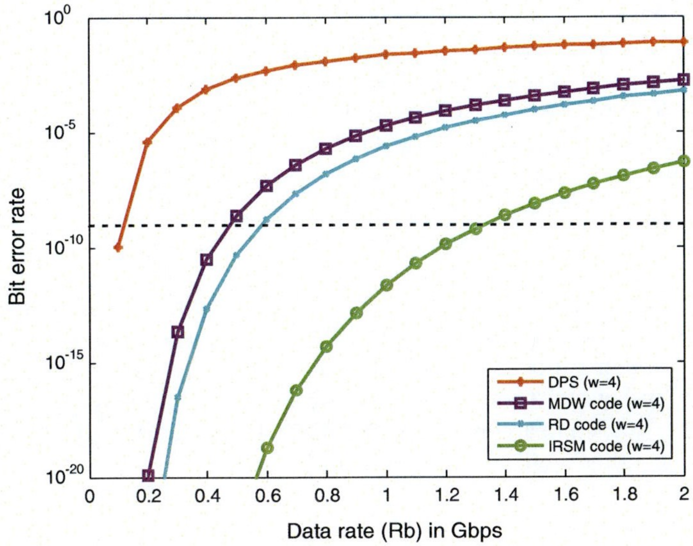

Fig. 2 BER against data rate for ($N_u = 60$)

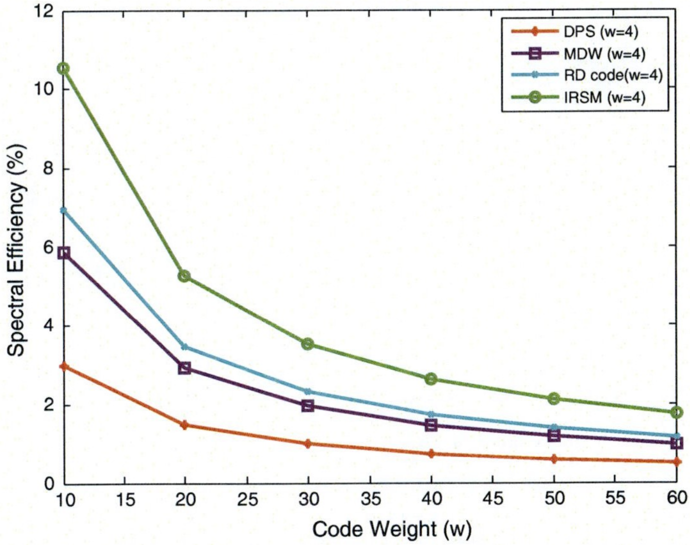

Fig. 3 Spectral efficiency against code weight

the IRSM code has the greatest SE compared to other encoding techniques including DPS, MDW, and RD codes. Observing that the relevance between "SE" and "w" is reversible proportional since the SE diminishes as the weight increases.

For minor considered code weight that is equal (10), IRSM code overcomes all compared codes in this study, which are DPS, MDW, RD, and IRSM codes whence SE percent by dint of 3, 5.88, 6.69, and 10.56%, respectively. In fact, these consequences are due to that IRSM code supplies 203 users, whereas DPS, MDW, and RD codes supply just 25, 48, 58, and 88 users, consecutively at a consent BER 1e-9 with taking account the same number of wavelength for all codes.

Additionally, Optisystem software ver. 7.0 has been utilized to examine the SAC-OCDMA system performance based on IRSM code whence two main parameters in optical communication network: Q-factor and BER. A plain graphic of 3 users is appeared in Fig. 4. This test was done with assist of the employed parameters referenced in Table 3 in accordance with ITU-T G.625 standard as well taking into account the non-linear effects in single mode fiber (SMF).

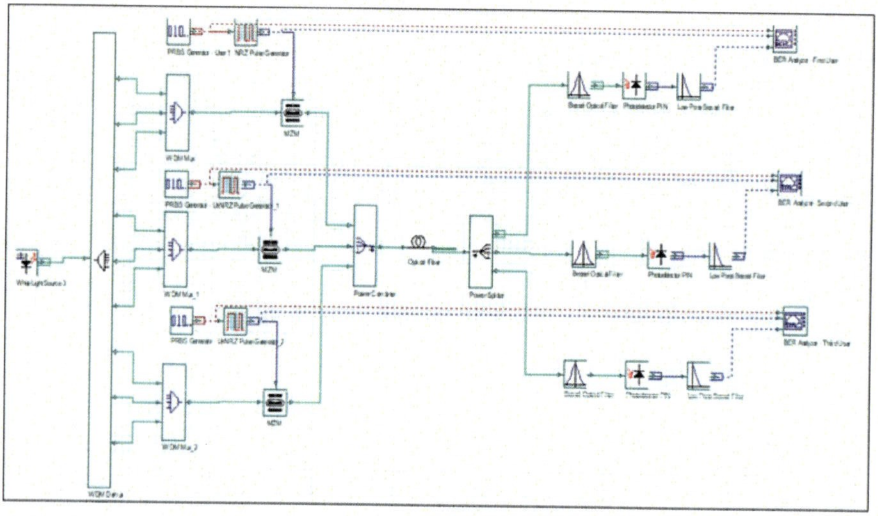

Fig. 4 IRSM code schematic block diagram for 3 users

Table 3 Used parameters in mimic test

Parameter	Value	Parameter	Value
Dark current	5×10^{-9} A	Data rate	622 Gbps
Distance	40 km	Attenuation	0.25 dB km^{-1}
Thermal noise	1.8×10^{-23}W/Hz	Cutoff frequency	466.5 MHz
Reference wavelength	1550 nm	Transmitted power	−115 dBm
Bandwidth of Bessel filter	0.3 nm	Dispersion	18 ps/nm/km

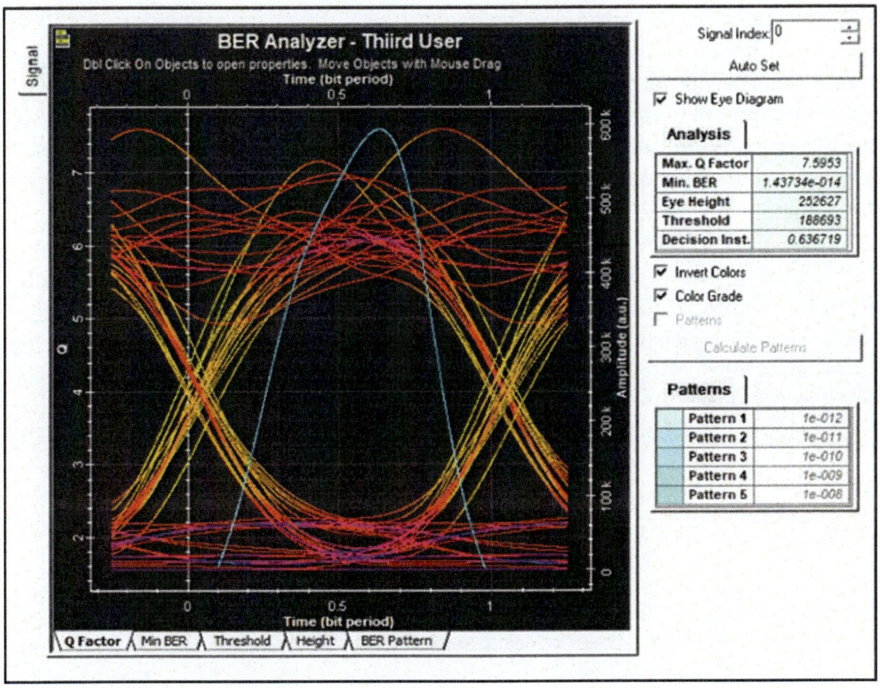

Fig. 5 Eye diagram of three users utilizing IRSM code

Based on Fig. 5, it exhibits the eye diagram of IRSM code at fiber length up to 40 km. It is perspicuously that IRSM code grants SAC-OCDMA system an agreeable performance represented consecutively, by 7.59 dB and 1.43×10^{-14} of Q-factor and BER when allowing for three users to be active. These results are overall consent in optical communication systems because Q-factor should to be at least 6 dB and the BER should to be at the most 10^{-9}. For that, it can deduce that our proposed code has the ability of satisfying optical communication demands.

5 Conclusion

In this research, a novel encoding technique has been proposed for SAC-OCDMA system and belongs to ZCC code families. This encoding technique is termed as IRSM code where its construction depends on easy five steps utilizing an identity matrix and shifting property. Also, the BER performance in different terms, of the IRSM code has been evaluated at high-speed SAC-OCDMA system relying the computational calculation as well as the mimic test using Optisystem software in order to more demonstration of viability of IRSM code. Results denote that IRSM code, based on

SDD technique, is able overcome symmetric encoding techniques proposed previously in different terms, for instance, system cardinality where IRSM code could fulfill an increment up to 252, 79, and 25% comparing with DPS, MDW, and RD codes, consecutively. Therefore, MAI phenomena has been restrained totally thanks to ZCC property for our proposed code permitting to ignore PIIN effect. Further, it has been founded that our code is convenient to meet optical communication requirements due to low BER and high quality factor although great length SMF. At the end, the SAC-OCDMA system capacity has been stopped at 88 with code length is very long and reaches 352, so it can extend this work in the future by proceeding an adjustment for IRSM code matrix or employing it as encoding–decoding pattern in various tow dimensional-OCDMA (2D-OCDMA) systems, seeking of the gain performance more satisfied than offered in this paper with reducing the long code length as possible.

References

1. Alayedi M, Cherifi A, Hamida AF, Bouazza BS, Aljunid SA (2021) Performance improvement of optical multiple access CDMA network using a new three - dimensional (spectral/time/spatial) Code. Wirel Personal Commun 118(4):2675–2698. https://doi.org/10.1007/s11277-021-08149-0
2. Alayedi M, Cherifi A, Hamida AF, Rahmani M, Attalah Y, Bouazza BS (2020) Design improvement to reduce noise effect in CDMA multiple access optical systems based on new (2-D) code using spectral/spatial half-matrix technique. J Opt Commun. https://doi.org/10.1515/joc-2020-0069
3. Cherifi A, Bouazza BS, Al-ayedi M, Aljunid SA, Rashidi CBM (2021) Development and performance improvement of a new two-dimensional spectral/spatial code using the Pascal triangle rule for OCDMA system. J Opt Commun 42(1):149–158. https://doi.org/10.1515/joc-2018-0052
4. Alayedi M, Cherifi A, Hamida AF, Rashidi CBM, Bouazza BS (2020). Performance improvement of multi access OCDMA system based on a new zero cross correlation code. In: IOP conference series: materials science and engineering, vol. 767. https://doi.org/10.1088/1757-899X/767/1/012042
5. Al-khafaji HMR, Ngah R, Aljunid SA, Rahman TA (2014) A new two-code keying scheme for SAC-OCDMA systems enabling bipolar encoding. J Mod Opt 5:62. https://doi.org/10.1080/09500340.2014.978914
6. Ahmed HY, Gharsseldien ZM, Aljunid SA (2016) An efficient method to construct diagonal permutation shift (DPS) codes for SAC OCDMA systems. J Theor Appl Inf Technol 94(2):475–484
7. Fadhil HA, Aljunid SA, Ahmad RB (2010) Design considerations of high performance optical code division multiple access: a new spectral amplitude code based on laser and light emitting diode light source. IET Optoelectron 4(1):29–34. https://doi.org/10.1049/iet-opt.2009.0010
8. Aljunid SA, Ismail M, Ramli AR, Ali BM, Abdullah MK (2004) A new family of optical code sequences for spectral-amplitude-coding optical CDMA systems. IEEE Photonics Technol Lett 16(10):2383–2385. https://doi.org/10.1109/LPT.2004.833859
9. Morsy MA (2018) Analysis and design of weighted MPC in incoherent synchronous OCDMA network. Opt Quant Electron 50(11):1–17. https://doi.org/10.1007/s11082-018-1657-z
10. Alayedi M, Cherifi A, Hamida AF (2019) Performance enhancement of SAC-OCDMA system using a new optical code. In: Proceedings—2019 6th international conference on image and

signal processing and their applications, ISPA, pp 1–4. https://doi.org/10.1109/ISPA48434.2019.8966912

11. El-Mottaleb SAA, Fayed HA, Ismail NE, Aly MH, Rizk MRM (2020) MDW and EDW/DDW codes with AND subtraction/single photodiode detection for high performance hybrid SAC-OCDMA/OFDM system. Opt Quantum Electron 52(5). https://doi.org/10.1007/s11082-020-02357-x
12. Alayedi M, Cherifi A, Ferhat Hamida A, Matem R, A. Abd El-Mottaleb A (2021) Improvement of SAC-OCDMA system performance based on a novel zero cross correlation code design. In: Proceedings of International Conference on Advances in Communication Technology, Computing and Engineering. RGN Publications
13. Kakaee MH, Essa SI, Seyedzadeh S, Mokhtar M, Anas SB, Sahbudin RK (2015) Proposal of multi-service (MS) code to differentiate quality of services for OCDMA systems. In: Proceedings of ICP 2014—5th international conference on photonics 2014, pp 176–178. https://doi.org/10.1109/ICP.2014.7002347
14. Bhanja U, Panda C (2020) Performance analysis of hybrid SAC-OCDMA-OFDM model over free space optical communication. CCF Trans Network 3(3–4):272–285. https://doi.org/10.1007/s42045-020-00039-6
15. Alayedi M, Cherifi A, Ferhat Hamida A, Bouazza BS, Aljunid SA (2021) Improvement of Multi Access OCDMA System Performance based on Three Dimensional-Single Weight Zero Cross Correlation Code. In: Proceedings - 2021 3rd International Conference on Computer and Information Sciences, ICCIS 2021
16. Alayedi M, Cherifi A, Ferhat Hamida A, Mrabet H (2021) A fair comparison of SAC-OCDMA system configurations based on two dimensional cyclic shift code and spectral direct detection. Telecommun Syst. https://doi.org/10.1007/s11235-021-00840-8
17. Sahraoui W, Aoudia H, Berrah S, Amphawan A, Naoum R (2020) Performances analysis of novel proposed code for SAC-OCDMA system. J Opt Commun. https://doi.org/10.1515/joc-2018-0125
18. Meftah K, Cherifi A, Dahani A, Alayedi M, Mrabet H (2021) A performance investigation of SAC - OCDMA system based on a spectral efficient 2D cyclic shift code for next generation passive optical network. Opt Quant Electron 53(10):1–28. https://doi.org/10.1007/s11082-021-03073-w

Information System Security Risk Priority Number: A New Method for Evaluating and Prioritization Security Risk in Information System Applying FMEA

Ismael Costa and Teresa Guarda

Abstract The emergence of the COVID-19 pandemic led several organizations around the world and in the most varied areas of activity, to move from the intention to implement a digital transformation in the medium/long-term, to an instant obligation to apply the digital transformation. The organizations' ability to adapt immediately meant their survival and even in some cases a positive evolution of their business. The digital transformation applied in an abrupt way has uncovered some critical factors for its success. One of the most relevant factors will be information security. Many of the digital systems put into operation more intensively during the pandemic, have shown to be highly fragile on issues related to information security. One relevant problem of the organizations is the low effectiveness and efficiency of financial, human, and material resources, allocated to the reduction or mitigation of the risks identified in their information systems. This study aims to offer a new method for prioritizing security risks. The new proposed method directs the organizations resources to more effectively and efficiently actions to reduce or mitigate the identified vulnerabilities of the information system.

Keywords Information system security analysis · Risk priority number method · System security risks · Failure mode and effect analysis

1 Introduction

The reality of the new normal revealed by the cultural, social, and economic impact of the COVID-19 virus, led to a radical transformation of the process of interaction and

---

I. Costa · T. Guarda
ISLA Santarém, Largo Cândido Dos Reis, 2000-241 Santarém, Portugal

T. Guarda (✉)
Universidad Estatal Península de Santa Elena, La Libertad, Ecuador

CIST—Centro de Investigación en Sistemas y Telecomunicaciones, Universidad Estatal Península de Santa Elena, La Libertad, Ecuador

Algoritmi Centre, Minho University, Guimarães, Portugal

A. Ullah et al. (eds.), *Proceedings of International Conference on Information Technology and Applications*, Lecture Notes in Networks and Systems 350,
https://doi.org/10.1007/978-981-16-7618-5_49

operation of people in their work process [1]. New forms of remote work were forcibly created using new digital communication platforms, and organizations were forced to reinvent their business models [2]. With an urgent need to implement remote work solutions, many organizations have been forced to make critical decisions for their operation and sustainability in a short period of time, without having an analysis, planning or effectively and efficiently solutions implementation for this purpose [3]. The obligation of social distance has highlighted the importance that online applications represented for the continuity of personal and professional needs [4].

In a digital society, an organization's main asset is information. Currently, this asset is more vulnerable than ever, taking constant risks. Attacks on information systems have an increasing impact on the normal functioning of organizations [5]. The intensive interconnection of various digital systems has created a great dependence on security aspects, affecting both the availability and the security of the entire information system [6]. The risk of information security is discussed in several articles through qualitative approaches, on the other hand, numerical approaches to quantify the risk have shown to be scarce [7].

Regardless of whether intrusions to the information system originate internally or externally [8], if the attacker identifies and exploits a failure, the entire security of the system is at risk [6]. It is therefore highly relevant to develop an efficient information security risk management. Determining the potential impact of a given risk by assessing the severity it will have if it occurs, assessing the likelihood of it occurring [8] and identifying which existing controls can detect the risk before it occurs, can be considered basic steps for risk assessment.

The main risk management strategy of an information system is the reduction or mitigation of risks and consequent negative impacts on the organization. This strategy is supported by the openness to investment in security technologies by the management of organizations [9].

The solution proposed in this article, intends to create a new method of quantitative evaluation of the Security Risk Priority Number. This new method is justified by the growing need of organizations to improve the effectiveness and efficiency of financial, human, and material resources, allocated to the reduction or mitigation of the risks identified in their information systems.

In the next chapter some concepts relevant to the topic of information security risk management are described, Chap. 3 presents the methodology used in the development of the proposed method for determining the Security Risk Priority Number. In Chap. 4 an example of application of the proposed method is presented in comparison with the already known method used by Failure Mode and Effect Analysis, followed by the last chapter with the presentation of the conclusions, limitations, and proposals for future work.

2 Background

The design of a functional safety system can be based on the IEC 61508 standard [10] where the entire life cycle of the system is covered. Regarding information security, the standard only refers to the need to carry out a threat analysis during the risk analysis process. An overview of information security management systems applicable to all types and dimensions of organization is described in ISO/IEC 27000 [11]. This document refers to the most common terms and definitions used in this family of standards. ISO/IEC 27001 [12] is a standard that contemplates policies for continuous improvement of the organization's information system. A basic security prerequisite is a cause of failure, which is comparable to a vulnerability. Vulnerability is defined by ISO/IEC 27002 [13] as "a weakness in an asset or group of assets that can be exploited by a threat".

Generally, two types of risk analysis methodology are used. One consists of a qualitative assessment of risk analysis that includes several non-technical questions that are easy to understand and simple to implement by managers. According to this method, it will not be necessary to quantify the frequency of occurrence of the identified threats [7]. Another methodology used is based on a quantitative analysis of risk, containing mathematical tools for risk assessment [14].

Vulnerabilities in information systems present in the industry or in other related areas are demonstrated at events such as Stuxnet [9] or Duqu [10], holistic analytical methods are needed to mitigate or reduce these risks. One of the tools used in the effective planning of the risk management process is the Failure Mode and Effect Analysis (FMEA). The growing increase in the security risks of information systems and the need for investments to combat this insecurity, is creating serious problems for organizations. To respond to this need, several studies, and methodologies for managing information security risks have been developed over time. However, few studies have been conducted to add the FMEA tool to the security of information [5].

Risk management based on the FMEA allows, at an early stage, to identify failures and recommend preventive or corrective actions to reduce or mitigate their occurrence [15]. To improve the reliability of military equipment, namely, the security aspects of mechanical equipment [6] and the United States Department of Defense developed the FMEA tool in the 1950s [16]. The basic concept of this tool has been extended to include existing and potential vulnerabilities and attacks related to the security of an information system. The FMEA tool provides a structured process for assessing failure modes and reducing or mitigating their effects through the implementation of corrective and predictive actions. This is a complex engineering analysis process that is also used to identify existing or potential failures, their root causes and impacts that they would cause to the organization, namely, to the success of business [17].

Generally, the prioritization of risk of failure modes in the FMEA is determined by calculating the Risk Priority Number (RPN), with three risk factors being evaluated: severity (S), occurrence (O), and detection (D). The RPN is determined by the following formula (1):

$$RPN = S \times O \times D \quad (1)$$

The severity factor represents the relevance of the impact that a given failure may have on the organization, the occurrence assesses the probability of the failure mode happening, and the detection quantifies the ability of the system to detect the failure before it can have a negative impact on the system [5].

There are several methodologies for evaluating factors O, S, and D, in this article, a 10-point linguistic scale will be considered, used in several FMEA applications [18].

3 Concept of the Security Risk Priority Number Method

3.1 *Structure of the Proposed Method*

In the present work, a new method for Security Risk Priority Number (SRPN) is proposed to direct the organization's resources, in a more effective and efficient way in the process of managing the security risks of the information system. The FMEA tool is the basis of the proposed model. Figure 1 illustrates the main steps of the proposed method. The steps of the proposed method are presented below:

Step 1. Create the evaluation team: Considering that the potential safety risks, the severity, occurrence, and detection values are estimated based on the experience of the evaluation group, this group should be multidisciplinary, containing specialists from various areas of the organization.

Step 2. Analyze and determine potential security risks: The risks, causes, effects of risks, and detection controls, are determined based on internal or external historical data, through research, literature, and specialized groups.

Step 3. Evaluate and determine the values to assign to severity (S), probability of occurrence (O), and detection (D).

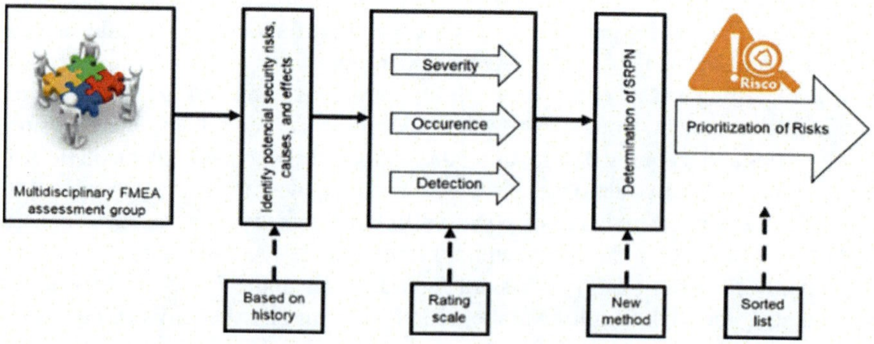

Fig. 1 The structure of the proposed new method for SPRN determination

Table 1 Severity rating scale adapted from [19]

Rating	Description	Definition
10	Extremely dangerous	Failure could cause total system breakdown, without any prior warning
9/8	Very dangerous	Failure could cause a serious system disruption with interruption in service, with prior warning
7/6	Dangerous	Failure could cause major system problems requiring major repairs or significant re-work
5	Moderate danger	Failure could cause a major system problem
4/3	Low to moderate danger	Failure could cause minor system problems that can be overcome with minor modifications to the system or process
2	Slight danger	Failure could cause little or no effect on the system
1	No danger	Failure causes no impact on the system

Step 4. Assign the SRPN value: In this step, the priority value to be assigned to each security risk identified in step 2 is determined. The SRPN value is determined according to the new method proposed in this article.

Step 5. Risk prioritization: Risk prioritization is achieved after reordering all risks using the proposed new method.

3.2 Evaluation of Severity (S), Occurrence (O) and Detection (D)

The severity factor assesses the severity of the organization's information system if a particular risk occurs. The occurrence values the existing probability that a risk may occur. The ability to identify a risk before it can have a significant impact on the organization's information system is measured by the Detection factor [5]. There are several methodologies for evaluating factors S, O, and D, however, in this study, a 10-point linguistic scale will be used [19], as it is simple to understand and widely applied. The used sales are described in Tables 1, 2 and 3.

3.3 New Method for SRPN

A cyber-attack is an example of an attack to the information system that may place the entire organization in total blockade. The assessment of information security risks, aims to respond to three main objectives: to identify the security risks of the systems, to prioritize the risks by their degree of criticality, and to justify the allocation of the organization's resources to reduce or mitigate the risks, which may jeopardize the smooth running of the organization.

Table 2 Occurrence rating scale adapted from [19]

Rating	Description	Definition
10	Certain probability of occurrence	Failure occurs at least once a day, or failure occurs almost every time
9	Failure is almost inevitable	Failure occurs predictably, or failure occurs every 3–4 days
8/7	Very high probability of occurrence	Failure occurs frequently, or failure occurs about once per week
6/5	Moderately high probability of occurrence	Failure occurs approximately once per month
4/3	Moderate probability of occurrence	Failure occurs occasionally, or failure occurs once every 3 months
2	Low probability of occurrence	Failure occurs rarely, or failure occurs about once per year
1	Remote probability of occurrence	Failure almost never occurs; no one remembers the last failure

Table 3 Detection rating scale adapted from [19]

Rating	Description	Definition
10	No chance of detection	There is no known mechanism for detecting the failure
9/8	Very remote/unreliable chance of detection	The failure can be detected only with a thorough inspection, and this is not feasible or cannot be readily performed
7/6	Remote chance of detection	The error can be detected with a manual inspection, but no process is in place, so that detection left to chance
5	Moderate chance of detection	There is a process for double-checks or inspections, but it is not automated and/or is applied only to a sample and/or relies on vigilance
4/3	High chance of detection	There is 100% inspection or review of the process, but it is not automated
2	Very High chance of detection	There is 100% inspection of the process, and it is automated
1	Almost certain chance of detection	There are automatic "shut-offs" or constraints that prevent failure

The degree of criticality of the risk is given by the value of RPN, which is usually calculated by the following Formula (2):

$$\text{RPN} = S \times O \times D \tag{2}$$

where S represents the value attributed to the implication that a certain risk of information security may originate to the organization. The O valuate the occurrence in which a certain risk may materialize, and D the value attributed to the probability of detecting a certain risk before it may cause any relevant damage to the organization.

The new method involves an interrelation between the three significant characteristics for the risk analysis of a security system, S, O, and D.

At a given risk of S equal to 1, O equal to 10, and D equal to 10, the value of the RPN would be given by the following expression (3):

$$\text{RPN} = S \times O \times D = 1 \times 10 \times 10 = \mathbf{100} \tag{3}$$

Considering a risk of S equal to 10, O equal to 1, and D equal to 5, the value of the RPN would be given by the following expression (4):

$$\text{RPN} = S \times O \times D = 10 \times 1 \times 5 = \mathbf{50} \tag{4}$$

The prioritization of risks in this calculation method is obtained after reordering all the risks in descending order of the RPN value. Using this method, a given risk which will not cause any impact on the organization, may be given a higher risk priority. On the other side, to another risk is given a lower risk priority which, although there is a low probability of occurrence, if it occurs would lead to interruption of the entire information system with serious temporary or permanent impacts for the organization. This method leads to a deficient allocation of resources by the organization, leading to inefficiently targeted costs and investments on security protection systems.

This study proposes a new methodology for assessing the degree of risk criticality and its respective prioritization. The new method considers the S that a given risk may cause to the information system as the most relevant factor. The O and D factors will be considered of identical relevance.

The risk prioritization value is given by the SRPN, which is determined by the following expression (5):

$$\text{SRPN} = S|O \times D \tag{5}$$

where S represents the value attributed to impact that a given information security risk can cause to the organization, followed by the calculation result achieved from the multiplication between the O and D factors.

The new method continues to include an interrelation between the three significant characteristics for the risk analysis of a security system, severity, occurrence, and detection, considering the severity attributed to the risk as the most relevant factor.

For the first case mentioned above, a certain risk of minimal severity ($S = 1$), with high occurrence ($O = 10$) and low probability of detection ($D = 10$), the value of the risk priority according to the new method would be given by the following expression (6):

$$\text{SRPN} = S\,|O \times D = 1|10 \times 10 = \mathbf{1|100} \tag{6}$$

Considering a risk of high severity ($S = 10$), with low occurrence ($O = 1$) and average probability of detection ($D = 5$), the value of the risk priority would be given by the following expression (7):

$$\text{SRPN} = S\,|O \times D = 10\,|1 \times 5 = \mathbf{10|5} \quad (7)$$

In the proposed new method, the prioritization of risks is obtained after doble reordering the risk table in descending order. The value of S as first order, and the multiplication factor of the occurrence by detection as second order. This method will always consider a higher risk priority to a given risk, which to happen will certainly have a high impact on the organization, to the detriment of another risk of high probability of occurrence and low detection, which if it occurs, will not cause significant impact on the organization's information system. In the examples mentioned above, the SRPN = 10|5, would be considered a higher priority than SRPN = 1|100, according to the new proposed method the organization should direct the available resources, first to the risk associated with SRPN = 10|5.

4 Example Application of the Usual Method Versus New Method

Table 4 presents the prioritization of risks determined by the multiplicative factor of S, O, and D. After reordering the risk list in descending order of the calculated RPN value, a higher risk priority is given to risks with a low impact to the organization ($S = 1$). On the other hand, risks with a high impact on the organization ($S = 10$) are considered of less importance in the order of prioritization.

Table 5 shows the risk prioritization determined using the new method proposed in this article (SRPN = $S|O \times$ D). After reordering the risk list in descending order of the determined value of SRPN, a priority is given to risks with high impact ($S = 10$) and a second priority to the interrelation between the occurrence and the detection ($O \times D$).

Compared to the usual prioritization of risks method, the proposed method ensures that the available resources are directed more effectively and efficiently, to reduce or mitigate the risks of the organization's information system.

5 Conclusions, Limitations and Further Work

This article presented a new method for valuing the Risk Priority Number, based on the attribution of a greater relevance to the impact that the security risk may have on the organization's information system. Application examples of the usual method of calculating risk prioritization according to FMEA procedures, through

Table 4 FMEA prioritization based in usual RPN calculation

Security information system failure mode and effects analysis (FMEA)								
Operation	Current or potential risk situation							
	Potential failure mode	Effect of failure	Severity	Cause of failure	Occurrence	Existing controls in the detection process	Detection	**RPN**
Internal network infrastructure	Network communication failure	Loss of productivity	1	Inefficient network infrastructure configuration	8	Nonexistent	10	**80**
Administrative workstation	Workstation without network connection	Loss of productivity	1	Interference caused by electromagnetic fields	10	Manual verification of electromagnetic fields	7	**70**
Mail server	Email system is inaccessible	Loss of timely information relevant to the organization's functioning	10	Hardware burned	6	Automatic shutdown system	1	**60**
ERP system	ERP system is blocked throughout the organization	Loss of irrecoverable data critical to the organization's functioning	10	Hacker attack	1	OS Firewall Software	5	**50**
3D CAD workstation	3D CAD software does not respond to technician commands	Delay in project delivery with some customer unsatisfaction	5	Hacker attack	1	OS Firewall Software	5	**25**

Table 5 FMEA prioritization with proposed method SRPN calculation

Security information system failure mode and effects analysis (FMEA)									
Operation	Current or potential risk situation								
	Potential failure mode	Effect of failure	Severity	Cause of failure	Occurrence	Existing controls in the detection process	Detection	**SRPN**	
Mail server	Email system is inaccessible	Loss of timely information relevant to the organization's functioning	10	Hardware burned	6	Automatic shutdown system	1	**10**	**6**
ERP system	ERP system is blocked throughout the organization	Loss of irrecoverable data critical to the organization's functioning	10	Hacker attack	1	Win10 Firewall Software	5	**10**	**5**
CAD 3D workstation	CAD 3D software does not respond to technician commands	Delay in project delivery with some customer unsatisfaction	5	Hacker attack	1	Win10 Firewall Software	5	**5**	**5**
Internal network infrastructure	Network communication failure	Loss of productivity	1	Inefficient network infrastructure configuration	8	Nonexistent	10	**1**	**80**
Administrative workstation	Workstation without network connection	Loss of productivity	1	Interference caused by electromagnetic fields	10	Manual verification of electromagnetic fields	7	**1**	**70**

RPN, and application of the proposed new method based on the SRPN prioritization value, were demonstrated. Prioritization based on the value of SRPN carried out through the new method will direct the financial, human, and material resources of organizations, to more effectively and efficiently actions to reduce or mitigate the identified vulnerabilities of the information system.

The new proposed method needs an evaluation through its application in an organizational environment, the lack of this evaluation is considered a limitation to the work performed.

Future work is necessary to carry out a reliable assessment of the new valuation method of the Security Risk Priority Number (SRPN), based on case studies thus reflecting the reality of the valuation methods in use by organizations versus the application of the new proposed method. It will also be relevant to carry out future studies related to the development of effectiveness and efficiency indicators in the allocation of resources to reduce or mitigate the security risks of information systems, using the new proposed method of enhancing the Security Risk Priority Number. Also, would be interesting to have future works applying the proposed method to information system project management or software development risk analysis.

References

1. Griffin D, Denholm J (2020) This isn't the first global pandemic, and it won't be the last. 2020. [Online]. Available: https://theconversation.com/this-isnt-the-first-global-pandemic-and-it-wont-be-the-last-heres-what-weve-learned-from-4-others-throughout-history-136231. Acedido em 13 5 2021
2. Carroll N, Conboy K (2020) Normalising the "New normal": changing tech-driven work practices under pandemic time pressure. Int J Inf 55
3. Ågerfalk PJ (2020) Artificial intelligence as digital agency. Eur J Inf Syst 1(29):1–8
4. Papagiannidis S, Harris J, Morton D (2020) WHO led the digital transformation of your company? A reflection of IT related challenges during the pandemic. Int J Inf Manage
5. Silva MM, Gusmão APHd, Poleto T, Silva LC, Costa APCS (2014) A multidimensional approach to information security risk management using FMEA and fuzzy theory. Int J Inf Manag 34:733–740
6. Schmittner C, Gruber T, Puschner P, Schoitsch E (2014) Security application of failure mode and effect analysis. In: International conference on computer safety, reliability, and security
7. Patel SC, Graham JH, Ralston PAS (2008) Quantitatively assessing the vulnerability of critical information systems: a new method for evaluating security enhancements. Int J Inf Manag 28(6):483–491
8. I. S. O. (ISO) e I. International electrotechnical commission, ISO/IEC 27005, Information technology—security techniques—information security risk management (2008)
9. Bojanc R, Blazic BJ (2008) An economic modelling approach to information security risk management. Int J Inf Manag 28:413–422
10. I. E. Commission, IEC 61508, Functional safety of electrical/electronic/programmable electronic safety-related systems (E/E/PE, or E/E/PES) (2010)
11. I. S. O. (ISO), ISO/IEC 27000 - Information technology — Security techniques — Information security management systems—overview and vocabulary. International Standardization Organization, 2018. [Online]. Available: https://www.iso.org/standard/73906.html. Acedido em 20 5 2021

12. I. S. O. (ISO) ISO/IEC 27001—Information Security Management, International Standardization Organization, 2013. [Online]. Available: https://www.iso.org/isoiec-27001-information-security.html. Acedido em 22 5 2021
13. I. S. O. (ISO) e I. E. C. (IEC), ISO/IEC:27002: information technology—security techniques—code of practice for information security management
14. Ozkan S, Karabacak B (2010) Collaborative risk method for information security management practices: a case context within Turkey. Int J Inf Manag 30(6):567–572
15. Abdullah K, Mohd Rohani J, Ngadiman M (2005) Development of FMEA information system for manufacturing industry. In: 3rd international conference on modeling and analysis of semiconductor manufacturing, Singapore
16. D. o. D. (US), MIL-P-1629: procedures for performing a failure mode, effects and Criticality analysis
17. McDemortt RE, Mikulak RJ, Beauregard MR (2009) The basics of FMEA (2nd). Taylor & Francis Group, New York
18. Lin Q-L, Wang D-J, Lin W-G, Liu H-C (2014) Human reliability assessment for medical devices based on failure mode and effects analysis and fuzzy linguistic theory. Saf Sci 62:248–256
19. Goodman S (1996) Design for manufacturability at midwest industries, Harvard: Lecture

Effect of Encryption Delay on FTP and VoIP Traffic Based on TCP/UDP

Muhammad Arif, Muhammad Asif Habib, Nasir Mahmood, Asadullah Tariq, and Mudassar Ahmad

Abstract With the rapid increase of E-commerce applications usage, the growth of data flow over the Internet is increasing these days. The negation of unnecessary data delays with data security is essential and ultimate requirement of organizations in terms of file transfer protocol (FTP) and voice over IP (VoIP) traffic in any network. This requirement is impossible to accomplish without the selection of the right protocols for data transmission. This paper provides brief modeling for two well-known transport layer protocols TCP and UDP. The core purpose of this research is to find out how TCP and UDP behave in the case of FTP and VoIP data flow over an IP cloud network (public network) with or without encryption delay. This paper estimates the performance by simulating the IP cloud network with the different scenarios for FTP and voice traffic with different parameters. For this voice and FTP Ethernet delays, FTP and voice packet sent and FTP traffic download/upload response time is simulated and tested on OPNET.

Keyword FTP · OPNET · PSTN · TCP · UDP · VoIP

1 Introduction

Data security has turned into a critical issue in the advanced world as the notoriety, and infiltration of the Web business sector and correspondence innovations has developed. This prevalence is a potential medium of security dangers. To overcome security dangers, present-day information interchanges utilize cryptography a viable,

M. Arif
Department of Software Engineering, The Superior University, Lahore, Pakistan
e-mail: arif@superior.edu.pk

M. A. Habib (✉) · N. Mahmood · M. Ahmad
Department of Computer Science, National Textile University, Faisalabad, Pakistan
e-mail: drasif@ntu.edu.pk

A. Tariq
Department of Computer Science, The Superior University, Lahore, Pakistan

A. Ullah et al. (eds.), *Proceedings of International Conference on Information Technology and Applications*, Lecture Notes in Networks and Systems 350,
https://doi.org/10.1007/978-981-16-7618-5_50

effective, and crucial part for the secure transmission of data by actualizing security parameters like confidentiality, authentication, accountability, and accuracy [1].

The Internet group concurs that security is one of the critical properties that ought to describe any information and communication technology (ICT) framework and application, with specific accentuation on those that depend on the Internet for their extremely nature, e.g., e-business. Outrageously, security is key to everything, and as a rule, safety and effectiveness are clashing prerequisites. While for an expansive class of Internet applications such an actuality has restricted impacts, there is a class of uses whose usefulness may be traded-off by security controls, specifically continuous applications, i.e., applications that force time requirements on parcel conveyance keeping in mind the end goal to imitate the first wellspring of data. For example, voice over IP and feature are conferencing. In the same way, it additionally has converse consequences for different applications like FTP, HTTP, and so on. Encrypt transferred data by using encryption keys are also presented using some standard algorithms. In TCP/IP network, 3DES, AES, and DES are standard encryption algorithms. Virtual private networking (VPN) that uses encrypted virtual tunnels is one of the best mechanisms for security. Here, in this study, we are going to take a brief analysis of the effect of encryption delay on TCP and UDP for voice and FTP-based applications (Traffic) over the network [2].

Peer-to-Peer (P2P) and Voice over Internet Protocol (VoIP) applications developed the essential correspondence benefits in the most recent couple of years, for organizations and people aiming to the voice and video quality, are high, and the calls are free for a direct association between two VoIP end clients. Virtual private network is characterized as a system that uses open network paths; however, keep up the security and insurance of private network [3].

Previously, the concept of VPN was associated and applied for public switched telephone network (PSTN) due to the fruitful results now it is related to IP-based networks (Internet). IP-based systems administration partnerships had consumed significant actions of assets and time to create intranets. These systems were introduced utilizing ATM, frame relay, and expensive rented line administrations to join remote clients [4]. For the portable specialists and littler destinations on the remote end, organizations added their systems with remote access servers or ISDN.

Small to medium-sized organizations utilized low-speed switched network services because they cannot afford to tolerate the cost of dedicated rented lines [5]. Organizations are now putting their extranet on to the Web because of the more availability of the net and more transfer speed. Today's VPN measures settled the security element utilizing sophisticated encryption strategies, astonishing tunneling protocols, information reliability, and protection are proficient.

Also, because these operations happen more than an open system, VPNs can cost altogether less to actualize than exclusive or rented administrations [6]. Even though early VPNs required broad ability to realize, innovation has developed to a level where sending can be a straightforward and reasonable answer for organizations of all sizes. Rapid growth in IT innovations, along with advancement in its features, is a unique piece of the pattern. Transfer of information from a server to the client is known as Gushing [7–11]. The concept of separation of duty on the basis of permissions is implemented to secure the medical information [12]. The information exchange rate and slip rate are the fundamentals for the QoS in gushed media. Conventional gushing uses TCP and UDP [8]. Cryptography is typically alluded to as "the investigation of the mystery." Encryption is the procedure of changing over typical content to an incoherent structure. Decoding is the procedure of changing over encoded content to standard content in the coherent structure [9]. The cryptography procedure can be ordered into two classifications symmetric key and asymmetric.

2 Related Work

The transmission control protocol (TCP) is the main element of the current Internet protocol suite. It supplemented the Internet protocol. same as TCP; user datagram protocol (UDP) is also the transport protocol of layer-4. UDP provides the simplest communication between two nodes. UDP interfaced with the network layer at the bottom and application layer at the upper layer in the TCP/IP suite.

The authors demonstrate voice streams, movement qualities show low burstiness, and it have little information rates [13]. VoIP has lessened loss rate and end-to-end delay necessities, and act as a brilliant constant voice correspondence. As UDP is considered more suitable for constant applications, VoIP keeps running on UDP. Most of the clients have seen delays over 150 ms, while deferments over 300 ms ordinarily reduce the discussion irritating.

Stream control transmission protocol (SCTP) is discussed in [7]. The deployment of SCTP is accomplished with network address translation (NAT) enable device [11]. Control chunks of SCTP are the primary concern for the development of NAT-enabled SCTP, which is used to deliver multi-homing and multi-streaming structures of SCTP. Multiplexing of SCTP control chunks can cause denial of service (DOS) attack, in the case of exposed SCTP-enabled NAT. This hazard for DOS attacks can lessen by the use of parameters processing limitations and SCTP chunks. In conventional IP routing, through tracking of multiple global IP addresses, lookup table conflicts can be passed up.

In [14], the authors discussed that SCTP combined the features of UDP and TCP. SCTP is considered a more reliable message-oriented protocol than TCP and UDP [10]. Except for fast recovery feature, SCTP uses the same congestion control feature used in TCP. Therefore, the performance of SCTP is much more efficient over Internet and satellite communication links as compared to UDP and TCP. An enhanced congestion control mechanism is presented in [14], which improved the performance of SCTP.

The author reviewed that SCTP has striking features such as multi-homing and multi-streaming [15]. The analysis of these above discussed depicts that these features have huge advantages over UDP and TCP protocols [16]. The dynamic separation of duty is used to secure and optimize communication parameters in the Internet of Things (IoT) [17, 18]. They appeared the finest number of streams of data when using multi-streaming and presented how multi-streaming affected the network performance. The promising solution is given after analyzing the drawbacks of multi-homing [19, 20].

Voice over IP (VoIP) is one of the swiftly rising communication technologies and sometimes, unable to retain its performance in the context of scalability. Session initiation protocol (SIP) is a lightweight protocol that was designed to overcome these limitations at the application layer protocol. But, SIP is a high vulnerability to security attacks [9, 10]. The data communication is made secure, efficient, and usable based on the permission-based implementation of dynamic separation of duty [21–24].

The proposed approach in this paper is tunneling utilization on a virtual private network (VPN) for secure and reliable communication. All the experiments were directed using a wireless VPN client (vpn01Client) that connected to the domain controller server (dc01Server) through the VPN server (vpn01Server).

3 Motivations

The main objective of this research is to measure the effect of encryption delay on voice and FTP traffic with TCP and UDP. The primary considerations for both applications will be:

- Impact of encryption delay for voice traffic I term of Ethernet delay, traffic sent, and IP packet drop concerning both TCP and UDP.
- Impact of encryption delay for FTP traffic in terms of Ethernet delay, FTP upload, and download response time and traffic sent an IP packet drop concerning both TCP and UDP.

The research problem on which we proposed a solution is of two types.

- Which protocol in TCP and UDP has efficient Ethernet delay, FTP upload and download response time, traffic sent, and IP packet drop with variant encryption delay?
- Which protocol in TCP and UDP has efficient Ethernet delay, traffic sent, and IP packet drop for VoIP traffic with variant encryption delay?

4 Proposed Methodology

In this section, we provide brief modeling for two well-known transport layer protocols TCP and UDP. The core purpose of our proposed methodology is to find out how TCP and UDP behave in the case of FTP and VoIP data flow over an IP cloud network (public network) with or without encryption delay. We followed the quantitative research approach. First, we have designed simulation scenarios in OPNET

14.5 Modeler, and then, we have settled down the attributes of the simulated environment according to our parameters. Statically, results for different settings are taken, and then, based on these results, conclusions are drawn.

4.1 Research Flow

TCP is a connection-oriented protocol that provides reliable data delivery with the concept of packet acknowledgment, so it would be better to consider TCP for less delay-sensitive traffic like FTP where UDP is connectionless and is favorable for delay-sensitive traffic like VoIP. Since we are dealing with the FTP and VoIP encrypted traffic under VPN, so we are mainly dealing here with both types (VoIP and FTP) of traffic, along with TCP and UDP as underlying protocols with different input and output parameters. Our primary considerations for both applications are as follows:

- Impact of encryption delay for voice traffic in terms of Ethernet delay, traffic sent, and IP packet drop concerning both TCP and UDP.
- Impact of encryption delay for FTP traffic in terms of Ethernet delay, FTP upload, and download response time and traffic sent an IP packet drop concerning both TCP and UDP.

As discussed above, a quantitative research approach is used for this research. We have compared the impact of transport layer protocols (TCP and UDP) for both FTP and VoIP encrypted traffic separately. And then, results are drawn in the form of graphs and tables based on output values. The following steps have been taken to complete this research:

1. Same input data size (in bytes), with varying encryption delay (0.5, 1, 1.5, 2, 2.5 ms) sent from the sender
2. To destination through a simulated network (default encryption of VPN).
3. Two main scenarios are created one for encrypted traffic with variant encryption delay, and the second is without encryption delay.
4. First scenario as shown in Fig. 1 is further divided into two sub-scenarios. The first one is for the FTP encrypted traffic against UDP and TCP. FTP data are sent to TCP and UDP separately, and the FTP response time will be noted against all the encrypted delays.
5. In another scenario, jitter will be calculated against all encrypted delays after sending voice encrypted traffic to TCP and UDP separately.
6. End-to-end packet delay is noted down.
7. Simulation outcomes will be used for comparison.
8. Interpretation of the compared results will result in the actual conclusion of the proposed scheme.

Figure 2 shows the sketch of the simulation scenario without applying encryption; in this scenario, two applications, i.e. FTP and VoIP, are taken. So, the FTP and

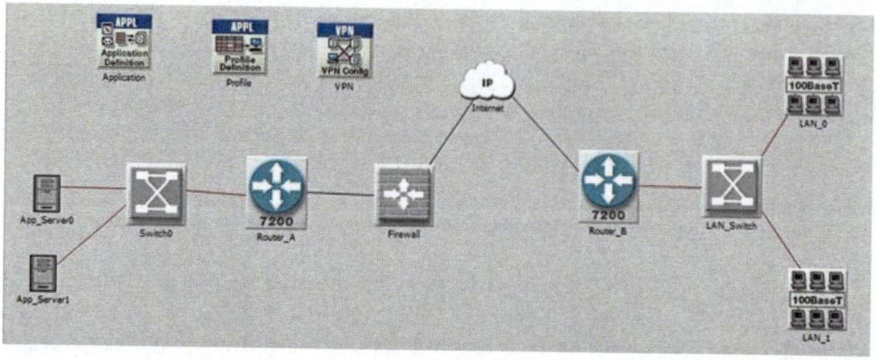

Fig. 1 TCP simulation scenario with encryption

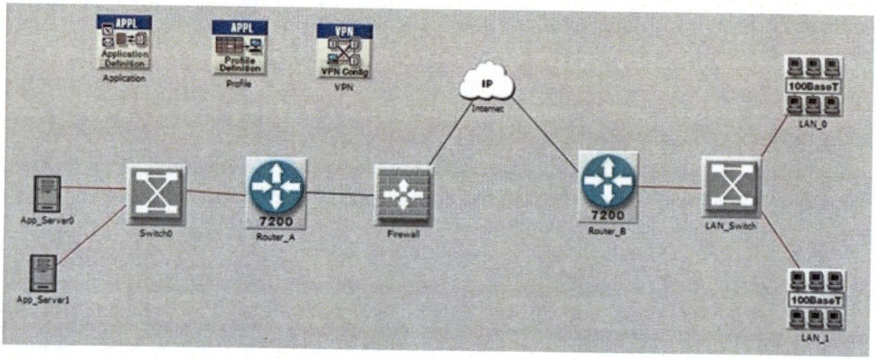

Fig. 2 Simulation scenario TCP without encryption

VoIP traffic will be routed over the network. The application server will select the traffic application and underlying transport layer protocol (TCP or UDP), while the application profile will set the remaining parameters like VoIP codec, no nodes, and encryption delays.

The above scenario in Fig. 1 represents the simulation with encryption. In application, both FTP and VoIP are settled, and their profile attributes are taken. Here, VPN is created between Router A and Router B, as shown in Fig. 1. Through VPN, all kinds of traffic can be sent and received securely over the public network. For security purposes, encryption/decryption is applied on both routes; by default, encryption algorithm used in VPN is DES. Figure 1 shows the encryption delay configuration and information about how varying encryption delay can be settled.

Table 1 Simulation parameters

Simulation details	Parameters
Simulation time	10 min (For each simulation)
No. of nodes	25 (in each LAN)
No. of applications	2 (FTP and VoIP)
No. of LAN	2
Performance measuring	Ethernet delay, FTP download/upload response
Parameters	Time, FTP traffic sent, voice traffic sent
Encryption delay	0.5, 1, 1.5, 2, 2.5 ms

4.2 Links Details

Links are the medium of communication between network devices, and it could be wired or wireless. In this simulation scenario, we are using a different kind of link, supporting different data rates to connect various devices of the network. Detail of interconnecting links is given as: PPP link/DS1 with maximum data rate support of 1.54 Mbps is used from Router A to Router B, where a secure VPN is created. While the Ethernet Base-100 link is used to connect switch0 and application servers on the other side of the network, it is used to connect switch1 and both LANs. The data rate support for this link is 100 Mbps.

The application server 0 (App Server0) in the scenario will entertain the FTP traffic request generated by the nodes, and application server 1 (App Server1) will consider the VOICE traffic. LAN 0 and LAN 1 are both local area networks with 25 nodes each. Both generate the FTP and VOICE traffic requests to both servers on the other side of the VPN network, where encryption delay can be varied according to Table 1, and the effect of encryption delay variation can be observed for FTP and VIOCE with TCP/UDP both [19].

5 Simulation Results

In this section, we are evaluating the results after simulations. Simulation parameters are described in Table 1. The behavior of UDP against FTP traffic received with all encryption delays has been represented by a graph in Fig. 3. It concludes that as encryption delay adds up, FTP traffic received in bytes also decrease with simulation time in all cases of encryption delays, as with Encryption delay-1, i.e., 0.5 ms it achieved the maximum of 5.9 KB wherewith encryption delay-5, i.e., 2.5 ms, it achieved the maximum of 5.1 KB. While FTP traffic received without any encryption is maximum of all, i.e., almost 8 KB with UDP. This shows

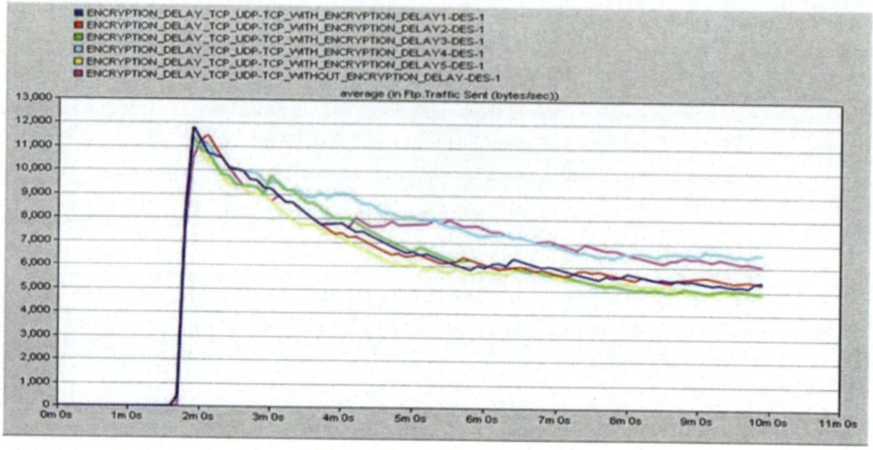

Fig. 3 Comparison of FTP traffic sent with all encryption delays and without encryption delay for UDP

UDP doesn't add any delay to FTP traffic received. It concludes that as encryption delay adds up FTP traffic sent in bytes also decrease with simulation time in all cases of encryption delays, as with encryption delay-1, i.e., 0.5 ms, it achieved maximum of 10.1 KB where with encryption delay-5, i.e., 2.5 ms it achieved minimum of all i.e., 5.1 KB. While FTP traffic sent without any encryption is maximum of all i.e., almost 8KB with UDP. This shows UDP doesn't add any delay to FTP traffic sent.

The graph in Fig. 4 shows a clear difference in sending the voice traffic for TCP and UDP. The traffic sent for TCP is observed much higher as compared to

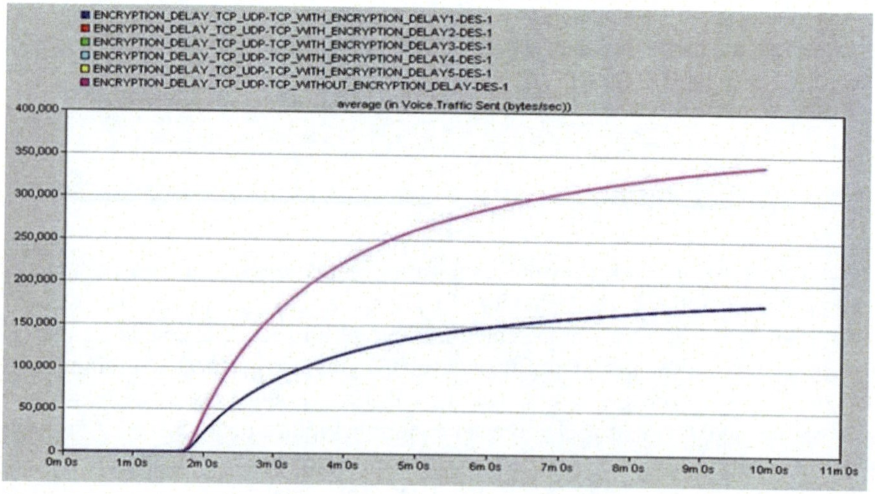

Fig. 4 Voice traffic sent in bytes/sec with respect to TCP and UDP

UDP for TCP. The graph shows the value of almost 340 Mbps, and for UDP, it is observed up to 20 Mbps approximately. Figure 4 shows voice traffic sent in bytes/sec with respect to TCP and UDP. The scope of this research can be enhanced in the future by examining the behavior of stream control transmission protocol (SCTP) and datagram congestion control protocol (DCCP) newly invented transport layer protocols for transportation of voice and FTP traffic in IP cloud environment (Public Network).

6 Conclusion

This research paper illustrates the in-depth examination and comparison of two well-known transport layer protocols for FTP and Voice over IP (VoIP) traffic flow. Network simulations scenarios have been prepared for both protocols using the OPNET network simulator to evaluate the performance against the selected parameters (FTP Ethernet delays, FTP, and voice packet sent, and FTP traffic download/upload response time). Considering Ethernet delay for both TCP and UDP, UDP is better without any encryption delay while considering the encryption delay TCP performs better. UDP is better for FTP download/upload response time without any encryption delay, while TCP is better with encryption delay. Moreover, UDP is better for FTP traffic sent if no encryption delay is added; TCP performs smoothly for FTP traffic sent if encryption delay is added. Whereas, for voice over IP traffic sent UDP and TCP almost act in the same manner. Based on the above discussion, this paper concludes and recommends that UDP is favorable to use when a user requires to send FTP data without any encryption delay on a network. While TCP will be appropriate if FTP Data is sent with encryption delay. Whereas, for VoIP traffic, any of both protocols can be used.

References

1. Nur Zincir-Heywood RAA (2015) Identification of VoIP encrypted traffic using a machine learning approach. J King Saud Univ–Comput Inf Sci 27
2. Snader JC (2015) VPNs illustrated: tunnels, VPNs, and IPsec: Addison-Wesley Professional
3. Verthein W, Taarud J, Little W, Telematics E, Zorn G (1999) Point-to-point tunneling protocol (PPTP) status of this memo this memo provides information for the internet community. It does not specify an Internet standard of any kind. Distribution of this memo is unlimited
4. Zorn G, Townsley WM, Rubens A, Palter B, Pall GS, Valencia AJ (1999) Layer two tunneling protocol. L2TP
5. Patel B, Aboba B, Dixon W, Zorn G, Booth S (2001) Securing L2TP using IPsec, 2070–1721
6. Kocher PC, The SSL protocol version 3.0, internet draft. Netscape Communications Corporation, 03/04/961996
7. Jaha AA (2015) Performance evaluation of remote access VPN protocols on wireless networks. Perform Eval, 4

8. Theodoro LC, Leite PM, de Freitas HP, Passos ACS, de Souza Pereira JH, de Oliveira Silva F et al (2015) Revisiting Virtual private network service at carrier networks: taking advantage of software defined networking and network function virtualization. ICN 2015, p 84
9. Duffield NG, Greenberg AG, Goyal P, Mishra PP, Ramakrishnan KK, Van der Merwe JE (2005) Virtual private network. ed: Google Patents
10. Hofmann P, An C, Loyola L, Aad I (2007) Analysis of UDP, TCP and voice performance in IEEE 802.11 b multihop networks. In: 13th European wireless conference, pp 1–4
11. Lee YHPJH (2010) Degraded quality service policy with bitrate based segmentation in a transcoding proxy. J Inf Commun Convergence Eng 8
12. Habib MA, Faisal CMN, Sarwar S, Latif MA, Aadil F, Ahmad M, Jabbar S, Khalid S, Chaudary J, Maqsood M (2019) Privacy-based medical data protection against internal security threats in heterogeneous Internet of Medical Things. Int J DistribSens Netw. https://doi.org/10.1177/1550147719875653
13. Thakur J, Kumar N (2011) DES, AES, and blowfish: symmetric key cryptography algorithms simulation based performance analysis. Int J Emerg Technol Adv Eng 1:6–12
14. Bruce S (1996) Applied cryptography, 2nd, Wiley and Sons, Inc.
15. Mandal PC (2012) Superiority of blowfish algorithm. Int J Adv Res Comput Sci Softw Eng 2
16. Wei X, Bouslimani Y, Sellal K (2012) VoIP based solution for the use over a campus environment. In: 2012 25th IEEE Canadian conference on electrical and computer engineering (sCCECE), pp 1–5
17. Habib MA, Ahmad M, Jabbar S, Khalid S, Chaudhry J, Saleem K, Joel JPC, Khalil MS (2019) Security and privacy based access control model for Internet of connected vehicles. Future Gener Comput Syst 97:687–696
18. Habib MA, Ahmad M, Jabbar S, Ahmed SH, Rodrigues JJ (2018) Speeding up the internet of things: LEAIoT: a lightweight encryption algorithm toward low-latency communication for the internet of things. IEEE Consum Electron Mag 7(6):31–37
19. Mahmood T, Nawaz T, Ashraf R, Shah SMA (2012) Gossip based routing protocol design for ad hoc networks. Int J Comput Sci Issues (IJCSI) 9(1):177
20. Tariq A, Rehman RA, Kim B, Forwarding strategies in NDN based wireless networks: a survey. In: IEEE communications surveys & tutorials (early Access). https://doi.org/10.1109/COMST.2019.2935795
21. Habib MA, Mahmood N, Shahid M, Aftab MU, Ahmad U, Faisal CMN (2014) Permission based implementation of dynamic separation of duty (DSD) in role based access control (RBAC). In: 2014 8th international conference on signal processing and communication systems (ICSPCS). IEEE, pp 1–10
22. Habib MA, Abbas Q (2012) Mutually exclusive permissions in RBAC. Int J Internet Technol Secur Trans 4(2–3):207–220
23. Habib MA (2011) Role inheritance with object-based DSD. Int J Internet Technol Secur Trans 3(2):149–160
24. Habib MA, Praher C (2009) Object based dynamic separation of duty in RBAC. In: 2009 international conference for internet technology and secured transactions, (ICITST). IEEE, pp 1–5

Malware Prediction Using LSTM Networks

Saba Iqbal, Abrar Ullah, Shiemaa Adlan, and Ahmad Ryad Soobhany

Abstract With a recent increase in the use of the Internet, there has been a rise in malware attacks. Malware attacks can lead to stealing confidential data or make the target a source of further attacks. The detection of malware has been posing a unique challenge. Malware analysis is the study of malicious code to prevent cyber-attacks and vulnerability assessment. This article aims for classification of malware using a deep learning model to obtain an accurate and efficient performance. The system proposed in this study extracts a number of features and trains the long short-term memory (LSTM) model. The study utilises hyper-parameter tuning to improve the accuracy and efficiency of the LSTM model. The findings revealed 99.65% accuracy using sigmoid function that outperforms other activation function. This work can be helpful in malware detection to improve security posture.

Keywords Malware · Security · Neural networks · Deep learning · LSTM

1 Introduction

There has been a continuous growth in the use and transmission of data over the Internet worldwide. There are rising concerns for the security of data at transfer and rest from various threats. To handle these threats, businesses are adopting novel technologies to detect and mitigate threats that can be compromise their security. Malware has been an ongoing security for many years. The name is derived from malicious

S. Iqbal (✉) · A. Ullah · S. Adlan · A. R. Soobhany
School of Mathematical and Computer Science, Heriot-Watt University Knowledge Park, Dubai, UAE
e-mail: si104@hw.ac.uk

A. Ullah
e-mail: a.ullah@hw.ac.uk

S. Adlan
e-mail: sia1@hw.ac.uk

A. R. Soobhany
e-mail: soobhany@hw.ac.uk

A. Ullah et al. (eds.), *Proceedings of International Conference on Information Technology and Applications*, Lecture Notes in Networks and Systems 350,
https://doi.org/10.1007/978-981-16-7618-5_51

code, and it comes in various forms such as viruses, worms, trojans, ransomware and rootkits having different functionalities. Malware differs in their behaviour, and once the system is compromised, the attacker gets unauthorised access without user's knowledge. Therefore, the requirement of conducting malware analysis is needed. It deals with malware functionalities along with their repercussions as discussed by Ellen [1]. There is a continuous growth in the use of machine learning and deep learning approaches for malware detection. Machine learning is a sub-discipline of artificial intelligence where deep learning (DL) is a sub-category of machine learning techniques [2]. Deep learning as also referred as deep neural networks that consists of multi-feed-forward layers in order to train the input data based on the type of the data set.

This paper uses static malware analysis along with deep learning techniques used to reduce the impact of malware attacks or nullify it. Sharp [3] emphasised the evolution of malware started since the beginning of the Internet world. Advanced malware leads to the inflexibility and inability of the system to detect these newly developed malwares. Even the anti-malware software are unable to detect these advanced malwares. Hence, an advanced model should be developed which can adapt constantly to these changing malware functionalities and detecting them effectively. The DL-based approaches help systems become more intelligent and adaptive to these advanced malwares, thereby detecting them based on their behaviour.

The contribution of this paper includes prediction of various types of malwares.

The study undertakes the following:

1. Limited malware classification analysis systems have been developed to date.
2. The PE data set of static malware analysis has been used in LSTM model for classification.
3. Comparative analysis of the LSTM performances using different hyperparameters.

2 Literature Review

This section explains the evolution of the malware, followed by Obfuscation techniques with brief instances. The section also discusses related work in the field of malware analysis using various machine and deep learning algorithms.

2.1 Camouflage Evolution in Malware

The malware designers use various techniques to circumvent security and penetrate through systems undetected. Therefore, malware analysis is essential to conduct. The various types of camouflage techniques are discussed below by Rad et al. [4].

Encryption: The earliest camouflage method and the first encrypted virus were 'Cascade' introduced in 1987.The antivirus scanner cannot detect the encrypted virus immediately because of its encrypted main body that has to be decrypted first to access the whole malicious code. However, the virus can be detected indirectly using the string signatures of the decrypted part of the virus.

Oligiomorphism: Advanced form of encryption, also known as semi-polymorphic in the concealment of malware comprising of a collection of decryptors randomly chosen for a new target. The antivirus consumes a longer time in detecting this type of malware because they have to decrypt all the decryptor first to access the main code for detection. The first of this type was 'Whale' a Dos virus introduced in 1990.

Polymorphism: It is a complex form of both encryption and oligiomorphism. It encrypts the main code in the same way, but the only difference between polymorphic and oligiomorphic/encryption viruses is that it creates an unlimited number of unique decryptors. Polymorphism works on a principle of modifying the appearance of the code whilst infecting the system thereby evading the detection by not leaving behind the permanent string signatures of its variants.

Metamorphism: Composed of mutation engine that mutates the whole code unlike polymorphism. The term was coined by Igor Muttik as 'Metamorphics are body-polymorphics'. Each variant as it duplicates differs in structure, code sequence, size and syntactic properties but have the same functionality.

2.2 Obfuscation Techniques

Obfuscation technique is a process to render unreadable source code or binary sequence of a malware whilst maintaining its functionality. The various obfuscation techniques are discussed below:

Junk/Dead Code Insertion: The binary sequence is modified using junk or dead instructions without any effect on the functionality or behaviour of the code. The instructions used in junk insertion does not change the value of CPU registers or the memory known to be no-operation (NOP).

Variable/Register Substitution: In variable/register substitution, the value of registers or memory variables is changed to different values in various versions of virus evading from signature-based detection. The first virus used this obfuscation method, that modifies the binary sequence of the malicious code, was W95. Regswap in 1998. This can be detected using wild card scanning. **Instruction replacement**: It is used to replace instructions with other instructions having the same meaning/functionality. The instructions, as illustrated below, have the same functionality of setting the register eax value to zero which makes it difficult to detect.

Subroutine Reordering: This method obfuscates a malicious code by randomly reordering its subroutine as shown below. It generates $n!$ of malware variants, n denotes to the number of subroutines. **Code transposition**: In code transposition, reordering sequence of instructions of the original malicious code can be done in two ways. The first method changes the order of the instructions which is then changed back to the original order using jumps or unconditional branches just before execution. The malware using this obfuscation method can easily be detected by removing the jumps and unconditional branches to get back the original malicious code. On contrary, the second method generates new malware variants by reordering independent instructions that do not have any effect on the other instructions. This method is difficult to implement causing a high detection cost.

Code Integration: In code integration technique, malware first disassembles the target programme, integrates itself to it thereby reassembling the new code into a new variant or generation. The first-ever malware to use this most sophisticated obfuscation technique was Zmist for Windows 95. The detection and recovery of the original malicious code are considerably difficult.

2.3 Deep Learning Techniques

Machine learning (ML) used in extracting features, pattern recognition and making predictions based on the input data set provided. ML divided into supervised and unsupervised learning. Deep learning frequently referred to deep neural networks (DNN). The traditional neural network only have 2–3 hidden layers, whereas in DNN it can go up to 150 hidden layers refer to the hidden multilayer as 'Deep'. These DNN models are trained using large labelled data set that learns by itself from the input data without human intervention. It works well with a huge amount of data. DNN algorithm learns higher-level features by combining lower-level features making the model learn features at multiple levels of abstraction. It learns complex functions mapping input to output and feature extraction without any human intervention [5].

Artificial neural networks (ANN) [6] are inspired by human brains. ANN similar to human brain has neurons or perceptrons, which are the most critical part of ANN. ANN architecture consists of an input layer, output layer and numerous hidden layers. It is also known as multi-layered perceptrons (MLP) where perceptrons are sigmoid neurons. Each neuron comprises of binary input weights, activation function and output. The predictions made by the neural network performed by the activation function, learning rate and multi-layered structure. The mechanism of neural network is that each neuron is fed using inputs by the input layer in order to generate different outputs for every single neuron based on weights and bias. The outputs of each neuron from this input layer are weighted and then fed to hidden layers consisting of multiple neurons which perform complex functions. The process of feeding output of one layer as an input for another layer is known to be a feed-forward neural network. The major drawback of this feed-forward neural network is that it does not hold memory for predictions to be made on sequences or time-series. Alternatively, to overcome this drawback, another neural network developed known to be recurrent neural network (RNN) having feedback loops that can store data in memory. Despite the advantage of using the RNN techniques, it has some disadvantages, such as gradient vanishing and exploding problems, and fails to process very long sequences when tanh or ReLU activation functions are used. RNN models are unable to store information for longer duration and incapable of handling long-term dependencies. Because of these drawbacks, it is difficult to train RNN models. These limitations can be addressed by using long short-term memory (LSTM) that has considerable functionality of storing information for a longer period of time.

LSTM is a type of RNN, consisting of additional functionality making it superior in predicting time-series data. Olah [7] stated that LSTM was introduced to overcome the drawback of short-term memory faced by standard RNN.

The deep learning models in the field of security is already progressing incredibly where malware analysis remains one of the major area providing better and efficient performances as compared to traditional methods. This section discusses the past researches that have been carried out for analysis of the malwares using deep and machine learning.

An LSTM framework consists of three gates shown in Fig. 1 that are input gate x_t/x_1, the output gate O_t, the forget gate f_t, immediate cell state C'_t, cell state C_t/C_1 and output state h_t/S_1 with their respective weights of unfolded single timestep LSTM architecture illustrated mathematically in below Eqs. 1–4

$$f_t = \sigma(W_t W_s S_{t-1} + W_f W_x x_t) \quad (1)$$

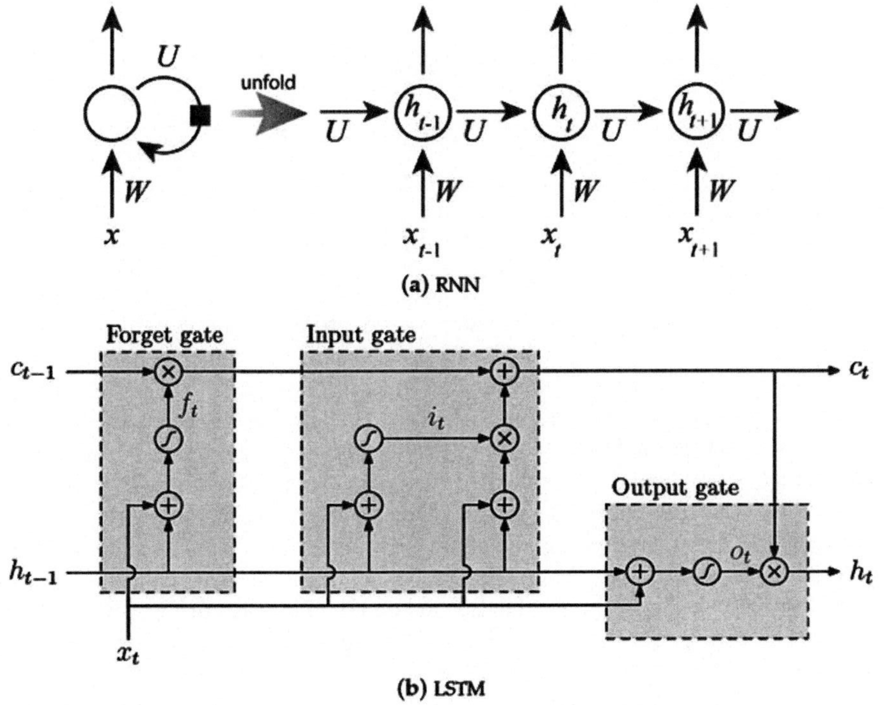

Fig. 1 **a** RNN structure and **b** LSTM structure source [8]

$$i_t = \sigma(W_i W_s S_t - 1 + W_i W_x x_t) \quad (2)$$

$$O_t = \sigma(W_o W_s S_{t-1} + W_o W_x x_t) \quad (3)$$

$$C_t' = \tanh(W_c W_s S_{t-1} + W_c W_x x_t) \quad (4)$$

Figure 1 concludes that forget gate f_t can be calculated when the input X_1 having some weight, and some previous output state S_0 is passed through the sigmoid function. Similarly, the input gate i_t and output gate O_t are calculated. Also, the intermediate cell state C'_t is calculated similarly.

The information being discarded depends on the forget gate value, which is combined with the previous cell state value. The result is obtained by multiplying the input gate, and the intermediate cell state decides about the information to be stored in the cell state. Finally, the new cell state for current input is calculated as below Eq. 5.

$$C_t = (i_t * C_t') + (f_t * C_{t-1}) \quad (5)$$

Hence, a new output state h_t is evaluated as from Eq. (6)

$$h_T = O_t * \tanh(C_t) \quad (6)$$

3 Related Work

D'Angelo et al. [9] have proposed their work of malware detection in a mobile Android environment using autoencoders-based artificial neural network. They provided the neural network with sparse matrices having two-dimensional representation of images named as API images which represent the behaviour of the mobile application extracted by using dynamic malware analysis. The encoders-based ANN is very effective when it comes to detecting malwares as it continuously draws the most useful and best features from the two-dimensional images provided. Vasan et al. [10] proposed the study of malware variants using image-based convolution neural network. Here, they converted the raw malware binaries into coloured images using finely tuned convolution neural network for the identification of the malwares. They trained the model using ImageNet data set and tuned it finely to cover the imbalance in the data set through the data augmentation technique to improve the performance. They concluded that the image-based CNN has better accuracy of 98.82% which is more compared to other algorithms, and the computational cost is lower than grey and coloured images analysis. Ren et al. [11] proposed two ends to end malware detection methods that are DexCNN and DexCRNN. The deep learning models are trained by the resampling the raw bytecodes from the classes.dex of Android applications. They used two resample methods on the classes.dex files to get it into fixed-size sequences required for the deep learning models as inputs. The accuracy for dexCNN achieved was 93.4% and accuracy for DexCRNN came out to be 95.8%. The pre-processed data set was divided into three sets that were 80% training set, 10% validation set and 10% test set.

Yakura et al. [12] wrote in his research about the importance of studying the malware in byte sequence which characterises its functionalities. He proposed deep learning model where the binary data of the malicious code is converted into image with an attention mechanism-based CNN. The main shortcomings of the proposed method were that the byte-sequence study becomes difficult if the packers use cryptographic methods like AES or RSA encryption. Andrade et al. [13] proposed large data sets for the classifications of malware which were previously limited to smaller data sets and were not made public. They trained their model based on long short-term memory neural network. Kang et al. [14] proposed a deep learning method using word2vec-based LSTM neural network for classification of malware and comparing it with one-hot encoding method. The malware is classified in a disassembled assembly source by extracting both the opcodes and API function thereby vectorising it using word2vec into fewer dimensions that reduces learning rate and increased classification rate. The vectorised results were then fed to LSTM to evaluate the classification results. Traditional use of either extracting opcodes or API function in malware classification had some limitations.

Sung et al. [15] also proposed a method of classifying malwares using an advanced version of word2vec model known as fast-Text model based on bidirectional LSTM (Bi-LSTM). The fast-Text model-based Bi-LSTM is used for classifying malwares using opcodes and API function for more accuracy and also this model includes words in sentences by N-gram algorithms for classification. This paper focuses on extracting features of the malwares and then pre-processing it. The fast-Text model embeds the word2vec and one-hot encoding files with the input files which was used to feed it to Bi-LSTM for the classification of files based on families.

In the proposed method, trained model of Bi-LSTM is verified using opcodes and API function where only test data set is used. It was concluded that API function names with opcodes used for malware classification had greater accuracy than using the opcodes or APIs alone. On comparing with word2vec and one-hot encoding method, the accuracy of the proposed method came to be 1.87% and 0.39% higher than one-hot encoding and word2vec simultaneously. The accuracy achieved by the proposed method came to be 96.76%.

This method requires to determine the size of Bi-LSTM and input vector along with the ability to detect malicious files where the input data set includes both normal and malicious files. Future work for classifying malicious or normal file from the randomly fed input data set will be carried out using the proposed model.

Table 1 critically analyses past research works based on malware analysis with machine learning and deep learning algorithms which also includes accuracy achieved by each model used in respective security domain. We have analysed that the accuracy achieved using both static and dynamic malware techniques performed

Table 1 Summary of related work using ML/DL techniques

Author	Focussed security domain	DL techniques	Year	Accuracy	Conclusion
[9]	Mobile security	AE	2019		The resulting framework can outperform more complex and sophisticated machine learning approaches in malware classification
[10]	IOT android mobile security	CNN	2020	Mailing data set-98.82% IoT-android data set-97.35%	More research is required for the improvement of the efficiency
[16]	Windows	TELM (Two hidden layer extreme learning)	2019	99.65%	Global dependencies of input samples needed to be extracted for both malware and botnet detection. Future work for improvement of ELM model is required
[17]	–	Multi-level different deep learning system using tree search	2018	–	Training with a larger data set can improve malware detection accuracy. For diverse data distributions, the traditional method cannot be fitted with deep learning model because of the complex malware data. Further research on using algorithms strategically to decrease computational should be made
[18]	IoT-environments	GLCM and machine learning techniques	2019	RF-95% Naıve Bayes-89% K-NN -80%	This model can be adopted in real-life applications for detecting known IoT malware using the learned pre-processed image features

(continued)

Table 1 (continued)

Author	Focussed security domain	DL techniques	Year	Accuracy	Conclusion
[11]		Dex-CNN and Dex-colour-inspiredneural network	2020	DexCNN-93.4% DexCRNN95.8%	The patterns learned by the proposed model for categorising benign or malicious area index files remain unclear that requires future research
[13]	Multi-class dataset	LSTM-based RNN	2019	90.63%	The multiclass dataset used was quite small for Deep learning and the resources required was high with an excellent result. Furthermore, research is required using fewer resources resulting in a good performance
[12]	Public malware dataset	Attention mechanism with CNN	2019	–	The proposed method has a higher classification accuracy than conventional methods. It reduces the workload of manually analysing the malware by extracting sequences of the malware

(continued)

Table 1 (continued)

Author	Focussed security domain	DL techniques	Year	Accuracy	Conclusion
[19]	Android	Monte-Carlo-based reinforcement learning	2020	–	The framework requires further research of using static analysis with dynamic analysis together as it limits meaningful features individually. A neural network can be used for the production of execution traces that have been trained with previously analysed data
[14]	Microsoft Malware Classification challenge dataset	Word2vec model-based LSTM	2019	97.59%	A requirement of high computing resources needed for using all the opcodes and API function. Word2vec model has better accuracy than one-hot encoding with lesser dimension
[15]	Drones and GCS (Microsoft dataset)	Fast-text model-based Bi-LSTM	2020	96.76%	This model requires further research on the size of input vector Bi-LSTM as well as classification of malware using a random dataset

better in malware predictions than using static or dynamic techniques individually. Further, study stated that using a larger relevant data set with API functions produces better results. It can be seen from the table that LSTM-based RNN model is frequently used because of its better performance than most of the deep learning and ML-based models. However, their performances can be improved using ensemble models where different algorithms are combined to build an effective ensemble model. Going further there has been survey research carried out by Aldweesh et al. [20] that focuses on usage of DL in IDSs giving a detailed review, its complete analysis and taxonomy of main DL architectures used in IDSs illustrated in Table 2. They also made a comparison of previous surveys based on cybersecurity DL to identify their drawbacks. They studied the DL solutions in detail that included input data, detection, deployment and evaluation strategies. After the survey, it was noticed that the earliest architectures were AE, DBN and RNN that were studied, whereas CNN is newly investigated in 2017. It mentions that the ensemble architectures need to be explored more. The survey further includes proposed IDSs based on deep learning discussing the data sets used and their performances achieved.

The main drawbacks discovered in the survey was that the proposed DL-based IDSs lacked the use of proper data sets. Since the proposed models used KD99 or NSL-KDD data sets containing old traffics without any novel attacks traffics without having any real-time properties. Coming onto CICIDS2017 IDS/IPS data set and NBaIoT data set containing traffic from stimulated environments overcame this issue. There need to be more researches done in the field of security using deep learning models. Table 1 shows the comparison made by Aldweesh et al. [20].

4 The Proposed Model

The main objective of this research is to create a relevant data set, and then, using the data set, a comparative study using deep learning technique is done to detect and classify malware. Figure 2 illustrates the proposed model architecture.

For static analysis, we have used indicators like file types, hashes, strings and file header data for determination of malicious files. Firstly, the filetype of the malware is found out helping to identify the targeted operating system, for instance PE32 file. The portable executable (PE) format is a file format for executable, object code, DLLs and others used in 32-bit and 64-bit versions of Windows operating systems. In addition to these, extracting strings from malicious files gives additional pieces of information and its functionalities.

We have extracted the PE headers of the benign and malware files which is then pre-processed where features are extracted from headers fields. The tasks in pre-processing step involves normalisation, removal of redundant features and labelling. Labelling is done either by signature-based antivirus or cloud-based service where multiple antivirus engine-based scan could be performed in parallel. We labelled each sample in both malware and benign group by cloud-based service Virustotal.

Table 2 Survey comparison of deep learning algorithms, Source [20]

DL Algorithm	Application		Dataset	Feature learning	Classification technique	Accuracy
2*AE	Wi-Fi intrusion (2016)		AWID	Stacked AE	Softmax	97.7%
	Wi-Fi (2018)	Sonation	AWID	3 SAE	K-means clustering	94.81%
3*DBN	Network (2014)	Intrusion	KDD99	LSTM	LSTM	93.94%
	Network (2016)	Intrusion	10% KDD99	DBN	Softmax regression	97.9%
	InVehicle (2016)	Network	Simulation of in vehicular network	DBN	Conventional stochastic gradient descent method	97.8%
4*RNN	Network (2015)	Intrusion	KDD99	LSTM	LSTM	93.82%
	Network (2016)	Intrusion	KDD99	LSTM	Softmax	96.93%
	Network (2018)	Intrusion	KDD99	–	Bi-directional LSTM	84.99%
	Attack detection in social networks (2018)		NSL-KDD	LSTM	LSTM	97.5%
3*CNN	Network (2017)	Intrusion	10% KDD99 + NSL-KDD	CNN	SVM + 1-NN	1-NN:96.19% + 86.74%SVM:95.27% + 77.68%
	Network (2018)	Intrusion	KDD99	CNN	Softmax	97.53%

(continued)

Table 2 (continued)

DL Algorithm	Application		Dataset	Feature learning	Classification technique	Accuracy
	Network (2018)	Anomaly	NSL-KDD + Kyoto Honeypot + MAWILab	–	CNN	Shallow CNN outperforms moderate and deep CNN
2*Ensemble	Network (2016)	Intrusion	10% KDD99	AE	DBN and BP for fine tuning	92.1%
	Network (2018)	Intrusion	KDD99	(1) None, (2) STL: sparse AE	(1) DNN (2) LSTM	DNN: 66% LSTM: 79.2%
Hybrid	Network (2018)	Intrusion	NSL-KDD	GAN	GAN	–

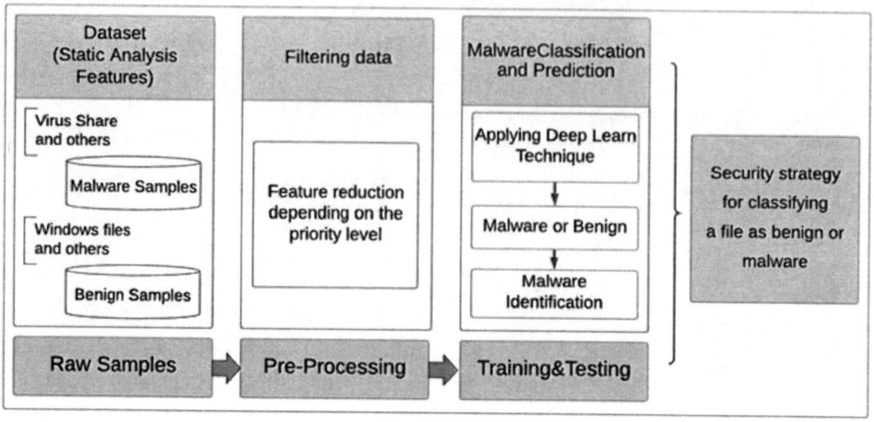

Fig. 2 Proposed model architecture

These samples were used in feature extraction where PE header fields are extracted to make a pre-processed data set for malware analysis. In the second phase, malware classification and prediction are performed, which involve applying DL techniques, malware benign classification and malware identification.

The data set is then fed to the LSTM network of deep learning technique having a single hidden layer and one input layer. Furthermore, the LSTM model is trained and tested using the training and testing set that has been divided in the pre-processing step in the ratio of 80:20. After implementing the aforementioned phases, the security strategy is precisely defined to classify the file as either benign or malware.

In this section, we describe the data set used in the analysis. Thereafter, the pre-processing, and exploration of the data are implemented using correlation technique.

4.1 Data Set Description

The data source used is a malware data set categorised into malicious and benign. The data set consists of 47,580 of malicious and benign files with their 1002 features or information. The data set is exported from the 'pe imports' elements of Cuckoo Sandbox reports where malware files were downloaded from virushsare.com and goodware files from portableapps.com and Windows7 x86 directories. It contains static analysis data that are the top-1000 imported functions. The data set has been validated first to verify its quality and usage required for the project use case. The data set consists of malware class represented as 0 for 'Goodware' and 1 for 'Malware'.

4.2 Data Pre-processing

The dataset used is the top 1000 features of PE imports as shared by Oliveira [21]. In this research, we have selected 29 features that define six categories comprising of benign programmes and malware programmes further categorised into 'Backdoor', 'Constructor', 'Email-worm', 'Hoax', 'Rootkit'. The five categories of malware have been defined based on the top 10 static API call information stated in Table 3

After the pre-processing phase, the data have to be split into train and test set before fed into the model, in order to be trained. In the experimented model, we have split the data into 80% train and 20% test. The key principle behind splitting this data is that the more samples in the train set, the lower the variance. The training set should be big enough to achieve low variance over the model parameters. Similarly, the test data should be small enough data to observe low variance amongst the performance results. The idea is to split the data to achieve low variance in both cases. If the data set is big enough to achieve lower variance on the training parameters; the increase of the training set is required, which simultaneously, will increase the training time.

4.3 Data Exploration

Figure 3 is used for visualising the correlation between the selected 29 features with malware class. So, the every square in the heat map helps us to visualise the correlation between the variables (features) on each axis. The sidebar shows the correlation range from -1 to 1. There are shades of colour from light to dark that defines the linear trend between two features or variables. The darkest shade of blue is $+1$, and the darkest shade of pink is -1. The correlation between two features is larger if the colour is darker. The diagonals that are all dark blue have values 1 because they are correlating each feature to itself.

4.4 Malware Prediction Using Deep Learning

This paper proposes a method for classifying malware. The proposed model is a deep learning model where it predicts whether a file is malicious or benign (0 for benign and 1 for malware). The data set originally having 1000 features belonging to malicious and benign files were collected using static analysis. Out of which, 29 features were selected depending on the priority level, as aforementioned. After pre-processing of the data set, it is fed to the LSTM model, in order to be trained. The LSTM model comprising of one input layer and one hidden layer.

Hyper-parameters tuning are performed using activation functions such as Sigmoid, ReLU along with different optimisers and epochs 10, 30, 50 and 100. In order to specify which works the best for the data set. Thereafter, a comparison

Table 3 Top 10 static features for malware families

Benign pro-grams	Backdoor	Constructor	Email-worm	Hoax	Rootkit
None	None	ExitProcess	GetProcAddress	ExitProcess	GetProcAddress
GetLastError	GetProcAddress	GetModuleFileNameA	LoadLibraryA	GetProcAddress	LoadLibraryA
GetProcAddress	LoadLibraryA	CIcos	GetModuleHandleA	LoadLibraryA	VirtualProtect
CloseHandle	GetModuleHandleA	adj fptan	ExitProcess	GetModuleFileNameA	ExitProcess
GetCurrentThreadId	VirtualAlloc	FreeLibrary	CloseHandle	RegCLoseKey	VirtualFree
GetCurrentProcess	ExitProcess	GetProcAddress	VirtualAlloc	CreateThread	VirtualAlloc
Sleep	VirtualFree	LocalAlloc	RegCLoseKey	CloseHandle	RegCloseKey
MultiByteToWideCha	r RegCLoseKey	vbaFreeVar	GetLastError	GetModuleHandleA	IsEqualGUID
EnterCriticalSection	GetModuleFileNameA	GetCommandLineA	GetModuleFileNameA	GetLastError	SetTimer
LeaveCriticalSection	CloseHandle	adj fdiv m64	WriteFile	WriteFile	InternetCrackUrlA

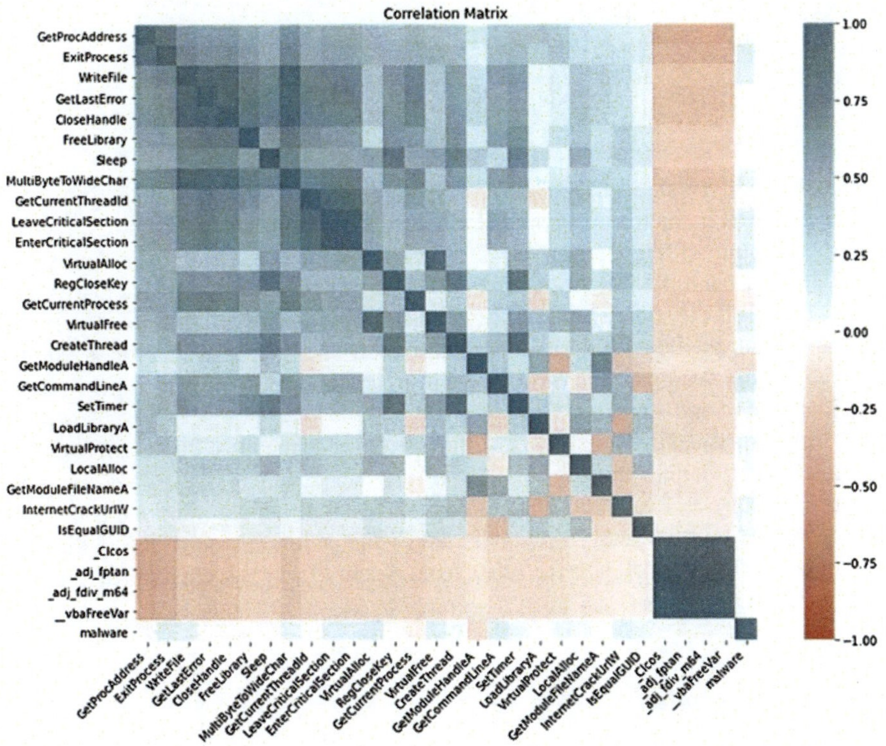

Fig. 3 Heat map of 29 features

analysis of the experiments was conducted between 29 and 1000 features. The output was a binary classification.

5 Results

This section presents the evaluation criteria used to assess the proposed model along with the results of the predictions.

Evaluation Criteria: The proposed malware prediction model is evaluated using three criteria root mean square root, mean absolute error and loss curve and accuracy curve.

1 RMSE- Root Mean Square Root

$$\text{RMSE} = \sqrt{\text{MSE} = \sqrt{\frac{1}{n}\sum_{j=1}^{n}(T_j - Y_j)^2}} \tag{7}$$

2 MAE- Mean Absolute Error

$$MAE = MAD = \frac{1}{n}\sum_{j=1}^{n}\left|T_j - Y_j\right| \tag{8}$$

In 7 and 8 equations, $T_j$is the actual value, whilst $Y_j$is the predicted value. 'n' is the no of predicted observations. So, MAE can be described as the average of the absolute difference between the actual and the predicted observations. On the other hand, RMSE can be defined as the square root of mean square error (MSE). It is the square root of the average of the squared difference between the actual and predicted observations. These metrics used for evaluation are in the range from 0-infinity and errors are within the +ve and −ve directions.

3 Loss Curve and Accuracy Curve

As explained by Karpathy [22], the loss curve is the most important to debug a model. It defines the learning rate of the neural network calculated by plotting training and validation loss. The model has a high learning rate if the loss decreases with increase in epochs, and it has a low learning rate if the loss increases with epoch. On the other hand, accuracy curve generally defines the overfitting scenarios.

5.1 LSTM Model

Tables 4 and 5 illustrate the RMSE and MAE values that are explained for training and testing set for epoch 10, 30, 50 and 100. There is not much MAE and RMSE difference between the training and testing set for 29 features. Whereas, there is a significant difference in RMSE between training and testing set for 1000 features.

Table 4 Evaluation of 29 features

	LSTM architecture parameters				
	Batch size = 400 activation function = sigmoid				
Epoch	Training MAE	Training RMSE	Testing MAE	Testing RMSE	Accuracy
10	0.03	0.17	0.03	0.17	98.97856
30	0.02	0.14	0.02	0.15	99.39680
50	0.02	0.12	0.01	0.14	99.51870
100	0.01	0.10	0.02	0.14	99.65111

Table 5 Evaluation of 1000 features

	LSTM architecture parameters				
	Batch size = 400 activation function = sigmoid features = 1000				
Epoch	Training MAE	Training RMSE	Testing MAE	Testing RMSE	Accuracy
10	0.00	0.06	0.02	0.13	99.82135
30	0.00	0.03	0.02	0.15	99.86548
50	0.00	0.03	0.03	0.17	99.89911
100	0.00	0.02	0.03	0.16	99.91172

5.2 *Analysis of Accuracy and Loss Curve*

Figure 4 demonstrates the accuracy curve and the loss curve for LSTM that describes the learning rate and types of overfitting cases. The graphs below are the result of

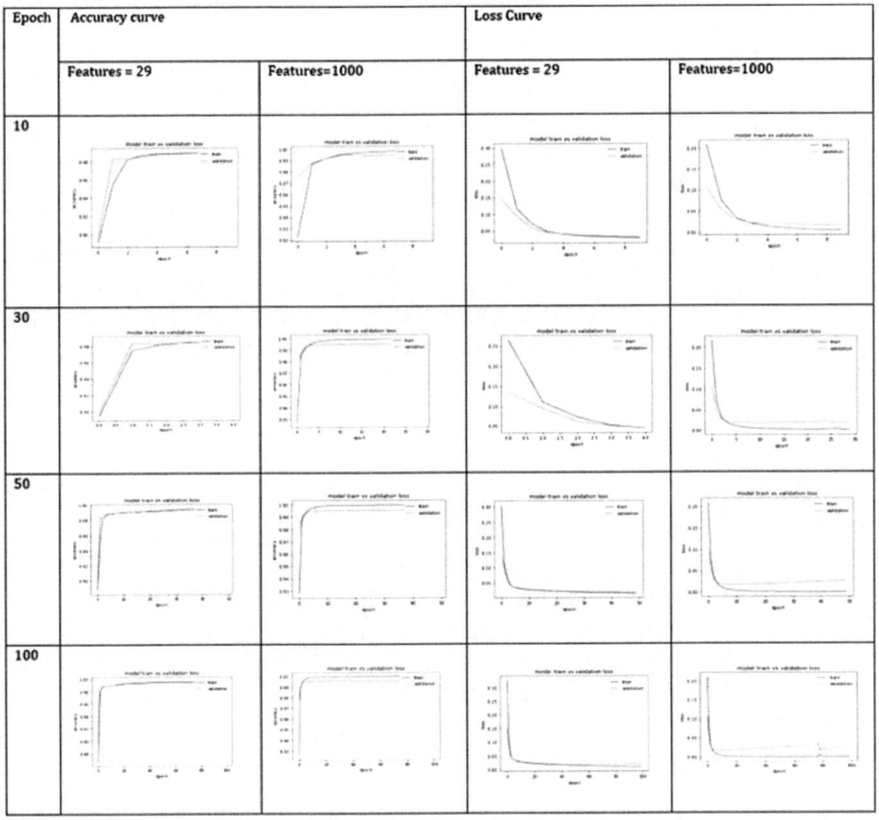

Fig. 4 Analysis of LSTM model

training and testing of LSTM model with epochs of 10, 30, 50 and 100. The epochs were used with 29 selected features as well as the complete 1000 features to compare the performance of LSTM as neural network model works better with a larger data set.

The accuracy curve is plotted between accuracy and epoch. Now, the 'accuracy' is calculated by training the model, and 'validation accuracy' is calculated by testing the model. The greater the difference between the training accuracy and the validation accuracy, there is a strong overfitting case, whereas if the difference is small, little overfitting case occurs.

The loss curve is plotted against the loss (error) and epochs. Again, the 'loss' is calculated, whilst training LSTM and also 'validation loss' is calculated whilst testing the model. The 'lower the loss the better is the learning rate of the model and vice versa.

The accuracy graph curve and loss graph curve in Fig. 4 illustrates the comparison between 29 and 1000 features, and it can be concluded after looking at the graphs that our LSTM model worked best for epoch = 50 and 100 with 29 features. There is no overfitting case in the accuracy curve as well as the difference in the loss is low meaning it higher learning rate.

The most important task after pre-processing is the input/output shape fed into the neural network (LSTM) model. This LSTM model works best when worked with a fewer features that consists of 29 features selected using weka but also gave better results when worked with the original data set consisting of 1000 features. The correlation matrix graph show us the correlation between each features and how much they are dependent on each other.

After carrying out the implementation and hyper-tuning of different parameters, we found that this model worked best with Sigmoid activation function, and the best predicted model came out to be at epoch 100 where accuracy is 99.65%. In other words that the model can predict the malicious file accurately with 99.65%.

6 Conclusion

This research is about classification of a given file into malicious or benign using deep learning technique by selecting static features for malware analysis. The model used for classification is LSTM model which is evaluated based on RMSE and MAE values. Various hyper-parameters were tuned during the experiments. Moreover, three different activation functions, namely ReLU, softmax and Sigmoid with different epochs were experimented. The LSTM model worked best with Sigmoid function resulting in high accuracy of 99% approximately. Whilst in ReLU activation function, the accuracy came down to 36–49% approximately when using optimizer as 'binary crossentropy', whereas the accuracy increased to 98% when using 'mse' optimizer. In softmax activation function, the loss come out to be constant which is a hypothetical case and does not work well with the data set.

Hence, ReLU activation function worked best with 'mse' optimizer, softmax activation function is not used because it's accuracy came out to be constant with the different hyper-parameters. In Conclusion, the best activation function that suits this research is 'sigmoid function with high accuracy between 98 and 99% using different parameters. The second most important factor that affected the accuracy was epoch. As the epoch increased, the accuracy slightly increased with increase in computational time. The epochs were chosen wisely for better performance of the model as computational time remains an important factor along-with the accuracy for malware detection.

This research can be further optimised in classifying the malware and their respective families. In addition, reducing computational time and effective analysing of these malwares can be blocked efficiently before the attacker cause a major damage to the system.

References

1. Ellen Z, What Is malware analysis? Defining and outlining the process of malware analysis. https://digitalguardian.com/blog/what-malware-analysis-defining-and-outlining-process-malware-analysis
2. In B, How does artificial intelligence work? https://builtin.com/artificial-intelligence
3. Sharp R (2009) An introduction to malware. Spring
4. Rad BB, Masrom M, Ibrahim S (2012) Camouflage in malware: from encryption to metamorphism. Int J Comput Sci Netw Secur 12(8):74–83
5. Brownlee J (2016) What is deep learning. Machine learning mastery 16
6. Nielsen MA (2015) Neural networks and deep learning, vol 25. Determination Press San Francisco, CA
7. Olah C (2015) Understanding lstm networks
8. Nait Aicha A, Englebienne G, Van Schooten KS, Pijnappels M, Kr¨ose B (2018) Deep learning to predict falls in older adults based on daily-life trunk accelerometry. Sensors 18(5):1654
9. D'Angelo G, Ficco M, Palmieri F (2020) Malware detection in mobile environments based on autoencoders and api-images. J Parallel Distrib Comput 137:26–33
10. Vasan D, Alazab M, Wassan S, Safaei B, Zheng Q (2020) Image-based malware classification using ensemble of CNN architectures (imcec). Comput Secur 92:101748
11. Ren Z, Wu H, Ning Q, Hussain I, Chen B (2020) End-to-end malware detection for android Iot devices using deep learning. Ad Hoc Netw 101:102098
12. Yakura H, Shinozaki S, Nishimura R, Oyama Y, Sakuma J (2019) Neural malware analysis with attention mechanism. Comput Secur 87:101592
13. Andrade EdO, Viterbo J, Vasconcelos CN, Gu´erin J, Bernardini FC (2019) A model based on Lstm neural networks to identify five different types of malware. Procedia Comput Sci 159:182–191
14. Kang J, Jang S, Li S, Jeong Y-S, Sung Y (2019) Long short-term memory-based malware classification method for information security. Comput Electr Eng 77:366–375
15. Sung Y, Jang S, Jeong Y-S, Hyuk J et al (2020) Malware classification algorithm using advanced word2vec-based bi-lstm for ground control stations. Comput Commun 153:342–348
16. Jahromi AN, Hashemi S, Dehghantanha A, Choo K-KR, Karimipour H, Newton DE, Parizi RM (2020) An improved two-hidden-layer extreme learning machine for malware hunting. Comput Secur 89:101655
17. Zhong W, Gu F (2019) A multi-level deep learning system for malware detection. Expert Syst Appl 133:151–162

18. Karanja EM, Masupe S, Jeffrey MG (2020) Analysis of internet of things malware using image texture features and machine learning techniques. Internet Things 9:100153
19. Sartea R, Farinelli A, Murari M (2020) Secur-ama: active malware analysis based on Monte Carlo tree search for android systems. Eng Appl Artif Intell 87:103303
20. Aldweesh A, Derhab A, Emam AZ (2020) Deep learning approaches for anomaly-based intrusion detection systems: A survey, taxonomy, and open issues. Knowl-Based Syst 189:105124
21. Oliveira A (2019) Malware analysis datasets: top-1000 PE imports. IEEE Dataport
22. Karpathy A et al (2016) Cs231n convolutional neural networks for visual recognition. Neural Netw 1(1)

Security Issues and Defenses in Virtualization

Rouaa Al Zoubi, Bayan Mahfood, and Sohail Abbas

Abstract Virtualization, the process of allowing efficient utilization of physical computer hardware, is the core of many new technologies. With this comes the importance of understanding the related security aspects to avoid the compromise of underlying resources and services. In this paper, we provide an overview on the two main virtualization architectures and the different types of virtualization approaches related to those architectures. We also review the literature for virtualization security requirements and security attacks. We highlight the latest security techniques proposed in the literature. Due to the growth of cloud computing in the industry, we also discuss virtualization security in the industry. As a result, we have found that the gap between academia and industry has become very small in this field, and more importance should be given to client and service provider responsibility awareness.

Keywords Virtualization · Virtual machine managers · Hypervisors

1 Introduction

The concept of virtualization was coined in the 1960s and has continued to evolve throughout the years. It was only until the 2000s that technology had reached a level that allows the complete support of modern virtualization, transforming the reality of the way networks, and applications are being developed [1]. Virtualization is considered as one of the main features of cloud computing. This technology is used for improving performance and increasing the efficiency of computing services. The main idea of virtualization is to share physical resources to utilize an actual machine's

R. Al Zoubi · B. Mahfood · S. Abbas (✉)
Department of Computer Science, College of Computing and Informatics, University of Sharjah, Sharjah, UAE
e-mail: sabbas@sharjah.ac.ae

R. Al Zoubi
e-mail: u20200419@sharjah.ac.ae

B. Mahfood
e-mail: bmahfood@sharjah.ac.ae

A. Ullah et al. (eds.), *Proceedings of International Conference on Information Technology and Applications*, Lecture Notes in Networks and Systems 350,
https://doi.org/10.1007/978-981-16-7618-5_52

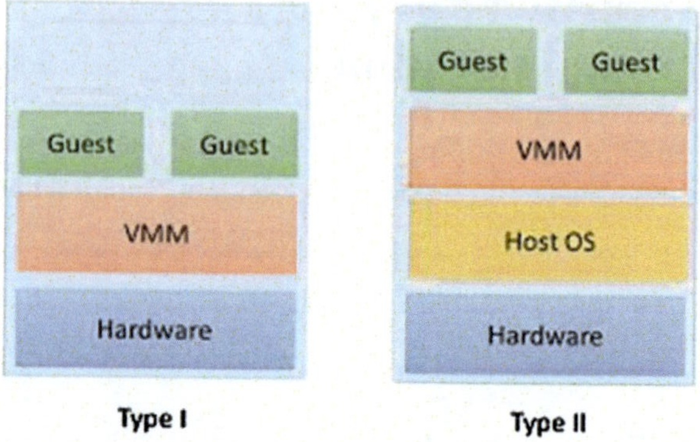

Fig. 1 VMM architectures

absolute limit by disseminating its abilities and resources among numerous clients or environments. With this rapid shift toward virtualization and cloud computing, it is very important to study the security threats they face, and the latest techniques are used to defend them. Virtualized environments require both traditional and specialized approaches to secure the different components of their intricate architecture. The two main components needed for virtualization are the virtual machine (VM) and the hypervisor, also known as virtual machine monitor (VMM). VMs are the emulation of a computer system that provide the functionality of a physical computer. They have their own processor, memory, network interface, and storage. The VMM on the other hand, acts as a monitor where it controls the creation of multiples guests (virtual machines) and distributes the resources of the actual physical hardware (the host) among them. It allows the VMs to run simultaneously without affecting performance, making it feel as if they are independent machines. Figure 1 shows the two basic architectures on which the VMM can interact with the host machine. In Type I (also called bare metal) the VMM runs directly on top of the system hardware, whereas in Type II, the VMM runs on top of the existing operating system (OS) [2].

Selecting security techniques also depends on the type of functionalities the virtualized environments provide. There are many types of virtualizations used in the industry for various purposes. Figure 2 shows the main types of virtualizations and provides a brief description of their functionality.

In this paper, an overview of virtualization technologies will be provided, and the most prominent traditional and specialized security threats that target virtualized environments will be discussed in addition to the state-of-the-art security solutions used to tackle such threats and vulnerabilities. The paper will also review virtualization security in the industry. The rest of the paper is organized as follows: Sect. 2 will address the security requirements of virtualization, attacks on virtualization, and the security solutions based on the main four layers of virtualization systems which consist of the service provider, hypervisor, virtual machines, and the guest image. Section 3 will discuss virtualization security in the industry, and the gap between security techniques studied in the literature and the techniques implemented in the industry. Section 4 will cover the use of security metrics to increase the quality of a

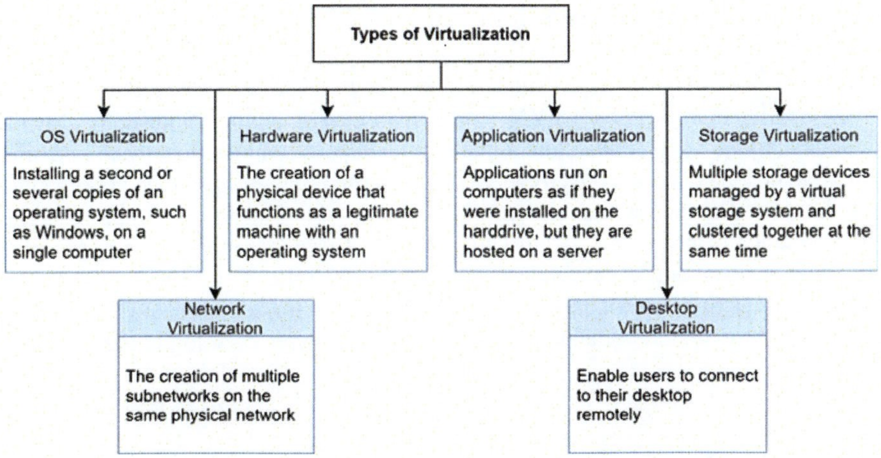

Fig. 2 Types of virtualization

cloud computing services. Finally, Sect. 5 will contain the conclusion and findings on this topic as well as future work.

2 Security Issues in Virtualization

2.1 Security Requirements

It's important to understand the security requirements of the virtualization environment before preparing and implementing the security of the system, as each part of the environment has to be secured and protected from any risks and attacks [3]. In this section, we will demonstrate the security requirements to prevent attacks on each layer of the virtualization architecture.

2.1.1 Service Provider Requirements

Service provider is a third-party company that offers cloud-based services. To protect the virtualization hardware, the cloud service provider must enable access to the hardware resources only to authorized users. Also, access control management has to be in place so that even the administrator can only access data that is assigned to him [3]. Users must also be provided with robust authentication mechanisms by the service provider where the service provider must adhere to security principles for the creation of trustworthy computing systems, such as the concept of least privilege [3].

2.1.2 Hypervisor Requirements

The hypervisor plays a major role in enabling the VM's to share the hardware resources, and it must preserve VM isolation and multiplexing of multiple VMs on a single hardware [4]. It must guarantee that no applications on the virtual machine can alter the source code of the hypervisor and virtual machines on the network. The hypervisor must also monitor the guest OS and applications in the virtual machine for suspicious behavior [5]. Additionally, the software that manages the hypervisor has to be secured and protected from any vulnerabilities, as well as access to the hypervisor has to be maintained and controlled by authorized groups. On the other hand, installing regular updates and keep track of the hypervisor logs can help make it more secure [6].

2.1.3 Virtual Machine Requirements

VMs have to be managed and protected in a way that each one operates independently from the others. In order to protect the operating system of the VM, regular patches and updates have to be done on the OS, in addition to the use of anti-virus software to maintain the system's protection from any malicious attacks [3]. Also, the VM's must communicate and share data between them; those communications have to meet the security requirements to prevent any potential attacks [7].

2.1.4 Guest Image Requirements

Hypervisor uses disk images to function like a disk driver for the virtual machine. Those disk images must be kept protected and secured against unauthorized access. Also, they must be reviewed and revised regularly following the requirements. Unnecessary images should not be made, and the existing ones should also be deleted from the system. If the VM is transferred from one computer to another, the image on the previous OS should be deleted [6]. In addition to that, the image of each VM has to be backed up and saved in a secure place. In the virtualization environment, there is a feature of taking a snapshot of the VM disk image in which this snapshot contains the state of the running image. This feature can be used to roll back the disk image to a safe state in case of any malicious attacks. However, snapshot access should only be granted to users who are authorized to access it [8].

2.2 *Attacks on Virtualization*

Each component in the virtualization environment may be a potential target for the attackers. However, attacks on various components can lead to security issues such as the compromise of the entire cloud infrastructure, data theft, and device hacking [3]. In this section, we will list some of the possible attacks on each component.

2.2.1 Service Provider Attacks

One of the main attacks occurs if the attacker gains access to the cloud hardware, wherein the attacker may install malicious software or code to harm the VMs by altering their source code and modifying their features. Cross-VM side-channel attacks can also be launched with physical access to the device. Similarly, CPU cache leakage is an attack that is used to test the load of other virtual Web servers on the network [9]. Furthermore, if access control is not properly enforced, various administrators, such as network and virtualization administrators, can gain access to customer data they are not allowed to see [3]. As a result of these operations, security risks such as data leakage and unauthorized traffic control will arise. The service provider must ensure that the software deployed on the cloud is written according to the best practices. Attacks like SQL injection, cross-site scripting, denial of service, and code execution can be caused by incorrect coding [3].

2.2.2 Hypervisor Attacks

A hypervisor attack can occur when an unknown cloud customer is allowed to rent a guest VM and use it to install a malicious guest OS that attacks and exploits the hypervisor by altering its source code and gaining access to memory contents (data and code) of other VMs in the system [5]. With more features in the hypervisor, the code size has also significantly grown, resulting in design and implementation flaws. Malicious hypervisors such as BLUEPILL rootkit, Vitriol, and SubVir are built on the fly to control the entire virtualization system, giving attackers host privileges to change and manage VMs [10]. Furthermore, an attack known as hyperjacking is a tactic used by malicious software to fully control the underlying operating system while being hidden from administrators and protection software [3, 11]. Another attack called VM escape is when a program running in one VM gains root access to the host computer. This is accomplished by crashing the guest OS and then running arbitrary code on the host OS. As a result, malicious VMs will take complete control of the host operating system [6, 12]. The VMs will communicate with the hypervisor and access other guest operating systems on the device after escaping the guest OS [3].

2.2.3 Virtual Machine Attacks

One type of attack that can be launched on virtual machines is running malicious programs on various VMs and obtaining the necessary permissions to monitor keystrokes and screen updates through virtual terminals, which attackers can use to steal sensitive data. Also, covert channels can be used for unauthorized communication with other VMs in the system if isolation is not adequately enforced [13]. Attackers can use Trojans, malware, and botnets to track traffic, steal sensitive data, and tamper with the functionality of guest operating systems. The Conficker worm, Zeus botnet, and command and control (CC) botnet communication operations are examples of such attacks, which result in data destruction, information collection, and backdoor creation by attackers [3]. The guest OS in VMs can be exploited by bugs in software, viruses, and worms. Zero-day attacks may also take advantage of unpatched VM operating systems [3]. VM migration is also a security threat which involves moving a virtual machine to another physical device without shutting it down [14]. This can happen when fault tolerance or load balancing is needed. The contents of the VM are exposed to the network during the conversion, which can result in data privacy and integrity being compromised [14]. As the VMs share the host resources, if an attacker uses one VM to obtain all of the resources of the host machine, the resources of other VMs would be disrupted, if not crashed, due to a lack of resources, resulting in a virtual denial of service attack [14].

2.2.4 Guest Image Attacks

If each image is not protected, unnecessary guest OS images in the cloud will result in various security issues. When a malicious guest OS image is transferred to another host, the other system is also compromised [6]. Furthermore, generating too many images and holding redundant images will deplete system resources, which an attacker can exploit to attack the system. When VMs are transferred from one physical machine to another, data from the VM images will remain on previous storage disks, which an intruder may access [3]. In the same way, attackers might be able to retrieve data from old, destroyed disks. Image backup protection is also a concern. The intruder can extract all information and data by gaining access to the backup images [3]. An attacker can access VM checkpoints on the disk, which contain VM physical memory contents and reveal sensitive VM state details. An intruder may build a new checkpoint and load it into the device to take the VM to any state they want. Details about previous VM states can be obtained by accessing all checkpoints in storage [8].

2.3 Security Solutions for Virtualization Components

Various security strategies can be used to address and counteract attacks on virtualization environments. Those strategies can mitigate, at the very least, the effect of those attacks on the system, if not prevented or detected.

2.3.1 Service Provider Security

One of the vital service provider security solutions is to prevent unauthorized individuals from gaining physical access to the system's virtualization hardware [3]. Another technique is ensuring each VM can be allocated access control that can only be set via hypervisor to protect VMs from unauthorized access by cloud administrators. Correctly, implementing the three fundamental concepts of access control known as identification, authentication, and authorization would prevent unauthorized access from system components. Likewise, if any administrator is involved in a security breach, cloud access control will identify that individual. Finally, installing an application layer firewall in front of Web applications and getting customer code tested for common vulnerabilities will help avoid Web application attacks [15].

2.3.2 Hypervisor Security

A well-known technique used for hypervisor security is the usage of hypersafe, a device that ensures the hypervisor's code integrity. By locking down the write-protected memory files, it protects the hypervisor implementation and prevents code alteration. It also protects the hypervisor's code from unauthorized entry, preventing control-flow hijacking attacks [16]. Only a local physical environment can be used to carry out a VM escape attack. As a result, insider attacks on the physical cloud system must be avoided. The interaction between guest machines and the host OS must also be appropriately configured [13]. Hypervisor isolation must be correctly enforced and maintained to prevent one VM from affecting other VMs. Furthermore, hardening the hypervisor will reduce the number of potential attack vectors. Other security techniques involve separating administrative roles, limiting hypervisor administrator access to alter, build, or remove hypervisor logs, and periodically monitoring hypervisor logs are some of these strategies [15].

2.3.3 Virtual Machine Security

In order to protect VMs, administrators must install software or applications that prevent VMs from using additional resources until they have been granted permission. Furthermore, a lightweight process that runs on a VM should be used to gather logs from VMs and track them in real time to detect any tampering [3]. In addition, best security practices must be used to harden the guest OS and the applications that run on it. Installing security software such as anti-virus, anti-spyware, firewall, host intrusion prevention system (HIPS), Web application defense, and log monitoring in the guest OS are other examples of VM security solutions [3]. Authors in [17] proposed the "Vigilant" scheme. It employs virtualization and machine learning techniques to control (VMs) via hypervisor without using a monitoring agent [17]. The advanced cloud protection system (ACPS) is a proposed system by the authors in [18] that controls and preserves the integrity of the OS in guest VMs. It also monitors activity of cloud components regularly using executable system files and employs virtual introspection techniques to embed a guest monitoring machine in the device without the intruder noticing it. As a result, any unusual behavior on the guest OS can be identified [18].

2.3.4 Guest Image Security

To ensure guest image security, virtualization users must have the policy to control image development, use, storage, and deletion. Image files must be checked for viruses, worms, spyware, and rootkits, which can hide from security software running on the guest OS [3]. The authors in [19] proposed an image management framework for effectively managing images in the cloud and detecting security breaches in images. Filters, virus scanners, and rootkit detectors are suggested as ways to defend against potentially infected images [19]. On the other hand, Nuwa [20] is a method for patching VM images in the cloud in an effective manner [3]. Nuwa rewrites patching scripts so that they can be implemented offline by analyzing patches. This results in the online patching installation scripts can be applied to images even offline [20]. Another essential security practice is double-checking that all data from previous or broken disks has been deleted when migrating VMs from one physical computer to another [3]. Cryptographic techniques such as encryption can be used to encrypt all backup images securing them [21]. When a VM is disconnected, it is important to remove any backups from the device as well. Further-more, to secure VM images from storage attacks, the cloud provider must encrypt the entire VM image when it is not in use. In case of checkpoint attacks, they can also be prevented by encrypting the checkpoint data [3]. SPARC is another tool for protecting checkpoints. This tool is used for dealing with security and privacy problems that arise due to VM checkpoints. SPARC allows users to choose which applications they want to checkpoint, ensuring that sensitive applications, and processes are not checked [3].

Table 1 presents a summary of the discussed security requirements for virtualization environments, the types of attacks they face as well as the security solutions used to protect each virtualization component.

3 Virtualization Security in the Industry

The field of virtualization security has been deeply researched over the past years. From finding vulnerabilities and studying the types of attacks faced to finding appropriate techniques for ensuring high levels of security, research allowed virtualization to reach the level that it is at today. Service providers such as Microsoft Azure, Google Cloud, and Amazon WorkSpaces (AWS) have become very advanced and are leading the major shift toward cloud computing. They all offer security services such as isolation, VMM integrity, platform integrity, restricted access, audit, intrusion detection, and much more. Through this review, we have found that most security techniques and services provided by service providers are compatible with the research being done reducing the gap between academia and industry. However, an important aspect that might still need studying is the roles and responsibilities of each of the client and the service provider according to the type of virtualization approach adopted. Often, security breaches occur due to the client being unaware of the security responsibilities they possess when migrating to the cloud. Usually, service providers secure the basic cloud infrastructure components, such as VMs, disks, and networks. The client on the other hand, oversees securing the operating system and software stack needed to run their applications as well as their data [22]. Clients must also ensure

Table 1 Summary of security aspects of virtualization described in paper

Category	Requirements	Attacks	Solutions
Service provider	1-Enable access to the hardware resources only to authorized people 2-Enable access control management 3-Provide users authentication mechanisms	1-Install malicious software on the cloud hardware 2-Cross-VM side-channel attacks 3-CPU cache leakage 4-Data leakage and unauthorized traffic control 5-SQL injection, cross-site scripting, denial of service, and code execution	1-Prevent unauthorized individuals from gaining physical access 2-Allocate access control for each VM 3-Correctly implement the three fundamental concepts of access control known as identification, authentication, and authorization
Hypervisor	1-Preserve VM isolation and multiplexing of multiple VMs on a single hardware 2- Guarantee that no applications on the virtual machine can alter the source code of the hypervisor and the virtual machines on the network 3- Monitor the guest operating system and applications in the virtual machine 4-Secure and protect the software that manages the hypervisor 5-Access to the hypervisor has to be maintained and controlled by authorized groups	1- BLUEPILL rootkit, Vitriol, and SubVir are malicious hypervisors built on the fly to control the entire virtualization system 2-Hyperjacking 3-VM escape	1-Use hypersafe 2-Appropriately configure the interaction between guest machines and host operating system 3-Maintain hypervisor isolation 4-Separate administrative roles

(continued)

Table 1 (continued)

Category	Requirements	Attacks	Solutions
Virtual machines	1-Each VM should operate independently from the others 2-Regular patches and updates have to be done on the OS 3- Use anti-virus software to maintain the system's protection from any malicious attacks 4- Secure communications between the VM's	1-Running malicious programs on various virtual machines and obtaining the necessary permissions to monitor keystrokes and screen updates 2-Covert channels 3-Use Trojans, malware, and botnets to track traffic and steal sensitive data 4-The Conficker worm and Zeus botnet 5-Bugs in software, viruses, and worms 6-Virtual machine migration	1-Install software that prevents VMs from using additional resources until they have been granted permission 2-Gather logs from virtual machines and track them in real time 3-Harden the guest OS and the applications that run on it 4-Install security software such as anti-virus, anti-spyware, firewall, and host intrusion prevention system 5-The use of"Vigilant" scheme
Guest images	1-Protect and secure disk images from unauthorized access 2-Delete any unnecessary images 3- The image of each VM has to be backed up and saved in a secure place	1-Holding redundant images 2-Attackers gain access to the backup images 3-Attacker access to VM checkpoints on the disk	1-Control image development, use, storage, and deletion 2-Check image files for viruses, worms, and spyware 3-Use an image management framework 4-Use filters, virus scanners, and rootkit detectors 5-Use cryptographic techniques to encrypt all backup images 6-Encrypt the entire VM image when it is not in use

the security of the endpoints used to access cloud services. Depending on the virtualization model, the cloud user might also be responsible for network security and, if required, communication encryption. If we consider real-attack attempts using information from MITRE ATTACK [23], the widely used knowledge base of adversary techniques extracted from real-world events, we can see that many of the recommended mitigation techniques are related to clients awareness and understanding of security threats and defense practices such as social engineering, password policies, multi-factor authentication, and many more.

4 Analyzing Virtualization Security

As more consumers (companies, organizations, and individuals) depend on cloud computing for their businesses and operations, concerns about the quality of such services have arisen. For that, performance, security, and reliability are the most important three categories for service measurement. Moreover, data protection and disaster recovery are very difficult factors to assess and cloud storage providers, and consumers will need a way to determine the service's efficiency. As a solution, security metrics can give users a better idea of what they're getting in terms of protection regarding their money. It also provides service providers with a numerical guide to increase the service's quality at their leisure. A security metric can involve three basic principles, which are: confidentiality, the prevention of unauthorized access to data, integrity, the prevention of unauthorized modification of data, and availability, which involves ensuring services are uninterrupted and are available when needed thereby preventing unauthorized destruction of data [24]. However, measuring the security of a cloud computing system or any information system is generally considered difficult. This is because many factors must be considered and there are many changing variables that can compromise the security of a system. Even with the difficulty of measuring the security, some research papers have addressed security metrics based on the CIA security standards. For example, the authors in [25] used the discrete event simulation (DES) method (a tool for simulating real-world structures that can be broken down into a series of logically distinct processes that evolve independently over time) and the Markov process (a stochastic model that describes a series of potential events where the probability of each event is solely determined by the state attained in the previous event) for measuring the availability. They concluded that DES is one of the most practical models for calculating availability, whereas Markov process is also a good alternative.

5 Conclusion and Future Work

Virtualization and cloud computing are currently very hot topics in the field of information technologies. Due to the need for various conventional and specialized security solutions, it is vital to continuously find new security techniques to defend against emerging threats. In this paper, we provided a brief background on virtualization which included the two types of architectures. In the first, the VMM runs on top of the existing operating system, while in the second, the VMM runs directly on top of the system hardware. Different types of virtualization approaches were also discussed along with their usage. This review focused on the security aspects of virtualization in literature discussing security requirements, types of attacks that target virtualized environments and highlighting the security techniques used for defense. With many companies turning to cloud computing, we also reviewed the virtualization security maintained by top service providers in the industry. We have concluded that most security techniques and services provided by service providers are indeed compatible with the research being done in this field, and the gap between academia and industry is becoming very small. As a future direction, more work should be done on client and service provider responsibility awareness, ensuring that both sides clearly understand their roles in securing the cloud infrastructure.

References

1. Yu FR, Liu J, He Y, Si P, Zhang Y (2018) Virtualization for distributed ledger technology (vdlt). IEEE Access 6:25019–25028
2. Sierra-Arriaga F, Branco R, Lee B (2020) Security issues and challenges for virtualization technologies. ACM Comput Surv 53(2):1–37
3. Kazim M, Masood R, Shibli MA, Abbasi AG (2013) Security aspects of virtualization in cloud computting. In: IFIP international conference on computer information systems and industrial management. Springer, pp 229–240
4. Szefer J, Keller E, Lee RB, Rexford J (2011) Eliminating the hypervisor attack surface for a more secure cloud. In: Proceedings of the 18th ACM conference on computer and communications security, pp 401–412
5. Szefer J, Lee RB (2011) A case for hardware protection of guest vms from compromised hypervisors in cloud computing. In: 2011 31st international conference on distributed computing systems workshops. IEEE, pp 248–252
6. Souppaya MP, Scarfone K, Hoffman P (2011) Guide to security for full virtualization technologies
7. Sabahi F (2012) Secure virtualization technology. Int J Comput Theory Eng 4(5):826
8. Gofman MI, Luo R, Yang P, Gopalan K (2011) Sparc: a security and privacy aware virtual machine check- pointing mechanism. In: Proceedings of the 10th annual ACM workshop on privacy in the electronic society, pp 115–124
9. Jin S, Ahn J, Cha S, Huh J (2011) Architectural support for secure virtualization under a vulnerable hypervisor. In: 2011 44th annual IEEE/ACM international symposium on microarchitecture (MICRO). IEEE, 272–283
10. Ibrahim AS, Hamlyn-Harris J, Grundy J (2016) Emerging security challenges of cloud virtual infrastructure, arXiv preprint arXiv:1612.09059
11. Compasti´e M, Badonnel R, Festor O, He R (2020) From virtualization security issues to cloud protection opportunities: an in-depth analysis of system virtualization models. Comput Secur 97:101905
12. Patil R, Modi C (2019) An exhaustive survey on security concerns and solutions at different components of virtualization. ACM Comput Surv (CSUR) 52(1):1–38
13. Reuben JS (2007) A survey on virtual machine security. Helsinki Univ Technol 2(36)
14. Chen L, Xian M, Liu J, Wang H (2020) Research on virtualization security in cloud computing. In: IOP conference series: materials science and engineering, vol 806(1). IOP Publishing, p 012027
15. Devi K, S G, R D (2018) Virtualization in cloud computing. IJARCCE 7(11):104–108
16. Wang Z, Jiang X (2010) Hypersafe: a lightweight approach to provide lifetime hypervisor control-flow integrity. In: 2010 IEEE symposium on security and privacy. IEEE, 380–395
17. Pelleg D, Ben-Yehuda M, Harper R, Spainhower L, Adeshiyan T (2008) Vigilant: out-of-band detection of failures in virtual machines. ACM SIGOPS Operati Syst Rev 42(1):26–31
18. Lombardi F, Di Pietro R (2011) Secure virtualization for cloud computing. J Netw Comput Appl 34(4):1113–1122
19. Wei J, Zhang X, Ammons G, Bala V, Ning P (2009) Managing security of virtual machine images in a cloud environment. In: Proceedings of the 2009 ACM workshop on Cloud computing security, pp 91–96

20. Zhou W, Ning P, Zhang X, Ammons G, Wang R, Bala V (2010) Always up-to-date: scalable offline patching of vm images in a compute cloud. In: Proceedings of the 26th annual computer security applications conference, pp 377–386
21. Tank DM, Aggarwal A, Chaubey NK (2021) Cyber security aspects of virtualization in cloud computing environments: analyzing virtualization-specific cyber security risks. In: Research anthology on privatizing and securing data. IGI Global, pp 1658–1671
22. Tank D, Aggarwal A, Chaubey N (2019) Virtualization vulnerabilities, security issues, and solutions: a critical study and comparison. Int J Inf Technol 1–16
23. "MITRE ATTCK®." [Online]. Available: https://attack.mitre.org/
24. Elsadig Abdalla Abdalla M (2020) Virtualization security issues: security issues arise in the virtual environment
25. Cueva-Parra L, Sahinoglu M (2009) Security metrics on cloud computing using statistical simulation and Markov process. In: 12th SDPS transdisciplinary conference proceedings on integrated systems, design and process science, Montgomery, Alabama

Malware Detection Using Machine Learning Algorithms for Windows Platform

Abrar Hussain, Muhammad Asif, Maaz Bin Ahmad, Toqeer Mahmood, and M. Arslan Raza

Abstract Windows is a popular Graphical User Interface-based Operating System that provides services like storage, run third-party software, play videos, network connection, etc. The purpose of such services can be demolished by targeting the availability of these services. Malware is one of the major security concerns for the Windows platform. Malware is any type of computer software that disturbs the availability of computer services. The traditional detection systems such as the intrusion detection/prevention system, Anti-Virus software cannot detect unseen malware due to the use of signature-based methods. So, there is a need to accurately detect such kind of malware in the Windows environment. In this work, a Machine Learning (ML)-based malware detection system is introduced which extracts features from the Portable Executable file's header to detect whether the executable is clean or malicious. After preprocessing the data, several ML models including Random Forest, Support Vector Machine (SVM), Decision Tree, AdaBoost, Gaussian Naive Bayes (GNB), and Gradient Boosting are applied to cope up with the malware. Moreover, a comparative analysis is conducted among ML models to select the appropriate one for the targeted problem. The experimental results show that the Random Forest outperformed the others with an accuracy level of 99.44\% for the detection of malware. This can be used to develop a desktop application for scanning the malware for the Windows platform with the added ability to customize the scanning process.

Keywords Machine learning · Portable executable · Malware · Random forest · SVM · Decision tree · Gradient boosting · GNB

A. Hussain · M. Asif (✉) · M. A. Raza
Department of Computer Science, Lahore Garrison University, Lahore, Pakistan
e-mail: drmuhammadasif@lgu.edu.pk

M. A. Raza
e-mail: marsalanraza@lgu.edu.pk

M. B. Ahmad
Karachi Institute of Economics and Technology (KIET), Karachi, Pakistan
e-mail: maaz@kiet.edu.pk

T. Mahmood
Department of Computer Science, National Textile University, Faisalabad, Pakistan

A. Ullah et al. (eds.), *Proceedings of International Conference on Information Technology and Applications*, Lecture Notes in Networks and Systems 350,
https://doi.org/10.1007/978-981-16-7618-5_53

1 Introduction

According to Stat Counter [1], the Windows Operating System (OS) is one of the popular OS worldwide. Windows OS usage has increased rapidly as it provides an easy Graphical User Interface (GUI) that may be used easily, even by a normal person. It provides several features such as storage for data, support for third-party and open-source software, entertainment provision, and communication services. On the other hand, attackers keep on building new types of malware for the Windows platform to exploit the computer services and confidential data of the user. A software program that is intentionally designed to harm the computer or network is termed malicious software or malware [2]. These seek to gain unauthorized entry to some automated machines such as computer systems or network devices. The purpose of this unauthorized entry is to get useful information and to damages the running services without the prior consent of the author. Sometimes, insiders deliberately insert malware into the system to sabotage the security mechanism of the organization, to gain administrative control, to collect confidential information, etc. [3, 5]. Malware exists in different forms such as viruses, worms, Trojan, spyware, ransomware, rootkit, backdoor, adware, and keyloggers [6].

Now a days, the Internet is the source of every piece of information and part of our everyday lives. As everyone is using the Internet, attackers usually use it to host infected applications on various sites. When a user installs such an application, it asks the user to accept the terms and conditions. Users usually do not bother it and accept the terms and conditions [7]. By doing so, they become the victim of the attack. The second most common way of spreading malware is through external devices, e.g. USB. Hence, there should be a method that protects confidential data from such types of threats.

Many researchers used static analysis [8] for malware detection and the rest of the researchers used dynamic analysis [9]. Static analysis is referred to as signature base analysis which examines the code without executing the application. The dynamic analysis is used to detect malware by examining the behavior of the application after running it in the sandbox. Dynamic analysis requires more resources and cost as these should be executed in an isolated environment. Static analysis is more reliable and safer than dynamic analysis. With the evolving technology, computers have now become more sophisticated and complex. Malware also has transitioned into a sophisticated and difficult-to-detect form. On a frequent and regular basis, some new 0-day malware (patch not available) is being launched by the attackers [10]. According to Tully [11], the current anti-malware systems like Intrusion Detection/Prevention System (IDS/IPS) and Anti-Virus (AV) software are unable to detect unknown malware (0-day malware) as most of these are based on signature-based methods. The signature-based systems extract the signature of the application then compare it with a well-known database of malicious signatures. If the signature is matched, the application is considered malicious. The attacker may also use packing, obfuscation, and other techniques to change the signature of the application to avoid detection [12]. So, there must be some intelligent detection methodology for the

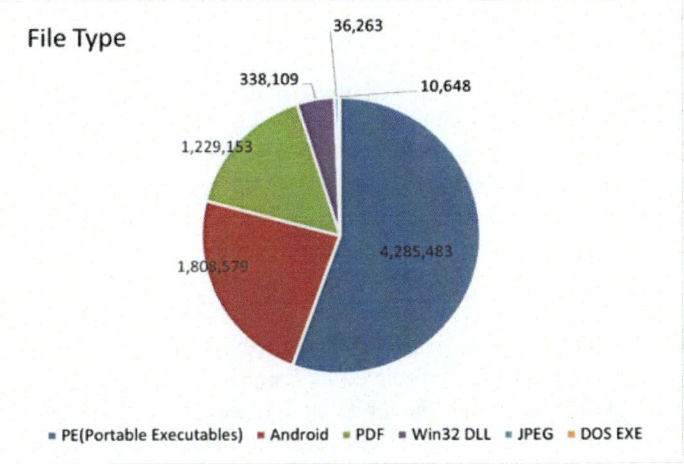

Fig. 1 Percentage of file type submitted to Virus total (29 July-4 August, 2020)

Windows platform to detect unseen malware. The scheme should be simple to use, efficient in terms of resource utilization, and capable enough to detect almost all possible malware, whether known or unknown.

In this work, a machine learning-based malware detection system is introduced (for Windows OS) in which the extraction of features from the Portable Executable (PE) file's header is used to detect whether the executable is clean or malicious. The system can also detect 0-day malware in the executable file. The 32-bit and 64-bit Windows platforms use PE file format for executable and Virus total statistics show that the most submitted file format is PE, as shown in Fig. 1.

The static analysis approach is used in the proposed model. The model diverges into four layers, i.e., data acquisition, preprocessing, prediction, and performance evaluation layers. The model detects malware without executing the executable file. A supervised machine learning strategy is used and binary classification is performed to classify the dataset into two main classes and determines whether the source executable file belongs to a malicious class or clean class. The proposed system's performance is evaluated by using the different performance evaluation metrics such as F1-score, precision, recall, accuracy, and support. The experimental results show that the Random Forest algorithm outperformed the others having an accuracy level of 99.44% in the detection of the malware.

The rest of the paper is organized into the following sections. Section 2 describes the related work of malware detection using different datasets and machine learning techniques. Section 3 presents the proposed system. Section 4 comprises the experimental and comparative analysis. Finally, Sect. 5 concludes the paper and discusses future directions.

2 Related Work

Many researches have been proposed until now on malware detection for the Windows platform. Naz and Singh [13] presented a comprehensive review related to the use of ML in malware detection for Windows. They introduced a static analysis method based on machine learning that used a classifier to train the model on extracted features of the PE file, but the dataset used was small that resulted in averaged accuracy. Darshan and Jaidhar [14] proposed a hybrid approach that comprised static and dynamic features of PE files and used a linear support vector classification approach to precisely find the unknown malware. The use of a small dataset for training the model caused problems in achieving high accuracy. A knowledge-domain analyzer for malware classification was introduced by Samantray and Tripathy [16]. It used dynamic features of files to train the model and to predict the malware. The dataset was collected from Kaggle, found the best relevant features using an extra tree classifier and k- best feature selection method. 95.53% accuracy was achieved using logistic regression.

Radwan [16] introduced a system where the static analysis approach was used to extract the PE file's features. Seven different machine learning classification models were used. Classification of Malware with PE headers (ClaMP) dataset from the Github repository was used for model training. Dataset was divided into two categories, one was raw features that comprised 53 features, and the second was the integrated features containing derived and expanded features. These summed up as 74 features. On 70/30 split, random forest outperformed other algorithms on integrated feature datasets having 93.23% accuracy and 97.56% accuracy on the raw dataset. With tenfold cross-validation, K-Nearest Neighbor (KNN) model performed well on an integrated dataset with 98.70% accuracy and Gradient Boosting (GB) tree performed well on the raw dataset with an accuracy of 93.55%. Shukla et al. [17] introduced a binary classification technique using a portable executable files dataset. The proposed technique used a Kaggle dataset which had PE header parameters of 14,599 malware files and 5012 benign files. The final model showed 97.2% accuracy by using the random forest algorithm.

Zhang et al. [18] proposed a static malware detection technique using the classification method. The method was applied to the EMBER dataset of extracted features of the PE file. The malware data is further re-labeled into multi-classes, based on their family using Virus total sha256 hash and K7 Antivirus Gateway (K7GW) detection results. Linear and ensemble models were applied but the ensemble-based random forest model outperformed the other one having the micro and macro average F1-score 0.96 and 0.89, respectively. Model performance could be increased by using the feature importance techniques. Sun et al. [19] introduced the malware family classification method based on static feature extraction. The dataset was collected from Virus-Share, used the Kaspersky engine and Exe-info PE to classify malware according to the malware family. The feature extraction technique is used to extract mainly three types of features, i.e., PE features byte-code features and assembler features. After that, eight classifier models were applied but only random forest performed well having a 93.56% F1-score. Low accuracy of detection rate may provide some false -negatives and false-positives results. Another classification system was introduced by Gandotra et al. [20]. The proposed system was desired to build for 0-day malware detection using static and dynamic features. The dataset was collected from an online repository of Virus-Share that contained only 3130 PE samples. The malware attributes were extracted manually by executing each sample file in the cuckoo sandbox. The dataset contained 18 features comprised of both static and dynamic features. Feature extraction was performed using Information Gain (IG) and ranker algorithm. After that, only 7 features were selected and used to build the classifier. Seven different classifier models were applied, i.e., Instance Base Learning (IBL) algorithms, Nave Bayes (NB), J48, Random Forest (RF), Bagging, Decision Table, and Multi-Layer Perceptron (MLP) from the library of WEKA. RF performed well with 99.97% accuracy of malware detection.

A static analysis method was presented by Mohammed et al. [21] comprised 1, 38,048 samples of malicious and benign executable files. Two of the models, i.e., decision tree and random forest were applied. RF precisely detected the malware with 99.43% accuracy. Moreover, a Web application was created for users to detect real-time detection of the executable files. Roseline and Geetha [22] proposed an intelligent malware detection technique by using the oblique RF paradigm. The ensemble-based decision tree-based models were used to train the model on three datasets, i.e., ClaMP dataset, the Antivirus dataset, and the Kaggle dataset. ClaMP contained 5210 samples of malicious and benign PE files headers. The antivirus dataset was collected from Virus-Share comprised 12,140 samples with 482 data features of malicious and benign API calls, registry system file systems, etc. The third dataset was collected from Kaggle that had hexdump-based features and disassembly-based features of nine malware families. It comprised 1805 features and 10,868 samples. Different decision tree models were used on each dataset. They claimed that RF outperformed other models in each dataset having accuracy levels of 99.14%, 99.23%, and 99.52%, respectively.

Gupta et al. [23] proposed architecture for malware detection. First, they prepared a large dataset comprised of clean files and malware sample files. Then performed automated malware analysis on the collected data by using the cuckoo sandbox, which consisted of the python script. They evaluated the performance using the following parameters: True Positive Rate, False Positive Rate, Precision, False Negative Rate, and Accuracy. The experimental results indicated that RF outperforms other models with an accuracy of 98.88%. Cho et al. [24] proposed a framework for malware detection. The classification process comprised behavior monitoring, sequence refining, sequence alignment, and similarity calculation. 150 malware samples were used in 10 different malware variant families. The classification was repeated 5 times and 87% accuracy was achieved. Burnap et al. [25] proposed an approach for the classification of self-organizing maps, which distinguished between malicious and benign files and reduced the over-fitting process in training of the samples. The classifiers used in this architecture were RF, NB, MLP, and SVM. The best result was achieved using the RF classifier having 98% accuracy.

Wang et al. [26] implemented a sandbox, feature extractor, and classifier. There were 3 stages in their work, i.e., collector, extractor, and classifier. The collector comprised a static analysis program and dynamic execution with the PinFWSandbox module. It recorded the dynamic and log file information and passed it to the extractor stage. Extractor performed static, dynamic, and system call feature extraction. The classifier was the final stage. It combined all the classifiers results and found that the dynamic op-code classifier gave better results for the F1-score, i.e., 96%. Makandar et al. [27] proposed classification of malware families using an artificial neural network. It converted malware binaries into gray-scale images and resized the images. Then, it extracted features of subband filtering on resized images through Gabor wavelet and GIST descriptor. It used feed-forward back-propagation neural network. The achieved accuracy was 96.35%. Devesa et al. [25] proposed an automatic detection system of malware with the use of logs. It used Qemu for emulating sandbox environment, and Wine for simulation purposes to extract features from behavior logs of malware samples. It implemented four classifiers, i.e., NB, RF, Sequential Minimal Optimization (SMO), and J48, to achieve better performance. The highest accuracy of 96.2% was observed by the RF classifier. The focus of most of the researchers is to develop an ML-based malware detection system using different datasets. It provided motivation to propose an ML-based malware detection system for Windows platforms.

3 Proposed System

This section presents the proposed machine learning-based malware detection system for Windows OS that extracts PE files header and uses them to detect whether the executable is clean or malicious. Figure 2 shows the architecture of the proposed system. It consists of four layers including the data acquisition layer, preprocessing layer, prediction layer, and performance evaluation layer. The following subsections describe these layers in detail.

3.1 Data Acquisition Layer

In this work, the data has been collected from an online dataset repository [29, 30]. Dataset consists of header features of PE. It is classified into two classes, i.e., clean and malicious. The clean and malicious classes are represented by 1 and 0 respectively in the legitimate attribute. The dataset contains 54 PE header features and 138,042 instances. Table 1 shows the PE header dataset description including some important features of the PE header, features contained in the dataset, and their description.

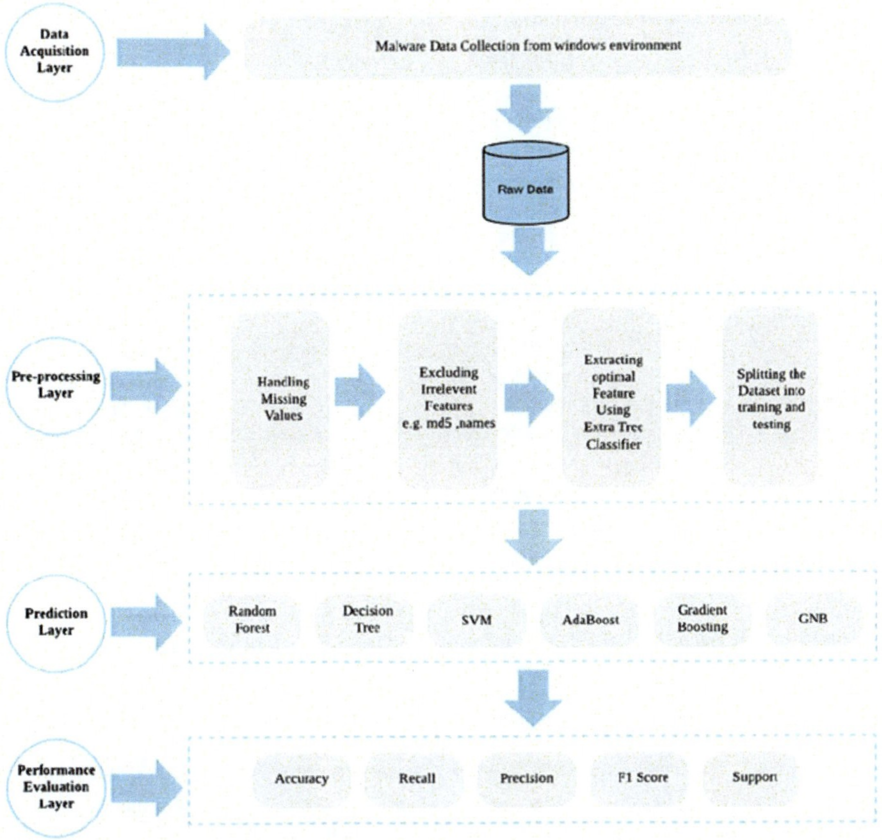

Fig. 2 Architecture of the proposed system

3.2 Preprocessing Layer

To avoid the impurity and biases in the results, data preprocessing is performed before applying the ML models. Following steps are performed to preprocess the data.

Handling Missing Values: First, the data is analyzed for missing values. There is no missing value found for this work.

Excluding Irrelevant Features: Some of the features like name, md5, etc. were not able to contribute to the prediction of malicious executables, so these are excluded from the dataset. These attributes include Name, Md5, LoaderFlags, SizeOfHeap-Reserve, ImportsNbDLL, ImportsNbOrdinal, and LoadConfigurationSize.

Extracting Optimal Features: After extraction of basic irrelevant features, the feature set is further reduced by extracting only the most contributed features in predicting the results. Feature selection is the method to select only important features and it can improve the accuracy and reduce the training time of the learning model. This is done by finding the correlation of all remaining features with the label attribute

Table 1 PE header dataset description

Dataset attributes	Description	Dataset attributes	Description
MajorImageVersion	Major version number of application	Md5	The unique hash of each executable
MinorImageVersion	Minor version number of application	Machine	Indicate what type of machine the executable was built, e.g., Intel × 86
MinorSubsystemVersion	Windowns NT Win32 subsystem minor version number	Subsystem	What type of window application, e.g., GUI, device drivers
MinorOperatingSystemVersion	Minor version number of Windows NT	ImageBase	Image first-byte ad dress when loaded in memory
NumberofSection	Number of section present in the file	Characteristics	What type of your file e.g. EXE, DLL, etc.
MajorOperatingSystemVersion	The major version number of the required operating system	DLL Characteristics	Flags used to indicate if a DLL image include entry point for process and thread initialization and termination
MajorSubsystemVersion	The major version number of the required window application	CheckSum	Used to validate the executable when loaded in memory
SizeOfOptionlHeader	Size optional header of Executable file	SizeOfImage	All header size in byte
SizeOfInitializedData	Size of initialized data	Name	Name of viruses
SizeOfStackReserve	Number of byte reserve for stack	BaseOfCode	Relative to the imagebase, pointer to the start of the code section
SizeOfUninitializedData	Size of uninitialized data	SizeOfCode	Size of executable code
SectionAlignment	Dictate minimum space required for each section	BaseOfData	Relative to the imagebase, pointer to the start of the data section

(continued)

Table 1 (continued)

Dataset attributes	Description	Dataset attributes	Description
SectionAlignment	Section alignment when loaded in memory, in byte	FileAlignment	Raw data section alignment in the image file, in bytes

which is the target attribute of this dataset. In the proposed system, an Extra Tree Classifier [31] is used to select the relevant and best features from the dataset. It is an ensemble-based learning technique in python which uses DT and RF concepts to find the relative importance of the attributes. Extra tree classifier uses some mathematical criteria for splitting like the Gini index, which is the hyperparameter in DT. In this work, the Extra Tree Classifier library of Scikit-learn in python [32] is used and it selected 15 features only out of 54 features from the dataset, based on their importance and correlation to the target attributes. Figure 3 shows the correlation of relevant features with target attributes.

Data Splitting: The preprocessed dataset is then divided into an 80:20 ratio, i.e., 80% for training and 20% for testing. There is a method (called Train-test split) that splits the dataset into two parts: training and testing. This method accurately divides the dataset as it takes the percentage of splitting. Classifiers are applied to the training dataset to train the machine learning model. The testing dataset is then used

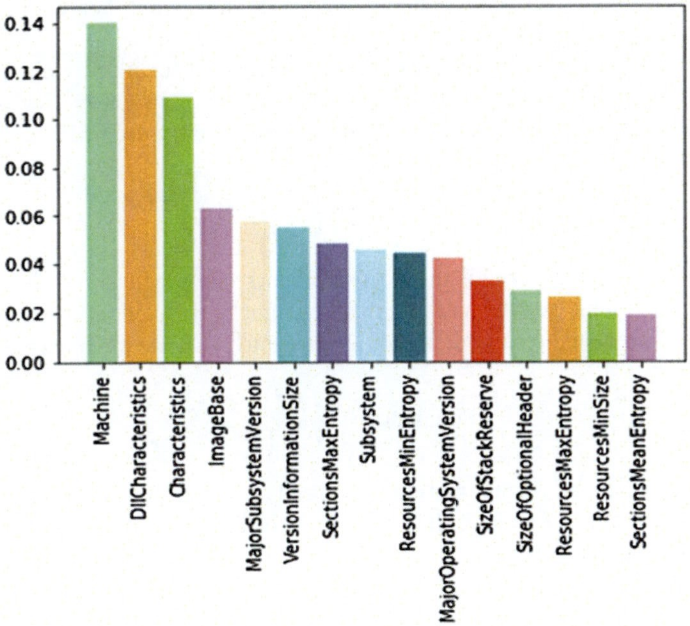

Fig. 3 Correlation of relevant features with target attribute

to test the performance of the trained classifiers. Dataset consisted of 138,042 data instances for training and testing purposes, where each segment contained around 27,610 instances.

3.3 *Prediction Layer*

The preprocessed and segmented dataset is now ready to be used to train and test ML techniques. At this layer, different ML techniques including Decision Tree, Random Forest, Support Vector Machine, AdaBoost, Gradient Boosting (GB), Gaussian Naive Bayes (GNB), and Adaboost are applied to perform malware detection.

3.4 *Performance Evaluation Layer*

In this works, classifiers' performance is evaluated based on accuracy, precision, recall, F1-score, and support metrics. The results of the performance evaluation layer lead to the decision whether the malicious executable is correctly identified or not. It helps in making an immediate decision for the needed precautions. The performance of each classifier is analyzed by applying the trained model on the test data set where the actual target attribute is already known and then comparing the outcomes of the model with known results.

Accuracy, Precision, Recall, F-Score, and Support: Accuracy or classification rate is the sum of true positive instances and true negative instances divided by the total number of instances. Equation 1 is used to calculate the accuracy.

$$\text{Accuracy} = \frac{\text{TP} + \text{TN}}{\text{TP} + \text{FP} + \text{FN} + \text{TN}} \tag{1}$$

The precision of the model is the fraction of correctly predicted results from the total predicted results, i.e., the measure that how much predicted results are relevant from the total predicted results. Equation 2 is used to calculate the precision.

$$\text{Precision} = \frac{\text{TP}}{\text{TP} + \text{FP}} \tag{2}$$

Recall (or sensitivity) is the fraction of correctly predicted results from the actual results, i.e., how much actual result is predicted correctly. The formula used for the calculation of recall is given in Eq. 3.

$$\text{Recall} = \frac{\text{TP}}{\text{TP} + \text{FN}} \tag{3}$$

F-measure or F1-score is the harmonic mean of precision and recall as shown in Eq. 4. The F-measure value near 1 indicates the perfect precision and recall of the ML model.

$$\text{F1 - Score} = \frac{2 \times \text{Precision} \times \text{Recall}}{\text{Recall} + \text{Precision}} \tag{4}$$

Support is defined as the number of instances of the true positives present in that class.

4 Experimental Analysis

For experimental analysis, Scikit learn python library [32] is installed on a machine having Intel 1.70 GHz Core i5 CPU, having 8 GB of RAM. Machine learning techniques are applied to perform malware detection on the PE dataset. The performance of models has been measured on preprocessed data by partitioning it into training and testing segments. For each classifier, values of model accuracy, recall, precision, F1-score, and support are calculated.

Table 2 lists the experimental results for all classifiers. It shows that the decision tree gives 99.1% accuracy, 99% recall, 99% precision, 99% F1-score, and support value 19,227. On the other hand, the Random Forest gives 99.4% accuracy (i.e., 0.6% error), 99% recall, and 0.99 F1-score. The Gradient Boosting accuracy is 98.9% which is lesser than the previous two models but it gives precision, recall, and F1-score of 99%. For AdaBoost and Support Vector Machine (SVM), accuracy is 98.5% and 97.6%, respectively, with 27,610 support instances. Gaussian NB (GNB) has the worst accuracy than other models, i.e., 92.9% and has 93% precision, recall, and F1-score.

The experimental results depict that the performance of GNB, SVM, AdaBoost, and GB algorithm is poor as compared to other algorithms. It is also observed that the accuracy of the DT is comparable with Random Forest. Moreover, the precision, recall, F1-score of DT and RF is the same. Based on these facts, it is can be deduced that the overall performance of the RF algorithm is better than DT, AdaBoost SVM,

Table 2 Comparison of machine learning techniques for malware detection in windows

Sr. No	Algorithms	Accuracy	Precision	Recall	F1-score	Support
1	Random Forest	99.44	0.99	0.99	0.99	27,610
2	Decision Tree	99.17	0.99	0.99	0.99	27,610
3	Gradient Boosting	98.97	0.99	0.99	0.99	27,610
4	AdaBoost	98.56	0.93	0.93	0.93	27,610
5	SVM	97.68	0.98	0.98	0.98	27,610
6	GNB	92.95	0.93	0.93	0.93	27,610

GB, and GNB for malware detection. The DT and RF both have almost the same learning rate. The accuracy of the RF (99.4%) is slightly higher than the DT (99.1%). In nutshell, keeping all factors in mind, it can be concluded that RF outperforms the rest of the machine learning algorithm for malware detection in the Windows environment. Generally, a machine learning algorithm is computationally intensive, it is deployed on the flask framework for real-time scanning of executables for detection. The model runs when the flask framework is running. So, overall execution time and computation power consumed by the Windows OS would reduce drastically, which may help to improve the overall performance.

5 Conclusion and Future Work

In this paper, a machine learning-based approach is introduced for detecting malware in the Windows OS. The proposed model extracts different features of the PE header file and compares it with the trained machine learning model and shows the result either as malicious or clean. In this work, six different machine learning algorithms are used for detecting malware in Windows executables. The results are compared based on accuracy, F1-score, recall, precision, and support. The experimental results show that Random Forest gives better performance as compared to other algorithms with 99.4% accuracy. In the future, the model can be deployed on the Flask framework to reduce the computational power in the Windows environment and enabled users to scan any executable file in real-time. Moreover, an optimal hybrid solution can be introduced using both dynamic and static analysis that should use more PE header features.

References

1. Statcounter: Global state: Operating System Market Share Worldwide [Online]. Available https://gs.statcounter.com/os-market-share. Accessed 19 June 2021
2. What is malware? [Online]. Available https://searchsecurity.techtarget.com/definition/malware. Accessed 19 June 2021
3. Ahmad MB, Fahad M, Khan AW, Asif M (2016) A first step towards reducing insider threats in government organizations. Int J Comput Sci Netw Secur 16(6):81–85
4. Ahmad MB, Fahad M, Khan AW, Asif M (2016) towards securing medical documents from insider attacks. Int J Adv Comput Sci Appl 7(8):357–360
5. Ahmad MB, Akram A, Asif M, Rehman SU (2014) Using genetic algorithm to minimize false alarms in insider threats detection of information misuse in windows. Environment 2014:1–12
6. What are the different types of Malware? [Online]. Available https://comtact.co.uk/blog/what-are-the-different-types-of-malware/. Accessed 19 June 2021
7. What is a cyber-attack? [Online]. Available https://www.ibm.com/services/businesscontinuity/cyber-attack. Accessed 19 June 2021
8. Anderson HS, Roth P (2018) Ember: an open dataset for training static PE malware machine learning models. arXiv preprint arXiv:1804.04637

9. Cabrera A, Calix RA (2016, October) On the anatomy of the dynamic behavior of pol-ymorphic viruses. In 2016 international conference on collaboration technologies and systems (CTS), Orlando, FL, USA, 31 October–4 November. IEEE, New York, USA, pp. 424–429
10. WWhat is zero-day (0day) exploit [Online] Available https://www.imperva.com/learn/application-security/zero-day-exploit/. Accessed 19 June 2021
11. Tully S, Mohanraj Y (2017) Mobile security: a practitioner's perspective. In: Mobile security and privacy, 2nd edn. Elsevier, pp 5–55
12. Hosseinzadeh S, Hyrynsalmi S, Leppnen V (2016) Obfuscation and diversification for securing the internet of things (IoT). Internet of Thing. ScienceDirect, pp 259–274
13. Naz S, Singh DK (2019, July) Review of machine learning methods for windows malware detection. In: 10th international conference on computing, communication and networking technologies (ICCCNT), Kanpur, India, 6–8 July. IEEE, New York, USA, pp 01–06
14. Darshan SLS, Jaidhar CD (2019) Windows malware detection system based on LSVC recommended hybrid features. J Comput Virol Hacking Tech 15(2):127–146 (Springer)
15. Samantray OP, Tripathy SN (2020) A knowledge-domain analyser for malware classification. In: 2020 international conference on computer science, engineering and applications (ICCSEA), Gunupur, India, 13–14 March. IEEE, New York, USA, pp 1–7
16. Radwan AM (2019, October) Machine learning techniques to detect maliciousness of portable executable files. In: 2019 international conference on promising electronic technologies (ICPET), Gaza, Palestine, 23–24 October. IEEE, New York, USA, pp 86–90
17. Shukla H, Patil S, Solanki D, Singh L, Swarnkar M, Thakkar HK (2019, December) On the design of supervised binary classifiers for malware detection using portable executable Files. In: 9th international conference on advanced computing (IACC), Tiruchirappalli, India, 13–14 December. IEEE, New York, USA, pp 141–146
18. Zhang S-H, Kuo C-C, Yang C-S (2019, August) Static PE malware type classification using machine learning techniques. In: International conference on intelligent computing and its emerging applications (ICEA), Tainan, Taiwan, 30 August–1 September. IEEE, New York, USA, pp 81–86
19. Sun B, Li Q, Guo Y, Wen Q, Lin X, Liu W (2017, December) Malware family classification method based on static feature extraction. In: 3rd IEEE international conference on computer and communications (ICCC), Chengdu, China, 13–16 Deccember. IEEE, New York, USA, pp 507–513
20. Gandotra E, Bansal D, Sofat S (2016, December) Zero-day malware detection. In Sixth international symposium on embedded computing and system design (ISED), Patna, India, 15–17 December. IEEE, New York, USA, pp 171–175
21. Mohammed AR, Viswanath GS, Babu KS, Anuradha T (2019, March) Malware detection in executable files using machine learning. In: International conference on E-Business and telecommunications. Springer, Berlin, pp 277–284
22. Roseline SA, Geetha S (2018, September) Intelligent malware detection using oblique random forest paradigm. In: International conference on advances in computing, communications, and informatics (ICACCI), Bangalore, India, 19–22 September. IEEE, New York, USA, pp 330–336
23. Gupta D, Rani R (2018) Big data framework for zeroday malware detection. Cybern Syst 49(2):103–121
24. Cho K, Kim TG, Shim YJ, Ryu M, Lm EG (2016) Malware analysis and classification using sequence alignments. Intell Autom Soft Comput 22(3):371–377
25. Burnap P, French R, Turner F, Jones K (2018) Malware classification using self-organizing feature maps and machine activity data. Comput Secur 73:399–410
26. Wang C, Ding J, Guo T, Cui B (2017, November) A malware detection method based on sandbox, binary instrumentation, and multidimensional feature extraction. In: International conference on broadband and wireless computing, communication and applications. Springer, Cham, pp 427–438
27. Makandar A, Patrot A (2015, December) Malware analysis and classification using artificial neural network. In: International conference on trends in automation, communications and computing technology (I-TACT-15), Bangalore, India (21–22 December). IEEE, New York, USA, pp 1–6

28. Devesa J, Santos I, Cantero X, Penya YK, Bringas PG (2010) Automatic behavior-based analysis and classification system for malware detection. ICEIS J 2(2):395–399
29. Dataset [Online]. Available https://www.kaggle.com/. Accessed 19 June 2021
30. Malware-DataSet [Online]. Available https://github.com/System-CTL/Malware-DataSet. Accessed 19 June 2021
31. Extra Trees Classifier [Online]. Available https://scikit-learn.org/stable/modules/generated/sklearn.ensemble.ExtraTreesClassifier.html. Accessed 19 June 2021
32. Scikit-learn [Online]. https://scikit-learn.org/stable/. Accessed 19 June 2021

An IoT-Based Remote Well Baby Care Solution

Leah Mutanu, Khushi Gupta, Jeet Gohil, and Abdihamid Ali

Abstract Regular infant health and growth monitoring are crucial during a baby's first 6 months of life to curb high infant mortality rates. However, in marginalized areas long distances to health centers, high cost of health care, limited health facilities, and numerous household responsibilities prevent babies from accessing this essential care. This paper presents a solution developed for remote postnatal care using IoT technology. Simulated laboratory tests are conducted, and the accuracy of the device evaluated by calculating the accuracy and error rates. The results show that the solution can take readings with acceptable accuracy levels and low error rates. Future studies will include reduction of error rates, additional measurements, and field tests.

Keywords Remote patient monitoring · IoT in healthcare · Infant postnatal care · Telehealth · Healthcare image processing

1 Introduction

Regular infant health and growth monitoring are critical during a baby's first 6 months of life to detect growth issues and the onset of preventable diseases that significantly contribute to high infant mortality rates. This is especially critical for preterm babies or babies with special needs who require close monitoring. Many mothers, especially in marginalized areas and developing countries do not take their babies for a health check-up until it is too late. The long distances to health centers, high cost of health care, limited health facilities, and numerous household responsibilities discourage mothers from infant postnatal care. Additionally, the recent COVID-19 health pandemic instilled fear of visiting health centers among parents [1]. These factors highlight the need for offering infant postnatal care remotely. While several innovations in Internet of Things (IoT) technology have found applications in healthcare, one area that has not received a lot of attention is infant postnatal care.

L. Mutanu · K. Gupta (✉) · J. Gohil · A. Ali
United States International University Africa, Nairobi, Kenya

L. Mutanu
e-mail: lmutanu@usiu.ac.ke

A. Ullah et al. (eds.), *Proceedings of International Conference on Information Technology and Applications*, Lecture Notes in Networks and Systems 350,
https://doi.org/10.1007/978-981-16-7618-5_54

2 Review of Related Work

A survey of IoT-based Healthcare Systems [2] describes healthcare as one of the industries that have benefited from innovations in IoT technology. Because this approach shifts the burden of patient monitoring from the provider end to the patient end, it should provide accurate data, be user friendly, flexible, and cost-effective. Previous studies focused on monitoring aspects such as the Cardiovascular and Respiratory System, Mobility related issues, Neurological Disorders, and Diabetes [3].

Our study focuses on the remote screening of infants during postnatal care. Growth monitoring is an intrinsic part of pediatric care around the world [4–7] and calls for observations of an infant's Weight, Standing Height or Supine length, Head circumference (often as a screen for malnutrition, but in some cases as a measure of brain development), and fine motor skills such as head or limp movement (0–4 months), or grasping at (4–8 months). Growth monitoring occurs alongside health monitoring, which assesses parameters such as body temperature, heart rate, and respiratory rate to detect the onset of illnesses [8]. Health workers often face difficulties plotting and interpreting growth charts to determine appropriate action [9] where measurements are taken manually using weighing scale or tape measures. Because of this, medical portable devices have been developed such as digital weighing scales, digital thermometers, and pulse oximeters. However, they remain as separate components with no way of automatically recording or transmitting data. The emergence of Wireless Sensor Networks has seen innovations such as a Sensory baby vest for monitoring of infant's respiration, heart rate, temperature, and humidity [10] and electronically transmitting the data. Most of these studies however focus on monitoring health [11–17] rather than growth parameters.

These solutions also do not focus on addressing challenges faced by marginalized users such as embedded, low cost, or low power requirements [3]. Such a system should utilize analytics that provides an early warning score (EWS) to detect the deterioration of a patient [18]. Our proposed solution makes use of predefined Early warning values (as guided by WHO [19] and other healthcare regulators) to detect outliers and generate reports and alerts for timely intervention. Monitoring systems should also go beyond reporting the current situation to predict risks using machine learning algorithms [20]. Web-based graphical interfaces that are simple to interpret can be used to provide reports generated from the prediction exercise [21]. In addition, the communication of alerts can be realized via SMS over Global System for mobile communication (GSM) or SMS over General Packet Radio Transmission (GPRS) [22, 23].

Several studies have also revealed the challenges presented by IoT-based health care solutions such as security, privacy, accuracy, data management, and energy consumption [24, 25]. The introduction of these devices in the community demands measures for addressing these challenges. Thus, the research objectives that steered this study were: (i) to develop a solution that can be used to remotely conduct infant growth and health monitoring, (ii) to develop a solution that factors in the nonfunctional requirements of users in marginalized communities, and (iii) to evaluate the prototype based on initial results conducted in a controlled environment. The rest of this paper presents the design and implementation of the solution in Section III, and the results obtained are discussed in Section IV. Section V provides concluding remarks and the outlook of future work in this area.

3 Solution Design

We named our proposed solution REWEBA, which stands for Remote Well Baby. In this section, we describe the System design and the implementation technology used.

3.1 System Design

Automation of infant monitoring starts with the input of patient details such as gender, location, and guardian contact, required during registration. After registration, the patient receives an identification number, required during subsequent measurements. During registration, the birth measurements (height, weight, skin condition, heart rate, and temperature) are taken using the IoT device. The Remote Well Baby software application (Reweba App) is hosted on a cloud server and includes a Web application, a data analytics module, and interfaces with image processing libraries and databases.

Growth and health measurements are taken using an IoT device and sent to a database for storage via the Azure IoT hub. Azure Cosmos DB stores the growth metrics while Azure blob storage stores the skin condition pictures. The numeric measurements are analyzed against predefined threshold values and alerts are generated where measurements fall outside expected ranges. Machine learning algorithms analyze the images taken for the patient's registration number, skin condition, and head circumference using Optical Character Recognition (OCR), Image classification techniques and Object detection, respectively. The Tesseract OCR engine was used to analyze the registration number image while OpenCV libraries were used for Object Detection. We used Azure Custom vision to develop a classification model for detecting common infant skin conditions such as eczema, measles, or chickenpox. Feedback is posted on a Webpage hosted on the cloud and via cell phone text messages using a third-party service called Twilio. Several checks are inbuilt to control the system workflow by invoking required procedures or logging errors. The overall picture showing the integrated components is illustrated in Fig. 1 and the Algorithm for the system workflow is illustrated in Fig. 2.

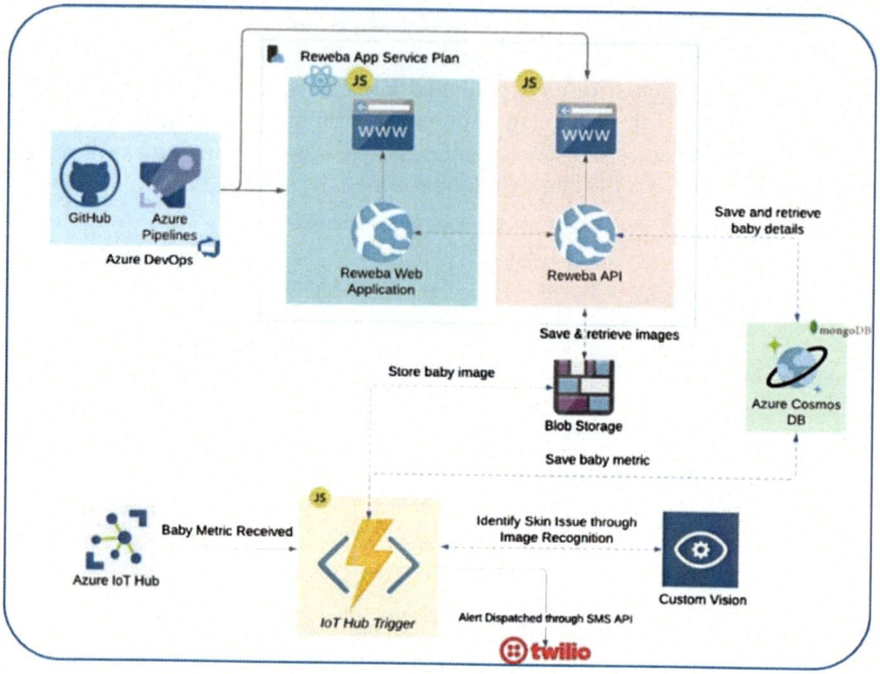

Fig. 1 Remote well baby (REWEBA) system design

3.2 *System Implementation*

The solution is implemented using a PI V4 Raspberry Pi microcontroller connected to a 10,000 mAh power bank supply and a minimum of 100Kbps Internet access. All the sensors used for measurements are embedded as shown on the circuit diagram in Fig. 3. A Backlight LCD 16X2 display is used to verify the measurements taken on the spot. In this section, we describe the technology used for implementing the measurements of demographic, growth, and health parameters.

Demographics Sensors. At the start of the monitoring process, identification of the baby's demographics (name, contact, and gender among others) is done through a registration number printed on the baby's registration card. An image of the baby's registration number is taken using an 8 Megapixel Resolution Raspberry Pi Camera module V2 and printed for scanning during subsequent measurements. The scanned image is analyzed using the Tesseract OCR engine to identify the registration number. Other demographic parameters taken include the location (latitude and longitude) of the baby using a Ublox NEO-6M GPS Sensor for location-based analytics. We also made use of a DHT11 temperature and humidity sensor to monitor the baby's environment. The room temperature and humidity are good indicators of environments that are not conducive for a baby and significantly play a role in the baby's health.

```
START
LOOP READ sensors measurements
        WRITE data to Local Storage for transmission
END LOOP
WHILE Connection Exists DO
        Initialize data transmission module
END WHILE
IF all data received THEN
        GET Threshold Values from CSV
        Parse Threshold metrics into JSON format
        IF Parameters out of range THEN
                Log an Issue Message
        END IF
        READ API Credentials
        Classify the image based on the threshold values
        IF error in image classification THEN
             Throw an Error
             SEND User SMS Error Alert
             STOP
        ELSE
             GET the class and classification score
             WRITE the score and class into the database
        END IF
ELSE
        Throw an Error
        SEND User SMS Error Alert
        STOP
END IF
IF no issues have been logged THEN
        SEND normal baby report via Web Interface
ELSE
        SEND via Web Interface and SMS with Logged issues
END IF
STOP
```

Fig. 2 Remote well baby workflow algorithm

Growth Monitoring Sensors. Typical Growth monitoring parameters measured during postnatal care include weight, height, head circumference, and fine motor skill milestones. To monitor the weight a Load cell sensor that uses a transducer to transform force or pressure into electrical output is used. The electrical signals generated by the HX711 Load cell are further amplified using the HX711 Weighing Sensor and fed into the microcontroller to derive the weight. An ultrasonic sensor is used to measure the baby's height using two ultrasonic transducers. One acts as a transmitter which converts electrical signal into ultrasonic sound pulses and the receiver listens for the transmitted pulses. The distance is then determined by the width of the output pulse. The ultrasonic sound waves are transmitted toward a wall and converts the reflected sound into an electrical signal as shown in Fig. 4a. To measure the head circumference, an image of the baby's head is taken from the top when the baby is lying down using the Raspberry Camera. To ensure that the largest circumference of the head is taken (occipito-frontal circumference), the tip of the nose and ears must be simultaneously visible. These images are transmitted to a server where Machine learning techniques are applied for image processing and object detection alongside a reference object for circumference calculations.

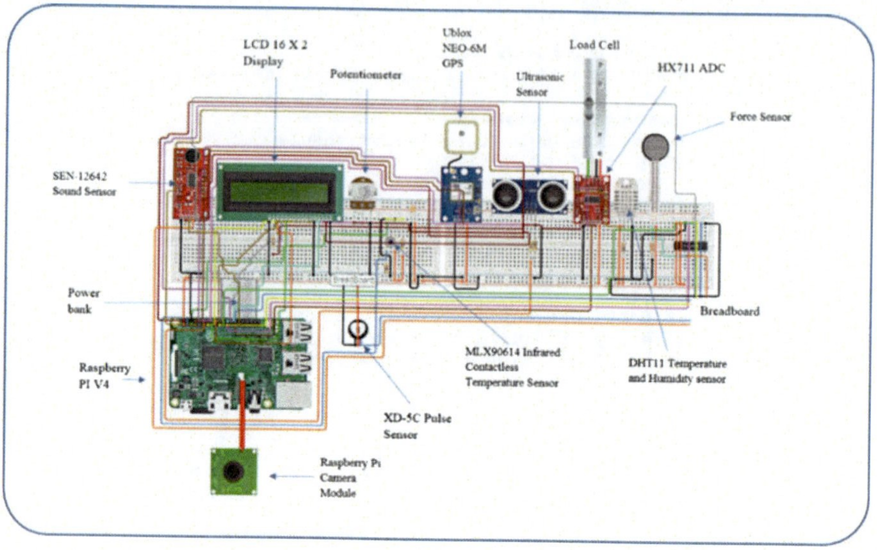

Fig. 3 Remote well baby IoT circuit diagram

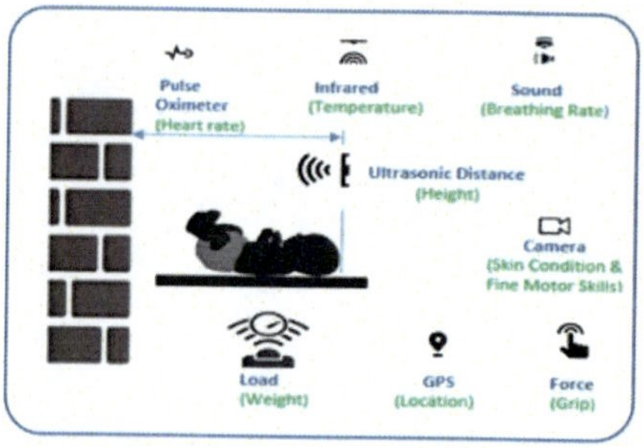

(a) Sensors Embedded in the Prototype

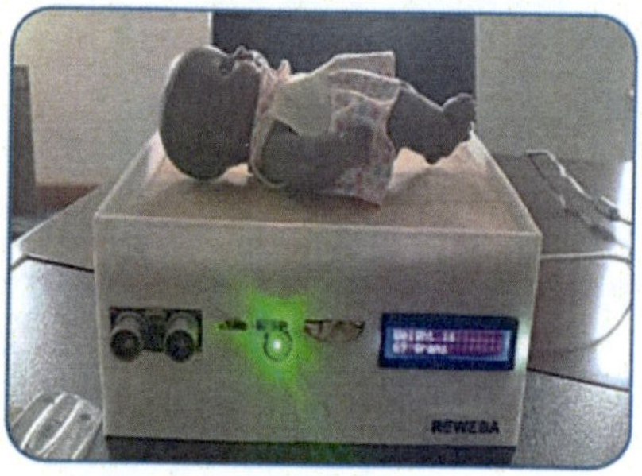

(b) Developed Prototype

Fig. 4 Remote well baby prototype

We made use of OpenCV the Open-Source Computer Vision Library [26] for image processing and implemented it using the Python Programming Language [27]. When taking readings, the presence of external objects needs to be minimized to reduce noise distortions. During postnatal care, a baby's fine motor skills are also monitored. We embedded a FlexiForce A401 Force resistive sensor in a squishy ball toy attached to the microcontroller for monitoring the baby's ability to grip items. The resultant drive voltage is recorded for analyzing grip motor skills.

Health Monitoring Sensors. The most popular way of detecting health concerns with a baby is by monitoring their temperature. We made use of an MLX90614 Contactless Infrared (IR) Digital Temperature Sensor which uses IR rays to measure the temperature of the object without any physical contact and communicates to the microcontroller using the I2C protocol. Our solution also made use of an XD-5C pulse sensor to measure the heart rate of the infant and a SEN-12642 sound sensor module to monitor the infants breathing rate. The sound sensor employs a microphone capable of determining noise levels within decibels at 3 kHz, 6 kHz frequencies. By placing it near the baby's nostrils, vibrations of the baby's breath are periodically taken and converted to a rate. Finally, the Raspberry camera is used to take an image of the baby's skin condition. This image can be taken from any part of the body such as the face or stomach. The image is transmitted to a server for classification using Machine Learning Neural Networks models. The implemented prototype is pictured in Fig. 4b.

Algorithms. Range frequency analytics were used to compare the input values to the WHO threshold values in the system and issue alerts to the doctor. To measure the head circumference, we used the level set method which has increasingly been applied to image segmentation because of its ability to achieve subpixel accuracy of object boundaries and contours for shape analysis. Lastly, to classify the skin disorders in children, we used k-fold cross-validation for image recognition to get the best results. With k-fold cross-validation, the data is distributed in a structured way so that the trained model is less overfitted to the training set. Because the number of publicly available images on infant skin conditions is limited, we made use of Transfer Learning; a machine learning technique that presents accurate results even with small datasets.

4 Results and discussion

The initial evaluation of the proposed system was conducted using laboratory simulated tests in a controlled environment. Evaluation of the functional requirements involved accuracy tests and the generation of reports and alerts. Evaluation of the nonfunctional requirements involved the assessments of cost, portability, security, and resource utilization of the developed solution. The results are discussed in this section.

4.1 Functional Requirements

Several measurements were taken using the device and transmitted to the server. Accuracy tests were conducted by comparing the readings with the expected values. The results presented show that the developed solution produces the desired output with acceptable accuracy levels.

Sensor Readings Accuracy tests. To evaluate the solution's accuracy, the measurements taken by the IoT device were evaluated against expected measurements taken using external devices. For example, a known weight (2Kg) is placed on the load cell and 10 separate readings are taken. The average weight is obtained

Table 1 Sensor accuracy tests

Sensor	Load cell (g)	Ultrasonic sensor (cm)	Room temp. sensor (°C)	GPS sensor (°S, °E)	Pulse sensor (BPM)	Room humidity sensor (%)	Sound sensor (Breaths/min)
Average readings	1980.75	50.11	25.47	−1.22463, 36.8725	71.1**	38.8	17.3**
Expected readings	2000	50	25	−1.2184, 36.8791	70	40	18
MAPE error	0.97%	0.22%	1.85%	0.57%, 0.02%	0.02%	0.03%	0.054%

** Test readings taken from an adult subject

and compared to the expected 2Kg weight to determine the accuracy of the weight sensor. Similar tests were conducted for the other sensors. The accuracy levels are calculated using the Mean Absolute Percentage Error (MAPE) equation shown in equation 1 [28], for each parameter.

$$\text{MAPE} = \frac{\sum \frac{|\text{Expected Value} - \text{Experimental Value}|}{\text{Expected Value}}}{N \text{ number of measurements}} \times 100\% \quad (1)$$

A sample of the accuracy results obtained in Table 1 show that the accuracy levels were within acceptable ranges. The MAPE error was below 2%, an indication that the solution can provide reliable information for monitoring infants remotely.

Image Recognition Accuracy Tests. Several images were uploaded to the classifier to classify the skin condition. We used a publicly available clinical resource website on dermatology and skin conditions known as DermNet [22]. The results show that the image classifier was able to work with different images irrespective of the body part or the skin tone as seen in the figure. Skin disorders can exhibit variability depending on the body part, keeping that in mind, we used a classification algorithm that factors in the location of the skin disorder. Additionally, we factor in the variability aspect by allowing the user to take multiple pictures for every body part. Lastly, the presence of noise in the images did not significantly hamper the classifier's accuracy.

The sample results obtained show that the model was able to classify the images accurately as shown in Table 2. The results provide a clear indication of the skin conditions. When the information is passed to a doctor, they verify by examining the transmitted images and request the mother to take the baby to the health center if any intervention is required. This eliminates the need for costly hospital trips where minor issues such as mild eczema or a heat rash are detected. In this way, the solution complements the provider by making healthcare more accessible.

Object detection Accuracy Tests. As described in the system design, the process of calculating the head circumference from images called for several image processing steps. These include resizing of the image, removal of noise, and detection

Table 2 Sample Image classification accuracy results

Sample 1 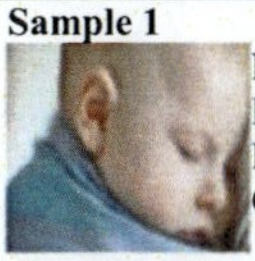 Normal: **96.4%** Eczema: 3.5% Measles: 0% Chicken pox: 0%	**Sample 2** 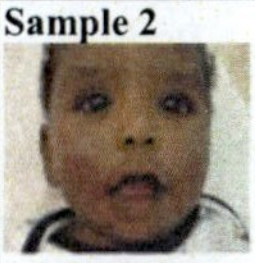 Normal: 0% Eczema: **99.9%** Measles: 0% Chicken pox: 0%	**Sample 3** 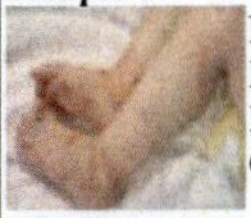 Normal: 0% Eczema: **99.9%** Measles: 0% Chicken pox: 0%
Sample 4 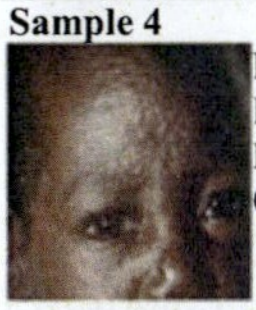 Normal: 1.2% Eczema: 0.1% Measles: **64%** Chicken pox: 34%	**Sample 5** 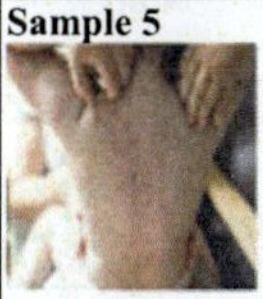 Normal: 0% Eczema: 0% Measles: 0% Chicken pox: **99%**	**Sample 6** 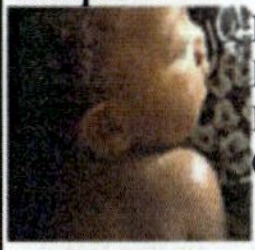 Normal: 0.8% Eczema: **79.2%** Measles: 17.9% Chicken pox: 2%

of the contour edges for measurements. We improved the accuracy of the results by performing a camera calibration to address the camera's extrinsic parameters (such as rotation and translation) and intrinsic parameters (such as focal length and optical center). For this experiment we used the checkerboard camera calibration approach and OpenCV chessboard calibration library to calibrate the camera used and generate undistorted images from any image taken with the camera. Additionally, we normalized the image to increase the contrast of the image and reduce the noise content from the image for improved image segmentation. An ellipse was then fitted in the resultant image and the perimeter was calculated to estimate the head circumference. The results of each process for a sample image and reference object are illustrated in Fig. 5.

To test the accuracy of the calculated head circumference, several images were taken on a doll, and the MAPE error calculated. To investigate if the results can be scaled to images of a different size, a bigger doll was used. The results presented in Table 3 show the MAPE error as 2.65% and 2.40% when two dolls with head circumference of 21.5 cm and 38 cm, respectively, were tested. As a novel idea, the concept has a lot of potential given the consistency of the results obtained in

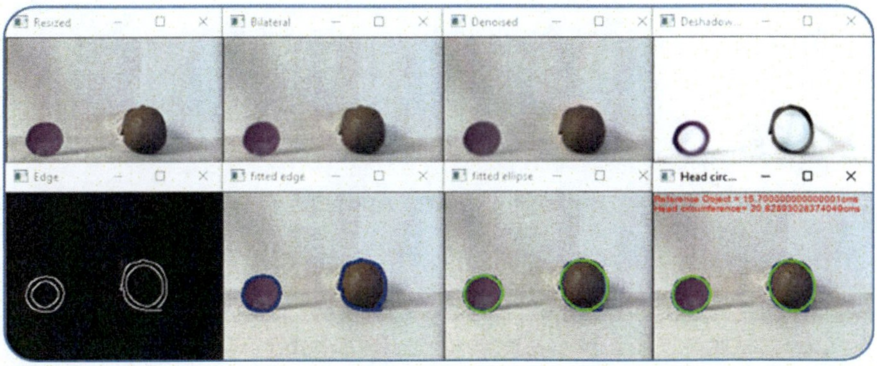

Fig. 5 Head circumference image processing steps

Table 3 Head circumference results

Image Samples	1	2	3	4	5	6	7	8	9	MAPE error
Circ. **(21.5 cm)**	21.11	20.93	21.04	20.63	21.01	20.95	20.83	21.16	20.86	**2.65%**
Circ. **(38 cm)**	37.26	37.08	36.38	37.21	36.63	37.83	36.95	37.58	37.12	**2.40%**

infant postnatal care. It provides a contactless option of automating the process of monitoring the head circumference digitally. The use of a reference object of known dimensions also complements the approach by providing a visual verification process for enhanced reliability. The enhanced accuracy of image recognition and object detection addresses the accuracy challenge encountered by previous IoT solutions as indicated by [24, 25] in section 2. We plan to reduce the error value further through enhanced image processing and noise cancellation techniques.

Report and Alerts. Following the accurate analysis of data, the system generates reports for each baby and are posted on the Website for authorized users as shown in Figure 6. Aggregated reports for all babies in a specific location are also generated for resource planning purposes by authorized government health representatives. Through individual reports, the provider can remotely monitor the key growth and health parameters of a baby. The frequency of monitoring can also be increased for preterm babies or babies with special needs without escalating costs or effort for the parents. Based on the results, the proposed solution meets the functional expectations.

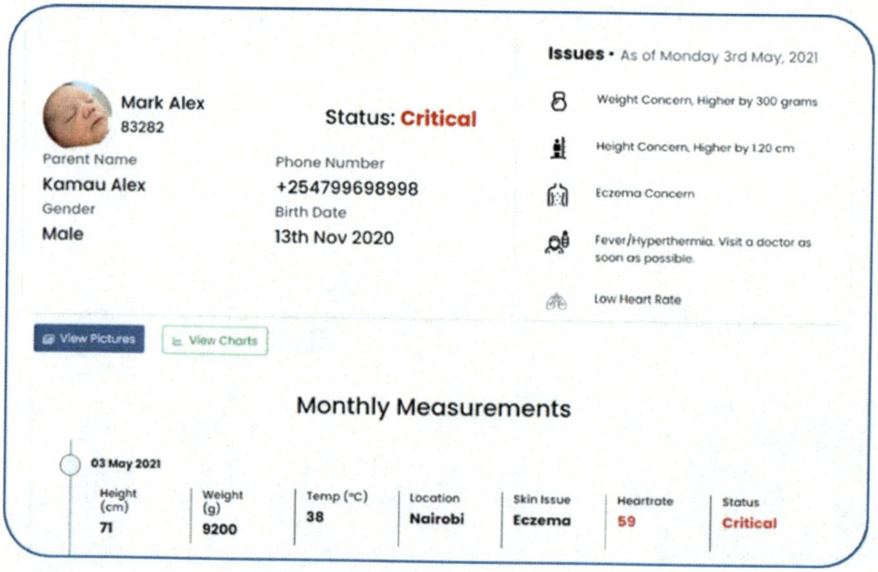

Fig. 6 Sample reports generated by the system

4.2 Nonfunctional Requirements

To evaluate the nonfunctional requirements, the operations of the system were examined. The results presented in this section show that the developed IoT solution has the potential of addressing the health access challenges encountered in marginalized communities.

Affordability. At the time of writing this paper, the total cost of the hardware components used to develop the solution was estimated as USD300, with an additional USD50 running costs (hosting, connection, and power charges) for the mandatory 6 months measurements. If shared within a closed community of about 20 users, each user will spend USD12.5 only for all postnatal care visits, which is often the average cost spent for transport and consultation fee during each postnatal visit in many marginalized communities. Existing automated solutions in the market such as baby monitors and other digital devices monitor specific parameters therefore multiple devices are required which costs more than our proposed solution. More comprehensive solutions in the market such as Neonatal incubators or are very costly as shown in Table 4. This is in line with the observations made by [3] thus, our solution, serves to bridge this gap.

Portability. The final device weighs 1 kilogram and has 23 cm × 23 cm × 10 cm dimensions, which is about the size of a lightweight laptop. Therefore, the device can easily be moved around for purposes of sharing. The device is packaged using a lightweight plastic case that protects the IoT device from dust and is easy to clean.

Privacy and Security. Security and privacy of healthcare data is a concern and therefore several security features are inbuilt into the solution. For authentication, we have included access controls for different users thereby enabling different stakeholders to access only what they are authorized to access. For example, a government health representative can only view aggregated charts without patient identification details, whereas the provider will be able to access details for a specific baby. We used the JavaScript library helmet.js to encrypt data at rest and in transit and cross-site scripting protection to secure the Web application interface. Users are also issued with

Table 4 Comparison of existing solutions

Automated existing solutions	Current costs (USD)	Comparison
Baby monitors Audio only Audio and camera Wearable (breathing and rollover)	20 40 100	Affordable Do not monitor height, head circumference, breathing rate, or skin conditions
Neonatal incubators (monitor temperature, humidity, oxygen, heart rate, breathing)	1000	Are costly and requires skills Manual monitoring of height, head circumference, and Skin conditions
Separate digital devices Infra-red thermometer gun Baby weighing scale Baby pulse oximeters	35 45 30	Affordable Do not monitor height, head circumference, breathing rate, or skin conditions

authentication tokens, which expire if the interface is left idle. Finally, the database layer enforces data integrity by ensuring that the passwords collected are stored in an encrypted format. To protect the child and offer a less intrusive solution, we designed our solution to make use of contactless sensors as much as possible. Therefore, this addresses the security challenges discussed by previous studies [24, 25].

Resource utilization. We have included a 16 GB SD card that temporarily stores data in JSON format if an Internet connection is not available. The data will be transmitted later when the connection resumes. The SD card can also be used to store measurements when the process is interrupted, for example when the baby becomes restless or if the baby falls asleep. The full set of measurements is approximately 2.5 Mbytes, which means that the SD card can hold measurements of about five babies before transmission is required.

The IoT device can be powered by a direct current or a 2600 mAh Lithium battery in the absence of electricity. To test how much power is consumed by the device we connected it to a power bank with a capacity of 10000 mAh and an output of 5 V and 2 A. We observed that after taking ten full sets of readings, the battery capacity only reduced by 5% ($\approx$500 mAh). Given that each reading takes about 2 minutes, the lithium battery can power the device for over 50 readings while the power bank can power it for over 150 readings. The device, therefore, is ideal for users in marginalized communities who face Internet connectivity and power challenges.

5 Conclusion and Future Work

This paper has described the process of developing a solution for infant growth and health monitoring. The solution factored in the nonfunctional requirements of users in marginalized communities. The results obtained revealed the feasible potential of using the proposed solution in infant postnatal care. We plan to enhance the solution to monitor additional parameters such as eye movement or head movements by analyzing video feeds. We are also working on enhancing the data capture and analysis process to reduce errors and hence improve the solution's accuracy.

References

1. Rao SP, Minckas N, Medvedev MM, Gathara D, Prashantha YN, Estifanos AS, Silitonga AC, Jadaun AS, Adejuyigbe EA, Brotherton H, Lawn JE (2021) Small and sick newborn care during the COVID-19 pandemic: global survey and thematic analysis of healthcare providers' voices and experiences. BMJ Global Health 6(3):e004347
2. Yassein MB, Hmeidi I, Al-Harbi M, Mrayan L, Mardini W, Khamayseh Y (2019) IoT-based healthcare systems: a survey. In: Proceedings of the second international conference on data science, E-learning and information systems, pp 1–9
3. Malasinghe LP, Ramzan N, Dahal K (2019) Remote patient monitoring: a comprehensive study. J Ambient Intell Human Comput 10:57–76

4. Neyzi O, Bundak R, Gökçay G, Günöz H, Furman A, Darendeliler F, Baş F (2015) Reference values for weight, height, head circumference, and body mass index in Turkish children. J Clin Res Pediatr Endocrinol 7(4):280
5. Wikland KA, Luo ZC, Niklasson A, Karlberg J (2002) Swedish population-based longitudinal reference values from birth to 18 years of age for height, weight and head circumference. Acta Paediatr 91(7):739–754
6. Cole TJ, Freeman JV, Preece MA (1998) British 1990 growth reference centiles for weight, height, body mass index and head circumference fitted by maximum penalized likelihood. Stat Med 17(4):407–429
7. Júlíusson PB, Roelants M, Nordal E, Furevik L, Eide GE, Moster D, Hauspie R, Bjerknes R (2013) Growth references for 0–19 year-old Norwegian children for length/height, weight, body mass index and head circumference. Ann Hum Biol 40(3):220–227
8. Davies P, Maconochie I (2009) The relationship between body temperature, heart rate and respiratory rate in children. Emerg Med J 26(9):641–643
9. De Onis M, Wijnhoven TM, Onyango AW (2004) Worldwide practices in child growth monitoring. J Pediatr 144(4):461–465
10. Linti C, Horter H, Osterreicher P, Planck H (2006, April) Sensory baby vest for the monitoring of infants. In: IEEE international workshop on wearable and implantable body sensor networks (BSN'06). https://doi.org/10.1109/bsn.2006.49
11. Ishak DNFM, Jamil MA, Ambar R (2017) Arduino based infant monitoring system. In: IOP conference series: materials science and engineering, vol 226, issue no 1, pp 1–6
12. Symon AF, Hassan N, Rashid H, Ahmed IU, Taslim Reza SM (2017) Design and development of a smart baby monitoring system based on raspberry Pi and Pi camera. In: 2017 4th international conference on advances in electrical engineering (ICAEE). https://doi.org/10.1109/icaee.2017.8255338
13. Sathya M, Madhan S, Jayanthi K (2018) Internet of things (IoT) based health monitoring system and challenges. Int J Eng Technol 7(1.7):175. https://doi.org/10.14419/ijet.v7i1.7.10645
14. De La Iglesia D, De Paz J, Villarrubia González G, Barriuso A, Bajo J (2018) A context-aware indoor air quality system for sudden infant death syndrome prevention. Sensors 18(3):757. https://doi.org/10.3390/s18030757
15. Shalannanda W, Zakia I, Sutanto E, Fahmi F (2020) Design of hardware module of IoT-based infant incubator monitoring system. In: 2020 6th international conference on wireless and telematics (ICWT). https://doi.org/10.1109/icwt50448.2020.9243665
16. Lobo C, Chitrey A, Gupta P, Sarfaraj, Chaudhari A (2020) Infant care assistant using machine learning, audio processing, image processing and IoT sensor network. In: 2020 international conference on electronics and sustainable communication systems (ICESC). https://doi.org/10.1109/icesc48915.2020.9155597
17. Hussain T, Muhammad K, Khan S, Ullah A. Lee MY, Baik SW (2019) Intelligent baby behavior monitoring using embedded vision in IoT for smart healthcare centers. J Artif Intell Syst 1(1):110–124. https://doi.org/10.33969/ais.2019.11007
18. Anzanpour A, Rahmani AM, Liljeberg P, Tenhunen H (2015) Internet of things enabled in-home health monitoring system using early warning score. In: Proceedings of the 5th EAI international conference on wireless mobile communication and healthcare, pp 174–177
19. WHO (2020) The WHO child growth standards https://www.who.int/childgrowth/standards/. Accessed 10th June 2020
20. Kaur P, Kumar R, Kumar M (2019) A healthcare monitoring system using random forest and internet of things (IoT). Multimedia Tools Appl
21. Enshaeifar S, Barnaghi P, Skillman S, Sharp D, Nilforooshan R, Rostill H (2020, April) A digital platform for remote healthcare monitoring. In: Companion proceedings of the web conference 2020, pp 203–206
22. Singh PP (2014, February) Zigbee and GSM based patient health monitoring system. In: 2014 international conference on electronics and communication systems (ICECS). IEEE, pp 1–5
23. Gupta S, Kashaudhan S, Pandey DC, Gaur PPS (2017) IOT based patient health monitoring system. Int Res J Eng Technol 4(3):2316–2319

24. De Michele R, Furini M (2019, September). IoT healthcare: benefits, issues and challenges. In Proceedings of the 5th EAI international conference on smart objects and technologies for social good, pp 160–164
25. Nazir S, Ali Y, Ullah N, García-Magariño I (2019) Internet of Things for healthcare using effects of mobile computing: a systematic literature review. Wirel Commun Mob Compu
26. Bradski G (2000) The opencv library. Dr Dobb's J Softw Tools 25:120–125
27. Van Rossum G, Drake FL Jr (1995) Python reference manual. Centrum voor Wiskunde en Informatica Amsterdam
28. Khair U, Fahmi H, Al Hakim S, Rahim R (2017, December) Forecasting error calculation with mean absolute deviation and mean absolute percentage error. J Phys Conf Ser 930(1):012002 (IOP Publishing)

Evaluation of Selective Reactive Routing Protocols of Mobile Ad-Hoc Network

Kashif Nazir, Muhammad Asif Habib, and Mudassar Ahmad

Abstract Mobile Ad-hoc Network (MANET) works without the help of any fixed infrastructure. As it has a dynamic topology thus, MANET can be created anywhere to use. MANET nodes are often mobile nodes, and they do not rely on any centralized infrastructure. Over the last few decades, many routing algorithms and routing protocols are designed to increase the efficiency and reliability of MANET; however, their productivity environment is always dependent. The most popular routing protocol introduced and used are DSR, AODV, and TORA. This Research evaluates the performance of these reactive routing protocols are DSR, AODV, and TORA. The performance parameters throughput, accumulated delay is also used despite the end-to-end delay, pause time, and routing. To implement a test-bed scenario, the most famous OPNET version 4.5 simulator is used. The concluded results show that AODV outperforms best in all the environments. The DSR takes 2nd position in performance, whereas TORA remained on the lower side as compared to the other two routing protocols. With the increase of network size and no mobile nodes, the end-to-end delay is more increased in TORA than in DSR. In terms of router discovery, the DSR protocols respond better than the other two protocols under discussion. The overall working of DSR was found good in changing environment.

Keywords End-to-end delay · MANET · Node · Performance · Throughput

1 Introduction

In the new technology era, the advents and usage of portable computing devices have changed the whole global village very rapidly. Governmental/semi-Government/none-Governmental organizations, Universities/Colleges/Schools, Armed forces, Companies, and some Agencies are currently using this type of new and efficient technology daily. The proliferation of innovative, powerful, reliable, compact, and mobile communication devices like cellular phones, laptops,

K. Nazir · M. A. Habib (✉) · M. Ahmad
Department of Computer Science, National Textile University (NTU), Faisalabad, Pakistan
e-mail: drasif@ntu.edu.pk

A. Ullah et al. (eds.), *Proceedings of International Conference on Information Technology and Applications*, Lecture Notes in Networks and Systems 350,
https://doi.org/10.1007/978-981-16-7618-5_55

Personal Digital Assistants (PDAs), and pagers that have tremendous computing and processing power changed the way of living standard of the person and increased the demand in the markets. Now the trend moved from the beginning of the first generation era to the Ubiquitous technology age. Every person utilizes several computing accessories with them, at the same time, through which they can access all the required information whatever and whenever they required. The increase in demand for computing devices makes the wireless networks more effective and the easiest solution for their interconnecting and, as a result, the wireless technology has been facing extraordinary production in the last few years [1]. Nowadays, most network connections between wireless products are designed using a centralized and fixed infrastructure [2, 3]. The network can increase from hundreds to thousands of mobile nodes using this technology that provides the nodes to take part in route maintenance, and route discovery process [4–6].

1.1 Characteristics of MANET

The main characteristics of MANET are the distributed operation, dynamic topology, multi-hop routing, shared physical medium, energy-constrained operation, autonomous terminal, light-weight nodes, and security.

1.2 MANET Applications

PAN is a localized and small-range network where different mobile nodes are generally associated with other known persons within a short-range. In this technology, the range of the network is very short such as in Bluetooth; the range is very short for communication between different mobile node devices like a mobile phone, pager, and laptop [7, 8].

1.3 MANET Challenges

Characteristics of MANET may face numerous problems like dynamic topology, limited bandwidth, routing overhead, quality of service, data losses, route changes, insecurity, and unreliability.

1.4 MANET Routing Protocols

Figure 1 describes the basic division of the routing protocols.

Reactive writing protocol is also known as on-demand routing protocol. Whenever a node wants to send data to some other node, it first initiates the route discovery for data propagation on an emergency basis. It follows the two steps; first, it initiates a route discovery, and after that, it maintains its route. The main parts of reactive routing protocol are route discovery and route maintenance. The main difference is that burden is higher as compared to the reactive routing protocol because of routing tables [9]. All the nodes within the segment use reactive routing protocol mechanism, and outside the segment, it uses table-driven protocol for data propagation [10].

2 Problem Background

The life cycle of MANET can be divided into three generations. Presently ad-hoc network falls into the third generation. The concept of the ad-hoc network can be started in 1970. Mainly PRNET uses the mixture of Carrier Sense Multiple Access (CSMA) and Areal Location of Hazardous Atmospheres (ALOHA) for vector routing and multiple accesses. The PRNET is then developed into the new and latest technology named Survivable Adaptive Radio Network (SURAN) in the early 1980s. SURAN provides more benefits as compared to PRENT by extending the radio efficiency and performance. It also holds the electronic threats. In 1980, the United States Department of Defense (DOD) continual support for multiple programs such as Near-Term Digital Radio (NTDR) and Globe Mobile Information System (GloMo). GloMo makes utilizes TDMA and CSMA/CA molds and also provides a self-healing and self-organizing network. The NTDR make also use of link-state routing and

Fig. 1 Classification of routing protocols

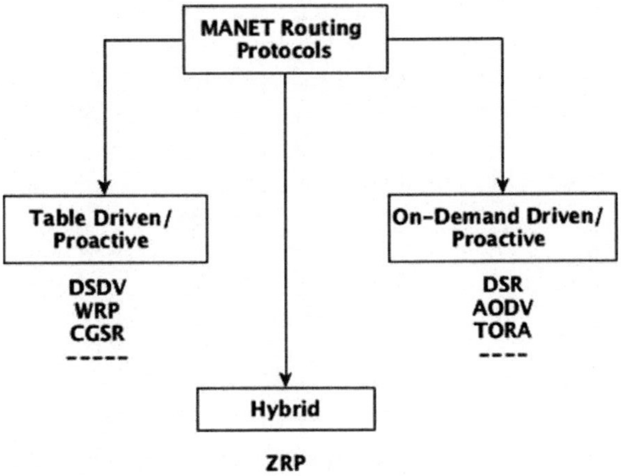

clustering an ad-hoc network. Unfortunately, the military damaged NTDR technology. At that, time this was the only genuine ad-hoc network that was in use. With time the growing interest in the mobile ad-hoc networks, a numberless great Internet Engineering Task Force (IETF) who work for consistent routing protocols for MANET and also increase in growth of different mobile devices like notebooks, PDA, palmtops, etc. [11–13].

3 Literature Review

I. Panda categorized the routing protocol into three strategies as a complete classification of MANET routing protocols [14]. The Research comprises of demand on routing protocol in which a routing table periodically updates every change and the second one is on-demand in which every node when it wants to establish a link and last one is a hybrid routing protocol that contains the qualities of both reactive and table-driven routing protocol [15]. The simulation result of the NS-2 simulator of some routing protocols of MANET is discussed. It is based on four performance metrics: Throughput, number of packets dropped, average delay, and routing overhead. M. V. Rajput et al., selected DSR and AODV from the on-demand, DSDV from table-driven, ZRP from the hybrid routing protocol. At the end of the result, AODV routing protocol performance in provisos of Throughput is better than other protocols. At the same time, DSDV showed poor performance in some cases, and also, its ratio of packets drop is higher than that of other protocols [16].

The routing protocols of MANET can be divided into three different types. The first one is proactive/reactive/hybrid. The second one is centralized/distributed based on some intermediate or gateway nodes that help deliver packets from one cluster to another cluster. Finally, the third one is dynamic/static based on its unique behavior [17].

The selection of a routing protocol for the quality of services is one of the fundamental problems in MANET. Therefore, the design of an optimal MANET routing protocol to achieve the quality of services parameters; is the most active and important research area in the MANET field. N. H. Saeed et al., provide a complete review of some of the already available MANET routing protocols that provide QoS support. Routing protocols are divided into three parts. The first part is based on route discovery divided into proactive, reactive, and combination of both routing protocols is the hybrid routing protocol. An example is OLSR, DSD, TORA, and Zone routing protocol. The second part is based on metrics that are further divided into single and multiple routing protocols, i.e., OLSR, FA, TDR, AQR, and ticket-based. Third and last part is based on a network architecture that is dividing into flat and hierarchal types. Examples of these types of protocols are OLSR, TDR, and Ticket based, etc. [18].

The routing protocols of MANETs make it more difficult to decide which protocol is appropriate for different networks because there are several protocols in routing not yet clear which protocol is suitable in some environments and which protocol is suitable in a different environment. At the time of result, R. Kaur and M. K. Rai described that in route discovery mechanism, the reactive routing protocol overhead is very low compared to the other two routing protocols. Still, its latency is very low compared to proactive, where the route is already always available in routing tables when needed. In a hybrid routing protocol, the latency inside the area is low because it uses a proactive routing technique inside the zone, and overhead is also decreased outside the area where it uses reactive routing mechanism [9].

A new routing algorithm is the on-demand Multicast Routing Protocol. It is used for routing unicast and multicast traffic via MANET. On-Demand Multicast Routing Protocol (ODMRP) creates some routes when some node wants to send data to another node. That's why delay can occur in this type of routing. However, there is no routing table information in this technology, so the overhead is low compared to other nodes. Experiments were performed based on a simulation that analyzed and studied the performance of ODMRP by using some parameters like End-to-End delay, average Throughput, and Packet Delivery Ratio (PRD). At the end of the results, a comparison is done between FSR and AODV routing protocols by changing the mobility and number of nodes. S. Goswami et al., can compare the result by using a simulator. ODMRP has better performance than FSR and AODV by using the parameters; average Throughput by changing the mobility models and change the number of nodes. Packet delivery fraction ratio of AODV gives better performance than FSR and ODMRP by changing the mobility models and changing the number of. Finally, at the end of the, et al., concluded that ODMRP for MANET shows best perform with compared to FSR and AODV [19].

4 Methodology

The simulator provides an environment for testing protocols similar to realistic scenarios. In the simulation technique, essential parameters are set to obtain the desired output results. Various simulators are available like OMNET++, GloMoSim, NS-2, OPNET, QualNet, and OPNET are well-developed commercial software products. OPNET Modelers 14.5 is selected for simulations because it is user-friendly. First, created the simulation scenarios and used various performance matrixes for the evaluation of routing protocols.

First of all, a set of performance metrics are selected, then a software platform is chosen, and then a simulation test-bed was designed. After that, the routing protocols can be evaluated by using different parameters. In this Research, three performance parameters are used for evaluating the routing protocols that affect the network performance. These parameters are Through- put, End-to-end delay, and routing overhead. The complete steps that are used in methodology are shown in Fig. 2. The brief discussion about the performance parameters/metrics is described as under.

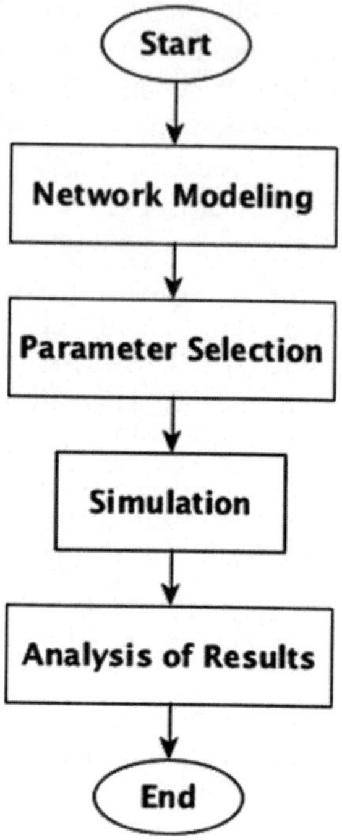

Fig. 2 Methodology steps

4.1 *Throughput*

It is defined as how much data can be sent from sender to receiver in a given time. Its unit is packets per second or bits per second. Many factors, such as unreliable and unidirectional communication, dynamic topology, and limited bandwidth and energy, affect the Throughput. Therefore, a network may require higher Throughput according to the situation.

4.2 *End-to-End Delay*

It is referred to as the time to transfer a packet across the network from sender to receiver. This time starts when the source generates packets until it receives at the receiver side, measured in seconds. The main reasons for delays are buffers queue, MAC controls, and some delay occurs due to the routing activities in MANET. Different applications have varying levels of end-to-end delay. For example, voice requires a low level of delay as compared to other data types like FTP may tolerate to delay up to some extent in a network because it is a delay-sensitive application. However, in MANETs, every node is mobile, and the signal between nodes is comparatively weak, so the delay is increased in the network.

Table 1 Methodology specification summary

Simulator	OPNET 14.5
Node provisions	15,30,45,60
Speed variation	25 m/s, 40 m/s
Pause time	100,300,500 s
Traffic load	FTP
Routing protocol	ODV, DSR and TORA
Observation	Throughput, end-to-end delay, routing-overheard

4.3 Routing Overhead

In reactive routing protocols, whenever a node wants to send data to another node, it initiates the route discovery for data propagation on an emergency basis. It follows the two steps; first, it initiates a route discovery, and after that, it maintains its route. Routing overhead is the total number of packets sent by a sender for route discovery, and it is expressed in packets per second or bits per second. MANET is developed to be scalable, so different routing protocols behave differently as the network size increase also the network traffic also increases as the network grow. When network traffic is increased, routing overhead is also increased as the numbers of nodes from source to destination are increased so routing, overhead is also increased. Some other causes of routing overhead are error packets route and network congestion. A summary of the methodology specifications is discussed in Table 1.

OPNET Modelers is used with version 14.5 in this simulation.

5 Results

The performance of AODV, DSR, and TORA, in terms of Throughput. First, we start from 15 nodes. After that, 30, 45, and 60 nodes are simulated. Throughput of AODV is better than DSR and TORA. Changing the input parameters like the speed of nodes and pause time does not affect the overall performance of AODV in 15 Node. Table 2 describes the relative Throughput of these three protocols.

Table 2 Relative effect of node number on throughput

Nodes	Relative AODV	Throughput TORA	DSR
10	Better	Better	Best
30	Best	Better	Low
45	Best	Better	Low
60	Best	Better	Low

Table 3 Relative performance of speed on protocol

Performance	variant due AODV	to speed DSR	Change TORA
25 m/s	Best	Better	Low
40 m/s	Best	Better	Low

After AODV, the performance of TORA is better than DSR in terms of Throughput. When we increase the speed of nodes at 25 m/s, it does not affect the overall performance of AODV. In that scenario, AODV outperforms best and then TORA and lastly DSR. When nodes move at a speed of 40 m/s, the result is the same as discussed above. By changing the pause time 100,300, and 500 s of the nodes, it performs the same function as AODV performs best and then TORA and lastly DSR in 15 nodes in terms of Throughput. The performance of TORA is better than the other two reactive routing protocols when we increased the speed of the nodes as described in Table 3.

As the network size increase, it affects the overall performance of the network. In this scenario, we change the numbers of nodes and two other input parameters like the mobility of nodes and the pause time of the nodes. In the case of 30 nodes. While considering Throughput using 45 nodes and with speed 25 and 40 m/s and pause times are 300 and 500 s.

In all scenarios, the performance of AODV is better. Secondly, TORA performs best, and lastly, DSR performs best in 45 nodes when the speed of the nodes is 25 and 40 and pause time is 100 and 300 in terms of Throughput. Finally, when the network is increased in terms of source, problems of hidden terminals, congestion, and network degradation come more into effect. So, these reasons cause the protocols to behave differently in the same environment by changing the parameters, and less delay causes more network throughput. From all the above figures, we conclude that the performance of DSR is low in the networks with 30, 45, and 60 nodes. From all the above reasons, we can observe that DSR outperforms TORA and AODV in the networks where nodes are less than ten, both in high-speed and mobility scenarios. In all remaining scenarios, the performance of AODV is best as compared to TORA and DSR. It outperforms DSR when the network size grows.

For end-to-end delay, considering the 15 nodes, the performance ratio of AODV and TORA is 1:2, and the ratio of AODV and DSR is 1:3. It means that DSR consumes three times greater than AODV for packet delivery and two times greater than TORA for sending packets to the receiver. In the case of end-to-end delay considering 30 nodes, all scenarios by increasing the number delay of AODV is less than previous scenarios. The delay in all cases is very low when the speed of nodes is 25 and 40. And delay is also low of AODV when we increase the pause time 300 and 500 s. On the other hand, the performance of TORA degrades in all scenarios. In all scenarios, the delay faced by TORA is very high as compared to the previous environment. The ratio between AODV and TORA is 1:15. DSR performs the same result that it performs in 15 nodes. In this environment, AODV outperforms best in all cases, and then DSR has a higher delay than AODV.TORA performs worst in all scenarios where mobility is high and pauses time increases in terms of End-to-end delay in 30 nodes.

In 45 nodes, all the scenarios show that AODV and DSR delay up to 0.02 s, but TORA's delay may rise to 14 minutes. In this environment, AODV has less delay as compared to DSR and TORA. DSR has more delay than TORA, and TORA performs worst in all scenarios by increasing the node mobility and pause time in terms of End-to-end delay. In the case of 60 nodes, end-to-end delay AODV has less delay by increasing the mobility and pause time. As the size of the network increase, all three on-demand protocols show higher delay due to route-cache burden and route discovery process.

The DSR has less overhead as compared to the other two routing protocols; it sent less amount of data packet for route discovery in the entire network; also, it is based on source routing, and all the path information are available to every node, so there is no need for nodes to send an unnecessary message for path discovery in the entire network. Overall, the performance of DSR is better than the other two reactive routing protocols in terms of Throughput. On the other hand, AODV sent a higher amount of data packets for route discovery than the DSR because all the nodes in AODV sent routing path information to sender and receiver for data sending, which increases the overhead of the network. TORA performs worst in routing overhead because it sends HELLO messages periodically to the entire network, increasing the routing overhead.

6 Conclusion

FTP high load traffic used all the nodes to send packets to the receiver. The performance matrixes of Research were Throughput End-to-end delay and routing overhead. The mobility of the nodes increased and increased the pause time of the nodes in the network in some scenarios. The application has been configured, profile and mobility configuration that run the simulation results. AODV performs best in all cases when the network size is high, increases the speed of moving nodes, changes the pause time, and TORA performs secondly and lastly DSR in terms of Throughput. In all the above environments, AODV has less delay as compared to DSR and TORA. DSR has more delay than TORA, and TORA performs worst in all scenarios by increasing the node mobility and pause time in terms of End-to-end delay. In the case of routing overhead, TORA sends the highest packets for route discovery into the entire network as compared to AODV and DSR. After TORA, AODV sends fewer packets for route discovery in-network, and lastly, DSR is sent with fewer amounts of data packets. This is valid for all environments where network size is consists of 15,30,45, and 60 nodes moving with the speed of 10,25, and 40 m/s with pause time 100,300, and 500 s. So, in terms of performance of these three routing protocols by using the parameters routing overhead, DSR performs best by sending the least amount of data packet for route discovery in the entire network after that AODV and then TORA sent a higher number of packets in the network. Hence, DSR performance is better than the other two reactive routing protocols. DSR depend on source routing in which every node is aware of the whole part of the network and the position of the nodes and the path followed by packet for data, so routing overhead is low compared to other nodes. In AODV, every node sent back a message for route discovery that increases the routing overhead. When some links are failed, multiple messages are sent to the sender and receiver nodes, increasing the network's burden and increasing routing overhead. TORA performs worst in routing overhead because it sends HELLO messages periodically to the entire network, increasing the routing overhead.

From this study, it is observed that there is no single protocol among shows overall higher performance. In some environments, one protocol may be shown best performance using the parameters Throughput, and others may be shown the best performance in Throughput and End-to-end delay. The choice of different routing protocols will depend on the environment, condition, and requirement used in the entire network.

As for future work, this research's direction will show the possibility of developing a new or altering the routing algorithm that will compensate for the MANET routing protocols' problems in this research area.

References

1. Conti M, Giordano S (2014) Mobile ad hoc networking: milestones, challenges, and new research directions. Commun Mag IEEE 52:85–96
2. Chlamtac I, Conti M, Liu JJ-N (2003) Mobile ad hoc networking: imperatives and challenges. Ad Hoc Netw 1:13–64
3. Macker JP, Corson MS (1998) Mobile ad hoc networking and the IETF. ACM SIGMOBILE Mobile Comput Commun Rev 2:9–14
4. Thouvenin R (2007) Implementing and evaluating the dynamic manet on-demand protocol in wireless sensor networks. University of Aarhus Department of Computer Science
5. Basagni S, Conti M, Giordano S, Stojmenovic I (2004) Mobile ad hoc networking: Wiley
6. Sun J-Z (2001) Mobile ad hoc networking: an essential technology for pervasive computing. In: 2001 international conferences onInfo-tech and Info-net. Proceedings. ICII 2001-Beijing, pp 316–321
7. Hoebeke J, Moerman I, Dhoedt B, Demeester P (2004) An overview of mobile ad hoc networks: applications and challenges. J Commun Netw 3:60–66
8. Rajabhushanam C, Kathirvel A (2011) Survey of wireless manet application in battle- field operations. Int J Adv Compu Sci Appl (IJACSA) 2
9. Kaur R, Rai MK (2012) A novel review on routing protocols in MANETs. Under-graduate Acad Res J (UARJ) 2278–1129
10. Kassim M, Rahman RA, Mustapha R (2011) Mobile ad hoc network (MANET) routing protocols comparison for wireless sensor network. In: IEEE international conference on system engineering and technology (ICSET), pp 148–152
11. Al-Omari SAK, Sumari P (2010) An overview of mobile ad hoc net- works for the existing protocols and applications. Int J Appl Graph Theor Wirel Ad hoc Netw Sens Netw 2
12. Ramanathan R, Redi J (2002) A brief overview of ad hoc networks: challenges and directions. IEEE Commun Mag 40:20–22
13. Kumar M, Mishra R (2012) An overview of MANET: history, challenges and applications. Indian J Comput Sci Eng (IJCSE). pp 121–125. ISSN 0976–5166

14. Panda I (2012) A survey on routing protocols of manets by using QoS metrics. Int J Adv Res Comput Sci Softw Eng 2
15. Pandey K, Swaroop A (2011) A comprehensive performance analysis of proactive, reactive and hybrid manets routing protocols. Int J Comput Sci Issues (IJCSI) 8
16. Rajput MV, Shah MN, Modi NK (2013) An analysis of multipath routing in mobile ad hoc networks. Int J Agric Innov Res 2
17. Garg P, Nagpal C, Bansal S (2013) Impact of random waypoint mobility model on hybrid routing protocols of scalable mobile Ad Hoc network. Int J Innov Res Dev 2
18. Saeed NH, Abbod MF, Al-Raweshidy HS (2012) MANET routing protocols taxonomy. In: International conference on future communication networks (ICFCN), pp 123–128
19. Goswami S, Joardar S, Das CB (2014) Reactive and proactive routing protocols performance metric comparison in mobile ad hoc networks NS 2. Int J Adv Res Comput Commun Eng 3

Detection and Identification of Malicious Node in Wireless Sensor Networks from Packet Modifiers and Droppers

Putty Srividya and Lavadya Nirmala Devi

Abstract The recent advancements in WSN are to produce economical and secure message broadcasting system. Packet droppers and modifiers are the common attacks in sensor networks. These attacks in WSN may interrupt the network and it is difficult to identify in multi-hop network. The present analysis aims to improve Quality of Service in Wireless Senor Network (WSN). This paper presents an effective method to identify the droppers and modifiers using ranking algorithms on the Directed Acyclic Graph (DAG) generated by the nodes. This also proposes a method that uses a mixture of broadcast cryptography technology and cluster key agreement to produce security to the information of the packets. This two-step authentication has improved the QoS in WSN. Provided better security than the existing /conventional methods.

Keywords Ranking algorithms · DAG · QoS · Packet droppers · Group key management · Public/secret key · Wireless sensor network

1 Introduction

The multi-hop sensor network modifies the secure transmission of packet-related data and common attacks that interfere with communication on the network. To retrieve a packet dropper, a detailed adoption step is Multiple Forwarding [1–5] at which each packet is transmitted on the different techniques of redundancy and therefore dropping packet approximately may not be tolerated. On the way to include modifiers of packet, utmost of the prevailing strategies [6–9] aims at filtering the messages changed and route over the variety of restricted hops. Those counter measures will then tolerate the dropping of packets and the attacks of modification. However, the intruders will remain offensive to the system while not be able to caught, for finding packet modifiers and droppers within the network system. It has been planned that the nodes could unceasingly screen the promoting behaviors of its respective neighbors [10–15] to see if their neighbors are misbehaving, and also the approach may be

P. Srividya (✉) · L. N. Devi
Department of ECE, University College of Engineering, Osmania University, Hyderabad, India

A. Ullah et al. (eds.), *Proceedings of International Conference on Information Technology and Applications*, Lecture Notes in Networks and Systems 350,
https://doi.org/10.1007/978-981-16-7618-5_56

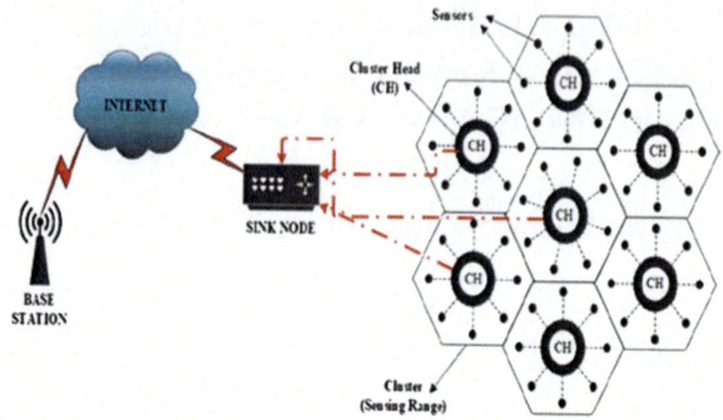

Fig. 1 Sensor network architecture

extended by victim nodes are misbehaving or not [15]. Wireless Network not solely satisfies the applying explicit requirements like timeliness, security, and irresponsibleness however additionally minimizes consumption of energy for extending life of the network. Although, previous studies have done to contemplate the trade-offs within the malicious attacker's presence. The mobile networks exceedingly monitor the atmosphere of node, which in turn observes the interested events that turns out collaborative and knowledge based over a sink, which might be an gateway in a well manner, node with memory, base station, or the user as shown in Fig. 1. A network sensing element was deployed usually on the associate unfriendly and unattended scenario for performing the observations and the knowledge assortment tasks. This will not offer a physical protection subject to uncompromising of the node. The opponent may employ various attacks for disrupting the communication. Among such attacks, two common measures like packets dropping and packets modifying, dropping or modify the packets by nodes that are meant to forward the data. In the hostile environment, as security is the major concern for mobile networks, the packet modifiers and droppers could be random. Investigations of such attacks are incredibly troublesome and typically not possible. This paper proposes a way to include security to the information given by source node packets by employing a cluster key management technique whenever the information is encrypted by cluster key and public/private key of a node.

2 Literature Review

The methods in detecting dropping packet attacks can be listed: neighbor monitoring approach, multipath forwarding approach, and acknowledgment approach. Multipath forwarding [1, 3] is a widely adopted countermove to relieve packet droppers, which is based on delivering redundant packets along multiple routes. Another procedure is to exploit the monitoring mechanism [4, 8]. The watchdog method was originally proposed to mitigate routing misbehavior in ad hoc networks [3]. It is then adopted to identify packet droppers in wireless sensor network [9, 10]. When the watchdog mechanism is deployed, each node monitors its neighborhoods promiscuously to collect the primary information on its neighbor nodes. A variety of esteem systems have been designed by interchange each node's firsthand observations, which are further used to quantify node's reputation. Based on the monitoring mechanism, the intrusion detection systems are proposed in [3, 4]. However, the watchdog method requires nodes to buffer.

3 Security for Data from Packet Droppers

The objective is to propose a simple effective theme to produce security to information. The information is conferred in source packets, so the data cannot be changed by attackers. The source packets will make use of the cluster key and public/private key of each node to write the information. Affected Node Draper/Modifier are considered as malicious nodes.

In this section the data formatting, node type topology is constructed by means of DAG (Directed Acyclic Graph). A tree routing is established from DAG and reports knowledge in turn to follow those routing trees. The tree routing is being employed for transmitting the packets to sink nodes. Each sender of packets to sink nodes adds a tiny low range of additional bits to the packets and jointly encrypts the packets victimization with the cluster key of the nodes. The cluster key are generated and maintained by the sink nodes. Public/private key is created by a fixed third party. When one round finishes, supported the additional carried bits in the packets received, the sink node turns a categorization rule of the node toward spotting nodes are considered to be dangerous i.e., the packet modifiers or droppers and the nodes at which the range of the node is suspiciously dangerous (i.e., supposed to be packet modifiers and droppers).

The reshaping of routing tree is made at each cycle. As a definite range of rounds of transmission has been approved, the decision can take enough information collected regarding the behavior of the nodes in several routing topologies. This information is employed to spot the dangerous nodes which may be droppers. In order to determine more number of the nodes which are in danger, the sink node runs the rule of heuristic ranking approach.

The next section describes a rule intended for DAG Association and packet transmission, by following the planned cluster key management rule. Initially, this focuses on the droppers of packet and does not override the node combination. Hence, it shows the approach for extending the theme conferred for handling the collusion of node and in turn observes various packet modifiers.

3.1 DAG Establishment and Transmission of Packet

The entire nodes within the network topology construct a DAG. The sink is aware of DAG and also the tree routing and in turn shares a novel key with every cluster. Every member of the cluster is aware of its cluster key. A trusty third party generates a public/private key shared by every node. When a node needs to forward/send the packet it gets attaches to the sequence of variety packets, cluster key generated by the sink, and also the public key of the receiver generated by the third party. Once the Associate Degree Transitional Innocent node obtains the packet it assigns an insufficient bit, forwarding route of the packet. On the other hand, misbehaving node receives as well drops the packets. The information within the source packets is still secure because the misbehaving node is unaware of the cluster key of the sender. By receiving the packet, sink node decrypts the packets by applying its public/private key and also the cluster key of the sender. Just in case collision knowledge cannot be altered because the misbehaving nodes cannot decrypt the cluster key of the sender. The sink identifies packet droppers supported the packet dropping magnitude relation to the information of the topology. Comprehensively, the theme includes the subsequent elements.

3.2 Transmission of Packets

A counter Cp is maintained by each node has track of the packets number such that, this will be sent thus so far. Once a u node features an information D item for reporting, this in turn composes and sends the packet subsequent to their parent nodes Pu: < Pu; at which Cp MOD Ns was that the number of packet sequence. Of the packet, Ru ($0 \leq \text{Ru} \leq \text{Np} - 1$) could be a number picked by node u randomly all over the formatting part system, and the metallic element is hooked up to the data for modifying the sink so as to search out trail on the packet that is forwarded. The term (X)Y signifies the results of X victimization encrypting key Y. Padding's padu, 0 and padu,1 square measure more to form all packets equal long. In the meantime, the sink will rewrite the packet still to search out the particular contented. So as to content those 2 objects at a same time, the square measure of padding's made as shown:

For the packets that were sent that is h hops away from sink, the padu length, 1 is $\log(\text{Np}) * (h - 1)$ bits which should be later represented, once a packet is forwarded for one hop, log(Np) info bits were excess and meanwhile, bits log(Np) are being shredded off.

Let the size of packet in maximum will be Lp bits, the node ID be lid bits and the data D be the LD bits. Padu; 0 must be $\text{Lp} - \text{Lid} * 2 - \log(\text{Np}) * h - \log(\text{Ns}) - \text{LD}$ bits, at which Lid * 2 bits were Pu and u fields of the packets, field Ru is the log (Np) long bits, filed padu, 1 is $\log(\text{Np})*(h - 1)$ bits long, and the Cp MOD Ns are log(Ns) long bits. On setting the padu, 0 for this values it ensures that the entire packets in network consists of similar Lp length.

3.3 Group Key Management

The agreement of Group key authorizes a user's collection to share a secret regular key through networks that are open and insecure. At that time, member could encrypt confidential message slightly at all by the common secret key and the cluster members entirely be capable to decrypt. Throughout this approach, an intimate intramural channel broadcast could also be recognized whereas not trying forward to a centralized key server to come back up with besides allocate secret keys to possible associates.

The security of implementation comprises three features. Initially, the secure transmission matter tends to formalize the cooperative remote groups, while at the same time the core must forcibly and efficiently transmit any hidden material from one to several channels. It could be impractical meant for a sender of United Nations remote agency is additionally in an extremely altogether different geographic region. This case is any worsened if the sender is dynamic or otherwise mobile. On the alternative hand, cryptography broadcast licenses senders to broadcast external to members of a non-cooperative planned cluster whereas not necessitating the dispatcher to turn with receivers before secret contents transmission, but it depends on a key server centralized to come back up with and allocate secret keys intended for each member of cluster. It means that (i) a confidential broadcast channel already is recognized, confidential varied channels unicast from server of the key to all receiver potential have to be bound to be formed, and (ii) the holding key server for each receiver can peruse the communications altogether and must be trustworthy altogether through a little possible sender and collectively the cluster members.

Members of the collection can be deleted or removed as follows. A member replacement might also be a portion of the sender through recovering the overall communal key of the user and introducing it to the overall public key chain of present receiver set. Through invoking the member repeatedly adding development, a dispatcher can combine couples of receiver sets interested in one cluster. Likewise, through appealing the operation of deletion member repeatedly, the sender could divide one set of receivers into a pair of collections. Every partitioning and merging might also be potency over. This unit displays the member deletion from the cluster receiver. At that time, the sender and the residual receivers jointly have to be bound to smear this alteration to their attendant cryptography and writing events.

Basic proposed system model:

We deliberate a group of N users composed, designated by $\{U1, \ldots, UN\}$. A sender wants to convey secret communications to a subset S receiver of the users N, wherever the S size is $n \leq N$. The problem is in what way to permit the dispatcher to securely and efficiently finish the transmission with the subsequent restraints:

1. This is hard to deploy authority of fully trusted key generation through users entirely and possible dispatchers in open network settings.
2. From the receivers the communication to sender is restricted, e.g., communication setting in the battlefield.

3. N should be large actually and up to millions, for example, in vehicular ad hoc networks.
4. Both the receiver and the sender groups are active because of ad hoc communication. Consistent with the application situations, some justifying structures might be subjected for resolving the problem:
5. Usually, N is a minor or average value, e.g., less than 256.
6. The receivers are communicated and cooperative through effectual local channels (broadcast).
7. An authority of partially trusted, e.g., infrastructure of a public key, is accessible to validate the receivers and the dispatchers.

We address the mentioned disadvantage by formalizing an improved key management paradigm cited as cluster key agreement primarily based on broadcast coding. The system style is illustrated in Figs. 4 and 5. The potential receivers area unit connected in conjunction with economical lay to rest cluster connections via communication infrastructures, they will boot connect with heterogeneous networks, each receiver contains a public/secret key attempt. The general public secret key certified by a certificate authority but the key is offered alone by the receiver. A remote sender can retrieve below certain constraints and tend to watch that the prevailing key management approaches do not offer effective solutions to the present disadvantage. On one hand, cluster key agreement provides an economical resolution to secure internal communication but, for a remote sender, it desires the sender to at constant time keep on-line with the cluster members and validates the credibleness of the general public key by checking its certificate that is significant (Fig. 2).

4 Algorithm

The formula in Fig. 3 specifies the secret writing procedure by victimization the ancient public/private key and cluster key management agreement. Within the ancient cluster, key management protocol the keys generated and maintained by a sure third party. Because the third party is aware of the key of all members, it decodes and bypasses messages transmitted over the network. However, getting the key to a fully secured third party can be problematic. The key area unit generated by the sync node in the Secret Write technique is because the routing tree is established by the sync node. Sync node generates cluster keys. The third party component of the technology used is fixed and is known about the common public key of each node and is implemented due to the Public Key Infrastructure (PKI) in open networks. Sync removes these nodes from the network to promote security; since the sync node manages the behavior of each node to remove dangerous nodes that could be droppers/modifiers the sink will delete these nodes from the network to boost the safety. The sink node will add or delete any node among the cluster. Within the planned technique, the sender has to be compelled to recognize solely the general public key of the receiver node from the third party and doesn't need any direct

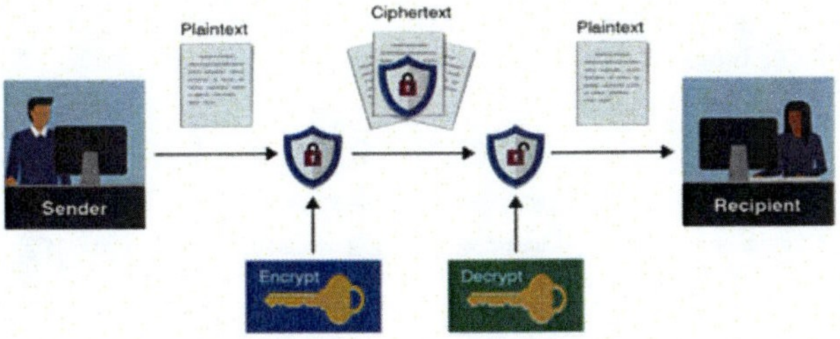

(a) Encryption and Decryption Architecture

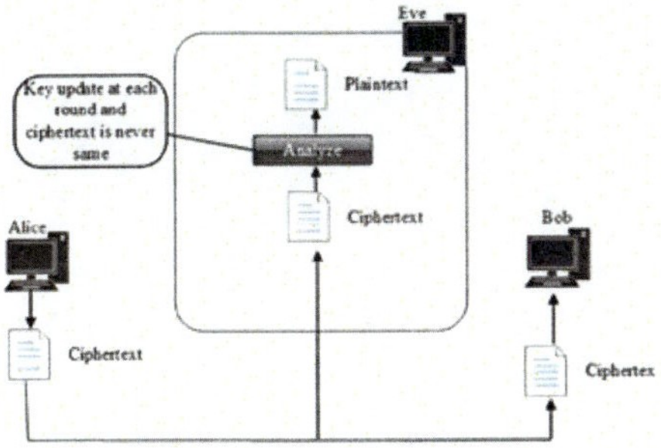

(b) Scenario of Cipher text

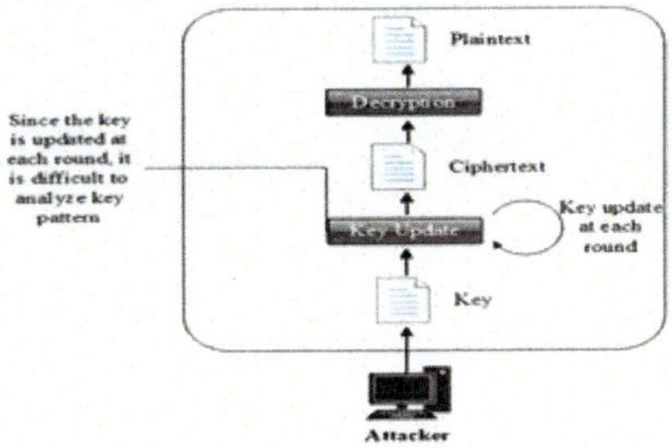

(c) Scenario of Key management technique

Fig. 2 **a–c** System architecture for encryption/decryption

KEYGEN (τ): generate a public-private key pair
1. Based on parameter τ security, this in turn estimates a triple element (q_1, q_2, *E*), at which *E* is elliptic curve points set that forms a group of cycle. The *E*, or (*E*) order is n at which n relates of the q1 and q2; q1 and q2 product are huge primes.
2. Fortuitous select two generators (i.e., base points) G, U, where ord(G) = ord(U) = n.
3 Compute point H= q_2 * U such that ord(H) = q_1 .
4. Select parameter T as the maximum plain text boundary and T < q_2
5. Write the known key: *PK*= (n, *E*, **G,** H, T).
6. Write the non-public key: *SK*= q_1.
ENC (*PK, M)*: Message encryption on *M* by public key *PK.*
a. Verify if message $M \epsilon \{0, 1\ldots T\}$.
b. Chance select $R \epsilon \{0, 1\ldots n\text{-}1\}$.
c. Make the cipher text *C* as $C = M * G + R * H$, where G, Hϵ *PK.*
d. Write *C.*
DEC (SK, *C)*: Decryption Message on *C* through private key *SK*
i. Calculate $\log_{\hat{U}}(q_1 * C) = \log_{\hat{U}}(q_1 *(M * G + R * H)) = \log_{\hat{U}}(M * q_1 * G) = M$ where $\hat{U} = q_1 * G$.
j. Write M.

Fig. 3 Encryption/decryption algorithm

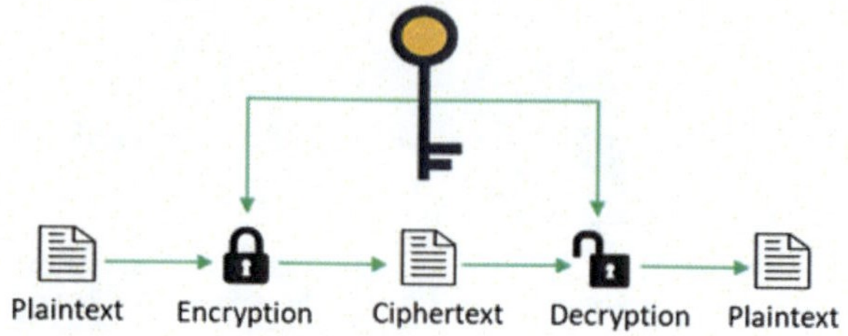

Fig. 4 Symmetric key encryption technique block diagram

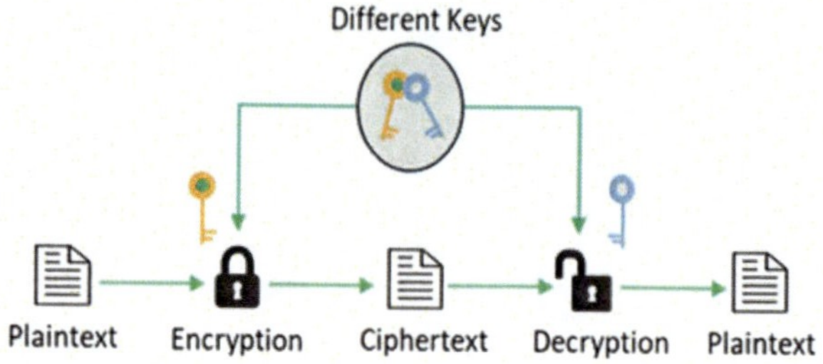

Fig. 5 Public key encryption technique block diagram

communication with the receiver, but rarely should send secret messages to any or all employees members. Figure 4 depicts the secret writing of information wherever every packet is appended with a bunch of keys and the public/private key of every node to get cipher text. Figure 5 depicts the decoding technique at the receiver, which uses the general public key of the sender obtained from the third party and its cluster key to decode the message, just in case of droppers or collisions no node will decode the info because the cluster secret is not shared with any third party.

5 Results and Discussions

The motivation of this paper is to produce increased security to the information from packet droppers/modifiers. This can be achieved by victimization of cluster key and public/private key secret writing technique. These keys are allotted by the sink node. The sink node is not solely responsible to blame the key generation and distribution. However, additionally within the construction of DAG and reconstruction/reshaping of the tree. This determines dangerous nodes which may be droppers/modifiers within the network and QoS support for network turnout, end-to-end latency, and information magnitude delivery relation with in the objectives. Extensive simulations are carried out using NS2 for the network with the thirty-five nodes with a transmission range of 150 m. Figure 6 shows the simulation setup for thirty-five nodes. A member replacement might also be a portion of the sender through recovering the overall communal key of the user and introducing it to the overall public key chain of the present receiver set. Through invoking the member repeatedly, a dispatcher can combine couple of receiver sets interested in one cluster. Likewise, through

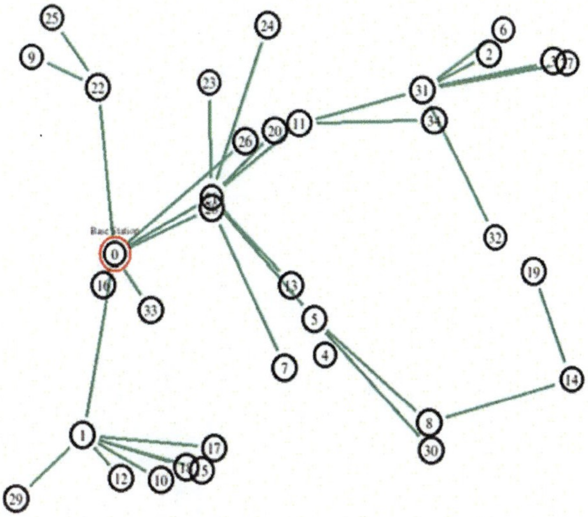

Fig. 6 Nodes tree routing

repeating the operation of deletion member repeatedly, the sender could divide one set of receivers into a different pair of collections. Every partitioning and merging might also be potency over. This unit displays the member deletion from the cluster receiver. At that time, the sender and the residual receivers jointly have to be bounded to fulfill this alteration to their consecutive encryption and decryption techniques.

Figure 6 offers the network produced by a structure of tree routing intended for 35 nodes where node 0 is measured as the base station. Figure 7 represents the packet modifiers and droppers that are founded by estimating PDR at each node and thus are treated as malicious nodes.

PDR is defined as the number of packets received at the sink node to the number of packets transmitted. From the above Fig. 9, it is observed that the time for PDR on an average is 2 s with a time interval of 10. Which indicates that the performance has been improved by eliminating malicious (Fig. 8).

The throughput is calculated for every 100 s with an interval of 5 cycles. From the above Fig. 10, it is observed that the throughput remains steady once the malicious nodes are eliminated.

The Fig. 11 gives the representation of time interval for estimating the end-to-end delay at each 50 s with 5 periodic intervals. In spite of overhead due to cryptography encryption and decryption with public/private key, the results show that delay is minimum.

Figures 12 and 13 represent detection rate and false positive rate of the malicious nodes with respect to the density of the network. Above simulation result shows that the results are quite better with the increase detection rate of malicious nodes and decrease in the false positive rate.

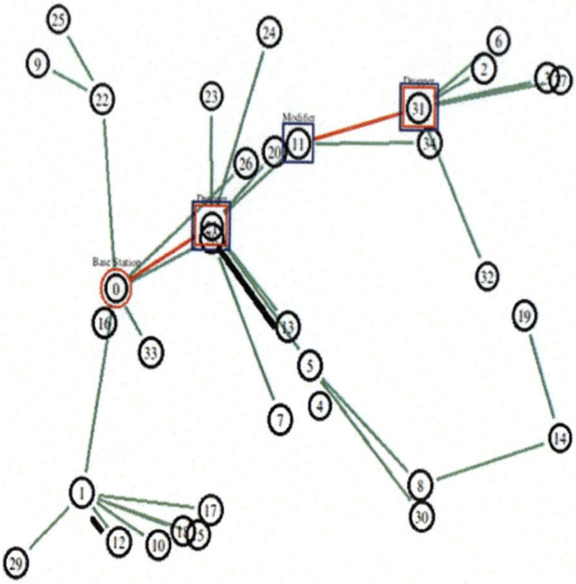

Fig. 7 Detection of packet droppers and modifiers

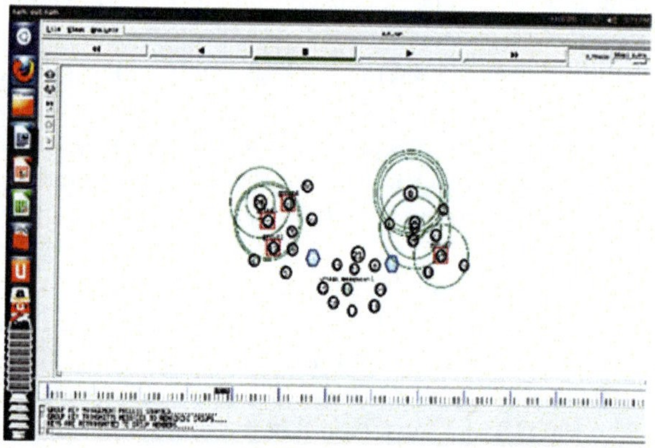

Fig. 8 Keys that are transmitted over the groups

6 Conclusion

The proposed technique provides a simple and effective method to identify packet droppers and modifiers. The malicious nodes are identified and removed from the network, routing tree structure dynamically to reconstruct each time in the network using heuristic ranking algorithm with DAG. To improve security significantly in the network, each packet is encrypted with cryptography and cluster key agreement technique between the nodes. Extensive simulations have been carried out to observe the effectiveness of this proposed method. Proposed method is outperforming the conventional methods in terms of security and QoS.

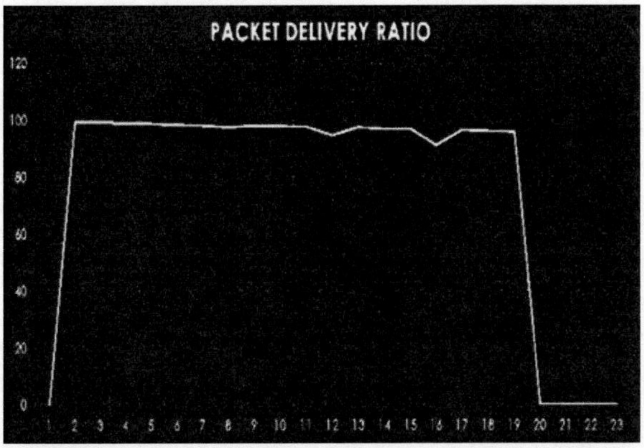

Fig. 9 PDR packet-delivery-ratio

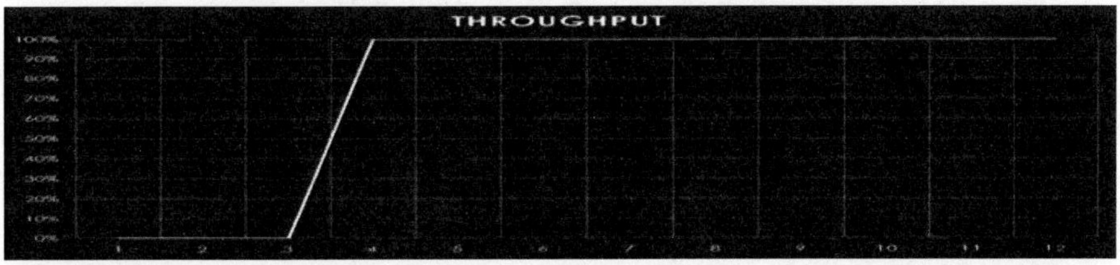

Fig. 10 Efficiency—throughput

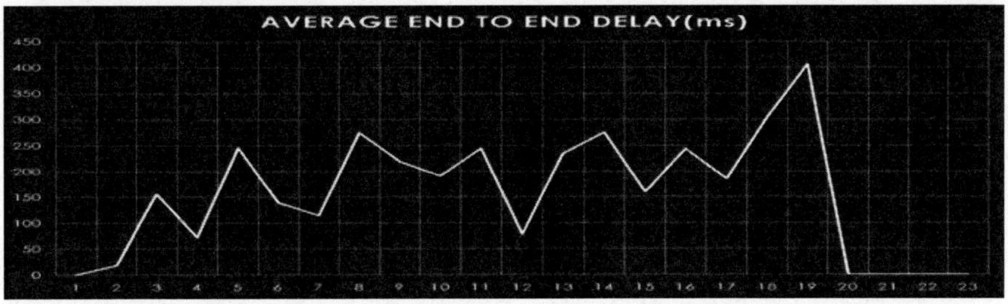

Fig. 11 End–end–delay average

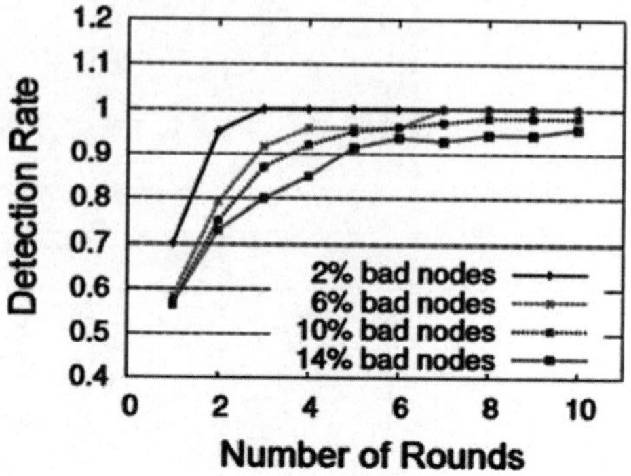

Fig. 12 Detection rate

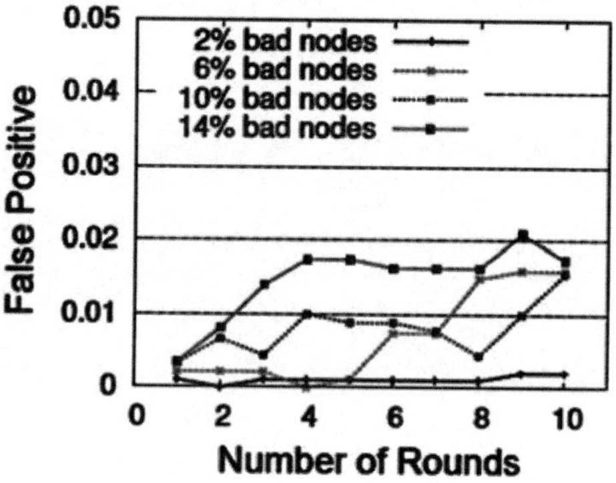

Fig. 13 False positive rate

References

1. Wang C, Feng T, Kim J, Wang G, Zhang W (2012) Catching packet droppers and modifiers in wireless sensor networks. IEEE Trans Parallel Distrib Syst 23(5):835
2. Chan H, Perrig A (2003) security and privacy in sensor networks. Computer 36(10):103–105
3. Bhuse V, Gupta A, Lilien L (2005) DPDSN: detection of packet-dropping attacks for wireless sensor networks. In: Proceedings of fourth trusted internet workshop
4. Kefayati M, Rabiee HR, Miremadi SG, Khonsari A (October, 2006) Misbehavior resilient multi-path data transmission in mobile ad-hoc networks. In: Proceedings of fourth ACM workshop security of ad hoc and sensor networks (SASN '06), pp 91–100. https://doi.org/10.1145/1180345.1180357
5. Khalil I, Bagchi S (2008) MISPAR: mitigating stealthy packet dropping in locally-monitored multi-hop wireless ad hoc networks. In: Proceedings of fourth international conference security and privacy in communication networks (SecureComm'08). Corpus ID: 9726406. https://doi.org/10.1145/1460877.1460913
6. Ye F, Luo H, Lu S, Zhang L (2004) Statistical en-route filtering of injected false data in sensor networks. In: Proceedings of IEEE INFOCOM IEEE JOURNAL ON SELECTED AREAS IN COMMUNICATIONS, vol 23, no 4, APRIL 2005, pp 839–850
7. Zhu S, Setia S, Jajodia S, Ning P (2004) An interleaved hop-by- hop authentication scheme for filtering false data in sensor networks. In: Proceedings of IEEE symposium security and privacy, ACM Journal Name, vol V, no N, Month 20YY, pp 1–32
8. Ganeriwal S, Balzano LK, Srivastava MB (2008) Reputation-based framework for high integrity sensor networks. ACM Trans Sens Netw 4(3):1–37

9. Li W, Joshi A, Finin T (2010) Coping with node misbehaviors in ad hoc networks: a multi-dimensional trust management approach. In: Proceedings of 11th International conference on mobile data management (MDM10). Corpus ID: 15932845. https://doi.org/10.1109/MDM.2010.57
10. Michiardi P, Molva R (2002) Core: a collaborative reputation mechanism to enforce node cooperation in mobile ad hoc networks. In: Proceedings of IFIP TC6/TC11 sixth joint working conf. command multimedia security: advanced communications and multimedia security. Corpus ID: 7925947. https://doi.org/10.1007/978-0-387-35612-9_9
11. Li Q, Rus D (2004) Global clock synchronization in sensor networks. In: Proceedings of IEEE INFOCOM. Corpus ID: 2747708 0-7803-8356-7/04/$20.00 (C) 2004 IEEE. https://doi.org/10.1109/INFCOM.2004.1354528
12. Sun K, Ning P, Wang C, Liu A, Zhou Y (2006) Tinysersync: secure and resilient time synchronization in wireless sensor networks. In: Proceedings of 13th ACM conference on Computer and Communications Security (CCS'06) CCS'06, October 30–November 3, 2006, Alexandria, Virginia, USA. Copyright 2006 ACM 1-59593-518-5/06/0010 ...$5.00
13. Song H, Zhu S, Cao G (2007) Attack-resilient time synchronization for wireless sensor networks. Ad Hoc Netw 5(1):112–125
14. Xiao B, Yu B, Gao C (2007) Chemas: identify suspect nodes in elective forwarding attacks. J Parallel Distrib Comput 67(11):1218–1230
15. Buchegger S, Le Boudec J (June, 2002) Performance analysis of the confidant protocol. In: Proceedings of ACM MobiHoc MobiHoc '02: Proceedings of the 3rd ACM international symposium on Mobile ad hoc networking & computing, pp 226–236. https://doi.org/10.1145/513800.513828

Digital Media and Education

E-learning Methodologies Involving Healthcare Students During COVID-2019 Pandemic: A Systematic Review

Carla Pires and Maria José Sousa

Abstract *Introduction* COVID-2019 pandemic has catalyzed the shift from presential education to online teaching. Study aim: to carry out a systematic review on e-learning methodologies involving healthcare students during COVID-2019 pandemic. *Methods* The Systematic Reviews and Meta-Analyses (PRISMA) checklist and flow diagram were followed. Keywords: (e-learning or "online classes" or "online education") and (COVID or SARS) and (medicine or nurse or pharmacist or physician or health) and (students or undergraduates or trainee). Databases: PubMed, SciELO, B-ON, DOAJ, and Cochrane Library. Inclusion criteria: review studies. *Results* From the 23 identified reviews, only 10 were selected: PubMed (n = 8), SciELO (n =2), B-on (n = 0), DOAJ (n = 0), and Cochrane Library (n = 0). *Discussion* All identified reviews were about online medical education, with a special focus on e-learning of anesthesia, surgery, and dentistry. The main obstacle to online e-learning for medical undergraduates was the impossibility of practical presential classes. *Conclusion* In general, diverse electronic platforms (e.g., social media platforms), telemedicine, flipped classroom, and multimodal systems may be applied to teach online medical students. Importantly, new innovative teaching strategies are emerging for the online training of clinical skills (e.g., online simulated patients/avatars). Online teaching is expected to become the new normal for activities that do not include interactions with patients. Medical curricula should be redesigned to address the needs of new social and scientific challenges.

Keywords Online medical education · e-learning · e-teaching · Health care students · New digital education strategies · COVID-2019

C. Pires (✉)
CBIOS - Universidade Lusófona's Research Center for Biosciences and Health Technologies, Campo Grande, 376, 1749-024 Lisbon, Portugal
e-mail: p5558@ulusofona.pt

M. J. Sousa
ISCTE-IUL—Instituto Universitário de Lisboa, Lisbon, Portugal
e-mail: maria.jose.sousa@iscte-iul.pt

A. Ullah et al. (eds.), *Proceedings of International Conference on Information Technology and Applications*, Lecture Notes in Networks and Systems 350,
https://doi.org/10.1007/978-981-16-7618-5_57

1 Introduction

Globally, schools were required to quickly switch from presential classroom teaching to online learning due to COVID-2019 pandemic. The creation of new online teaching formats is demanding, time-consuming, and requires the existence of an appropriated Information Technology (IT) infrastructure. Training opportunities and psychological support to students and faculty members should be offered, since the acquisition of new digital skills (e.g., training in didactic methods and/or IT skills) or controlling anxiety or fear of COVID-2019 may be necessary in some cases [8, 10, 12].

A national survey about online training covered 2721 medical students from 339 medical schools of UK (May 4, 2020 to May 11, 2020). According to the findings of this survey, the main identified benefit of online training was the flexibility of platforms and the main barriers were family distractions and problems with internet connectivity [8]. Communication interactions, student assessment, use of technology tools, online experience, pandemic-related anxiety or stress, time management, and technophobia were identified as the main challenges of online health training in other study [26].

In general, medical teaching involving direct contact to patients was compromised during pandemic, since health disciplines require profound knowledge and practice skills. For instance, presential activities students/residents of oncologic surgery or dentistry were disrupted [11, 18]. Classic training with face-to face interactions (lectures, tutorials, laboratory sessions, patient contact in clinical settings, and/or rotations) were replaced by online learning. Additionally, the education of undergraduates of pharmacy or nursing faced a significant reduction of experimental rotations (e.g., practices in hospitals or clinics) [3, 12, 13, 21, 27]. Thus, technological developments or clinical and surgical simulations and virtual reality are needed for the online training of medical students, such as new digital substitutes of patients (e.g., simulated patients or avatars) [4, 11, 12]. Avatars were successfully applied to train the disclosure of a cancer diagnosis to a patient or to check if patients authorize (or not) a certain nursing procedure. Communication competences of health students were increased in both studies [4, 9].

Co et al. [5] developed a new web-based surgical skill learning session (intervention group), which was compared to a face-to-face teaching group (control group). Students' performance of surgical skills was comparable between both groups, which support the potential successful capacity of online teaching methodologies for presential clinical activities [5]. Simulation-based training tools for surgical skill acquisition were also applied in an ophthalmology curriculum [20].

The importance of integrating online platforms on educational practices after pandemic is recognized by diverse studies [8, 26]. For instance, pharmacy educators have suggested the expansion and accreditation of e-learning/teaching facilities in future, the introduction in pharmacy curricula of computer science and data science courses to strengthen undergraduates' competencies and skills in information and communication technologies (ICT), and the inclusion of laboratory work in virology to study actual and emerging virus [16].

Challenges to online health education should be evaluated and discussed, considering the actual pandemic scenario, as well as the new paradigm of health education after the COVID-2019 pandemic [20, 26]. In this sense, the study objective was to carry out a systematic review on e-learning methodologies involving healthcare students during COVID-2019 pandemic.

2 Methods

A systematic review on e-learning methodologies involving healthcare students during COVID-2019 pandemic was carried out. The Systematic Reviews and Meta-Analyses (PRISMA) checklist and flow diagram were followed [23].

2.1 Research Question

What type of new strategies on e-learning and/or e-teaching of health students were adopted during COVID-2019 pandemic?

2.2 Timeframe and Keywords

Positively, the search was carried out without time limitations on 22-4-2021. Overall, 13 keywords were selected. The searched stream was [(e-learning or "online classes" or "online education") and (COVID or SARS) and (medicine or nurse or pharmacist or physician or health) and (students or undergraduates or trainee)]. This online search strategy was based on [23] requisites [23]. Purposively, synonyms or related terms to each keyword were used aiming at selecting the maximum number of studies. The applied search stream is compliant with the searching strategies of Cochrane Training, i.e., "Search strategies should avoid using too many different search concepts, but a wide variety of search terms should be combined with OR within each included concept" [17].

2.3 Searched Databases

Five scientific databases were conveniently selected: PubMed, SciELO, B-ON, DOAJ, and Cochrane Library. These databases comprise a significant number of updated and peer-reviewed papers, which have justified their selection [2, 6, 7, 25, 28]. Gray literature and other registers or resources were not considered to avoid the inclusion of potential inconsistencies or less robust studies.

2.4 *Inclusion Criteria*

Only, review studies, systematic reviews, or meta-analysis on e-learning methodologies involving healthcare students during COVID-2019 pandemic were selected. The inclusion of review studies aimed at evaluating and discussing the maximum number of studies about the present topic. Reviews written in English, Portuguese, French, or Spanish were fully analyzed. Regarding the reviews written in other languages (e.g., Chinese), only the English abstracts were considered to avoid translation costs.

Inclusion criteria followed PICOS, i.e., participants, interventions, comparisons, outcomes, and study design criteria (PICOS) [19]: participants (P) were health undergraduates (e.g., medicine, pharmacy, or nursing students) and the respective teaching tutors, interventions (I): any online learning and/or teaching strategy; comparisons (C): studies with similar objectives, research designs or methodologies were compared, the outcome (O): any students' benefit and study design (S): review studies, including all types of study designs were considered (e.g., qualitative, quantitative, experimental, etc.).

2.5 *Exclusion Criteria*

All qualitative, quantitative, or experimental studies were excluded, since the present work is a systematic review of reviews (e.g., systematic reviews, meta-analysis, or other type of reviews) on e-learning methodologies involving healthcare students during COVID-2019 pandemic. Reviews on other topics were excluded.

3 Results

Twenty-three reviews were identified: PubMed ($n = 13$), SciELO ($n = 3$), B-on ($n = 7$), DOAJ ($n = 0$), and Cochrane Library ($n = 0$), with only 10 reviews selected (8 reviews in PubMed and 2 reviews in SciELO; $n = 10$) (Fig. 1).

Particularly, systematic reviews or meta-analysis were not identified and/or selected, i.e., only non-structured/narrative reviews were selected ($n = 10$). This may have happened because the number of studies on the present topic is still limited, since COVID-2019 only was declared by World Health Organization (WHO) on December 31, 2019 [22].

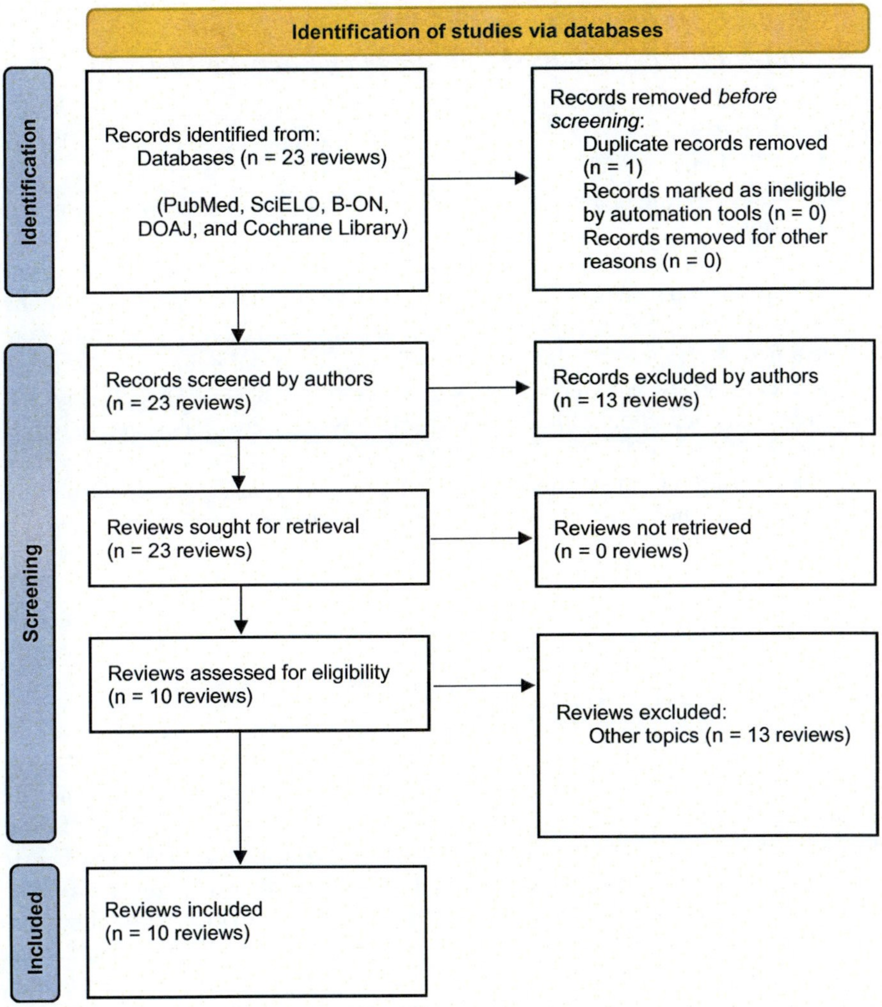

Fig. 1 PRISMA 2020 flow diagram for new systematic reviews: selection of review studies on e-learning methodologies involving healthcare students during COVID-2019 pandemic [24, 23]

3.1 *Classification of the Selected Reviews Per Studied Topic*

The topics of the 10 selected reviews on e-learning methodologies involving health undergraduates were, as follows: 4 (medical education) [3, 15, 22, 27], 1 (anesthesia students) [29], 2 (surgery students) [1, 11], and 3 (dental education) [10, 14, 18]. Interestingly, only medical students were enrolled in the selected reviews, i.e., reviews involving other type of health professions, such as pharmacy or nursing were not found.

3.2 Characteristics of the Selected Reviews: Country and Number of Analyzed Studies Per Each Review

The 10 selected reviews were, respectively, from Brazil ($n = 3$) [3, 18, 27]; USA ($n = 2$) [14, 15], India ($n = 2$) [1, 22], collaborations between diverse countries (UK, Ireland, USA, Australia, Hong Kong, Santiago, Chile, and South Africa) ($n = 1$) [29], Pakistan ($n = 1$) [10] and Germany ($n = 1$) [11].

The number of evaluated papers per each review was estimated based on the number of discussed papers, since systematic reviews or meta-analysis on e-learning methodologies involving healthcare students during COVID-2019 pandemic were not identified. Averages and standard deviations (SD) of the number of discussed papers in each review per country: USA (average = 53; SD = 41); Germany (27; only 1 review); international collaborations (21; only 1 review); Pakistan (20; only 1 review); Brazil (average = 19.3; SD = 10.8); and India (average = 18.3; SD = 7.4). These findings seem to support that the number of reviews and studies on the present topic is limited, which reinforce the relevance of the present systematic review.

4 Discussion

4.1 Online Medical Education

Diverse topics were discussed in the 4 identified reviews, such as challenges of health education in this new pandemic scenario [22, 27], the most used e-learning platforms and their advantages [3] or the application of social media platforms in classes [15].

Suddenly, teachers were required to select appropriate e-learning platforms, develop new educational materials, and/or plan online activities. Disciplines on how to manage a pandemic were missing in medical curricula (e.g., ventilatory therapy). Anatomical teaching with cadavers and clinical practices were interrupted, while digital and online resources started to be applied to teach medical procedures and surgical techniques. Some studies reported the integration of medical students in hospitals during COVID-2019 pandemic [27]. Diverse challenges of e-learning methodologies involving medical students were identified, as follows: (1) lack of skills (e.g., engagement of teachers); (2) time management (e.g., training on the implementation of online classes); (3) lack of infrastructure/resources (e.g., low-income countries); (4) poor communication (e.g., establishment of teams at medical schools); (5) negative attitudes (e.g., institutional culture); and (6) student engagement/concentration (e.g., rotations between synchronous and asynchronous modes; chats and forums) [22].

Zoom® and Google® (Google Meet) were the most frequent used e-learning platforms. All the evaluated e-learning platforms were available in different devices (laptops, smartphones, and tablets). Synchronous and asynchronous classes were conducted (e.g., recorded lectures or live interactions). Recorded lectures present some advantages, such as flexible attendance or repeated consultation by students. A well-integrated trained team is required to provide quick solutions and/or support with regard to the use of digital tools [3].

Social media platforms showed potential to be used as relevant medical education tools: Facebook (e.g., communication between faculties and students), Twitter (e.g., conferences), Instagram (e.g., sharing of images as training resource), YouTube (e.g., explanation of complex concepts), WhatsApp (e.g., discussion of clinical cases), and/or podcasts (e.g., presentation of topics from different medical specialties). Some medical schools are already applying these platforms in both formal and informal education settings. Social media platforms seems to offer unique communication capabilities, but should be carefully introduced in medical education. Guidance on the application of social media platforms is lacking (e.g., supervision of students learning outcomes during their implementation) [15].

4.2 Anesthesia Students: e-learning Adoption

Only one review on online training of anesthesia students was identified. In a similar way to what happened with other medical specialities, e-learning, and digitalization were adopted (e.g., online trainee meetings or tutorials; problem-based learning groups; presentation of anatomical models or simulations). Learning outcomes of anesthesia students were negatively impacted in practical classes (e.g., reduced caseload, diminution of sub-specialty experience, and supervised procedures). Positively, National training bodies and medical regulators have demonstrated flexibility and innovation, regarding the new e-learning adopted methodologies [29].

4.3 Surgery Students: e-learning Adoption

Online teaching opportunities in surgery settings were analyzed in two reviews [1, 11]. Among the identified online teaching opportunities were the application of (1) social media (e.g., Facebook), (2) flipped classroom (e.g., pre-recorded video lectures), (3) telemedicine as a substitute for didactic clinic, and (4) multimodal systems for learning basic surgical skills (e.g., visual, auditory, and tactile components to teach suturing) [1].

Despite residents of surgical oncology may practice on simulation labs (e.g., robotic simulators) or see web videos, diverse disadvantages of online training were identified, such as no hands-on experience; no cadaver dissection; no training of real life or complicated situations; less time spend in operating room; lack of intraoperative experience and practical skills; or lack of surgical exposure. In order to overcome these obstacles, videos of surgeries may be posted in online platforms (e.g., VISTA), with the reception of peer feedback and coaching. In addition, web-based platforms and virtual learning resources were helpful tools, such as Websurg®, deutsche Gesellschaft für Chirurgie (DGCH), or Journal of Medical Insight (JOMI) [11].

4.4 Online Dental Education

Three reviews on dental online education were selected. Classroom activities in faculties were suspended, with the exception of dental emergencies. All dental facilities had to be equipped with appropriate protective equipment due to the generation of aerosols and droplets in routine dental procedures during COVID-2019 pandemic. The main challenges of online dental education were the preclinical simulations of laboratory activities (e.g., use of mannequins), cybersecurity and the application encrypted systems [18].

Besides the common online education strategies (asynchronous and synchronous teaching; blended learning, flipped classrooms or problem-based learning), other supportive technologies were used, such as VoiceThread (VT) (a cloud-based program that integrates videos and PowerPoint in a presentation), quizzes (e.g., EDpuzzle), WebEx, or the administration of online exams (e.g., Canvas, ExamSoft, ProctorU, Honorlock, Respondus Monitor, and Examity) [14]. Diverse free or paid e-learning platforms were identified in the field of dental education, such as mobile apps (e.g., prosthodontics images), video conferencing networks and social media: Instagram® (e.g., lives), Facebook®, WhatsApp® or email (e.g., oral diagnoses), Telegram®, YouTube® or other (Pinterest®), Moodle®, Zoom®, Jitsi®, WebEx®, Microsoft Teams® or Google Classroom®, and Hangout® [10, 18].

New emergent online teaching strategies were identified in dentistry, such as virtual reality/augmented reality (VR/AR)-based simulation devices (e.g., Simodont, DentSim, Periosim, etc.) or haptic technology [10]. Virtual reality systems and haptic technology are not portable, which limit their application online. The investment in these technologies is recommended to improve students' psychomotor skills [14]. Additionally, other electronic resources may be available to all students and professors, such as the application of an automatic digital slide scanner to collect photos of dental pathologies, virtual patient (VP)-based learning (e.g., simulated patients, since online simulation with dental mannequins is extremely difficult), automatic digital slide scanner or sharing of clinical and/or histopathological images [18].

E-health infrastructures and/or services (e.g., teledentistry) were incipient in dental education [18]. Finally, the introduction or reinforcement of some topics was recommended in dentistry curricula, such as training in teledentistry (e.g., triaging of patients), crisis management, inter-professional education, or extramural rotations [10].

4.5 Limitations of the Selected Reviews

The main drawback of online e-learning was the impossibility of practical classes for medical education, with a limited number of studies purposing innovative solutions. The suitability of e-materials (e.g., readability) or the duration of the online classes were not analyzed. No study evaluated the e-literacy of professors and students. Controlled, longitudinal, and multicentric studies on the present topic were almost inexistent. Review studies were focused on medical students. In a similar way to the findings of Santos et al. [27], the restrictions on net access, the adoption of digital tools in developing countries, and the impact of digital classes on the relationship between medical students and professors or patients were not evaluated. Importantly, interpersonal relationships are essential to medical training, which seems to have been devaluated [27].

4.6 Study Strengths

This is the first systematic review of reviews on e-learning methodologies involving healthcare students during COVID-2019 pandemic, as far as we know. The number of published reviews on the present topic is limited. Additionally, the selected reviews discussed a restricted number of studies, which is indicative of a lack of research on the present topic. Diverse studies confirmed that e-learning methodologies will be more frequently used after pandemic. These findings support the relevance of the present systematic review.

4.7 Future Research

Representative and longitudinal studies are recommended. E-learning strategies enrolling different health professionals and collaborative practices are suggested (e.g., pharmacy or nursing students). Guidelines and good practices on e-learning and e-teaching should be developed since studies were heterogeneously developed and implemented. E-learning platforms may be optimized through usability tests. 5 G may be applied to teach practical medical classes (e.g., augmented reality). The adoption of digital tools in developing countries or the impact of digital classes on the relationship between health students and professors (interpersonal relationships) should also be investigated.

5 Conclusions

COVID-2019 pandemic has catalyzed the shift from presential education to online teaching, with the suppression of previous barriers and resistances to the implementation of this model. Staff and students should be kept motivated by school administrations and trained to perform online learning. Cybersecurity and encrypted systems are crucial for online training [18].

In general, in-person education activities were interrupted in medical schools. The online shift from presential to online education was feasible. For instance, diverse electronic free or paid platforms were identified as adequate for e-learning activities (e.g., social platforms); school administrators were required to ensure the continuity of education with respect for regulations and the safety of staff and students (e.g., social distancing and bio safety); and telemedicine, and multimodal systems were used to teach medical students online. Common online education strategies were applied in medical education (e.g., asynchronous, and synchronous teaching; blended learning, flipped classrooms, or problem-based learning) [14].

Besides, e-learning platforms may be successfully applied in medical teaching, in-person clinical educational practices (e.g., surgery) cannot be fully substituted [1, 18, 29]. New innovative e-learning strategies are emerging to teach clinical skills [14, 18], such as automatic digital slides, international video platforms, simulated patients (e.g., avatars), simulated robotic training, or haptic technology. These technologies are very expensive and usually are not available in developing countries [4, 10, 11]. Despite, online education may be a poor substitute for in-person clinical activities, such as anesthesia, surgery, or dentistry (all hands-on speciality), e-learning may become the new normal in future for the activities that do not include interactions with patients [29].

Medical curricula should be updated and rethought to meet the needs of new social and scientific challenges (e.g., virology training) [16]. Schools are required to plan the return to clinical activities (e.g., investment in new infrastructures) [18]. The sharing of resources between schools are recommended (e.g., digital images, videos, or didactic materials) [14]. These findings are likely to be applicable in other healthcare educational settings (e.g., nursing or pharmacy).

References

1. Agarwal PK (2020) A combined approach in prolonged COVID-19 pandemic to teach undergraduate surgery students-future primary care physicians. J Family Med Prim Care 9(11):5480–5483. https://doi.org/10.4103/jfmpc.jfmpc_1129_20
2. B-on (2021) Biblioteca do Conhecimento online. [Online knowledge library]. Available online: https://www.b-on.pt/. Accessed on 4 May 2021
3. Camargo CP, Tempski PZ, Busnardo FF, de Martins MA, Gemperli R (2020) Online learning and COVID-19: a meta-synthesis analysis. Clinics 75:e2286. Epub November 6, 2020. https://doi.org/10.6061/clinics/2020/e2286

4. Carrard V, Bourquin C, Orsini S, Schmid Mast M, Berney A (2020) Virtual patient simulation in breaking bad news training for medical students. Patient Educ Couns 103(7):1435–1438. https://doi.org/10.1016/j.pec.2020.01.019
5. Co M, Chung PHY, Chu KM (2021) Online teaching of basic surgical skills to medical students during the COVID-19 pandemic: a case–control study. Surg Today (2021). https://doi.org/10.1007/s00595-021-02229-1
6. Cochrane Library (2021) The Cochrane database of systematic reviews. Available online: https://www.cochranelibrary.com/. Accessed on 4 May 2021
7. DOAJ. (2021). Directory of Open Access Journals (DOAJ). Available online https://doaj.org/. Accessed on 1 May 2021
8. Dost S, Hossain A, Shehab M, Abdelwahed A, Al-Nusair L (2020) Perceptions of medical students towards online teaching during the COVID-19 pandemic: a national cross-sectional survey of 2721 UK medical students. BMJ Open 10(11):e042378. https://doi.org/10.1136/bmjopen-2020-042378
9. Hara C, Goes F, Camargo R, Fonseca L, Aredes N (2021) Design and evaluation of a 3D serious game for communication learning in nursing education. Nurse Educ Today 100:104846. https://doi.org/10.1016/j.nedt.2021.104846
10. Haroon Z, Azad AA, Sharif M, Aslam A, Arshad K, Rafiq S (2020) COVID-19 era: challenges and solutions in dental education. J Coll Physicians Surg Pak JCPSP 30(10):129–131. https://doi.org/10.29271/jcpsp.2020.supp2.129
11. Hau HM, Weitz J, Bork U (2020) Impact of the COVID-19 pandemic on student and resident teaching and training in surgical oncology. J Clin Med 9(11):3431. https://doi.org/10.3390/jcm9113431
12. Häusler M, Bosse HM, Fischbach T, Graf N, von Kleist-Retzow JC, Kreuder J (2020) Alice im digitalen Wunderland: pädiatrische Lehre in der COVID-19-Pandemie: Eine Umfrage und Stellungnahme der AG Lehre der Deutschen Gesellschaft für Kinder- und Jugendmedizin (DGKJ) [Alice in the digital wonderlandpediatric teaching during the COVID-19 pandemic]. Monatsschr Kinderheilkd 1–8. https://doi.org/10.1007/s00112-020-01076-7
13. Hsieh HY, Hsu YY, Ko NY, Yen M (2020). Nursing education strategies during the COVID-19. Epidemic Nurs 67(3):96–101. https://doi.org/10.6224/JN.202006_67(3).13
14. Iyer P, Aziz K, Ojcius DM (2020) Impact of COVID-19 on dental education in the United States. J Dent Educ 84(6):718–722. https://doi.org/10.1002/jdd.12163
15. Katz M, Nandi N (2021) Social media and medical education in the context of the COVID-19 pandemic: scoping review. JMIR Med Educ 7(2):e25892. https://doi.org/10.2196/25892
16. Kawaguchi-Suzuki M, Nagai N, Akonoghrere RO, Desborough JA (2020) COVID-19 pandemic challenges and lessons learned by pharmacy educators around the globe. Am J Pharm Educ 84(8):ajpe8197. https://doi.org/10.5688/ajpe8197
17. Lefebvre C, Glanville J, Briscoe S, Littlewood A, Marshall C, Metzendorf M-I, Noel-Storr A, Rader T, Shokraneh F, Thomas J, Wieland LS (2021) Chapter 4: searching for and selecting studies. In: Higgins JPT, Thomas J, Chandler J, Cumpston M, Li T, Page MJ, Welch VA (eds) Cochrane handbook for systematic reviews of interventions version 6.2 (updated February 2021). Cochrane, 2021. Available from www.training.cochrane.org/handbook
18. Machado RA, Bonan P, Perez D, Martelli Júnior H (2020) COVID-19 pandemic and the impact on dental education: discussing current and future perspectives. Braz Oral Res 34:e083. https://doi.org/10.1590/1807-3107bor-2020.vol34.0083
19. McKenzie JE, Brennan SE, Ryan RE, Thomson HJ, Johnston RV, Thomas J (2021) Chapter 3: defining the criteria for including studies and how they will be grouped for the synthesis. In: Higgins JPT, Thomas J, Chandler J, Cumpston M, Li T, Page MJ, Welch VA (eds) Cochrane handbook for systematic reviews of interventions version 6.2 (updated February 2021). Cochrane, 2021. Available from www.training.cochrane.org/handbook
20. Mishra K, Boland MV, Woreta FA (2020) Incorporating a virtual curriculum into ophthalmology education in the coronavirus disease-2019 era. Curr Opin Ophthalmol 31(5):380–385. https://doi.org/10.1097/ICU.0000000000000681

21. Moreau C, Maravent S, Hale GM, Joseph T (2021) Strategies for managing pharmacy experiential education during COVID-19. J Pharm Pract 34(1):7–10. https://doi.org/10.1177/0897190020977730
22. Nimavat N, Singh S, Fichadiya N, Sharma P, Patel N, Kumar M, Chauhan G, Pandit N (2021) Online medical education in India—different challenges and probable solutions in the age of COVID-19. Adv Med Educ Pract 12:237–243. https://doi.org/10.2147/AMEP.S295728
23. PRISMA (2020) Preferred reporting items for systematic reviews and meta-analyses (PRISMA) checklist and flow diagram. Available online http://www.prisma-statement.org/. Accessed on 1 May 2021
24. Page MJ, McKenzie JE, Bossuyt PM et al (2021) The PRISMA 2020 statement: an updated guideline for reporting systematic reviews. BMJ (Clin Res Ed 372: n71. https://doi.org/10.1136/bmj.n71
25. PubMed (2021) PubMed.gov US National Library of Medicine National Institutes of Health. Available online https://pubmed.ncbi.nlm.nih.gov/. Accessed on 1 May 2021
26. Rajab MH, Gazal AM, Alkattan K (2020). Challenges to online medical education during the COVID-19 pandemic. Cureus 12(7):e8966. https://doi.org/10.7759/cureus.8966
27. Santos BM, Cordeiro MEC, Schneider IJC, Ceccon RF (2020). Educação Médica durante a Pandemia da Covid-19: uma Revisão de Escopo. Revista Brasileira de Educação Médica, 44(Supply 1):e139. Epub October 2, 2020. https://doi.org/10.1590/1981-5271v44.supl.1-20200383
28. SciELO (2021) Scientific electronic library online. Available online https://scielo.org/. Accessed on 4 May 2021
29. Sneyd JR, Mathoulin SE, O'Sullivan EP, So VC, Roberts FR, Paul AA, Cortinez LI, Ampofo RS, Miller CJ, Balkisson MA (2020) Impact of the COVID-19 pandemic on anaesthesia trainees and their training. Br J Anaesth 125(4):450–455. https://doi.org/10.1016/j.bja.2020.07.011

The Teaching of Technical Design in Technical Courses in Computer Graphics at Federal Institutes of Education, Science, and Technology

Eliana Paula Calegari, Luciana Aparecida Barbieri da Rosa, Raul Afonso Pommer Barbosa, and Maria Jose de Sousa

Abstract Computer graphics are a tool for the development of projects in technological areas, as they allow the representation and virtual visualization of spaces, construction systems, products, objects, and environments. For graphic representation through computer graphics, technical drawing knowledge is required. Thus, the objective of this work was to analyze how the contents of technical design are addressed in the curricula of Technical Courses in Computer Graphics offered by Federal Institutes. For this, a survey was carried out of the Campuses that have Technical Courses in Computer Graphics and verified the existence of disciplines that address contents of the technical design area in the curricula of the courses, from that, a content analysis was made on the syllabus on the approach to technical design in these disciplines. As a result, it was found that there are four offers of the Technical Course in Computer Graphics at Federal Institutes. It was found that the approach of the technical design area in the disciplines is focused on digital projects and technologies, such as Computer-aided Design (CAD) programs. This approach is considered essential for the Technical Course in Computer Graphics and is in line with the course proposal established in the National Catalog of Technical Courses, prepared by the Ministry of Education.

Keywords Federal Institute · Professional and technological education · Technical course in computer graphics · Pedagogical course project · Technical drawing

E. P. Calegari (✉)
Instituto Benjamin Constant, Rio de Janeiro, Brazil

L. A. B. da Rosa
Instituto Federal de Rondônia, Vilhena, Brazil

R. A. P. Barbosa
Fundação Getúlio Vargas, Rio de Janeiro, Brazil

M. J. de Sousa
Instituto Universitário de Lisboa, Lisbon, Portugal

A. Ullah et al. (eds.), *Proceedings of International Conference on Information Technology and Applications*, Lecture Notes in Networks and Systems 350,
https://doi.org/10.1007/978-981-16-7618-5_58

1 Introduction

The Federal Institutes of Science and Technology Education were established in Brazil by Law No. 11.892 of December 29, 2008. This law establishes the offer of Professional and Technological Education (EFA), at different levels and modalities, with emphasis on local, regional, and national socioeconomic development. Thus, the EFA is now understood as the educational process that seeks to solve the demands of society, of integration and verticalization of education, in order to strengthen the local productive, social and cultural arrangements, besides constituting a specialized center in offering science education, develop extension programs and conduct applied research, cultural production, entrepreneurship, cooperativism, and also produce knowledge aimed at preserving the environment [3].

The Ministry of Education (MEC) makes available the National Catalog of Technical Courses, in which the areas of the EFA high school courses are defined, as well as the profile of the graduates. Among the courses described in this catalog is the Technical Course in Computer Graphics, in the Information and Communication area. The professional profile is described as follows: The Computer Graphics Technician participates in the elaboration and development of computer graphics projects in two or more dimensions, using modeling, illustration, animation, audio, and video tools, and also works in the development of digital simulators and electronic models. The field of work of this professional is quite broad, involving advertising agencies, design studios, communication companies, architecture firms, video producers, digital games, and companies developing content for the Internet [5].

According to Costa et al. [5] in recent decades it is notorious the contribution of computers to the development of projects, especially in relation to graphic representation, due to the accuracy and ease in drawing and redesigning. The computer graphics resources used for graphic representation are CAD systems, which stand out for their easy manipulation and visualization, besides providing representations in three-dimensional space. Thus, computer graphics offers tools, such as drawing software that help in the development of projects, especially in the stages of creating virtual models and prototypes.

According to Silva et al. [12], the technical drawing is a form of communication through graphic expression that has the objective of representing the form, considering the dimensions and position of objects according to the needs required by technological areas, such as engineering, architecture, and design. In this way, technical drawing is considered a universal graphic language for industry, because it provides precise information for the planning, conception, and construction of a given object.

Therefore, for the elaboration of projects in technological areas, technical drawing is essential for graphic representation. Therefore, the teaching of contents in the area of technical drawing is essential for the development of projects with the use of computer graphics resources. Considering that the knowledge about technical drawing is essential for the formation of students of the Technical Course in Computer Graphics, as well as the lack of investigations on how the contents of technical drawing are mobilized in the curricula of this course, this study aims to analyze how the contents of technical drawing are addressed in the curricula of Technical courses in Computer Graphics offered by the Federal Institutes of Education, Science, and Technology.

2 Methodology

This work is configured as a research of exploratory and documentary nature, characterized as to the qualitative approach. According to Gil [6], the exploratory research aims to provide greater familiarity with the problem, in order to make it more explicit.

Regarding its operationalization, it was carried out from internet research, with the aim of analyzing how the contents of technical drawing are addressed in the curricula of Technical courses in Computer Graphics offered at the Federal Institutes of Education, Science, and Technology. The data survey was carried out in October 2020. Figure 1 shows a summary of the methodological aspects of this work.

The first stage of the research, survey 1, consisted of identifying all the Campi of the Federal Institutes (http://redefederal.mec.gov.br/instituicoes). The data collected on the Federal Network website were listed according to the state where the institution's Campi are located.

In the second stage of the research, survey 2, an internet search was conducted on the websites of the campuses of the Federal Institutes for the verification of the Technical courses in Computer Graphics, currently offered, where they were organized according to the state and the campuses of the Federal Institute.

In the third stage of the research, survey 3, a mapping was performed through the websites of the campuses of the Federal Institutes aiming to collect the Pedagogical

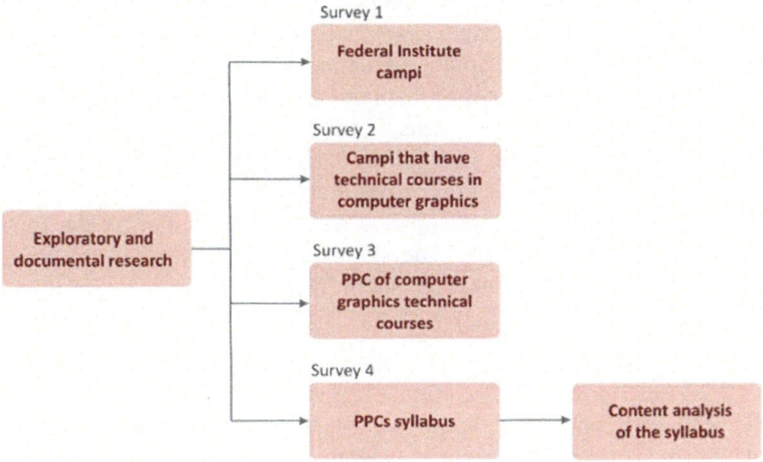

Fig. 1 Summary of methodological aspects. *Source* Prepared by the authors

Course Projects (PCP) of the Technical Course in Computer Graphics. In the PCPs of the Technical Computer Graphics courses, the disciplines that address content about technical drawing were verified. In the PCPs, data were collected regarding the modality of the course (face-to-face or distance learning) and the form in which it is offered (integrated, concomitant, or subsequent), the name of the discipline, the period of the course in which it is offered and the workload of the discipline.

The fourth step, survey 4, consisted in the investigation of the disciplines syllabus that deal with the contents of technical drawing in the PCP of the Technical courses in Computer Graphics, and these were analyzed through content analysis. According to Bardin [1, p. 46], content analysis consists of an operation of "classification of constituent elements of a set by differentiation and then by regrouping according to gender (analogy), with previously defined criteria," which allowed the condensation and simplified representation of the raw data found.

The RQDA program (R Package for Qualitative Data Analysis) will be used to process the data. After collecting the data, the process of classifying it into subcategories will begin. The subcategories are words or small groups of words that are defined from the frequency in which they appear in the data, in addition to the greater representativeness [1]. From the grouping of the data, it will be possible to interpret the results and characterize the approach of technical drawing in the Technical Courses in Computer Graphics of the Federal Institutes.

3 Results and Discussion

A search was conducted on the Ministry of Education's website to identify the campuses of the Federal Institutes in the Brazilian regions. Thus, in the 7 states of the region there are 69 campuses of Federal Institutes. In the Northeast region there are 199 campuses of Federal Institutes distributed in the 9 states of this region. In the Center-West region there are 55 campuses of Federal Institutes distributed in the 3 states of that region. In the Southeast region there are 165 campuses of Federal Institutes distributed in the 4 states of that region. In the South region there are 103 campuses of Federal Institutes distributed in the 3 states of that region.

After surveying the campuses of the Federal Institutes in all states of the 5 Brazilian regions, a search was made on the website of each campus to verify if they offer the Technical Course in Computer Graphics. We identified 4 offers of the Technical Course in Computer Graphics in the Federal Institutes. One offer in the North region, in IFRO, two offers in the Northeast region, in IFCE and IFCE, and one offer in the Southeast region, in IFTM.

3.1 *The Pedagogical Projects of the Technical Courses in Computer Graphics*

A search was conducted on the website of each campus that offers the Technical Course in Computer Graphics to collect the PCP. In the PCP, information was verified about the modality of the course, the form of the offer, and the total workload of the course. Three Technical courses in Computer Graphics take place in the face-to-face modality (IFPE, IFCE, and IFTM) and one course in the EaD modality (IFRO). According to the PCP of the Technical Course in Computer Graphics of IFRO, 49% of the course hours are in person. It was probably chosen the face-to-face modality or with part of the workload in person, due to the nature of the course being practical, with the need for classes in computer labs with the use of specific programs [7].

Regarding the form in which the Technical Course in Computer Graphics is offered, two courses (IFPE and IFCE) are offered in the subsequent form, which is that in which the student must have completed high school. One course is offered in the concomitant form (IFRO), which is intended for students who have completed elementary school, and who are regularly enrolled and attending the 1st or 2nd year of high school, from the public or private education network, including the EJA modality, and that have been selected in a public selection process, and another course in the integrated form (IFTM), where the student takes the high school and the technical course at the Federal Institute.

The total workload of the Technical courses in Computer Graphics varies from 1185–3200 h. The course with the lowest total workload is IFPE's 1185 h, and the course with the highest workload is IFTM's 3200 h. The biggest workload in IFTM's course occurs because besides the technical subjects, high school subjects are also accounted for.

3.2 *Survey of Course Syllabuses that Address the Teaching of Technical Drawing*

In the syllabus of Technical courses in Computer Graphics offered by IFRO, IFPE, IFCE and IFTM, we identified the disciplines that have content in the area of technical drawing, the course hours and in which period of the course it is offered.

In two PCPs (IFRO and IFTM), the content about technical drawing is addressed in disciplines that have the name of the area "Technical Drawing," in another PCP (IFPE) the name of the discipline is "Architecture (CAD 2D)" and in the other (IFCE) is "CAD 2D/3D." The names of the disciplines that do not have the words "technical drawing" are related to computer-aided design (CAD). According to Costa (2018), the knowledge of technical drawing is essential for the elaboration of computer-aided designs. The course workload varies from 40 to 80 h, and in two PCPs (IFRO and IFPE) the subject is offered in the first semester, in another PCP (IFCE) in the second semester of the course, and in another PCP (IFTM) in the last year of high school. It can be stated that probably, the offering of the disciplines in the initial semesters of the course occurs because they are basic content for the realization of computer-aided designs [9].

3.3 *Syllabus Analysis*

The syllabus of the disciplines that address technical drawing contents were analyzed through the content analysis technique. After reading, classification and data analysis, subcategories were elaborated and from the common aspects they were grouped into macro-categories. Figure 2 illustrates the categories that were built from the analysis performed.

As can be seen in the categories developed from the analysis of the syllabus, the approach to technical drawing in the courses is related to design issues and CAD tools. Regarding the macro-category "Project," the following subjects are addressed: floor plan, architectural project, and mechanical project. Thus, it can be inferred that the emphasis of the syllabus is directed to projects in architecture and engineering. According to Costa (2018), technical drawing is essential for the preparation of projects in the areas of architecture and engineering, because it allows to communicate clearly and accurately the elements that are part of the project. Ribeiro et al. [11] state that technical drawing is a way of graphic expression that aims at the

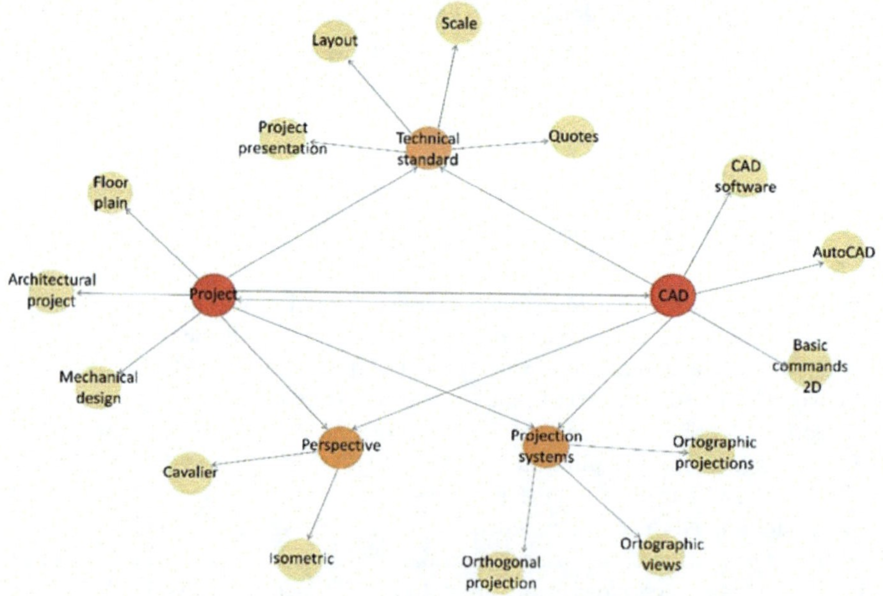

Fig. 2 Data categorization. *Source* Image generated by the software RQDA

representation, dimension, and positioning of objects according to the needs that are required by architecture and by the various engineering modalities.

Regarding the CAD macro-category, the following aspects are addressed: CAD Sofware, AutoCAD, and basic commands 2D. Thus, it can be stated that CAD tools, such as CAD programs, are addressed in the syllabus, AutoCAD being a widely used program for graphical representation through technical drawing. However, there are many other CAD programs that have resources and tools for the development of projects in technological areas, some proprietary, such as: 3D Slash, Fusion 360, SolidWorks, and others free, such as: NanoCAD, BricsCAD, ProgeCAD, FreeCAD, and LibreCAD [10].

According to Costa (2018), computer-aided design (CAD) systems are currently unavoidable tools and cannot fail to be considered in the context of technical design. The advancement of technology has led to the emergence of computers and the development of increasingly sophisticated programs with high capacity to process data, thus, technical design has appropriated these CAD tools to improve and optimize the preparation of projects. In this sense, knowledge of CAD tools is indispensable for projects in engineering, architecture, product design, or any other area that uses drawings for the graphic representation of projects, because today most industrial sectors develop parts and products through fully automated processes.

The macro-categories "Project" and "CAD" are directly interconnected with the subcategories "Technical Standards," "Projection System," and "Perspective." These subjects are the basis for the construction of projects and for the elaboration of computer-aided design (CAD). Thus, to prepare a project it is necessary to know the Technical Standards of technical drawing established by ABNT, in addition to understanding the Projection System for the graphic representation of the project through the construction of orthographic views, and to know the theoretical basis for the construction of perspective drawing, because it allows to represent an object as close as possible to reality.

Regarding the technical standards, Leake and Borgerson [8] state that they are used to standardize the elements used for the representation of technical drawing, and thus facilitate its writing and interpretation. In this sense, it was agreed to standardize the language of representation of technical drawing, that is, the strokes and symbols, through technical standards. The technical standards that were cited in the syllabus are related to the following subjects: dimension, scale, layout, and project presentation. These subjects are part of the following technical standards: NBR 10,068—Drawing sheet—Layout and dimensions, NBR 8196—Employment of scales and NBR 10,126—Dimensioning in Technical Drawing.

In the subcategory "Projection Systems" the following contents are associated: orthographic projections, orthographic views, and orthogonal projection. The representation of objects in technical drawing occurs through the system of orthographic projections. The method of representation by this system is based on the descriptive method devised by Gaspar Monge. The basic operation of this method is the orthogonal projection that aims to represent the true magnitude of the figures in space that are parallel to the projection plane [4]. Thus, this method aims to represent three-dimensional (3D) objects on a two-dimensional (2D) plane. Thus, the representation of objects in technical drawing is made from orthogonal projections resulting in the construction of orthographic views. According to NBR 10,647 (1989), orthographic views are figures resulting from orthogonal projections of the object on conveniently chosen planes, whose purpose is to accurately represent its shape and details. Given the above, the projection system is the fundamental theoretical basis for the preparation of technical drawings through orthographic views.

In the subcategory perspective the following subjects are related: isometric and cavalier. The perspective technique allows to graphically represent the three dimensions of an object (width, height, and depth) in a single plane, for example, on a sheet of paper that has two usable dimensions (width and height). Perspective is the representation closest to the visual experience, and generally corresponds to a global view of the object. Thus, in technical drawing, perspective is used to communicate information about the form and function of objects. The graphic representation through perspective can be divided into three types, each of them showing the object in a different way: conic, orthogonal axonometric: isometric, dimetric, and trimetric, and oblique axonometric: cavalier.

Generally, the graphic representation through perspective is used in the initial phases of a project, and the representation through orthographic views are suitable for the final phase of the project, because they have the characteristic of communicating the object unequivocally. The graphic representations through the techniques of perspective drawing are of great relevance for the development of projects in technological areas since they allow to represent an object as close as possible to reality.

4 Conclusions

The objective of this study was to analyze how the contents of technical drawing are addressed in the curricula of the Technical courses in Computer Graphics offered by the Federal Institutes. Thus, a survey was conducted about the existence of the Technical Course in Computer Graphics on the campuses of the Federal Institutes. From a total of 571 campuses of Federal Institutes, 4 campuses offer this course, they are: IFRO Porto Velho Zona Campus, IFPE Olinda Campus, IFCE Advanced Jaguarana Campus, and IFTM Uberaba Campus.

The IFRO offers the Technical Course in Computer Graphics in distance education modality in concomitant form, while IFPE and IFCE offer it in face-to-face and in subsequent form, and the IFTM offers it in face-to-face in integrated form. Thus, it was verified that there is an offer of this course in the North region, two offers in the Northeast region and one offer in the Southeast region.

In the PCP of Technical Courses in Computer Graphics offered by Federal Institutes, it was possible to verify the disciplines that have contents about technical drawing. In the curriculum of the Technical Course in Computer Graphics IFRO, there is a specific subject of technical design that has the nomenclature "Technical Drawing," already in the curriculum of the mentioned course offered by IFPE there is the subject called "Architecture (CAD 2D)" that addresses content in technical drawing.

In relation to the curriculum of the Technical Course in Computer Graphics offered in IFCE, there is the subject "CAD 2D/3D" that has contents about technical drawing, and finally, in the curriculum of that course offered by IFTM there is the subject "Technical Drawing and CAD" that also covers contents of technical drawing area. In this sense, it was found that in all curricula of Technical Courses in Computer Graphics offered by Federal Institutes there are disciplines that deal with contents in technical drawing.

Through content analysis, it was verified the approach of the contents about technical drawing in the disciplines of the Technical Courses in Computer Graphics offered by the Federal Institutes. Thus, the contents about technical drawing are approached through projects, covering projects in architecture and engineering, and CAD, in which contents about CAD programs are treated.

As a basis for the development of projects and for computer-aided design (CAD), contents related to technical standards, the projection system, and perspective are covered. Regarding the technical standards, the focus is on the contents about dimension, scale, layout, and projects presentation, which have specific technical standards. About the projection system, the subjects' syllabus includes contents about orthographic projections, orthographic views, and orthogonal projections. And, in relation to the contents about perspective, there is the approach of contents related to isometric and cavalier perspectives.

Thus, the approach to technical drawing in the course syllabus is focused on projects and digital technologies, such as CAD programs. This approach is extremely important for the Technical Course in Computer Graphics because it is in line with the course proposal established in the National Catalog of Technical Courses. This catalog establishes that the professional profile of the technician in computer graphics is focused on the elaboration of projects in two or more dimensions, in the areas of architecture, engineering, design, digital games, video production companies, and others, using digital tools for drawing, modeling, illustration, animation, audio, and video. In this sense, from the analysis and conclusions made it is understood that this work offers subsidies for the elaboration or reformulation of PCPs of technical courses in Computer Graphics.

References

1. Bardin L (2011) Análise de conteúdo. Edições 70, Lisboa
2. Brasil (1996). Lei nº 9.394 de 20 de dezembro de 1996: Estabelece as diretrizes e bases da educação nacional. Disponível em http://www.planalto.gov.br/ccivil_03/leis/l9394.htm. Acesso em 20 de setembro de 2020
3. Brasil (2008) Lei nº 11.892, de 29 de dezembro de 2008: Institui a Rede Federal de Educação Profissional, Científica e Tecnológica, cria os Institutos Federais de Educação, Ciência e Tecnologia, e dá outras providências. Disponível em http://www.planalto.gov.br/ccivil_03/_ato2007-2010/2008/lei/l11892.htm. Acesso em 20 de setembro de 2020

4. Bornancini JC, Petzold N, Orlandi JH (1987) Desenho técnico básico: fundamentos teóricos e exercícios a mão livre. 4ª ed. Sulina, Porto Alegre
5. de Costa CBM, de Silva EP, Santos M, dos Santos MM, de Pereira MNL, Schroeder N, Branchine SM (2020) Catálogo nacional de cursos técnicos. 2014. Disponível em: http://portal.mec.gov.br/index.php?option=com_docman&view=download&alias=77451-cnct-3a-edicao-pdf-1&category_slug=novembro-2017-pdf&Itemid=30192. Acesso em 11 de setembro de 2020
6. Gil AC (2010) Como elaborar projetos de pesquisa. Atlas, São Paulo
7. IFRO (2020) Resolução No 13/Reit-Cepex/Ifro, de 01 de julho de 2020: Dispõe sobre a aprovação da Reformulação do Projeto Pedagógico do Curso Técnico em Computação Gráfica, Concomitante ao Ensino Médio, EaD, do IFRO. Disponível em https://portal.ifro.edu.br/zona-norte/cursos/2679-tecnico-compt-1. Acesso em 18 de setembro de 2020
8. Leake J, Borgerson J (2013) Manual de desenho técnico para engenharia: desenho, modelagem e visualização. LTC, Rio de Janeiro
9. Marques JC (2017). O Ensino do Desenho Técnico e suas relações com a História da Matemática, da Arquitetura e a Computação Gráfica. XXI EBRAPEM, Encontro Brasileiro de Estudantes de Pós-Graduação em Educação Matemática, Pelotas
10. Parsekian GA, Achon CL, de Oliveira EP, De Paula N (2014). Introdução ao CAD. EdUFSCar, São Carlos
11. Ribeiro CA, Peres MP, Izidoro N (2011) Apostila de desenho técnico mecânico. Instituto Federal de Educação, Ciência e Tecnologia de São Paulo
12. Silva A, Ribeiro CT, Dias J, Sousa L (2013) Desenho Técnico Moderno. LTC, Rio de Janeiro

Digital Learning Technologies in Higher Education: A Bibliometric Study

Andreia de Bem Machado, Maria José Sousa, and Gertrudes Aparecida Dandolini

Abstract In a society mediated by technological tools, which interconnect several continents, digital education is part of higher education institutions. In this context, these institutions have to assume the role of being connected in time and space to the changes resulting from the impacts to which society is subjected, whether from a pandemic or for political reasons, in short, in different contexts of transformation. Thus, the aim of this article is: to analyze, through a bibliometric study, the main technologies, practices, and contexts used to promote the digital development in higher education. Therefore, bibliometrics were carried out, based on a systematic search in the Scopus and Web of Science databases, to answer the following problems: What are the main technologies for digital higher education? What are the main practices and contexts of digital education? As a result, it was identified that the highest productivity in the Scopus database was in the years 2018 and 2020 and in the Web of Science database in the years 2019 and 2020. It was found that Australia is the country that stands out in publications, in both databases, and research is concentrated in the areas of knowledge, social sciences, information technology, engineering, and education. The contributions of this survey are to verify whether technologies and technologies practices they are tools for transforming education. In addition, it was found that the technologies and educational strategies adopted in digital education are a fundamental instrument to facilitate justice and inclusive access to education, eliminating barriers to learning, expanding the view of teachers in order to qualify the process student learning process.

A. de Bem Machado (✉) · G. A. Dandolini
Engineering and Knowledge Management Department, Federal University of Santa Catarina, Florianópolis, Brazil
e-mail: andreia.bem@ufsc.br

G. A. Dandolini
e-mail: gertrudes.dandolini@ufsc.br

M. J. Sousa
Political Science and Public Policy Department, ISCTE—University Institute of Lisbon, Lisbon, Portugal
e-mail: maria.jose.sousa@ISCTE-IUL.pt

A. Ullah et al. (eds.), *Proceedings of International Conference on Information Technology and Applications*, Lecture Notes in Networks and Systems 350,
https://doi.org/10.1007/978-981-16-7618-5_59

Keywords Higher education · Digital development · Technologies · Pedagogical practices

1 Introduction

In a digital world, the possibilities of increasingly interactive resources have revolutionized the way people communicate and share knowledge through innovative technologies. Communication-related technology adds access to knowledge, which has been expanded through digital communication networks [1]. The countless paths taken by the innovation made possible that technology points to different realities and orientations in the communication process originated by social media. These are requirements for twenty-first-century education that makes use of communication technologies in learning contexts supported by mobile technologies, applications for tablets, and smartphones becoming increasingly popular in the educational environment [2]. Education through these tools moves to a digital teaching and learning process [3]. This new education format becomes interactive in which learning content is available online using digital media.

There are numerous benefits for the learning process with the use of ICT in classrooms, one of them is the use of open education platforms in a complementary way to improve students' academic results [4]. Collaborative learning on these platforms, called digital learning environments, has encouraging effects in increasing knowledge, competence, satisfaction, and problem solving skills [5].

Digital learning and the use of virtual learning environments heralds a new era in higher education [6]. In this digital context, transformation technologies happen at the speed of megabytes with digital resources that advance the cultural structure, visualized in social relations, man versus man, and man versus machine. Thus, the practices of using technologies for digital culture permeate knowledge that manifests in a network, being that in higher education institutions, are mediated by the teacher, assuming this important role in education.

In this context, and with the aim of answering the following questions: RQ1: Which are the main technologies for digital higher education? RQ2: Which are the main digital education practices and contexts? In this article, a bibliometric study was carried out, through a systematic search in two databases. Its applications helps in understanding new issues and identifying trends for future researches Scientific mapping allows investigating a global image of scientific knowledge from a statistical perspective. It mainly uses the three structures of knowledge to present the structural and dynamic aspects of scientific research [7].

Thus, the present article is organized as follows: the next topic presents the method used in the study, in the third topic the results. Lastly, the discussions and conclusions are presented.

2 Methodology

To increase knowledge, measure, and analyze scientific literature publications on trust in the field of digital education, bibliometric analysis was performed from a search in Clarivate Analytics' Scopus and Web of Science (WoS) database. The study was developed using a strategy composed of three phases: execution plan, data collection, and bibliometry. To assess the results in a more in-depth way for the bibliometric analysis, this result was exported to a bibliographic management software called EndNoteWeb. These data provided the organization of relevant information in a bibliometric analysis, such as temporal distribution; leading authors, institutions, and countries; type of publication in the area; key words and the most referenced works[1] [8].

2.1 Data Collection and Research Strategy

Considering research problems: Which are the main technologies for digital higher education? Which are the main digital education practices and contexts? It was delimited, still in the planning phase, the search terms, i.e., "digital technolog*" and "digital learning" and "higher education." The use of the truncator (*) occurred intending to potentiate the result by seeking technologies and their written variations presented in the literature. And, as a basic principle for the search, we chose to plan to search for the use of the terms in the "title, abstract and keyword" fields, without delimiting temporal, language, or other restrictions that may limit the result. The research was carried out on May 30, 2021.

3 Results

From research planning to data collection, a total of 28 documents were retrieved in both the Scopus database and the Web of Science database. Eligible articles in the Scopus database were published between 2001 and 2021. The Web of Science eligible database articles were from 2014 to 2021. In the Scopus database, it was found that the highest productivity was in 2018 and 2020, with a total of six documents in each of the years. In the Web of Science database, the highest productivity was in 2019 with 8 publications and in 2020 with seven publications. The first publication in the Scopus database was in 2001, entitled "DISA: Insights from an African model for digital library development" [9], while in the Web of Science database it was in 2014

[1] Software-based on the *web* that contributes to the work of the researcher during the writing process of hissa. Bibliographic reference management artifact produced by *Thomson Scientific*. allows you to search databases *online*, organize references, extension files,.pdf as well as create and organize the bibliography in a text editor. source: http://www.endnote.com.

entitled "Teaching and learning transformatives in Higher Education: Using Social Media in a Team-Based Learning Environment" [10].

When analyzing the country that published the most in the area, it is clear that Australia stands out both in the Scopus database and in the Web of Science database, with an average of 19% of the total publications, totaling 6 in the first and 4 in the second. Secondly, in the Scopus database, with 6% of the appearances, there are China, Ireland, Romania, the Russian Federation, and the United States of South Africa, with two documents from each of these countries in the Scopus database. In the Web of Science database are Russia, Spain, and Ukraine with 10% of the publications, that is, three articles published in the Web of Science database. Figure 1 and Fig. 2 show the countries with at least two publications, seven from the Scopus database and six from the Web of Science.

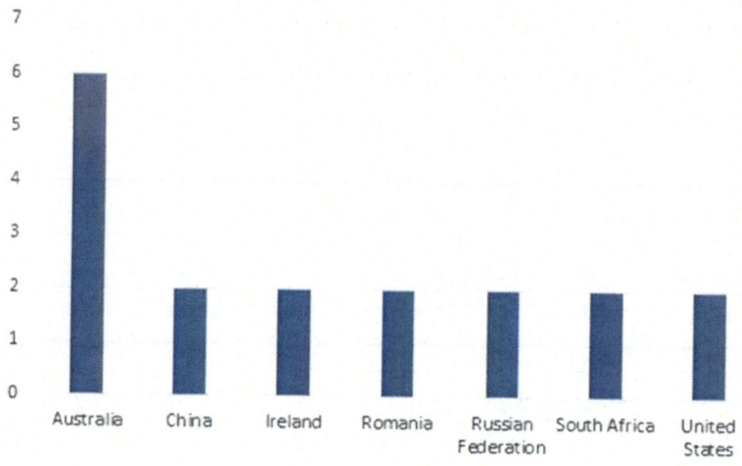

Fig. 1 Publications distribution by country of work (Scopus)

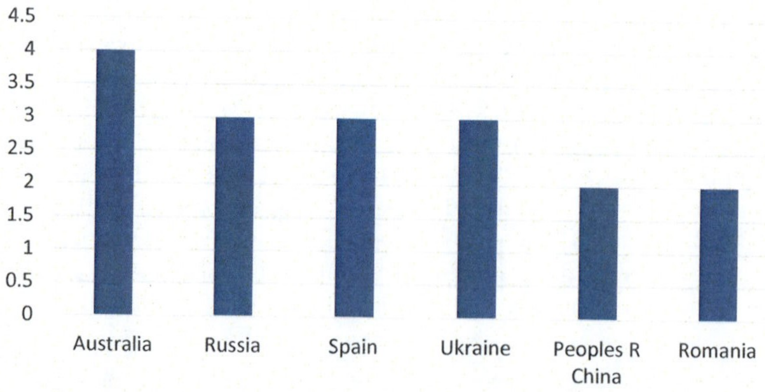

Fig. 2 Publications distribution by country of work (Web of Science)

The VOSviewer program was chosen to visualize the authors' network as it employs a uniform structure of mapping and grouping [11]. VOSviewer is a network building and visualization software program that focuses on graphical representation and is useful for interpreting huge bibliometric maps. These networks can be built based on citation relationships, bibliographic combination, co-citation, or co-authorship and can include journals, authors, or institutions. The circles in the views reflect the items under investigation related to each denomination. The greater the weight of the item in the net, the larger the circle. The distance between items reflects the degree to which they are related. The longer the link, the thicker the related lines will be. Color and location are two methods of grouping. Thus, the analysis regarding the identification of authors was carried out, it was observed that there are no reference authors in the Scopus and Web of Science databases on the topic of Digital Learning Technologies in Higher Education. The first database contains 77 authors with one publication in this area, shown in Fig. 3.

In the Web of Science database, the highlighted author is Elena Fleaca, from the Polytechnic University of Bucharest, Romania, with two publications. The other 83 authors who publish in this area have only one publication, as shown in Fig. 4.

It is concluded that the country that most publishes in the two databases is Australia, but the affiliation that publishes is located in Romania, which appears in the distribution of countries by publication in fourth place in the Scopus database and sixth in the Web of Science database.

From the general survey, it was also possible to analyze the type of document research in the area of digital learning technologies in higher education. It is noticed that publications focus on journal articles in the two databases surveyed with 35% of the total number in the Scopus database and 46%in the web of Science database.

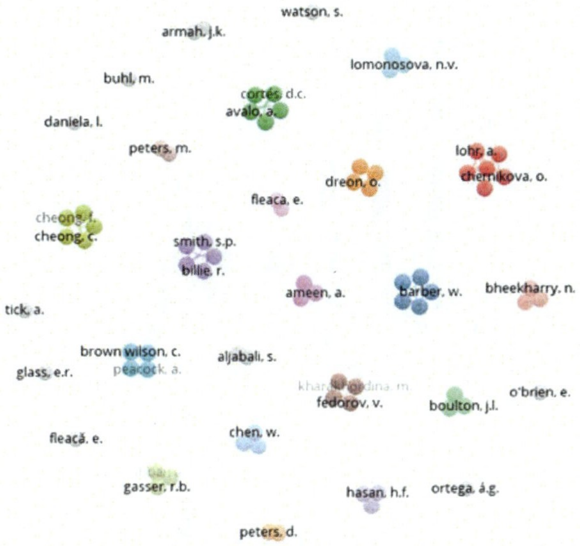

Fig. 3 Scopus coauthors

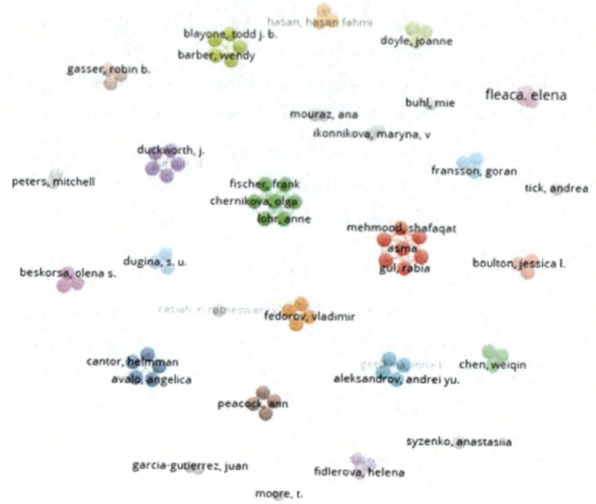

Fig. 4 Web of science coauthors

Regarding the areas of concentration that of publications that are highlighted in the Scopus database, 35% concentrates in the area of Social Sciences, 24% in the area of Computer Science, and 18% Engineering. As shown in Fig. 5.

In the Web of Science database, the publications that have the highlight concentrates on education with 45%, according to Fig. 6.

Based on the bibliometric analysis, and on the work group retrieved, there were 242 occurrences hosted on the Scopus databases. Then, it was found that nine words

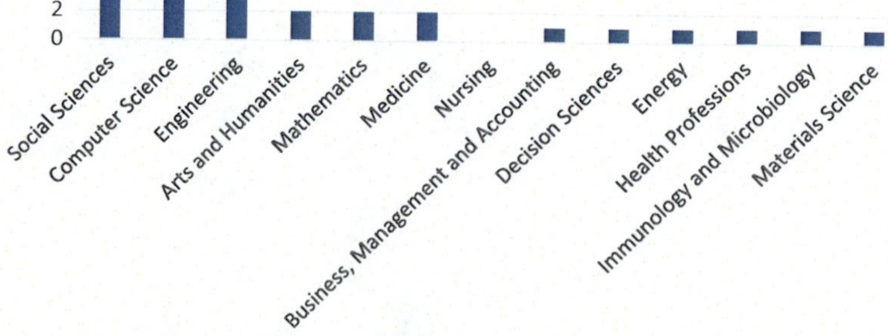

Fig. 5 Scopus publications areas of concentration

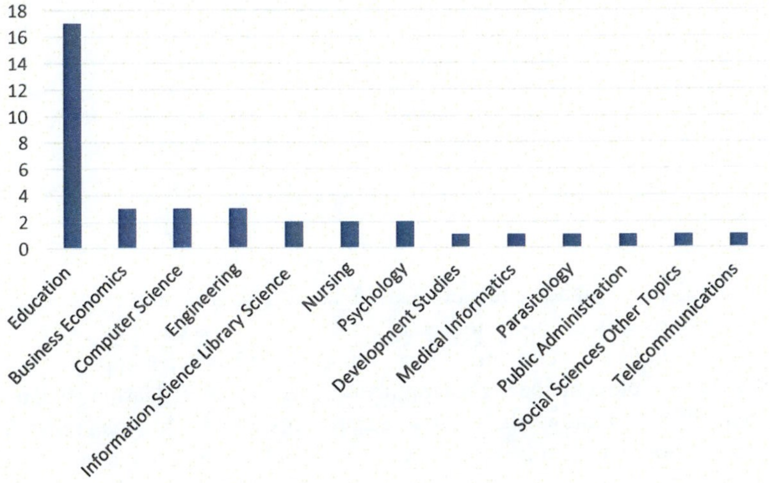

Fig. 6 Web of Science publications areas of concentration

are highlighted in the Scopus database, which are: "E-learning," digital technologies, students, teaching, education, higher education, computer-aided instruction, learning, and human, from the recovered works, showing the keywords, as shown in Fig. 7.

In the Web of Science database, 168 occurrences were found, it was uncovered that five words are highlighted in the database: digital learning, higher education,

Fig. 7 Tag cloud Scopus

Fig. 8 Tag cloud Scopus

digital technologies, higher education, and technology, from the retrieved works, showing the keywords as shown in Fig. 8.

After analyzing the documents, it was found that 13 articles appear in the two databases. The article with the highest number of citations is "E-learning and nursing assessment skills and knowledge—An integrative review" [12], which ranks first in the Scopus database and second most cited works in the Web of Database Science.

4 Discussion and Conclusions

Through a bibliometric study, we sought to understand the academic production on the subject of digital learning technologies in higher education. In both databases, Scopus and Web of Science, analyzed in the research, 28 papers were retrieved. In the Scopus database, the works were registered between the years 2001 and 2021 and in the Web of Science database between the years 2014 and 2021. It was found that the country that most publishes in the two databases is Australia, but the affiliation with the most publications in both databases, is located in Romania, ranking fourth in the Scopus database, with four publications, and sixth in the Web of Science database, with two publications.

It turned out that they present interdisciplinary characteristics, involving areas of knowledge related to Social Sciences in the Scopus database and education in the Web of Science database. Furthermore, the analysis of the most used keywords demonstrates that digital learning technologies appear as a topic related to the words "E-learning," digital technologies, students, teaching, education, higher education, computer-aided instruction, learning, and human. And in the Web of Science database the related words are associated with digital learning, higher education, digital technologies, higher education, and technology. Concluding, the theme is relevant in the process of digital teaching and learning in higher education, but there are no authors referenced in the Scopus database. In the Web of Science database there is an author, Elena Fleaca, from the Polytechnic University of Bucharest, Romania, with two publications, which are highlighted on this basis, concluding that the topic lacks studies.

As limitations, the method presented here is not capable of qualitatively identifying the theme of digital learning technologies in higher education and, therefore, it recommends carrying out integrative literature reviews that allow for broadening and deepening the analyzes carried out here.

Acknowledgements This research is part of the Athena Project, I would like to register here my thanks to the Athena Project. I thank the Federal University of Santa Catarina and the department of Engineering and Knowledge Management where I am doing my post-doctorate for the opportunity to do this study.

References

1. de Machado AB, Souza MJ, Catapan AH (2019) Systematic review: intersection between communication and knowledge. J Inf Syst Eng Manag 4(1). https://doi.org/10.29333/jisem/5741
2. Sousa MJ, Rocha Á (2020) Learning analytics measuring impacts on organizational performance. J Grid Comput 18(3):563–571
3. Sousa MJ, Sousa M (2019) Policies to implement smart learning in higher education. In: Proceedings of the 18th European conference on e-learning. ACPI
4. Sousa MJ., Rocha Á (2018) Digital learning in an open education platform for higher education students. In: 10th international conference on education and new learning technologies (EDULEARN)
5. Männistö M, Mikkonen K, Kuivila H-M, Virtanen M, Kyngäs H, Kääriäinen M (2020) Digital collaborative learning in nursing education: a systematic review. Scand J Caring Sci 34(2):280–292
6. Virtanen MA, Haavisto E, Liikanen E, Kääriäinen M (2018) Ubiquitous learning environments in higher education: a scoping literature review. Educ Inf Technol 23(2):985–998
7. Sweileh WM, Al-Jabi SW, AbuTaha AS, Sa'ed HZ, Anayah FM, Sawalha AF (2017) Bibliometric analysis of worldwide scientific literature in mobile-health: 2006–2016. BMC Med Let Me know Decis Mak 72
8. Morris SA, Van der Veer Martens B (2008) Mapping research specialties. Annu Rev Inf Sci Technol 42(1):213–295
9. Peters D, Pickover M (2001) DISA: insights of an African model for digital library development recovered: el 2 de junio de 2021, de Dlib.org website http://www.dlib.org/dlib/november01/peters/11peters.html
10. Rasiah RRV (2014) Transformative higher education teaching and learning: using social media in a team-based learning environment. Procedia Soc Behav Sci 123:369–379
11. Van Eck NJ, Waltman L (2010) Software survey: VOSviewer, a computer program for bibliometric mapping. Scientometrics 84(2):523–538
12. McDonald EW, Boulton JL, Davis JL (2018) E-learning and nursing assessment skills and knowledge—an integrative review. Nurse Educ Today 66:166–174

Personalised Combination of Multi-Source Data for User Profiling

Bruno Veloso, Fátima Leal, and Benedita Malheiro

Abstract Human interaction with intelligent systems, services, and devices generates large volumes of user-related data. This multi-source information can be used to build richer user profiles and improve personalization. Our goal is to combine multi-source data to create user profiles by assigning dynamic individual weights. This paper describes a multi-source user profiling methodology and illustrates its application with a film recommendation system. The contemplated data sources include (*i*) personal history, (*ii*) explicit preferences (ratings), and (*iii*) social activities (likes, comments, or shares). The MovieLens dataset was selected and adapted to assess our approach by comparing the standard and the proposed methodologies. In the standard approach, we calculate the best global weights to apply to the different profile sources and generate all user profiles accordingly. In the proposed approach, we determine, for each user, individual weights for the different profile sources. The approach proved to be an efficient solution to a complex problem by continuously updating the individual data source weights and improving the accuracy of the generated personalised multimedia recommendations.

Keywords User modeling · Multi-source · Profiling · Recommender systems

F. Leal (✉)
REMIT, Porto, Portugal
e-mail: fatimal@upt.pt

B. Veloso · B. Malheiro
INESC TEC, Porto, Portugal
e-mail: brunov@upt.pt

B. Malheiro
e-mail: mbm@isep.ipp.pt

B. Malheiro
School of Engineering, Polytechnic Institute of Porto, Porto, Portugal

B. Veloso · F. Leal
Universidade Portucalense, Porto, Portugal

A. Ullah et al. (eds.), *Proceedings of International Conference on Information Technology and Applications*, Lecture Notes in Networks and Systems 350,
https://doi.org/10.1007/978-981-16-7618-5_60

1 Introduction

Consumers are overloaded with multi-source information, which surpasses human processing capabilities. In this context, recommendation systems became essential to help users find products, services, or people since they can predict the user behavior based on historical data. With the ubiquitous Information and Communication Technologies (ICT) support, users simultaneously became contributors and consumers of shared data. This user-shared data is present in numerous online services. Recommendation systems have used multi-source information to build a thorough profile and, thus, generate meaningful, personalised recommendations.

This paper contributes with a multi-source user profiling approach that aims to improve recommendation accuracy. Therefore, the challenge is creating a weighted user profile using heterogeneous data sources and generating recommendations for personalised media content. Hence, the proposed platform includes (*i*) a multi-source Profiler; and (*ii*) a film Recommender.

In the Profiler, our method assigns dynamic individual weights to combine data from different profile components: (*i*) personal history, (*ii*) explicit preferences (ratings), and (*iii*) social activities (likes, comments, and shares). Furthermore, such information is collected from multi-sources: (*i*) personal history encompasses YouTube media content simulating a set of films watched by a user; (*ii*) social activities use information from social networks (Facebook and Twitter); and (*iii*) films and series are classified by IMDb features (genres, actors, and places). The Recommender integrates a Content-based and Collaborative Filter and provides personalised film recommendations using those heterogeneous profiles.

The experiments, which were performed with MovieLens dataset, compare the proposed method with standard approaches. In the standard approach, we calculate the best global weights to apply to the different profile sources and generate all user profiles accordingly. In the proposed approach, we determine, for each user, individual weights for the different profile sources to combine the available data and build the user profile. The results show the benefits of using weights in multi-source profiling approaches.

The remaining paper is organised as follows. Section 2 presents the related work on multi-source user profiling. The proposed method, including the profiler, recommender, and evaluator, is in Sect. 3. Section 4 describes an experimental evaluation and the results. Finally, Sect. 5 contains the conclusions and a discussion.

2 Related Work

User profiling is a collection of information that identifies user behavior [1]. The constant technological development and the emergence of new online services increase user-generated information in different data sources. Therefore, the user-related information available online is enormous, unstructured, and from multiple data sources. The multi-source profiling methodologies present in related work use weight vectors, semantic technologies, and probabilistic analysis to address user modeling in its respective systems.

Ma et al. propose a weighted fusion approach of user interests from multiple data sources on the Web (Twitter, Facebook, and LinkedIn) using, simultaneously, semantic reasoning to discover implicit interests [2].

Smailovic et al. propose an advanced user profiles approach for the SmartSocial Platform using reasoning upon multi-source user data [3]. This approach includes analysis of telecom operator (calls or text messaging), context-aware information (motion, environment, and position), and Internet services (Facebook, LinkedIn, and Twitter).

Sun et al. exploit multi-source user-related information for personalised restaurant recommendations [4]. The recommendations are based on user ratings, social factors, region topic, and mobility factors. In this scenario, the solution comprises a Bayesian graphical model considering the collaborative filtering of latent factors—the social and mobility factors—to learn the user preferences. Then, a probabilistic analysis fuses the heterogeneous information.

Aghasaryan et al. present a technology that gathers user-related data from different multimedia services [5]. The profiling process is based on keyword inference which translates the sets of weighted keywords of each data model into user model concepts.

Heitmann proposes a multi-source, cross-domain personalization framework with semantic interest graphs [6]. The framework is composed of a conceptual architecture for recommender systems and a cross-domain personalization approach. In this context, it was developed the SemStim—an unsupervised, graph-based recommendation algorithm that uses information from DBpedia to provide recommendations.

In terms of datasets that include heterogeneous data sources, a lack of knowledge hinders evaluating the recommendation systems and their user profiles. Farseev et al. refer that the research efforts on multi-source user profiling are relatively sparse, and it is not easy to find an appropriate large-scale dataset for user profile assessment [7].

Despite these efforts, the problem of multi-source profiling in recommendation systems is still insufficiently explored, especially in terms of assessment due to the lack of user-related data from multiple data sources. Therefore, facing this related work, we presented a Media Recommendation System with a personalised combination of multi-source data for user profiling using weighted vectors and semantic technologies for profiling and the MovieLens dataset for assessing the generated user profiles.

3 Proposed Method

This paper proposes a weighted multi-source dynamic profiling to generate a user profile from multiple data sources, improving personalised recommendations. We integrated the proposed profiling approach into a multimedia recommendation system illustrated in Fig. 1.

The recommendation platform includes two main modules: (*i*) multi-source profiler to create the global/individual user profile; and (*ii*) recommender using content-based and collaborative filtering.

The provided Graphical User Interface (GUI) allows the user to rate the generated recommendations and to connect his/her social networks, which will be used in the multi-source profiling. The experiments were conducted using MoviLens dataset, which simulated multi-source data.

3.1 Multi-Source Profiler

The profiler module collects and structures information from multiple sources for building the user profile. Specifically, the profiler module integrates two components: (*i*) personal profile; and (*ii*) social profile.

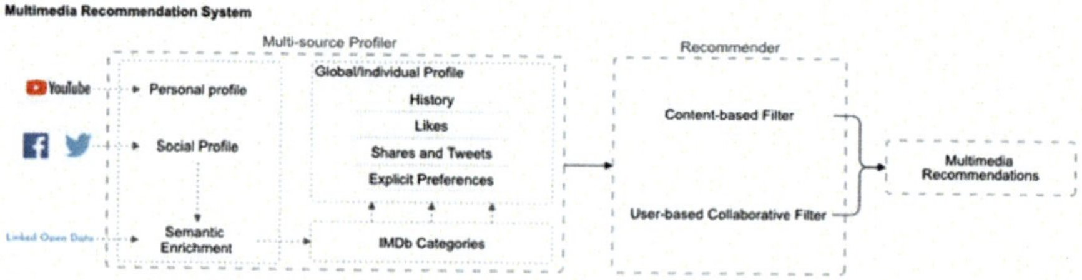

Fig. 1 Multimedia Recommendation System

Personal Profile includes historical data (YouTube) and explicit preferences (ratings to the recommendations). The system uses YouTube to simulate viewed films and, thus, build historical information.

Social Profile is based on social activities collected from Facebook and Twitter. The social activities encompass shares, likes, and tweets. In like-related activities, Facebook categorizes the pages that the user has liked. Concerning the shares and tweets, it was required an additional semantic enrichment to obtain the same categorization. This enrichment relies on Linked Open Data (LOD) for categorizing social preferences. Therefore, the social profile comprises two categories: Facebook likes and social data (tweets and shares).

User Profile encompasses the *Personal Profile* and *Social Profile* using four vectors of categories: (*i*) YouTube history; (*ii*) Facebook likes; (*iii*) social data; and (*iv*) explicit preferences. Each vector position represents a category, including a percentage concerning the number of times the user has viewed, liked, and shared content about that category. We separated Facebook likes in a particular vector since Facebook provides specific categories for its pages. The remaining social data (shares and tweets) need to be semantically enriched to categorize social preferences. The vector of explicit preferences includes the recommendation ratings. For standardisation, all profile vectors adopt IMDb film categories.

Algorithm 1 depicts the user profile composition. The system looks for user-related data (line 1) and uses the OMDB API[1] (line 2) to get the corresponding film features. Using those inputs, the Global Profile is built as follows:

- If the user data is an instance of history, each film genre feature is added to a partially user history profile (line 3–5).
- If the user data is an instance of ratings for each film genre feature is added to a partially user rating profile modeled by the normalised rating (line 6–8).
- If the user data is an instance of like for each film genre feature is added to a partial user-like profile (line 9–11).
- If the user data is an instance of shares and tweets, the system uses the LOD to get semantic information about the comment. In addition, the module applies a semantic match for looking at similar interest points between the semantic film genre and the semantic information collected for the user comment. If it has a positive matching, each film genre feature is added to a partially user tag profile (line 12–16).
- At the end, the four components of the user profile are returned.

[1] omdbapi.com.

Algorithm 1: User Profile Composition

```
UD ← GetUserData()
MF[n] ← GetMovieFeatures()
if UD = History then
    for f = 0; f < n; f++ do
        UP_u[0][f] += MF[f];
if UD = Rating then
    for f = 0; f < n; f++ do
        UP_u[1][f] += MF[f] * rating;
if UD = Like then
    for f = 0; f < n; f++ do
        UP_u[2][f] += MF[f];
if UD = Tag/Comments then
        SemCpts = GetLODConcepts()
        MovCpts = GetMovieConcepts()
if SimMatch(SemCpts, MovCpts) then
        for f = 0; f < n; f++ do
            UP_u[3][f] += MF[f];
return UP_u[c]
```

The final user profile depends on the selected multi-source profiling approach: (*i*) Global User Profiling with equal weights for all users; or (*ii*) Individual User Profiling with personalised weights for each user. The Global User Profile combines the multi-source user data to generate the final user profile vector by applying the determined global weights (35% for YouTube history, 25% for Facebook likes, 15% for shares and tweets and 25% for explicit preferences provided by recommendation ratings). The Individual User Profile applies personalised weights, based on the user activity, to integrate the user-related multi-source data. Algorithm 2 generates the final user profile according to the selected multi-source data profiling. First, it retrieves the weight combination to employ (line 1). Then, applies the retrieved weights to the individual profile components (history, likes, ratings, tags) to create the current user profile (line 4). Next, applies the selected recommendation filter and calculates the evaluation metrics (line 5). Finally, in case of the individual profiling, if the current results improve the previous one, it updates the optimal weights of the user (line 6–8).

Algorithm 2: Calculate Individual/Global Profile and Weights

$WC_u[c] \leftarrow GetWeights(u)$
for $c = 0; c < 4; c{+}{+}$ **do**
  for $f = 0; f < 20; f{+}{+}$ **do**
    $UP_u[c, f] = UP_u[c, f] * WC_u[c]$
  $EvMetrics[m] \leftarrow CalcRec(UP_u[c])$
  if IP **then**
    if $EvMetrics[m]_t$ **better than** $EvMetrics[m]_{t-1}$ **then**
      $WC_u[c] \leftarrow UpdateWeights(u)$
return $WC_u[c]$

3.2 Recommendation Service

The developed recommendation system is based on the Content-based Filtering (CbF) and memory-based Collaborative Filtering (CF). The CbF algorithm uses the Keyword Vector Space Model (VSM) to build the vector categories (user profile and media items) and Collinearity and Proximity Similarity described in Veloso et al. for matching the media items with the user profile [8]. The user-based collaborative filtering (CF) implements the *k*-Nearest Neighbours algorithm together with Pearson Correlation coefficient. It predicts items which a user may choose based on the information shared by other users with similar behaviour. The system recommends IMDb films. The films are organised by category (descending order taking into account the weight category in the user profile). Here, the user can rate the recommendations via star rating, which will be used to refine recommendations a posteriori.

3.3 Graphical User Interface

For interacting with the system, the user needs to register on the platform. GUI allows the register (Context Profile), the social networks aggregation, to see social activities (Social Profile), access to YouTube history (Content-based Profile), and receive recommendations. After registration, the user is invited to aggregate the multiple data sources to build his user profile. On the one hand, for aggregating the social networks to the platform, the user needs to introduce the credentials of the respective social network. On the other hand, for accessing YouTube history, the user must associate his Google account.

With all accounts connected, the GUI shows and interlinks the social activities performed by users on Facebook and Twitter (shares, likes, and tweets) and the YouTube history. The YouTube history contains the last 50 videos viewed. Therefore, GUI can be used only to consult the multiple data sources activities.

Whenever the user accesses the recommendation page, the automatic client starts, and the information from social networks, YouTube, and explicit preferences are processed, creating the corresponding user profile. Finally, after matching the user profile and IMDb films, the user can access and rate the recommendation.

3.4 Evaluation Procedure

The evaluation procedure of recommendation systems can be divided into predictive accuracy and classification metrics. The predictive accuracy metrics measure the error between the predicted rating and the rating assigned by the user. There are two important metrics: (*i*) the Mean Absolute Error (MAE), which measures the average absolute deviation between the predicted rating and the rating assigned by the user; and (*ii*) the Root Mean Square Error (RMSE), which emphasizes the largest errors [9]. Equation 1 displays the GMAE, which is the weighted average of the MAE for all users.

$$GMAE = \frac{1}{u}\sum_{u=1}^{m}\left(\frac{1}{n} \times \sum_{i=1}^{n}\left|p_{i,u} - r_{i,u}\right|\right) \tag{1}$$

where $p_{i,u}$ represents the predicted rating for user u to the item i, $r_{i,u}$ is the rating applied by the user u to the item i, and u represents the number of users.

Equation 2 displays the GRMSE, the weighted average of the RMSE for all users, and where u represents the number of users.

$$\text{GRMSE} = \frac{1}{u}\sum_{u=1}^{m}\left(\sqrt{\frac{1}{n} \times \sum_{i=1}^{n}\left(p_{i,u} - r_{i,u}\right)^2}\right) \tag{2}$$

where $p_{i,u}$ represents the predicted rating for user u to the item i, $r_{i,u}$ is the rating applied by the user u to the item i and u represents the number of users.

The classification accuracy metrics measure the frequency with which the recommendation system provides correct recommendations. There are three important metrics: (*i*) Recall, (*ii*) Precision, and (*iii*) F-measure. The Recall determines the number of relevant items selected from the total number of relevant items available. The Precision defines the number of relevant items selected from the total number of items. Finally, the F-measure combines Recall and Precision into a single metric [10]. Precision, Recall, and F-measure metrics range between 1 (best) and 0 (worst). The TP is the number of relevant items recommended by the system, i.e., true positive results, FN is the number of relevant items not recommended by the system, *i.e.*, the false-negative results, and FP corresponds to the number of irrelevant items recommended by the system, *i.e.*, the false-positive results.

The Global F-measure (GF) was used to evaluate the performance of the different filters. Equation 3 displays the GF, which is the weighted average of the F-measure for all users.

$$GF = \frac{1}{n}\sum_{i=1}^{n}\left(2 \times \frac{\text{Precision} \times \text{Recall}}{\text{Precision} + \text{Recall}}\right) \tag{3}$$

4 Experimental Evaluation

To evaluate the system, we have performed a set of experiments to analyse the accuracy of the recommendations supported by the proposed multi-source profiling approach. For each recommendation filter, we apply the multi-source user profile module which is composed of two different approaches: Global User Profiling (GP) and the Individual User Profiling (IP).

4.1 Dataset

For assessing the proposed multi-source profiling approach for Media Recommendation System, we have limited datasets due to multi characteristics that compose the Global Profile (history, ratings, likes, and comments). However, MovieLens dataset contains ratings that simulate likes and tags, which simulate the shares and tweets. The MovieLens dataset was selected and adapted to assess our approach by comparing the recommendations generated with the standard methodologies. This dataset belongs to GroupLens Research.[2] The dataset was chosen due to its lower data sparsity, size, and by its multimedia origin. It contains 100 023 ratings and 2488 tag applications across 8570 films. This information was provided by 706 users between April 02, 1996 and March 30, 2015. Each user is represented by an ID and has rated at least 20 films.

The dataset was divided into a training set (80%) and test set (20%), ensuring that, regardless of the number of ratings, tags, and likes, each user in the training set has 80% of their explicit data and the remaining 20% are in the testing set.

4.2 Evaluation Results

The offline performance of the recommendation service was determined by testing each implemented filter separately. To assess the user profile generated by the system,

[2] grouplens.org/datasets/movielens/

Table 1 GF and GMAE offline filter results (Best results are highlighted with bold)

Tests	GF	GMAE	GRMSE
CbF	0.571	–	–
CbF-IP	**0.590**	–	–
CbF-GP	0.396	–	–
CF	**0.667**	0.760	0.934
CF-IP	**0.634**	**0.557**	**0.688**
CF-GP	0.628	**0.542**	**0.673**

Best results are highlighted with bold

our evaluation contemplates three different tests for each recommender filter: (*i*) the use of recommender filter without the multi-source user profile construction; (*ii*) the use of recommender filter with the GP; and (*iii*) the use of recommender filter with IP. The training data set was submitted to each filter, and the individual user's recommendations were generated. Then, these user's recommendations were evaluated using the test data set for calculating the global evaluation metrics. Table 1 displays the GF and GMAE results for the implemented filters, using a 60% similarity threshold and a 3.5 rating threshold.

Regarding CbF-IP recommendations, GF increased 3.32%. The results show that for CbF the recommendations improve with multi-source individual profiling approach. In the case of CF recommendations, it was also possible to calculate the rating accuracy from the predicted and observed user ratings. The results show that GMAE and GRMSE decrease both with CF-IP and CF-GP, displaying higher predictive accuracy. GF decreases, indicating lower classification accuracy.

5 Conclusion

The rapid growth of different platforms has caused the distribution of user-related information. The collection of this information from multi-sources enables the creation of novel user profiling methods. This paper describes a multimedia recommendation system which integrates multi-source user data for user profiling. Specifically, the proposed recommendation platform integrates information from YouTube, Facebook, and Twitter, and allows the classification of recommendations. The proposed method includes two multi-source profiling approaches: (*i*) global profiling; and (*ii*) individual profiling. While the global profile uses the calculated global weights to combine the multiple data sources, the individual profile dynamically personalises such weights to improve the final recommendations. To evaluate the proposed multi-source profiling method, we have conducted several experiments using the MoviLens dataset. The offline tests show that the content-based filter with individual profiles performed better than its counterparts. In the case of the collaborative filters, the multi-source profiles display better predictive accuracy.

As future work concerning the recommendation service, we plan to adopt a hybrid recommender filter to use the multi-source data generated by the users in a more efficient manner. In terms of user modeling, the focus will be on building an algorithm that needs to be more scalable, precise, and faster to find the optimal weights to the user profile. Finally, we intend to evaluate the recommendation system with real multi-source viewer data.

References

1. Hasan O, Habegger B, Brunie L, Bennani N, Damiani E (2013) A discussion of privacy challenges in user profiling with big data techniques: the eexcess use case. In: Big Data (Big Data Congress), 2013 IEEE International Congress. pp 25–30. IEEE
2. Ma Y, Zeng Y, Ren X, Zhong N (2011) User interests modeling based on multi-source personal information fusion and semantic reasoning. In: Zhong N, Callaghan V, Ghorbani A, Hu B (eds) Active media technology, Lecture notes in computer science, vol 6890, pp 195–205. Springer, Berlin, Heidelberg
3. Smailovic V, Striga D, Podobnik V (2014) Advanced user profiles for the smart social platform: reasoning upon multi-source user data. ICT Innovations 2014, Web Proceedings ISSN 1857-7288, pp 258–268
4. Sun J, Xiong Y, Zhu Y, Liu J, Guan C, Xiong H (2015) Multi-source information fusion for personalized restaurant recommendation. In: Proceedings of the 38th International ACM SIGIR Conference on Research and Development in Information Retrieval. pp 983–986. ACM
5. Aghasaryan A, Betg´e-Brezetz S, Kodialam M, Mukherjee S, Senot C, Toms Y, Wang L (2009) Multi-source user profiling and keyword inference for personalized application enablement. In: Intelligence in next generation networks, 2009. ICIN2009. 13th International Conference. pp 1–5. IEEE
6. Heitmann B, Hayes C (2014) Semstim at the lod recsys 2014 challenge. In: Semantic web evaluation challenge, pp 170–175. Springer
7. Farseev A, Nie L, Akbari M, Chua TS (2015) Harvesting multiple sources for user profile learning: a big data study. In: Proceedings of the 5th ACM on International Conference on Multimedia Retrieval, pp 235–242. ACM
8. Veloso B, Malheiro B, Burguillo JC (2015) A multi–agent brokerage platform for media content recommendation. Int J Appl Math Comput Sci 25(3):513–527
9. Herlocker JL, Konstan JA, Borchers A, Riedl J (1999) An algorithmic framework for performing collaborative filtering. In: Proceedings of the 22nd Annual International ACM SIGIR Conference on Research and Development in Information Retrieval, pp 230–237. ACM
10. Basu C, Hirsh H, Cohen W et al (1998) Recommendation as classification: using social and content-based information in recommendation. In: AAAI/IAAI. pp 714–720

Metrics and Indicators of Online Learning in Higher Education

Maria José Sousa and Teresa Mourão

Abstract This article aims to present research regarding the effectiveness of online learning within the domain of Higher Education, as well as to suggest means of measuring it within other contexts of online learning. The research's purpose consists on identifying and analyzing different environments and practices of online learning, as well as how to proceed with its measurements, with the intent of involving its different stakeholders in decision-making processes for quality improvement within such learning practices. This study is based on a bibliographic review with the purpose of providing solid answers to the following research questions: (Q1) What are the Higher Education contexts in which online learning can occur? (Q2) What are the key metrics that allow us to measure the effectiveness of online learning in Higher Education? The results that were found from the literature review identify that the contexts of online learning constitute themselves as innovative pedagogical models that empower students, facilitating and promoting their learning process (Q1). Furthermore, the main metrics regarding the different Pedagogical and Technological Dimensions, in respect to the domain of Higher Education, and the realization of the importance of its measurement within various dimensions, and for the multiple elements inherent to learning processes, have also been identified (Q2). The dynamics of digital learning in Higher Education are framed by the contexts and processes in which they take place, they are also mobilized by its actors, and become measurable, through indicators and metrics which constitute themselves as core moderating elements for the success of online education in this domain.

Keywords Higher education · Learning technologies · Learning processes · Online learning contexts · Online teaching metrics · Learning analytics

---

M. J. Sousa (✉)
Political Science and Public Policy Department, ISCTE—University Institute of Lisbon, Lisbon, Portugal
e-mail: maria.jose.sousa@ISCTE-IUL.pt

T. Mourão
ISCTE—University Institute of Lisbon, Lisbon, Portugal

A. Ullah et al. (eds.), *Proceedings of International Conference on Information Technology and Applications*, Lecture Notes in Networks and Systems 350,
https://doi.org/10.1007/978-981-16-7618-5_61

1 Introduction

In the last decade, online education has been characterized by rapid and profound advances resulting from social and economic changes with major impacts on education systems, specifically in Higher Education. The emergence of online learning methodologies and tools is happening at a vertiginous rate, where knowledge assumes a central role of flexible access in terms of space and time.

At the same time, digital learning assumes a multitude of forms, which need to be identified, classified and measured, and also encompasses a range of actors, from students and teachers, to scientists, designers, software developers, institutional leaders, and government policymakers alike.

As such, it is against this background that, in the present study, we seek to discover how the effectiveness of online learning in Higher Education can be measured. We propose to do so by presenting several metrics/indicators that emerged from a previous literature review, which resulted in the following research questions: (Q1) What are the Higher Education contexts in which online learning can occur? (Q2) What are the key metrics that allow us to measure the effectiveness of online learning in Higher Education?

In conceptual terms, the literature review is established from basic concepts, transversal to the thematic of online education: Digital Learning, as an educational approach that integrates multiple contexts, modalities, and digital tools and, Digital Learning Analytics, which promotes data analysis regarding education in general, obtained from virtual learning environments, for educational improvement and its corresponding decision making processes, combining the management of learning processes with the creation of more effective environments. This study continues with the presentation of results, highlighting context analysis proposals and metrics related to digital learning in Higher Education, from which will be highlighted, among others—in the Discussion session—the emerging concepts of m-learning (mobile learning) and Social Network Learning, as the current methods used by universities for data collection and analysis, based on Web Analytics and Social Network Analytics. The study is, finally, closed by demonstrating the potential that online learning in Higher Education holds in increasing the academic performance of students, teachers, and the quality of teaching practiced in these institutions. Furthermore, critical dimensions that affect data collection and analysis in online education have also been listed.

2 Methodology

The methodological approach of the research was qualitative and the main technique to collect and analyze data was content analysis from the literature review of papers on Pedagogical and Technological Dimensions and Digital Learning Analytics. For this paper, a bibliometric research was performed using B-On, which is a research resource that allows access to scientific texts included in Elsevier, WOS, Sage and Springer databases, among others.

3 Presentation of Results

As a result of the conducted systematic literature review, the following table that systemizes several dimensions, that need to be addressed in order to facilitate the implementation of online education metrics analysis, is presented. It is important to note that digital learning pedagogies and technologies supports digital learning, using technology with the aim to improve the quality of teaching and engaging participants in the learning process, as shown in Table 1, which also presents online learning metrics, arising from the literature review, across the following dimensions listed.

All of the tools listed in Table 1, when used in learning environments supported by innovative pedagogical models, can contribute to facilitating student learning and improving academic performance.

4 Discussion

From the literature review, we were able to verify that the number of existing articles on the subject of online education, its effectiveness, efficiency and quality, is a trend. Many authors have proposed the systematization of concepts and practices related to Digital Learning, Learning Analytics, and Learning Analytics Contexts.

A common starting point in this bibliographic review can be observed, which is expressed in the idea that in order to assess the quality of online education, it is essential to first identify the various forms in which it can occur, since this type of education includes a wide range of practices and can be configured from anywhere at any time, involving multiple factors including cognitive-behavioral factors, such as intention, awareness, and the expectations that will influence the results of the learning process.

Creating a synthesis framework that would help identify and understand the methodologies and tools for digital learning was then the first step of this study, in order to answer our initial research question: What are the Higher Education contexts in which online learning can occur?

In order to achieve this objective, we proceed with a proposal for the systematization of the different Dimensions for Pedagogies and Technologies in Higher Education (Table 1), based on a theoretical framework linked to concepts such as Digital Learning and Digital Learning Methodologies—adapted from previous publications of our own—of which resulted four domains: (1) Collaborative Digital Learning Contexts, as spaces, facts or learning environments that support innovative pedagogical models and that are linked to collaborative work and communities of

Table 1 Metrics/indicators of online learning in higher education based in literature review

Pedagogical and technological dimensions	Metrics	Authors
Collaborative digital learning contexts (target students): Collaborative Communities; Cooperative learning; Collaborative learning; Network participation	New knowledge. Learning outcomes. Assessments Number of active participations Number of nodes in the network Number of students in each node of the network Number of students in each community	Abdulmajed, H., Park, Y.S., Tekian, A. (2015), Alhajri, S. (2016), Al-Jaber and Al-Ghamdi [1]; Amory, A. (2014); Barber, et a;l. [2]; Batez [3], Chen et al. [4], Curwood, P., Scott, J., Carvalho, L., and Simpson, A. (2015), Epure, M. and Mihães, L.C. (2017), Friend and Militello [5], Guerra, W.J.G., Hernández, M.A.M., and Pírez, L.E.R. (2014), Kocaman-Karoglu, A. (2016), Kosonen et al. [6], Lau [7]; Le [8], Liyanagunawardena et al. [9]
Learning environments: Learning Management Systems (LMS); Social Network Learning; Mobile learning; YouTube; Facebook; Instagram; Wikipedia; LinkedIn; Google; Websites eLearning; Mobile learning; Learning	YouTube Analytics. Google Analytics. AdWords Social network analytics	
Object repository; Blended learning; Blackboard; Moodle Learning Manager; Twitter; Videoconferencing; MOOC—massive open online courses		Mantri, A. (2014), Martin-Garcia et al. [10], Masterman [11], McMahon et al. [12], McNaughton et al. [13], Moorefield-Lang and Hall [14], Muñoz González, J. M., García, S. R., and Pichardo, I. C. (2015), Nielsen and Hoban [15], Nourkami-Tutdibi et al. [16], Obaid et al. [17], Persky et al. [18], Piñero Martín et al. [19], Rai et al. [20], Rudow and Sounny-Slitine [21], Salmon et al. [22], Sohrabi and Iraj [23], Stansbury and Earnest [24], Stewart [25], Sungkur et al. [26]; Tena et al. [27], Trotskovsky and Sabag [28]
Digital learning processes: Flipped classroom using digital media; Experiential online development; Open educational practice; Online learning environments; Technology integrated teaching methods; Digital storytelling; Educational games; Augmented reality; Web-based video; Digital video; Webinars	Feedback course start/course completion Test/exam results. Skills levels. Academic performance assessments Course access points. Time in system. Number of clicks and accesses Number of game wins. Number of views. Number of webinars/videos/games Number of participants in webinars/games Number of accesses to Open Education Platforms Number of new digital learning experiences	
Digital learning facilitators:	Evaluation score	

(continued)

Table 1 (continued)

Pedagogical and technological dimensions	Metrics	Authors
Project based-learning; problem based-learning; active learning; gamification; simulation; narrated stop-motion animation	Number of simulations Number of problems solved Number of projects designed Number of implemented projects Number of companies/institutions involved in the pedagogical practices Involvement of students in learning activities Involvement of students through educational resources or tools Involvement of students in discussion activities	Unger et al. [29], van der Keylen et al. [30], Wood and Bilsborow [31], Xu [32]

Source Sousa et al. [33, 34] adapted

practice that empower students, while, at the same time, facilitating and promoting learning processes; (2) Learning Environments as systems that help the management of learning processes by using interactive features such as online discussions, videoconferencing, and discussion forums in order to enhance the learning outcomes of university students—for instance, the current Mobile Learning; Social Media Learning and, the more traditional, Moodle and MOOC—Massive Open Online Courses; (3) Digital Learning Processes: Computer-based tools that use technological means, such as the internet, to facilitate learning processes through students' motivational and involvement procedures, examples of which are Flipped Classrooms using Digital Media, Open Educational Practices, and Webinars; (4) Digital Learning Facilitators: Participatory teaching approaches, which refer to student involvement from the resolution of problems of practical and concrete nature, different from the traditional teaching methods, Project/Problem-based Learning, Gamification, Simulation and Authentic Learning.

Furthermore, it is important to emphasize that, from this synthesis, many of the emerging practices that have been presented have come to revolutionize the Higher Education Teaching System and its respective online learning processes. We refer to their use, both by students and teachers alike, be it in the praxis of the academy or its surroundings. Collaborative digital resources and learning facilitators, whose teaching approaches allow students to better explore, discuss, build models, engage in projects and enhance their problem solving skills—through a set of innovative educational techniques and tools, guided by teachers and experts—enable, according to the literature review, a teaching–learning dynamic with greater motivation, preparation and an enhancement in its respective results, not only in university, but also in the students' own daily lives (e.g., Authentic Learning Methodologies, Mobilization of Learning in Life Skills).

On the other hand, learning systems based on m-learning or Social Network Learning have, as a strong point, the power of motivating new generations of students who are familiar with this type of devices and tools—especially as leisure tools—by enabling them to use such means, in a broader context, as facilitators of learning processes within an academic environment, namely due to their multifunctionality, accessibility, flexibility, and personalization.

In addition, it is also worth mentioning the importance of online platforms, whose use has expanded, as an alternative to physical contexts, and as an interactive teaching–learning space that connects the teacher-student dyad and each of them with their peers.

Having identified the Pedagogical and Technological Dimensions, we then proceed to the definition of online learning metrics and indicators in Higher Education (Table 1), in order to answer our second research question: What are the key metrics that allow us to measure the effectiveness of online learning in Higher Education?

Our research points to the importance of measuring the various dimensions/elements of the learning process in Higher Education in regard to their respective contexts, for which we defined the following criteria—whose metrics are specified in Table 1—which will now be explained: (1) Collaborative Digital Learning Contexts (Target: Students), in this dimension we focused specifically on university students and on the impact that Digital Collaborative Pedagogical Models—promoted by the academy—have on the performance and quality of online learning carried out by these actors. The dynamics promoted through these environments are essential for the participation of students, with the additional moderation done by teachers, leading to the promotion of debate, exchange of ideas and stimulus for the production of answers to the considered learning questions, as well as knowledge construction. For this purpose, we fundamentally used the adherence criteria, participation, and results; (2) Learning Environments appear in our analysis, in generic terms, linked to Learning Management Systems, as support tools for the creation and management of environments, in this case, related to Higher Education. They enhance online learning through the use of platforms designed from pedagogical methodologies for the production and dissemination of components, resources, and materials present in this type of education. The resources made available by these structures allow for the planning, management, and control of online university education and have, as their main objective, to intervene and stimulate in an innovative, complete, and dynamic way these learning processes. Our proposal for effectiveness, efficiency, and quality of the identified structures was also presented in a generic way, through Big Data quantifiers present in Web Analytics, namely Google Analytics; AdWords; YouTube Analytics, among others. These types of platforms allow us to use a full set of features for the collection and systematization of information and it will be up to each entity/organization of Higher Education to select its data matrix for proper analysis according to its quality metrics. This topic also discusses other online learning resources/systems used in academia such as Mobile Learning and Social Network Learning. According to the conducted literature review, the main methods used by institutions for data collection and analysis used by universities are Web Analytics and Social Network Analytics; Web Analytics being the collection of data from the use of web pages; and Social Network Analytics being the collection of data by the use of social networks; (3) Digital Learning Processes, in this scope, were mainly presented online methodologies in the learning processes, based on technological means and open educational resources, such as Experiential Online Development/Open Educational Practice/Online Learning Environment and Webinars, in which the involvement of students is appealed, being called to intervene in such online environments, from an active participation, based on constructivism, for new knowledge development, as well as for the progression of their autonomy and independence. Here, the teacher assumes the role of a moderator/guide, which is why the teacher's feedback, toward the students, is of the utmost privilege. The suggested metrics focus, essentially, on the adherence of students to these proposals, their permanence and participation (accesses/time spent and clicks), quality of their interventions, assessment of skills (test/exams evaluations and academic performance). Some computer-based tools and innovative technological solutions were also presented, which aim to integrate such resources through the introduction of different learning styles, motivational, creative, and participatory involvement of students—either through Digital Storytelling, from technological supports with integration of Augmented Reality, Virtual Reality, 360° Cameras, among others—or even Videos, Games, Web-Based Video/Digital Video and Educational Games. The criteria are also innovative in respect to the traditional ones, and are based on the adhesion, participation and performance (Number of game participants; Number of videos/ games; Number of game wins and Number of views); (4) Digital Learning Facilitators: In this domain, we found participative-motivational approaches, which direct the online learning process of university students, adapted to concrete conditions, toward the practical resolution of real problems. The first methodologies presented are generally projects, either intrinsic to the academy itself, or linked to its surroundings, namely with a connection to the business domain. In this way, students work on a project for a significant period of time (e.g., a semester), involving the resolution of a given problem, with significant impact in motivational terms, which results in a public presentation of a report or, even, of a specific product. These participatory teaching approaches, which refer to the involvement of students by solving practical and concrete real-life problems, are different from the traditional teaching methods: Project/Problem-based Learning; Gamification; Simulation and Active/Authentic Learning. For this type of proposal, students are tutored by the teacher and the skills to be assessed are related to learning content obtained through know-how, creativity, critical thinking development, mutual aid, and communication skills (vd. Table 1). We found yet another set of increasingly used motivational and engagement methodologies: Gamification, which consists of applying components normally associated with games to academic subjects and facilitating the learning process through multimedia environments, and Simulation, an experimental educational experience developed from digital immersive games, which involve students in learning experiences linked to risk-free simulations, close to its real counterparts. These practices represent a greater stimulus, with greater involvement and effectiveness in student learning outcomes as well. The metrics identified have to do mainly with the number of games/simulations and its corresponding obtained scores.

In summary, through the literature review, our research questions were answered by identifying the teaching–learning contexts and respective metrics and indicators for measuring the quality of online education in Higher Education. It is, however, up to each entity to identify the metrics and domains that best suit them, and, from there forward, to proceed with the use of analytical tools responsible for the processing of data collection, analysis, use, and dissemination, in the form of detailed descriptions and reports, in order to provide guidance with the purpose of facilitating decision making processes for the quality, effectiveness, and efficiency of their academic institutions.

Although it has been proven that online education allows for the enhancement of learning potential, inevitably increasing the quality of Higher Education and academic performance of students, teachers, managers, and other agents involved in this environment, it is nevertheless necessary to take into consideration some of the reported obstacles/ difficulties, from the literature review, regarding this teaching modality, namely for its actors–students (lack of technological resources to access this type of education, lack of understanding on how to properly use the proposed virtual tools and difficulties in time management); teachers (lack of specific training regarding methodologies, difficulties in conducting online teaching strategies, difficulty in assuming their multifunctional role as teacher/tutor/guide/mediator/facilitator and limited support from the educational institutions); managers and university educational institutions (difficulties related to the complex nature of these systems, as well as the management of large volumes of data and their working time availability; difficulty in the coverage of technical support given to the involved agents).

Furthermore, it is also worth highlighting, from this systematic literature review, the contribution of Wolfgang Greller and Hendrik Drachsler, who propose six critical dimensions that affect online education analysis: Internal Limitations, External Constraints, Instruments, Objectives, Data, and Stakeholders. For each dimension, several interpretations can be provided. Internal Limitations may include academic skills offered by Universities; External Constraints may impede the collection and analysis of online education data. These are, for instance, legislation, standards, and conventions. The instruments used to collect and perform data analysis can also result in positive or negative impacts on the obtained results, them being the basis for decision making. In addition, there is also influence on the type of data and the purpose for which this data is used—the Learning Objectives. Data sets can be open (public) or closed (accessible to a limited number of people) and, therefore, used for reflective or predictable processes. The final critical dimension is that of the stakeholders involved in the process of analyzing the multiple dimensions of online education: the Higher Education institutions, its students, the government, society in general, and others.

The collection and analysis of learning metrics, considering multiple methods and factors, are, therefore, of dynamic and complex nature. Examples of this are the many ethical and legal issues of analyzing student learning data and the possibility of unauthorized entities to access this data, a topic worth discussing in another knowledge forum.

5 Conclusions

At a time when online education is experiencing multiple and profound advances, becoming an everyday practice, the present article sought to conduct a survey regarding the effectiveness of online education in Higher Education. Our contribution was to systematize the scientific information made available by various studies on this topic, bridging the lack of understanding about Digital Learning Analytics associated with the different Pedagogical and Technological Dimensions. To this end, the research resulted from a literature based on the concepts of Pedagogical and Technological Dimensions, teaching approaches that integrate multiple contexts, modalities, digital tools, and Learning Analytics—which promote the analysis of data on education, obtained in virtual learning environments, for the improvement of teaching and its decision processes, combining the management of learning processes with the creation of even more effective environments. To achieve the proposed objectives for this study, we started from the following research questions: (Q1) What are the Higher Education contexts in which online learning can occur? (Q2) What are the key metrics that allow us to measure the effectiveness of online learning in Higher Education? These questions helped us to operationalize this research and we proceeded to systematize the different Pedagogical and Technological Dimensions for Higher Education, through the following categories: Collaborative Digital Learning Contexts (innovative spaces for collaborative work that enhance learning processes); Learning Environments (learning process management systems); Digital Learning Processes (computer-based tools that use technological resources to facilitate learning processes) and Digital Learning Facilitators (participatory teaching approaches, based on problem-solving approaches through multimedia and playful environments). For each of these dimensions in Higher Education, the key metrics were identified and the importance in measuring the various pedagogical and technological elements inherent in the learning processes in Higher Education has also been highlighted. However, some limitations of this research worthwhile to point out, namely the time frame of the articles search for the literature review, and the keywords used to make the search, which could include other dimensions besides technologies and pedagogies for digital Higher Education. To summarize, the dynamics of digital learning in Higher Education are framed by the contexts and processes in which it takes place, it is also mobilized by its actors, and becomes measurable, through indicators and metrics which are the core moderating elements for the success of online education in this domain.

References

1. Al-Jaber MA, Al-Ghamdi SG (2020) Effect of virtual learning on delivering the education as part of the sustainable development goals in Qatar. Energy Rep 6:371–375
2. Barber W, King S, Buchanan S (2015) Problem based learning and authentic assessment. Electronic Journal of e-Learning 13(2):59–67
3. Batez M (2021) ICT skills of university students from the faculty of sport and physical education during the COVID-19 pandemic. Sustainability (Switzerland) 13(4):1–13
4. Chen L, Chen TL, Chen NS (2015) Students' perspectives of using cooperative learning in a flipped statistics classroom. Australas J Educ Technol 31(6):621–640
5. Friend J, Militello M (2015) Lights, camera, action: advancing learning, research, and program evaluation through video production in educational leadership preparation. J Res Leadersh Educ 10(2):81–103
6. Kosonen K, Ilomäki L, Lakkala M (2015) Using a modeling language for supporting university students' orienting activity when studying research methods. J Interact Media Educ 1(1):1–15
7. Lau KH (2014) Computer-based teaching module design: principles derived from learning theories. Med Educ 48(3):247–254
8. Le AT (2021) Support for doctoral candidates in Australia during the pandemic: the case of the University of Melbourne. Stud High Educ 46(1):133–145
9. Liyanagunawardena TR, Lundgvist K, Williams SA (2015) Who are with us: MOOC learners on a Future Learn course. Br J Edu Technol 46(3):557–569
10. Martin-Garcia A, Serrano MJ, Gomez M (2014) Fases y clasificación de adoptantes de blended learning en contextos universitarios. Aplicación del análisis CHAID. Revista Española de Pedagogía 259:457–476
11. Masterman E (2016) Bringing open educational practice to a research-intensive university: prospects and challenges. J Article 14:31–42
12. McMahon CJ, Tretter JT, Faulkner T, Krishna Kumar R, Redington AN, Windram JD (2020) Are e-learning Webinars the future of medical education? An exploratory study of a disruptive innovation in the COVID-19 era. Cardiol Young 1–10
13. McNaughton S, Westberry N, Billot J, Gaeta H (2015) Exploring teachers' perceptions of video-conferencing practice through space, movement and the material and virtual environments. Int J Multiple Res Approaches 8:87–99
14. Moorefield-Lang H, Hall T (2015) Instruction on the go: reaching out to students from the academic library. J Libr Inf Serv Dist Learn 9:57–68
15. Nielsen W, Hoban G (2015) Designing a digital teaching resource to explain phases of the moon: a case study of preservice elementary teachers making a slowmation. J Res Sci Teach 52(9):1207–1233
16. Nourkami-Tutdibi N, Hofer M, Zemlin M, Abdul-Khaliq H, Tutdibi E (2021) Teaching must go on: flexibility and advantages of peer assisted learning during the covid-19 pandemic for undergraduate medical ultrasound education—perspective from the "sonobystudents" ultrasound group. GMS J Med Educ 38(1):1–7
17. Al-Youbi AO., Al-Hayani A, Bardesi HJ, Basheri M, Lytras MD, Aljohani NR (2020) The King Abdulaziz University (KAU) pandemic framework: a methodological approach to leverage social media for the sustainable management of higher education in crisis. Sustainability (Switzerland) 12(11)
18. Persky AM, Fuller KA, Jarstfer M, Rao K, Rodgers JE, Smith M (2020) Maintaining core values in postgraduate programs during the covid-19 pandemic. Am J Pharm Educ 84(6):697–702
19. Piñero Martín ML, Esteban Rivera ER, Rojas Cotrina AR, Callupe Becerra SF (2021) Trends and challenges of latin american graduate programs in COVID-19 contexts. Rev Venez Gerencia 26(93):123–138
20. Rai SS, Gaikwad AT, Kulkarni RV (2014) A research paper on simulation model for teaching and learning process in higher education. Int J Adv Comput Res 15(4):582–587
21. Rudow J, Sounny-Slitine MA (2015) The use of web-based video for instruction of GIS and other digital geographic methods. J Geogr 114(4):168–175

22. Salmon G, Gregory J, Dona KL, Ross B (2015) Experiential online development for educators: the example of the Carpe Diem MOOC. Br J Edu Technol 46(3):542–556
23. Sohrabi B, Iraj H (2016) Implementing flipped classroom using digital media: a comparison of two demographically different groups perceptions. Comput Hum Behav 60:514–524
24. Stansbury JA, Earnest DR (2017) Meaningful gamification in an industrial/organizational psychology course. Teach Psychol 44(1):38–45
25. Stewart B (2015) Open to influence: what counts as academic influence in scholarly networked Twitter participation. Learn Media Technol 40(3):287–309
26. Sungkur RK, Panchoo A, Bhoyroo NK (2016) Augmented reality, the future of contextual mobile learning. Interactive Technology and Smart Education 13(2):123–146
27. Tena R, Almenara J, Osuna J (2016) E-Learning of Andalusian University's Lecturers. Turkish Online J Educ Technol 15:25–37
28. Trotskovsky E, Sabag N (2015) One output function: a misconception of students studying digital systems—a case study. Res Sci Technol Educ 33(2):131–142
29. Unger D, Kulhavy D, Busch-Peterson K, Hung IK (2016) Integrating faculty led service learning training to quantify height of natural resources from a spatial science perspective. Int J High Educ 5(3):104–116
30. van der Keylen P, Lippert N, Kunisch R, Kühlein T, Roos M (2020) Asynchronous, digital teaching in times of covid-19: a teaching example from general practice. GMS J Med Educ 37(7):1–8
31. Wood D, Bilsborow C (2014) 'I am not a person with a creative mind': facilitating creativity in the undergraduate curriculum through a design-based research approach. Electron J e-Learning 12(1):111–125
32. Xu H (2016) Faculty use of a learning object repository in higher education. J Inf Knowl Manag Syst 46(4):469–478
33. Sousa MJ, Rocha A (2018) Corporate digital learning—proposal of learning analytics model. In: Rocha A, Adeli H, Reis LP, Costanzo S (eds) Trends and advances in information systems and technologies WorldCIST'18. Advances in intelligent systems and computing. Springer, Berlin, p 745
34. Sousa MJ, Cruz R, Martins JM (2017) Digital learning methodologies and tools—a literature review. Edulearn17 Proc 5185–5192

Youth and Adult Education (YEA) and Distance Education in the Web of Science (WoS) Database from 2000 to 2020: Bibliometry

Alcione Félix do Nascimento, Luciana Aparecida Barbieri Da Rosa, Raul Afonso Pommer Barbosa, Maria Carolina Martins Rodrigues, Larissa Cristina Barbieri, and Maria Jose de Sousa

Abstract This study seeks to analyse the characteristics of publications related to Youth and Adult Education (YEA) and Distance Education in the Web of Science (WoS) database from 2000 to 2020. The main thematic areas, year publications, research field, institutions, authors, title of sources, countries, languages; research areas and finally, a relationship between authors with more publications and citations. For this, the methodology of quantitative bibliometric study was adopted. The results of the research are presented, highlighting the main characteristics of scientific production in the WoS database related to the terms education of young people and adults and distance education in the period of study, resulting in 42 articles, there was also a growth at from 2015, highlighting the importance of the themes for teaching students' learning. 316 citations were identified, of the 10 (ten) most cited articles on youth and adult education and distance education, in the last twenty years, in the research carried out on the Web of Science. The conclusions show productive trajectories for development for research in the area.

Keywords DL · Bibliometry · YAE

A. F. do Nascimento · L. A. B. Da Rosa · L. C. Barbieri
IFRO, Vilhena, Brazil

R. A. P. Barbosa (✉)
Fundação Getulio Vargas FGV-EAESP, São Paulo, Brazil

M. C. M. Rodrigues
Universidade de Algarve, Faro, Portugal

M. J. de Sousa
ISCTE, Lisbon, Portugal

A. Ullah et al. (eds.), *Proceedings of International Conference on Information Technology and Applications*, Lecture Notes in Networks and Systems 350,
https://doi.org/10.1007/978-981-16-7618-5_62

1 Introduction

1.1 Introductory Aspects and Issues of the Study

The different metamorphoses that have occurred in the environment in the last decades have impacted the teaching–learning process of the students. It is known that the school environment in regular education and the educational institutions face some challenges. Negreiros et al. [23] refer to the emergence of YAE as a process that challenges the exclusion aspects of the Educational System due to the history of failures or dropouts.

Saenger et al. [29] highlight distance learning as a teaching modality related to educational processes today, providing opportunities for students at all social levels, including those who need to work and study. In this sense, the opportunity to complete elementary and high school emerges, thus allowing social inclusion and a rescue of citizenship in the teaching of Youth and Adult Education modality.

Still in this context, it is noteworthy that the YAE education modality has some specificities that encompass other views, i.e. the teaching–learning process of these students should be promoted differently from the regular education modality.

The Youth and Adult Education (YAE) consists of the opportunity of schooling for this public who, for other reasons, could not follow their studies at the age considered normal for their age group. For this reason, when the student seeks a schooling certification, he or she is faced with the chance of acquiring new knowledge and its expansion, as well as to resize their interpersonal relationships resulting from this movement that takes place in the school environment (Brasil 2005).

Therefore, YAE students sometimes belong to the same social class, have low or no purchasing power, have only the basics for their subsistence and their main means of information and leisure is limited to television programmes, which results, in a way, in a scenario of social maladjustment and limitation, where the search for school occurs through personal perspectives, motivated by an expectation of changing their reality [3].

In this sense, seeking to broaden the discussion, the theme distance education comes to meet as a perspective of adding in Youth and Adult Education the digital inclusion of these students having as a tool in the process the use of ICTs helping in the social integration founded.

Technological advances arise as a stimulant for teaching innovation, presenting as a result the remodelling in the ways of learning, thus, the several modifications imposed to the individual, coming from the Information and Communication Technologies (ICTs), which provide and dynamize the global connection within the didactic model of education. Today, different studies positively approach concepts of ICT application in the educational process [7]. In this same direction, this is observed as the main challenge, given that the quality of higher education courses is directly linked to the qualification of the teachers involved [14].

Distance Education (DL) also acts as a complement to YAE having an important role in the teaching–learning process, in this sense the DL teaching modality necessarily arises with the advance brought by new technologies that present an understanding in the teaching–learning process and as well as several tools necessary for the development of communication between students and teachers, being a basic requirement for interaction in DL, for this reason the student will have to develop a basic skill in contact with technology, especially the internet [21, 24].

Based on the assumptions listed above, the central question of this study emerges:

What are the contributions of publications on youth and adult education and distance education in the database Web of Science in the period from 2000 to 2020? In this sense, the general objective of the study is to analyse the characteristics of publications related to youth and adult education and distance education in the Web of Science database in the period from 2000 to 2020. It has as specific objectives: (1) to identify the characteristics of the state of the art of youth and adult education and distance education; and (2) to verify the publications at national and international level regarding countries, institutions, authors and language;

2 Theoretical Background

2.1 Conceptions About Professional and Technological Education

The distinctive governmental strands focused on Vocational and Technological Education EFA happened in the 1990s, being through the edition of Law no. 8. 948, of December 08, 1994, being established the System and the National Council of Technological Education, in the same way that allowed the development and expansion of the Federal Network of Technological Education resulting from the potential transformation of all ETFs at the time in CEFETs, with the intention of spreading the offer of Higher Courses of Technology (CSTs), and thus enabled the formation of a workforce potentially qualified and recognized for companies [2].

Through the bewildering legislation generated by the Brazilian public spheres focused on the EFA standards, it is noteworthy that it was a period of advances and setbacks. In the period of Lula's government (2003–2011) new commitments and changes took place aiming to ensure better living conditions to the Brazilian population regarding education. In this sense, Frigotto and Ciavatta [9] criticize that these changes would have as their main character the large companies that would have qualified employees.

In this sense, Bastos [4] emphasizes that there is a gap in this argument, since this same State is also constituted by social actors that defend the interest in maintaining a mitigated type of education, aimed at professionals with low salaries, since linked to qualification, contradictorily, the capitalist mode of production requires the production of a reserve army.

The results of these public policies for EFA have not stood out since the 1990s, passing through different governments. However, this reconstruction did not happen, and the Lula government decided to create focused programmes, such as the Factory School, Integration of Vocational Education to Secondary Education in the form of Youth and Adult Education (PROYAE) and Youth Inclusion (PROJOVEM) [10].

Thus, it is possible to highlight the need for the development of new professional skills that are required in the world of work among several innovations we have, the encouragement of autonomy, teamwork, creativity and innovation, for this reason the institutions of professional education have placed the research as a pedagogical basis as well as the work as one of its essential principles within education [8].

In this same direction, we can understand the importance of fostering the training of professional education teacher, developing diverse skills, taking into account the teaching modality employed, being necessary to the effectuation of strategies and practices that integrate scientific, technological, social and humanistic knowledge, which make up the core of general and universal knowledge, so that it takes advantage of educational opportunities that each modality comes to offer [17].

The next topic will present the different optics of the legislation on the vocational education bias.

2.2 *Youth and Adult Education—YAE and Its Aspects*

In the historical trajectory of YAE, one realizes that its beginning occurred in the period still with the teachings of the Jesuit priests during the development of the catechization work to teach the first letters still in the Empire period. In this way, we can attribute that the literacy of adults had as main interest for reading and the study of catechism, as well as the development of small tasks of everyday life [25].

In this sense, the consolidation of the public system in Brazil, which took place during the 1930s, highlights the fact that Youth and Adult Education started to have its importance in the country, which was going through major changes resulting from the industrialization and urbanization process, among which was the offer of free public education, reaching increasingly different social sectors [27].

In this perspective, the Constitution of 1934 instituted the creation of a National Education Plan, which recognized adult education as an obligation and duty of the State, including in its rules, the provision of full primary education also free, but with the condition of compulsory attendance, extended to adults (Brasil 2000).

Still in this context, YAE denotes a modality aimed at young people and adults who have not had access to regular education [22]. Viegas and Moraes (2017). They complement themselves of the importance in the teaching support of the YAE public since it is an education with archable and adapted for this public.

Haddad and Di Pierro [12] point out that the LDB was the beginning of discussions about the guarantee of citizens' rights, as well as the duties of the political spheres, identifying the educational assignment in the political-historical environment, seeking reflection on the diversity of the distinct individuals of the YAE.

In a special way, basic education is assumed by the LBD as a public policy in the national education system. In this sense, YAE is strategically inserted as a government project focused on specific literacy and schooling levels, promoting the full development of the individual.

In this vein, Machado [17] argues that the approval of the new LDB is a decisive act and a key point in the so-called reconfiguration of the field since it gives it magnitude in the premises of the citizens' right to the school environment. It is important to point out that the supply of YAE is the government's responsibility, enabling these individuals to take wings to a better qualification and quality of life.

Ventura [31] explains that based on the records and historical approaches on Youth and Adult Education (YAE) within the context of Brazilian education, it was possible to understand that the modality was conducted for decades based on a compensatory, utilitarian, emergency and discontinuous vision, with the predominance of weak policies from the institutional point of view, and light in terms of its quality in the educational process.

The LDB had its relevance, being considered as a legal landmark regarding the beginning of a differentiated educational perception for a public that has a diversified characteristic, either with its life history often disregarded by the rights already constituted in the Constitution, defining itself as a conception of youth and adult education, which reserves inferences to the development of people in a full way [13].

Youth and Adult Education is a teaching modality that has undergone major transformations in various aspects, having in its trajectory important changes to achieve advances and thus allow a model that takes into account the reality and experiences of this public that is differentiated, so the experiences, age and education level are key features for the survey of a diagnosis and thus develop a methodological process that complements this demand within the teaching modality.

Otherwise, we will find a space with an authoritarian conception with no relation of exchange between the students and the knowledge, which will result in the inexistence of contents and objectives with structured mechanisms, using methods that do not take into consideration the several experiences that brought experiences that can be worked on to complement and enrich the knowledge in the school space. Therefore, the two themes have been primordial in what concerns the teaching–learning process. Distance learning and YAE have been implemented in educational institutions.

In this sense, evidence of these themes will be presented below.

2.3 Distance Education and Its Importance in the Teaching–Learning Process

The emergence of writing provided a more effective communication, so that the transmitter and receiver did not need to be physically present at the same place and time. Distance Education (DL) provides respect for the student's learning pace, its cultural aspects disseminating and enabling new knowledge in a more comprehensive way, especially for people who did not have access to studies [20].

Giolo [11] highlights a study related to the accelerated growth of Distance Education in Brazil, having as a landmark its regulation still in 2005. This study covers the entire historical context of EaD Public Policies and with that a legal framework from the normative of its legislation allowed an acceleration in spreading this modality in Brazil.

However, Arruda [1], from a historical analysis considers that in Brazil the modifications in the school environment, would be the least affected in the relationship of the development of digital technologies, especially when we observe the changes and infrastructure of these spaces, as well as in the training of teachers.

For [15], the nature of the distance learning process has for years been presented as a stepping-stone in the educational environment, however, in a short period of time, it will become indispensable and far-reaching.

The understanding of DL seeks to reduce and overcome geographical obstacles that make it impossible to access learning and interaction between individuals in the educational environment and offers advantages when compared to the systematically traditional/presencial model of education. Distance learning forces the student to develop his/her own study routine, providing communication between students/teacher/tutors in a synchronous or asynchronous way, besides the emergence of new opportunities for interaction between groups of diverse regionalisms, because it is possible the interaction of students and teachers from several states (Preti 2000).

Maia and Meirelles [18] point out that the DL mode has been considered as a structured system, rooted in efficient methodological and evaluative procedures that provide the teaching–learning process of the students. However, it is necessary to analyse the student's profile, to list autonomy as the most important element, since the need for the development of activities is the level that needs to be developed independently. Thus, the increasing volume of knowledge that must be assimilated in the process of school education requires more rational organization of teaching methods [16].

It is noteworthy to highlight that this new educational model of ODL will require new skills from teachers so that they have an efficient teaching–learning process demanding mainly from teachers in continuous training, as well as quality in the level of teaching transmitted [26].

From this perspective, Marchisotti et al. [19] argue that the primordial age of the insertion of remote teaching with face-to-face teaching are important factors, in addition to teacher training for this dialogic relationship between individuals. In addition, DL will have to endure obsolete, coercive teaching, taking advantage of a plastered culture with already established standards that have an impact on learning. With this, we can say that computer networks provide favourable support so that the horizontal organization of learning can function in a multifaceted way with community creation and methodologies aimed at creativity and social connectivity [30].

In this sense, it is noted that the transformation goes through the scope of change related to human interdependence in the face of formative mobility for the reestablishment of dialogue, EaD even being mediated by technologies cannot be an educational space that distances the subjects, it is necessary to the sensitization, approach and recognition so that it puts knowledge into circulation, through digital culture, providing the opportunity to create citizens of the current or virtual world, even knowing that along the pedagogical trajectory the educator is recognized as a mere mediator of the processes of educating [6].

Thus, distance education in current times presents itself in front of a new scenario, where the training space of distance education gains great relevance in this period of social isolation, the development of various tasks that were previously performed only in person now provides an important space for advances and new learning processes in various areas and technological support and with the development of new methods and modalities of studies that enable this atypical period. This new reality is at a national and international level, so it is important to note the considerations of other authors who provide the basis for this confrontation and transformations.

Xiao and Yi [32] discuss this new scenario, faced initially in China, especially the direct impact that educational institutions, as well as the subjects that incorporate them, had to adapt. The authors corroborate the new reality that the pandemic brought when it imposed the closure of the entire classroom school system abruptly, such decision brought several discussions of a renewal of thought related to the dynamics of technological mediation at all levels of education in Chinese society.

Such institutional reconfiguration of the educational system was gradually reflected worldwide. In Brazil, such transformations in the school environment directly impact other educational modalities at all levels. We can exemplify the teaching modality of Youth and Adult Education (YAE), which attends to a public within a differentiated teaching context, and from this new reality it was necessary to add YAE to DE with learning mediated by digital technologies.

Next, the methodological aspects of the study will be presented.

3 Methodology

Nature of the research

The research is characterized by a bibliometric study of a quantitative nature, which analysed the state of the art in articles published between the years 2000 and 2020 about youth and adult education and distance education. Rousseau [28] and Camps et al. [5] define bibliometrics as a discipline that allows the quantitative study of scientific production, by analysing its very nature and the transfer of a science in each period.

The analysis of the researched themes was carried out in the Web of Science (WOS) database of the Institute for Scientific Information (ISI), considered to be one of the main multidisciplinary databases for scientific publication.

Graph 1 Document Types. *Source* Elaborated by the author (2020)

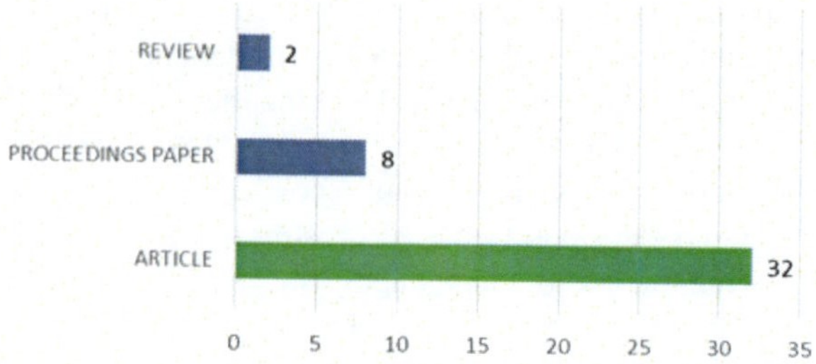

In the search field the keywords youth and adult education and distance education were entered, and thus it was possible to identify the state of the art on these subjects. Among the characteristics analysed, the main authors, year of publication, journals, institutions, countries, research areas, and the ten most cited articles in the twenty-year period were searched.

Study scope

In WOS there is a citation index, where it is mentioned for each article, the documents it cited and the documents that cited it (Graph 1).

The terms youth and adult education and distance education were searched in the Web of Science in the period 2000 and 2020, resulting in a total of 42 documents that cover the themes object of this study, thirty-two articles.

The research sought to analyse the main issues: main years of publication of the articles, field of research, educational institutions, authors, title of the sources, main countries, languages, areas of knowledge and the relationship between authors with more publications and the most cited.

Data collection

This study was divided into two stages; the terms youth and adult education and distance education were used in the WoS search field, delimiting the period between 2000 and 2020 and then a survey of the main characteristics of the publications was carried out.

In the second step, the most cited publications were compared with the authors who published the most in the same period. Figure 1 shows the stages of the research.

In this sense, the bibliometric analysis of this study is presented below.

4 Research Results and Analysis

Results and Discussion

Next, we present the results of the research, highlighting the main characteristics of the scientific production in the WoS database related to the terms youth and adult education and distance education in the period from 2000 to 2020, resulting in 42 articles.

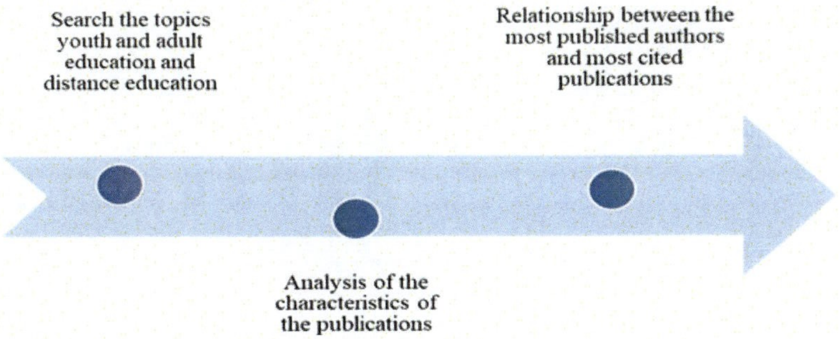

Fig. 1 Research stages. *Source* Elaborated by the author

General characteristics of publications on youth and adult education and distance education in the Web of Science

This item will present the main general characteristics of the publications related to the themes: 1. year of publications; 2. field of research; 3. main institutions; 4. main authors; 5. title of the sources; 6. main countries; 7. languages; 8. research areas and finally, 9. Relationship between authors with the most publications and the most cited.

Graph 2 shows the number of Web of Science articles related to the topic, published between the years 2000 and 2020. In this sense, there has been an increase since 2015, evidencing the importance of the themes for the teaching and learning of the students.

The research fields with the highest representation of publications were research in education (12), occupational environmental public health (6), followed by geography (3) and sociology (3), according to Graph 3. From this perspective, it is possible to infer, as listed above, that these themes still have a greater representation in the field of education, but there is multidisciplinary when we analyse the other areas that have studies on the subject.

The ten institutions that stood out the most and had research on the subject of youth and adult education and distance education are presented in Graph 4. Thus, the institutions that published the most were: University of California System (3) and University of California San Francisco (2), both based in the USA and Autonomous

Graph 2 Publication of years. *Source* Elaborated by the authors

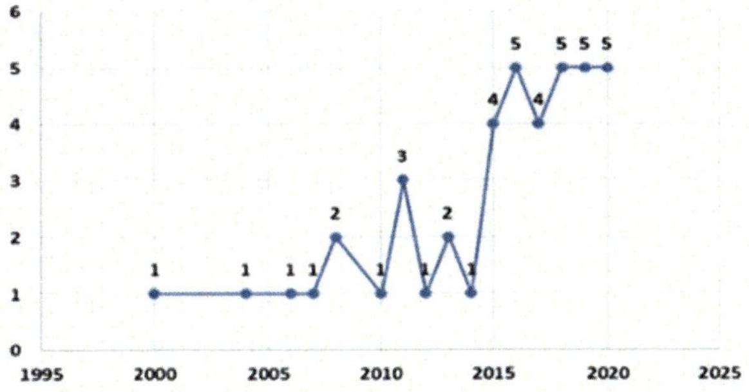

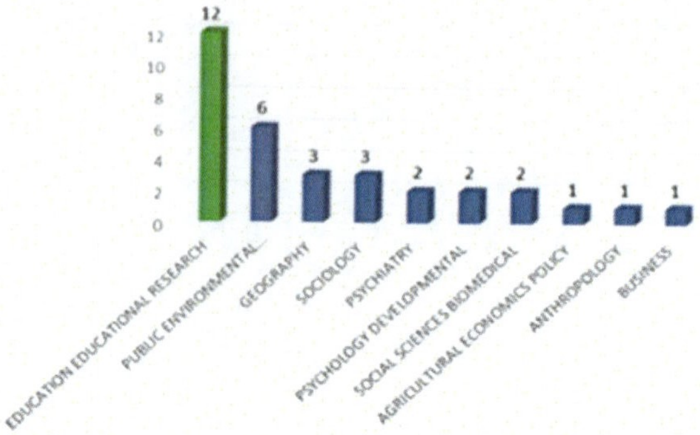

Graph 3 Research field. *Source* Elaborated by the authors

University of Barcelona (1) based in Spain. It is important to note that Brazil had no university that published on this theme when analysing the 20 years of publication in the Web of Science database.

Graph 5 shows the main authors who published articles in the analysed period. It was verified that among the ten authors who published the most on the theme, there was only one article per author during the period of analysis. Thus, it is also verified that there is no researcher who stands out when the themes of youth and adult education and distance education are analysed.

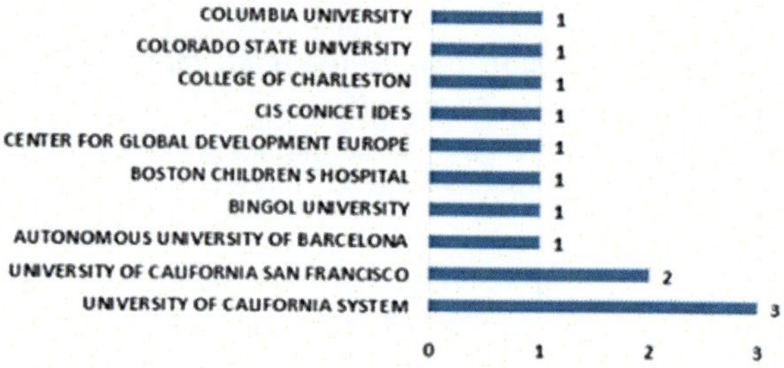

Graph 4 Main institutions. *Source* Elaborated by the authors

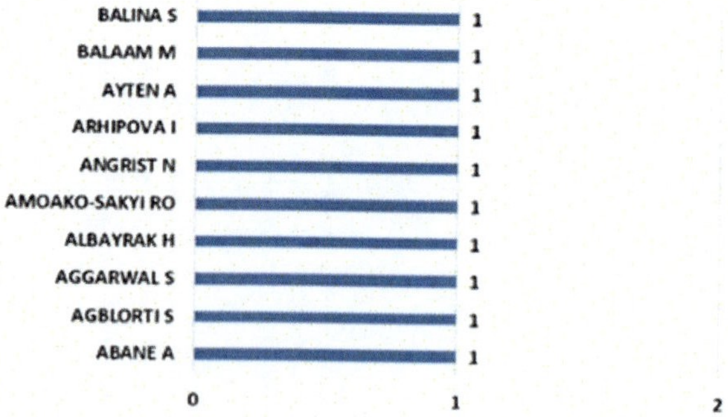

Graph 5 Main authors. *Source* Elaborated by the authors

Graph 6 shows the main sources of publications and number of published articles related to the investigated themes.

The journals and conference in the ranking of the ten that published the most articles involving the theme were: Childrens Geographies (2); Olhres Journal (2) and 12th International Conference of Education Research and Innovation (ICERI-2019) (1). It is important to note that the first three sources that published the most in the journals of the area belong to the area of education, highlighting the Journal Olhres which is from the Department of Education of Unifesp. In addition, there is also in the same amount publications in journals in distinct multidisciplinary areas.

The quantity of articles distributed by the main countries is shown in Graph 7. From the data above, it is noteworthy that the countries that have published the most were: USA (12), followed by Brazil (8), England (5) and Canada and Spain (2). In second place is Brazil, with eight articles published in the Web of Science on the subject, which reveals that this theme is still underdeveloped.

Regarding the languages of the published papers in the study area, 33 are published in English, as shown in Graph 7, corresponding to 78% of the published papers. However, it is worth noting that Brazil is in this rank (6) published papers, totalling 15%.

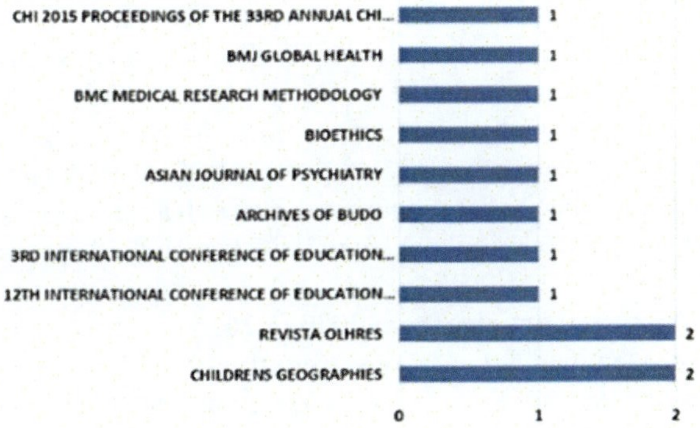

Graph 6 Number of articles per source. *Source* Elaborated by the authors

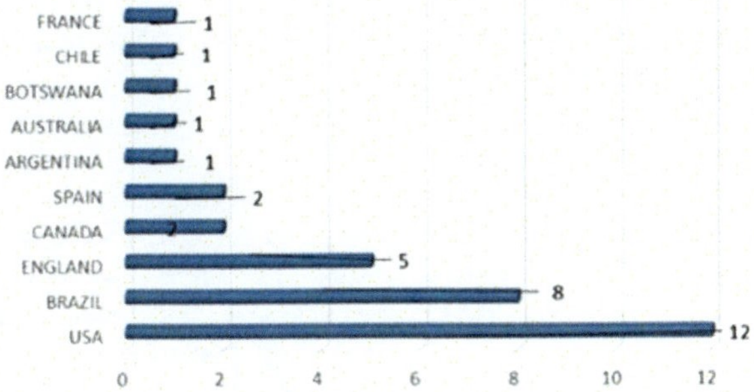

Graph 7 Articles per country. *Source* Elaborated by the authors

Graph 8 Quantity of articles per language. *Source* Elaborated by the authors

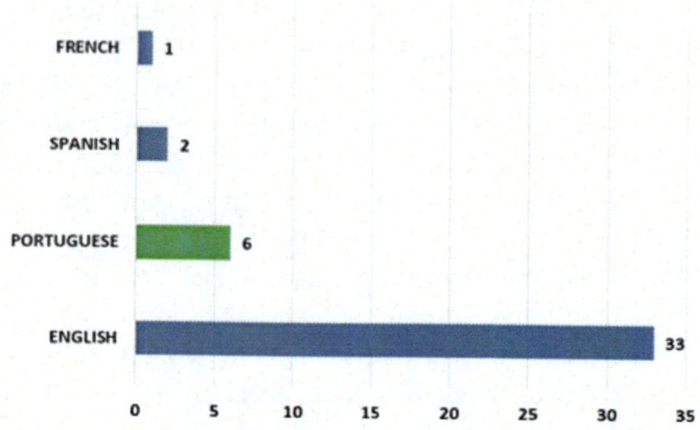

The research areas with the highest number of publications were Education Educational Research (13), Public Environmental Occupational Health (6) and Psychology (4) according to Graph 8.

Thus, it is considered that the analysis of the study developed had its research conducted from the Web of Science (WoS) database of the Institute for Scientific Information (ISI), considered one of the main multidisciplinary bases of publication in the scientific field, resulting in a total of 42 documents that cover the themes object of this study, being thirty-two articles (Graph 9).

Finally, the study quantified bibliometrically the data evidenced in the search field where it inserted the keywords youth and adult education and distance education, and thus it was possible to identify the state of the art on these subjects. Among the characteristics analysed, it searched for the main authors, year of publication, journals, institutions, countries, research areas, and the ten most cited articles in the twenty-year period that we will analyse in the following chart.

Most cited articles from 2000 to 2020

We identified 316 citations, from the 10 (ten) most cited articles on youth and adult education and distance education, from 2000 to 2020, in the Web of Science search, presented in Chart 1 (Table 1).

Graph 9 Research area. *Source* Elaborated by the authors

Table 1 List of the most cited publications in the period (2000–2020)

WOS	Title	Author	Journal	Year
66	Preventing youth violence through the promotion of community engagement and membership	Zeldin, S.	Journal of Community Psychology	2004
51	Labelling of mental disorders and stigma in young people	Wright et al.	Social Science & Medicine	2011
41	Trails and Physical Activity: A Review	Starnes et al.	Journal of Physical Activity & Health	2011
28	Mobility, education and livelihood trajectories for young people in rural Ghana: a gender perspective	Porter et al.	Childrens Geographies	2011
25	The effect of an intervention combining self-efficacy theory and pedometers on promoting physical activity among adolescents	Lee et al.	Journal of Clinical Nursing	2012
21	The regional migration of young adults in England and Wales (2002–2008): a 'conveyor-belt' of population redistribution?	Smith, D.P.; Sage, Joanna	Childrens Geographies	2014
16	Transnational youth transitions: becoming adults between Vancouver and Hong Kong	Tse, Justin K. H.; Waters, Johanna, L.	Global Networks-a Journal of Transnational Affairs	2013
13	Multiple psychoactive substance use (alcohol, tobacco and cannabis) in the French general population in 2005	Beck, F.; Legleye, S.; Spilka, S.	Presse Medicale	2008
11	The level of bravery and aggressiveness of the sports activity organisers for the youth-simulation research	Klimczak et al.	Archives of Budo	2016

(continued)

Table 1 (continued)

WOS	Title	Author	Journal	Year
9	Relationship of muscular strength on work performance in high school students with mental retardation	Smail, Karen M.; Horvat, Michael	Education and Training in Developmental Disabilities	2006

Source Elaborated by the authors

The construction of Chart 1 was developed with the purpose of relating the most cited publications in the last twenty years period to the proposed theme, relating the authors who were most cited in the same period. In view of this comparative chart, it is possible to evidence that although Zeldin, S. appears as the author of the article that was most cited, where of the 316 citations presented, 66 were cited by him, even so, Zeldin is not listed among the ten authors who most published on the subject under study in the database indicated in Chart 4.

In addition, Abane Albert and Amoako-Sakyi authors are listed in fourth place with 28 citations out of the 316 presented in Chart 1. Even with a lower number of citations, these two authors are listed in Chart 4—which presents the main authors who published articles in the analysed period, each author with a single spublished article, so we can find two authors who published the most in the period and despite not having the most cited articles in the Web of Science even so they won the fourth place with 28 citations among the researched subjects.

Next, the final considerations of the study will be presented.

5 Conclusions

This study aimed to analyse the characteristics of publications related to youth and adult education and distance education in the Web of Science database from 2000 to 2020. To achieve the objectivity of the proposed study, a bibliometric research of a quantitative nature was carried out.

As for the perspective of the study, it is possible to observe that these themes have greater representation in the field of education, but there is multidisciplinary when we analyse the other areas that have studies on the subject.

The institutions that published the most research on the themes of youth and adult education and distance education were University of California System and University of California San Francisco, both based in the USA, and Autonomous University of Barcelona, based in Spain. It is important to note that Brazil had no university that published on this theme when analysing the 20 years of publication in the Web of Science database.

About the authors who have published the most on the subject, there is a varie-ty in the period of analysis, with only one article per author in the period analysed. In relation to the journals and conferences in the ranking of the ten that had most published articles involving the theme were: Childrens Geographies (London), Re-vista Olhres (Brazil) and 12th International Conference of Education Research and Innovation (ICERI-2019). It is important to note that the first three sources that published the most in the journals of the area belong to the area of educa-tion, highlighting the Journal Olhres which is from the Department of Education of Unifesp.

Regarding the articles per country, that is, the quantity of articles distributed by the main countries that published the most were: USA, followed by Brazil, England and Canada, and Spain. It is noteworthy that in second place we have Brazil with eight articles published in the Web of Science, which reveals that this theme is still little worked on.

It is also worth mentioning that in relation to the languages of the papers published in the study area, thirty-three are published in English, corresponding to 78% of the published papers. However, it is worth noting that Brazil comes sixth in the rank totalling 15% of publications.

Still in relation to the articles, we verified 316 citations of the ten most cited articles on youth and adult education and distance education, in the last 20 years it is possible to verify that Zeldin, S. author of the article that was most cited during the study does not appear among the ten authors who have most published on the subject. However, we have two authors who have published the most Albane, A. and Amoako-Sakyi are authors who have published the most on the topic and have their articles ranked fourth with twenty-eight citations in the Web of Science in the period from 2000 to 2020.

The contribution of this study is to present the state of the art on the themes and to help future studies. The limitation is the inexistence of a researcher who stands out in relation to the analysis of the themes of Youth and Adult Education (YAE) and distance education (DL).

It is suggested that future studies should deepen and broaden these studies, and that there is a productive trajectory for the development of more research on the respective themes studied.

References

1. Arruda E (2018). Implementação das tecnologias digitais nos currículos das escolas de Educação Básica dos países membros da OCDE. In: SIQUEIRA, Ivan Claudio Pereira (org) Subsídios à elaboração da BNCC: estudos sobre temas estratégicos da parceria CNE e Unesco. Moderna, São Paulo
2. Azevedo LA (2011) De CEFET a IFET: cursos superiores de tecnologia no Centro Federal de Educação Tecnológica de Santa Catarina: gênese de uma nova institucionalidade? 2011. 192 f. Tese (Doutorado em Educação)–Programa de Pós-Graduação em Educação, Universidade Federal de Santa Catarina, Florianópolis
3. Barreto V (2006) Alunos e Alunas da YAE. Trabalho com educação de jovens e adultos. Brasília

4. Bastos MHC (2005) O Ensino de História da Educação em Perspectiva Internacional. EDUFU, Uberlândia, pp 95–130
5. Camps D, Recuero Y, Ávila RE, Samar ME (2006) Estudio bibliométrico de un volumen de la revista Archivos de Medicina. Archivos de Medicina 2(3):4
6. Carvalho JS (2015) Educação cidadã à distância: uma perspectiva emancipatória a partir de Paulo Freire. Tese (Doutorado em Educação)–Faculdade de Educação, Universidade de São Paulo, São Paulo
7. Castells M (1999) A sociedade em rede. Paz e Terra, São Paulo
8. Cordão FA, Moraes F(2017) Educação profissional no Brasil: síntese histórica e perspectivas. Senac, São Paulo
9. Frigotto G, Ciavatta M (2003) Educação básica no Brasil na década de 1990: subordinação ativa e consentida à lógica do mercado. Educação Soc Campinas 24(82):93–130
10. Frigotto G, Ciavatta M, Ramos M (2005) A política de educação profissional no governo Lula: um percurso histórico controvertido. Educação Soc Campinas 26(92):1087–1113
11. Giolo J (2018) Educação a Distância no Brasil: a expansão vertiginosa. Revista Brasileira de Política e Administração da Educação-Periódico científico editado pela ANPAE 34(1):73–97
12. Haddad S, Di Pierro MC (2000) Escolarização de Jovens e Adultos. Rev Bras Educação 14:5–130
13. Julião EF (2015) A diversidade dos sujeitos da educação de jovens e adultos. In: Medeiros CC, Gasparello A, Barbosa JL (eds) Educação de jovens, adultos e idosos: saberes, sujeitos e práticas. UFF/Cead, Niterói, pp 157–170
14. Karpinski JA, Del Mouro NF, Castro M, Lara LF (2017) Fatores críticos para o sucesso de um curso em EAD: a percepção dos acadêmicos. Avaliação, Campinas; Sorocaba, SP 22(2):440–457
15. Lévy P (2000) Cibercultura, 2 edn. Editora, São Paulo, p 34
16. Luria A (2006) Vigotskii. In Linguagem, desenvolvimento e aprendizagem. Ícone, São Paulo, pp 21–38
17. Machado LRS (2019) Formação Docente para a Educação Profissional: limites e possibilidades de institucionalização. Belo Horizonte, MG. Cadernos de Pesquisas 26(4)
18. Maia MC, Meirelles FS (2002) Educação a Distância: O caso Open University. ERA eletrônica 1(1):1–15
19. Marchisotti GG, Oliveira FB, Lukosevicius AP (2017) The social representation of distance education from a brazilian perspective. Ensaio: Avaliação e Políticas Públicas em Educação, Rio de Janeiro 25(96):743–769
20. Maturana H (1999) Emoções Trannsformación em laconvivencia. Dolmen Ediciones S.A. Santiago, Chile
21. Moraes MC (2008) Educação à distância e a ressignificação dos paradigmas educacionais: fundamentos teóricos e epistemológicos. In: Rocha A (Org) MORAES, Maria Cândida; PESCE, Lúcia; BRUNO, Pesquisando fundamentos para novas práticas na educação online. São Paulo: RG Editores
22. Nascimento SM (2013) Educação de Jovens e Adultos YAE, na visão de Paulo Freire. 2013. 45 f. Monografia (Especialização em educação: métodos e técnicas de ensino)-Diretoria de Pesquisa e Pós-Graduação, Universidade Tecnológica Federal do Paraná, Paranavaí
23. Negreiros F, Silva CFC, Sousa YLG, Santos LB (2017) Análise psicossocial do fracasso escolar na Educação de Jovens e Adultos. Revista Psicologia em Pesquisa 11(1)
24. Oliveira FA, dos Santos AMS (2020) Construção do Conhecimento na Modalidade de Educação a Distância: Descortinando as Potencialidades da EaD no Brasil. EaD Em Foco 10(1)
25. Paiva VP (1987) Educação popular e educação de adultos. -5ª edição-São Paulo-Edições Loyola–Ibrades
26. Preti O (org) (2012) Educação a Distância: construindo significados. NEAD: IE–UFMT, Plano, Cuiabá; Brasília. Redpath L (2000) Confronting the bias against on-line learning in management education. Academy of Management Learning & Education 11(1):125–140
27. Ribeiro VMM (1997) Educação de jovens e adultos: proposta curricular para o 1° segmento do ensino fundamental. Ação Educativa, São Paulo

28. Rousseau R (2001) Indicadores bibliométricos y econométricos en la evaluación de instituciones científicas. ACIMED 9:23–29
29. Saenger, P. H., Fontoura Teixeira, M. do R. (2020). Educação a distância e os estudantes da YAE. Revista Eletrônica Científica Ensino Interdisciplinar, 6(17).
30. Santaella L (2015) Flusser ressignificado pela cultura digital. In: Hanke M, Ricarte É, (orgs) Do conceito à imagem: a cultura da mídia pós-Vilém Flusser. EDUFRN, Natal, RN, pp 12–24
31. Ventura J (2011) A trajetória histórica da educação de jovens e adultos trabalhadores. In: Tiriba L, Ciavatta M (orgs) Trabalho e Educação de Jovens e Adultos. Liber, Brasília. Livro. Editora UFF, 276p, pp 57–97
32. Xiao C, Yi L (2020) Analysis on the influence of epidemic on education in China. In: Das V, Khan N (eds) Covid-19 and student focused concerns: threats and possibilities. American Ethnologist website

Empowering Learning Using a Personal Portfolio Application in an Undergraduate Information Technology Micro-Subject

Anthony Chan and David Tien

Abstract Creating a new undergraduate information technology subject to allow students to bring together a record of their learning experiences outside the classroom requires an understanding of opportunities, student capabilities, and workplace scenarios. The challenge is more difficult when a regional/rural university is located where not many technology opportunities exist. Subject design will have to think outside the box to give students a fair opportunity to collate a portfolio of their learning. The challenge is compounded with an industry who prefers outsourcing and hiring of short-term contractors as opposed to full-time employment. With the pandemic and many organizations preferring now to work from home, student outlook for how a unique subject should operate is covered and student preferences recorded. This enables other institutions who may decide to introduce such a subject and use the same approach in the creation and design of a subject where the teaching staff becomes a facilitator.

Keywords Student portfolio · Micro-subject · Authentic learning · Self-reflection · Subject development · Yarning · Working from home

1 Introduction

The opportunity for regional and rural students to own a portfolio documenting their experiences is limited given a lower population and the remoteness from events held in the capital cities of Australia. About 28% of the Australian population live in rural and remote Australia [23]. The education levels of people living in rural and remote areas are also influenced by factors such as decreased study options, the skill and educational requirements of available jobs, and the earning capacity of jobs in these communities [13, 15].

Charles Sturt University is a regional university located in six regional and rural cities of New South Wales, Australia. Most of our online students come from remote

A. Chan (✉) · D. Tien
Charles Sturt University, Wagga Wagga, New South Wales, Australia
e-mail: achan@csu.edu.au

A. Ullah et al. (eds.), *Proceedings of International Conference on Information Technology and Applications*, Lecture Notes in Networks and Systems 350,
https://doi.org/10.1007/978-981-16-7618-5_63

locations, small towns, and regional cities in Australia. Our Bachelor of Information Technology program underwent a change process in which two-unit micro-subjects (courses) were created. Two subjects were created to assist our students start off an early professional development portfolio and to encourage our students to explore and discover the opportunities further afield.

2 Portfolio and Opportunities

A student's portfolio is valued as a practice from the mid-nineties in the nursing profession as a tool to address the theory and practice divide and to provide these nursing students with skills that will enable them to maintain their professional profile for registration purposes [7]. Paulson [18] states that the portfolio is a purposeful collection of student work that exhibits the student's efforts, progress, and achievements in one or more areas and the benefit of such a collection must include student participation in selecting contents, the criteria for selection, the criteria for judging merit, and evidence of student self-reflection. Portfolios can become a window into the students' heads, a means for both staff and students to understand the educational process at the level of the individual learner. A portfolio is where the student is a participant in, rather than the object of, assessment [18].

Students' perceptions from an early experience in the industry are likely to have the student seeking employment in that industry, the region or country preferred, and the functional area or sector which is most attractive to the student [12]. However, a study indicated that [hospitality] students become less interested in selecting that career choice after exposure to the subject and industry. Hence, hospitality students need to master a second language and be familiar with international developments of the industry. The students should be encouraged to take part in an exchange program and undertake a period of work abroad. This has been seen to help enhance the student's commitment to the industry [1]. Clarke et al. [5] discovered that students who have a very narrow and limited knowledge of the range of computing careers available get disappointed when the world of work does not match that stereotype. Therefore, the feature of a wide offering of jobs and career paths that employ computing skills should be highlighted and expanded.

Zafarino [32] stated that the traditional full-time employment in IT companies has now been overtaken by the outsourcing. A tech career site, dice.com, noted that there are about 30,000 contract positions which represents one-third of the total available tech jobs. Survey data from TEKsystems that 26% of IT hiring managers expect to increase headcount of contingent workers [2]. With the move of database systems to the cloud, the introduction of faster mobile networks, and the ability to outsource talent from international markets, the need to have an in-house team to be physically present at a location to maintain computer database servers or cabled network cabinets was no longer necessary [19]. Song [26] reported that "private 5G networks are now in companies which drastically enhances and improves workflow from a variety of connected devices." The SARS-CoV-2 worldwide pandemic has further driven companies and organizations into the "offsite office" or more popularly known as "work from home" concept [10]. Companies such as Optus have accelerated their "experts at home" program which is about providing flexibility, and NBN Co [Australia's national broadband company] has commented that digital dependence will be much higher than before the crisis [8]. In 2019, independent contractors of Australia recorded 10.9% owner-managers without employees and 5.9% owner-managers with employees for the month of August 2019. The total number of employees is 2,154,200. The changing definition from "independent contractors" to the current "owner-managers" makes it difficult to compare older statistics to the newer definition today. But comparing generally to 2009, the percentage of independent contractors was at 9.1% [11].

Even before the pandemic, highly skilled temporary and contract staff are certainly becoming the new normal in workplaces especially with non-routine jobs [17]. Jobs in the tech industry are classified as a combination of creativity and the possession of a set of complex manual skills. Unlike commodities or retail service industries of general businesses, there is a decreasing number of physical locations that IT students can go to for the typical work experience. Most IT work is done virtually [at home] and hosted in the cloud; hence, the concept of a personal portfolio becomes even more relevant as proof of self-education, responsibility about updating skills, and personal time management [28].

3 The Subject Design

The subject is a two-point value requiring students to spend a total of 35–40 h. The subject was designed following the familiar 100 point system as in the verification of personal identity in Australia. The same system is also used in the accumulation of points to be eligible for a post-study Australia visa for international students. Two major activities were constructed to match the needs of today's information technology landscape based on this portfolio system. The first compulsory item was an introduction to the entire subject, and the second, an introduction of roundtable sharing of experiences. This also forms a debrief. The subject author based this on "the opportunity to habitus to transpose and hence transform practices" [20]. The First Nations practice of yarning circles Yarning [31] conducted online seemed a perfect fit to the students who may be either studying on campus or online. Yarning has been used in the field of science, technology, engineering, and mathematics (STEM): in the field of medical treatment [22], research discourse [9] as well as in Indigenous orality, connection with family and mobile phones [3]. The actual activities that were constructed for the students to pick, based on their personal interest, were drawn from the university ethos and values [16].

Based on the university value of **insightful**, students are encouraged to attend presentations by researchers, guest speakers, and topical experts to understand people and the world. Students will be knowledgeable, and the university will be able to attract the best and brightest and helps to retain talent [27]. Being stronger together under the value of **inclusive**, computing students will join any professional computing or technology association to participate in their meetings, discussions, and events [24]. To alleviate conditions where students and professionals may disagree [5], students will also be able to volunteer with any registered charities or not-for-profit organization registered with the Australian Charities and Not-For-Profits Commission. Students will learn to be easy, warm, and welcoming [25]. The value of **impactful** will see the students being able to pick up, understand new areas in the technology arena from other organizations, and broaden their perspective of thinking through Massive Open Online Courses (MOOC) and be outcome driven to achieve a certificate of completion [4, 6, 30]. Lastly, the value of **inspiring** will allow our students to lead for the future by being exposed to the latest technology on display in exhibitions, be creative in what they do, and update themselves outside the classroom setting [21]. There is no registration requirement for information technology students to practice in the industry. However, the exposure to these events and participation with organizations would be very beneficial in bringing about a hybridization of teaching and learning, bring distance education to campus, and will bring campus-level educational practices to distance education [29].

4 Stakeholder Responses and Implementation

Based on a statistical significance between evaluation of students through the model of the student's portfolio and evaluation in the traditional way resulted in favor of the portfolio model of assessment in a similar activity [14], three group interviews (total $n = 28$) were held to gauge student interest in such a subject. All were undergraduates who have completed at least six months of university studies. This authentic assessment was then presented, students were asked three simple open-ended questions, and the top three of their responses were as follows:

Q1. What would you like to collect in your portfolio?

A variety of experiences to exhibit my skills (26%), wide choice in activities that I can participate (14%) and emails, testimonials, and certificates (12%).

Q2. What kind of support would you like to help you build this portfolio? Activities (approved and recommended) by the university (17%), assigned mentor (13%), no examination (6%).

Q3. How would you like to put this together? Personal ownership of documents and artifacts (29%), ability to build continuously: add-on (or replaced) (16%), free lifetime portal (8%).

Other significant comments of value included opportunities for group study or study discussion around topics of interest, individual contribution to the knowledge space of selected topics, and a role in the student chapter of the professional association. Three paths of authentic assessment methods will form the subject: firstly, a student diary, also known as journal for students to construct online and place copies of their documents; secondly, a method of self-reflection and recording of thoughts to be entered after an event; this will record personal reflection and an authentic learning experience; and finally, a portfolio for submission for formal evaluation of learning experience and the opportunity of open sharing with peers. This will be done through a group process which has been called "digital yarning" by the university First Nations teaching group.

5 Future Direction

The most valuable aspect of this subject seems to be an extended duration so that full participation can take place. Currently, a typical study session of 14 weeks is too short, and it would be beneficial to extend to 28 weeks. Another higher-level subject could also be constructed so that university undergraduates can begin to contribute expertise, take on leadership roles, and provide guidance to junior members. This would increase interest and expand knowledge about studies in the computing and information technology career areas. The professional associations could also assist with student-level memberships and activities targeted at university undergraduates. With this support, a detailed longitudinal study of participants can be done to determine if such an effort will lead to a longer career path and sustained interest in the industry. The industry could benefit from a view toward education funding, and government could decide to contribute more funding into self-education expenses for contractors and itinerant IT workers.

6 Conclusion

Subject development such as this one can contribute positively to new experience and provide knowledge to students enrolled in university micro-subjects. Students will also benefit greatly from a program that is truly flexible.

Acknowledgements The authors acknowledge and appreciate the consultation of the First Nations teaching group of the university in understanding and application of First Nations yarning method in this subject.

References

1. Barron P, Maxwell G (1993) Hospitality management students' image of the hospitality industry. Int J Contemp Hospital Manage 5(5). https://doi.org/10.1108/09596119310046961
2. Bednarz A (2017) Life as an IT contractor. NetworkWorld. https://www.networkworld.com/article/2824528/life-as-an-it-contractor.html

3. Brady FR, Dyson LE, Asela T (2008) Indigenous adoption of mobile phones and oral culture. Cult Attitudes Towards Technol Commun 384–398. https://opus.lib.uts.edu.au/handle/10453/10883
4. Chen Y-H, Chen P-J (2015) MOOC study group: facilitation strategies, influential factors, and student perceived gains. Comput Educ 86:55–70. https://doi.org/10.1016/j.compedu.2015.03.008
5. Clarke VA, Joy Teague G (1996) Characterizations of computing careers: students and professionals disagree. Comput Educ 26(4):241–246. https://doi.org/10.1016/0360-1315(96)00004-8
6. Cohen A, Shimony U, Nachmias R, Soffer T (2019) Active learners' characterization in MOOC forums and their generated knowledge. Br J Edu Technol 50(1):177–198. https://doi.org/10.1111/bjet.12670
7. Dolan G, Fairbairn G, Harris S (2004) Is our student portfolio valued? Nurse Educ Today 24(1):4–13. https://doi.org/10.1016/j.nedt.2003.08.002
8. Fernyhough J (2020) Optus staff to work from home permanently. Finan Rev. Retrieved 17 April from https://www.afr.com/companies/telecommunications/optus-staff-to-work-from-home-permanently-20200417-p54kro#:~:text=Telecoms%20giant%20Optus%20says%20it,the%20offshore%20call%20centre%20model
9. Geia LK, Hayes B, Usher K (2013) Yarning/Aboriginal storytelling: towards an understanding of an Indigenous perspective and its implications for research practice. Contemp Nurse 46(1):13–17. https://doi.org/10.5172/conu.2013.46.1.13
10. Hern A (2020) Covid-19 could cause permanent shift towards home working. The Guardian. https://www.theguardian.com/technology/2020/mar/13/covid-19-could-cause-permanent-shift-towards-home-working
11. Independent Contractors: How Many? (Australia) (2019) Retrieved 10 January from https://www.selfemployedaustralia.com.au/Research/How-Many/independent-contractors-how-many
12. Kevin Jenkins A (2001) Making a career of it? Hospitality students' future perspectives: an Anglo-Dutch study. Int J Contemp Hosp Manag 13(1):13–20. https://doi.org/10.1108/09596110110365599
13. Lamb S, Glover S (2014) Educational disadvantage and regional and rural schools. In: Research Conference 2014, Melbourne. https://research.acer.edu.au/cgi/viewcontent.cgi?article=1228&context=research_conference
14. Mahasneh OM, Murad OS (2014) Suggested model (related to the student portfolio) used in evaluation the students in university courses. Higher Educ Stud 4(3):72–81. https://eric.ed.gov/?id=EJ1075607
15. National Regional, Rural and Remote Education Strategy: Final Report (2020) https://www.dese.gov.au/reviews-and-consultations/national-regional-rural-and-remote-education-strategy
16. Our Values (2021) Charles sturt university. Retrieved 15 September from https://about.csu.edu.au/our-university/ethos/our-values
17. Pash C (2017) Contract and temp jobs are on the rise in Australia: Here's where they are. Business Insider Australia. https://www.businessinsider.com.au/contract-and-temp-jobs-are-on-the-rise-in-australia-heres-where-they-are-2017-8
18. Paulson FL, Paulson PR, Meyer CA (1991) What makes a portfolio a Portfolio? Educational Leadership. https://web.stanford.edu/dept/SUSE/projects/ireport/articles/e-portfolio/what%20makes%20a%20portfolio%20a%20portfolio.pdf
19. Pollitt E (2019) Tech jobs: it's a contractor's world. Australian Comput Soc. Retrieved 13 May from https://ia.acs.org.au/article/2019/tech-jobs--it-s-a-contractor-s-world.html
20. Radoll P (2009) The emergence of the indigenous field of practice: factors affecting Australian Indigenous household ICT adoption. In: Proceedings of the 21st annual conference of the australian computer-human interaction special interest group: design: Open 24/7, Melbourne, Australia. https://doi.org/10.1145/1738826.1738885
21. Rinallo D, Borghini S, Golfetto F (2010) Exploring visitor experiences at trade shows. J Bus Ind Mark 25(4):249–258. https://doi.org/10.1108/08858621011038207

22. Ristevski E, Thompson S, Kingaby S, Nightingale C, Iddawela M (2020) Understanding aboriginal peoples' cultural and family connections can help inform the development of culturally appropriate cancer survivorship models of care. JCO Glob Oncol 6:124–132. https://doi.org/10.1200/JGO.19.00109
23. Rural and Remote Health (2020) Australian institute of health and welfare. Retrieved 1 October from https://www.aihw.gov.au/reports/australias-health/rural-and-remote-health
24. Slack M, Murphy J (1995) Faculty influence and other factors associated with student membership in professional organizations1. Am J Pharm Educ 59. https://www.researchgate.net/publication/255578256_Faculty_Influence_and_Other_Factors_Associated_with_Student_Membership_in_Professional_Organizations1
25. Smith K, Holmes K, Haski-Leventhal D, Cnaan RA, Handy F, Brudney JL (2010) Motivations and benefits of student volunteering: comparing regular, occasional, and non-volunteers in five countries. Canadian J Nonprofit Soc Econ Res 1(1):65–81. https://doi.org/10.22230/cjnser.2010v1n1a2
26. Song J (2020) How 5G will change the Enterprise. Retrieved 22 October from https://www.forbes.com/sites/forbestechcouncil/2020/10/22/how-5g-will-change-the-enterprise/?sh=4eced7ed5692
27. Sutton H (2015) Build a strong community of adult learners to benefit the whole campus. Enroll Manage Rep 19(8):1–5. https://doi.org/10.1002/emt.30110
28. Syzdykova Z, Koblandin K, Mikhaylova N, Akinina O (2021) Assessment of E-portfolio in higher education. Int J Emerg Technol Learn (iJET) 16(2):120–134
29. Waddoups G, Howell S (2002) Bringing online learning to campus: the hybridization of teaching and learning at brigham young university. Int Rev Res Open Distance Learn 2. https://doi.org/10.19173/irrodl.v2i2.52
30. Wang X, Yang D, Wen M, Koedinger K, Rose CP (2015) Investigating how student's cognitive behavior in MOOC discussion forums affect learning gains. In: 8th international conference on educational data mining, Madrid, Spain
31. Yarning Circles (2020) Queensland government. https://www.qcaa.qld.edu.au/about/k-12-policies/aboriginal-torres-strait-islander-perspectives/resources/yarning-circles
32. Zafarino S (2019) Full-time or contract tech professionals? when to hire tech contractors. CIO. Retrieved 3 October from https://www.cio.com/article/3405570/full-time-or-contract-tech-professionals-when-to-hire-tech-contractors.html

Author Index

A. Ullah et al. (eds.), *Proceedings of International Conference on Information Technology and Applications*, Lecture Notes in Networks and Systems 350,
https://doi.org/10.1007/978-981-16-7618-5

D

E

F

G

H

I

J

K

L

M

N